A joint report by the
twenty-three UN agencies
concerned with freshwater

Water for People Water for Life

The United Nations World Water Development Report

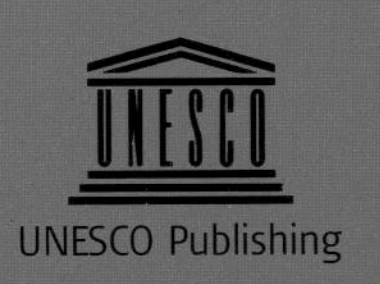

Published in 2003 jointly by the **United Nations Educational, Scientific and Cultural Organization (UNESCO)**, and **Berghahn Books**.

This Report has been elaborated on behalf of the partners of the United Nations World Water Assessment Programme (WWAP) with support from the following countries:
Bolivia; Estonia; France; Germany; Guinea; Hungary; Japan; Mali; Mauritania; Netherlands; Peru; Russian Federation; Senegal; Sri Lanka; Thailand; Turkey; United Kingdom.

United Nations Funds and Programmes
United Nations Centre for Human Settlements (Habitat)
United Nations Children's Fund (UNICEF)
United Nations Department of Economic and Social Affairs (UNDESA)
United Nations Development Programme (UNDP)
United Nations Environment Programme (UNEP)
United Nations High Commissioner for Refugees (UNHCR)
United Nations University (UNU)

United Nations Regional Commissions
Economic and Social Commission for Asia and the Pacific (ESCAP)
Economic Commission for Africa (ECA)
Economic Commission for Europe (ECE)
Economic Commission for Latin America and the Caribbean (ECLAC)
Economic Commission for Western Asia (ESCWA)

Secretariat of United Nations Conventions and Decades
Secretariat of the Convention on Biological Diversity (CBD)
Secretariat of the Convention to Combat Desertification (CCD)
Secretariat of the International Strategy for Disaster Reduction (ISDR)
Secretariat of the United Nations Framework Convention on Climate Change (CCC)

Specialized UN Agencies
Food and Agriculture Organization (FAO)
International Atomic Energy Agency (IAEA)
International Bank for Reconstruction and Development (World Bank)
United Nations Educational, Scientific and Cultural Organization (UNESCO)
United Nations Industrial Development Organization (UNIDO)
World Health Organization (WHO)
World Meteorological Organization (WMO)

Library of Congress Cataloging-in-Publication Data
A catalogue record for this book is available from the Library of Congress

British Library Cataloguing in Publication Data
A catalogue record for this book is available from the British Library

ISBN UNESCO: 92-3-103881-8
ISBN Berghahn: 1-57181-627-5 (cloth), 1-57181-628-3 (paperback)

UNESCO Publishing: http://upo.unesco.org/
Berghahn Books: www.berghahnbooks.com

Printed in Barcelona

Design & production
Inside cover montages: Emmanuel Labard and Estelle Martin, Atelier Takavoir, Paris.
Design and production: Andrew Esson, Baseline Arts Ltd, Oxford.
Index: Paul Nash, Perth.

Contents

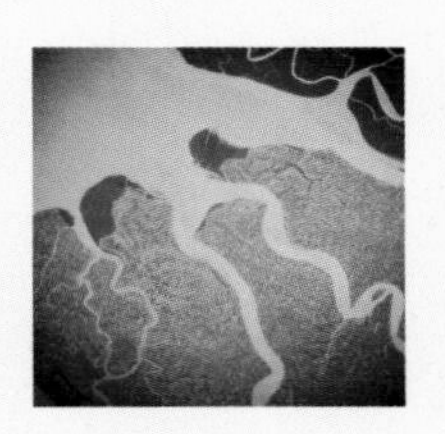
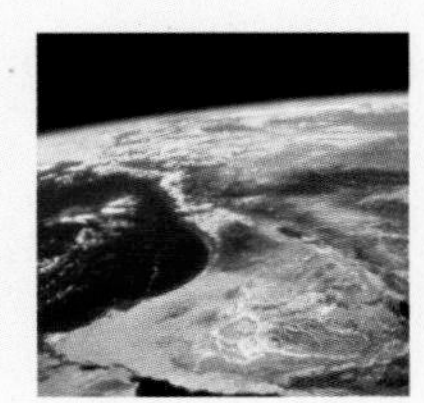

Figures

Maps

Boxes

Boxes by Region

Africa

Asia

Europe

Latin America and the Caribbean

North America

Oceania

Global

Tables

Acknowledgements

The World Water Assessment Programme would like to thank the following people for their gracious help and support in preparing the *World Water Development Report*. All those whom we have inadvertently omitted from this list, please accept our heartfelt apologies, along with our thanks.

Abdelsalam Ahmed, Abdalla
Abderrahman, Walid
Abdulrazzak, Mohamed
Abegunawardane, M.H.
Abewickrema, Nanda
Abeysinghe, A.M.H.
Abrams, Len
Adebayo, Yinka
Adeel, Zafar
Adriaanse, Martin
Aekaraj, Sukontha
Affeltranger, Bastien
Aggarwal, Pradeep
Ahamed, Aziz
Alabaster, Graham.
Alcamo, Joseph
Aliakseyeva, Natallia
Almeida, Georgina
Alpan Atamer, Sema
Al-Rashed, Muhammad
Al-Weshah, Radwan
Amarasinghe, Upali
Amunugama, Sarath
Amarathunga, A.P.
Amdur, Richard
Amezal-Benureau, Aîcha
Andersson, Ingvar
Aparicio, Javier
Appelgren, Bo
Aranya, Fuangawasdi
Arata, Akio
Aratani, Keita
Arico, Salvatore
Ariyabandu, R. de S.
Ariyaratne, B.R
Arulvijitkul, Pongsak
Asai, Takashi
Ashley, Sara
Askew, Arthur
Atannon, Medard
Atapattu, N.K.
Athukorala, Kusum
Aureli, Alice
Baba, Toshio
Baer, A.
Bailey, Ariane
Baker, F.W.G.
Ballayan, Dominic
Bandusena, S. B.
Barg, Uwe
Barker, Randy
Baron, J.
Barraqué, Bernard
Barret, Philippe
Bartram, Jamie
Basoberry, Antonio
Battikha, Veronica
Bauer, Peter
Bazza, Mohamed
Beano, J.E.
Belhadi, Ahmed
Belin, Solange
Berbalk, Dagmara
Bernardt, Cristoph
Berghahn, Marion
Bhattarai, Madhusadan
Björklund, Gunilla
Blix, Bozena
Bloch, Julia
Bloxham, Martin
Blyth, Simon
Boedeker, Gerold
Bogardi, Janos
Bommelaer, Olivier
Bonell, Mike
Bories, Jacques
Bos, Robert
Bosnjakovic, Branko
Boulharouf, Rajeb
Bourg, Anna-Kay
Bowen, Thomas
Bresser, Ton
Brewster, Marcia
Briceño, Salvano
Bridgewater, Peter
Briffault, Robinia
Briscoe, John
Bruck, Stephan
Bruinsma, Jelle
Bruno, Rudolf
Budarin, Vladimir F.
Burapatanin, Somboon
Burchi, Stefano
Burke, Jacob
Burke, Niamh
Carmovas, Alvaro
Carr, Richard
Casal, François
Casman, Elizabeth
Castelein, Saskia
Chadwick, Matthew
Chane, Bayou
Chantanintorn, Sanong
Chantrel, Pierre
Charriet, Bertrand
Cheret, Ivan
Cheriez, Jean-Francis
Clua, Alexandra
Cofino, William Peter
Coles, Peter
Collins, Terry
Cosgrove, William J.
Courtecuisse, Arnaud
Cowan, Donald
Crespo Milliet, Alberto
Curtin, Fiona
Curtis, Ian
Daler, Dag
Daley, Ralph
Das Gupta, Ashim
Davy, Thierry
Dayaratne, W.J.M.
De Fraiture, Charlotte
De La Rocque, Jacques
De Santis, Giorgia
De Sherbinin, Alex
De Silva, K.S.R.
De Silva, Vijitha
De Vaulx, Maurice
Delli Priscolli, Jerome
Demissie Yadeta, Meseret
Dengo, Manuel
Denzer, Ralf A.
Dharmaratne, G.H.P.
Diagne, Babacar
Diop, Salif
Dissanayake, I.
Dissanayake, J.B.
Dodge, Terry
Donkor, Stephen Maxwell
Donzier, Jean-François
Dorn, Ronald
Dowlatabadi, H.
Drammeh, Halifa
Droogers, Peter
Dukhovny, Victor
Dzikus, Andre
Edwards, Mark
Elakanda, D.C.S.
Eliasson, Åse
Enderlein, Rainer
Facon, Thierry
Fadjouwin, Grégoire
Fahmi, Ahmed
Fahmy, Hussam
Faurès, Jean-Marc
Fekete, Balazs
Fekri, Hassan
Felici, Eleana
Fernando, M.J.J.
Feuillette, Sarah
Flörke, Martina
Florin, Reto
Foster, Stephen
Fournier, Fred
Fraser, Andrew
Frijters, Ine
Frithsen, Jeff
Fritsch, Jean-Marie
Frogé, Eric
Fugl, Jens
Fukumoto, Shinobu
Fukushima, Tamiya
Gajanayake, P.S.
Galbraith, Hector
Galloni, Susanna
Gangopadhyay, Subuhrendu
Garnier, Olivia
Gertman, Isaac
Gilbrich, Wilfred
Giordano, Meredith
Gleick, Peter
Glidden, Stanley
Goitia, Julio Sanjinez
Gourcy, Laurence
Goya, Yoshiyuki
Grabs, Wolfgang
Green, Pamela
Groombridge, Brian
Guettier, Philippe

Guinchard, Françoise
Gupta, Rajiv
Gurkok, Cahit
Guttman, Cynthia
Haddadin, Munther
Haenel, Jeanette
Hall, Alan
Hall, David
Haller, Laurence
Hallifax, Peter
Handawela, James
Hansen, Eggret
Harding, John
Hartvelt, Frank
Hashizumi, Hiroki
Hassan, Fekri
Hattori, Tsukasa
Heller, L.
Hellmuth, Molly
Hemakumara, H.M.
Henderson, Mark
Henrichs, Thomas
Henry, Rosamund
Herrod, Susan
Hipel, Keith
Hirai, Yasuyuki
Hiroki, Kenzo
Hislop, Lawrence
Holaas, Olve
Holt McCluskey, Alyssa
Hong, Eunkyoung
Hoogeveen, Jippe
Hoque, Azm Fazlul
Huber-Lee, Annette
Hueb, Jose
Huidobro, Pablo
Iglesias, Ana
Iijima, Ikuko
Ikebuchi, Shuichi
Imbulana, K.A.U.S.
Imura, Takashi
Inoki, Hiromasa
Inoue, Tomoo
Intrawityanunt, Pitak
Ito, Seiji
Iwasaki, Yoshihisa
Jansky, Libor
Jaskiewicz, E.
Jayasekara, A.M.
Jayasekara, H.B.
Jayasinghe, Ananda
Jayasinghe, U.G.
Jayasundera, B.K.
Jayatillake, H.M.
Jayatissa, W.
Jayawardane, D.S.
Jayaweera, Upali
Jezeph, David
Jinadasa, W.P.
Jinapala, K.
Jingu, Makoto
Jonch-Clausen, Torkil
Jones, Patricia
Jorgensen, Dorthe
Kaewkulaya, Jesda
Kalyan, Ray
Kaneki, Makoto
Karam, Rosanna
Karunaratne, A.D.M.
Karunaratne, G.R.R.
Karunatilake, W.A.K.
Kasiwatta, Ruwan
Katavetin, Parichart
Kavvas, Levent
Kawaguchi, Daiji
Kawasaki, Hideaki
Kawashima, Tadao
Kaysavawong, Pavich
Kemp-Benedict, E.
Kemper, Karin
Kendrick, Shawn
Kikuchi, Ryosuke
Kimura, Masayoshi
Kirshen, Paul
Kite, Jeffery
Klohn, Wulf E.
Kobayashi, Takanobu
Kobori, Iwao
Koda, Kazuhisa
Kodituwakku, K.A.W.
Koga, Katsutoshi
Konaka, Takayuki
Konradsen, Flemming
Koreeda, Nobukazu
Korsjukov, Margus
Koshinaga, Kenji
Koujo, Friedrich
Kovacs, Yves
Kularatne, Gamini
Kulkarni, K. M.
Kulshreshtha, Suren
Kuylenstierna, Johan
Laje, Cristina
Lantagne, D.
Larsen, Henrik
Lawrence, J. Gary
Le Marechal, Dominique
Lemaire Drinkwater, Vanessa
Lenton, Roberto
Lentvaar, Jan
Leogardo, Vincent
Liiv, Harry
Lisbjerg, Dennis
Loh, Eloise
Lopez, Alexander
Lord, Marcia
Lorenz, Carolin M.
Lorenz, Frederick M.
MacDevette, David
MacQueen, Anna
Macomber, Marcia
Märker, Michael
Major, David
Makin, Ian W.
Mandalia, Lalji
Marques, Marcia
Matoba, Yasunobu
Matsuda, Wakako
Matsukawa, Noriyuki
Matsuura, Tadashi
Matthews, Geoffrey
Maurer, Thomas
Mawere, Gilbert
Mayfield, Colin
McDonald, David
Medagama, Jaliya
Medrano, Juan Carlos
Meigh, Jeremy
Mendis, B.P.N
Merzoug, Mohamed Salem O
Milburn, Anthony
Miller, John
Millward, Michael
Min, Qingwen
Mirabzadeh, Parastu
Misaki, Takahito
Missoten, Robert
Mitsuhashi, Hisashi
Mitchell, Bruce
Miwa, Junji
Mmayi, Patrick
Mohammed, Waleed A.
Mokuoane, E.M.
Molden, David
Molle, Françoise
Moller, Birgitte
Moore, Roger
Morris, Neil
Mostert, Eric
Muckleston, Keith
Mulongoy, Jo
Murakashi, Manami
Murase, Masahiko
Mushiake, Katsumi
Mutagamba, Maria Lubega
Muthukuda, P.S.
Mutuwatte, Lal
Nakamura, T.
Nakashima, Douglas
Nakayasu, Masaaki
Nandalal, K.D.W.
Natchtnebel, Peter
Ndiaye, Tamsir
Neto, Fred
Nelson, H.G.P.
New, James
Newton, Adrian
Nicol, Alan
Niida, Hiroshi
Nishida, Yuji
Niyangoda, S.M.S.B.
Noguchi, Hiroshi
O'Connell, Patrick Enda
Obes de Lussich, Angelica
Oda, Hideaki
Ogawa, Mamiko
Ojima, Satoshi
Okazumi, Toshio
Oki, Taikan
Olivera, Fancisco
Ongchotiyakul, Pacharee
Otchet, Amy
Otte, Alexander
Ouezzin Coulibaby, Jeannette
Ouriaghli, Catherine
Ozbilen, Vedat
Pacheco, Anibal
Pacini, Nicola
Panabokke, C.R.
Panyachatraksa, Prapat
Pasztor, Janos
Pathirana, P.P.S.R.K.
Pattanee, Surapol
Perera, Asoka
Perera, Gamini Jayawickrema
Pilon, Paul
Pimpunchat, Narongsak
Piper, David
Pizarro, Roberto
Poinsot, Claire
Poulisse, Jan
Prasad, K.
Pregassame, Radja
Price, Roland
Priest, John
Propersi, Federico
Proussevitch, Alexander
Pruess, Annette
Puri, Shammy
Rapport, David
Ratnayake, Ranjith
Ratwatte, Anuruddha
Ray, Kalyan
Rees, Judith
Reichert, Peter
Renault, Marie
Reuss, Martin
Richmond, Mark
Richts, Andrea
Rijsberman, Frank
Robarts, Richard

Robertson Vernhes, Jane
Roche, Pierre-Alain
Rodda, John C.
Rogers, Peter
Roll, Gulnara
Romero, Ricardo
Rosegrant, Mark
Rosen, Dov
Rubasinghe, Sirisoma
Rudolf, Bruno
Saad, Bahaa
Safar-Zitoun, Mahamed
Salih, Abdin
Samarasekara, Mahinda
Samarasinghe, S.A.P.
Samaraweera, P.
Samir Farid, M.
Samsoen, Nicolas
Sanders, Sylvie
Satterthwaite, David
Schaff, Thomas
Scheraga, Joel
Schneegans, Susan
Selvarajah, S.
Senaratne, P.C.
Senaratne, Sunidha
Seneviratne, Lakshman
Seneviratne, P.W.
Seta, Masaaki
Sethaputra , Sacha
Shahin, Mamdouh
Shamir, Uri
Shamir, Yona
Shanahan, Peter
Shanmugarajah, C.K
Shiklomanov, Igor
Shindo, Soji
Shirai, Minoru
Shrestha, Dinesh Lal
Sieber, Jack
Sighomnou, Daniel
Simachaya, Wijarn
Simonovic, Slobodan
Sinou, D.
Siripornpibul, Chaiporn
Siriwardane, Nihal
Skinner, Andrew
Smith, David
Sokolov, Vadim
Solanes, Miguel
Somaratne, P.G.
Somatilaka, H.S.
Sonoda, Toshihiro
Sora, Shuichi
Sorenson, Per
Soussan, John
Sprey, Melvin
Sreenath, Narasingarao
Sreenath, Sree
Stephan, RayaMarina
Storm, Borge
Struckmeier, Wilhelm
Strzepek, Kenneth
Sullivan, Caroline
Suzuki, Motoyuki
Swayne, David
Szentmartoni, Erzsebet
Szöllösi-Nagy, Andras
Takahashi, Teiji
Takara, Kaoru
Takashima, Hatsuhisa
Takeuchi, Kuniyoshi
Talayssat, Camille
Tan-Torres Edjer, T.
Tanaka, Michiko
Taylor, Richard
Tejada-Guibert, Alberto
Thaikar, Pipat
Thanalerdsomboon, Prakit
Ti, Lehuu
Tiard, Ferline
Timmerman, Jos G.
Tobin, Vanessa
Tolb, Richard
Tollan, Arne
Toyama, Masato
Treves-Habar, Janine
Tromp, Dik
Trongkarndee, Nitiphan
Tropp, Hakan
Trottier, Julie
Tsujisaka, Ginko
Tsukamoto, Shigemitsu
Tsuneyama, Shuji
Tsutsui, Seiji
Turrel, Hugh
Uematsu, Ryuji
Unt, Peeter
Vallée, Domitille
Van Beek, Eelco
Van Koppen, P.
Van Vierssen, Wim
Vandeweerd, Veerle
Vander Zaag, Pieter
Venkat, Rao
Vasilenko, Olga
Vassolo, Sara
Vinogradov, Sergei
Vlachos, Evan
Vogel, Richard M.
Vörösmarty, Charles
Vrba, Jaruslav
Wallace, Jim
Wallace, Pamela
Wallin, B.G.K.
Wang, Jinxia
Warnakulasooriya, H.V.
Watanabe, Yukio
Weerakoon, D.W.R.
Weerasinghe , K.D.N.
Wiberg, David
Wickramage, M.
Wickramanayake, Mangala
Wickramaratne, Rohith
Wickremarachchi, M.S.
Wijesinghe, M.W.P.
Wilkinson, Christine
Williams, Robert
Williams, Sue
Wirasinha, Ranjith
Witharana, Prabath
Witt, Ronald
Wolf, Aaron
Wolter, Hans
Wood, Rohan
Woodward, Alistair
Wouters, Patrricia
Wright, Albert
Wu, Jisong
Yamada, Yasuyuki
Yamaguchi, Hiroshi
Yamauchi, Hiroshi
Yasuda, Goro
Yates, David
Yohe, Gary
Yokota, Taeko
Yoneyama, Ken
Yoshida, Hitoshi
Yoshinaga, Kenji
Yoshitani, Junichi
Zacharopoulou, Catherine
Zatlokal, Barbara
Zhu, Zhonping
Zimmer, Dani

Foreword

THE CENTRALITY OF FRESHWATER IN OUR LIVES CANNOT BE OVERESTIMATED. Water has been a major factor in the rise and fall of civilizations. It has been a source of tensions and fierce competition between nations that could become even worse if present trends continue. Lack of access to water for meeting basic needs such as health, hygiene and food security undermines development and inflicts enormous hardship on more than a billion members of the human family. And its quality reveals everything, right or wrong, that we do in safeguarding the global environment.

But if the water problems facing our world are sometimes a cause of tension and concern, they can also be a catalyst for cooperation. Two thirds of the world's major rivers are shared by several states. More than 300 rivers cross national boundaries. Increasingly, countries with expertise in the management of watersheds and flood plains, or with experience in efficient irrigation, are sharing that knowledge and technology with others. Scientists from many nations and disciplines are pooling their efforts, assessing risks and working to bring about a much-needed 'blue revolution' in agricultural productivity. Policy-makers can and should draw on these experiences, which have generated a rich inventory of lessons and 'best practices'.

With these issues in mind, the nations of the world have established a comprehensive and demanding water resources agenda. In the Millennium Declaration adopted by the General Assembly in 2000, world leaders resolved 'to halve, by the year 2015, the proportion of the world's people who are unable to reach, or to afford, safe drinking water', and 'to stop the unsustainable exploitation of water resources'. Water resources also figured prominently at the World Summit on Sustainable Development in Johannesburg in 2002. The Plan of Implementation adopted there reiterated the Millennium Development Goal on water, set a new target of halving the proportion of people who do not have access to basic sanitation by 2015, and recognized the key role of water in combating poverty and in the realms of agriculture, energy, health, biodiversity and ecosystems.

This first edition of the *World Water Development Report, Water for People, Water for Life*, is the main outcome of the World Water Assessment Programme, a long-term project started in response to decisions of the General Assembly and the Commission on Sustainable Development. A joint project involving twenty-three United Nations specialized agencies and other entities, it provides a comprehensive view of today's water problems and offers wide-ranging recommendations for meeting future water demand. This coincides with the International Year of Freshwater, which is being observed throughout 2003. Finally, it shows the United Nations at work, helping the world to confront current and impending water crises. I recommend this publication to the widest possible audience.

Kofi Annan
Secretary General, United Nations

Prologue

IT GIVES ME GREAT PLEASURE TO INTRODUCE this first edition of the *World Water Development Report, Water for People, Water for Life*, which, like the World Water Assessment Programme itself, is a product of many hands working in close collaboration. The Report emphasizes the critical importance of freshwater in every region of the world. It shows how human development is stifled without an adequate supply of water of suitable quality. It shows how peace and harmony between and within nations are threatened where, for whatever reason, there is insufficient water to meet human and environmental needs. And it shows how the value of freshwater exceeds narrow economic calculations by encompassing a whole range of social, cultural and ethical considerations.

Water is a common element in the lives of all peoples and societies. Water has been the foundation, and sometimes the undoing, of many great civilizations and today is essential for the agricultural, economic and industrial activity that helps societies to develop.

As a drop of water makes its way from the mountain top to the sea, it can be used many times over: to alleviate thirst, sustain crops, facilitate industry, help generate electricity, remove wastes and, at the same time, support the natural environment. As an element with multiple uses, water links activities together and, in so doing, it links people together. For these reasons, it must be managed wisely through an integrated approach that takes account of all uses and users.

The achievement of an integrated management of freshwater resources places a premium on cooperation. Within in any one country, this requires the many ministries involved in water management to come together in common purpose, especially when demand for the resource outstrips supply. When freshwater is subject to several national jurisdictions, as is often the case with major river systems and lakes, the need for integrated international approaches and well-designed mechanisms for the amicable resolution of disputes is imperative.

Through the World Water Assessment Programme, many United Nations agencies and other entities concerned with freshwater issues have come together in common effort to monitor and assess this vital resource. I am proud that UNESCO has been playing the role of coordinator and catalyst for this system-wide process by hosting the Secretariat of the World Water Assessment Programme. As a key outcome of this collaborative work, the Report is an important contribution to the International Year of Freshwater and the Third World Water Forum in Kyoto in March.

I commend and welcome the *World Water Development Report* as a basis for informed discussion on freshwater issues. I am sure that its findings and analyses will be invaluable for placing public debate, policy formation and decision-making on a sound footing, grounded upon evidence and cogent argumentation.

The world is full of uncertainty and it is difficult to predict the future. By working together, however, we can help to shape the world and make it a better place – for people, for life.

Koïchiro Matsuura
Director-General, UNESCO

Preface

AS WE ENTER THE TWENTY-FIRST CENTURY A GLOBAL WATER CRISIS IS THREATENING the security, stability and environmental sustainability of all nations, particularly those in the developing world. Millions die each year from water-related diseases, while water pollution and ecosystem destruction grow. Again, those in developing countries are hardest hit. In its Millennium Declaration, the United Nations (UN) called on the nations of the world to 'halve by 2015 the proportion of people who are unable to reach, or to afford, safe drinking water', and to 'stop the unsustainable exploitation of water resources by developing water management strategies at the regional, national and local levels, which promote both equitable access and adequate supplies'.

Action Now!

Current thinking accepts that the management of water resources must be undertaken using an integrated approach, that assessment of the resource is of fundamental importance as the basis for rational decision-making and that national capacities to undertake such assessments must be fully supported. Management decisions to alleviate poverty, to allow for economic development, to ensure food security and the health of human populations as well as to preserve vital ecosystems, must be based on our best possible understanding of all relevant systems. Hence the need for comprehensive assessments.

The World Summit on Sustainable Development held in Johannesburg in August/September 2002 suggested that there are five priority areas that need immediate action: water and sanitation, energy, health, agriculture and biodiversity. Water is at the heart of sustainable human development.

There have been many assessments in the past, but up until now there has been no global system in place to produce a systematic, continuing, integrated and comprehensive global picture of freshwater and its management.

The United Nations Response: The World Water Assessment Programme

The need for a more people-oriented and integrated approach to water management and development has been more fully accepted as a result of a number of major conferences and international events. The UN has responded by undertaking a collective system-wide continuing assessment process, the World Water Assessment Programme (WWAP).

The UN system has the mandate, credibility and capacity to take on the task of systematically marshalling global water knowledge and expertise to develop, over time, the necessary assessment of the global water situation, as the basis for action to resolve water crises.

Building on the achievements of the many previous endeavours, WWAP focuses on the evolving freshwater situation throughout the world. The results of this assessment are to be published at regular intervals in the *World Water Development Report* (WWDR). The programme will evolve with the WWDR at its core. There is thus a need to include data compilation (geo-referenced meta-databases), interpretation and dissemination, information technologies, comparative trend analyses, methodology development and modelling. The recommendations from the WWDR include capacity-building to improve country-level assessment, with emphasis on developing countries. This includes the capacity-building in education and training, in monitoring and database science and technology. The programme identifies situations of water crisis and thus provides guidance for donor agencies. It develops knowledge and understanding necessary for further action.

WWAP focuses on terrestrial freshwater, but links with the marine near-shore environments and coastal zone regions as principal sinks for land-based sources of pollution and sedimentation; these are areas where the threat of flooding and the potential impact of sea-level rise is particularly acute.

WWAP is undertaken by the concerned UN agencies, aided by a Trust Fund, donors providing support in cash and in kind. The generosity and foresight of the government of Japan initiated the programme and allowed production of the first edition of the WWDR. The United Nations Educational, Scientific and Cultural Organization (UNESCO) currently provides and manages the Trust Fund and hosts the WWAP Secretariat at its headquarters in Paris.

The programme serves as an 'umbrella' for coordinating existing UN initiatives within the freshwater assessment sphere. In this regard it will link strongly with the data and information systems of UN agencies.

WWAP consists of the following coordinated elements:

- the *WWDR component*, involving the preparation of the triennial report and resultant advice, when requested, to governments;
- a *Water Information Network and Water Portal* comprising a global-scale meta-database, knowledge management systems to facilitate the assessment and dissemination of information, an online library, web site and newsletter. The network will allow communication with governments and water-related non-government groups, facilitate capacity-building and raise awareness about water;
- a *capacity-building component*, the prime purpose of which is to promote the ability of governments to conduct their own assessments through human resource development, education and training, provision of methodologies, institution and infrastructure development and development of data and information networks; and
- a series of specific applications (for example on conflict resolution).

The *World Water Development Report*

The report is a periodic review, continuously updated, designed to give an authoritative picture of the state of the world's freshwater resources and our stewardship of them. The WWDR builds upon past assessments and will constitute a continuing series of assessments in the future.

The WWDR is targeted to all those involved in the formulation and implementation of water-related policies and investments, and aims to influence strategies and practices at the local, national and international levels. While a broad, global picture is given, particular emphasis is placed on developing-country situations, where management capacities are likely to be weaker, with the intention of identifying areas in particular need of attention. It lays the foundations for efficient and effective capacity-building in areas where stewardship challenges are greatest.

The involvement of national governments, as the prime beneficiaries of the process, is actively sought. This is considered a precondition for high-quality and continuity of data collection and analysis, and for the subsequent credibility and sustainability of the report.

As a UN-led exercise, the preparation of the WWDR is a joint effort of the UN and its member states to collect and prepare reliable data in a harmonized and meaningful manner. Data and information used in the report are from official sources such as national authorities and basin agencies, or equivalents. National and local governments, institutions and universities, user associations, the private sector, non-governmental organizations and national consultants are also involved.

The first edition of the WWDR offers an inaugural assessment of progress since the Rio Summit. Using seven pilot case studies, it suggests some possible directions for developing appropriate assessment methodologies, to be further explored in subsequent reports. In this way there will be a progressive build-up of knowledge and understanding.

This report represents an action-oriented and people-centred product. If we are to address the water crisis, we must act now!

Gordon Young
Coordinator of WWAP

‘The world is full of uncertainty and it is difficult to predict the future.

By working together, however, we can help to shape the world and make it a better place – for people, for life.’

Part I: Setting the Scene

Water is an essential element of daily life for each and every one of us. Throughout this book, it is explored in all its facets.

Part I presents the background, starting with an introduction to the water crisis in its many shapes and forms. It then provides a glimpse of the milestones on the long policy road that has brought us to where we stand today. Finally, the chapter on indicators proposes some tools to help us assess our progress towards building a better future.

1 The World's Water Crisis

Table of contents

When the planet herself sings to us in our dreams, will we be able to wake ourselves, and act?

Gary Lawless, *Earth Prayers from around the World*

WE ARE IN THE MIDST OF A WATER CRISIS THAT HAS MANY FACES. Whether concerning issues of health or sanitation, environment or cities, food, industry or energy production, the twenty-first century is the century in which the overriding problem is one of water quality and management. Water management has evolved, but in 2003 some 25,000 people are still dying every day from malnutrition and 6,000 people, mostly children under the age of five, are dying from water-related diseases. It is a real-world crisis that numbers alone can dehumanize. The months of writing this text have seen headlines of millions facing malnutrition in southern Africa, millions affected by floods in Bangladesh, floods throughout central and eastern Europe, and hundreds killed by Nile fever. But the silent deaths of millions of others do not make daily headlines, nor does the plight of those poor and powerless people who are still deprived of a basic human right. Yet these terrible losses, with the waste and suffering they represent, are preventable.

We know the problem: it is one of management, and we have agreed on targets for improvements to be made by 2015. But will we honour these commitments? Will we muster the political will to meet our goals? To do so we must provide more than a quarter of a million individual people with improved water supply and hygiene each and every day. We must act now.

In this chapter we look at the general context in which these events and human dramas are unfolding. What are the forces at work? Who are the major actors? What are the stakes? And how do the complex and often subtle interactions between the different actors and their environments affect the water situation?

THE FACT THAT THE WORLD FACES A WATER CRISIS has become increasingly clear in recent years. Challenges remain widespread and reflect severe problems in the management of water resources in many parts of the world. These problems will intensify unless effective and concerted actions are taken, as is made clear in the *World Water Vision* (Cosgrove and Rijsberman, 2000, p. xxi):

> ***This increase in water withdrawals implies that water stress will increase significantly in 60% of the world, including large parts of Africa, Asia and Latin America. Will this lead to more frequent and more serious water crises? Assuming business as usual: yes.***

Water, People and Sustainable Development

The 'business as usual' qualification is important. We cannot carry on as we do, and many aspects of water resources management must change. This is recognized in the United Nations (UN) Millennium Declaration (2000), which again called upon all members of the UN

> *to stop the unsustainable exploitation of water resources by developing water management strategies at the regional, national and local levels which promote both equitable access and adequate supplies.*

Water is essential for life. We are all aware of its necessity, for drinking, for producing food, for washing – in essence for maintaining our health and dignity. Water is also required for producing many industrial products, for generating power, and for moving people and goods – all of which are important for the functioning of a modern, developed society. In addition, water is essential for ensuring the integrity and sustainability of the Earth's ecosystems.

None of these facts are in dispute. And yet, we all too often take the availability of water for granted, as if there existed an abundance of the resource. This assumption has now been challenged and found to be untenable. In recent years the availability of and access to freshwater have been highlighted as among the most critical natural resource issues facing the world. The UN environmental report *GEO 2000* states that global water shortage represents a full-scale emergency, where 'the world water cycle seems unlikely to be able to adapt to the demands that will be made of it in the coming decades' (UNEP, 1999). Similarly, the World Wide Fund for Nature (WWF) emphasizes that 'freshwater is essential to human health, agriculture, industry and natural ecosystems, but is now running scarce in many regions of the world' (WWF, 1998).

Complacency is not an option. Water consumption has almost doubled in the last fifty years. A child born in the developed world consumes thirty to fifty times the water resources of one in the developing world (UNFPA, 2002). Meanwhile, water quality continues to worsen. The number of people dying from diarrhoeal diseases is equivalent to twenty fully-loaded jumbo jets crashing every day, with no survivors. These statistics illustrate the enormity of the problems facing the world with respect to its water resources, and the startling disparities that exist in its utilization.

This book assesses the world's water situation. The water crisis that exists is set to worsen despite continuing debate over the very existence of such a crisis. For many years over the past decades, 6,000 people, and mainly children under five, have died every day. Descriptions more severe than 'a crisis' have been associated with events in which 3,000 people have lost their lives in a single day. What phrase can be used for the recurrence of higher loss of life every day of every year over decades? That the world is in a water crisis is undeniable, and the time to take action is now.

What are the forms of this water crisis, what difference will it make to people's lives, what forces are causing it and what can we do about it? This chapter paints the picture: it examines the importance of water in people's lives, identifies main concerns and trends in water resources and their uses, and discusses the main factors that are causing changes to the availability and use of this most vital of resources. It is certain that the water crisis is a crisis that manifests itself in the everyday lives of billions of people. In different ways and in different places, the nature of the water crisis is a crisis of lost lives and lost livelihoods.

Chapter 18 of Agenda 21 (UN, 1992, p. 275), adopted at the Earth Summit in Rio de Janeiro, defined the overall goal of water policy developments:

> *Water is needed in all aspects of life. The general objective is to make certain that adequate supplies of water of good quality are maintained for the entire population of this planet, while preserving the hydrological, biological and chemical functions of ecosystems, adapting human activities within the capacity limits of nature and combating vectors of water-related diseases.*

The task for water policy-makers thus becomes a part of the wider challenge of achieving sustainable development. We must keep a focus on the first principle of the Rio Declaration:

> *Human beings are at the centre of concerns for sustainable development. They are entitled to a healthy and productive life in harmony with nature.*

It is clear that water is integral to sustainable development, and is related in some way to each of the five theme areas elaborated at the World Summit on Sustainable Development (WSSD) in Johannesburg, August/September 2002. These include water and sanitation, energy, health, agriculture and biodiversity. Furthermore, as discussed below, water is relevant to all three strands of development – social, economic and environmental.

Health, hygiene and social development

Water is most obviously related to the issue of social development through its impacts on health. Without safe drinking water, humans – not to mention animals and plant life – cannot survive. Water-related diseases are amongst the most common causes of illness and death, and the majority of people affected by them live in developing countries. Good sanitation facilities and hygienic practices can significantly reduce diarrhoeal and infectious diseases and prevent worm infections. Water for washing prevents scabies and trachoma. One important aspect of water quality is the avoidance of changes in its chemical composition. Water resource management also has an impact on malarial infection rates by preventing mosquito breeding grounds. Furthermore, since adequate water resources are essential for food production, they have an impact on people's health through the prevention of malnutrition, thus enabling people to more readily recover from illness and lead healthier lives.

Improved sanitation facilities can impact remarkably on people's lives, in terms of safety, privacy, convenience and dignity, especially with regard to the lives of women. In fact, the provision of water schemes often has a greater impact on the lives of women as in most societies the responsibility for domestic water and sanitation is theirs. However, most decisions affecting communities are taken by men. Well-planned water and sanitation schemes have been shown to be a good way of breaking this gender demarcation, allowing women to exercise authority within a community and empowering them to make decisions affecting the community and beyond. There has been a trend in recent years towards local management of water supply schemes and water resources. This is empowering communities to work together for the betterment of their societies.

Water is often an initial starting point for community initiatives, as the essential nature of the issues means they are widely understood. Many communities, once empowered in this way, continue to work together on subsequent initiatives. Sanitation is also a good starting point for addressing long-term poverty issues in a community. Often this can be done by focusing on children as they are the ones to suffer most immediately from ill-health and are primary agents of change, increasing the pace at which necessary behavioural changes are adopted by communities.

Water and poverty reduction

Much of sustainable development is focused on getting people out of poverty. People privileged enough to live in more prosperous parts of the world, along with the better-off in many developing countries, rarely have to confront the consequences of water scarcity. For many of the world's poor however, the story is very different. Inadequate access to water forms a central part of people's poverty, affecting their basic needs, health, food security and basic livelihoods. Improving the access of poor people to water has the potential to make a major contribution towards poverty eradication.

Poverty is no longer seen as a simple lack of income or, at the national level, low per capita Gross National Product (GNP). It is today recognized to be a complex, multifaceted situation that involves both the material and non-material conditions of life. Many international organizations have put forward new approaches to poverty reduction in recent years, which have important implications for the development of all aspects of life, including key areas of natural resource management such as water. These approaches are leading to a rethinking of many water policies and laws, with the emphasis on new institutional and management frameworks that more explicitly target the needs and opportunities of poor people. One of the earliest new approaches springs from the United Nations Development Programme's (UNDP) 'Human Poverty Index' introduced in the 1997 *Human Development Report*, which views poverty in terms of a lack of basic human capabilities. The index consists of five key indicators: literacy, life expectancy, access to safe water, availability of health services and the proportion of underweight children aged five and under. Income poverty is also recognized, with extreme poverty defined as the lack of income needed to satisfy basic food needs, and overall poverty as the lack of income needed to satisfy a range of basic needs including food, shelter, energy and others.

The World Bank initiated a broad electronic debate on the meaning of poverty through their web site, which gave fruit to the *2000 World Development Report*. Key elements of poverty are given, such as the inability to satisfy basic needs, lack of control over resources, lack of education and skills, poor health, malnutrition, lack of shelter and lack of access to water supply and sanitation, vulnerability to shocks and a lack of political freedom and voice. As self-evident as the statement that 'poverty is a situation people want to escape' may seem, it reflects the important and often underestimated point that poverty is dynamic and people move into and out of poverty as the conditions of their lives change. This approach is reflected in the World Bank/

International Monetary Fund (IMF) Poverty Reduction Strategy Papers (PRSPs), which call for multidimensional assessments of poverty that reflect specific local conditions. Other international financial institutions, such as the Asian Development Bank, have also developed new poverty-based policies in recent years that are guiding major changes in their approach to the assistance provided to developing countries.

The Organization for Economic Cooperation and Development (OECD) Development Assistance Committee (DAC) has produced Poverty Guidelines (2001) that recognize the need for a sharper and more explicit focus on poverty reduction. In these, 'poverty, gender and environment are mutually reinforcing, complementary and cross-cutting facets of sustainable development', so that any poverty reduction strategy must focus on gender and environmental issues. Poverty itself is defined as being rooted in the lack of economic, human, political, socio-cultural and protective capabilities.

In a joint contribution to the WSSD preparatory process on linking poverty and environmental management, the government of the United Kingdom, the European Commission, UNDP and the World Bank also emphasized the material and non-material aspects of poverty including lack of income and material means, poor access to services, poor physical security and the lack of empowerment to engage in political processes and decisions that affect one's life. They focused on livelihoods, health and vulnerability as three key dimensions of poverty reduction.

The 'livelihoods approach', a complex and dynamic model, has been developed by UNDP and others (Carney, 1998; Rennie and Singh, 1996). The core of this approach is that poverty reflects poor access to livelihood assets (natural, social, human, financial and physical capital in the Department for International Development [DFID] model) and vulnerability to external shocks and trends in society, the economy and the environment such as market price movements, natural disasters and political change. All of these new approaches are based on a far more explicit poverty reduction agenda than has been evident in the past. They reflect the international consensus that other needs and priorities, including environmental protection and peace and stability, are unlikely to be realized in a world in which the poverty of so many is found alongside the affluence of so few.

One of the main characteristics of poverty is now seen as vulnerability: the extent to which people are vulnerable to the harmful impacts of factors that disrupt their lives and which are beyond their immediate control. This includes both shocks (sudden changes such as natural disasters, war or collapsing market prices) and trends (for example, gradual environmental degradation, oppressive political systems or deteriorating terms of trade). Many such vulnerabilities related to water resources (for example, health threats, droughts or floods, cyclones and pollution). The need to integrate vulnerability reduction into water policies (and in particular the links between water policies, disaster mitigation and climate change) is being increasingly considered. Recognition of vulnerability as a key issue is also expressed in the growing interest in impact assessment as a way of identifying vulnerable individuals or communities who may carry a disproportionate negative burden resulting from development, including water resource development. Early determination of possible environmental, social and health impacts of water resources development provides ample opportunities for environmental management plans, health promotion and protection and social safeguards to be implemented to optimum effect.

The idea of Integrated Water Resources Management (IWRM) is widely accepted as the starting point for water policies, but along with it there is the increasing recognition of the need to adapt IWRM to the specific needs of the poor. There is also a need to ensure that developing integration is not done at the expense of meeting such pressing needs that may arise in one particular aspect of water management such as drinking water supply, improved irrigation or protection of threatened ecosystem functions. Effective and immediate actions to meet these needs are important if water is to be prioritized over other areas of policy.

Partnerships between different stakeholders at all levels (international, national and local) are stressed in most new policy approaches, with the recognition that solutions to water problems cannot be achieved by one organization or even one segment of society. In particular, the inclusion of civil society organizations and of local community groups is emphasized in most new approaches and policies. Indeed, changing institutional mandates is central to new water policies and laws around the world.

International targets and the Millennium Development Goals

The integral role of water in international development has been recognized over the last two decades, with several international agreements specifying targets on water supply and sanitation dating back to the United Nations Children's Fund's (UNICEF) 1980 International Water Supply and Sanitation Decade (IWSSD), which established the target of universal coverage of a safe water supply and sanitation by 1990. While significant numbers of people gained access to improved drinking water and better sanitation over the decade, the target was not met due to population growth. It was, however, readopted as a target for the year 2000 at the World Summit for Children in 1990. More recently targets have been established by the Water Supply and Sanitation Collaborative Council (WSSCC) as part of the process leading up to the Second World Water Forum in The Hague in March 2000. The targets were presented in the report *Vision 21: A Shared Vision for Hygiene, Sanitation and Water Supply and a Framework for Action* (see box 1.1, WSSCC, 2000).

Box 1.1: *Vision 21* – water supply and sanitation targets

- To reduce by 2015 by one-half the proportion of people without access to hygienic sanitation facilities.
- To reduce by 2015 by one-half the proportion of people without sustainable access to adequate quantities of affordable and safe water, which was also endorsed by the United Nations Millennium Declaration.
- To provide water, sanitation and hygiene for all by 2025.

Source: WSSCC, 2000.

Upon review of the events of the IWSSD, the use of such global targets was criticized as failing to focus on the changes that contribute progressively to health and development, and targets were held to be too simplistic, dividing the world into those who 'have' and those who 'have not'. The *Vision 21* report stressed the indicative nature of these targets and the need to consider them in a local context. However, it is still the view of many that such targets remain helpful in assessing the magnitude of the task ahead in meeting the water and sanitation needs of the poor.

In this wider context, the United Nations General Assembly Millennium meeting in the year 2000 established a number of Millennium Development Goals that have become the key international development targets of the modern era. Only one directly relates to water (the Millennium Development Goal on environmental sustainability) but improved water management can make a significant contribution to achieving all of the goals. The relationships, both direct and indirect, between the Millennium Development Goals and water are listed in table 1.1. The Millennium Development Goals provide a context within which wider issues linking water, sustainable development and poverty reduction can be understood. Table 1.1 illustrates the importance of thinking about water in relation to a wider context: the ways in which it can contribute to the overall reduction of poverty and the development of people and nations.

Water and economic development

The economic well-being of society has so far exerted the greatest demand on the world's water resources. The major economic role of water lies in its relationship with agriculture. This is certainly true at a national level, where food security issues and national economic performance are related, albeit in a complex way. But it is certain that irrigation and the control of crop timing can equally affect the macro-economics of a country or region. At a local level, agriculture is the mainstay of many rural communities, and the availability of adequate water allows production of food for household nutrition and for sale at local markets. In addition, the availability of irrigation water enables more crops to be grown per year, and the economics involved in the selling of produce, in irrigation and in year-round farming increases employment opportunities, which has direct economic benefits on a local community.

Water is an essential raw material in many industries that have a major influence on economic performance at the national level, but also at local and household levels. Water also plays a large role in power generation in many countries whether through cooling, or directly through hydroelectricity generation. Water transport is also important in many parts of the world, allowing access to markets as well as generating its own economy.

Improved access to water and sanitation plays a huge indirect role in local communities, insofar as the time taken for these basic tasks is time made unavailable for economic activity. It can take some people hours to collect water, and, in areas without sanitation facilities, seeking privacy for defecation can also be time-consuming. In addition, illness as a result of a water-related disease, not to mention the expense of medication or looking after someone who is ill, prevents many people from carrying out economically active work. The time, energy and resources saved by improved water and sanitation can very often be used on productive economic activity.

Many poor people in urban areas buy their water from private vendors, often at a rate well in excess of piped water supply. This means that a significant proportion of household expenditure is spent on water. Reduced water prices would have a major impact on the economic status of such people, and, with money being available for other things, may effect economic growth.

Biodiversity, environmental sustainability and regeneration

Water is an essential part of any ecosystem, in terms of both its quantity and quality. Reducing the availability of water for the natural environment will have devastating effects, as will the pollution from domestic, industrial and agricultural wastewaters. Just as the environment is integrally tied up with the social, health and economic impacts of water use, ensuring environmental sustainability and regeneration will also have positive effects on these areas.

Damage to the environment is causing a greater number of natural disasters. Flooding occurs in areas where deforestation and soil erosion prevent the attenuation of flood waters. Climate change, which, it is suggested, is fueled both by emissions and by degradation of the world's natural environment, is blamed for the increasing number of floods and droughts. The environment is also a source of many resources – food (agriculture, fisheries and livestock) and raw materials from forests.

Table 1.1: Water, poverty and the Millennium Development Goals

	Millennium Goals	How water management contributes to achieving goals	
		Directly contributes	**Indirectly contributes**
Poverty:	to halve by 2015 the proportion of the world's people whose income is less than $1/day	• Water as a factor of production in agriculture, industry and other types of economic activity • Investments in water infrastructure and services act as a catalyst for local and regional development	• Reduced vulnerability to water-related hazards reduces risks in investments and production • Reduced ecosystems degradation boosts local-level sustainable development • Improved health from better quality water increases productive capacities
Hunger:	to halve by 2015 the proportion of the world's people who suffer from hunger	• Water as a direct input into irrigation, including supplementary irrigation, for expanded grain production • Reliable water for subsistence agriculture, home gardens, livestock, tree crops • Sustainable production of fish, tree crops and other foods gathered in common property resources	• Ensure ecosystems integrity to maintain water flows to food production • Reduced urban hunger by cheaper food grains from more reliable water supplies
Universal primary education:	to ensure that, by 2015, children everywhere will be able to complete a full course of primary schooling		• Improved school attendance from improved health and reduced water-carrying burdens, especially for girls
Gender equality:	progress towards gender equality and the empowerment of women should be demonstrated by ensuring that girls and boys have equal access to primary and secondary education		• Community-based organizations for water management improve social capital of women • Reduced time and health burdens from improved water services lead to more balanced gender roles
Child mortality:	to reduce by two thirds, between 1990 and 2015, the death rate for children under the age of five years	• Improved quantities and quality of domestic water and sanitation reduce main morbidity and mortality factor for young children	• Improved nutrition and food security reduces susceptibility to diseases
Maternal mortality:	to reduce by three quarters, between 1990 and 2015, the rate of maternal mortality	• Improved health and reduced labour burdens from water portage reduce mortality risks	• Improved health and nutrition reduce susceptibility to anaemia and other conditions that affect maternal mortality
Major diseases:	to halve, by 2015, halt and begin to reverse the spread of HIV/AIDS, the scourge of malaria, the scourge of other major diseases that affect humanity	• Better water management reduces mosquito habitats and malaria incidence • Reduced incidence of range of diseases where poor water management is a vector	• Improved health and nutrition reduce susceptibility to HIV/AIDS and other major diseases
Environmental sustainability:	to stop the unsustainable exploitation of natural resources and to halve, by 2015, the proportion of people who are unable to reach or to afford safe drinking water	• Improved water management, including pollution control and sustainable levels of abstraction, are key factors in maintaining ecosystems integrity • Actions to ensure access to adequate and safe water for poor and poorly serviced communities	• Development of integrated management within river basins creates conditions where sustainable ecosystems management is possible and upstream-downstream impacts are mitigated

This table shows that improving water management can make a significant contribution to achieving all of the Millennium Development Goals established by the UN General Assembly Millennium meeting in 2000.

Source: Soussan, 2002.

While the environment does have the capacity to cope with certain kinds of pollution, where exceeded, pollution can result in watercourse contamination beyond use, which can in turn result in costly treatment procedures as well as the loss of people's livelihoods (e.g. loss of fish stocks and other aquatic life).

Another aspect of environmental sustainability is its 'unfair' effect on the poor. It is the poor people who often have to live in 'undesirable', marginal areas, more at risk from floods, etc. Furthermore, the poor often live in closer relationship with the environment, and do not have an alternative open to them as do the wealthy, when, for example, a fish stock is depleted. The world's poor suffer disproportionately as a result. These aspects and more are further explored throughout this chapter.

Water Resources in Crisis

Water resources can only be understood within the context of the dynamics of the water cycle. These resources are renewable (except for some groundwater), but only within clear limits, as in most cases water flows through catchments that are more or less self-contained. Water resources are also variable, over both space and time, with huge differences in availability in different parts of the world and wide variations in seasonal and annual precipitation in many places. This variability of water availability is one of the most essential characteristics of water resource management. Most efforts are intended to overcome the variability and to reduce the unpredictability of water resource flows.

Both the availability and use of water are changing. The reasons for concern over the world's water resources can be summarized within three key areas: water scarcity, water quality and water-related disasters. Each is discussed briefly here and expanded on throughout this report.

Water scarcity

The precipitation that falls on land surfaces is the predominant source of water required for human consumption, agriculture and food production, industrial waste disposal processes and for support of natural and semi-natural ecosystems. The fate of this water is either to be 'taken up' by plants and the soil and then eventually returned to the atmosphere by evapotranspiration, or to drain from the land into the sea via rivers, lakes and wetlands. Our primary source of water is runoff diverted by humans for use in irrigated agriculture, in industry and in homes (rural and urban); for consumption of various kinds; and for waste disposal. It is the water of evapotranspiration that mainly supports forests, rainfed cultivated and grazing land, and a variety of ecosystems. Despite a withdrawal of only 8 percent of total annual renewable freshwater resources, it has been estimated that 26 percent of annual evapotranspiration and 54 percent of accessible runoff is now appropriated by humans (Shiklomanov, 1997). As the per capita use increases due to changes in lifestyle (leisure and domestic practices) and as population increases, the proportion of appropriated water is increasing. This, coupled with spatial and temporal variations in water availability, means that the water to produce food for human consumption, industrial processes and all the other uses described above is becoming scarce.

It has been estimated that today more than 2 billion people are affected by water shortages in over forty countries: 1.1 billion do not have sufficient drinking water and 2.4 billion have no provision for sanitation (WHO/UNICEF, 2000). The outcome can mean increases in disease, poorer food security, conflicts between different users and limitations on many livelihood and productive activities. Current predictions are that by 2050 at least one in four people is likely to live in countries affected by chronic or recurring shortages of freshwater (Gardner-Outlaw and Engelman, 1997). At present many developing countries have difficulty in supplying the minimum annual per capita water requirement of 1,700 cubic metres (m^3) of drinking water necessary for active and healthy life for their people (see map 1.1). The situation is particularly grave in many of the cities of the developing world. This is worrying given predictions of a 60 percent world urban population by 2020. At present, half the population of developing countries live in water poverty.

Flows of water are also essential to the viability of all ecosystems. Unsustainable levels of extraction of water for other uses diminish the total available to maintain ecosystems integrity. As land is cleared and water demand grows for agriculture and other human uses at the expense of natural ecosystems, the appropriation of evapotranspiration moisture by humans looks set to continue. This will inevitably lead to the further disturbance and degradation of 'natural' systems and will have profound impacts upon the future availability of water resources. Actions to ensure that the needs of the environment are taken into account as a central part of water management are critical if present trends are to be reversed.

This situation is aggravated by the fact that many water resources are shared by two or more countries. Currently there are 263 river basins that are shared by two or more nations and that are home for roughly 40 percent of the global population. In the majority of cases, the institutional arrangements needed to regulate equity of resource use are weak or missing.

Water quality

Even where there is enough water to meet current needs, many rivers, lakes and groundwater resources are becoming increasingly polluted. The most frequent sources of pollution are human waste (with 2 million tons a day disposed of in

Map 1.1: Internal renewable water resources generated within a country on a per capita basis, circa 1995

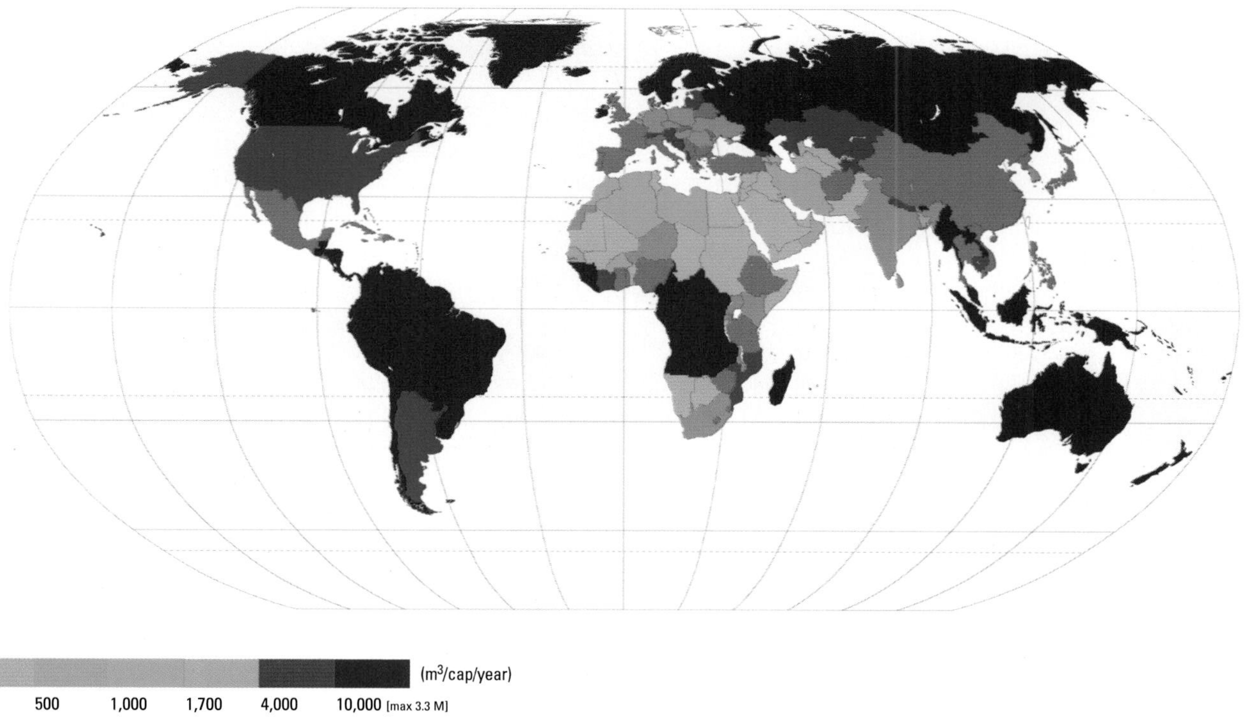

(m³/cap/year)

<500 500 1,000 1,700 4,000 10,000 (max 3.3 M)

This map shows the per capita total internal renewable water availability by country, i.e. the fraction of the country's water resources generated within the country.

Source: Map prepared for the World Water Assessment Programme (WWAP), by the Centre for Environmental Research, University of Kassel, based on Water Gap Version 2.1 D, 2002.

watercourses), industrial wastes and chemicals, and agricultural pesticides and fertilizers. It has been estimated that half of the population of the developing world is exposed to polluted sources of water that increase disease incidence. Key forms of pollution include faecal coliforms, industrial organic substances, acidifying substances from mining aquifers and atmospheric emissions, heavy metals from industry, ammonia, nitrate and phosphate pollution from agriculture, pesticide residues (again from agriculture), sediments from human-induced erosion to rivers, lakes and reservoirs and salinization. The situation is particularly bad in developing countries where institutional and structural arrangements for the treatment of municipal, industrial and agricultural waste are poor. Levels of suspended solids in rivers in Asia have risen by a factor of four over the last three decades. Asian rivers also have a biological oxygen demand (BOD) some 1.4 times the global average, as well as three times as many bacteria from human waste as the global average. They also include rivers that have twenty times more lead than that in surface waters of OECD countries. A report on the state of India's rivers concluded that:

> *India's rivers, especially the smaller ones, have all turned into toxic streams. And even the big ones like the Ganga are far from pure. The assault on India's rivers – from population growth, agricultural modernization, urbanization and industrialization – is enormous and growing by the day.... Most Indian cities get a large part of their drinking water from rivers. This entire life stands threatened (CSE, 1999, p. 58).*

Such a statement holds true for many other rivers in Asia and around the world.

Such degradation of water resources is both a national and international problem, exacerbated by the failure of national and regional institutions to protect downstream users from upstream polluters.

The poor, many of whose livelihood systems depend directly or indirectly on water resources, feel the impacts of such pollution disproportionally. In many countries fishing is a key livelihood activity of the poor, and even where this is not the case, fish frequently provide the bulk of animal protein in the diet.

The destruction of fish and their habitats through pollution can have devastating impacts upon these poor communities.

Water-related disasters

Between 1991 and 2000 over 665,000 people died in 2,557 natural disasters, of which 90 percent were water-related events. The vast majority of victims (97 percent) were from developing countries (IFRC, 2001). Growing concentrations of people and increased infrastructure in vulnerable areas such as coasts and floodplains and on marginal lands mean that more people are at risk (Abramovitz, 2001). While poor countries are more vulnerable, in every country it is the very poor, the elderly, and women and children who are especially hard hit during and after disasters. After such events national statistics of infrastructural damage and loss of life are available but rarely is it possible to determine the effect on the livelihood systems of the population. The 'failure to focus on [the impact of] disasters on livelihoods indicates and perpetuates misplaced aid priorities' (IFRC, 2001, p. 35). Asia has fared particularly badly, with roughly 40 percent of all disasters taking place on the continent. Each event frequently leaves thousands of communities more vulnerable to the next disaster, with both individual and state barely able to recover from one disaster before the next catastrophe strikes. Worldwide, floods were the most reported disaster event; the year 2000 saw 153 flood events alone, some of the worst taking place in Mozambique and along the length of the Mekong River (South-East Asia), while in terms of loss of life, droughts claimed the greatest number of victims. Such events can and must drive policy changes. For example, in Bangladesh, the 1988 flood and 1991 cyclone brought with them a determination from disaster-related institutions that they would not meet a similar threat unprepared again. As a result, Bangladesh initiated changes that, while painful to develop, have now provided it with robust cyclone and flood preparedness and management strategies, all of which have been severely tested since. Consequently, the significance of disasters as a driver of water resource management should not be underestimated. What is important is thus not just the specific impact of disasters, but the way in which they interact with other aspects of water management, and the ways in which vulnerable people adjust their resource management to take account of the risks.

Changes Affecting Water

It is important to set the issues surrounding water in a global context. The world is changing at an ever increasing rate. Many of these changes are having an impact on how we, as humans, utilize the world's water. This section recaps and describes a number of the changes that have taken and are taking place, and the effect these are having on the water situation. Of course, none of these issues are isolated and most are interdependent. Nevertheless, focusing on the reasons why the world is now facing a water crisis is helpful in understanding what factors will worsen the crisis, and where developments outside of the water sector can be brought to bear to improve the situation.

Geopolitical changes

The last half-century has seen major changes in the political make-up of many countries. While this provides a backdrop for many of the following issues, a number of specific issues do arise from the changing political scene. Many countries that were once colonies have gained independence, and have assumed the ability for, and responsibilities of, self-governance. The rise of communism after the Second World War, and the Cold War that followed, impacted on how water resources were managed. The command economies that focused on agriculture resulted in the construction of many large irrigation schemes, some with severe environmental implications (e.g. the Aral Sea in central Asia, which has been desiccated due to intensification of irrigation in the region). The fall of communism and the rise of democracy across the world, both in previous communist states and military dictatorships, has changed the way water resources are managed. This has allowed a greater public awareness of water issues, and enabled local groups to take care of their local water resources. However, many of these new democracies are having to deal with inherited, old ways of working and past environmental impacts. The changing economic structure in many countries has resulted in there being less money available for investment in water management.

Population growth

Rapid growth of the world's population has been one of the most visible and dramatic changes to the world over the last hundred years. Population growth has huge implications for all aspects of resource use, including water. Although water is a renewable resource, it is only renewable within limits; the extent to which increasing demands can be met is finite. As population increases, freshwater demand increases (see figure 1.1) and supplies per person inevitably decline.

Per capita water supplies decreased by a third between 1970 and 1990 and there is little doubt that population growth has been and will continue to be one of the main drivers of changes to patterns of water resource use. Future projections of worldwide population growth have been revised downward in recent years, primarily as a result of significant declines in birth rates. Although there are differences of opinion, most projections expect this slow-down of growth rates to continue and for the world's population to stabilize at about 9.3 billion people (still over 50 percent higher than the 2001 population of 6.1 billion) somewhere in the middle of the twenty-first century (UNFPA, 2002).

Figure 1.1: World population and freshwater use

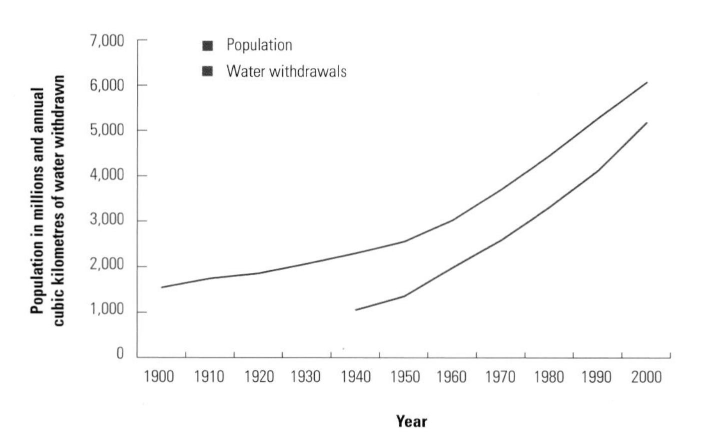

There is a direct correlation between population growth and the increase in freshwater consumption.

Source: Gardner-Outlaw and Engelman, 1997.

Figure 1.2: World population prospects

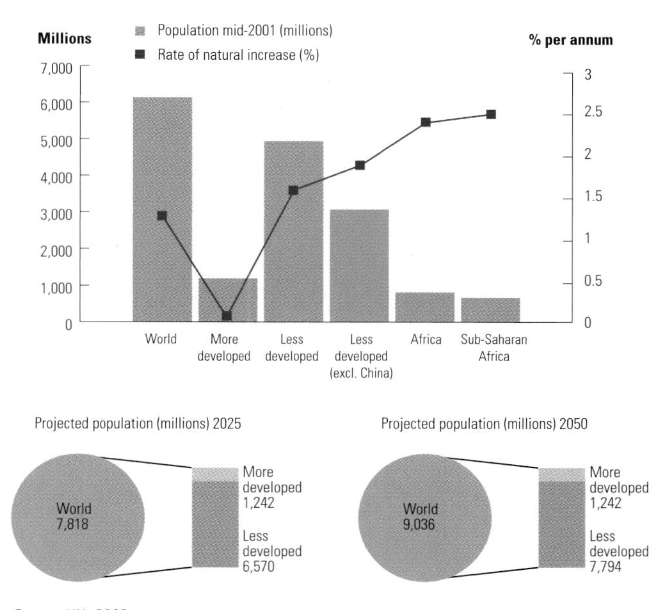

Source: UN, 2002.

Although population growth rates are reducing and global population will eventually stabilize, the increase in numbers of people will still be a major driver of water resource management for at least another fifty years. A number of scenarios have been developed based on the most recent UN population projections (see figure 1.2). Based on these projections, the future for many parts of the world looks bleak. The most alarming projection suggests that nearly 7 billion people in sixty countries will live water-scarce lives by 2050. Even under the lowest projection, just under 2 billion people in forty-eight countries will struggle against water scarcity in 2050 (Gardner-Outlaw and Engelman, 1997).

Agricultural demand

Population growth not only leads to greater demand for water for domestic supply but also impacts on the majority of other uses for water. The demand for food increases with population, and hence does the water required for agricultural production. The area of irrigated land more than doubled in the twentieth century.

In some parts of the world, such as South Asia, most of the recent expansion in irrigated areas has been in wells through private investments to exploit groundwater. This has allowed countries such as Bangladesh and many states in India to expand dry-season agriculture. There are worrying signs in some areas that groundwater resources are being over-exploited, with groundwater levels falling, which could develop into a crisis as the situation is not sustainable and threatens the food security, not to mention the water supply, of millions of people.

However much irrigation expands, most of the world's farmlands will continue to be watered through rainfall in the future. Water management is particularly at issue for rainfed areas where the rainfall is limited or erratic (and consequently a constraint upon production) and the communities do not have the means to supplement it by storage, groundwater use or other sources.

Energy requirements

In addition to rising agricultural demand, increasing population also necessitates increased energy demand. The most obvious use of water for energy production is through the operation of hydroelectricity facilities. The storage required may have serious health implications for the surrounding human population in terms of the incidence of water-related diseases such as malaria, dengue fever and bilharzia. Interestingly, evaporation from the surface of reservoirs represents the greatest consumptive use of water from hydroelectricity generation.

However, hydropower has formed the basis for major national and regional development with large benefits for all, including the poor, in many countries. While industrialized nations have tapped most of their economically feasible hydropower (about 70 percent for Europe and North America), developing countries as a whole have vast untapped hydro resources (with only about 15 percent currently developed overall). These opportunities for hydro-based development are particularly important for a group of the world's poorest nations (such as Nepal, Lesotho, Lao and Tajikistan) for whom water and terrain represent their greatest natural resources.

Additionally, modern hydropower plants increasingly provide major benefits to local populations. Hydro is also, relative to other forms of major energy production, environmentally benign. As water scarcity increases these facilities will increasingly become the focus of attention.

Substantial amounts of water are used for cooling and in chemical processes. The majority is returned to the watershed, with relatively little loss to contamination or evaporation, though the change of temperature can have important ecological consequences, which are discussed in chapter 10.

Urbanization

In addition to general population growth, the changing demographics are affecting how water resources are managed. At the beginning of the twentieth century, only a small percentage of the population lived in cities in most regions of the world, but as the world population has increased, so has the proportion that live in urban areas. The urban population rose greatly throughout the twentieth century and is projected to reach 58 percent of the world population by 2025 (UNFPA, 2002). In the next thirty years, the greatest urban growth will occur in Asia (see figure 1.3). UN (2002) estimates show that in real terms the urban population of the less developed world is expected to nearly double in size between 2000 and 2030 from a little under 2 billion to nearly 4 billion people (see figure 1.4). Between 2015 and 2020, urban population will exceed rural for the first time, and will continue to escalate sharply while rural numbers remain more or less static.

As the population in these centres grow, so do their demands for resources; reflecting both the high concentrations of people and the very different lifestyles and aspirations of city dwellers. Among the consequences of this urban influx are the overloading of water supply and sanitation infrastructure – a situation made worse by the geographical location of some of these cities. The problems of water supply have forced many urban authorities to over-exploit fragile sources, such as aquifers, and there are many examples of falling water levels in many cities (e.g. in Manila in the Philippines). Deterioration of water supplies and sanitation leads to a progressive decline in urban living conditions – water shortages, pollution and unsanitary water conditions all of which contribute to an urban water and health crisis. Many poor people in cities also pay very high prices for their water from private vendors, and agitation and even riots over poor water supplies (especially during droughts) are far from unknown. Inadequate coverage and decline in urban infrastructure hits the poorest hardest as wealthier households tend to have access to urban water supplies or can afford tubewells if the supply is unreliable or of poor quality.

Figure 1.3: Actual and projected urban population in different regions of the world in 1950, 2000 and 2030

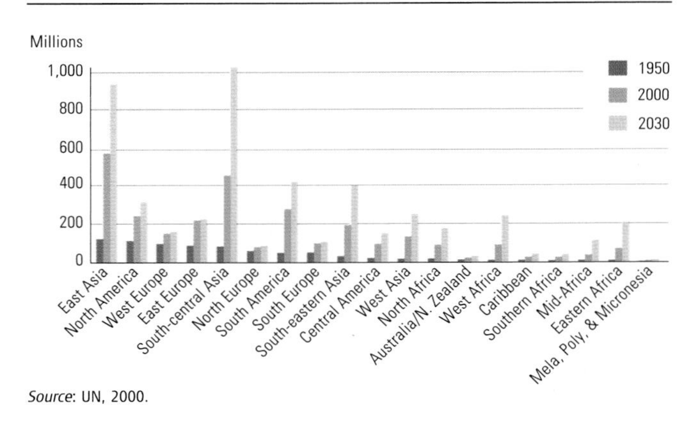

Source: UN, 2000.

Figure 1.4: Urban and rural population, less developed countries, 1950–2030

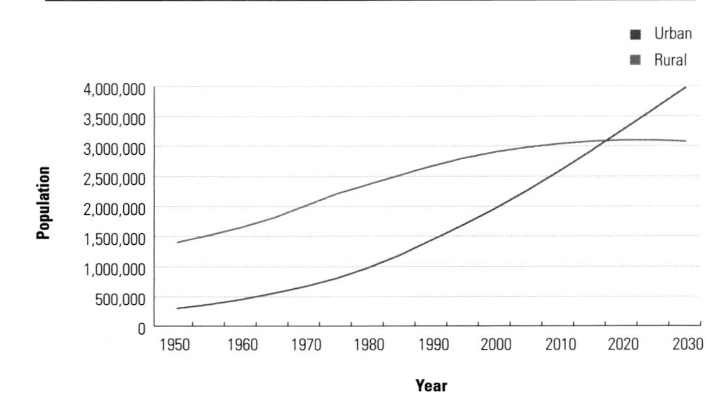

The urban population rose greatly throughout the twentieth century and is projected to reach 58 percent of the world population by 2025. Meanwhile, the rural population is expected to stabilize from 2010.

Source: UN, 2000.

Economic growth and industry

The twentieth century saw unprecedented economic growth. Much of this growth, which has provided the wealth enjoyed by so many people in the Western world, was dependant on water consumption, as industries (and their demand for water) have been growing at a very fast rate.

Besides the pressure exerted on water resources by increasing demand, industrialization poses a great threat to water quality (table 1.2). This is often centred around major urban centres that serve as foci for industrial development. Like urban population growth, many of these industries are growing at a rapid rate. For example, the paper and steel industries, which rank as some of the most important industrial sources of water pollution in Latin America, have been growing twice as fast as the economy of these countries as a whole (Gleick, 1993).

Industrial wastewater, like municipal sewage, often contains suspended solids that silt up waterways, suffocate bottom dwelling organisms and impede fish spawning. Wastes such as organic material use up oxygen, limiting its availability for other aquatic organisms, while others pose a direct threat to human health. Cadmium, lead and mercury are particularly dangerous because they can interfere with hormones and reproduction. Copper and zinc are less dangerous to humans but are toxic to aquatic life (Stauffer, 1998; Gleick, 1993).

The threat of pollution of water resources comes not only from the regular operation of the industries but also from the risk of accidents. For example, the disastrous fire in the Schweizerhalle (Switzerland) pesticide chemical plant in 1986 led to a serious pollution of the Rhine, which, for several days, caused fishing activity and drinking water supplies to be stopped even 1,000 kilometres (km) downstream, in the Netherlands. In the United Kingdom in 1988, the accidental dumping of 20 tons of concentrated aluminium sulphate solution into the water treatment plant at Camelford, Cornwall, resulted in a drop in the pH level from 7–8 to 3.5–4.2 in the Camel River, resulting in the death of most of the fish in the river. Consumers were exposed for three days to drinking water of pH levels 3.9 to 5.0, raising fears over health impacts among the local community.

Globalization

Globalization is related to the economic growth experienced in recent years, and is a growing feature of almost every aspect of our world. We are frequently reminded that we live in an increasingly interconnected world. Many global brands advertise that new lifestyles are changing demands and aspirations around the world. Changes to production technologies and transport opportunities have created an increasingly international market. Policy decisions on one side of the world can affect people on the other.

Many developing countries have to deal with the more hazardous industries, such as those producing dyes, asbestos and pesticides. For example in Bangladesh, Sri Lanka and Pakistan, textile manufacturers and tanneries are located in peri-urban areas where they place tremendous pressures on local water resources, both through the demand required for production and pollution from waste disposal. Chemical and pharmaceutical production around Indian cities, such as New Delhi and Ahmedabad, are leading to pollution so severe that it is contaminating groundwater aquifers. It is not just industrial production that is responding to the globalized economy, but also the agricultural sector. Many areas within easy reach of airports in cities such as Nairobi, Kenya, produce large quantities of vegetables, fruit and flowers (often in greenhouses) that are air-freighted to markets in Europe, Japan and North America, placing great demands on limited water resources. Clearly, under such conditions, special measures need to be taken. Understanding the full impacts of globalization on the world's water resources is challenging, but what is clear is that the world is irrevocably changing, with tremendous implications (good and bad) for the management of water resources, as for every other aspect of life.

Table 1.2: Water pollutants by industrial sector

Sector	Pollutant
Iron and steel	Organic residues, oil, metals, acids, phenols and cyanide
Textiles and leather	Organic residues, suspended solids, sulphates and chromium
Pulp and paper	Organic residues, solids, chlorinated organic compounds
Petrochemicals and refineries	Organic residues, mineral oils, phenols and chromium
Chemicals	Organic chemicals, heavy metals, suspended solids and cyanide
Non-ferrous metals	Fluorine, suspended solids
Micro-electronics	Organic residues, organic chemicals
Mining	Suspended solids, metals, heavy metals, acids, salts

Industries and their demand for water have grown at a very fast rate from the twentieth century up to the present, posing a great threat to water quality.

Source: Stauffer, 1998.

Technological changes

The last century and in particular the last few decades have witnessed an acceleration of major and highly significant technological changes, many of which have had direct impact on water resources and their management. However, the application of these advances has not been uniform and thus the benefits have been biased towards the more prosperous nations. A few illustrative examples are given here.

Surveillance and monitoring of the resource itself have been greatly affected by advances in remote sensing, giving us much better appreciation of spatial and temporal variation in many aspects of the resource. New instrumentation allows more precise, more efficient and more effective monitoring of precipitation, energy balances, river flows and water quality. Advances in data and information transmission, coupled with great increases in the ability to store information have allowed information systems to expand exponentially. Data and information can be shared more easily, increasing our knowledge base, which can be made more readily available to all. Systems of analysis allowing better diagnosis of problems and prediction and forecasting of future scenarios are being developed apace.

Exploitation and use of the resource is being made more effective by technological changes such as more efficient ways of boring wells and extracting groundwater, better systems of water transmission, from piped systems to tanker transport, and better systems for producing freshwater through desalination techniques.

Management of demand is also benefiting from technological advances. More efficient systems of irrigation – drip irrigation instead of spray irrigation – more efficient toilets and shower-heads, recycling techniques and new wastewater technologies are allowing water to be conserved more effectively.

However, while of potential benefit to all humankind, these techniques are being preferentially applied within the more affluent countries and sections of societies. Cultural backgrounds and political situations as well as ability to pay affect the rate at which new technologies are applied. The applications and thus the benefits are not even and it is the aspiration of many organizations to help redress this imbalance.

Lifestyle

Because water is so integral to many of life's needs and behaviours, increasing prosperity heightens pressures on all resources, including water resources. Water is often viewed as a right, and many, particularly in the Western world, believe that they can use it in unlimited quantities. For example, we expect to be able to have food crops all year round, requiring irrigation. The water required for the production of many consumable goods is significant. Acquiring a refrigerator or a television requires electricity, and more electricity places demands on water. While this issue is often difficult to quantify due to its diverse nature, its impact on the changing face of the global water resources should not be underestimated.

Recreation and tourism

One such change in lifestyle that deserves special mention is the explosive increase in tourism in the last three decades. During the 1970s, only one person in thirteen from industrial countries had travelled to a developing country as an international tourist. By the end of the 1990s it was one in five (Honey, 1999). Cuba has seen a fivefold increase in tourists since 1990 (Figueras, 2001 quoted in Mastny, 2002a). For Saint Lucia, and Antigua and Barbuda, tourism receipts now account for nearly 50 percent of the GDP; for the Maldives it is nearly 90 percent. Tourism is the only sector in which developing countries consistently run a trade surplus.

This boom in tourism has multiple impacts. There are undoubtedly economic benefits at a national level due to the increased revenue available, but the development also requires the use of disproportionate shares of local natural resources, of which water is often the most crucial. Much of this water, when used, is disposed of without adequate treatment in ways that impact irrevocably on the surrounding water resources and their ecosystems. In a 1994 study for the Caribbean Tourist Organization, it was disclosed that 80 to 90 percent of sewage from hotels and associated facilities was released in coastal waters and was thought to have adverse effects on coral reefs and mangrove swamps. Hotels and their guests consume vast quantities of water. In Israel, water use by hotels along the River Jordan is thought to be contributing to the drying up of the Dead Sea where the water level has dropped 16.4 metres since 1977 (Gertman and Hecht, 2002). Golf tourism has an enormous impact on water withdrawals – an eighteen-hole golf course can consume more than 2.3 million litres a day. In the Philippines, water use for tourism threatens paddy cultivation. Tourists in Grenada generally use seven times more water than local people and this discrepancy is common in many developing tourist areas (Mastny, 2002b). Tourism is, however, vital to the economic well-being and the reduction of poverty in many developing countries. Since natural resources are a powerful part of the attraction of this industry, it provides added incentive for resource preservation. In many cases though, tourism leaves an undeniable ecological footprint. Countries that depend on tourism are making major efforts to simultaneously maintain their tourism industries and reduce the environmental impact (including water use) of the industry.

Recreation is a major use of and a major issue in the planning of water resources in all parts of the world. The use of lakes and reservoirs for sailing, fishing and water skiing is an important consideration even in the prosperous countries of Europe and North America. It can add significant economic benefits to these resources, but also has implications for water quality in ecological terms.

Climate change

The previous discussion has identified water scarcity, water quality and water-related disasters as the main challenges facing the modern world. As if these were not bad enough, the news is that things are likely to get worse. Global circulation models of the atmosphere suggest that the increased carbon dioxide and other greenhouses gases are likely to cause changes to the global climate. It is generally agreed that more precipitation can be expected from 30° North and 30° South because of increased evapotranspiration. In contrast, many tropical and subtropical regions are expected to receive lower and more erratic precipitation in the future. Indeed, the 2001 Intergovernmental Panel on Climate Change report suggests that this may already be happening, and that 'natural systems are vulnerable to climate change, and some will be irreversibly damaged' (McCarthy et al., 2001, p. 4). The effect of climate change on streamflow and groundwater recharge varies regionally, but generally follows projected changes in precipitation. Impacts on ecosystems and on the availability of water resources for ecological and human needs will follow these trends.

That there will be climate change in the future is no longer in doubt. What is far from clear is exactly what changes will take place, and the pace of change. All projections of future trends at regional and subregional levels carry strong warnings with respect to the uncertainties that lie ahead. Despite this, there is a general consensus that the many parts of the world already experiencing water stresses (and likely to experience greater stress in the future even if rainfall patterns do not change) are the very ones where rainfall will be lower and more variable as climate change really takes hold. These include arid and semi-arid regions of the developing world that are already poor and already have great problems in water resource management. The impact of climate change is likely to make all of these problems worse.

Climate change is also likely to lead to increased magnitude and frequency of precipitation-related disasters – floods, droughts, mudslides, typhoons and cyclones. The 1990s saw a string of 'natural' disasters. The 1998 monsoon season in South-East Asia brought with it the worst flood in living memory to Bangladesh, placing some 65 percent of the country underwater, while Hurricane Mitch wreaked devastation across Central America in the same year. In Venezuela in December 1999, following torrential rain that saw two years' worth of rainfall in just two days, 15 million metres of mud, rocks and trees detached themselves from the mountain side and descended into the urban areas below killing 30,000 people. The event also caused approximately US$2 billion in damage (IFRC, 2001). The same year saw the coast of Orissa in India devastated by one of the worst cyclones ever recorded, while early 2000 witnessed devastating floods in Mozambique. These headline-catching events are just the tip of the iceberg: 'The total number of disasters (not just "great" ones) has also been on the rise, with the year 2000 setting a new record' (Abramovitz, 2001, p. 8).

Flows in rivers and streams in many places are likely to decrease at low flow periods as a result of increased evaporation. It is also predicted that climate change will degrade water quality through increased pollutant concentrations and loads from runoff and overflows of waste facilities and due to increased water temperatures. A recent study estimates that climate change actually accounts for about 20 percent of the global increase in water scarcity, the remaining 80 percent accounted for by population growth and economic development. Countries that already suffer from water shortages, such as Pakistan, India, Mexico, northern China and the countries of the Middle East and sub-Saharan Africa will be hardest hit (Vörösmarty et al., 2000).

Interestingly, climate change has focused attention on the use of energy, and the intensity with which various forms of energy generation emit greenhouse gases. Hydropower is able to make a significant contribution to reducing the emission of such gases in energy production. Currently, about 19 percent of the world's electricity is produced from hydropower.

The Development of International Water Policies

The issues discussed in the previous sections have been gradually coming into prominence over the past number of years, and attention has been given to them in the international debates on water policies and management issues that have taken place over the last decade or so, particularly since the Earth Summit in Rio de Janeiro. The history could even go back further, to the Mar del Plata Action Plan of 1977, but perhaps the best starting point is the Dublin Conference of 1992, from which emerged the Dublin Statement on Water and Sustainable Development that was a contribution to the preparation of the Earth Summit in Rio. This statement contains much of merit, including the four Dublin Principles that have become the cornerstone of much debate on international approaches to water policies:

- Freshwater is a finite and vulnerable resource, essential to sustain life, development and the environment.
- Water development and management should be based on a participatory approach, involving users, planners and policy-makers at all levels.
- Women play a central part in the provision, management and safeguarding of water.

- Water has an economic value in all its competing uses and should be recognized as an economic good.

The focus of these principles, and of the action plan, on issues of environment, gender, governance and sustainability are still relevant today. They are taken up in Chapter 18 of Agenda 21, prepared at Rio, which states that:

> *The holistic management of freshwater as a finite and vulnerable resource, and the integration of sectoral water plans and programmes within the framework of national economic and social policy, are of paramount importance for action in the 1990s and beyond.*

This same document approved seven programme areas for action at the national and international levels:

1. Integrated water resources development and management.
2. Water resources assessment.
3. Protection of water resources, water quality and aquatic ecosystems.
4. Drinking water supply and sanitation.
5. Water and sustainable urban development.
6. Water for sustainable food production and rural development.
7. Impacts of climate change on water resources.

Despite the content of Chapter 18, water resources were not a particularly prominent issue at Rio, with issues such as forests, climate change and biodiversity having a far higher profile. The balance has, to a great extent, been redressed since then through the importance given to freshwater issues by the Commission on Sustainable Development (CSD) in its second (1994), sixth (1998) and eighth (2000) sessions and in the 1997 UN General Assembly Special Session. All contained a call for a concerted effort to develop more integrated approaches to water management and for a stronger focus on the needs of poor people and poor nations. Actions to protect ecosystems and to ensure better participation by women, the poor and other marginalized groups in the governance of water were identified as specific priorities. The importance of policies that create an enabling environment, protect the weak and create better governance conditions were particularly recognized.

The CSD 2 meeting in May 1994 set the tone. It emphasized that, if existing trends continued, 35 percent of the world's population would be living in conditions of water scarcity or stress by the year 2025, up from 6 percent in 1990. Problems with water quality, the threat of water-related disasters and water-related health, food security and environmental deterioration were identified as deepening this acute situation:

> *'While in the past there was a tendency to regard water problems as being local or regional in nature, there is a growing recognition that their increasingly widespread occurrence is quickly adding up to a crisis of global importance'* (CSD, 1994, p. 3).

This statement reflects the increasing prominence given to water resource issues in international policy processes on environment and sustainable development in the period following Rio in 1992. The CSD 2 report on freshwater also drew out the strong link between water resources and poverty.

The CSD 6 session was of particular importance in the development of international approaches to water policies, building on the detailed discussions held in an expert group meeting on strategic approaches to freshwater management held in Harare (Zimbabwe) in January 1998. The Harare meeting represented a gathering of key stakeholders in international water policy development and concluded that water stresses were of global significance and that water was a key resource in sustainable development. The need for fundamental changes in the dominant approaches to water management was recognized, with a move away from technical and sectoral approaches and towards integrated approaches in which the social dimension of water management was central. This is reflected in the overall CSD 6 report (CSD, 1998, p. 5):

> *it is important that consideration of equitable and responsible use of water become an integral part in the formulation of strategic approaches to integrated water management at all levels, in particular addressing the problems of people living in poverty.*

A growing trend in international policy debates is emphasized here: that IWRM on its own is not enough; it must be focused on the specific causes of water stress and the needs of those people without water security.

The report of the Harare expert group meeting goes more deeply into the rationale for this policy refocusing. While progress had been made in some areas,

> *overall progress has been neither sufficient nor comprehensive enough to reduce general trends of increasing water shortages, deteriorating water quality and growing stresses on freshwater ecosystems. There is a compelling case for integrating these approaches to freshwater management into national economic frameworks as key elements in policies for sustainable development and poverty alleviation (CSD, 1998, p. 5).*

This statement is important in that it further qualifies the concept of IWRM: integration is not just between different parts of water management, it also relates to how water is integrated into wider processes of sustainable development, environmental management and poverty reduction. The Harare meeting recommended the following issues as keys to policies for water management: sustainability, capacity-building, information management, environment and development, economics and finance, participation and institutions and, finally, international cooperation.

CSD 8 met in the spring of 2000. The overall focus was particularly on issues of rural development and sustainable agriculture. Links between agriculture, land and water were considered within this context, emphasizing that agriculture was the overwhelmingly dominant water user in most parts of the world. Natural limits to water availability and deficiencies in many aspects of water management were highlighted and the need to realize potential increases in the productivity of water use in both rainfed and irrigated agriculture were recognized as key policy priorities. This represents a further move away from supply-side policies (where the key was seen to be to increased water availability) and towards demand management approaches. Again this theme has parallels in other areas and has been further emphasized since.

The year 2000 saw several of these policy trends come together through a number of international events. These included the Millennium Session of the United Nations General Assembly. The UN Millennium Declaration specifically states in the targets set for 2015 (paragraph 19):

> *We resolve further to halve, by the year 2015, the proportion of the world's people whose income is less than one dollar a day and the proportion of people who suffer from hunger and, by the same date, to halve the proportion of people who are unable to reach or to afford safe drinking water.*

This resolution is quoted in full because it demonstrates the link, in one paragraph, between poverty, hunger and water security. This link is significant in policy terms, as it defines, for the global community, the overriding policy priority for water resource management. The principal crisis is one of the governance barriers that prevent the poor from having sustainable access to water resources. This means that the global community must first and foremost ensure that national and international water resources policies prioritize the reduction and eventual eradication of poverty. Through the sustainable use of water resources we can begin to meet basic needs, reduce vulnerabilities, improve access and empower poor people to control the water resources upon which they depend.

These links were demonstrated in table 1.1, which looked at the contribution that improved water management can make to each of the different Millennium Development Goals. Progress in realizing these goals is extremely hard to measure, and separating out the specific contribution of water management to the progress made is difficult, but essential to action. The available indications suggest that there have been some notable achievements in many parts of the world and prospects for achieving many of the goals are good. This is particularly true in South America and parts of Asia. The picture for Africa and, to a lesser extent, South Asia is not as encouraging. If we focus on the world's prospects for achieving the Millennium Development Goal concerned with water supply, a similar picture emerges with the exception that South Asia is on track for achieving the target ahead of schedule. Prospects for Africa and parts of East Asia and the Pacific are again, less optimistic, with some progress made but from a low base, and little hope for achieving the 2015 goal if present trends continue.

There has, in consequence, been active development towards refining the approach to water resources within the CSD and in the UN system in the years since Rio. There have also been parallel developments of great significance, perhaps the most important of which was the preparation of the *World Water Vision*, launched at the World Water Forum in The Hague in March 2000, and the Ministerial Declaration on Water Security in the 21st Century, affirmed by the representatives at the parallel Ministerial Conference in The Hague.

The *World Water Vision*'s subtitle, 'making water everybody's business', sums up the intention and need to build a consensus on the importance of water and the ways forward in the future rather than coming up with radically new approaches. The *Vision* itself was built around a series of scenarios. The 'business as usual' scenario was proved not to be a viable option, as the projection into the future of present trends in water use and resource degradation would rapidly become unsustainable in many parts of the world, greatly deepening emerging crises in these regions. The *Vision* did define a sustainable scenario in which the key needs of all people were met and ecological integrity was maintained. It emphasized that realizing this sustainable future would require 'that people's roles and behaviours must change to achieve sustainable water resources use and development' (Cosgrove and Rijsberman, 2000, p. xiii).

The Hague Ministerial Declaration represented the political response to the *Vision* and the emergence of an international consensus on the importance of water in sustainable development. It identified seven challenges for the global community, challenges that provide the basis for the policy issues discussed below and that make up the thematic chapters of this report.

1. **Meeting basic needs:** recognizing that access to safe and sufficient water and sanitation are basic human needs and are essential to health and well-being, and to empower people, especially women, through a participatory process of water management.

2. **Securing the food supply:** enhancing food security, particularly of the poor and vulnerable, through the more efficient mobilization and use of water and the more equitable allocation of water for food production.
3. **Protecting ecosystems:** ensuring the integrity of ecosystems through sustainable water resources management.
4. **Managing risks:** providing security from floods, droughts, pollution and other water-related hazards.
5. **Sharing water resources:** promoting peaceful cooperation and developing synergies between different uses of water at all levels, whenever possible, within and – in the case of boundary and transboundary water resources – between concerned states, through sustainable river basin management or other appropriate approaches.
6. **Valuing water:** managing water in a way that reflects its economic, social, environmental and cultural values in all uses, with a move towards pricing water services to reflect the cost of their provision. This approach should account for the need for equity and the basic needs of the poor and the vulnerable.
7. **Governing water wisely:** ensuring good governance, so that the involvement of the public and the interests of all stakeholders are included in the management of water resources.

The seven challenges from The Hague represent a major turning point in the development of water policies, but they are not the final word. Indeed, work has continued since The Hague in further defining the key challenges that face water policy-makers, and will continue over the coming years. Work undertaken within the preparation of this report has identified a further four challenges for the future.

8. **Water and cities:** acknowledging that urban areas are increasingly the focus of human settlements and economic activities, and that they present distinctive challenges to water managers.
9. **Water and industry:** focusing on industry needs and the responsibility to respect water quality and take account of the needs of competing sectors.
10. **Water and energy:** recognizing that water is vital for all forms of energy production, and that there is a need to ensure that energy requirements are met in a sustainable manner.
11. **Ensuring the knowledge base:** reflecting that good water policies and management depend upon the quality of knowledge available to decision-makers.

The report does not follow the order of these challenges, instead opting to concentrate on two main issues in the water arena – 'needs, uses and demands', and 'management'. The challenges are reorganized within these two parts. Taken together, the eleven challenges highlight the elements essential to defining a compelling policy agenda. Now it is everybody's business to turn these challenges into specific policies and actions that reflect their differing needs and priorities, and the potential available to them in different places at different times.

There has been active development of these basic principles since the meeting in The Hague in March 2000. The German government hosted an International Conference on Freshwater in December 2001, again with widespread participation and active debates on key water issues. The Ministerial Declaration produced at this meeting stressed the contribution that water management can make to reaching the Millennium Development Goals. Issues of finance, governance, gender and capacity development were emphasized. The Conference as a whole identified five 'keys': actions that were seen as essential to moving the water debate forward. These keys relate to the water security of the poor, decentralization, partnerships, sharing water and governance. The Ministerial Declaration also emphasized the importance of mainstreaming water-related issues in the preparation of the 2002 WSSD in Johannesburg.

This theme was again taken up during the preparations for the Summit. Kofi Annan, the Secretary General of the United Nations, stated his conviction that water should be one of the key issues debated at the Summit. The Netherlands Crown Prince, Willem Alexander, an influential voice in the debate on water, produced a paper called 'No Water, No Future' as his contribution to the WSSD. The Prince's introductory statement underscores the links between water resources and sustainable development:

> *The World Summit on Sustainable Development should reaffirm the importance of achieving water security and adopt targets and actions that will allow us to meet this challenge jointly. In this context, I would even daresay that if nations cannot manage their water resources, Sustainable Development remains a faraway dream.*

The Fourth Prepcom for the WSSD in Bali in May–June 2002 defined the context within which issues such as water management must be considered:

> *Poverty eradication, changing unsustainable patterns of production and consumption and protecting and managing the natural resource base of economic and social development are overarching objectives of, and essential requirements for, sustainable development. (CSD, 2002, p. 1)*

The WSSD reaffirmed the existing Millennium Development Goal on water provision and agreed on a new goal on improved sanitation. This is of great significance, as it is the first time that the key issue of sanitation has been specifically recognized as a focal point for international action. The WSSD Plan of Implementation also reaffirmed the links between water resources, poverty reduction, disaster management and other issues such as health and food security. As such, it reflected the emerging international consensus demonstrated in the Hague and Bonn Declarations and elsewhere.

Similar conclusions are being reached elsewhere. The World Bank has drafted a new Water Resources Sector Strategy (World Bank, 2002) that emphasizes the strong links between water development and poverty reduction, the need for both better management development of water resources in most developing countries, and the importance of the global community supporting these developing nations by sharing the risks of water investment, using both investment and guarantee instruments. The Global Environment Facility (GEF) summarized discussions at the Ministerial Roundtable on Financing Environment and Sustainable Development held in Monterrey and Bali at the preparatory conference for the WSSD (GEF, 2002).

The prominence of water issues at the Monterrey meeting reflects the increasing recognition of the importance of financing investments in water management. Many water investments, and especially large-scale infrastructure, are expensive and risky, with uncertain rates of return and long pay-back periods. Some, such as large dams, are also very controversial and can have high associated social and environmental costs. Others, including many water supply schemes built by governments for low-income communities, are unlikely to ever recover their costs. But the need for such investments is clear if many water-related problems are to be overcome. There are uncertainties over exactly how much money is needed to provide all the world's people with a minimum level of water security, but the costs undoubtedly run into many billions a year. The issue of how to finance these investments is emerging as a major policy issue for the future.

At the regional level, key statements have been made by groups of ministers in both Asia and Africa. A joint statement by ministerial delegations from ten Asian countries (Cambodia, Indonesia, Lao PDR, Malaysia, Myanmar, Nepal, Philippines, Sri Lanka, Thailand and Viet Nam) in May 2002 agreed upon the vital importance of sustainable water management for their countries. The statement recognized water as a basic human need that should be given a special social valuation and called on governments to ensure that all people have affordable access to safe water for basic needs. The statement also recognized the importance of water for food security, the need for participation in water management and the links between good governance and good water management.

The context for discussion of water issues in Africa has been redefined by the development of the New Partnership for Africa's Development (NEPAD), the launching of the African Ministerial Conference on Water and the formation of a consensus on water through meetings of the majority of African countries in Accra and Abuja in April 2002. NEPAD is a vision for a new approach to development in Africa based on partnership between Africa and the rest of the world. It calls for a programme of action to build integrated development that balances social, economic and political issues. It is recognized that water will play a key role in many aspects of this new development trajectory, a conclusion reflected in the Accra Declaration on Water and Sustainable Development. This called for policies, strategies and real commitments to implementation in six key areas, with all undertaken in a manner designed to protect the environment:

- improved access to potable water services and sanitation;
- water use to address food security and income generation;
- IWRM in national and shared water basins;
- water-related disaster prevention, mitigation and management;
- empowerment and capacity-building focuses on improving equity and gender sensitivity; and
- pro-poor water governance and water policies.

These key statements from ministers from Africa and Asia reflect the integration of new approaches based on the social and environmental values of water into mainstream policy development. Links to poverty reduction are particularly notable in both, as are the role that water management plays in disaster mitigation and environmental sustainability.

It is consequently clear that there is a strong momentum in the international community to recognize the importance of water management in the wider processes of poverty reduction and sustainable development. But to do so necessitates changes to policies and laws as well as new management practices. Such changes are happening in many places, though this is a long-term process and conservative forces often resist them. Actions to support future reform through enhanced international cooperation in particular will be a key issue for future water management. There are a number of good examples of reform, some of which are considered later in this book. The future is likely to see a continuation of the types of changes to policy approaches that have emerged in the years since Rio, with in particular a consensus on the need for more integrated approaches, stronger partnerships and a more effective focus on poverty reduction and sustainable development in water policy processes.

References

Abramovitz, J. 2001. ***Unnatural Disasters.*** Washington DC, Worldwatch Paper 158, Worldwatch Institute.

Abu-Zeid, M. and Abdel-Dayem, S. 1989. Egypt programmes and policy options for facing the low Nile flows. Cairo. Proceedings of the International Seminar on Climatic Fluctuations and Water Management, 11 December.

Carney, D., ed. 1998. *Sustainable Rural Livelihoods: What Contribution Can We Make?* London, Department for International Development.

Clarke, R. 1993. *Water: The International Crisis.* London, Earthscan Publications.

Cosgrove, B. and Rijsberman, F.-R. 2000. *World Water Vision: Making Water Everybody's Business.* London, World Water Council, Earthscan Publications Ltd.

CSD (Commission on Sustainable Development) 2nd Session, 1994. p. 3.

——. 6th Session, 1998. p. 5.

——. 10th Session, 2002. p. 1.

CSE (Centre for Science and the Environment). 1999. *The Citizen's Fifth Report* p. 58. Delhi.

Drakakis-Smith, D. 1997. *Third World Cities.* London, Routledge.

Dublin Statement. 1992. Official outcome of the International Conference on Water and the Environment: Development Issues for the 21st Century, 26–31 January 1992, Dublin. Geneva, World Meteorological Organization.

EEC (European Economic Community). 2000. Framework Directive in the Field of Water Policy (Water Framework). Directive 2000/60/EC of the European Parliament and of the Council of 23 October 2000, establishing a framework for Community action in the field of water policy [Official Journal L 327, 22.12.2001].

EIA (Environmental Impact Assessment). 1992. *International Energy Annual 1990.* Washington DC, Department of Energy.

Federal Ministry for the Environment, Nature Conservation and Nuclear Safety, And Federal Ministry for Economic Co-operation and Development. 2001. *Ministerial Declaration, Bonn Keys, and Bonn Recommendation for Action.* Official outcomes of the International Conference on Freshwater, 3–7 December 2001, Bonn.

Gardner-Outlaw, T. and Engelman, R. 1997. *Sustaining Water, Easing Scarcity: A Second Update.* Washington DC, Population Action International.

Gertman, I. and Hecht, A. 2002. The Dead Sea Hydrography from 1992 to 2000. Journal of Marine Systems, vol. 39/3–4, pp. 169–81.

Gleick, P.-H. 1993. *Water in Crisis: A Guide to the World's Fresh Water Resources.* New York, Oxford University Press.

Honey, M. 1999. *Ecotourism and Sustainable Development: Who Owns Paradise?* Washington DC, Island Press.

IFRC (International Federation of Red Cross and Red Crescent Societies). 2001. *World Disasters Report 2001.* Geneva.

IPCC (Intergovernmental Panel on Climate Change). 2000. *Emissions Scenarios – A Special Report of Working Group III of the Intergovernmental Panel on Climate Change.* Cambridge, Cambridge University Press.

——. 2001. *Third Assessment Report.* Cambridge, Cambridge University Press.

Lohmar, B.; Wang, J.; Rozelle, S.; Huang, J.; Dawe, D. 2001. 'Investments, Conflicts and Incentives: The Role of Institutions and Policies in China's Agricultural Water Management'. In: E. Biltonen, and I. Hussain (eds.), *Managing Water for the Poor.* Colombo, International Water Management Institute.

Mastny, L. 2002a. *Travelling Light: New Paths for International Tourism.* Washington DC, Worldwatch Paper 159.

Mastny, L. 2002b. 'Redirecting International Tourism'. In: L. Starke (ed.), *State of the World 2002.* New York, Norton.

McCarthy, J.J.; Canziani, O.F.; Leary, N.A.; Dokken, D.J.; White, K.S. 2001. *Climate Change 2001: Impacts, Adaptation, and Vulnerability.* Cambridge, Cambridge University Press.

Mehta, L. 2000. Water for the twenty-first century: challenges and misconceptions. Working paper 111, Brighton, Institute of Development Studies.

Ministerial Declaration of The Hague on Water Security in the 21st Century. 2000. Official outcome of the Second World Water Forum, 3–7 December 2001, The Hague.

Molden, D. and Rijsberman, F. 2002. *Assuring Water for Food and Environmental Security.* Colombo, International Water Management Institute.

Ohlsson, L. 1996. *Hydropolitics.* London, Zed Books.

Petrella, R. 2001. *The Water Manifesto.* London, Zed Books.

Postel, S. 1993. 'Water and Agriculture'. In: P.H. Gleick (ed.), *Water in Crisis: A Guide to the World's Fresh Water Resources.* New York, Oxford University Press.

Rennie, J. and Singh, N. 1996. *Participatory Research for Sustainable Livelihoods: A Guidebook for Field Projects.* Manitoba, International Institute for Sustainable Development.

Rogers, P. and Aitchinson, J. 1998. *Towards Sustainable Tourism in the Everest Region of Nepal.* Kathmandu, The World Conservation Union.

Seabright, P. 1997. *Water: Commodity or Social Institution?* Stockholm, Stockholm Environment Institute.

Shiklomanov, I.-A. (ed.) 1997. *Comprehensive Assessment of the Freshwater Resources of the World.* Stockholm, Stockholm Environment Institute.

Soussan, J.-G. 2002. 'Poverty and Water Security'. Concept paper prepared for the Water and Poverty Initiative, Asian Development Bank Manila.

Stauffer, J. 1998. The Water Crisis: Constructing Solutions to Freshwater Pollution. London, Earthscan Publications.

UN (United Nations). 2002. *World Population Prospects: The 2000 Revision.* New York, Population Division, Department of Economic and Social Affairs.

——. 2000. *World Urbanization Prospects: The 1999 Revision.* New York.

——. 1992. Agenda 21. Programme of Action for Sustainable Development. Official outcome of the United Nations Conference on Environment and Development (UNCED), 3–14 June 1992, Rio de Janeiro.

UNEP (United Nations Environment Programme). 1999. *Global Environmental Outlook 2000.* London, Earthscan Publications.

UNFPA (United Nations Population Fund). 2002. *The State of the World Population 2001.* New York.

UN-Habitat. 2001. *The State of the World's Cities Report 2001.* Nairobi.

Vörösmarty, C.-J.; Green, P.; Salisbury, J.; Lammers, R.-B. 2000. 'Global Water Resources: Vulnerability from Climate Change and Population Growth'. Science 289.

Whittington, D. and Choe, K. 1992. Economic Benefits Available for the Provision of Improved Potable Water Supplies. WASH (Water, Sanitation and Hygiene for All) Technical Report no. 77, Washington D.C, The United States Agency for International Development (USAID).

WHO/UNICEF (World Health Organization/United Nations Children's Fund). 2000. *Global Water Supply and Sanitation Assessment 2000 Report.* Geneva.

WRI (World Resources Institute). 1986. *Water Resources 1986.* New York, Basic Books.

WSSCC (Water Supply and Sanitation Collaborative Council). 2000. *Vision 21: Water for People – A Shared Vision for Hygiene, Sanitation and Water Supply and A Framework for Action.* Geneva.

WWF (World Wide Fund for Nature). 1998. *Living Planet Report 1998: Over-consumption is Driving the Rapid Decline of the World's Natural Environments.* Gland, Switzerland.

2 Milestones

Study the past if you would divine the future.
Confucius

The road to sustainable management of water resources has been paved with thirty years of international conferences and decisions. This chapter gives the highlights of these major worldwide efforts.

This table is available on the WWAP web site, with hyperlinks to the full text of the outcomes: http://www.unesco.org/water/wwap/milestones

Dates	Events	Outcomes	Quotations
1972	**UN Conference on the Human Environment, Stockholm** Main issues: preservation and enhancement of the human environment	Declaration of the UN Conference on the Human Environment	'A point has been reached in history when we must shape our actions throughout the world with a more prudent care for their environmental consequences.' (6. Declaration of the UN Conference on the Human Environment)
1977	**UN Conference on Water, Mar del Plata** Main issues: assessment of water resources, water use and efficiency	Mar del Plata Action Plan	'relatively little importance has been attached to water resources systematic measurement. The processing and compilation of data have also been seriously neglected.' (Recommendation A: Assessment of water resources, Mar del Plata Action Plan)
1981–1990	**International Drinking Water and Sanitation Decade**		'The goal of the Decade was that, by the end of 1990, all people should possess an adequate water supply and satisfactory means of excrete and sullage disposal. This was indeed an ambitious target as it has been estimated that it would have involved the provision of water and sanitation services to over 650,000 people per day for the entire ten year period. Although major efforts were made by government and international organisations to meet this target, it was not achieved.' (Choguill, C.; Francys, R.; Cotton, A. 1993. Planning for Water and Sanitation.)
1990	**Global Consultation on Safe Water and Sanitation for the 1990s, New Delhi** Main issues: safe drinking water, environmental sanitation	New Delhi Statement: 'Some for all rather than more for some'	'Safe water and proper means of waste disposal ... must be at the center of integrated water resources management' (Environment and health, New Delhi Statement)
	World Summit for Children, New York Main issues: health, food supply	World Declaration on the Survival, Protection and Development of Children	'We will promote the provision of clean water in all communities for all their children, as well as universal access to sanitation.' (18. World Declaration on the Survival, Protection and Development of Children)
	Beginning of the International Decade for Natural Disaster Reduction (1990–2000)	Recognition of the increased general vulnerability of people and property to natural disasters	'to reduce through concerted international action, especially in developing countries, the loss of life, property damage and social and economic disruption caused by natural disasters...' (Resolution 44/236 of the UN General Assembly)

Dates	Events	Outcomes	Quotations
1992	**International Conference on Water and the Environment, Dublin** Main issues: economic value of water, women, poverty, resolving conflicts, natural disasters, awareness	Dublin Statement on Water and Sustainable Development	Principle 1: 'Fresh water is a finite and vulnerable resource, essential to sustain life, development and the environment' Principle 2: 'Water development and management should be based on a participatory approach, involving users, planners and policy-makers at all levels' Principle 3: 'Women play a central part in the provision, management and safeguarding of water' Principle 4: 'Water has an economic value in all its competing uses and should be recognized as an economic good' (Guiding principles. *The Dublin Statement on Water and Sustainable Development*)
	UN Conference on Environment and Development (UNCED Earth Summit), Rio de Janeiro Main issues: cooperation and participation, water economics, drinking water and sanitation, human settlements, sustainable development, food production, climate change	Rio Declaration on Environment and Development	'establishing a new and equitable global partnership through the creation of new levels of cooperation among States, key sector societies and people' (Rio Declaration)
		Agenda 21	'The holistic management of freshwater ... and the integration of sectoral water plans and programmes within the framework of national economic and social policy, are of paramount importance for action in the 1990s and beyond.' (Section 2, Chapter 18, Agenda 21)
1994	**Ministerial Conference on Drinking Water Supply and Environmental Sanitation, Noordwijk** Main issues: drinking water supply and sanitation	Programme of Action	'To assign high priority to programmes designed to provide basic sanitation and excreta disposal systems to urban and rural areas.' (Action Programme)
	UN International Conference on Population and Development, Cairo	Programme of Action	'To ensure that population, environmental and poverty eradication factors are integrated in sustainable development policies, plans and programmes.' (Chapter III – Interrelationships between population, sustained economic growth and sustainable development, C- Population and Environment, Programme of Action)

Dates	Events	Outcomes	Quotations
1995	**World Summit for Social Development, Copenhagen** Main issues: poverty, water supply and sanitation	Copenhagen Declaration on Social Development	'To focus our efforts and policies to address the root causes of poverty and to provide for the basic needs of all. These efforts should include the provision of ... safe drinking water and sanitation.' (Chapter I – Resolutions adopted by the Summit, Commitment 2.b., Copenhagen Declaration)
	UN Fourth World Conference on Women, Beijing Main issues: gender issues, water supply and sanitation	Beijing Declaration and Platform for Action	'Ensure the availability of and universal access to safe drinking water and sanitation and put in place effective public distribution systems as soon as possible.' (106 x, Beijing Declaration)
1996	**UN Conference on Human Settlements (Habitat II), Istanbul** Sustainable human settlements development in an urbanizing world	The Habitat Agenda	'We shall also promote healthy living environments, especially through the provision of adequate quantities of safe water and effective management of waste.' (10. The Habitat Agenda, Istanbul Declaration on Human Settlements)
	World Food Summit, Rome Main issues: food, health, water and sanitation	Rome Declaration on World Food Security	'To combat environmental threats to food security, in particular, drought and desertification ... restore and rehabilitate the natural resource base, including water and watersheds, in depleted and overexploited areas to achieve greater production.' (Plan of Action, Objective 3.2, Rome Declaration)
1997	**First World Water Forum, Marrakech** Main issues: water and sanitation, management of shared waters, preserving ecosystems, gender equity, efficient use of water	Marrakech Declaration	'to recognize the basic human needs to have access to clean water and sanitation, to establish an effective mechanism for management of shared waters, to support and preserve ecosystems, to encourage the efficient use of water.' (Marrakech Declaration)
1998	**International Conference on Water and Sustainable Development, Paris**	Paris Declaration on Water and Sustainable Development	'to improve co-ordination between UN Agencies and Programmes and other international organizations, to ensure periodic consideration within the UN system ... [To] emphasize the need for continuous political commitment and broad-based public support to ensure the achievement of sustainable development, management and protection, and equitable use of freshwater resources, and the importance of civil society to support this commitment.' (Paris Declaration)

Dates	Events	Outcomes	Quotations
2000	**Second World Water Forum, The Hague** Main issues: water for people, water for food, water and nature, water in rivers, sovereignty, interbasin tranfer, water education	World Water Vision: Making Water Everybody's Business	• 'Involve all stakeholders in integrated management; • Move to full-cost pricing of water services; • Increase public funding for research and innovation; • Increase cooperation in international water basins; • Massively increase investments in water' (Vision Statement and Key Messages, World Water Vision)
	7 challenges: Meeting basic needs, Securing the food supply, Protecting ecosystems, Sharing water resources, Managing risks, Valuing water, Governing water wisely	Ministerial Conference on Water Security in the 21st Century	'We will continue to support the UN system to re-assess periodically the state of freshwater resources and related ecosystems, to assist countries where appropriate, to develop systems to measure progress towards the realization of targets and to report in the biennial World Water Development Report as part of the overall monitoring of Agenda 21.' (7.B, Ministerial Declaration)
		UN Millennium Declaration	'We resolve ... to halve, by the year 2015 ... the proportion of people who are unable to reach or to afford safe drinking water.' (19, UN Millennium Declaration)
	End of the International Decade for Natural Disaster Reduction (1990–2000)		
2001	**International Conference on Freshwater, Bonn** Water – key to sustainable development	Ministerial Declaration	'Combating poverty is the main challenge for achieving equitable and sustainable development, and water plays a vital role in relation to human health, livelihood, economic growth as well as sustaining ecosystems.' (Ministerial Declaration)
	Main issues: governance, mobilizing financial resources, capacity-building, sharing knowledge	Recommendations for action	'The conference recommends priority actions under the following three headings: • Governance • Mobilising financial resources • Capacity building and sharing knowledge' (Bonn Recommendations for Action)

Dates	Events	Outcomes	Quotations
2002	**World Summit on Sustainable Development, Rio+10, Johannesburg**	Johannesburg Declaration on Sustainable Development	'We recognize that poverty eradication, changing consumption and production patterns, and protecting and managing the natural resource base for economic and social development are overarching objectives of, and essential requirements for sustainable development.' (Para. 11, Declaration on Sustainable Development)
		Plan of Implementation	'The provision of clean drinking water and adequate sanitation is necessary to protect human health and the environment. In this respect, we agree to halve, by the year 2015, the proportion of people who are unable to reach or to afford safe drinking water (as outlined in the Millennium Declaration) and the proportion of people who do not have access to basic sanitation ...' (II.7, Plan of Implementation)
			'Develop integrated water resources management and water efficiency plans by 2005, with support to developing countries, through actions at all levels to: a) Develop and implement national/regional strategies, plans and programmes with regard to integrated river basin management, b) Employ the full range of policy instruments, including regulation, monitoring, voluntary measures, market and information-based tools. c) Improve the efficient use of water resources.' (IV.24, Plan of Implementation)
2003	**International Year of Freshwater**		'Water is likely to become a growing source of tension and fierce competition between nations, if present trends continue, but it can also be a catalyst for co-operation. The International Year of Freshwater can play a vital role in generating the action needed – not only by governments but also by civil society, communities, the business sector and individuals all over the world.' (UN Secretary-General, Kofi Annan)
	Third World Water Forum, Kyoto		

Signing Progress: Indicators Mark the Way

By: UNECE (United Nations Economic Commission for Europe

Table of contents

EN IRAPUATO
EL CAMBIO ES COMPROMISO DE TODOS
SE CONSIGNARA A LA PERSONA QUE
SE SORPRENDA DEPOSITANDO
BASURA EN ESTE LUGAR
RECUERDA: LA LIMPIEZA ES COMPROMISO DE TODOS
DENUNCIA CUALQUIER ANOMALIA A. LOS TELS.: 61929/61186/60099 EXT. 146

The woods are lovely, dark and deep,
But I have promises to keep,
And miles to go before I sleep,
And miles to go before I sleep.

Robert Frost, *Stopping by Woods on a Snowy Evening*

THE EXPRESSION 'EVERYTHING IS RELATIVE' is often heard, but is difficult to counter as there is some truth to the argument. Indicators, however, change this status quo and establish a common ground, enabling us to compare and draw conclusions from areas as vast as entire regions to the smallest drainage basins. They are the tools that allow us to know where we are, to measure what we have accomplished, and what yet needs to be achieved.

The importance of indicators cannot be overestimated, as it is they that will allow for informed, rational decisions and actions to be taken. Indicators are thus at the heart of our book's mission – taking stock for people and planet.

This chapter offers an overview of the concept of indicators and of the World Water Assessment Programme's (WWAP's) construction of a base upon which future indicators can be further developed. We are, however, only in the initial stages of this important project – the journey of a thousand miles begins with a single step.

INTERNATIONALLY ACCEPTED INDICATORS on different aspects of water management need to be developed. These should include indicators for the relevant targets in the UN Millennium Declaration and for other relevant national and international goals. These indicators should be developed through participatory processes, including stakeholders from different levels and around the world. The World Water Assessment Programme [WWAP] should take a lead role in the development of these indicators (Bonn Freshwater Conference, 2001).

The Challenge

Development of indicators goes hand-in-hand with WWAP's mandate to monitor progress towards meeting the targets and goals set out in Agenda 21, Commission on Sustainable Development (CSD) processes, and other important intergovernmental forums, such as the Ministerial Meeting of the Second World Water Forum. It was resolved in the planning stages of the *World Water Development Report* (WWDR) that to measure progress – or the lack of it – specific tools would have to be developed. As a result, one of WWAP's tasks is the development and implementation of a comprehensive set of indicators, which would structure and provide the information required for assessing the state of water resources and help in monitoring progress towards sustainable management.

Indicators are used to simplify, quantify, communicate and create order within complex data. They provide information in such a way that both policy-makers and the public can understand and relate to it. They help us to monitor progress and trends in the use and management of water resources over time and space. Similarly, indicators can help us to compare results in different areas or countries and examine potential links between changing conditions, human behaviour and policy choices. Because 'good' indicators are easy to understand, they offer a tool for raising awareness about water issues that cuts across every social and political group.

Developing 'good' indicators is not an easy task, however, and involves collection, collation and systematization of data. The need for clarity and ease of understanding means that indicators often condense large volumes of data into brief overviews and reduce the complexities of the world into simple and unambiguous messages. The need for scientific validity, on the other hand, requires that indicators must simplify without distorting the underlying patterns or losing the vital connections and interdependencies that govern the real world. They must therefore also be transparent, testable and scientifically sound.

According to the World Health Organization (WHO), indicators have to meet a whole set of criteria, which both 'condition and limit the types of indicators that can be developed and the ways in which they may be constructed and used' (WHO, 1999). Because the same indicator has to satisfy often conflicting but equally important social, political, financial and scientific objectives, deriving indicators becomes an objective maximization exercise constrained by available time, resources and partnership arrangements.

The solution lies in identifying or developing denominators common to as many cases as possible, so that comparisons may be made. If data can be gathered according to commonly agreed or standardized norms, then lessons can be drawn that may be transposable from one case to another.

Laying the foundation: the first edition of the WWDR

Attempts to develop indicators are not new. Since the early 1960s, efforts have been underway to develop a meaningful set of indicators and indices for water resources. Against this backdrop, the WWDR has been mandated to develop indicators and adopt a methodology for their development by learning from previous initiatives. This has proven to be a slow process in which each milestone provides the way – or ways – forward. Amidst such complexity and trade-offs, the WWAP secretariat agreed on a methodological approach, identified some indicators, carried out limited testing and developed a better understanding and appreciation of the problems inherent to indicator development. These are significant achievements and they move the WWDR, as a reporting mechanism, towards where it aims to be in the future.

In this book, indicators are much in evidence. In each of the eleven challenge chapters that make up parts III and IV, for example, various figures and statistics are presented. These are indicators – tools for monitoring changes through time and space. A comparison of the 'before' and 'after' pictures allows us to assess progress, as does a comparison of the present state of things with the desired 'target'. What are the major drivers exacerbating freshwater availability in the world? How many people have access to drinking water? What are the signs and markers of malnutrition, declining biodiversity, and transboundary water as a medium for cooperation? These are a few of the major questions surrounding freshwater management issues. The report offers numerous such indicators that might help us to assess issues and better tackle them in the future.

The seven pilot case studies that make up part V also present indicators for measuring progress against targets. In a few instances, the indicators developed in one river basin are the same, or at least similar, to those in another basin. But on average, differences from one case study to the next outweigh the similarities. This lack of common measurement is one of the great challenges for the WWDR.

Bangkok and Tokyo are both urban basins, but are not subject to the same pressures and do not occupy similar natural environments or enjoy the same socio-economic status. The Senegal River basin, Lake Titicaca and Lake Peipsi are all transboundary water bodies, but they are not comparable. This book highlights such variability while reporting on existing commonalities.

Nonetheless, the task at hand is far from complete. Gathering harmonized data and developing indicators that are universally applicable are complex and delicate tasks, not best done in a hurry.

Introduction to Indicators

Indicators are our link to the world (IISD, 1999).

Indicators help to reflect and communicate a complex idea. They are everywhere and are part of our everyday lives. We use them to observe, describe, and evaluate actual states, to formulate desired states or to compare an actual with a desired state. These simple numbers, descriptive or normative statements can condense the enormous complexity of the world around us into a manageable amount of meaningful information. For example, a farmer in North India can determine the suitability of field preparation to sow wheat based on the soil's ability to form a ball when pressed, while her counterpart in Canadian prairies would need to take a moisture meter reading of the soil. Although the purpose of both acts is the same, the former relies on the soil's ability to form a ball as an indicator and the latter on the value in her moisture meter.

Indicators can therefore be either descriptive or normative; they may indicate quantitative or qualitative information and they may (or may not) be applicable to varying temporal and spatial dimensions. Probably the most essential criterion for an indicator is that it should enlighten its user and provide a sense of the issue being examined. An indicator can be a simple data point or variable, or can be a simplified value derived using a complex mathematical algorithm.

Indicator criteria

The usefulness of an indicator is most important; and for an indicator to be useful the complexity employed in developing indicators seldom matters. What matters is that the selection of the variable or the mathematical aggregation of several variables gives a clear picture of the issue being assessed, referenced or monitored. An E-coli bacteria count in a water source may be accurate but it will be of no use to a mother portering water from an improvised Johad (well) in the Thar Desert in India. However, a photograph or an image of a child suffering from disease after drinking contaminated water can provide her with a concept of better protection of water during its collection and storage.

Indicators can tell a different story or message according to specific contexts, for particular purposes and for specified target groups, and therefore resist universal application. Both the design and the use of indicators involve many personal and negotiated decisions, explicit and implicit assumptions, normative and subjective judgement, disciplinary and method-specific rules. They are based on beliefs, internalized values and norms and on one's perception of 'reality'. For example, the grey cloud may bring happiness to a rice farmer in Sri Lanka, but the same can mean disaster to a potter in Viet Nam. In essence, if we have learned to watch the relevant indicators, we can understand and cope with our dynamic environment.

To indicate, according to the dictionary, is 'to point out, to announce, to give notice to, to determine or to estimate'. Many definitions of indicator and index can be found in literature, but the following are used in this report.

- An indicator, comprising a single data (a variable) or an output value from a set of data (aggregation of variables), describes a system or process such that it has significance beyond the face value of its components. It aims to communicate information on the system or process. The dominant criterion behind an indicator's specification is scientific knowledge and judgement.

- An index is a mathematical aggregation of variables or indicators, often across different measurement units so that the result is dimensionless. An index aims to provide compact and targeted information for management and policy development. The problem of combining the individual components is overcome by scaling and weighting processes, which will reflect societal preferences.

- The aggregation of variables into an indicator may be likened to combining the annual withdrawal of surface and groundwater into the total annual withdrawal. An index combines for example, the withdrawal with the availability of water to indicate water stress. The emphasis in an index is not on scientific justification, but on responding to the societal needs. Figure 3.1 shows the difference between variables, indicators and indices.

- A variable is an observed datum derived by using basic statistics or monitoring, such as amount of rainfall or runoff, or number of diarrhoeal cases. Indicators are derived when the basic variables or observed data are aggregated using objective and scientific methods; for example mathematical aggregation of evaporation and transpiration data provide an indicator on evapotranspiration. When such core indicators are lumped together we get an index, for example, the aridity index of an area, which can be constructed by aggregating different indicators on water availability and use (see box 3.1).

The main functions of indicators are thus simplification, quantification, communication and ordering. Indicators can relate and integrate information and allow comparison of different regions and different aspects. Most significantly, indicators:

- provide information on the system or process under consideration in an understandable way and are therefore a means of communication to policy-makers and the public;
- evaluate the effect of performed policy actions and plans (after the fact) and help to develop effective, new actions (beforehand); and
- help to translate information need into data that have to be collected and to translate collected data into policy relevant information.

Purpose and use of indicators

Clearly, the growing interest in the use of indicators and indices is closely connected to the increasing complexity of policy problems and the large amount of available data. In the water sector, beyond their face value, indicators can provide various types of information, as listed below.

- **Descriptive**: describing the state of the resource is the most common use of indicators. An example of the descriptive use of indicators is given by figure 3.2, presenting the values of four indicators, namely available water resources, water demand, internal renewable water resources, and water supply on a global scale. These maps clearly show global differences in resources, demand and supply of water. However, without reference to a specific location and context, they cannot be used to make management decisions.
- **Showing trends:** regular measurement of indicators provides time series, which can show trends that may provide information on the system's functioning or its response to management. Figure 3.3 presents the development of consumable fish as a percentage of the minimum biomass needed for a sustainable renewal of the fish population. The indicator gives information on both the aquatic ecosystem and the long-term viability of the fisheries sector.
- **Communication:** indicators can be an instrument to communicate policy objectives and results to the public. Such indicators help promote action. The United Nations' Children's Fund (UNICEF) offers an example to illustrate the use of simple health-related indicators to communicate the relationship between inadequate access to water and sanitation and the health of children of the world. A commonly asked policy question is: why are children most vulnerable to unsafe water and environmental

Figure 3.1: Translation of an information need into policy-oriented information using variables, indicators and indices

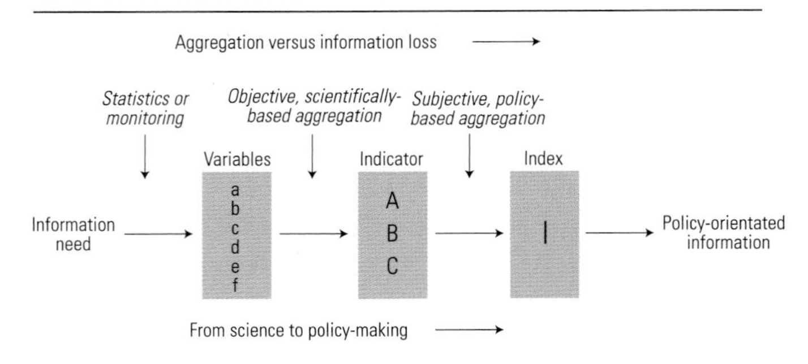

This figure shows the difference between variables, indicators and indices, which all represent different stages of information collation. Indicators take variables and condense them into manageable information sets, which are then further condensed by indices. These can then be translated into policy-oriented information.

Source: Lorenz, 1999.

Box 3.1: The Water Poverty Index as an illustration of the difference between variables, indicators and index

The purpose of the WPI is to provide decision-makers with 'an evaluation tool for assessing poverty in relation to water resource availability'. The occurrence of poverty reflects the conditions of life upon which a person depends and the existence of a combination of circumstances or 'functioning' which gives rise to capabilities upon which an individual can build. The five key components of the WPI are:

1. resource: physical availability of surface and groundwater;
2. access: the extent of access to this water for human use;
3. capacity: the effectiveness of people's ability to manage water;
4. use: the ways in which water is used for different purposes; and
5. environment: the need to allocate water for ecological services.

In the calculation of the WPI, these components (indicators) are each made up of a collection of subcomponents (variables).

Source: Sullivan. Prepared for the World Water Assessment Programme (WWAP), 2002.

Figure 3.2: Calculated available water resources, water demand, internal renewable water resources and water supply

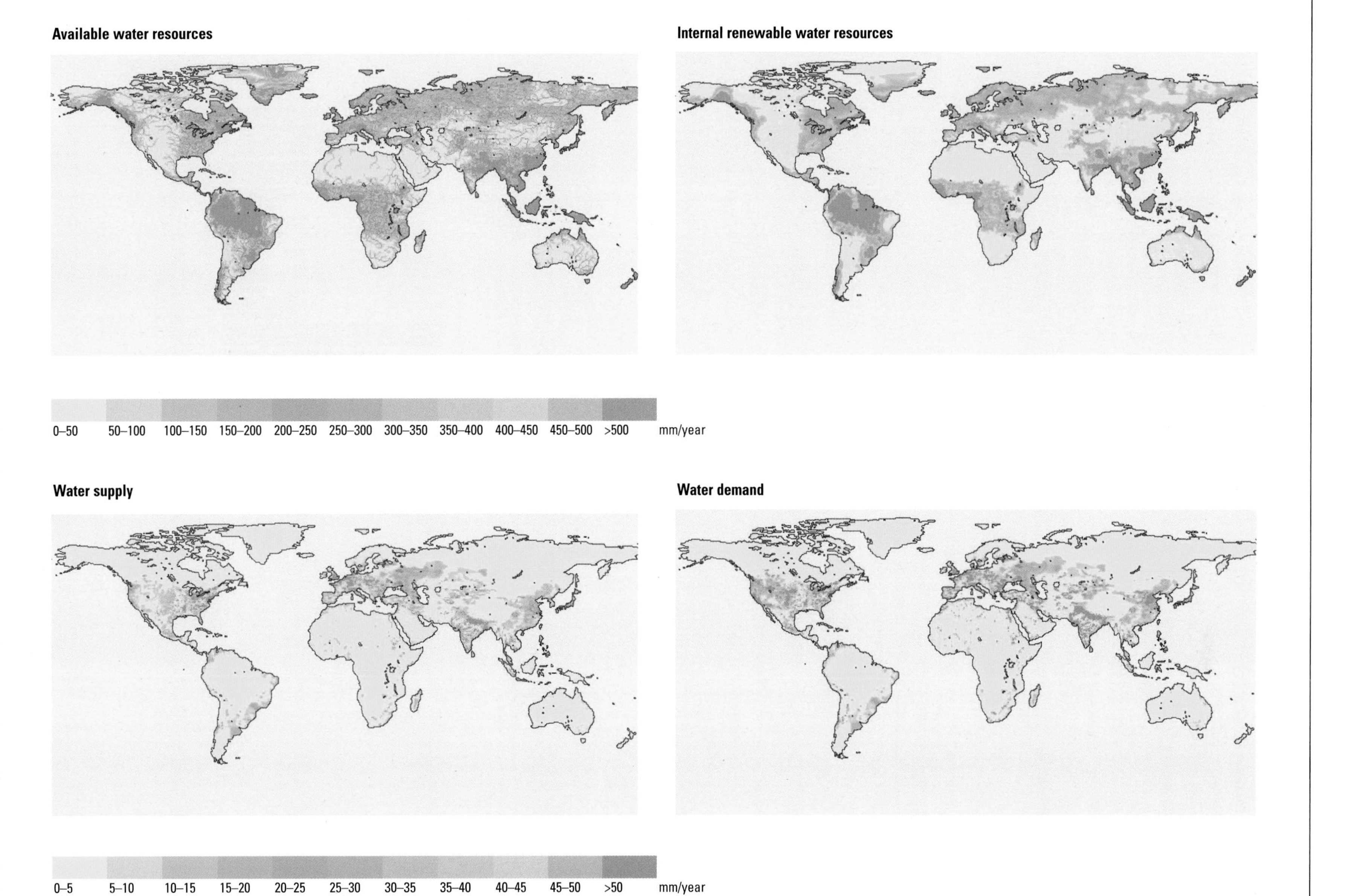

The four indicators represented in the above maps clearly show differences in global water resources, supply and demand. However, they are not usable as indicators at a local level because the scale is too large.

Source: Watermodel, presented at the Second World Water Forum.

risks? A simple indicator-based response to this question is that in 1998, 2.2 million people died from diarrhoeal diseases of which 1.8 million were children under five years of age. The statement clearly demonstrates that poor access to water and sanitation primarily affects children under five years of age since this age group represents almost 85 percent of total deaths. The indicator response also demonstrates, however, the risks of simplification, because it overlooks the fact that a large proportion of diarrhoeal diseases are food-borne rather than linked to contaminated drinking water.

- **Assessment:** an indicator value can also be compared to a reference condition that represents some desired state. Specification of a reference condition depends strongly on the purpose of the indicator, its users, and the issue being reflected. Reference conditions include policy targets, sustainability criteria and historical states. Box 3.2 demonstrates that existence of a certain mammal, fish or bird species can be a simple reference indicator to assess the quality of the ecosystem.
- **Predicting the future**: when models are linked to indicators, a time series can be extended into an (estimated) future. Alternative scenarios can be assessed in terms of how well each one moves the system towards a desired state. Figure 3.4 shows an example of this use.

Box 3.2: Use of species as indicators for ecosystem quality

Species can be very good indicators for ecosystem quality. In the first place, the presence of a certain species provides information on ecological processes. For example, the occurrence of migrating fish indicates that a river can be safely navigated by fish. Mammals are species from the higher levels of the food chain. Their presence indicates that the underlying food web is functioning well enough to provide the mammal with sufficient food. Furthermore, they need relatively large areas of suitable habitat. Therefore, the presence of mammals is an indicator for the quality and size of an ecosystem.

Beyond their face value, species serve a function of communication as well, particularly species of the higher trophic levels, such as fish, mammals and birds that appeal to the public. Therefore, the return of a popular species is often stated as an objective of ecological restoration measures to increase social acceptance of these measures (e.g. salmon in the Rhine, beaver in the Elbe).

Figure 3.3: Development of consumable fish as a percentage of the minimum biomass needed for a sustainable renewal of the fish population

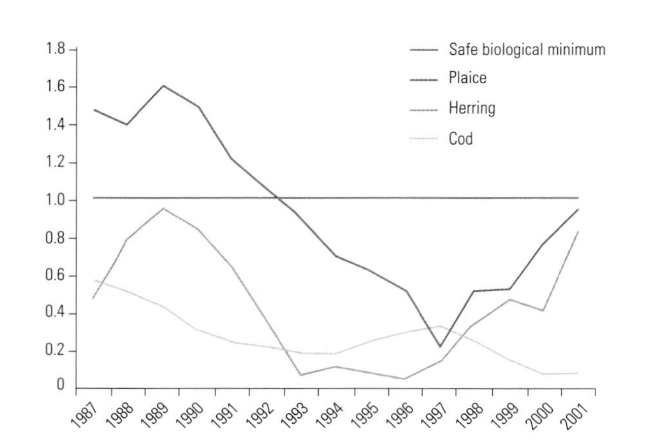

This indicator gives information on both the condition of the aquatic ecosystem and the long-term viability of the fisheries sector. There was a trend of rapid decline in the development of consumable fish in the early 1990s, but their populations have been steadily increasing again since 1997.

Source: CIW, 2000.

Figure 3.4: Annual freshwater demand in different regions of the world for the current situation and for two alternative projections for the year 2025

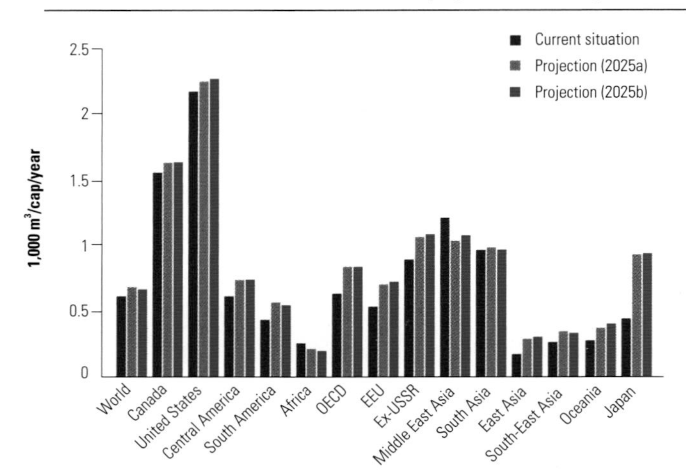

This graph represents the indicator of future annual freshwater demand. Projections show that demand is likely to rise in most areas of the world, excluding Africa. Indicators can therefore be used to make projections into a possible future.

Source: Watermodel, presented at the Second World Water Forum.

Indicators and Sustainable Development

Integrated Water Resources Management (IWRM)

> *Indicators for monitoring progress towards sustainable development are needed in order to assist decision-makers and policy-makers at all levels and to increase focus on sustainable development* (UN Sustainable Development web site).[1]

The sustainable development, management, protection and use of freshwater resources act as guiding principles for indicator development and assessments. Sustainable development in particular has received much attention since its position at the forefront of the World Commission on Environment and Development (WCED) in 1987. The WCED defined sustainable development as 'the development that meets the needs of the present without compromising the ability of future generations to meet their own needs'. A key factor in the elaboration of sustainable development is the integral view taken of the central concepts: that the interests of people, society, economy and environment need to be seen as an interconnected whole and trade-offs respecting all interests need to be made. Economic development has to be viable from a social and environmental point of view, social development has to be viable in the light of economy, and environment and environmental policies have to be attuned to social and economic development. The trade-offs are ultimately a societal and political choice.

Integrated Water Resources Management (IWRM) can be regarded as the vehicle that makes the general concept of sustainable development operational for the management of freshwater resources. IWRM adopts a holistic approach, which implies that information is needed on the state of the economy, society and water resources, and their mutual relationships. It also invokes the need for greater participation, which means that there must be tools for effective communication between different groups of stakeholders – e.g. policy-makers, the public and scientists. Indicators can help simplify information on IWRM and establish effective communication between various stakeholders.

1. Citation taken from http://www.un.org/esa/sustdev/isd.htm posted circa 2000.

Indicator development models

> *The search for indicators is evolutionary. The most important process is one of learning* (WHO, 1999).

Two main issues stand out as the main drivers for indicator development: the need to present complex phenomena in meaningful, understandable, comparable and objective numbers to decision-makers and the public, and the need to establish objective benchmarks to analyse changes over time and space. Although tracking weather, rainfall and other hydrological variables has been carried out for a long time, focused indicator development work appears to have begun only in the early 1980s, when the need arose for combining different variables to produce an aggregate value (Spreng and Wils, 2000). The effort gained additional impetus during the late 1980s, when sustainable development became a global issue. This necessitated the development of indices, concepts and frameworks that enabled putting these indicators and indices into cause-effect relationships, thereby integrating the environmental, economic and social domains of development. Considerable progress seems to have been made since the early 1990s. Figure 3.5 summarizes the history of indicator development.

In the literature, different approaches exist on how an indicator or a set of indicators can be built (Bossel, 1999; Meadows, 1998; Van Harten et al., 1995). The major indicator development models appear to have been shaped by four approaches, namely the bottom-up approach where the logic goes from data to parameters to indicators, the top-down approach, which follows the logic down from vision to themes to actions to indicators, the systems approach, which bases indicators on a comprehensive analysis of system inflows and outputs, and the cause-effect approach (commonly known as the Pressure-State-Response [PSR] approach or the Driving force-Pressure–State-Impact-Resource [DPSIR] or the Driving Force-Pressure-State-Exposure-Effect-Action [DPSEEA]), which subscribes to the logic of indicators denoting various causes and effects.

1. **The bottom-up approach** uses an information pyramid, wherein the logic is to aggregate available primary data along several hierarchical levels into indicators using intuitive and mathematical approaches. Water resource specialists tend to be critical of this approach as being too reductionist. It is reasoned that the lumping of information not only reduces the 'internal variability' of the system, but also loses the relational issues to other resources and processes. Finding data alone without building models to test hypotheses in a real environment is of little use. Once data are available, very often in abundance, there is some danger that the bottom-up approach has to go through a bureaucratic compression process. With such compression,

Figure 3.5: Overview of the history and development of sustainability indicators

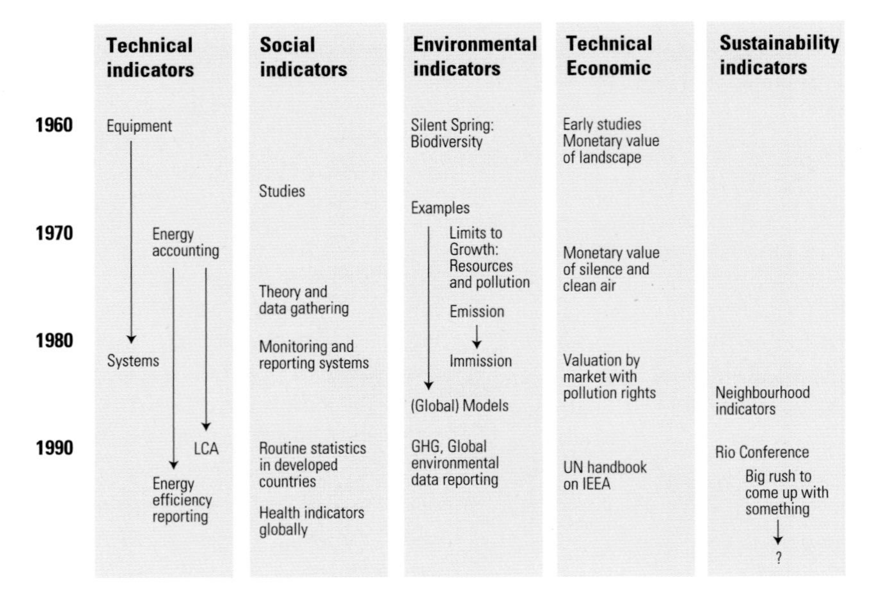

LCA: Life Cycle Analysis, GHG: Global Hydrological Gauging, IEEA: Integrated Environmental and Economic Accounting.

Source: Spreng and Wils, 2000.

creative ideas may be abandoned. It is essential to understand whether one piece of data is of significance beyond the measured quantity and beyond their immediate first use and can be used as an indicator. This approach is widely used in data-rich situations.

2. **The top-down approach** draws from the Logical Framework (log frame) approach, which is a programme management tool that serves purposes in both design and monitoring within programme cycle management. A log frame follows a generalized structure where the goal of an intervention is structured according to its purpose, outputs and specific activities. A generalized log frame approach would commence with defining an overarching goal (typically the attainment of an institutional or organizational outcome in the real world). Recognizing the breadth of such goals, many different interventions may be required to attain the goal, each with a unique and clearly defined purpose. Achieving this purpose requires interventions to accomplish outputs, through a particular set of activities. Risks and assumptions – those externalities that could undermine outputs securing the purpose or the goal – need to be internalized during design and implementation.

The Logical Framework approach, in contrast to some other approaches that employ indicators, is essentially a 'top-down' approach. The starting point is the definition of a real-world outcome to be attained – the goal – backed by an objectively verifiable indicator. Action is then designed around the interventions needed to attain the goal. The distinction between purpose, output and activity often varies, but is a valuable step in distinguishing between different levels of intervention. For example, 'reduce the number of lives lost to floods by 2010' is most appropriate as a purpose, backed by outputs such as 'regular cyclone warning received by national disaster management agencies by 2005', an output for which any number of activities may be needed.

Indicators, whether quantitative, qualitative or time-based, are in all cases set at all levels of the log frame from goal down to activity. Because of the diversity of starting points for action, purposes to be attained, and types of intervention, unanimity on universal indicators assumes much less importance at lower levels of the Logical Framework.

Unanimity is vital, however, at the upper levels where agreement is needed on a common goal and on how attainment of that goal is to be measured. With goals in water having been set at an international level, for example the Millennium Development Goals, the indicators of these goals have become accepted as international development targets. These targets therefore represent a unique set of high-profile indicators.

Table 3.1 presents a draft version of the Logical Framework of the Global Water Partnership (GWP) Framework for Action

Table 3.1: Draft Logical Framework of the Global Water Partnership Framework for Action

Intervention logic	Targets
Goal: Economic well-being and social development under environmental sustainability and regeneration improved	*International development targets met, in particular:* 1. The proportion of people living in extreme poverty in developing countries should be reduced by at least one-half by 2015 (Copenhagen) 2. The death rate for infants and children under the age of five years should be reduced in each country by two thirds of the 1990 level by 2015 (Cairo) 3. There should be a current national strategy for sustainable development in the process of implementation in every country by 2005, so as to ensure that current trends in the loss of environmental resources are effectively reversed at both global and national levels by 2015 (Rio) 4. Reduce by half the number of undernourished people on the earth by 2015 (Rome).
Purpose: Global water security provided through efficient, equitable and sustainable management and use of water	*Global water security targets achieved:* 1. Comprehensive policies and strategies for IWRM in process of implementation in 75% of countries by 2005 and in all countries by 2015 2. Proportion of people not having access to hygienic sanitation facilities reduced by half by 2015 3. Proportion of people not having sustainable access to adequate quantities of affordable and safe water reduced by half by 2015 4. Increase water productivity for food production from rainfed and irrigate farming by 30% by 2015 5. Reduce the risk from floods for 50% of the people living in floodplains by 2015 6. National standards to ensure the health of freshwater ecosystems, established in all countries by 2005, and programmes to improve the health of freshwater ecosystems implemented by 2015
Outputs: Political will to mobilize people, and resources secured	1.1 Complete targets and log-frame for water security by August 2000 1.2 Regional and National Programmes for Action completed by August 2001 1.3 Programmes for Action discussed at the Bonn Conference (Dublin+10) 1.4 Programmes for Action and national targets prepared by governments before Rio+10 meeting in mid 2002 1.5 Third World Water Forum (on major water issue arising from Second World Water Forum) held in March 2003 1.6 First edition of *World Water Development Report* published March 2003
Effective water governance for IWRM realized	2.1 IWRM mainstreamed in policy and strategy implementation processes in all countries by 2005 2.2 Cooperation mechanisms between riparian states in all major river basins developed and strengthened by 2005, and shared waters agreements formulated by 2015 2.3 The economic value of water recognized and reflected in national policies and strategies by 2005, and mechanisms established by 2015 to facilitate full cost pricing for water services 2.4 GWP Toolbox of options for water management developed by 2001
Effective water wisdom generated	3.1 Water Awareness initiatives instigated in all countries by August 2001 3.2 Capacity for informed decision-making at all levels and across all stakeholders increased by 2005 3.3 Investment in research on water issues increased by August 2001 3.4 Hygiene education in 80% of all schools by 2010
Solutions to urgent water priorities prepared: protecting the resource, enhancing crop productivity per drop, improving sanitation, urban upgrading, improved flood management	4.1 Programmes to tackle urgent priorities formulated, resourced and under implementation in all countries by 2005 4.2 Action programmes to protect surface and groundwater resources prepared and in process of implementation by 2003, and defined standards achieved by 2010 4.3 Task force on food-water security reports by end 2001 and action programmes for enhancing crop per drop prepared and in process of implementation by 2003 4.4 Action programmes for sanitation formulated and in process of implementation, and knowledge/information about good hygiene practices made universal by 2003 4.5 Action programmes to integrate water needs (supply and waste) with spatial planning and social and economic needs prepared and in process of implementation by 2003 4.6 Action programmes for flood preparedness and protection formulated and under process of implementation by 2003
Investment needs for water security identified and agreed upon	5.1 Investment needs for closing the resources gaps identified and (indicative) investment plans developed in all countries by 2002 5.2 Mechanisms for mobilizing new financial resources identified and under process of implementation by 2003 5.3 Investments committed to the water domain doubled by 2005 5.4 Private sector-led International Research Foundation established by 2002
	Activities: Detailed activities will be developed as part of the continuing work to complete the framework for action.

Source: Based on GWP, 2000.

(2000), with indicators at goal level reflecting international targets, the purpose reflecting the outcome of the *World Water Vision*, and outputs reflecting the future products of the Framework for Action process.

3. **The systems approach** completely analyses the inflows, stock and outflows of an issue before defining indicators. It draws from the concept of system dynamics and offers a way forward in understanding the behaviour of the system over time. The approach adheres to the notion that 'all systems depend to some degree on the resource-providing and waste-absorbing capacities of their environment, and argues that

   - most systems interact with other systems that are essential to their viability;
   - many interactions are hierarchical, with subsystems contributing to the functioning of a system, which contributes to the functioning of a suprasystem, and so on; and
   - 'The viability of the total system depends on the viability of many but not necessarily all of its subsystems' (IISD, 1999).

   The systems approach has been applied in developing sustainability indicators and relies on specific indicators dealing with human systems (including social and individual development and governance), support systems (including economics and infrastructure) and natural systems (including resources and environment). Although the approach is seen as very promising, it is complex and often considered at a stage of development where it still is 'too academic' to solve real-world problems. Similarly, the core definition of the system itself tends to be too vague to allow development into a meaningful indicator development exercise.

4. **The cause-effect** approach is the most widely used approach to indicator development. Also considered a milestone, the pressure-state-response (PSR) conceptual framework was first introduced by the Organization for Economic Cooperation and Development (OECD) in 1994. This enabled trade-offs and the linking of environmental, economic and social indicators (OECD, 1994). Following the PSR framework of the OECD, several cause-effect classifications have been developed:

   - The Driving force-Pressure-State-Impact-Response (DPSIR) framework was used by the European Environmental Agency (Hettelingh et al., 1998), United Nations Environment Programme (UNEP) (Swart and Bakkes, 1995; Bakkes et al., 1994) and the World Resource Institute (Hammond et al., 1995).
   - The Driving force-State-Response (DSR) framework of the United Nations Commission on Sustainable Development was used for the indicators of Agenda 21 (DPCSD, 1996).
   - The Pressure-State-Impact-Response (PSIR) framework is mostly used in the Netherlands (Hoekstra 1998; Van Harten et al., 1995; Rotmans et al., 1994; Van Adriaanse, 1993).
   - The Driving force-Pressure-State-Exposure-Effects-Action (DPSEEA) framework is used in the burden of disease studies of the WHO.

Although this is the most applied approach and offers a very promising guideline for indicator development, it all too often fails to take the entire system into consideration because of the subjectivity involved in understanding the pressure, state and responses. Similarly, although the policy concerns are nested within the pressure, state and responses, the approach also appears to lack any explicit linkage to policy in the development of indicators.

Assessment of the previous approaches in relation to indicator development shows that a substantive effort is currently being made to derive indicators to measure success or failure of efforts towards sustainable development. Efforts such as developing the Genuine Progress Indicator, or achieving IISD indicators appear to have been initiated to provide a better measure for development than those provided by Gross Domestic Product/Gross National Product (GDP/GNP) and, more recently, by the Human Development Index (HDI). These efforts aim to derive an index similar to the HDI but with added environmental indicator/index values. Most of these efforts are still to be recognized. Since 1997 (Rio+5), the United Nations Development Programme (UNDP) has also been trying to integrate environmental indicators into the broad framework of the HDI.

Several other approaches have also been initiated to assess ecosystem integrity or health. The ecological footprint evaluation and environmental sustainability indicators have been developed exclusively to measure the quality of an area's environment. There are also a few water resource indicators being developed and used. The most utilized thus far are the fifteen basin scale indicators developed and applied by the Water Resources Institute (WRI) in 1998.

There are several UN-agency-led initiatives that report progress on water resource development, such as the Aquastat and FAOSTAT initiative of FAO, and the Global Resource Information Database (GRID) and Global International Waters Assessment (GIWA) initiatives of UNEP. While a few of these are developed and maintained to fulfil sectoral objectives, others try to fulfil a broader objective. WHO's efforts to estimate global burden of disease, expressed in Disability-Adjusted Life Years (DALY), are not limited to water-related diseases. This indicator aims not only to measure the health status of populations, but also to provide an approach to analysing the cost-effectiveness of different health interventions and, in a revised version, the quality of the health services in member states. Initially it was developed as a tool for use in the health sector, but with the development of comparative risk assessment methods, it is rapidly becoming an indicator for cross-sectoral health issues.

The most significant and inclusive initiative could prove to be the CSD-led initiative on sustainability indicators, which aims to report progress on the implementation of the recommendations of Agenda 21. However, these indicators have yet to be applied. Box 3.3 lists a few major indicator development exercises which used one of the above approaches.

WWDR Indicator Development

Obviously, there are strengths and weaknesses to each approach. One of the WWDR's tasks is to evaluate these and to learn from them. The most common way forward is to clearly identify from the outset within each approach the aim and intended uses of the indicators. All approaches recommend that the development of indicators be based on a thorough understanding of the system or process under consideration and preferably on a conceptual model. More importantly, they urge that a participatory approach be taken in order to gather different, often contradicting, viewpoints, both to legitimize the result and to enhance learning.

Clearly, development of indicators requires a common understanding of the issues at hand between the UN agencies responsible for the assessments and those information users who have asked for the indicator-based report. Such a process necessitates a large number of choices that unavoidably reflect the knowledge and values of their developers (Bossel, 1999). Involvement of the right institutions and people is therefore essential (see the United Nations Economic Commission for Europe [UNECE] guidelines on monitoring and assessment of transboundary rivers, 2000).

In developing indicators for the first WWDR, the Secretariat has tried not only to combine the four approaches reviewed, but also to take into consideration the lessons learnt by other indicator development efforts. The process was as inclusive as possible, representing both information providers and users, such as representatives of the pilot case studies, experts from the UN agencies, scientists, representatives from major water-related non-governmental organizations (NGOs) and a few members of the interested public. As outlined in chapter 15 on governing water wisely, the WWDR must provide:

- decision-makers with a means to understand the importance of water issues so as to involve them in promoting effective water governance;
- water specialists with a way to step 'outside the water box', so they learn to take into account broader, social, political and economic issues into the water equation, which would then require
- transparent and mutually communicable strategies for decision-makers and water specialists, in such a way that both have a clear understanding of the state of progress in terms of a global desire to achieve water-related goals and targets through effective implementation of policies and related actions.

Thus, an essential part of WWDR indicator development is ensuring participation of different stakeholders, particularly the member states involved in the WWAP process. This important element is illustrated in box 3.4, which depicts the efforts made by the Greater Tokyo case study in developing indicators of flood risks and water quality. The approach taken for the development of indicators in this issue of the WWDR can be summarized in six steps as outlined in figure 3.6.

Figure 3.6: Schematic representation of a guideline for indicator development

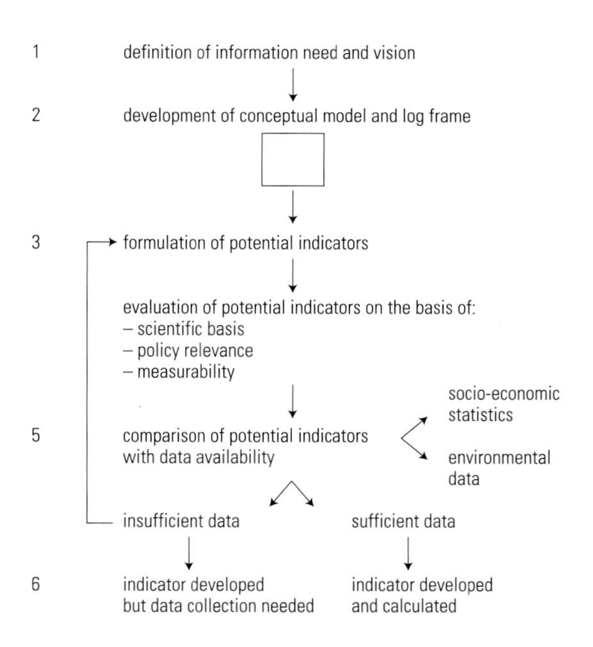

This scheme shows the different steps to achieving developed and calculated indicators from the definitions of needs.

Source: Based on Lorenz, 1999.

Box 3.3: Overview of major indicator development efforts

- **Human Development Index** of the United Nations Development Programme (UNDP – 1990): a composite index, which combines indices on gender-related development, a gender empowerment measure and human poverty.
- **Human Poverty Index (HPI) of UNDP**: measures the level of deprivation in three essential elements of human life – longevity, knowledge and decent living standards.
- **Indicators of Sustainable Development** of the United Nations Commission on Sustainable Development (CSD – 1997): aims at assisting decision-makers to track progress towards sustainable development. A total of seven water-related indicators are drawn, based on Chapter 18 of Agenda 21.
- **Environmental Indicators** of the Organization for Economic Cooperation and Development (OECD – 1994): development of key environmental indicators endorsed by OECD Environment Ministers and the broader OECD core set of environmental indicators.
- **European System of Environmental Pressure Indices (EPI)**: the EPI initiative is financed by the European Commission's Environment Directorate and aims at a comprehensive description of the most important human activities that have a negative impact on the environment. The EPI is based on the Driving force-Pressure-State-Impact-Response (DPSIR) model, and water-related indicators are included in the initiative.
- **Environmental Sustainability Index (ESI) and Environmental Performance Index (EPI)**: the ESI combines measures of current conditions, pressures on those conditions, human impacts and social response, and is a measure of overall progress towards environmental sustainability. The ESI scores are based on a set of twenty core 'indicators', each of which combines two to eight variables. The EPI permits national comparisons of various efforts made to manage some common policy objectives, primarily air and water quality, climate change and ecosystem protection.
- **Genuine Progress Indicator (GPI)**: developed in 1994 as a new measuring tool for the performance of the economy as it actually affects people's lives. The GPI draws from the Gross Domestic Product (GDP) the financial transactions that are relevant to the well-being of households and adjusts them for aspects of the economy that the GDP ignores. Access to water supply, sanitation and health benefits are considered in the GPI.
- **International Institute for Sustainable Development (IISD) Measurement and Indicators**: has conceptually defined indicators. Pilot indicators and indices to assess sustainability have been calculated (http://www.iisd.org/measure/default.htm).
- **Ecological Footprint:** developed by Earth Day Network, it measures the amount of nature's resources that an individual, a community or a country consumes in a given year. Water resources is one of the core indicators in the accounting.
- **Capital Theory** of the World Bank (Serageldin and Steer, 1994): illustrates sustainable development with four types of capital – natural capital, human-made capital, human capital, and social capital. For sustainable development, the total amount of capital in a society should be non-declining at any time. Flows between these capitals will change the amount of each capital. Sustainability indicators should describe the amount of each capital and the flows between capitals.
- **UN System Water Resource Indicator-Based Reporting System:** UN agencies are involved in publishing sectoral indicators that provide information on the state of the world's water. A few notable ones are those elaborated for the burden of disease led by WHO, UNICEF and WSSCC. Other efforts include the FAO's publications based on FAOSTAT and Aquastat, and UNEP's publications based on GRID statistics.
- **Specific Water Availability:** water stress indicator based on specific water availability, which is the water per capita that is left after agricultural, domestic and industry uses (Shiklomanov, 1997).
- **Standard Indicator:** deals with water availability in a country based on the people living in that country (Falkenmark et al., 1989). The indicator assumes it is likely that there will be fewer problems because of water shortages if more water is available per capita.
- **Water Indicators** developed by the World Resources Institute (WRI – 1998): indicator-based assessment of watersheds and freshwater systems on the basis of fifteen global indicators that characterize watersheds according to their value, current condition and vulnerability to potential degradation.
- **UN/ECE Task Force on Monitoring and Assessment Guidelines (2000):** development of guidelines for monitoring and assessment that integrate indicator development in the monitoring cycle.

Box 3.4: Indicators for flood risk and water quality in Greater Tokyo

Water quality indicators

As people become more concerned with improving the environment and more involved in the management process, this process must be made as transparent and understandable as possible. New water quality indicators have therefore been developed, based on the increased and more diverse needs in water management. In 1998, a study was conducted in the five rivers of Greater Tokyo to develop easily understandable indicators to monitor water quality conditions. The river administrator proposed indicators through the Internet and collected opinions from the public. In this study, comprehensibility of indicators is considered important. The study emphasized and proposed indicators based on three aspects: recreation, biodiversity and drinking water (*Source:* Kanto Regional Development Bureau, MLIT 2002).

Flood risk indicators

As a countermeasure to the high flood damage in Greater Tokyo, an easily comprehensible indicator was developed and introduced to the public, showing the degree of flood damage risk. The degree of safety from flood damage can be expressed by the combination of flood frequency and inundation level. Each of these criteria can be expressed by a colour and a height, respectively. In the accompanying diagram developed for Greater Tokyo, the green represents low flood frequency, the red, high flood frequency.

(*Source:* Yasuda and Murase, 2002).

Six criteria for indicators

In establishing indicators, the criteria must be absolutely clear. The following six criteria are proposed in the Greater Tokyo case study (Yasuda and Murase, 2002):

- Relevance: the numerical value of an indicator should represent the degree of 'what should be measured' directly.
- Clarity: ambiguity and arbitrariness should be excluded from measuring with an indicator.
- Cost: the cost of the evaluation by an indicator should be affordably low.
- Continuity: availability of coherent data both in historical and regional scope should be respected.
- Comprehensibility: definition/expression of an indicator should be intuitively/easily comprehensible to users.
- Social benefit: net social benefit that an indicator yields, as it is applied, should be maximized.

Step 1: Defining the information need

To define the information need it is important to distinguish possible uses of information.

- Information used for policy development;
- information used for policy implementation and evaluation;
- Information used for planning water resources development;
- Information used for operational water management; and
- Information used for communication with the public.

As outlined earlier in this chapter, the WWAP and WWDR have to assess the success of strategic approaches to the sustainable development, management, protection and use of freshwater resources in achieving the goals described in Chapter 18 of Agenda 21.

Within this context, the aims of indicators in the WWDR are:

- to provide a simple yet meaningful description of the complex water resource phenomena and management issues as a basis for action by decision-makers and the public;
- to provide insight into problems and potentials for integrated water resources management on a global scale;
- to keep track of developments regarding the state of the water resources and the effectiveness of the global response in solving problems;
- to assess the impact of water resources development on economic, social, health and environmental conditions; and
- to keep track of progress in meeting the set targets and goals.

The WWDR is centred around the eleven challenge areas (chapters 5–15), each of which requires appropriate information. For each challenge area, one UN agency or a group of agencies has taken the lead in supervising the preparation of a position paper describing the issues at stake. These position papers specify the information need required in each case and have driven the development of the appropriate indicators.

Step 2: Developing a conceptual model

A conceptual model is a verbal or visual abstraction of a part of the world from a certain point of view. Information on the system, its spatial and temporal scales and the cause and effect chain can be put into the conceptual model, representing the problem to be solved. The theory or concept that is important to a subject and that needs to be described in the conceptual model is not an objective fact, but depends on cultural background. In the conceptual model, values are expressed, worldviews are at play and theories on system functioning are developed and shared (implicitly or explicitly). Development of a conceptual model implies that a minimum amount of knowledge is available on the *systems* under consideration.

Integrated water resources management requires trade-offs between society, economy and the environment. Therefore, information is needed on cause-effect relationships and socio-economic and environmental effects of policy measures. In this first edition of the WWDR, a combination of log frame and DPSIR concept has been chosen as a working model. In this framework, the log-frame (as presented in table 3.1) provided a conceptual basis reflecting the policy needs, and the DPSIR provided the cause-effect relationships between human activity, environmental effect and societal response. These were manifested as indicators representing driving forces, pressure, state, impact and response (Hettelingh et al., 1998; Hammonds et al., 1995; Swart and Bakkes, 1995; Bakkes et al., 1994).

- Driving force indicators describe the driving forces of water use, such as those listed in chapter 1 on the world's water crisis: poverty, population growth, urbanization, globalization, industrial expansion, agricultural development, energy production and use, recreation and tourism.
- Pressure indicators describe the pressure on water systems as a result of human activities (e.g. use of natural resources, discharges of waste).
- State indicators describe the quality/quantity change in the 'state' of water as a result of the pressure.
- Impact indicators describe the impacts on ecosystems, resources, human health, social conditions, materials and amenities caused by the change in state.
- Response indicators describe the societal response to these changes and coping mechanisms, which are reflected in institutions, environmental, economic and sectoral policies. The response can be directed at different parts of the cause-effect chain (e.g. driving force, pressure, state or impact).

To give a practical example of the use of the DPSIR framework, we have tried to relate it to the components of the Water Poverty Index (see table 3.2).

Step 3: Formulating potential indicators

The potential indicators are the variables that describe dominant processes and characteristics in the conceptual model. For the WWDR, the process adopted mostly sectoral strategies and a list of indicators was selected for initial evaluation.

Step 4: Evaluation of potential indicators on the basis of selection criteria

The selection criteria relate to the scientific and policy requirements that indicators should fulfil. A list of commonly used criteria is included in table 3.3. Certain criteria are contradictory as they are linked to different information needs, therefore a potential indicator cannot fulfil all of the criteria listed in the table (e.g. 'specific for a certain stress or effect' and 'broadly applicable to many stressors and sites, usable in different regions'). For the information needed, a selection of criteria needs to be made. For example, the concentration of chlorophyll in the water is a generally accepted and scientifically robust indicator for the algal biomass. If communication is the major aim of the indicators, the selection criterion 'simple, easily interpretable and appealing to society' is more important than strong scientific robustness. This holds, for example, for indicators developed in interaction with the local community within the framework of the local Agenda 21.

Table 3.3 thus presents a list of the important criteria for the WWDR indicators. In general terms, they are well established and give meaningful and accurate information on the state or quality of the subject and are adequately documented. The pilot case studies have these adopted criteria and attuned them to their regional needs and conditions.

The lead UN agencies have used this preliminary set of indicators to establish the state of the resource and report progress against the set targets in various challenge areas. The wish lists of indicators have also been included, and will be developed and included in subsequent editions of the WWDR (see table 3.4).

Step 5: Assessing data availability

The potential indicators need to be assessed with regard to availability of data. Data have to be available in order to construct the indicator. If the data are not available, they have to be collected.

Table 3.2: Example of the use of the DPSIR framework on the basis of the components of the Water Poverty Index (WPI)

DPSIR	WPI indicator	WPI component
Driving force	% of households receiving a pension/remittance or wage	capacity
(e.g. population density, poverty)	expenditure measured by ownership of durable items	capacity
Pressure	domestic water consumption rate	use
	agricultural water use	use
	water use for livestock	use
	industrial water use	use
State	surface water assessment	resource
	groundwater assessment	resource
	reliability of resources	resource
	water quality assessment	resource
Impact on environment	people's use of natural resources	environment
	crop loss	environment
	% of households reporting erosion of land	environment
Impact on people	access to clean water	access
	reports of conflict over water use	access
	access to sanitation	access
	% of water carried by women	access
	time spent in water collection	access
	access to irrigation coverage	access
	under-five mortality rate	capacity
	% of households reporting illness due to water supplies	capacity
Response	education level of population	capacity
	membership in water users associations	capacity

Source: WWAP Secretariat, based on information from WPI, Sullivan et al., 2002a.

Table 3.3: Criteria for the selection of indicators

Scientific requirements

- robust, well-founded basis in scientific knowledge
- representativeness, describing the state or quality of an issue or subject, giving significant and precise information
- clearly and consistently defined, so as to be unambiguous or lend themselves to various interpretations, or to give inconsistent results in different situations
- be developed within an agreed-upon conceptual and operational framework
- quantitative expression
- be sensitive insofar as a small change to be measured should result in a measurable change in the indicator
- anticipatory, early warning, capable of indicating of degradation or risk before serious harm has occurred
- stability, low natural variability in order to separate stress caused effects from random fluctuations
- specific for a certain stress or effect
- broadly applicable to many stresses and sites, usable in different regions
- uncertainties need to be specified
- transformable (intelligent)

Policy requirements

- tailored to the needs of the primary users
- have ownership by users
- problem has to be manageable, thus cause-effect chain of indicator has to be known to enable tackling of the problem
- having a target or threshold against which to compare or an explicit scale ranging from undesirable states to desirable states (along with specific weightings) in order to assess the significance of the information
- recording either changes in the means recommended by policy or changes in the development impact attributable to policy
- lend itself to be linked to models, forecasting and information systems
- simple, easily interpretable and appealing to society in order to ease communication between policy makers and society
- matching with national and international policy plans and indicating the progress of policy
- availability of historical data in order to show trends over time
- data should be readily collectable and, thereby, lowering the technical and collection costs
- normalized to make things comparable and providing a basis for regional, national and international comparison

Sources: Information collected from the WWDR indicator workshop; Hoon et al., 1997; Van Harten et al. 1995; De Zwart 1995; Hendriks, 1995; Swart and Bakkes, 1995; OECD, 1994; Kuik and Verbruggen, 1991; Liverman et al., 1988.

In that case, the investment in data collection should be in balance with the information gained by the indicator. Data on socio-economic activities can be found in socio-economic databases and statistics. Data on the quantity and quality of resources may be obtained from literature, monitoring programmes, surveys and special projects.

For an integrated use of socio-economic and environmental indicators, the data have to be converted to the appropriate spatial level. Socio-economic statistics tend to be collected for administrative regions and aggregated to larger spatial levels (e.g. the national level), whereas environmental data are collected at the level of a water body or river basin. These differences in scale complicated the integrated use of the different indicators in the adopted framework. More discussion on data is provided in the next section of this chapter.

Step 6: Developing indicators

In the last step the indicators are developed. If sufficient data are available, an indicator can be calculated. If not, the data should be collected by means of monitoring or surveys. If data collection is not possible, new or alternative indicators will need to be considered by returning to the step of formulating potential indicators.

It should be noted that indicator development is an ongoing process, which requires adherence to the notion of learning by doing. This means that a set of indicators is never definite and the selected list of indicators is no exception. With time, the indicators will be adapted and improved on the basis of changing information needs, progress in policy development and scientific insights and experiences gleaned from the use of the indicators. The countries involved in the process of developing and proposing indicators must have full ownership of those indicators, and not be restricted to a preconceived list. The seven pilot case studies of this first WWDR are a good example of the broadening of indicator horizons.

Table 3.4: List of indicators used in the first edition of WWDR and indicators that will be developed in the future

Challenge area	Indicators used	Future indicators
Drivers	• Consumption: developing vs. developed world • Incidence of worms, scabies, trachoma, diarrhoea • Human Poverty Index: 5 indicators • Withdrawals: % of total annual renewable freshwater • Evapotranspiration: % used by humans • Runoff: % used by humans • Water shortages: no. of people and countries affected, no. countries unable to supply minimum drinking water • Wetlands: % threatened • Transboundary rivers: % population dependent on Internal renewable water resources • Polluted water: % of population exposed to pollution indicators: coliforms, industrial substances, acid, heavy metals, ammonia, nitrates, pesticides, sediments, salinization • Natural disasters: deaths from water-related causes, number of floods • Lifestyle indicator: summer watering of gardens, availability of certain foods year round • Precipitation changes in north and south • Precipitation-related disasters: death and property damage • Mentions of water in international agenda, WB, GEF, WSSD • Ministerial statements mentioning water	• National and subnational indicators of major drivers
Water cycle (chapter 4)	• Long-term average water resources • Global hydrological network • Country data on water resources • Water availability versus population • Mean annual precipitation • World maximum point rainfalls for different durations • World's largest groundwater systems • Groundwater use for agricultural irrigation • Annual flows to the world's oceans • Largest rivers in the world by mean annual discharge with their loads • Countries using the largest quantities of desalinated water and treated wastewater	• Use/yield {Yield = f(Q, variability in both space and time, storage)} • PDSI or aridity index (moisture index) • Organic pollutants load • Biological contaminants (E. coli/thermo-tolerant coliform) • UNESCO/IAEA/IAH/ECE Groundwater Index • Water stress threshold map • Naturally occurring inorganic contaminants fluor and arsenic
Promoting health (chapter 5)	• Distribution of unserved people – water supply • Distribution of unserved people – sanitation • Actual and total water supply coverage, global, urban and rural breakdown • Actual and total sanitation coverage, global, urban and rural breakdown • Incidence of cholera in the world	• Burden of water-associated diseases (expressed in DALYs) with Comparative Risk Assessment • Fraction of the burden of ill-health resulting from nutritional deficiencies, attributable to water scarcity impacts on food supply • Access to improved drinking water sources and extension of piped water supply • Investment in drinking water supply and sanitation • Percentage of Health Impact Assessments (HIA) of water resources development and compliance with HIA recommendations
Protecting ecosystems (chapter 6)	• Degree of river fragmentation • Terrestrial Wilderness Index • Land converted to agriculture • Area of wetland drained • Hydrological indicators (flow, etc.) • Emissions of water pollutants by sector • Compliance with water quality standards for key pollutants • Biological water quality (based on community response) • Biological assessment (perturbation from reference condition)	• Uptake of strategies/legislation uptake for environmental protection • Reporting procedures in place at national level • Sites/species afforded protection by legislation • Restoration schemes • Formation and empowerment of regulatory or other institutions

Table 3.4: continued

Challenge area	Indicators used	Future indicators
Protecting ecosystems (chapter 6... cont'd)	• Numbers or presence/absence of non-native (alien) species • Levels of endemism • Rapid Biodiversity Inventory – Conservation International/Field Museum AquaRAP • Numbers/proportion of threatened species (critically endangered) • Living Planet Index • Commercial or other fisheries catch • Food production trends	
Water and cities (chapter 7)	• Megacities around the world • Proportion of urban populations with access to 'improved' water supply and sanitation • Water supply: access to 'improved' water supply – %, type by %; house connection, yard tap, public tap, unserved • Water supply cost per litre • Water supply: unaccounted for water – % of distribution input, community-supported upgrading programmes: numbers of people/households in programme • Water consumption levels: Domestic: litres per capita per day (lpcpd), water metre tariff (punitive structure aimed at reducing undue consumption), % of water recycled, type of water source (groundwater, river, mix, etc.) • Industry and commercial: m^3 per day • Child mortality rates: deaths per 1,000 live births • Children < 5 years: diarrhoeal diseases linked to inadequate water and sanitation • Sanitation: access to 'improved' sanitation – %, sanitation: sewer connections – %, solid waste collection – % • Water source (river) distance from demand centre: % > 8 km, inter-basin transfer: % • Water impounding reservoirs (dams): supply volume m^3 per year	• Regional and country level breakdown of values • Access to 'safe, sufficient' water supply – % • Access to 'safe, convenient' sanitation – % • Urban ecological footprint
Securing the food supply (chapter 8)	• Average food price (whole world) • Average per capita food consumption (whole world and regions) • Lending for irrigation and drainage (whole world) • Area equipped for irrigation vs. total arable land by country • Irrigated area (regions) • Agricultural water use by country • Number of chronically hungry people by country • Average grain yields (whole world) • Area of arable land (whole world) • Cropping intensity (whole world) • Consumption of livestock products (regions) • Fish consumption (marine, inland and aquaculture, whole world) • Water used for irrigation (net and gross, whole world)	• Water used for irrigation (net and gross, groundwater and surface water), informal (supplemental, spate, local water harvesting), surface irrigation, sprinkler irrigation, drip irrigation • Productivity: $ or vol./m^3, efficiency, jobs per drop • Proportion of crops marketed at government controlled prices • Agricultural subsidies • Breakdown of food consumption broken down into cereals, oil crops, livestock and fish • Total investment (private, state, development agencies) in irrigation and drainage • Food imports/exports between regions
Industry (chapter 9)	• Economic value (in US$) obtained annually by industry per cubic metre of water used • Competing water uses for main income groups of countries • Contribution of main industrial sectors to BOD production in high income OECD countries and in low income countries • Industrial water efficiency	• Industrial use of water per capita by total developed water per capita • Reuse/recycling • Pollution from industry

Table 3.4: continued

Challenge area	Indicators used	Future indicators
Energy (chapter 10)	• Distribution of households with access to electricity in forty-three developing countries • World's electricity production • Deployment of hydropower • Installed hyrdocapacity	• Efficiency/Productivity (output per m^3) • Use of water in thermal towers and competition with other uses • Access to electricity rural and urban coverage for the whole world • Per unit cost of renewable and non-renewable energy sources
Risks (chapter 11)	• List of severe natural disasters since 1994 • Major drought events and their consequences in the last century • Trends in causes of food emergencies, 1981–99 • Trends in great natural catastrophes	• Losses – country and basin level data, by region and globally, in human life (number/yr), in real and relative social and economic values (total losses, % of GNP, growth, investments and development benefits) • Population exposed to water-related risk (number of people/yr, income groups) • Other than water-related risks (% of losses from seismic, fire, industrial and civil stability risk) • Number of people living with 100–year flood. Vulnerability map based on the proportion of land within 1 km of river with slope of less than 1 degree • Legal and institutional provisions for risk-based management (established/not established) • Budget allocation for water risk mitigation (total and % of total budgets/yr) • Risk reduction in flood plains (% of total flood plain populations) • Risk reduction and preparedness action plans formulated (% of total number of countries) • Risk-based resource allocation (country, international organizations, [yes/no])
Sharing water (chapter 12)	• Basins of high/medium water stress (abstraction as proportion of river flow) • Dependence of country's water resources on inflow from neighbouring countries (inflow as ratio of total water availability) • Number of international basins • Number of treaties/cooperative events for international rivers • Shared aquifers – number/resource volume/conflicts relating to changes that might suggest international basins where there is a requirement for greater cooperation. Indicators of these types of changes are: 1. Newly internationalized basins 2. Basins with unilateral projects and lack of institutional capacity (treaties/bodies/positive relations) 3. International basins where non-water related hostility exists between states	• Demand changes (sectoral) and distribution • Mechanisms for sharing within country (allocations/priorities) both routinely and at times of resource shortage • Proportion of water use by industry, agriculture and domestic sector • Existence of law for judicious distribution of water • Water Policy accounts and statements
Valuing water (chapter 13)	• Annual investment in water for agriculture, water supply and sanitation and environment and industry • Sources of investment funds • Annual investment in urban and rural sanitation • Level of cost recovery for water supplies for agriculture • Level of cost recovery for urban water supplies • Price of water from municipal water supply systems • Comparison of the price of water from the public utilities and informal water vendors	• Price of water charged to farmers for irrigation • Average price of water in rural water supply systems • Sewerage charges

Table 3.4: continued

Challenge area	Indicators used	Future indicators
Ensuring knowledge (chapter 14)	• Gross primary school enrolment • Illiteracy rate • Density hydrological monitoring stations worldwide, by region • Research and Development expenditure for selected countries • Number of television sets and radio receivers per 1,000 people • Number of telephone lines per head • Expenditure on ICT • Number of hydrological monitoring stations, by World Meteorological Organization (WMO) regions	• No. of water resource institutions • No. of water resource scientists • Newspaper circulation, per 1,000 inhabitants • Water topics in school curriculum • Number of web sites with available water resource countries information
Governing wisely (chapter 15)	• Existence of institutions (water resources authorities) responsible for management (including issuing abstraction and discharge licences), which are independent of water users. Percentage of land area covered by such institutions. Number of water authorities and average area covered by each • Existence of water quality standards, for effluent discharges, minimum river water quality targets • Existence of defined water rights	• Water quality in rivers, lakes etc. • Numbers of instances when water service providers experience a raw water shortage • Existence of legislation advocating Dublin principles • Institutional strengthening and reform (post-1992) • Defined roles of government (central and local) • Existence of participatory framework and operational guidelines • Private sector involvement and stakeholders' responsibility established and implemented • Asset ownership properly defined • Financial commitment for IWRM adoption

This list of indicators gives only major indicators used in the report. An appropriate database system is being developed under the guidance of the Data Advisory Committee. Similarly, efforts are already underway to further develop 'proto-indicators'.

Observational Challenges for Indicator Development

After decades of intensive experimental and modelling studies, our understanding of hydrology at the local scale is arguably well in hand. In contrast, knowledge of the complex function of drainage basins, especially in relation to anthropogenic challenges, is far less developed. The indicators developed for the WWDR must be able to track the sources, transport, and fates of water as well as waterborne constituents through large heterogeneous watersheds. On the one hand, there is the challenge of quantifying elements of the land-based hydrological cycle and moving progressively towards these larger scales; on the other hand, the dynamism associated with human-water interactions are all too often difficult to understand. The former has been the focus of major international observational and scientific coordinating programmes (e.g. WMO-HYCOS, Coordinated Enhanced Observing Period [CEOP], Committee on Earth Observation Satellites [CEOS], Global Terrestrial Network on Hydrology [GTN-H], UNESCO-HELP [Hydrology, Environment, Life and Policy]), but for the latter there are only a few monitoring systems in place (CSD process through inputs from the United Nations Administrative Committee on Coordination Subcommittee on Water Resources [UN ACC/SCWR] now reconstituted as UN-Water, GWP, World Water Council [WWC]). Similarly, while the deteriorating state of current monitoring networks has been noted in the former case, a substantial build-up is also required to make them capable of capturing the system's dynamism. In both, substantial improvements are needed to overcome the observational challenges.

Hydrologic information base

The need for hydrologic information that can be used to develop indicators has never been more timely. With the advent of high quality biophysical data sets, including those from remote sensing and operational weather forecasts, the scientific community is rapidly approaching a situation in which the hydrological cycle can be monitored over large regions and in near-real time. Paradoxically, the trend of reduction in stations that routinely monitor hydrographic variables may greatly limit the usefulness of these emerging high technology tools, since they ultimately require calibration against known ground-based standards in order to prove their reliability.

Our ability to monitor the terrestrial water cycle using traditional discharge-gauging stations – the mainstay of water resource assessment – continues to deteriorate rapidly across much of the world. A time series plot (made in 1999 but little different today) showing global station holdings from the WMO Global Runoff Data Centre (GRDC) and earlier UNESCO hydrologic data banks amply demonstrates the difficulty at hand. Declines can be found all over the world, even in otherwise well-monitored countries such as the United States and Canada. Marked declines are most dramatic in the developing world. Insofar as these regions are, by their very nature, subject to the direct and immediate effects of rapid anthropogenic change, it is of great concern that the infrastructure to adequately monitor water resources is substantially lacking. A major effort will be required to rehabilitate these atrophied monitoring systems. Without a sustained international commitment to baseline monitoring, WWAP's indicator development will suffer a major setback.

Several factors contribute to the loss of hydrologic reporting (Ad Hoc Group on Global Water Data Sets, 2001; NRC, 1999; Kanciruk, 1997). Data collection is now highly project-oriented, yielding poorly integrated time series of short duration, restricted spatial coverage, and limited availability. In addition, there has been a legal assault on the open access to basic hydrometeorological data sets, aided in large measure by commercialization and fears surrounding the piracy of intellectual property. Delays in data reduction and release (up to several years in some places) are also prevalent.

Ironically, global analysis of hydraulic engineering is limited by a shortage of water data. Existing dam registries give no information on river discharge, either into or out of impoundments. The registries also fail to give complete information on storage volumes and area. Relationships linking water level to surface area and volume are not supplied, yet such information is essential for computing accurate impoundment residence times and in predicting hydrograph distortion. The volume of groundwater seepage loss is also not documented, even for the largest reservoirs, and the hydrology of millions of farm ponds and rice paddies is, at best, educated guesswork.

Information based on human impacts and stewardship

A critical requirement for water resource assessment across all scales includes a broad set of socio-economic variables to help quantify the utilization of water. The conjunction of these variables can produce two fundamental quantities, that is, the rate of water withdrawal and/or consumption which can then be compared to water supply. This comparison produces the important indicator of relative water use, which is a measure of the capacity of the water resource system to provide services to a community of local and regional users. However, the situations are not that simple to observe. Ethno-engineering solutions to water conveyance in the Gilgit area of Pakistan and in arid Egypt have proved common understanding of hydraulics to be wrong. Similarly, while the total contribution of people's participation may have been guessed at and accounted for, calculating losses due to declining sources of traditional wisdom have never been attempted. Such issues are important and need to be captured in indicators, and were visible even in the pilot-case-study countries where the national governments worked with the WWAP secretariat to produce reports.

These are some of the problems that WWAP has had to confront. There has been a steady realization that these challenges must be examined as part of an integrated set of problems, where everything is linked to everything else (though some links are of greater strategic importance than others). While the combination of DPSIR and policy-driven log frames (upon endorsement of lead UN agencies) can structure the desired indicators, this process must be looked upon as switching on the engine before putting the car in gear and placing one's foot on the pedals. There is a whole chain of required actions: identification and development of global water data sets, in-country capacity-building and technology transfer, statistical analysis, modelling, data interpretation, comparative trend analysis, and rapid data dissemination. Although in this report a large number of indicators have been presented, the longer-term aim of the WWDR is to present most indicators in a geospatial format.

Limitations, Caveats and Discussion Points

The previous section introduced the approach adopted in developing indicators both used and proposed through the WWDR process. In this section, some of their major limitations are highlighted.

The issue of the right scale

Scale is an important aspect in indicator development and use. This paragraph discusses four issues of scale. Firstly, indicators are often targeted to a certain spatial scale. As the information needs may differ at local, regional and global levels, indicators developed for a certain spatial scale may not be useful for another. For instance, data for an indicator at a high spatial level cannot always be obtained by aggregating the data of a lower spatial level. This hampers the calculation of indicator values on the global level. For example, the WHO/UNICEF Joint Monitoring Programme (JMP) publicizes indicator values on water supply and sanitation at a global scale as the percentage of people with access to improved drinking water resources and access to improved sanitation facilities (see table 3.5 for a selection of countries). The aim of the indicator is to compare water supply and sanitation over the world. The information need at the regional level, however, will be different; at that scale the distribution of access over the regions or the type of technology in a certain region are more interesting factors. The regional data can only be aggregated to a national indicator value if the indicators and data collection are the same between the

Table 3.5: Sample of indicator values per country on water supply and sanitation

	% of population with access to improved drinking water sources in the year 2000			% of population with access to improved sanitation facilities in the year 2000		
	Total	**Urban**	**Rural**	**Total**	**Urban**	**Rural**
Afghanistan	13	19	11	12	25	8
Algeria	94	88	94	73	90	47
American Samoa	100	100	100	–	–	–
Andorra	100	100	100	100	100	100
Angola	38	34	40	44	70	30
Argentina	79	85	30	85	89	48
Australia	100	100	100	100	100	100
Austria	100	100	100	100	100	100

Source: WHO/UNICEF, 2000.

different regions of a country, one of the challenges the JMP is working on.

The problem of different information needs hinders the use of available regional data for comparison at a global scale, such as in the WWDR. A second example is GNP, which is an indicator for economic activity measuring the total monetary value of all final goods and services produced for consumption in society during a particular time period. GNP as an average national number is not representative of regional variations between communities within a country. However, although GNP is useful to compare the economic activity between different countries at a global scale, it cannot be directly related to welfare as in large countries there may be great internal variations, in most cases, however, certainly when considering health, the economic indicators go hand-in-hand with welfare (rural poverty normally equals rural ill-health). Recent studies have shown that if economic indicators are linked to societal equity indicators, the association between economy and welfare becomes much more solid. However, the welfare aspects (e.g. health, education, security, environmental quality) cannot be directly related to the GNP of a country. In table 3.6 an overview is given of a number of indicators, their aim and the correct spatial scale of their use.

A second important issue is selecting the optimal spatial scale to aggregate and present the values of the indicator or index. This scale depends on the information need and aim of the indicator. Figure 3.7 shows how the spatial scale modifies the information provided by an indicator. The Water Stress Index (WSI) calculates the percentage of water demand that cannot be satisfied without taking measures. The figure's four spatial scales present different scores of the Water Stress Index for almost each part of the world. For example, Brazil scores white (WSI 0–25 percent) on a grid cell scale, mostly yellow on a watershed scale (WSI 25–50 percent), white on

Table 3.6: Overview of a number of indicators, their aims and the correct spatial scale of their use

Indicator	Information provided	Aim	Spatial scale
Gross National Product	Economic activity per country	Comparison of economic activity between countries	Global
Water Stress Index	% of water demand that cannot be satisfied without taking measures	Indication of areas suffering from water stress	Low spatial scale: grid cell or watershed
Water Poverty Index	Index based on five components: resource availability, access to water, capacity of people to manage water, use of water, environment	Providing information on water and poverty-related issues	Different scales possible depending on the aim: communities and regions for comparison within a country or countries for comparison on a global scale
Indicator species	Presence or abundance of the species	Indication of the ecosystem quality	The scale of one ecosystem or comparable ecosystems located in the same climate range

Figure 3.7: Values of the Water Stress Index (WSI) presented on four different spatial scales: country, grid cell, regional and watershed

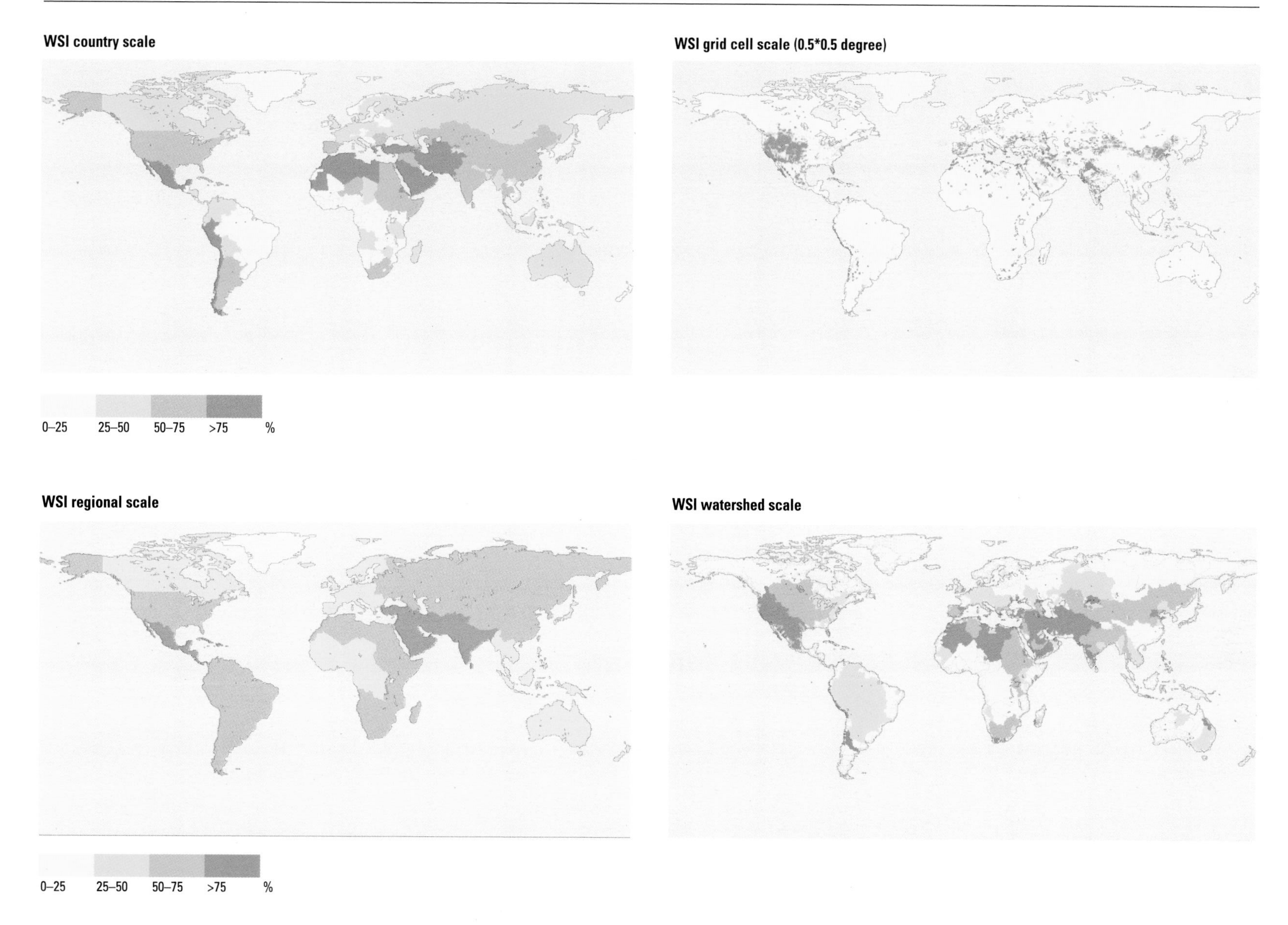

The grid cell is the optimal and most reliable scale to point out water-stressed areas. Comparison of the four different types of spatial scales used in the above maps shows that aggregation of the Water Stress Index on a large scale leads to information loss.

Source: Second World Water Forum, 2000.

a country scale and orange on a regional scale (WSI 50–75 percent). The hot spot of red colour (WSI >75 percent) in the West Coast of the United States in the figure on grid cell scale has turned into an orange colour in the figures on country and regional scales. Comparison of the four spatial scales leads to the conclusion that large-scale aggregations of the WSI result in too much information loss. The grid cell is the optimal scale to detect the areas of water stress in the world because it allows more detailed data to be shown rather than grouping all data together to produce an average.

Thirdly, selection of the right temporal scale is important for indicator aggregation and presentation as well. For example, the availability of water depends strongly on the seasons. A mean value calculated over one year can hide a shortage of water in dry periods and floods during wet times. In this case a seasonal-based mean or a minimum and maximum value over a year will provide more relevant information. The temporal scale of an indicator is also dependent on the point in time and the period of data collection. For example, water-quality monitoring is often based on a measurement frequency of once a month. Consequently, information on the minimum oxygen concentration in a water body will not be provided accurately by this measurement frequency. The minimum oxygen level is however important for the survival and development of aquatic fauna.

The last scale-related discussion point is the level of aggregation of the data. During the process of indicator or index development a trade-off has to be made between aggregation versus loss of detail. Which trade-off is made in a certain situation depends on the aim, the user, the system under consideration, knowledge of the system, data availability and the available financial resources. Too much detail can even lead to information loss, as the overall picture fades and becomes less clear. The initial selection of indicators is scientifically based, but if the need is policy-driven, a trade-off between scientific completeness and simplification for management will be inevitable.

Figure 3.8 provides an interesting example (EEA, 2001) based on an extensive knowledge and information base encompassing different institutes and scientists in countries bordering the Rhine River. Scientists have carried out studies in different regions on various time scales, looking into patterns, trends and data distribution profiles in order to understand and monitor the processes taking place. Other scientists are undoubtedly interested in the detailed information and in data interpretation. All this information has been aggregated into just two indicators, representing the entire river basin, completely discarding spatial features. The figure provides, however, a clear message to decision-makers and the public: oxygen content and biodiversity were poor in the 1970s, but have greatly improved since. Scientists have responsibility to ensure that the conclusions drawn from the figure are correct, despite the simplifications and need to provide additional information (e.g. benchmarks) which enable both decision-makers and the public to classify the present condition in terms of 'good', 'acceptable 'or 'poor'. In general, the search for the right balance between the policy aim of indicators and their scientific foundation requires an ongoing dialogue between scientists and policy-makers in order to improve and focus the indicator set. Proper documentation of the aggregation procedure and the original data enables retrospective testing, verification of the approach and increased transparency. While aggregation aims to reduce multidimensional information to a single dimension, visualization may present multidimensional information at a glance. Depending on what is to be visualized a specific design can be chosen (e.g. tables, diagrams, line charts and maps). Visualization techniques offer powerful means for knowledge transfer and communication.

It is essential to note that scale is a critical issue but it is very much linked to the issue of boundaries. Discrepancies between natural and administrative boundaries make indicator interpretation difficult, especially since rivers may often serve as administrative boundaries between countries or between provinces/states within countries. So collecting water-associated data in two administrative areas separated by a river often covers up the impact of the aquatic environment on health and risks, unless the data collected are segregated for respective distances from the river.

Figure 3.8: A time series of the oxygen concentration and living organisms in the Rhine River since 1900

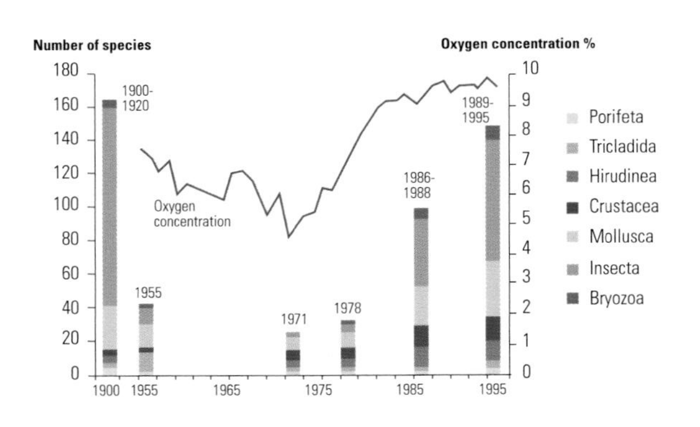

This figure provides a clear message to policy-makers and to the public. Although oxygen content and biodiversity were poor in the 1970s, they have since then strongly improved. This conclusion has been drawn from extensive data, which were then aggregated into easily understandable indicators.

Source: EEA, 2001.

Geospatial presentation of indicators

A related issue is the visual presentation of indicators to users. With a wide spectrum of users foreseen, calculated indicators must be presented as simply as possible while retaining their scientific rigour. In order to achieve this objective, offset the scale and aggregation problems and respect spatial variability in socio-economic and hydrological situations, WWAP is working towards presentation of indicators using geospatial tools. Although the difficulties treated above have a practical implication for water resource assessments at regional, continental, and global scales, existing monitoring mechanisms can be aligned to produce basic datasets to calculate various indicators.

WWAP has noted these concerns and has already undertaken, on a pilot basis, spatial representation of a set of water resource data from Africa. The Data Synthesis System (DSS) for Africa is an operational, digital information system for water resource assessment cast within a geographic information system framework accessible via the World Wide Web. The system includes a broad suite of spatial and statistical data encompassing point scale and gridded socio-economic and biogeophysical products for data exploration and download. This data is organized according to water indicator themes and is presented in the spatial context of the river basin to analyse the changing nature of water in relation to human needs and activities at the global, regional and case study scales. Figures 3.9 and 3.10 provide examples of how indicators are being synthesized and presented for the Nile basin.

Implications for tracking changes through time

Indicator time series show trends, which may provide information on the system's functioning or its response to management practices. The availability of extended time series can provide information on conditions throughout history, which could be used to derive a historically based reference condition. A comparison in time requires consistency in data collection and indicator construction. Changing indicators and data collection resulting from changing information needs means disruption of time series. While upgrading is necessary, it is a balance between better information to be gained at that moment or information gained on trends from extended time series of data (being less than perfect). For example, in figure 3.8 the number of species and the distribution over the species groups between 1900 and 1920 could be used as a historical reference, showing that the present number of species is approaching the historical reference.

Misinformation and misinterpretation

Indicators can voluntarily misinform or involuntarily be misinterpreted due to subjective elements inside the indicator or index, inadequate definition of the indicator, mathematical problems in the process of aggregation or the use of unreliable data.

1. Aggregating a number of indicators to one index involves the various steps of selection, scaling (transforming indicators into dimensionless measures), weighting (valuation), aggregation and presentation. These steps will require some combination of expert judgement, multicriteria analysis, public opinion polls, value-based decisions and modelling experiments. As they are greatly dependent on subjective perceptions and subject to change, these steps must be taken with caution. To avoid the problem of misinformation and misinterpretation, a clear description of the subjective elements in the indicators and indices should be given, such as the reference condition, the measuring rod, the weighting factors and the aggregation method.

2. Indicators can also misinform due to imprecise description, and it is therefore necessary to have a mechanism for cross-referencing and validating them. It may be useful for a proposed indicator to be broad enough to allow it to be deconstructed to its root variables so that distortions, if any, can be properly traced.

3. In most cases, the mean or median of a dataset has to be calculated in some step of the aggregation process. The distribution of data underlying the indicator or index may be very complex, that is, may contain outlying observations, have a heavy tailing or otherwise be heavily skewed, or be multimodal in nature. The straightforward calculation of mean or median may be misleading in such cases – expert judgement and proper documentation are required.

4. A fourth possibility of misinformation by indicators is the use of unreliable data. If the data collection or processing has been carried out in an inadequate manner, the indicators based on these data can provide wrong information. Measuring hydrological and environmental variables needs to be conducted applying the principles of quality assurance in the field (sampling, field measurements) and in laboratories. Methodology needs to be well validated, taking into consideration aspects such as suitability of purpose (are the performance characteristics, e.g. measurement uncertainty, suitable for the objectives at hand), robustness and traceability. Also the collection of socio-economic statistics needs to be quality assured (e.g. getting a representative random sample, asking the right questions in a survey). Most indicator reports do not provide detailed

Figure 3.9: Nile basin map showing climatic moisture index

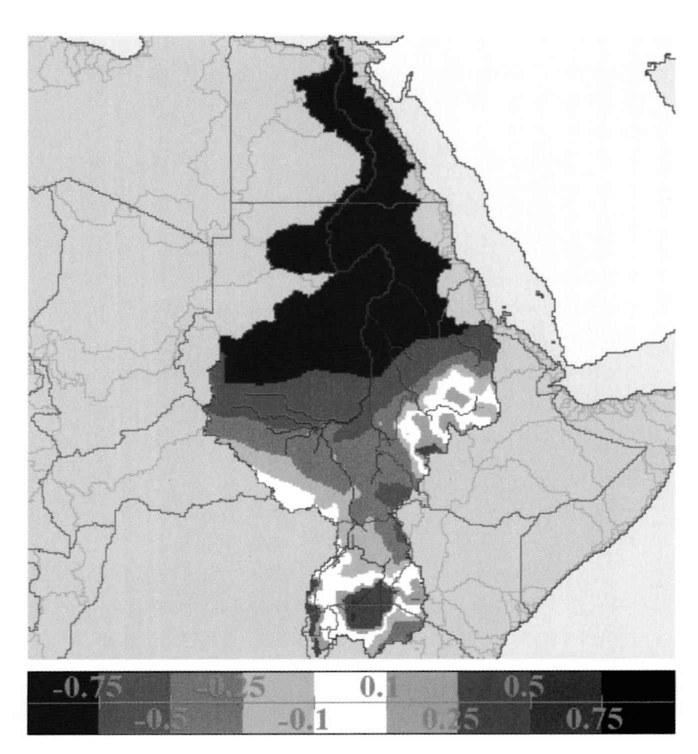

The method of Willmott and Feddema (1992) was applied relating potential evaporation (PET) to precipitation (PPT) to generate a mapping of relative water scarcity from a climatic perspective. Climatic Moisture Index is calculated as follows:

If PPT < PET then CMI = PPT/PET - 1
If PPT > PET then CMI = 1 - PET/PPT
If PPT = PET then CMI = 0

Negative values (shown in red on map) depict areas where the ecological demand for water (PET) exceeds the amount of available water in the form of precipitation (PPT). Positive values (shown in blue on map) depict areas where the amount of available water in the form of precipitation (PPT) is sufficient to meet the ecological demand for water (PET).

Source: Data Synthesis System for World Water Resources, UNESCO/IHP Contribution to the World Water Assessment Programme (WWAP), University of New Hampshire/UNESCO, 2002.

Figure 3.10: Nile basin map showing indicator on people with drought stress threshold

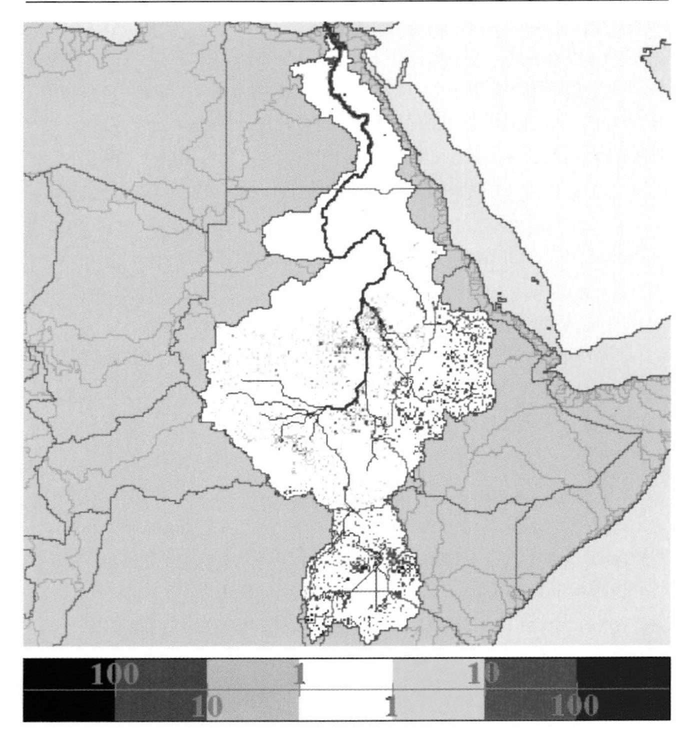

Thousands of people above (red) and below (blue) the 0.4 Demand per Discharge Stress Threshold for 30-year recurrence drought.

These two figures provide examples of how indicators are being synthesized and presented.

information on the quality of data. Reliable information is, however, necessary for proper assessment, especially when far-reaching decisions have to be made on the basis of indicators.

Data availability and implications

The dependence of indicator development on data can lead to the situation in which data availability drives the selection of indicators, which, in turn, reinforces the collection of the same data. Water quality monitoring systems have in the past been an example of the 'data-rich, but information-poor syndrome', in which plenty of data are produced but not tailored to information needs (Ward et al., 1986). Comparison with data availability may lead to modifying the indicator set, but should also feedback to more precise specification of data needs with more efficient design of monitoring programmes. An important element of the latter is coordinating socio-economic and environmental data collection and harmonizing associated spatial and temporal scales. Developing indicators and measuring them by monitoring programmes is a continuous, dynamic process, as information needs and measurement techniques can change over time (Cofino, 1995). Our efforts thus far have provided us with the following critical observations.

- Although efforts have been made, the adoption of earth systems modelling data into water resource assessment has been slow. The usefulness of doing so is nicely exemplified by studies aimed at understanding the impact of greenhouse warming on regional water systems.

- The capacity of the earth systems science community to generate data sets from high technology observing systems and modelling is enormous and growing. These products emerge as either remote sensing data sets or value-added products from climate, numerical weather prediction, or ecosystem simulations. It has been estimated that by 2010 the constellation of forty-five major planned satellite systems will provide 52 billion remotely sensed observations per day of the continental land mass, oceans, and atmosphere, up four orders of magnitude from year 2000 (Lord, 2001). Major efforts need to be initiated to effectively 'drink from the fire hose' in order to sensibly exploit information contained in these data streams. Collectively, these represent an enormously important contribution to the water resources community, a contribution that has yet to be productively tapped.

- With the recent advent of these spatially discrete and high-resolution earth systems data sets, the community finds itself poised to assemble a truly global picture of progressive changes to inland water systems and a potential means to monitor water availability worldwide. These newly developed digital products are often global in domain, are spatially and temporally coherent, and provide a consistent, political 'boundary-free' view of major elements defining the terrestrial water cycle. The UN agencies have already constituted an expert panel to assess the relevancy of all such potentially useful data sets. In the next phase, merging these data resources into integrated water assessment frameworks will pose a central challenge for WWAP.

- A coherent view of the contemporary state of global water resources will be impossible without a more fruitful interchange between the physical science and socio-economic communities (Young et al., 1994). The review in earlier sections of this chapter point to the explicit need for a common language in order to bridge the conceptual and practical gaps now separating these disciplines. Indeed, a more interdisciplinary approach is required, merging knowledge gained from years of case study experience with a growing technical capacity to monitor the changing state of the global hydrosphere. A systematic assessment of required data sets and a formal programme to collect this information worldwide is critically needed.

- There is a fundamental need to more fully document the role that humans play in the terrestrial water cycle. The science of hydrology is distinct from engineering hydraulics, but in some sense drainage basin dynamics are now more closely tied to the character of water engineering than to the behaviour of natural fluvial systems, at least in heavily managed part of the world. The important, yet crude, back-of-the-envelope calculation of Postel et al. (1996) showing a 54 percent control of the 35 percent of global runoff to which we have access is surely an understatement for many regions of the world. Humans are now an intrinsic part of the water cycle and strongly determine the availability of water resources. Feedbacks to the atmosphere and changes in runoff associated with widespread land conversion and management also are critical. At the same time that their influence on water supplies will grow, humans will increasingly be placed into competition for scarce water supplies and exposed to flood and drought, pollution, public health problems and economic stress, many such problems of our own making.

- The continental and global scales are also an important organizing framework, testing the capacity of state-of-the-art monitoring technologies to gain a synoptic view of the state of the terrestrial hydrosphere. A move towards a near real-time and operational capacity is sought. This capacity is important since the condition of water availability – from drought through flood – changes so rapidly. Not only is this important for preparedness in the face of weather-related vulnerability but also for securing a more accurate picture of the temporal and spatial aspects of chronic water stress.

Box 3.5: Achievements

The following methodological sketch has been agreed upon to further develop the indicators used and reported by UN agencies across the different chapters and case studies:

- Data and databases are important aspects of the exercise. For this, a conceptual framework on data collection and metadata preparation has been agreed upon, and major databases have been identified.
- The role of member countries in the process has been duly recognized. There is an understanding that the indicators will have the participation and thus the ownership of involved countries and UN system agencies.
- Based on the conceptual framework on information network efforts for data improvement, activities have been devised for implementation.
- Indicator development will be an inclusive exercise in which other major efforts will play a role in methodological refinement. WWAP will simply be taking a lead in the exercise.

- A logical coupling of global-regional-local scale is a must. While the effort of the WWDR will focus on presenting a global outlook, such understanding will build upwards from critical testing of methodologies at the regional and local scale first.

- World Wide Web-based technologies (including modelling, GIS, World Wide Web and metadata search engines) are essential to the provision of timely biogeophysical data. At the behest of UN agencies involved in WWAP's Data Advisory Committee, the programme has already received guidance on the way forward in terms of how these technologies may be used for data management.

Much work is therefore required to collect and prepare the biogeophysical and socio-economic data sets for use in future water resource assessments. In addition to the geography of water supply, issues of economic and technological capacity to provide water services, along with population growth, levels of environmental protection and health services, investment in water infrastructure, including irrigation and other hydraulic engineering works, must be included in future such analyses.

Conclusions

Indicators are thus vital instruments for the assessments embodied in the WWDR. The publication of this first edition is just the beginning of a long-term effort to develop a comprehensive set of indicators and their 'most friendly' presentation. In the years to come, we will be building upon the material presented in this chapter. We will have to improve the indicator development process (see box. 3.5) and involve more stakeholders from wider cultural backgrounds. Indicators identified and used in this report must be qualified using both scientific and political processes. While the scientific issues underlined in this chapter would have to be carefully taken into consideration, the more important point is to offer a suitable platform for the member states to become more involved in developing the scientific basis and the resulting indicators. It is essential that both the conceptual framework for indicator development and data gathering be subject to further scrutiny. We look forward to being able to demonstrate the progress that has been achieved in the second edition of the *World Water Development Report.*

References

Ad Hoc Group on Global Water Data Sets. 2001. 'Global Water Data: A Newly Endangered Species'. Co-authored by Vörösmarty, C. (lead); Askew, A.; Barry, R.; Birkett, C.; Döll, P.; Grabs, W.; Hall, A.; Jenne, R.; Kitaev, L.; Landwehr, J.; Keeler, M.; Leavesley, G.; Schaake, J.; Strzepek, K.; Sundarvel, S.-S.; Takeuchi, K.; Webster, F. An op-ed piece to *EOS Transactions American Geophysical Union*, Vol. 82, pp. 54, 56.

Ahmad, Y.J.; El Serafy, S.; Lutz, E., eds. 1989. *Environmental Accounting for Sustainable Development.* Washington DC, World Bank.

Alcamo, J.; Heinrichs, T.; Rösch, T. 2000. *World Water in 2025.* World Water Series Report #2. Kassel, Centre for Environmental Systems Research, University of Kassel.

Bakkes, J.-A.; Van Den Born; G.-J.; Helder, J.-C.; Swart, R.-J.; Hope, C.-W.; Parker, J.-D. 1994. *An Overview of Environmental Indicators: State of the Art and Perspectives.* UNEP/EATR.94-01; RIVM/402001001. Nairobi Environmental Assessment Sub-Programme, United Nations Environment Programme.

Bosch, P. 2001. *Guidelines for the Data Collection of the Kiev Report.* Technical report 66, European Environment Agency.

Bossel, H. 1999. *Indicators for Sustainable Development: Theory, Method, Applications.* A report to the Balaton group. Winnipeg, International Institute for Sustainable Development.

Bouni, C.; Narcy, J.-P. 2001. *The French Model for Water Management Examined in Terms of Governance: the Example of the Seine-Normandy Basin. Summary.* Seine Normandy Water Agency.

CIW (Commissie Integraal Waterbeheer). 2000. *Water in beeld 2000: Voortgangsrapportage over het waterbeheer in Nederland.* The Hague, Commissie Integraal Waterbeheer (Commission of Integrated Water Management).

Cobb, C.; Glickman, M.; Cheslog, C. 2001. 'The Genuine Progress Indicator 2000 Update'. In: *Redefining Progress Issue Brief.*

Corvalán, C.; Briggs, D.; Zielhuis, G., eds. 2000. *Decision Making in Environmental Health: From Evidence to Action.* London, E&FN, Spon.

Corvalán, C.; Kjellstrom, T.; Smith, K. 1999. 'Health, environment and sustainable development: identifying links and indicators to promote action'. *Epidemiology* Vol. 10, pp. 656–60.

Cofino, W.-P. 1995. '*Quality Management of Monitoring Programmes'.* In: M Adriaanse; J. Van der Kraats; P.-G. Stoks; R.-C. Ward (eds.), *Proceedings of the International Workshop Monitoring Tailor-made I, 20–23 September 1994.* Beekbergen.

CSD (Commission on Sustainable Development) 1998. Official Records of the Economic and Social Council, 22 December 1997 and 20 April 1 May 1998, Supplement No. 9. New York, UN (United Nations).

DPCSD (Department for Policy Coordination and Sustainable Development). 1996. *Indicators of Sustainable Development: Framework and methodologies.*

De Zwart, D. 1995. 'Biomonitoring'. *Monitoring Water Quality in the Future.* Vol. 3. National Institute of public health and the Environment. Bilthoven.

Desai, M. 1995. *Poverty, Famine and Economic Development.* Aldershot, Edward Elgar.

EEA (European Environment Agency). 2001. *Environmental Signals 2001.* Copenhagen, European Environment Agency regular indicator report.

Esty, D.; Peter, C. 2002. *Environmental Performance Measurement: The Global Report 2001–2002.* New York, Oxford University Press.

Falkenmark, M.; Lundqvist, J.; Widstrand, C. 1989 (November). 'Macro-scale Water Scarcity Requires Micro-scale Approaches: Aspects of Vulnerability in Semi-arid Development'. *Natural Resources Forum.* Vol. 13, No. 3, pp. 258–67.

Falkenmark, M. 1998. 'Dilemma when Entering the 21st Century – Rapid Change but Lack of a Sense of Urgency'. *Water Policy*, Vol. 1, pp. 421–36.

Federal Ministry for the Environment, Nature Conservation and Nuclear Safety, And Federal Ministry for Economic Cooperation and Development. 2001. Ministerial Declaration, Bonn Keys, and Bonn Recommendation for Action. Official outcomes of the International Conference on Freshwater, 3–7 December 2001, Bonn.

Gleick, P.-H. 2000a. 'Water Futures: A Review of Global Water Resources Projections'. In: F.-R. Rijsberman (ed.), *World Water Scenario.* London, Earthscan Publications.

——. 2000b. *Water: The Potential Consequences of Climate Variability and Change for Water Resources of the United States.* Pacific Institute for Studies in Development, Environment, and Security. Oakland, CA, Pacific Institute.

——. 1998. The World's Water: 1998–1999: The Biennial Report of Freshwater Resources. Washington DC, Island Press.

GWP (Global Water Partnership). 2000. *Towards Water Security: A Framework for Action.* Sweden, Global Water Partnership Secretariat.

Hammond, A.; Adriaanse, A.; Rodenburg, E.; Bryant, D.; Woodward, R. 1995. *Environmental Indicators: a Systematic Approach to Measuring and Reporting on Environmental Policy Performance in the Context of Sustainable Development.* Washington DC. World Resource Institute.

Hendriks, J. 1995. *Concentration of Microcontaminants and Response of Organisms in Laboratory Experiments and Rhine Delta Field Surveys.* RIZA nota 95.035. Lelystad, the Netherlands, Institute for Inland Water Management and Waste Water Treatment.

Hettelingh, J.P.; De Hann, B.J.; Strengers, B.J.; Klein Goldewijk, C.G.M.; Van Woerden, J.W.; Pearce, D.W.; Howarth, A.; Ozdemiroglu, E.; Hett, T.; Capros, P.; Georgakopolous, T.; Cofala, J.; Amann, A. 1998. *Integrated Environmental Assessment of the baseline scenario for the EU State of the Environment.* 1998 Report. The Netherlands, RijksInstituut voor Volksgezondheid en Milieu, or the National Institute of Public Health and the Environment.

Hoekstra, A.Y. 1998. 'Perspectives on Water'. Thesis, Technical University of Delft.

Hoon, P.; Singh, N.; Wanmali, S.S. 1997. 'Sustainable Livelihoods: Concepts, Principles and Approaches to Indicator Development, A Draft Discussion Paper'. Poverty and Sustainable Livelihoods, Social Development and Poverty Eradication Division, Bureau for Development Policy, United Nations Development Programme (prepared for the Workshop on Sustainable Livelihoods Indicators, UNDP. New York, 21 August 1997).

ICOLD (International Commission on Large Dams). 1994. *Dams and the Environment: Water Quality and Climate.* Paris.

IISD (International Institute for Sustainable Development). 1999. 'Beyond Delusion: A Science and Policy Dialogue on Designing Effective Indicators for Sustainable Development'. Workshop Report.

Kanciruk, P. 1997. 'Pricing Policy for Federal Research Data'. *Bull. Am. Meteor. Soc.* Vol. 78, pp. 691–2.

Kanto Regional Development Bureau. 2002. *The Proposal of New Water Quality Indicators. (Atarashi Suishitsu Sihyo no Tei-an).* Japan, Ministry of Land, Infrastructure and Transport.

Kjellstrom, T.; Corvalán, C. 1995. 'Framework for the Development of Environmental Health Indicators'. *World Health Statistics Quarterly*, Vol. 48, p. 2.

Kuik, O. and Verbruggen, H. 1991. *In search of indicators for sustainable development.* Dordrecht, Kluwer Academic Publishers.

Lenoir, T. and Gumbrecht, H.-U., eds. 1996. 'Writing Science'. In: P. Galison and D.-J. Stump (eds.), *The Disunity of Science: Boundaries, Contexts, and Power.* Stanford, CA, Stanford University Press.

Levy, M.-A. 2002. 'Measuring Nations Environmental Sustainability'. In: C. Daniel and P. Esty (eds.), *Environmental Performance Measurement: The Global Report 2001–2002.* New York, Oxford University Press.

Liverman, D.; Hanson, M.E.; Brown, B.-J.; Meredith, R.-W. 1988. 'Global Sustainability: Towards Measurement'. *Environmental Management.* Vol. 12, No. 2, pp. 133–43.

Liverman, D.; Moran, E.-F.; Rindfuss, R.-R.; Stern, P.-C. eds. 1998. *People and Pixels: Linking Remote Sensing and Social Science.* Washington DC, National Research Council.

Lord, S. 2001. 'NOAA Data Assimilation Activities'. Presented at the Coordinated Enhanced Observation Program International Workshop. February. Greenbelt MD, NASA/Goddard Space Flight Center.

Lorenz C.M. 1999. *Indicators for Sustainable Management of Rivers.* Thesis. Vrije Universiteit Amsterdam.

Meadows, D. 1998. 'Indicators and Information Systems for Sustainable Development'. In: 'A report to the Balaton group'. Hartland Four Corners, the Sustainability Institute.

NRC (National Research Council). 1999. *A Question of Balance: Private Rights and the Public Interest in Scientific and Technical Databases.* Washington DC, National Academy Press.

OECD (Organization for Economic Cooperation and Development). 1994. *Environmental indicators: OECD core set.* Paris.

Pereira-Ramos, L. 2002. *The SEQ-EAU – the French System for Evaluating Water Quality in Rivers.* Paris, Seine Normandy Water Agency.

Pigou. 1920. *The Economics of Welfare.* Oxford, Transaction Publishing.

Postel, S. 1997. *Last Oasis: Facing Water Scarcity.* New York, W.W. Norton.

Postel, S.-L.; Daily, G.-C.; Ehrlich, P.R. 1996. 'Human appropriation of renewable fresh water'. *Science*, Vol. 271, pp. 785–8.

Rotmans, J.; Van Asselt, M.B.A.; De Bruin, A.J.; Den Elzen, M.J.G.; De Greef, J.; Hilderink, H.; Hoekstra, A.Y.; Janssen, M.A.; Koester, H.W.; Martens, W.J.M.; Niessen, L.W.; De Vries, H.J.M. 1994. *Global Change and Sustainable Development, a Modelling Perspective for the Next Decade.* Global Dynamics and Sustainable Development

Programme GLOBO Report Series no. 4. Bilthoven, National Institute of Public Health and Environmental Protection.

Scheller, A. 1999. *Researchers' Use of Indicators.* Interim Report of The Indicator Project supported by the Alliance for Global Sustainability Centre for Energy Policy and Economics. Zurich, Swiss Federal Institute of Technology.

Seine-Normandy Water Agency. 2002. *Wetlands in the Seine-Normandy Basin: Situation and Water Agency Policy.* Paris.

Sen, A.-K. 1999. *Development as Freedom.* Oxford, Oxford University Press.

——. 1995. *Mortality as an Indicator of Economic Success and Failure.* Discussion paper 66, London School of Economics and Political Science.

——. 1981. *Poor, Relatively Speaking.* Oxford Economic Papers, No. 35.

Serageldin, I. and Steer, A. 1994. 'Epilogue: expanding the capital stock'. In: I. Serageldin and A. Steer (eds.), *Making Development Sustainable: from concepts to actions* Environmentally Sustainable Development Occasional Paper Series N°2. Washington DC, The World Bank.

Shiklomanov, I.-A. (ed.) 1997. 'Assessment of Water Resources and Water Availability in the World'. In: *Comprehensive Assessment of the Freshwater Resources of the World.* Stockholm, Stockholm Environment Institute.

——. 1996. *Assessment of Water Resources and Water Availability in the World: Scientific and Technical Report.* St. Petersburg, Russia, State Hydrological Institute.

Spreng, D. and Wils, A. 2000 [1996]. *Indicators of Sustainability: Indicators in Various Scientific Disciplines, Alliance for Global Sustainability.* AGS Report.

Sullivan, C.-A.; Meigh, J.-R.; Fediw, T. S. 2002a (May). *Derivation and Testing of the Water Poverty Index, Phase 1.* Final Report. Wallingford UK, Center for Ecology and Hydrology.

Sullivan, C.A.; Meigh, J.R.; O'Regan, D. 2002b. *Evaluating Your Water, a Management Primer for the Water Poverty Index.* Wallingford, Center for Ecology and Hydrology.

Swart, R.-J and Bakkes, J.-A., eds. 1995. *Scanning the Global Environment: A Framework and Methodology for Integrated Environmental Reporting and Assessment.* UNEP/EATR.95-01; RIVM 402001002 Environmental Assessment Sub-Programme. Nairobi, United Nations Environment Programme.

Townsend, P. 1979. *Poverty in the UK.* Harmondsworth, Penguin.

UNCSD (United Nations Common Supply Database). 2002. *Indicators of Sustainable Development.*

UNECE (United Nations Economic Commission for Europe). 2000 (March). *Guidelines on Monitoring and Assessment of Transboundary Rivers.* UNECE Task Force on monitoring and assessment. Lelystad, Rijksinstituut voor Integraal Zoetwaterbeheer en Afvalwaterbehandeling (Institute for Inland Water Management and Waste Water Treatment)

USEPA (United States Environmental Protection Agency). 1996. *Environmental Indicators of Water Quality in the United States.* EPA 841-R-96-002. Office of Water, 4503F. United States.

Van Adriaanse, A. 1993. *Environmental Policy Performance Indicators.* Amsterdam, SDU Publishers.

Van Dreicht, G. and Knoop, J.-M. (in prep.). *Water Stress Assessment and Forecast at the Global Scale.* Development of a computer programme for simulation of water demand and water availability for global-scale analysis of water stress. ARiBaS 1.07. Rapport No. 402001016. Bilthoven, RijksInstituut voor Volksgezondheid en Milieu (National Institute of Public Health and the Environment).

Van Harten, H-.A-.J.; Van Dijk, G.-M.; De Kruijf, H.-A.-M. 1995. *Waterkwaliteits indicatoren: overzicht, methodologie ontwikkeling en toepassing.* RIVM report 733004001. Bilthoven, Rijksinstituut voor Volksgezondheid en Milieuhygiëne (National Institute for Public Health and Environment Hygiene).

Vörösmarty, C. 2002. *Global Water Assessment: Potential Contributions from Earth Systems Science.* Prepared for the WWAP.

Vörösmarty, C. and Sahagian, D. 2000. 'Anthropogenic Disturbance of the Terrestrial Water Cycle'. *BioScience,* Vol. 50, pp. 753–65.

Vörösmarty, C.; Fekete, B.-M.; Tucker.; B.A. 1996. *River Discharge Database, Version 1.0* (RivDIS v1.0), Volumes 0 through 6. A contribution to IHP-V Theme 1. Technical Documents in Hydrology Series. Paris, United Nations Educational, Scientific and Cultural Organization.

Vörösmarty, C.; Fekete, B.-M.; Meybeck, M;. Lammers, R. 2000. 'A Simulated Topological Network Representing the Global System of Rivers at 30-minute Spatial Resolution (STN-30)'. *Global Biogeochemical Cycles,* Vol. 14, pp. 599–621.

Ward, R.-C.; Loftis, J.-C.; Bride, G.-B. 1986. 'The 'Data-rich But Information-poor' Syndrome in Water Quality Monitoring'. *Environmental Management* Vol. 10 No. 3, pp. 291–7.

WHO (World Health Organization). 1999. *Environmental Health Indicators: Framework and Methodologies.* Prepared by David Briggs. University College Northampton, Nene Centre for Research.

——. 1996. *Linkage Methods for Environment and Health Analysis – General Guidelines.* Document WHO/EHG/95.26. Geneva.

WHO/UNICEF (World Health Organization/United Nations Children's Fund). 2000. *Global Water Supply and Sanitation Assessment 2000 Report.* Geneva, World Health Organization/New York, United Nations Children's Fund.

Willmott, C.-J. and Feddema, J.-J. 1992. 'A more rational climatic moisture index'. *Professional Geographer,* Vol. 44, No. 1. pp. 84–88.

World Bank. 2002. World Development Report 2002: Building Institutions for Markets. Washington DC.

——. 2001a. *African Development Indicators 2001.* Washington DC.

——. 2001b. *World Development Report 2001.* Washington DC.

World Economic Forum. 2002a. *2002 Environmental Sustainability Index.* An Initiative of the Global Leaders of Tomorrow Environment Task Force, Annual Meeting 2002. Yale Center for Environmental Law and Policy, Yale University, Center for International Earth Science Information Network, Columbia University.

——. 2002b. *Pilot Environmental Performance Index.* An Initiative of the Global Leaders of Tomorrow Environment Task Force, Annual Meeting 2002. Yale Center for Environmental Law and Policy. Yale University, Center for International Earth Science Information Network, Columbia University.

WRI (World Resource Institute). 1998. *Watersheds of the World: An Assessment of the Ecological Value and Vulnerability of the World's Watersheds.* Washington DC.

Yasuda, G. and Murase, M. 2002 (February). 'Flood Risk Indicators'. Paper presented at the workshop on indicator development for the World Water Development Report. Rome.

Young, G.-J.; Dooge, J.-C.-I.; Rodda, J.-C. 1994. *Global Water Resource Issues.* Cambridge, Cambridge University Press.

ii

Part II: A Look at the World's Freshwater Resources

We all depend on the same vital element: water. Diverse by its very nature, it is solid, vapour and liquid; it is in the air, on the Earth's surface and within its ground – ever-changing, and giving shape to a dramatic range of natural ecosystems.

For the Earth's inhabitants, diversity of the resource also means great disparities in well-being and development. As we degrade the quality of our water and modify the natural ecosystems on which people and life depend, we also threaten our own survival.

Before we further explore the complex relationship linking water and people, this part provides a brief look at the current state of the finite but dynamic and wonderful resource that is freshwater.

4 The Natural Water Cycle

By: UNESCO (United Nations Educational, Scientific and Cultural Organization) / WMO (World Meteorological Organization)

Table of contents

We made from water every living thing.

The Koran (Sura 21:30)

WATER IS THE MOST WIDELY OCCURRING SUBSTANCE on this planet. Globally distributed by the hydrological cycle, driven by the energy cycle, the circulation of water powers most of the other natural cycles and conditions the weather and climate. Water has shaped the Earth's evolution (Dooge, 1983) and continues to fashion its progress, in marked contrast to those bodies in the solar system without water. While the greater part of the water within the Earth's hydrological cycle is saline, it is the lesser volume of freshwater within the land-based phase which provides a catalyst for civilization. This is the water precipitated from the atmosphere onto land, where it may be stored in liquid or solid form, and can move laterally and vertically and between one phase and another, by evaporation, condensation, freezing and thawing. On the land surface, this water can travel at widely differing velocities usually by predictable pathways (Young et al., 1994) which can slowly change with time. These pathways combine to form stream networks and rivers within river basins, the water flowing by gravity from the headwaters to the sea. Some basins, such as the Amazon, are massive, others minute. Depending on the nature of the geology, soils and land cover within the basin, a varying proportion of this water may infiltrate to recharge the underlying aquifers, some recharge re-emerging later to sustain river flows.

This groundwater combined with surface water forms the world's freshwater resource; renewable but also finite and vital not only to human systems but also to the terrestrial environment.

The existence of springs and other sources of water have played a major role in determining human settlement. Rivers and lakes provided routes for transport of goods and people, later supplemented by canals. Falling water provided and continues to provide power for industry. Today, water is employed for a wide variety of different purposes: desalination, recycling and reuse of wastewater, rainwater harvesting and similar non-conventional methods provide or add to the resource in certain localities. The value of water, its cost, competition between different uses including its cultural and aesthetic aspects, all add further dimensions to the consideration of water resources.

STUDIES OF THE WORLD WATER BALANCE commenced in the late nineteenth century, and examples of these and later studies are listed by Lvovitch (1970) and Baumgartner and Reichel (1970). Korzun (1978) and more recently Shiklomanov (forthcoming) refined estimates of the global budget and its regional variations. Table 4.1 shows the approximate volumes for the globe in the different phases of the hydrological cycle, the annual volumes recycled with their average replacement periods. This table highlights the enormous disparity between the huge volume of saltwater and the tiny fraction of freshwater and, in addition, the long residence time of polar ice and groundwater, as opposed to the brief period for which water remains in the atmosphere. Some 96.5 percent of the total volume of the world's water is estimated to exist in the oceans and only 2.5 percent as freshwater, but these and similar estimates lack precision. Nearly 70 percent of this freshwater is considered to occur in the ice sheets and glaciers in the Antarctic, Greenland and in mountainous areas, while a little less than 30 percent is calculated to be stored as groundwater in the world's aquifers. Again there is the disparity between these large volumes and the much smaller estimated volumes of water stored in rivers, lakes and reservoirs and in the soil, together with the water in plants and the atmosphere.

Measuring Water Resources

Even more important are the wide variations in the distribution of water in time and space across the globe and the problems such variations pose for the reliability of assessments of water resources. These assessments depend largely on hydrological data obtained from measurements made by ground-based networks of instruments and surveys, from sensors in satellites and from a number of other sources. These assessments are essential prerequisites for successful water resource development and management (WMO/UNESCO, 1997).

Although hydrological measurements have been made in Egypt and China for several thousand years, networks of hydrological instruments, as we know them, were started in Europe and North America in the eighteenth and nineteenth centuries. Today, most national networks consist of stations where variables such as precipitation, evaporation, soil moisture, ice, sediment, water quality, groundwater level and river water level and discharge are measured, continuously in some cases, or daily, monthly or less frequently in others (WMO, 1994). There have been major improvements in monitoring groundwater and in knowledge of groundwater as a result of the extension of geological mapping across the world and the hydrogeological interpretation of these maps. In some countries, ground-based weather radar systems are employed for determining the distribution of precipitation, while remotely sensed data from satellites are used for estimating the extent of lying snow, precipitation, soil moisture and certain other variables. However, many national networks are still composed of instruments and sensors that were first introduced in the nineteenth century. Many of these devices suffer from inherent errors; they lack maintenance and they are not calibrated regularly. The characteristics of the network and of its development vary from country to country: in some there are measurements of the whole range of hydrological variables and systems for accessing remotely sensed data, in others there are only rudimentary networks sampling just a few of the variables. Figure 4.1 shows the number of monitoring stations that make up the global hydrological network, by type and percentage of the total (WMO, 1995). Of course, these networks are not only used for assessing water resources, but also for flood and drought forecasting and prediction, for pollution protection, water conservation, groundwater protection, inland navigation and for a host of other purposes.

Knowledge of water resources is only as good as the available data, but the various assessments of water resources that have been conducted, together with other surveys, invariably indicate that hydrological data, including hydrogeological data, are lacking in many parts of the world. Indeed, it is a twin paradox that those areas with most water resources, namely mountains, have the least data and that the nations of Africa, where the demand for water is growing fastest, have the worst capabilities for acquiring and managing water data. This lack of data applies to surface water and groundwater, and to quantity and quality. Indeed, with the exception of Latin America, the reliability and availability of data have declined sharply since the mid-1980s particularly in Africa, in eastern Europe (Rodda, 1998) and around the Arctic (Shiklomanov et al., 2002), largely because national hydrological and allied networks have been degraded by lack of investment. There are many countries with no data on water chemistry, productivity, biodiversity, temporal changes and similar biological expressions of the state of the aquatic environment. Systems for storing, processing and managing these data and using them for assessing water resources and other purposes, such as flood forecasting, are often rudimentary.

Since the 1960s, both the United Nations Educational, Scientific and Cultural Organization (UNESCO) and the World Meteorological Organization (WMO) have been pursuing collaborative programmes that are designed to improve national hydrological capabilities, particularly for nations in need. These are the International

Table 4.1: The distribution of water across the globe

Location	Volume, (10^3 km^3)	% of total volume in hydrosphere	% of freshwater	Volume recycled annually (km^3)	Renewal period years
Ocean	1,338,000	96.5	–	505,000	2,500
Groundwater (gravity and capillary)	23,400[1]	1.7		16,700	1,400
Predominantly fresh groundwater	10,530	0.76	30.1		
Soil moisture	16.5	0.001	0.05	16,500	1
Glaciers and permanent snow cover:	24,064	1.74	68.7		
Antarctica	21,600	1.56	61.7		
Greenland	2,340	0.17	6.68	2,477	9,700
Arctic Islands	83.5	0.006	0.24		
Mountainous regions	40.6	0.003	0.12	25	1,600
Ground ice (permafrost)	300	0.022	0.86	30	10,000
Water in lakes:	176.4	0.013	–	10,376	17
Fresh	91.0	0.007	0.26		
Salt	85.4	0.006	–		
Marshes and swamps	11.5	0.0008	0.03	2,294	5
River water	2.12	0.0002	0.006	43,000	16 days
Biological water	1.12	0.0001	0.003		–
Water in the atmosphere	12.9	0.001	0.04	600,000	8 days
Total volume in the hydrosphere	1,386,000	100	–		
Total freshwater	35,029.2	2.53	100		

[1] Excluding groundwater in the Antarctic estimated at 2 million km^3, including predominantly freshwater of about 1 million km^3.

This table shows great disparities: between the huge volume of saltwater and the tiny fraction of freshwater; between the large volumes of water contained by the glaciers and the water stored in the aquifers; and between the amount of groundwater and the small volumes of water in rivers, lakes and reservoirs.

Source: Shiklomanov, forthcoming.

Figure 4.1: The global hydrological network by type

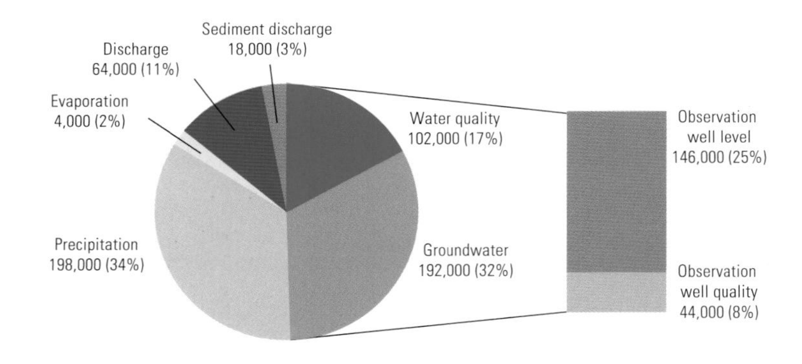

This figure shows the number of monitoring stations that make up the global hydrological network.

Source: WMO, 1995.

Hydrological Programme (IHP) and the Hydrology and Water Resources Programme (HWRP). On the global scale, WMO's World Hydrological Observing System (WHYCOS) aims to stimulate hydrological data collection and management in near real-time in a number of data-poor regions. Starting from a European base and progressing region by region, the UNESCO Flow Regimes from International Experimental and Network Data (FRIEND) Project (Gustard and Cole, 2002; van Lanen and Demuth 2002; Gustard, 1997) continues to improve the archiving of data and its use for assessing water resources, flood prediction and many other purposes in Europe, Africa, Asia and Latin America. FRIEND also involves the International Association of Hydrological Sciences (IAHS), which has been a partner in the IHP since it started in 1965. Hydrogeological interpretation of geological maps and other areas of hydrogeology have been stimulated by the IHP through cooperation with the International Association of Hydrogeologists (IAH) (Struckmeier and Margat, 1995). Another example of a collaborative programme is the UNEP/WHO/UNESCO/WMO Global Environment Monitoring Systems (WHO, 1991)[1] which is endeavouring to improve global water quality data. The GEMS archive contains some 1.6 million data points, but coverage is poor for Africa, Central Asia and for river mouths. Unfortunately similar biological equivalents seem to be lacking. However, there are other international initiatives, such as the Global Terrestrial Observing System (GTOS, 2002), which are likely to generate improved monitoring of aquatic ecosystems. UNESCO and WMO and other United Nations (UN) bodies have mounted a series of national and basin-wide technical assistance projects since the 1960s, to help developing countries assess and manage their water resources more effectively.

There are also a number of advances stemming from the use of remote sensing that provide increasing potential for monitoring a growing number of hydrological variables and overcoming the difficulties of determining meaningful spatial patterns from ground-based observations (Schultz and Engman, 2001). Data provided by Geographical Information Systems (GIS) along with digital terrain models are also becoming very important. For example, Food and Agriculture Organization (FAO)/UNESCO prepared exercises intended for self-learning in the application of GIS to hydrologic issues in West Africa using Arcview (Maidment and Reed, 1996). Data produced by tracer techniques are also proving very useful in quantifying sources of streamflow, residence times and exploring flow paths. The advent of global data centres, such as the Global Runoff Data Centre (GRDC), has eased the problem of access to world and national datasets. The Global Precipitation Climatology Centre (GPCC) collects precipitation data and there are proposals for a centre for world groundwater to be established in the Netherlands, under the auspices of UNESCO and WMO. Another example is the World Glacier Monitoring Service (WGMS) that collects data on the fluctuations of selected glaciers and has been publishing these data since 1967 (Kasser, 1967). Now the application of Landsat imagery, GIS and digital terrain modelling in certain parts of the world allows for the rapid analysis of glacier changes (Paul, 2002). The International Geosphere Biosphere Programme (IGBP) of the International Council for Science (ICSU) has also stimulated several initiatives involving the collection of global datasets, some concerned with hydrology, and it has assisted ongoing programmes such as the Global Network for Isotopes in Precipitation (GNIP) (Gat and Oeschger, 1995), which has, since 1961, provided monthly time series of isotope data from over 550 stations managed by the International Atomic Energy Agency (IAEA) and WMO. Indeed, IAEA has spearheaded efforts for the application of isotopes in hydrology, such as by improving the understanding of aquifers in many developing countries by collecting and analysing data on rates, sources of recharge and the age of groundwater. Unninayar and Schiffer (1997) provided a compendium of the systems designed to observe the globe's atmosphere, hydrosphere and land surface, while the IAHS Global Databases Metadata System gives a metadata listing of key water-related datasets. The Internet is a prime tool in accessing these data, such as by FAO's AQUASTAT or UNESCO's Latin American and Caribbean Hydrological Cycle and Water Resources Activities Information System (LACHYCIS). Table 4.2 and figure 4.2 present an overview of the world's available water resources.

Progress in understanding water resources has developed considerably over the last twenty to thirty years, particularly through advances in modelling. Now a very wide range of models

Figure 4.2: Water availability versus population

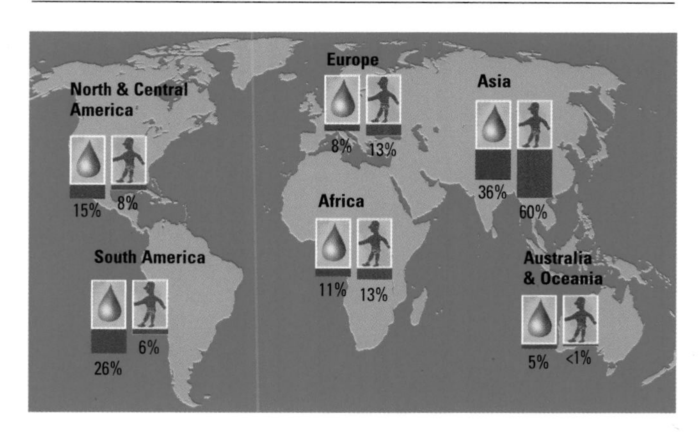

The global overview of water availability versus the population stresses the continental disparities, and in particular the pressure put on the Asian continent, which supports more than half the population with only 36 percent of the world's water resources.

Source: Web site of the UNESCO/IHP Regional Office of Latin America and the Caribbean.

1. GEMS Water, a combined effort of the United Nations Environment Programme, World Health Organization, World Meterological Organization.

Table 4.2: Water availability per person per year

Ranking	Continent	Country	Water resources							Land	Population	
			Total internal renewable water resources (km³/year)[1]	Groundwater: produced internally (km³/year)[2]	Surface water: produced internally (km³/year)[3]	Overlap: Surface and groundwater (km³/year)[4]	Water resources: total renewable (km³/year)*	Water resources: total renewable per capita (m³/capita year)	Dependency ratio (%)	Land area (km²)	Population in 2000 (1000 inh)	Population density in 2000 (inh/km²)
1	N C America	**Greenland**	603.00	–	–	–	603.00	**10,767,857**	0	341,700	56	0
2	N C America	**United States, Alaska**	800.00	–	–	–	980.00	**1,563,168**	18	1,481,353	627	0.4
3	South America	**French Guiana**	134.00	–	–	–	134.00	**812,121**	0	88,150	165	2
4	Europe	**Iceland**	170.00	24.00	166.00	20.00	170.00	**609,319**	0	100,250	279	3
5	South America	**Guyana**	241.00	103.00	241.00	103.00	241.00	**316,689**	0	196,850	761	4
6	South America	**Suriname**	88.00	80.00	88.00	80.00	122.00	**292,566**	28	156,000	417	3
7	Africa	**Congo**	222.00	198.00	222.00	198.00	832.00	**275,679**	73	341,500	3,018	9
8	Asia	**Papua New Guinea**	801.00	–	801.00	–	801.00	**166,563**	0	452,860	4,809	11
9	Africa	**Gabon**	164.00	62.00	162.00	60.00	164.00	**133,333**	0	257,670	1,230	5
10	Oceania	**Solomon Islands**	44.70	–	–	–	44.70	**100,000**	0	27,990	447	16
11	N C America	**Canada**	2,850.00	370.00	2,840.00	360.00	2,902.00	**94,353**	2	9,220,970	30,757	3
12	Oceania	**New Zealand**	327.00	–	–	–	327.00	**86,554**	0	267,990	3,778	14
13	Europe	**Norway**	382.00	96.00	376.00	90.00	382.00	**85,478**	0	306,830	4,469	15
14	N C America	**Belize**	16.00	–	–	–	18.56	**82,102**	14	22,800	226	10
15	Africa	**Liberia**	200.00	60.00	200.00	60.00	232.00	**79,643**	14	96,320	2,913	30
16	South America	**Bolivia**	303.53	130.00	277.41	103.88	622.53	**74,743**	51	1,084,380	8,329	8
17	South America	**Peru**	1,616.00	303.00	1,616.00	303.00	1,913.00	**74,546**	16	1,280,000	25,662	20
18	Asia	**Laos**	190.42	37.90	190.42	37.90	333.55	**63,184**	43	230,800	5,279	23
19	South America	**Paraguay**	94.00	41.00	94.00	41.00	336.00	**61,135**	72	397,300	5,496	14
20	South America	**Chile**	884.00	140.00	884.00	140.00	922.00	**60,614**	4	748,800	15,211	20
21	Africa	**Equatorial Guinea**	26.00	10.00	25.00	9.00	26.00	**56,893**	0	28,050	457	16
22	N C America	**Panama**	147.42	21.00	144.11	17.69	147.98	**51,814**	0	74,430	2,856	38
23	South America	**Venezuela**	722.45	227.00	700.14	204.69	1,233.17	**51,021**	41	882,050	24,170	27
24	South America	**Colombia**	2,112.00	510.00	2,112.00	510.00	2,132.00	**50,635**	1	1,038,700	42,105	41
25	South America	**Brazil**	5,418.00	1,874.00	5,418.00	1,874.00	8,233.00	**48,314**	34	8,456,510	170,406	20
26	Asia	**Bhutan**	95.00	–	95.00	–	95.00	**45,564**	0	47,000	2,085	44
27	South America	**Uruguay**	59.00	23.00	59.00	23.00	139.00	**41,654**	58	175,020	3,337	19
28	Africa	**Central African Rep.**	141.00	56.00	141.00	56.00	144.40	**38,849**	2	622,980	3,717	6
29	N C America	**Nicaragua**	189.74	59.00	185.74	55.00	196.69	**38,787**	4	121,400	5,071	42
30	Asia	**Cambodia**	120.57	17.60	115.97	13.00	476.11	**36,333**	75	176,520	13,104	74
31	Africa	**Sierra Leone**	160.00	50.00	150.00	40.00	160.00	**36,322**	0	71,620	4,405	62
32	Oceania	**Fiji**	28.55	–	–	–	28.55	**35,074**	0	18,270	814	45
33	South America	**Ecuador**	432.00	134.00	432.00	134.00	432.00	**34,161**	0	276,840	12,646	46
34	Europe	**Russian Federation**	4,312.70	788.00	4,036.70	512.00	4,507.25	**30,980**	4	16,888,500	145,491	9

Table 4.2: *continued*

35	N C America	**Costa Rica**	112.40	37.30	75.10	0.00	112.40	**27,932**	0	51,060	4,024	79
36	Africa	**Guinea**	226.00	38.00	226.00	38.00	226.00	**27,716**	0	245,720	8,154	33
37	Asia	**Malaysia**	580.00	64.00	566.00	50.00	580.00	**26,105**	0	328,550	22,218	68
38	Asia	**Brunei Darussalam**	8.50	0.10	8.50	0.10	8.50	**25,915**	0	5,270	328	62
39	Africa	**Guinea-Bissau**	16.00	14.00	12.00	10.00	31.00	**25,855**	48	28,120	1,199	43
40	Oceania	**Australia**	492.00	72.00	440.00	20.00	492.00	**25,708**	0	7,682,300	19,138	2
41	Africa	**Congo, Dem. Rep.**	900.00	421.00	899.00	420.00	1,283.00	**25,183**	30	2,267,050	50,948	22
42	Europe	**Croatia**	37.70	11.00	27.20	0.50	105.50	**22,669**	64	55,920	4,654	83
43	South America	**Argentina**	276.00	128.00	276.00	128.00	814.00	**21,981**	66	2,736,690	37,032	14
44	Asia	**Myanmar**	880.60	156.00	874.60	150.00	1,045.60	**21,898**	16	657,550	47,749	73
45	Europe	**Finland**	107.00	2.20	106.80	2.00	110.00	**21,268**	3	304,590	5,172	17
46	Africa	**Madagascar**	337.00	55.00	332.00	50.00	337.00	**21,102**	0	581,540	15,970	27
47	Europe	**Yugoslavia**	44.00	3.00	42.40	1.40	208.50	**19,759**	79	102,000	10,552	103
48	Europe	**Sweden**	171.00	20.00	170.00	19.00	174.00	**19,679**	2	411,620	8,842	21
49	Africa	**Cameroon**	273.00	100.00	268.00	95.00	285.50	**19,192**	4	465,400	14,876	32
50	Europe	**Slovenia**	18.67	13.50	18.52	13.35	31.87	**16,031**	41	20,120	1,988	99
51	Africa	**Sao Tome and Principe**	2.18	–	–	–	2.18	**15,797**	0	960	138	144
52	N C America	**United States, Hawaii**	18.40	13.20	5.20	0.00	18.40	**15,187**	0	16,636	1,212	73
53	N C America	**Honduras**	95.93	39.00	86.92	29.99	95.93	**14,949**	0	111,890	6,417	57
54	Europe	**Latvia**	16.74	2.20	16.54	2.00	35.45	**14,642**	53	62,050	2,421	39
55	Africa	**Angola**	184.00	72.00	182.00	70.00	184.00	**14,009**	0	1,246,700	13,134	11
56	Asia	**Mongolia**	34.80	6.10	32.70	4.00	34.80	**13,739**	0	1,566,500	2,533	2
57	Europe	**Ireland**	49.00	10.80	48.20	10.00	52.00	**13,673**	6	68,890	3,803	55
58	Asia	**Indonesia**	2,838.00	455.00	2,793.00	410.00	2,838.00	**13,381**	0	1,811,570	212,092	117
59	Europe	**Albania**	26.90	6.20	23.05	2.35	41.70	**13,306**	35	27,400	3,134	114
60	Asia	**Georgia**	58.13	17.23	56.90	16.00	63.33	**12,035**	8	69,700	5,262	75
61	Africa	**Mozambique**	99.00	17.00	97.00	15.00	216.11	**11,814**	54	784,090	18,292	23
62	Asia	**Viet Nam**	366.50	48.00	353.50	35.00	891.21	**11,406**	59	325,490	78,137	240
63	N C America	**United States**	2,818.40	–	–	–	3,069.40	**10,837**	–	9,158,960	283,230	31
64	Europe	**Hungary**	6.00	6.00	6.00	6.00	104.00	**10,433**	94	92,340	9,968	108
65	Africa	**Namibia**	6.16	2.10	4.10	0.04	17.94	**10,211**	66	823,290	1,757	2
66	Africa	**Zambia**	80.20	47.00	80.20	47.00	105.20	**10,095**	24	743,390	10,421	14
67	N C America	**Guatemala**	109.20	33.70	100.70	25.20	111.27	**9,773**	2	108,430	11,385	105
68	Europe	**Austria**	55.00	6.00	55.00	6.00	77.70	**9,616**	29	82,730	8,080	98
69	Europe	**Romania**	42.30	8.30	42.00	8.00	211.93	**9,445**	80	230,340	22,438	97
70	Europe	**Bosnia and Herzegovina**	35.50	–	–	–	37.50	**9,429**	5	51,000	3,977	78
71	Africa	**Botswana**	2.90	1.70	1.70	0.50	14.40	**9,345**	80	566,730	1,541	3
72	Europe	**Slovakia**	12.60	1.73	12.60	1.73	50.10	**9,279**	75	48,080	5,399	112
73	Europe	**Estonia**	12.71	4.00	11.71	3.00	12.81	**9,195**	1	42,270	1,393	33
74	Asia	**Nepal**	198.20	20.00	198.20	20.00	210.20	**9,122**	6	143,000	23,043	161
75	Africa	**Mali**	60.00	20.00	50.00	10.00	100.00	**8,810**	40	1,220,190	11,351	9
76	Asia	**Bangladesh**	105.00	21.09	83.91	0.00	1,210.64	**8,809**	91	130,170	137,439	1,056
77	Europe	**Switzerland**	40.40	2.50	40.40	2.50	53.50	**7,462**	24	39,550	7,170	181

Table 4.2: *continued*

Ranking	Continent	Country	Water resources							Land	Population	
			Total internal renewable water resources (km³/year)[1]	Groundwater: produced internally (km³/year)[2]	Surface water: produced internally (km³/year)[3]	Overlap: Surface and groundwater (km³/year)[4]	Water resources: total renewable (km³/year)*	Water resources: total renewable per capita (m³/capita year)	Dependency ratio (%)	Land area (km²)	Population in 2000 (1000 inh)	Population density in 2000 (inh/km²)
78	N C America	**United States, Conterminous**	2,000.00	1,300.00	1,862.00	1,162.00	2,071.00	**7,407**	3	7,663,984	279,583	36
79	Europe	**Luxembourg**	1.00	0.08	1.00	0.08	3.10	**7,094**	68	2,586	437	169
80	Europe	**Greece**	58.00	10.30	55.50	7.80	74.25	**6,998**	22	128,900	10,610	82
81	Africa	**Reunion**	5.00	2.80	4.50	2.30	5.00	**6,935**	0	2,500	721	288
82	Europe	**Portugal**	38.00	4.00	38.00	4.00	68.70	**6,859**	45	91,500	10,016	109
83	Asia	**Kazakhstan**	75.42	6.10	69.32	0.00	109.61	**6,778**	31	2,699,700	16,172	6
84	Europe	**Lithuania**	15.56	1.20	15.36	1.00	24.90	**6,737**	38	64,800	3,696	57
85	Asia	**Thailand**	210.00	41.90	198.79	30.69	409.94	**6,527**	49	510,890	62,806	123
86	Asia	**Philippines**	479.00	180.00	444.00	145.00	479.00	**6,332**	0	298,170	75,653	254
87	Africa	**Gambia**	3.00	0.50	3.00	0.50	8.00	**6,140**	63	10,000	1,303	130
88	Europe	**Netherlands**	11.00	4.50	11.00	4.50	91.00	**5,736**	88	33,880	15,864	468
89	Europe	**Belarus**	37.20	18.00	37.20	18.00	58.00	**5,694**	36	207,480	10,187	49
90	Africa	**Chad**	15.00	11.50	13.50	10.00	43.00	**5,453**	65	1,259,200	7,885	6
91	Asia	**Turkmenistan**	1.36	0.36	1.00	0.00	24.72	**5,218**	97	469,930	4,737	10
92	Africa	**Côte d'Ivoire**	76.70	37.70	74.00	35.00	81.00	**5,058**	5	318,000	16,013	50
93	Africa	**Swaziland**	2.64	–	–	–	4.51	**4,876**	41	17,200	925	54
94	N C America	**Mexico**	409.00	139.00	361.00	91.00	457.22	**4,624**	11	1,908,690	98,872	52
95	Africa	**Mauritania**	0.40	0.30	0.10	0.00	11.40	**4,278**	96	1,025,220	2,665	3
96	Africa	**Senegal**	26.40	7.60	23.80	5.00	39.40	**4,182**	33	192,530	9,421	49
97	Asia	**Kyrgyzstan**	46.45	13.60	44.05	11.20	20.58	**4,182**	0	191,800	4,921	26
98	N C America	**El Salvador**	17.78	6.15	17.60	5.97	25.26	**4,024**	30	20,720	6,278	303
99	Africa	**Benin**	10.30	1.80	10.00	1.50	24.80	**3,954**	58	110 620	6,272	57
100	Asia	**Azerbaijan**	8.12	6.51	5.96	4.35	30.28	**3,765**	73	86 600	8,041	93
101	N C America	**Jamaica**	9.40	3.89	5.51	0.00	9.40	**3,651**	0	10,830	2,576	238
102	Asia	**Korea, Dem. People's**	67.00	13.00	66.00	12.00	77.14	**3,464**	13	120,410	22,268	185
103	Asia	**Turkey**	227.00	69.00	186.00	28.00	229.30	**3,439**	1	769,630	66,668	87
104	Europe	**France**	178.50	100.00	176.50	98.00	203.70	**3,439**	12	550,100	59,238	108
105	N C America	**Cuba**	38.12	6.48	31.64	0.00	38.12	**3,404**	0	109,820	11,199	102
106	Asia	**Japan**	430.00	27.00	420.00	17.00	430.00	**3,383**	0	364,500	127,096	349
107	Europe	**Italy**	182.50	43.00	170.50	31.00	191.30	**3,325**	5	294,110	57,530	196
108	Asia	**Iraq**	35.20	1.20	34.00	0.00	75.42	**3,287**	53	437,370	22,946	52
109	Africa	**Togo**	11.50	5.70	10.80	5.00	14.70	**3,247**	22	54,390	4,527	83
110	Europe	**Macedonia, The Fmr Y.**	5.40	–	5.40	–	6.40	**3,147**	16	25,430	2,034	80

Table 4.2: *continued*

111	Africa	**Niger**	3.50	2.50	1.00	0.00	33.65	**3,107**	90	1,266,700	10,832	9
112	Asia	**Afghanistan**	55.00	–	–	–	65.00	**2,986**	15	652,090	21,765	33
113	N C America	**Trinidad and Tobago**	3.84	–	–	–	3.84	**2,968**	0	5,130	1,294	252
114	Asia	**Pakistan**	248.00	55.00	243.00	50.00	418.27	**2,961**	41	770,880	141,256	183
115	Africa	**Uganda**	39.00	29.00	39.00	29.00	66.00	**2,833**	41	197,100	23,300	118
116	Europe	**Ukraine**	53.10	20.00	50.10	17.00	139.55	**2,815**	62	579,350	49,568	86
117	Europe	**Spain**	111.20	29.90	109.50	28.20	111.50	**2,794**	0	499,440	39,910	80
118	Asia	**Armenia**	9.07	4.20	6.27	1.40	10.53	**2,780**	14	28,200	3,787	134
119	Africa	**Ghana**	30.30	26.30	29.00	25.00	53.20	**2,756**	43	227,540	19,306	85
120	Europe	**Moldova, Republic of**	1.00	0.40	1.00	0.40	11.65	**2,712**	91	32,910	4,295	131
121	Europe	**Bulgaria**	21.00	6.40	20.10	5.50	21.30	**2,680**	1	110,550	7,949	72
122	Asia	**Sri Lanka**	50.00	7.80	49.20	7.00	50.00	**2,642**	0	64,630	18,924	293
123	Asia	**Tajikistan**	66.30	6.00	63.30	3.00	15.98	**2,625**	17	140,600	6,087	43
124	Africa	**Tanzania, United Rep. of**	82.00	30.00	80.00	28.00	91.00	**2,591**	10	883,590	35,119	40
125	Africa	**Nigeria**	221.00	87.00	214.00	80.00	286.20	**2,514**	23	910,770	113,862	125
126	N C America	**Dominican Republic**	21.00	11.70	21.00	11.70	21.00	**2,507**	0	48,380	8,373	173
127	Europe	**United Kingdom**	145.00	9.80	144.20	9.00	147.00	**2,465**	1	240,880	59,634	248
128	Asia	**China**	2,879.40	891.80	2,715.50	727.90	2,896.57	**2,259**	1	9,327,420	1,282,437	137
129	Africa	**Sudan**	30.00	7.00	28.00	5.00	64.50	**2,074**	77	2,376,000	31,095	13
130	Asia	**Uzbekistan**	16.34	8.80	9.54	3.00	50.41	**2,026**	77	414,240	24,881	60
131	Asia	**Iran, Islamic Rep. of**	128.50	49.30	97.30	18.10	137.51	**1,955**	7	1,622,000	70,330	43
132	Africa	**Mauritius**	2.21	0.68	2.03	0.50	2.21	**1,904**	0	2,030	1,161	572
133	Asia	**India**	1,260.54	418.54	1,222.00	380.00	1,896.66	**1,880**	34	2,973,190	1,008,937	339
134	Europe	**Germany**	107.00	45.70	106.30	45.00	154.00	**1,878**	31	356,680	82,017	230
135	N C America	**Puerto Rico**	3.40	–	–	–	7.10	**1,814**	0	8,870	3,915	441
136	Europe	**Belgium**	12.00	0.90	12.00	0.90	18.30	**1,786**	34	30,230	10,249	339
137	Africa	**Ethiopia**	110.00	40.00	110.00	40.00	110.00	**1,749**	0	1,000,000	62,908	63
138	N C America	**Haiti**	13.01	2.16	10.85	0.00	14.03	**1,723**	7	27,560	8,142	295
139	Africa	**Eritrea**	2.80	–	–	–	6.30	**1,722**	56	101,000	3,659	36
140	Africa	**Comoros**	1.20	1.00	0.20	0.00	1.20	**1,700**	0	2,230	706	317
141	Asia	**Syrian Arab Republic**	7.00	4.20	4.80	2.00	26.26	**1,622**	80	183,780	16,189	88
142	Europe	**Poland**	53.60	12.50	53.10	12.00	61.60	**1,596**	13	304,420	38,605	127
143	Africa	**Zimbabwe**	14.10	5.00	13.10	4.00	20.00	**1,584**	30	386,850	12,627	33
144	Africa	**Somalia**	6.00	3.30	5.70	3.00	13.50	**1,538**	56	627,340	8,778	14
145	Africa	**Malawi**	16.14	1.40	16.14	1.40	17.28	**1,528**	7	94,080	11,308	120
146	Asia	**Korea, Republic of**	64.85	13.30	62.25	10.70	69.70	**1,491**	7	98,730	46,740	473
147	Africa	**Lesotho**	5.23	0.50	5.23	0.50	3.02	**1,485**	0	30,350	2,035	67
148	Europe	**Czech Rep.**	13.15	1.43	13.15	1.43	13.15	**1,280**	0	77,280	10,272	133
149	Asia	**Lebanon**	4.80	3.20	4.10	2.50	4.41	**1,261**	1	10,230	3,496	342
150	Africa	**South Africa**	44.80	4.80	43.00	3.00	50.00	**1,154**	10	1,221,040	43,309	35
151	Europe	**Denmark**	6.00	4.30	3.70	2.00	6.00	**1,128**	0	42,430	5,320	125
152	Africa	**Burkina Faso**	12.50	9.50	8.00	5.00	12.50	**1,084**	0	273,600	11,535	42

Table 4.2: *continued*

			Water Resources							Land	Population	
Ranking	Continent	Country	Total internal renewable water resources (km^3/year)[1]	Groundwater: produced internally (km^3/year)[2]	Surface water: produced internally (km^3/year)[3]	Overlap: Surface and groundwater (km^3/year)[4]	Water resources: total renewable (km^3/year)*	Water resources: total renewable per capita (m^3/capita year)	Dependency ratio (%)	Land area (km^2)	Population in 2000 (1000 inh)	Population density in 2000 (inh/km^2)
153	Asia	**Cyprus**	0.78	0.41	0.56	0.19	0.78	**995**	0	9,240	784	85
154	Africa	**Kenya**	20.20	3.00	17.20	0.00	30.20	**985**	33	569,140	30,669	54
155	Africa	**Morocco**	29.00	10.00	22.00	3.00	29.00	**971**	0	446,300	29,878	67
156	Africa	**Egypt**	1.80	1.30	0.50	0.00	58.30	**859**	97	995,450	67,884	68
157	N C America	**Antigua and Barbuda**	0.05	–	–	–	0.05	**800**	0	440	65	148
158	Africa	**Cape Verde**	0.30	0.12	0.18	0.00	0.30	**703**	0	4,030	427	106
159	Africa	**Rwanda**	5.20	3.60	5.20	3.60	5.20	**683**	0	24,670	7,609	308
160	N C America	**Saint Kitts Nevis**	0.02	0.02	0.004	0.00	0.02	**621**	0	360	38	106
161	Africa	**Burundi**	3.60	2.10	3.50	2.00	3.60	**566**	0	25,680	6,356	248
162	Africa	**Tunisia**	4.15	1.45	3.10	0.40	4.56	**482**	9	155,360	9,459	61
163	Africa	**Algeria**	13.90	1.70	13.20	1.00	14.49	**478**	4	2,381,740	30,291	13
164	Africa	**Djibouti**	0.30	0.02	0.30	0.02	0.30	**475**	0	23,180	632	27
165	Asia	**Oman**	0.99	0.96	0.93	0.90	0.99	**388**	0	212,460	2,538	12
166	N C America	**Barbados**	0.08	0.07	0.01	0.002	0.08	**307**	0	430	267	621
167	Asia	**Israel**	0.75	0.50	0.25	0.00	1.67	**276**	55	20,620	6,040	293
168	Asia	**Yemen**	4.10	1.50	4.00	1.40	4.10	**223**	0	527,970	18,349	35
169	Asia	**Bahrain**	0.004	0.00	0.004	0.00	0.12	**181**	97	690	640	928
170	Asia	**Jordan**	0.68	0.50	0.40	0.22	0.88	**179**	23	88,930	4,913	55
171	Asia	**Singapore**	0.60	–	–	–	0.60	**149**	–	610	4,018	6,587
172	Europe	**Malta**	0.05	0.05	0.00	0.00	0.05	**129**	0	320	390	1,219
173	Asia	**Saudi Arabia**	2.40	2.20	2.20	2.00	2.40	**118**	0	2,149,690	20,346	9
174	Africa	**Libyan Arab Jamahiriya**	0.60	0.50	0.20	0.10	0.60	**113**	0	1,759,540	5,290	3
175	Asia	**Maldives**	0.03	0.03	0.00	0.00	0.03	**103**	0	300	291	970
176	Asia	**Qatar**	0.05	0.05	0.001	0.00	0.05	**94**	4	11,000	565	51
177	N C America	**Bahamas**	0.02	–	–	–	0.02	**66**	0	10,010	304	30
178	Asia	**United Arab Emirates**	0.15	0.12	0.15	0.12	0.15	**58**	0	83,600	2,606	31
179	Asia	**Gaza Strip (Palestine)**	0.05	0.05	0.00	0.00	0.06	**52**	18	380	1,077	2,834
180	Asia	**Kuwait**	0.00	0.00	0.00	0.00	0.02	**10**	100	17,820	1,914	107
181	Africa	**Seychelles**	–	–	–	–	–	**–**	0	450	80	178
182	Asia	**West Bank**	0.75	0.68	0.07	0.00	0.75	**–**	0	5,800	–	–

[1] 2+3-4* Aggregation of data can only be done for internal renewable water resources and not the total renewable water resources, as that would result in double counting of shared water resources.

(–) No data available

Sources: Water resources: FAO: AQUASTAT 2002; land and population: FAOSTAT, except for the United States (Conterminous, Alaska and Hawaii): US Census Bureau.

exists and more are being developed including: rainfall-runoff models, aquifer models, ecosystem models and catchment models, many including monitoring of water quality. There are process models, hydroecological models and management models backed up by decision support systems and expert systems. There are stochastic and deterministic models with a complexity that ranges from simple lumped and black-box models to very sophisticated physically-based models with a high resolution of the land surface, including the surface-soil-vegetation-atmosphere interface and the processes operating there. Satellite data are being employed in a number of different types of models and they are proving valuable in assessing the water quality of large basins.

However, there are also studies which demonstrate that the sophistication and likeness to reality of a model are no guide to its predictive success (Naef, 1981). The additional problem of scale exists when the results of a limited experiment carried out over distances of tens of metres have to be extrapolated to kilometres by modelling. Scale is also a problem that has to be addressed when different types of models are to be coupled, meteorological and hydrological models for example, but the increase in computing power is one of the factors easing this difficulty. Some of these techniques have been employed to estimate water resources on a global or continental scale grid, producing maps showing variations with time (McKinney et al., 1998).

Improving knowledge of hydrological processes is essential to the understanding that permits water resources to be safeguarded and managed. The physical processes operating at the surface of the ground where the atmosphere, soil and vegetation meet are important to runoff generation and infiltration and also to the climate models developed for atmospheric studies, such as for work on climate change. Likewise, studies of the interaction of water with the biotic environment are necessary for a number of practical applications, the control of algal blooms and the maintenance of fish stocks, for example. Hydrological processes operating in bodies of surface water are a major factor in the complex and seasonally dynamic groundwater flow fields associated with them; consequently the representation of these processes has to be adequately captured in models which aim to portray these systems.

Global Hydrology and Water Resources

Variability over a wide range of scales in space and time is the most obvious feature of the global pattern of the hydrological cycle and of its component parts which determine water resources (see map 4.1).

Map 4.1: The long-term average water resources according to drainage basins

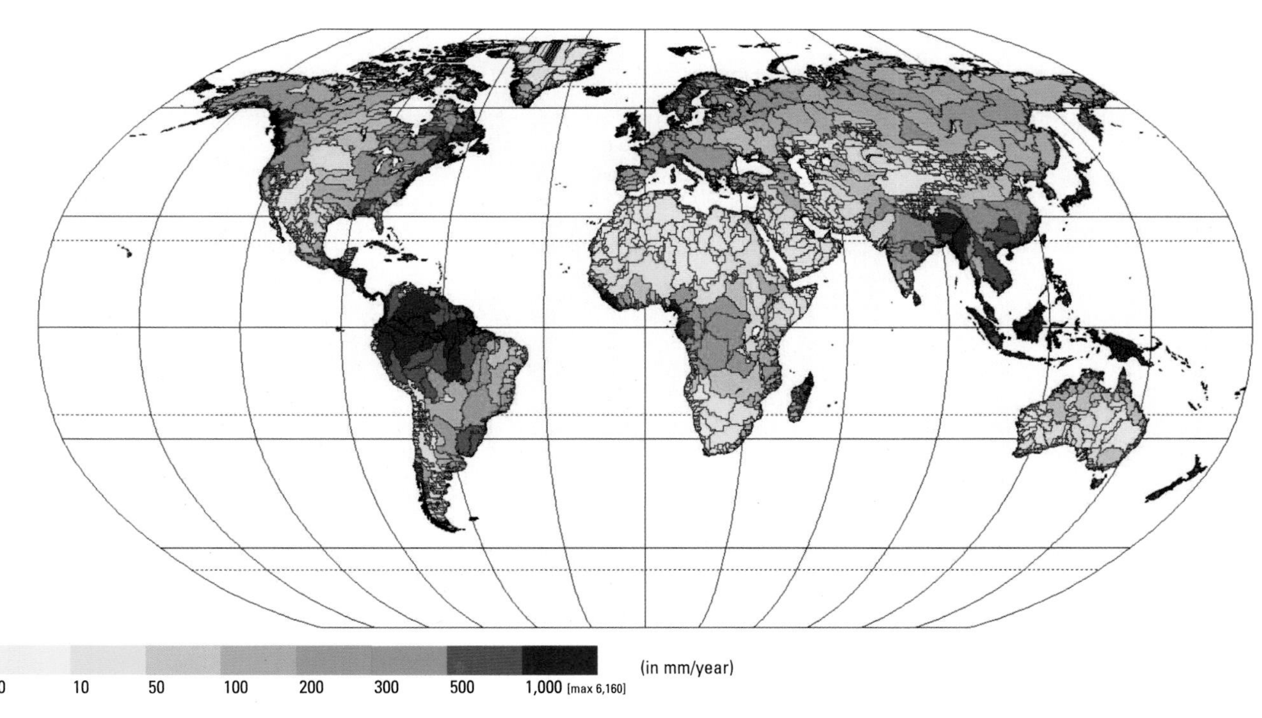

The long-term average of water resources by drainage basin is used as an indicator of water available to the populations in the basin. The use of the drainage basin as the basic unit sharpens the contrast between adjacent water-rich and water-poor countries, compared to map 4.4, based on a grid scale.

Source: Map prepared for the World Water Assessment Programme (WWAP) by the Centre for Environmental Research, University of Kassel, based on Water Gap Version 2.1.D, 2002.

But space and time deal very unfairly with certain parts of the world: while some regions and periods experience water scarcity, others are replete with water resources. In addition to the variations from year to year and beyond, there is the progress of the seasons which gives a regular rhythm to a number of these patterns and consequently to water resources over much of the globe outside the tropics. Precipitation provides the input to the land phase of the cycle. Evaporation, transpiration and sublimation return a large part of this water to the atmosphere, while much of the remainder is temporarily stored in the soil and in aquifers, glaciers and ice sheets. What is left runs off to the sea over a much shorter time span, with lakes and reservoirs holding some of the flow. Rivers transport most of this runoff from the land but some groundwater reaches the sea directly.

Climate change

Climate change is a natural, ongoing process. Thanks to the progress made in a number of techniques, the variability of the climate during the last 500,000 years has been well documented. Analysis tells us that a difference of a few degrees in annual average temperature of the earth can lead to massive impacts on glacier extension, sea level, precipitation regimes and distribution, and patterns of biodiversity.

The different assessments carried out by the Intergovernmental Panel on Climate Change (IPCC) have shown with increasing evidence that the emissions of greenhouse gases released in the atmosphere since the nineteenth century – which will continue for the coming decades, even if the rate is reduced or stabilized – will lead to a 'global warming' of the earth over the period 1990–2100, with an expected increase of the average annual temperature in the range of 1.4°C to 5.8°C. The projected rate of warming is very likely to be without precedent during at least the last 10,000 years. Among the associated effects are rises in the ocean level (in the range of 0.09 to 0.88 metres for the same period) and, as a consequence of the availability of more energy in the climate system, an intensification of the global hydrological cycle. In some areas, this will lead to changes in the total amount of precipitation, in its seasonal distribution pattern and in its frequency and intensity. Together with changes in evapotranspiration, these new conditions may directly affect the magnitude and timing of runoff, the intensity of floods and drought and have significant impacts on regional water resources, affecting both surface water and groundwater supply for domestic and industrial uses, irrigation, hydropower generation, navigation, in-stream ecosystems and water-based recreation. Hydrological sciences have underscored 'non-linearity' and 'threshold effects' in hydrological processes, which means that the terrestrial component of the hydrological cycle amplifies climate inputs. The regional drought that struck the African Sahel during the 1970s and the 1980s provides an illustration of these concepts: while the decrease in precipitation over this region during the two mentioned decades was in the magnitude of 25 percent compared to the 1950–69 period, the major rivers flowing through the region have experienced reductions in annual flows of a magnitude of 50 percent (Servat et al., 1998). In others words, what can be considered as a minor change in the total or in the temporal pattern of precipitation may nevertheless have tangible effects on water resources.

As a consequence of sea level rise, a calculable effect on water tables is that the interface between freshwater and brackish water will move inland, which may have significant impacts on the development and the life of people in coastal regions and in small islands.

Most numerical simulations have shown that an intensification of the hydrological cycle will not simply result in a smooth drift towards new conditions, but will most probably be associated with an increased variability of rainfall patterns at different time scales (interannual, seasonal, individual storm event). Thus, climate change will have to be taken into account in managing the temporal variability of water resources, and in managing the risks of water-related disasters (floods and droughts).

For water resource managers, the impacts of climate change are still considered as minor compared to the problems they are facing with the present climate variability. However, as it is likely that the variability may increase due to climate change, the impacts of the latter might become a real concern for water managers. In fact, coping with present-day climate variability while applying principles of Integrated Water Resources Management (IWRM) that take due account of risk, is certainly the most sound option for coping with climate change in the future.

Precipitation

The world pattern of precipitation (see map 4.2) shows large annual totals in the tropics (2,400 millimetres [mm] and more), in the mid-latitudes and where there are high mountain ranges (Jones, 1997).

The monsoon, tropical cyclones and mid-latitude frontal and convective storm systems are important mechanisms controlling precipitation, while orographic lifting is another. Towards the poles and with increasing altitude, a greater proportion of the precipitation occurs as snow. The annual snowfall over the earth is estimated to be about 1.7×10^{13} tons (Shiklomanov et al., 2002), covering an area that varies from year to year between 100 and 126 million km^2.

Small annual precipitation totals (200 mm and less) occur in the subtropics, the polar regions and in areas furthest from the oceans. There are also rain shadows in the lee of mountains, such as in the valleys east of the Sierra Nevada in the western United States

Map 4.2: Mean annual precipitation

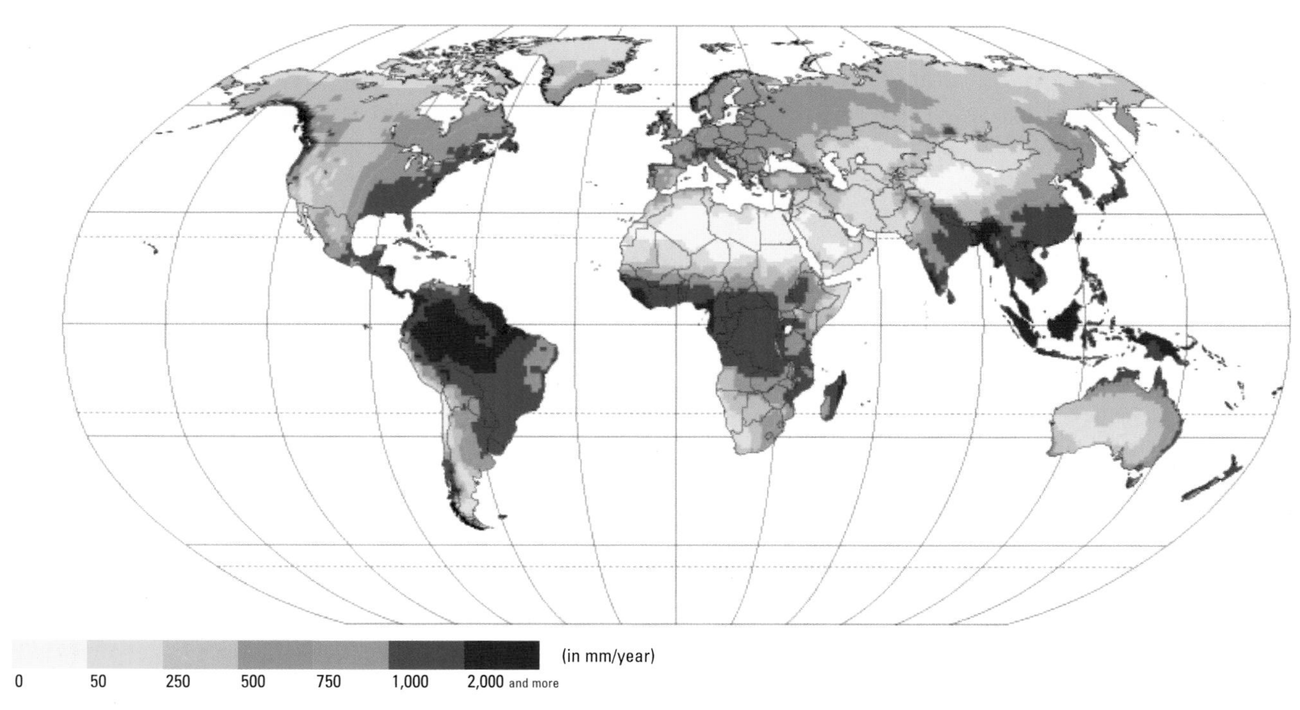

The world pattern of precipitation shows strong disparities between large annual rainfall in the tropics (some areas get in excess of 10,000 mm), and semi-arid and arid regions (such as the Sahara Desert). Differences within the African continent are particularly significant.

Source: Map prepared for the World Water Assessment Programme (WWAP) by the Centre for Environmental Research, University of Kassel, based on an Analysis by the Global Precipitation Climatology Centre (GPCC) (data extracted from the GPCC website in 2002 and Rudolf et al., 1994).

where totals are small. The world's deserts and semi-deserts are located in these areas, some vast such as the Sahara Desert, others very local in nature. In certain arid regions there may be no rain for several years, in marked contrast to locations where heavy precipitation occurs virtually every day and annual totals are enormous, for example in Hawaii, United States (11,000 mm). Such extremes in precipitation give rise to floods on the one hand and droughts on the other, with few parts of the globe left immune: in fact, deserts can experience flash floods, while humid areas may suffer from prolonged droughts.

Shiklomanov (1998a) estimates the total precipitation on the land surface to be 119,000 cubic kilometres (km^3) per year with other estimates ranging from 107,000 to 119,000 km^3.

Evaporation

The pattern of evaporation is conditioned by the availability of water to evaporate. Where water is readily available, such as in an open water surface, evaporation is at the potential rate and only restricted by atmospheric conditions. Where water is in limited supply, in an arid area for example, the actual rate of evaporation from the land surface is much lower than the potential. In general terms, potential evaporation rates are highest in the arid subtropics (more than 2,000 mm per annum), decreasing poleward to about 500 mm at latitude 50° and also decreasing with altitude. Actual rates are highest in the tropics and in mid-latitudes where large precipitation totals ensure a plentiful supply of soil moisture. Evaporation from the land surface is estimated by Shiklomanov (1998a) to be 74,200 km^3 a year, with the lowest of other estimates being 70,000 km^3.

Soil moisture

The soil acts as a significant reservoir when it is well developed, partitioning precipitation between runoff and infiltration and releasing water for plant growth. Soil moisture storage is dependent on a number of factors in addition to precipitation and evaporation: factors such as soil type, soil depth, vegetation cover and slope. The consequence is that even within a small basin, the pattern of soil moisture can be very heterogeneous. Consequently, the best guide to the global distribution of soil moisture storage may be the balance between precipitation and evaporation. This balance has a marked seasonal pattern over much of the world within the top part of the soil profile. This dries during the summer and returns to a wet state in the winter. Korzun (1974) postulated that the active soil water occurs mainly in the top two metres of the soil – within the rooting depth of most vegetation. On this basis he estimated the globe's total volume of soil moisture to be approximately 16,500 km^3. This figure assumes that soil moisture is 10 percent of the two-metre layer, and that the area of soil containing moisture covers 55 percent of the land area or 82 million km^2.

Groundwater

Since earliest antiquity, the human race has obtained much of its basic requirement for good-quality water from subterranean sources. Springs, the surface manifestation of underground water, have played a fundamental role in human settlement and social development. But for many millennia, capability to abstract this vital fluid was tiny in comparison to the available resource.

Heavy exploitation followed major advances in geological knowledge, well drilling, pump technology and power development, which for most regions dates from the 1950s (Foster et al., 2000). Today, with a global withdrawal rate of 600–700 km^3/year (Zektser and Margat, forthcoming), groundwater is the world's most extracted raw material, and, for example, forms a cornerstone of the Asian 'green agricultural revolution', provides about 70 percent of piped water supply in the European Union, and supports rural livelihoods across extensive areas of sub-Saharan Africa.

The science base: from maps to models

The surface extension of aquifers is now reasonably well known in most parts of the world, as a result of major improvements in geological mapping and hydrogeological interpretation over the past ten to thirty years, which have been stimulated by the IHP and facilitated by IAH (Struckmeier and Margat, 1995).

Hydrogeological interpretation involves building up a conceptual model of how the groundwater system functions, through identification of recharge processes, three-dimensional flow regime, discharge areas and the relationship with surface water. This forms the essential scientific basis for water resource management and protection, increasingly via numerical modelling of aquifers. But, especially in the developing world, conceptual (and thus numerical) models of the groundwater flow system for mapped aquifers cannot always be established with sufficient confidence or in adequate detail as a result of:

- lack of knowledge of three-dimensional geology;
- inadequate monitoring of groundwater levels; and
- insufficient data on hydraulic head variations with depth, that control flow patterns from recharge to discharge areas.

A vast reservoir of freshwater

Groundwater systems (aquifers and in some cases interbedded aquitards) unquestionably constitute the predominant reservoir and strategic reserve of freshwater storage on planet Earth – probably about 30 percent of the global total and as much as 98 percent of the fraction in liquid form (Shiklomanov, 1998a). Certain aquifers (such as those in table 4.3 and map 4.3) extend quite uniformly over very large land areas and have much more storage than all the world's surface reservoirs and lakes. In sharp contrast to surface water bodies, they hardly lose any of their stored water by direct evaporation.

Nonetheless, calculation of the total volume of global groundwater storage is by no means straightforward, and the precision and usefulness of any calculation will inevitably be open to question. Actual estimates, which are always massive, range from 7,000,000 km^3 (Nace, 1971) to 23.4 million km^3 (Korzun, 1974), but all are subject to major assumptions about the effective depth and porosity of the freshwater zone.

The boom in groundwater resource exploitation

Rapid expansion in groundwater exploitation occurred between 1950 and 1975 in many industrialized nations, and between 1970 and 1990 in most parts of the developing world. Systematic statistics on abstraction and use are not available, but globally groundwater is estimated to provide about 50 percent of current potable water supplies, 40 percent of the demand of self-supplied industry and 20 percent of water use in irrigated agriculture (Zektser and Margat, forthcoming). These proportions, however, vary widely from one country to another. Moreover, the value of groundwater to society should not be gauged solely in terms of relative volumetric abstraction. Compared to surface water, groundwater use often brings large economic benefits per unit volume, because of ready local availability, drought reliability and good quality requiring minimal treatment (Burke and Moench, 2000).

The list of major cities with considerable dependence on this resource is long (Foster et al., 1997; Burke and Moench, 2000). It assumes even greater importance for the supply of innumerable medium-sized towns: it is believed that more than 1.2 billion urban dwellers worldwide depend on well, borehole and spring sources. As regards groundwater use for irrigated agriculture, FAO has certain country-level data on its AQUASTAT Database for the 1990s (table 4.4 provides some data for selected countries).

The case of India is worthy of specific mention, since groundwater directly supplies about 80 percent of domestic water supply in rural areas, with some 2.8 to 3.0 million hand-pump boreholes having been constructed over the past thirty years. Further, some 244 km^3/year are currently estimated to be pumped for irrigation from about 15–17 million motorized dugwells and tubewells, with as much as 70 percent of national agricultural production being supported by groundwater (Burke and Moench, 2000; Foster et al., 2000).

Table 4.3: Some large aquifers of the world

No.	Name	Area (million km²)	Volume (billion m³)	Replenishment time (years)	Continent
1	Nubian Sandstone Aquifer System	2.0	75,000	75,000	Africa
2	North Sahara Aquifer System	0.78	60,000	70,000	Africa
3	High Plains Aquifer System	0.45	15,000	2,000	North America
4	Guarani Aquifer System	1.2	30,000	3,000	South America
5	North China Plain Aquifer Systems	0.14	5,000	300	Asia
6	Great Artesian Basin	1.7	20,000	20,000	Australia

The largest aquifers occur in Africa, where they represent a very precious resource, since rainfall is almost non-existent. However, a wise exploitation of this resource is necessary.

Sources: Margat, 1990a, 1990b.

Map 4.3: Groundwater resources of the world

The map clearly shows that the conditions of groundwater storage vary from area to area. While some regions are underlined by aquifers extending over large areas, others have no groundwater, except for the floodplain alluvial deposits usually accompanying the largest rivers. In mountainous regions, groundwater generally occurs in complexes of jointed hard rocks.This global map is based on important hydrogeological mapping programmes that have been carried out on all continents except Antarctica. It forms the first step of a worldwide hydrogeological mapping and assessment programme (WHYMAP) recently started by UNESCO, IAH, the Commission for the Geological Map of the World (CGMW), the International Atomic Energy Agency (IAEA) and the German Federal Institute for Geosciences and Natural Resources (BGR). In this programme, a series of groundwater-related global maps will be produced and provided in digital format.

Source: Map prepared for the World Water Assessment Programme (WWAP) by the Federal Institute for Geosciences and Natural Resources (BGR)/Commission for the Geological Map of the World (CGMW)/International Association of Hydrogeologists (IAH)/United Nations Educational, Scientific and Cultural Organization (UNESCO), 2002.

Table 4.4: Groundwater use for agricultural irrigation in selected nations

Country	Irrigated area (M ha)	Irrigation use (km³/year)	% of groundwater
India	50.1	460	53
China	48.0	408	18
Pakistan	14.3	151	34
Iran	7.3	64	50
Mexico	5.4	61	27
Bangladesh	3.8	13	69
Argentina	1.6	19	25
Morocco	1.1	10	31

Groundwater is more widely used for irrigation than surface water in countries such as India, Bangladesh and Iran. Note that arid countries such as Saudi Arabia are not listed in the table despite their using almost 100 percent groundwater from irrigation.

Sources: Burke and Moench, 2000; Foster et al., 2000.

Aquifer replenishment - controls and uncertainties

Groundwater is in slow motion from areas of aquifer recharge (which favour the infiltration of excess rainfall and/or surface runoff) to areas of aquifer discharge as springs and seepages to watercourses, wetlands and coastal zones (Zektser, 1999). The large storage capacity of many aquifers over long time periods (see table 4.3) transforms a highly variable recharge regime into a much more constant discharge regime.

The contemporary rate of aquifer recharge (replenishment through deep infiltration) is often used as an indicator of groundwater resource availability. However, average aquifer recharge rate is not necessarily a constant parameter, and is also frequently subject to considerable uncertainty (Foster et al., 2000), since it varies considerably with:

- changes in land use and vegetation cover, notably introduction of irrigated agriculture with imported surface water, but also with natural vegetation clearance, soil compaction, etc.;
- changes in surface water regime, especially diversion of river flow;
- lowering of the water table, by groundwater abstraction and/or land drainage, leading to increased infiltration; and
- longer-term climatic cycles, with considerable uncertainty remaining over the impacts on groundwater systems of the current global warming trend.

Figure 4.3: Typical groundwater flow regimes and residence times under semi-arid climatic conditions

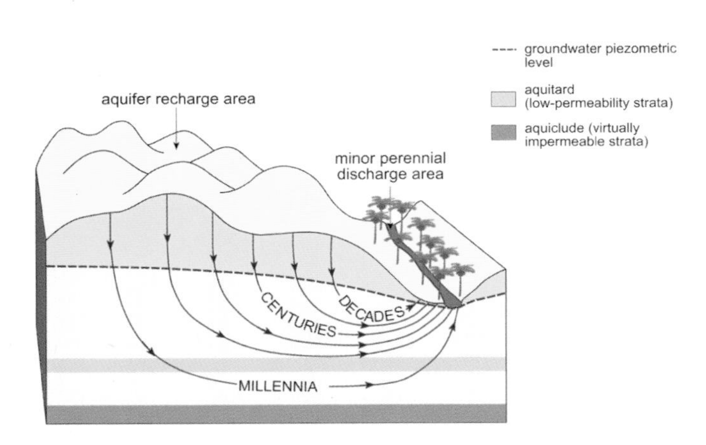

As a result of the very large storage capacity and very low flow velocity of groundwater systems, aquifer residence times can often be counted in decades or centuries, and sometimes in millennia.

Source: British Geological Survey.

These variations mean that groundwater recharge estimates have always to be treated with caution.

As a result of the very large storage of groundwater systems, aquifer residence times can often be counted in decades or centuries and sometimes in millennia (figure 4.3). Evidence of this, and the major influence on aquifer recharge of climate change during Quaternary history, has been revealed through analyses of environmental isotopes.

Development and application of these techniques, promoted by the IAEA-Isotope Hydrology Section, has demonstrated that much of the deeper groundwater in large geological basins and thick coastal deposits originated as recharge infiltrating during wetter epochs, often 10,000 or more years ago. In some more arid regions this 'fossil groundwater' may be the only resource, and thus should be used judiciously.

Groundwater development: the risk of unsustainability

The rapid expansion in groundwater exploitation has led to major social and economic benefits, but is also encountering significant problems (see table 4.5). For example, it has been estimated that mining of groundwater storage is occurring at a rate of about 10 km³/year on the North China Plain within the Hai He basin, and about 5 km³/year in the 100 or so recognized Mexican aquifers.

This abstraction is not physically sustainable in the longer term. In both cases most of the consumptive use of the pumped groundwater is by irrigated agriculture, but there is also competition with urban water supply abstraction coupled with inadequate attention to wastewater in general and to opportunities that integrated planning could provide.

A significant fraction of total aquifer replenishment is commonly required to maintain dry-weather river flows and/or to sustain some types of aquatic and terrestrial ecosystems (Foster et al., 2000; Alley, 1999). Groundwater abstraction reduces (in some cases

Table 4.5: Groundwater exploitation and associated problems

Socio-economic benefits	Sustainability problems
• economical provision of good-quality urban water supply	• inefficient resource utilization on a very widespread basis
	• growing social inequity in the access to groundwater in some regions
• low-cost development of drought-reliable rural water supplies	• physically unsustainable abstraction rates in more arid regions
	• reduction in dry weather baseflow in some downstream watercourses
• accessible and reliable water supply for irrigated crop cultivation	• irreversible aquifer damage locally due to saline intrusion/upconing
	• localized land subsidence due to aquitard compaction
• improved drainage and salinity alleviation in some areas	• damage to some groundwater-dependent ecosystems

Sources: Alley, 1999; Foster et al., 2000.

seriously) natural aquifer discharge to the aquatic environment, and resource development involving consumptive use of groundwater (or export from the sub-basin concerned) has the greatest impact. This should be an important consideration in resource planning and environmental management, but it is one that has been all too widely overlooked in the past.

The integrity of the soil layer overlying aquifers plays a key role in allowing groundwater recharge to take place. Anthropogenic influences can be highly significant in this context. For example, there is mounting evidence from across the African Shield that clearance of natural vegetation has led to soil erosion and compaction. As a consequence, infiltration and aquifer recharge and discharge have been reduced, leading to the fall of dry-weather flow in many smaller rivers which are vital to human survival and livelihood.

Natural groundwater quality problems

While the quality of unpolluted groundwater is generally good, some groundwater naturally contains trace elements, dissolved from the aquifer matrix, which limit its fitness for use (Edmunds and Smedley, 1996). These elements can be troublesome for domestic use (iron) or pose a public health hazard (fluoride, arsenic). With the introduction of more systematic and comprehensive analysis of groundwater supplies, supported by hydrogeochemical research, detailed knowledge of their origin and distribution is steadily increasing with the hope that associated problems can either be avoided or treated on a sound footing in the future.

There are significant areas of the globe where serious soil and groundwater salinization are present or have developed as a result of:

- rising groundwater table, associated with the introduction of inefficient irrigation with imported surface water in areas of inadequate natural drainage;

- natural salinity having been mobilized from the landscape, consequent upon vegetation clearing for farming development with, in these cases, increased rates of groundwater recharge; and

- excessive disturbance of natural groundwater salinity stratification in the ground through uncontrolled well construction and pumping.

Such situations always prove costly to remedy (Foster et al., 2000).

Vulnerability of aquifers to pollution

Aquifers are much less vulnerable to anthropogenic pollution than surface water bodies, being naturally protected by the soil and underlying vadose (unsaturated) zone or confining strata. But, as a result of large storage and long residence times when aquifers become polluted (see figure 4.4), contamination is persistent and difficult to reverse (Clarke et al., 1996).

Some aquifers are more vulnerable than others, and can be affected by a wide range of pollutants discharged or leached at the land surface. Moreover, most aquifers will (sooner or later) be affected by relatively persistent contaminants (such as nitrate, salinity and certain synthetic organics), if widely leached into groundwater in aquifer recharge areas.

The more spectacular groundwater pollution incidents, with large plumes of high concentration, are associated with industrial point sources from accidental spillage or casual discharge in vulnerable areas. However, more insidious and persistent problems are associated with diffuse pollution sources generated through intensification of agricultural cultivation or from unsewered urban and industrial development. The compilation of maps of aquifer vulnerability provides land use managers with a valuable tool for the establishment of preventive and protective measures (Vrba and Zaporozec, 1994).

Certain clear tendencies, including widespread quality deterioration of shallow aquifers in areas of rapid urbanization and agricultural intensification, have been identified (Foster and Lawrence, 1995). However, it is not possible to make reliable estimates of the proportion of active replenishment or of groundwater storage affected by pollution, because few nations have adequate groundwater quality monitoring networks set up for this purpose. This will, however, be a major thrust of the recently launched European Union Water Framework Directive.

The future: management and monitoring needs
A major challenge for the future is to stabilize aquifers exhibiting serious hydraulic imbalance and, where feasible, reinstate some discharge to the surface water environment. This can be achieved only by implementing realistic balance-of-demand management measures and supply or recharge enhancement. This variously requires (Foster et al., 2000):

- an institutional framework of appropriate style and scale;
- a sound system of groundwater abstraction and use rights;
- adequate financial investment in water-saving technology;
- active groundwater user and broader stakeholder participation;
- economic instruments to encourage reduced water consumption; and
- incentives to increase water harvesting and aquifer recharge.

Groundwater recharge enhancement requires good planning, design and operation, with appropriate monitoring to ensure that the selected method is effective and sustainable. A range of potentially cost-effective methods is available for 'banking' excess rainwater, surface runoff and reclaimed wastewater in aquifers (Bouwer, 2002).

The principal problems that have arisen in relation to groundwater in urban development (Foster et al., 1997) result from the common failure by urban water and environmental managers to identify and manage potentially negative interactions between wastewater elimination and groundwater supply, and to recognize the association between groundwater abstraction and urban drainage and infrastructure in low-lying cities.

In relation to groundwater pollution threats, the major management task is one of protection. This requires sustained institutional action to identify 'hazardous activities' and 'vulnerable areas', broadcasting the latter so as to 'make groundwater more visible' to stakeholders and the broader public, and thereby mobilizing their participation in pollution control.

A key requirement in many countries will be to transform the role of the national or local government departments responsible for groundwater from exclusively 'supply' development to primarily 'resource-custodian' and 'information-provider' development (Foster et al., 2000).

For the most part, monitoring of aquifer abstraction and use, water level fluctuation and recharge quality is far from adequate for water-resource management needs. This deficiency has also reduced the current ability to present a comprehensive and well-substantiated statement on the global status of groundwater resources and the role of groundwater in some processes of global change.

Glaciers and ice sheets
The world's largest volume of freshwater is stored within the ice caps and glaciers, about 90 percent of it in the Antarctic and much of the remainder in Greenland (see table 4.1). Despite plans for towing icebergs to coastal locations in lower latitudes, this water is considered inaccessible and not available for use. Most of the available water contributing to water resources drains from the smaller ice sheets and glaciers in North and South America, Europe and Asia. The present glaciation is estimated to cover an area of about 16.2 million km^2 and the total water volume of ice across the globe is considered to exceed 24 million km^3 (Korzun, 1974). This smaller volume that is distributed across the continents and from which melt water is released sustains river flows and contributes to seasonal peaks. Without further precipitation, the 74 km^3 of water stored within Swiss glaciers is estimated to be sufficient to maintain river flows for about five years (Bandyopadhyay et al., 1997).

In addition, underground ice resides in the areas of permafrost extending over north-east Europe, northern Asia, including the Arctic islands, northern Canada and the fringes of Greenland and Antarctica, as well as in the higher parts of South America. The total area of permafrost is about 21 million km^2, some 14 percent of the land area. The depth of permafrost ranges from 400 to 650 metres. Korzun (1974) has estimated the volume of ice to be 300,000 km^3. However, this water makes a very limited contribution to water resources.

Lakes and reservoirs
There are 145 large lakes across the globe with an area of at least 100 km^2, holding some 168,000 km^3 of water (Korzun, 1974). This is estimated to be about 95 percent of the total volume of all the world's lakes, which number some 15 million, giving a total volume of lake water of 176,400 km^3 (see table 4.1). Of this total, about 91,000 km^3 is freshwater and 85,400 km^3 is saline. However, these figures should be treated with caution, as the hydrology of about 40 percent of the world's large lakes has not been studied and their volumes are approximations (Shiklomanov, forthcoming). Most of the world's lakes are situated in the Northern Hemisphere and located in glaciated areas: for example, Lakes Superior, Huron, Michigan, Erie and Ontario in the United States lie behind moraines left by the receding ice. Some lakes are found in large tectonic depressions (Baikal in the Russian Federation, Victoria and Nyassa in eastern Africa), others in valleys blocked by landslides (Teletskoye in the Altai Mountains, Russian Federation). There are also some lakes of volcanic origin, some created by wind action and those resulting

from the collapse of particular strata. The world's major lakes include the Caspian Sea, which contains some 91 percent of the world's inland saline waters and Lake Baikal, which represents 27 percent of the world's lake freshwater.

According to the World Commission on Dams (WCD, 2000) there was a world total of 47,655 large dams in 1998 and an estimated 800,000 smaller ones (Hoeg, 2000). A large dam by the definition of the International Commission on Large Dams (ICOLD) has a height of more than 15 metres, or has a dam of above 5 metres holding a reservoir volume of more than 3 million cubic metres (Mm^3). These include some dams that have been constructed to increase the capacity of existing lakes, for example the Owen Falls Dam on the Nile below Lake Victoria. Dams have been built across river valleys to create reservoirs and some create cascades of reservoirs, for example, on the Nile and Colorado. Reservoirs have been constructed alongside rivers and are filled by pumping. Together these dams hold back a large volume of water and contribute a significant amount of storage globally (see table 4.6).

The first dams were constructed some 5,000 years ago, but the commissioning of large dams peaked between the 1960s and the 1980s particularly in China, the United States, the former USSR and India. However, some 300 dams over 60 metres high were listed as under construction in 1999, and authorities are claiming that many more will be needed in the future to meet the burgeoning demand for water. Cosgrove and Rijsberman (2000) maintain that a further 150 km^3 of storage will be required by 2025 to support irrigation alone and 200 km^3 more to replace the current overconsumption of groundwater. Of course, reservoirs are also built to satisfy various needs: flood control, drinking water supply, recreation and so on.

Vörösmarty et al. (1997) estimated that there are 633 large reservoirs with capacities of over 0.5 km^3 storing a total volume of nearly 5,000 km^3. It is considered that this represents 60 percent of the total global capacity which can then be calculated as a figure in excess of 8,000 km^3. These large reservoirs regulate about 40 percent of the earth's total runoff, increasing residence times by nearly fifty days and retaining about 30 percent of the sediment transported by the rivers where they are located. These reservoirs also cause increased evaporation – since they create a larger total surface area exposed to evaporation – representing some 200 km^3 a year, according to Cosgrove and Rijbersman (2000).

River flows

Although the volume of water in rivers and streams is very small by comparison with those in the other components of the world water balance (see table 4.1), in many parts of the world this water constitutes the most accessible and important resource.

Map 4.4 shows how the pattern of river flows reflects the global distribution of precipitation, with zones of large flow in the tropics and middle latitudes and small flows over much of the remainder. In fact about 40 percent of the total runoff enters the world's oceans between 10° N and 10° S. But not all rivers reach the ocean; there are a number of areas of inland drainage which are not connected to it including: the Caspian Sea basin, most of middle and central Asia, the Arabian Peninsula, much of north Africa and central Australia. Together they cover about 30 million km^2 (20 percent of the total land area), but they produce only 2.3 percent (about 1,000 km^3 per year) of the runoff (UNESCO, 1993). In these areas, groundwater is particularly important, although it is very difficult to assess its contribution to the resource. In one study for Africa, FAO (1995) estimated the renewable groundwater resource to be 188 km^3/year for the continent as a whole, or 5 percent of the volume of runoff. There is also the water lost by evaporation from the large rivers that cross these arid and semi-arid areas and from the reservoirs and

Table 4.6: The world's largest reservoirs

Reservoir	Continent	Country	Basin	Year of filling	Dam height (m)	Full volume (km^3)
Owen Falls (Lake Victoria)	Africa	Uganda, Kenya, United Republic of Tanzania	Victoria-Nile	1954	31	204
Bratskoye	Asia	Russian Federation	Angara	1967	106	169
Nasser	Africa	Egypt	Nile	1970	95	169
Kariba	Africa	Zambia, Zimbabwe	Zambezi	1959	100	160
Volta	Africa	Ghana	Volta	1965	70	148
Daniel Johnson	North America	Canada	Manicouagan	1968	214	141
Guri	South America	Venezuela	Caroni	1986	162	136
Krasnoyarskoye	Asia	Russian Federation	Yenisey	1967	100	73
Vadi-Tartar	Asia	Iraq	Tigris	1956–1976	–	72
WAC Bennett	North America	Canada	Peace	1967	183	7

Source: Shiklomanov, forthcoming.

Map 4.4: Long-term average runoff on a global grid

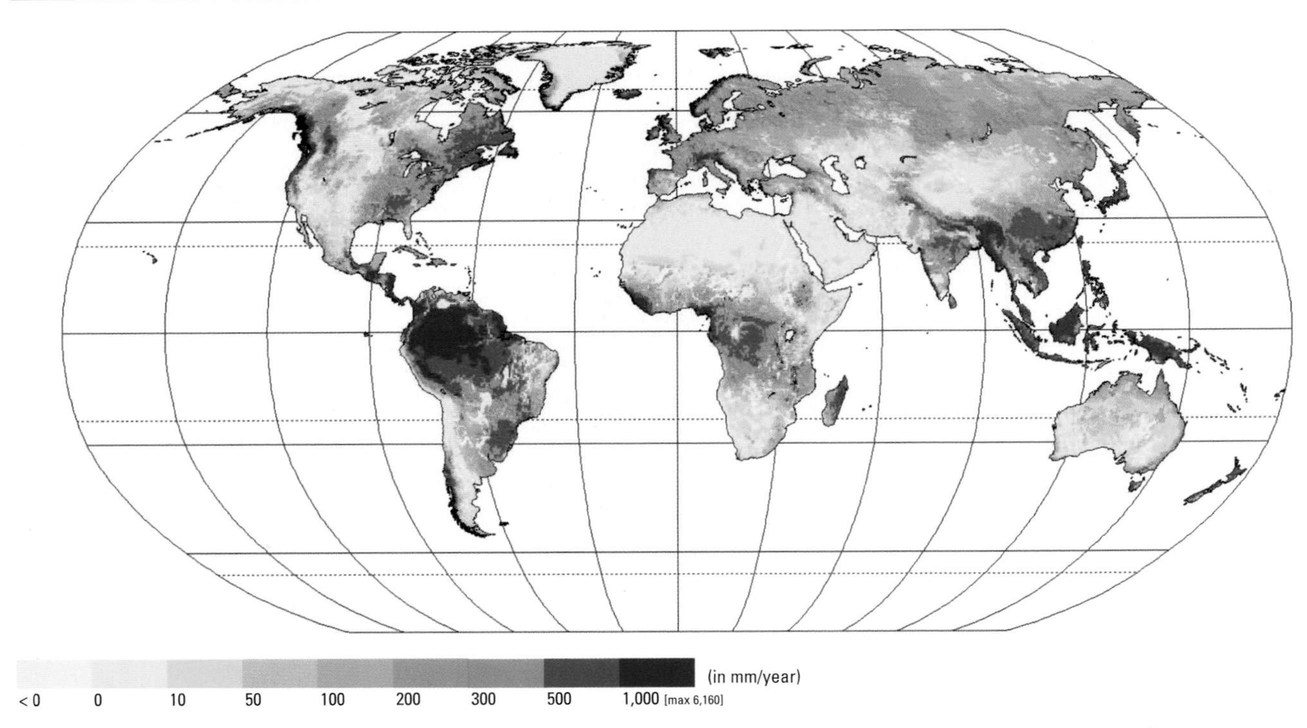

The enormous variation in climate around the Earth leads to great variability in the streamflow, which is in line with the rainfall. This map shows similar patterns to map 4.1.

Source: Map prepared for the World Water Assessment Programme (WWAP) by the Centre for Environmental Research, University of Kassel, based on Water Gap Version 2.1.D, 2002.

swamps associated with them, for example from the Indus (Pakistan), the Niger (West Africa), the Nile (eastern Africa) and the Colorado (Argentina). About 1,100 km^3 of runoff per year is lost in this way (Shiklomanov, forthcoming). For the large number of rivers with a groundwater component, this is included in the determination of flow, even though some groundwater flows to the oceans directly.

The world's largest river, the Amazon in Latin America, contributes some 16 percent of the global total annual runoff, while the five largest river systems (Amazon, Ganges with the Brahmaputra in India, Congo in Central Africa, Yangtze in China and Orinoco in Venezuela) together account for 27 percent (see table 4.7).

These figures are derived from the forthcoming study by Shiklomanov who collected and analysed flow records from the world hydrological network divided between twenty-six homogeneous and comparable regions covering the globe (Shiklomanov, 1998a). The 2,500 most suitable records were selected from this network and adjusted to the period 1921 to 1985. This adjustment was necessary because although a few of the records were for longer periods of observation, many were for shorter periods, many records had gaps and some had to be estimated from precipitation totals. From this study, the average total flow per year from the land surface to the ocean was estimated to be 42,800 km^3 with slight variations from year to year.

There are groups of wet years and dry years, but no trend over the sixty-five-year period. In terms of water resources the variability from one year to the next is very important but this variability is masked by the averaged data. This distorted view pertains particularly to arid and semi-arid regions, where the coefficient of variation (C_v) of annual discharges is in excess of 0.7 and where the driest years can experience a discharge of less than 10 percent of the long-term average. For wet regions (with average annual rainfall greater than 1,000 mm), annual variability is benign and the coefficients of variation are smaller (typically between 0.15 and 0.3) and the driest year is rarely less than 40 percent of the long-term average. So, where river flows are lowest around the world, the year-to-year variability is highest. Smaller rivers show greater annual variability than larger rivers. Runoff is unevenly distributed through the year for most regions of the globe with 60 to 70 percent occurring in the spring and early summer and 2 to 10 percent in the three driest months. For example, in Russia and Canada between 55 to 70 percent of the runoff occurs between May and August, while 47 to 65 percent of the runoff in India and China is between July and September. Floods contribute a large proportion of the flows during these periods when they transport the major part of the annual load of sediment and of materials in solution (see table 4.8). A number of severe floods have occurred in recent years, for example those on the

Table 4.7: The largest rivers in the world by mean annual discharge with their loads

River	Basin area (km²)	Mean annual discharge (m³/sec)	Maximum discharge (m³/sec)	Minimum discharge (m³/sec)	Runoff (mm/year)	Volume (km³)	Suspended solids (million tons/year)	Dissolved solids (million tons/year)
Amazon (South America)	4,640,300	155,432	176,067	133,267	3,653	4,901	275	1,200
Congo (Central Africa)	3,475,000	40,250	54,963	32,873	1,056	1,296	41	43
Orinoco (Venezuela)	836,000	31,061	37,593	21,540	1,172	980	32	150
Yangtze (China)	1,705,383	25,032	28,882	21,377	463	789	247	478
Brahmaputra (India)	636,130	19,674	21,753	18,147	975	620	61	540
Yenisei (Russian Federation)	2,440,000	17,847	20,966	15,543	231	563	68	13
Lena (Russian Federation)	2,430,000	16,622	19,978	13,234	216	524	49	18
Parana (Argentina)	1,950,000	16,595	54,500	4,092	265	516		
Mississippi (United States)	3,923,799	14,703	20,420	10,202	118	464	125	210
Ob (Russian Federation)	2,949,998	12,504	17,812	8,791	134	394		

The world's largest river, the Amazon, contributes by itself some 16 percent of the global total annual stream water flow, and the Amazon and the other four largest river systems (Congo, Orinoco, Yangtze, Brahmaputra) combined account for 27 percent.

Sources: GRDC, 1996; Berner and Berner, 1987.

Oder (Germany) in July 1997, those generated by the supercyclone in Orissa (India) in October 1999 and those from cyclone Eline, which affected Mozambique and neighbouring countries in February 2000 (Cornford, 2001).

Water quality

The quality of natural water in rivers, lakes and reservoirs and below the ground surface depends on a number of interrelated factors. These factors include geology, climate, topography, biological processes and land use, together with the time the water has been in residence. However, over the last 200 years human activities have developed to such an extent that there are now few examples of natural water bodies. This is largely due to urban and industrial development and intensification of agricultural practices, combined with the transport of the waste products from these activities by surface water and groundwater and by the atmosphere. The scale and intensity of this pollution varies considerably. Table 4.8 shows some of the chemical determinants of the world's average rivers, both natural and polluted.

There are global problems such as heavy metals, regional problems like acid rain and much more localized ones – groundwater contamination, for example. In many places groundwater has become

Table 4.8: The chemical composition of average river water (concentration in milligrams/litre)

	Calcium (Ca^{++})	Magnesium (Mg^{++})	Sodium (Na^{+})	Potassium (K^{+})	Chlorine (Cl)	Sulphate (SO_4)	Bicarbonate (HCO_3)	Silicon dioxide (SiO_2)	Total Dissolved Solids (TDS)
Actual	14.7	3.7	7.2	1.4	8.3	11.5	53.0	10.4	110.1
Natural	13.4	3.4	5.2	1.3	5.8	8.3	52.0	10.4	99.6

The difference shown in this table between the natural and the actual chemical composition of river water highlights the importance of the water pollution all over the world.

Source: Meybeck, 1979.

Table 4.9: The world's major water quality issues

Issue scale	Water bodies polluted	Sector affected	Time lag between cause and effect	Effects extent
Organic pollution	rivers ++ lakes + groundwater +	aquatic environment	<1 year	local to district
Pathogens	rivers++ lakes + groundwater +	health ++	<1 year	local
Salinization	groundwater ++ rivers +	most uses aquatic environment health	1–10 years	district to region
Nitrate	rivers + lakes + groundwater ++	health	>10 years	district to region
Heavy metals	all bodies	health aquatic environment ocean fluxes	<1 to >10 years	local to global
Organics	all bodies	health aquatic environment ocean fluxes	1 to 10 years	local to global
Acidification	rivers ++ lakes ++ groundwater +	health aquatic environment	>10 years	district to region
Eutrophication	lakes ++ rivers +	aquatic environment most uses ocean fluxes	>10 years	local
Sediment load (increase and decrease)	rivers + lakes	aquatic environment most uses ocean fluxes	1–10 years	regional
Diversions, dams	rivers ++ lakes + groundwater ++	aquatic environment most uses	1–10 years	district to region

+ Serious issue on a global scale
++ Very serious issue on a global scale

Pollutants of many kinds eventually find their way into water bodies at all levels. Although it may take some years for problems to become evident, poor water quality affects both human health and ecosystem health.

Source: WHO/UNEP, 1991.

contaminated as a result of leakage from storage tanks, mine tailings and accidental spillages (Herbert and Kovar, 1998). This contamination highlights the dimension of time; because groundwater systems are almost impossible to cleanse and many contaminants are persistent and remain a hazard for long periods even at low concentrations. There are also parts of the world where naturally occurring trace elements are present in groundwater, the most prevalent being arsenic and fluoride. These cause serious health effects. Indeed, health is an important factor in many of the world's major water quality problems listed by WHO/UNEP (1991) (see table 4.9).

Arsenic is widely distributed throughout the earth's crust: it occurs in groundwater through the dissolution of minerals and ores. Long-term exposure to arsenic via drinking water causes cancer of the skin, lungs, urinary bladder and kidney, as well as other effects on skin such as pigmentation changes and thickening. Cancer is a late expression of this exposure and usually takes more than ten years to develop. A recent study (BGS and DPHE, 2001) suggests that Bangladesh is grappling with the largest mass 'poisoning' in history, potentially affecting between 35 and 77 million of the country's 130 million inhabitants. Similar problems with excessive concentrations of arsenic in drinking water occur in a number of other countries. Excessive amounts of fluoride in drinking water can also be toxic. Discoloration of teeth occurs worldwide, but crippling skeletal effects caused by long-term ingestion of large amounts are prominent in at least eight countries, including China where 30 million people suffer from chronic fluorosis. The preferred remedy is to find a supply of drinking water with safe levels of arsenic and fluoride but removal of the concentrations may be the only solution.

Organic material from domestic sewage, municipal waste and agro-industrial effluent is the most widespread pollutant globally (UNEP, 1991). It is discharged untreated into rivers, lakes and aquifers, particularly in the more densely populated parts of Asia, Africa and South America and to a varying extent around certain urban centres over the remainder of the world. Its volume has risen over the last hundred years and this rise is likely to continue into the future as the pace of development accelerates. It contains faecal materials, some infected by pathogens, which lead to increased rates of morbidity and mortality among populations using the water.

This organic material also has high concentrations of nutrients, particularly nitrogen and phosphorus, which cause eutrophication of lakes and reservoirs, promoting abnormal plant growth and depleting oxygen. Nitrogen levels have also risen because of the increased use of nitrogenous fertilizers in agriculture, in both developed and developing countries. There is concern because nitrate concentrations in large numbers of sources of surface water and groundwater exceed the WHO guideline of 10 milligrams per litre. In many parts of the world trends in many heavy metal concentrations in river water have risen due to leaching from waste dumps, mine drainage and melting, to the extent that they can reach five to ten times the natural background level (Meybeck, 1998). Concentrations of organic micropollutants from the use of pesticides, industrial solvents and like materials have also increased. There is anxiety about the health effects of these and other pollutants, but the consequences of exposure to these substances is often not clear.

For the developed world, acidification of surface water was a serious problem in the 1960s and 1970s, particularly in Scandinavia, western and central Europe and in the north-east of North America, but since then sulphur emissions have decreased and the acid rain problem has diminished. The main impact was on aquatic life which generally cannot survive with pH levels below 5, but there are also health problems because higher acidity raises concentrations of metals in drinking water. Acidification is likely to continue in countries and regions with increasing industrialization, such as India and China.

For the developing world, increasing salinity is a serious form of water pollution. Poor drainage, fine grain size and high evaporation rates tend to concentrate salts in the soils of irrigated areas in arid and semi-arid regions. Salinity affects large areas, some to a limited extent, others more severely. In some cases natural salinity is mobilized from the landscape by the clearance of vegetation for agriculture and the increased infiltration this may cause. Shiklomanov (forthcoming) estimates that some 30 percent of the world's irrigated area suffers from salinity problems and remediation is seen to be very costly (Foster et al., 2000).

Most rivers carry sediment in the form of suspended load and bed load, in some cases the latter is charged with metals and other toxic materials (see map 4.5). This sediment load is adjusted to the flow regime of the river over time, and changes to this regime accompanied by increases or decreases in the load can cause problems downstream. These include the progressive reduction of reservoir volumes by siltation, the scouring of river channels and the deposition of sediment in them, threatening flood protection measures, fisheries and other forms of aquatic life. River diversions, including dams, can produce some of these effects on sediment, but in addition they may alter the chemical and biological characteristics of rivers, to the detriment of native species. The world total of suspended sediment transported to the oceans is reported to be as high as 51.1 billion tons per year (Walling and Webb, 1996).

Despite regional and global efforts to improve the situation since the 1970s, knowledge of water quality is still incomplete, particularly for toxic substances and heavy metals (Meybeck, 1998). In addition, there appear to be no estimates of the world total volume of polluted surface water and groundwater, nor the severity of this pollution. Shiklomanov (forthcoming) provides estimates of the volume of wastewater produced by each continent, which together gave a global total in excess of 1,500 km^3 for 1995. Then there is the contention that each litre of wastewater pollutes at least 8 litres of freshwater, so that on this basis some 12,000 km^3 of the globe's water resources is not available for use. If this figure keeps pace with population growth, then with an anticipated population of 9 billion by 2050, the world's water resources would be reduced by some 18,000 km^3.

Human impacts on water resources

Preceding paragraphs have discussed various aspects of the influence of human activities on the hydrological cycle and on water resources. In turn there are also those aspects of land use which influence the hydrological cycle. Wetlands, for example, can have profound effects, many beneficial to humankind, including flood storage, low flow maintenance, nutrient cycling and pollutant trapping (Acreman, 2000). This view forms a key component of the policies stemming from the Convention on Wetlands (Davis, 1993) and those of many national initiatives, some concerned with the economic value of wetlands (Laurans et al., 1996).

Studies of the hydrological impacts of changing land use have a long and well documented history (Swanson et al., 1987; Blackie et al., 1980; Rodda, 1976; Sopper and Lull, 1967). The techniques comprising the paired basin approach were developed in Switzerland in the 1890s, with subsequent research following in Japan and the United States between 1910 and 1930. Similar basins, usually small and contiguous with the same land use, were instrumented in order to measure their water balances to quantify the effects of change,

Map 4.5: Sediment load by basin

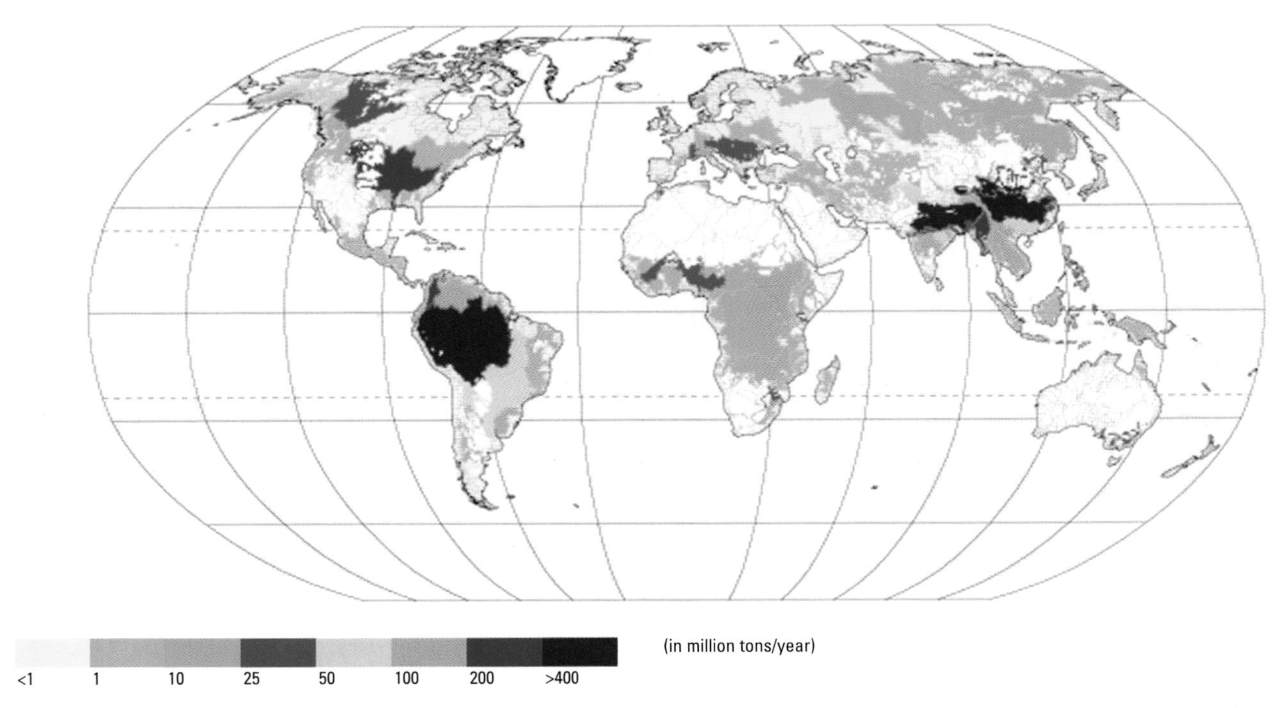

Changes in sediment yield reflect changes in basin conditions, including climate, soils, erosion rates, vegetation, topography and land use. It is influenced strongly by human actions, such as in the construction of dams and levees (see high sediment load in China and the Amazon basin, where large dams have been implemented), forest harvesting and farming in drainage basins.

Source: Syvitski and Morehead, 1999.

particularly with respect to runoff. After a calibration period, land use was altered in one basin and the differences in the hydrological behaviour of the pair were quantified during the ensuing period. Clear-cutting of forest, the effects of fire, different cropping practices, grazing and afforestation were among the changes studied. Coweeta in the United States, Valdai in Russia and Jonkershoek in South Africa were among those locations where these experiments were launched in the 1920s and 1930s.

A large number of studies of this type were started in different parts of the world in the 1950s and 1960s and the majority contributed to the programme on representative and experimental basins, which formed a prominent part of the International Hydrological Decade (IHD, 1965–74) of UNESCO. Research practice was strengthened (Toebes and Ouryvaev, 1970) and results compared (IAHS/UNESCO, 1970) and as time went on, greater interest was given to water quality matters. The results from many representative and experimental basins were harnessed to the FRIEND Programme, to gain better knowledge of the effects of human activities on the hydrological cycle on a regional scale and to upgrade the assessment of water resources. The European Network of Experimental and Representative Basins (ERB), which was launched in 1986, developed a study of methods of hydrological basin comparison (Robinson, 1993). Maksimovic (1996) provided a review of methods for investigating urban hydrology. At the start of the twenty-first century with the IHP broadened towards the social, political and environmental dimensions of water and water resources, it is fitting that the Hydrology, Environment, Life and Policy Programme (HELP) is returning to a network of selected river basins across the world as the basis for their study.

There are some global overviews of the results from these various studies of representative and experimental basins, for example in Falkenmark and Chapman (1989), but few recent publications. One reason may be that conditions vary to such an extent that the conclusions from one set of basin studies may not hold in another set sited in a different climatic zone. There is also the problem that there are many results for temperate latitudes and relatively few for other regions of the globe, despite increased attention to key regions such as the humid tropics (Bonell et al., 1993). Nevertheless, Ellis (1999) summarized the findings of fifty-two studies mainly concerned with the impact of urban areas on the hydrological cycle. He concluded that:

- infiltration and recharge to aquifer systems are reduced and surface runoff is increased in both volume and rate due to growth in the impermeable surface areas leading to increased downstream flooding;
- declining water levels and land subsidence may occur due to groundwater mining;
- pollutant loads to water courses and surface water bodies increase from surface runoff and sewage outfalls especially during storms in urban areas;
- leakage to groundwater occurs from old and poorly maintained sewers;
- extensive soil and groundwater are contaminated by industrial leakages, spills of hazardous industrial chemicals and poorly planned solid and liquid waste disposal practices;
- increased artificial surface water infiltration and recharge from source control devices lead to poor groundwater quality; and
- habitats and the diversity of species are reduced in receiving waters.

Hibbert (1967) surveyed the results from thirty-nine studies, mainly in the United States, on the effects of altering the forest cover on basin water yield. He showed, in general, that reduction in forest area increases yield and that reforestation decreases yield. The increase was a maximum in the first year of complete felling, with an upper limit equivalent to a depth of 4.5 mm a year for each percent reduction in forest cover. As the forest regrows, the increased streamflow declines in proportion to the rate of forest recovery. Dunne and Leopold (1978) reached similar conclusions from these and other findings and added that the effect of reducing forest cover may be far less important in arid regions. Results from ninety-four paired basin studies in different parts of the world were reported by Bosch and Hewlett (1982). They found that for pine and eucalyptus forest there was an average of 40 mm change in yield per 10 percent change in cover, while the corresponding figures for hardwood and scrub were 25 and 10 percent. Later research has generally agreed with these results, but the emphasis has switched from relatively simple studies of water quantity to those of the processes involved (Kirby et al., 1991), including basin hydrobiogeochemistry, in attempts to understand the mechanisms in operation. But it seems that the further the functioning of a basin is unravelled, the more complex and detailed the processes appear and the greater the number of questions and uncertainties generated (Neal, 1997).

Desalinated water resources

With population growth and concerns about water scarcity growing, several countries, especially in the Near East region, are developing desalination plants to convert saline water (e.g. sea-water, brackish water or treated wastewater) into freshwater.

The deterioration of existing groundwater resources, combined with the dramatic decline in costs, has given new impetus to this old technology, once considered an expensive luxury. The global market for desalination currently stands at about US$35 billion annually, and could double over the next fifteen years.

The process of desalting has a great deal to contribute to relatively small-scale plants providing high-cost water for domestic consumption in water-deficient regions. For irrigation, however, costs do constitute a major constraint. Therefore, except in extreme situations, desalinated seawater has not been used for irrigation, and the contribution of desalinated seawater on a global scale to total resource availability is very small.

In 2002 there were about 12,500 desalination plants around the world in 120 countries. They produce some 14 million m^3/day of freshwater, which is less than 1 percent of total world consumption. The most important users are in the Near East (about 70 percent of worldwide capacity) – mainly Saudi Arabia, Kuwait, the United Arab Emirates, Qatar and Bahrain – and North Africa (6 percent), mainly Libya and Algeria. Among industrialized countries, the United States (6.5 percent) is a big user (in California and parts of Florida). Most of the other countries have less than 1 percent of worldwide capacity. It is expected that the demand for desalinated seawater will increase in those countries that already apply this option, and will also appear in other regions and countries as their less expensive supply alternatives become exhausted. However, safe disposal of generally toxic chemical by-product of desalination is still a concern.

Among the various desalination processes, the following are the most interesting for large-scale water production: Reverse Osmosis (RO), Multi-Effect Distillation (MED) and Multi-stage Flash Distillation (MSF). The latter is used mainly in the oil producing countries of the Middle East. Currently, RO offers the most favourable prospects, as it requires less energy and investment than other technologies. A lot of energy is needed to desalinate water, although the form and amount of energy input depends on the process used. For RO, for example, about 6 kilowatts per hour (kWh) of electricity is required for each m^3 of drinking water produced. For distillation processes such as MED and MSF, the energy input is mainly in the form of heat (70°C to 130°C hot water or steam). For MED, specific heat consumption is in the range of 25–200 KWh/m^3 and for MSF, 80–150 kWh/m^3.

Gulf States such as Saudi Arabia, the United Arab Emirates and Kuwait use dual-purpose power and desalination plants on a grand scale. Jordan, Israel and the Palestinian Authority are increasingly seeing a viable and economic solution to ensuring future water supplies in desalted water. The North African sea resort of Tunis is investigating this alternative to major water transport schemes. Despite the costs, desalination plants are currently underway in Italy, Spain, Cyprus, Malta, South Africa, Algeria, Morocco, the Canary Islands, the Republic of Korea and the Philippines. In western America, where drought and water shortages are common occurrences, cities such as Santa Barbara in California are investing alongside water utilities to supplement their drinking water supply with desalinated water.

The Regional Dimension[2]

An examination of the state of freshwater resources at the continental level reveals clear disparities that a global examination can hide, since it is based on averages. In order to get an unbiased view of the global state of freshwater resources, it is therefore important to address this regional dimension of the overall picture. In turn, this regional and global appreciation is not complete without being informed by data from the local level, such as those provided in the case studies section of this report.

Africa

The African continent occupies an area of 30.1 million km^2 spanning the equator. It has a rapidly growing population of well over 700 million, many living in some of the world's least developed countries. The hydrological network is also the world's least developed with sparse coverage and short fragmentary records, except for the Nile Basin and in certain countries in the north and south of the continent. Most of Africa is composed of hard Precambrian rocks forming a platform with some mountainous areas, mainly on the fringes of the continent and where the rift valley crosses east Africa. The climate is much more varied than the relief with the hottest of deserts and the most humid of jungles – the amount and distribution of precipitation in space and time being paramount. Annual totals vary from 20 mm a year over much of the vastness of the Sahara to 5,000 mm towards the mouth of the Niger (see map 4.2 showing mean annual precipitation). With large amounts of solar radiation and high temperatures, African evaporation rates are high. The deserts that cover about one third of the continent in the north and south have little surface water but large volumes of groundwater. The Congo (central Africa) is the world's second largest river and the Nile the longest (6,670 km), but the Orange (South Africa), Zambesi (southern Africa), Niger and Senegal (West Africa) rivers are also important. The average annual flow from Africa for the 1921–1985 period was about 4,000 km^3. Many of the rivers have a considerable hydropower potential that is already exploited by certain dams and power plants such as those at Kariba (Zambia) and Aswan (Egypt). These schemes have aided development but some have serious hydroecological and social effects. Parts of the large areas under irrigation suffer from high water tables and increased salinity. Over the past ten years, Africa has experienced nearly one-third of all water-related disaster events (in this case, floods and droughts) that have occurred worldwide, with nearly 135 million people affected (80 percent by drought).

2. This section draws heavily on Shiklomanov (forthcoming) and on the expertise of the UNESCO Regional Hydrology Advisers.

Asia

The Asian continent has an area of 43.5 million km^2 occupying one third of the land surface of the globe and supports a population of 3,445 million people. It is a continent of great contrasts – contrasts in relief, climate, water resources, population density and standard of living, for example. There are also contrasts in the hydrological network: countries fringing the Pacific and Indian Oceans, such as Japan and Malaysia, have networks with high levels of capability and they contrast with those towards the centre of the continent where networks are generally deficient. Asia's geology and relief are very complicated and the climate is extremely varied, the monsoon dominating the south and east. Climatic differences are intensified by high mountain chains and plateaux, disrupting the pattern of precipitation that, in general, decreases from south to north and from east to west (see map 4.2 on mean annual precipitation).

High rates of evaporation occur across the southern half of the continent with areas of desert in the west and centre. Some of the world's largest rivers drain Asia to the Arctic, Pacific and Indian Oceans (see table 4.7): the Ganges and the Brahmaputra (India); Yangtze (China); Yenisey, Lena, Ob, Amur (Russian Federation); and Mekong (South-East Asia) for example, but there are large areas draining to the Aral and Caspian Seas and further areas of inland drainage in western China. The mean annual runoff from Asia for the period 1921 to 1985 is estimated to be 13,500 km^3, about half of which originated in South-East Asia, in contrast to the Arabian Peninsula with an estimated 7 km^3. There are large aquifers and many lakes in Asia, such as Lake Baikal in the Russian Federation. China, India, Russia and Pakistan have a large number of reservoirs, primarily used for irrigation. The continent faces serious flood problems and sedimentation problems, particularly in China, as well as pollution of surface water and groundwater in densely populated areas. However, the water problems of the basin of the Aral Sea are the most acute.

Europe

Occupying an area of 10.46 million km^2, Europe is one of the world's most densely populated and developed regions. It has a dense hydrological network containing a number of stations with 200 years or more of records. This network is best developed in the west and least in the east of the continent. Precipitation, in general, increases westwards to the Atlantic rim from about 400 mm in Russia and Poland to over 4,000 mm a year in Norway and Scotland (see map 4.2 on mean annual precipitation). It also increases with altitude such as in the Alps. However, parts of the south, Spain and Italy, for example, receive low rainfall amounts and experience high rates of evaporation that cause water resource problems. Although there are some large rivers, such as the Volga (250 km^3), the Danube (225 km^3) and the Rhine (86 km^3), the majority are relatively small, many with steep courses. The mean annual runoff from Europe, for the period of 1921 to 1985, is estimated at 2,900 km^3 per year, most from northern Europe and least from the southern part of eastern Europe. The number of lakes and reservoirs is large and there are extensive aquifers. Over the last 200 years, industry, energy generation, agriculture and urban development have changed the pattern of runoff from the continent and altered its quality characteristics. Many rivers and lakes were badly polluted by discharges of untreated sewage, mine wastes and agricultural effluents.

Latin America and the Caribbean

South America has an area of 17.9 million km^2 and a population of less than 400 million, about 6 percent of the world total, but produces about 26 percent of the world's water resources. It has a modern hydrological network with 6,000 or so stations, some with records longer than fifty years. Precipitation averages over South America are about 1,600 mm a year, with a mean of about 2,400 mm across the Amazon basin. Totals can be as low as 20 mm a year in the Atacama Desert and over 4,000 mm in the Andes in southern Chile. Evaporation rates are high across much of the continent, and with the variability of the precipitation in certain areas, such as north-east Brazil, drought can be a frequent problem. The Amazon is the world's largest river, but the Rio de la Plata, Orinoco, San Francisco and Paranaiba rivers are also very important. The average runoff from South America for the 1921–1985 period was calculated to be about 12,000 km^3 per year. There are large and productive aquifers, lakes and reservoirs, but the high density of population in certain areas and the untreated sewage resulting causes water pollution problems and there are similar problems due to agricultural effluents and mine wastes in some parts.

Central America has a surface area of 807,000 km^2 and a population of 35 million. Various factors have put substantial pressure on the water resources, in spite of their abundance. The annual per capita water availability exceeds 3,000 m^3/year, but only 42 percent of the rural population and 87 percent of the urban population have access to drinking water. Two thirds of the population live in areas with drainage to the Pacific ocean, while 30 percent of its water discharges into this water body. The other third of the population is located in the Caribbean basin, which generates 70 percent of the 'isthmus' water. This uneven distribution puts stress on the region's water resources.

The Caribbean has a surface area of 269,000 km^2. Countries differ in size, population and economic conditions. The temperature varies between 24°C in February and 31°C in August, also presenting a wide rainfall variation throughout the region, from 500 mm/year in the Netherlands Antilles to 7,700 mm/year in the Dominican Republic.

The region has sufficient water but the availability of safe water is becoming a major socio-economic issue. Population growth has notably increased water demand. Water quality is a generalized issue in the region due to the degradation caused by agricultural toxic substances and the mismanagement of solid waste as well as mining and industrial activities.

North America

North America, including its adjacent islands to the north and south, has an area of 24.25 million km^2 and a population of some 450 million, more than half living in the highly developed United States and Canada. These countries have the most advanced hydrological networks of the world, with routine use made of radar and satellite data. The relatively simple structure of the continent, with high mountains in the western third and vast plains extending to the east towards lower mountains, allows the arctic and the tropics to influence the climate through their weather extremes – the hurricanes, for example, which track across the south. Precipitation roughly follows the same pattern as the relief. Along the Pacific rim rainfall can reach 3,000 mm and more at higher altitudes; considerable variations occur within the western mountains and plateaux, while annual totals of between 500 and 1,500 mm occur eastwards. To the south there are very dry areas shared by Mexico and the United States. The north is dominated by the Great Lakes and a large number of others, while there are many reservoirs and extensive aquifers over much of the continent. The Mississippi-Missouri is the principal river system followed by the St. Lawrence, Mackenzie, Columbia and Colorado. These rivers and the multitude of smaller ones carried an annual average of runoff of about 7,900 km^3 to the surrounding seas over the period from 1921 to 1985. Throughout the nineteenth and twentieth centuries, human activities considerably changed the natural pattern of runoff in the majority of drainage basins and the position of the water table in many aquifers. The flow in most rivers is regulated, while numerous abstractions and discharges are made for a variety of different purposes. Large interbasin transfers take place in Canada to assist power generation. Agriculture, particularly irrigated agriculture in the west, causes resource problems and pollution arises from industry and mine wastes. There have been serious flood problems in the Mississippi basin in recent years and there are recurring floods in countries fringing the Caribbean in the wake of hurricanes.

Oceania

Australia is the smallest continent with an area of 7.6 million km^2 and a population approaching 20 million. However, the islands of Tasmania and New Guinea, those of New Zealand and those comprising the remainder of Oceania add a further 1.27 million km^2 and some 10 million additional people. Australia and New Zealand are developed countries with advanced hydrological networks but they are less advanced over the rest of the region. Australia is a large ancient plateau, raised along its eastern fringe, but the structure and geology of much of the remainder of Oceania is more varied and recent. Australia is the driest continent with a mean annual precipitation of 200 to 300 mm over much of the country, with totals rising to 1,200 mm and more along the eastern fringe and to 1,000 mm in the south-west corner. By way of contrast, many of the islands have much higher rainfalls: for example along the west coast of New Zealand's South Island 5,000 mm a year is recorded. Evaporation rates are high over Australia and over the rest of the region. The rivers of Oceania are short and fast-flowing and produce an average annual runoff of about 2,000 km^3. The average runoff from Australia is only some 350 km^3 a year: except for the Murray Darling, most of the rivers are short and drain the eastern coast. There are considerable quantities of groundwater, but there are problems of salinity, some induced by irrigation. There are relatively few lakes in Australia and many of them are ephemeral.

Conclusions

The natural water cycle is spatially and temporally complex. Humans require a stable water supply and have developed water engineering strategies dating back several thousands of years. Anthropogenic control of continental runoff is now global in scope. Increasingly, humans are a significant actor in the global hydrological cycle, defining the nature of both the physics and chemistry of hydrosystems. Runoff distortion through water engineering and land management will complicate our ability to identify climate change impacts on water systems and, hence, seriously affect water supplies. Pressure on inland water systems is likely to increase, together with population growth, economic development and potential climate change. Critical challenges lie ahead in coping with progressive water shortages, water pollution and our slow movement forward in providing universal supplies of clean water and sanitation. The situation is paradoxical: although we have succeeded in meeting certain challenges, the solutions have, in most cases, also created new problems.

Significant progress has been made in establishing the nature of water in its interaction with the biotic and abiotic environment. Eco-hydrology, dealing with ecologically sound water management and the functioning of ecosystems, is rapidly becoming a very active discipline. The results gained from understanding the basic hydrological processes have played a considerable part to date in the successful harnessing of water resources to the needs of humankind, reducing risks from extremes and so on. Our modelling capability has significantly improved with the rapid advances in computing and GIS technologies. As a result, estimates of climate change impacts on water resources are improving.

However, there are still many uncontrolled or unknown elements that hinder understanding, such as the following.

- Variability in space and time:
  - On the surface and on large spatial scales, there are huge contrasts between the very dry deserts and the rain forests; on smaller spatial scales, contrasts exist between one side of a mountain range and the other, e.g. the south and north flanks of the Himalayan mountains. There is variability too in the availability of groundwater, some areas having ready access to plentiful supplies, other areas being almost devoid of easily accessible and renewable groundwater resources.
  - On time scales from hours to decades, there is often high variability, from high-intensity precipitation events of short duration, through marked differences between seasons in precipitation to interannual and interdecadal variation. The evidence is that all these types of variability are becoming more intense as climate changes.
- Lack of adequate database and data collection in various parts of the world is now well established since Agenda 21. In spite of progress in some national water data infrastructure, our ability to describe status and trends of global water resources is declining. We still do not know the behaviour of some of the hydrological parameters in the humid tropics, highlands and flatlands, so the need for research and capacity-building is both relevant and important.

References

Acreman, M. 2000. 'Wetlands and Hydrology'. *Conservation of Mediterranean Wetlands*, No. 10. Arles, MedWet, Tour du Valat.

Alley, W.-A. 1999. *Sustainability of Groundwater Resources*. US Geological Survey Circular 1186.

Bandyopadhyay, J.; Rodda, J.-C.; Kattlemann, R.; Kundzewicz, Z.W.; Kraemer, D. 1997. 'Highland Waters: A Resource of Global Significance'. In: B. Messerli and J.-D. Ives (eds.), *Mountains of the World: A Global Priority*. London and New York, Parthenon Publishing.

Baumgartner, A. and Reichel, E. 1970. *Preliminary Results from a New Investigation of World's Water Balance, Symposium on the World Water Balance*. Vol. 3, No. 93. International Association of Hydrological Sciences.

Berner, E.-K. and Berner, R.-A. 1987. *The Global Water Cycle: Geochemistry and Environment*. Englewood Cliffs, Prentice Hall.

BGS (British Geological Survey) and DPHE (Department of Public Health Engineering). 2001. *Arsenic contamination of groundwater in Bangladesh*. Four volumes. BGS Technical Report WC/00/19. Keyworth, British Geological Survey.

Blackie, J.-R.; Ford, E.-D.; Horne, J.-E.-M.; Kinsman, D.-J.-J.; Last, F.-T.; Moorhouse, P. 1980. *Environmental Effects of Deforestation: An Annotated Bibliography*. Occasional Publication No. 10. Cumbria, Freshwater Biological Association.

Bonell, M.; Hufschmidt, M.-M.; Gladwell, J.-S. (eds.). 1993. *Hydrology and Water Humid Tropics*. Cambridge, United Nations Scientific, Educational and Cultural Organization, Cambridge University Press.

Bosch, J.-M. and Hewlett, J.-D. 1982. 'A review of catchment experiments to determine the effect of vegetation changes on water yield and evapotranspiration'. *Journal of Hydrology*, Vol. 55, pp. 3–23.

Bouwer, H. 2002. 'Artificial recharge of groundwater: hydrogeology and engineering'. *Hydrogeology Journal*, Vol. 10, pp. 121–42.

Burke, J.-J. and Moench, M.-H. 2000. *Groundwater and society: resources, tensions and opportunities*. United Nations Publication ST/ESA/205.

Clarke, R.; Lawrence, A.-R.; Foster, S.-D.-D. 1996. *Groundwater – a threatened resource*. Nairobi, United Nations Environmental Programme Environment Library 15.

Cornford, S.-G. 2001. 'Human and Economic Impacts of Weather Events in 2000'. *WMO Bulletin*, Vol. 50, pp. 284–300.

Cosgrove, W.-J. and Rijsberman, F.-R. 2000. *World Water Vision: Making Water Everybody's Business*. London, World Water Council, Earthscan Publications Ltd.

Davis, T.-J. 1993. *The Ramsar Convention Handbook*. Gland, Ramsar.

Dooge, J.-C.-I. 1984. 'The Waters of the Earth'. *Hydrological Sciences Journal*, Vol. 29, pp. 149–76.

——. 1983. 'On the Study of Water'. *Hydrological Sciences Journal*, Vol. 28.

Dunne, T. and Leopold, L.-B. 1978. *Water in Environmental Planning*. New York, Freeman and Company.

Edmunds, W.-M. and Smedley, P.-L. 1996. 'Groundwater, Geochemistry and Health'. In: J.D. Appleton, R. Fuge and G.J.H McCall (eds.), *Environmental Geochemistry and Health with Special Reference to Developing Countries*. London, The Geological Society Publishing House.

Ellis, J.-B. 1999. 'Preface'. In: B.-J. Ellis (ed.), *Impacts of Urban Growth on Surface Water and Groundwater Quality*. International Association of Hydrological Sciences. Pub. No. 259.

Falkenmark, M. and Chapman, T. (eds.) 1989. *Comparative Hydrology: An Ecological Approach to Land and Water Resources*. Paris, United Nations Educational and Cultural Organization.

FAO (Food and Agriculture Organization). 1997. *Water Resources of the Near East Region: A Review*. Rome.

——. 1995. *Water Resources of African Countries: A Review*. Rome.

Foster, S.-D.-D.; Lawrence, A.-R.; Morris, B.-L. 1997. *Groundwater in Urban Development: Assessing Management Needs and Formulating Policy Strategies*. New York, World Bank Technical Paper 390.

Foster, S.-D.-D.; Chilton, P.-J.; Moench, M.; Cardy, W.-F.; Schiffler, M. 2000. *Groundwater in Rural Development: Facing the Challenges of Supply and Resource Sustainability*. New York, World Bank Technical Paper 463.

Foster S.-D.-D. and Lawrence, A.-R. 1995. 'Groundwater Quality in Asia: An Overview of Trends and Concerns'. *UNESCAP Water Resources Journal*. Series C, 184, pp. 97–110.

Garmonov, I.-V.; Konplyantsev, A.-A.; Lushnikova, N.-P. 1974. 'Water Storage in the Upper Part of the Earth's Crust'. In: V.V. Korzun (ed.), *The World Water Balance and Water Resources of the Earth*. St Petersburg, Gidrometeoizdat.

Gat, J. and Oeschger, H. 1995. *GNIP: Global Network for Isotopes in Precipitation*. International Atomic Energy Agency, PAGES, World Meteorological Organization, International Association of Hydrological Sciences.

GEMS (Global Environment Monitoring System). 1992. *UNEP/WHO/UNESCO/WMO GEMS Water Operational Guide*. Nairobi, United Nations Environment Programme.

GRDC (Global Runoff Data Centre). 1996. *Freshwater Fluxes from Continents into World Oceans*. Report No. 10. Koblenz, Federal Institute of Hydrology.

GTOS (Global Terrestrial Observing System). 2002. *Biennial Report 2000–2001*. Rome.

Gustard, A. and Cole, G.-A. 2002. *FRIEND – A Global Perspective 1998–2002*. Wallingford, United Nations Educational, Scientific and Cultural Organization/International Hydrological Programme V, Centre for Ecology and Hydrology.

Gustard, A. (ed.). 1997. *FRIEND'97 – Regional Hydrology: Concepts and Models for Sustainable Water Resource*, Proceedings of the Postojna, Slovenia Conference, September–October. International Association of Hydrological Sciences Publication No. 246.

Herbert, M. and Kovar, K. (eds.). 1998. *Groundwater Quality: Remediation and Protection*. International Association of Hydrological Sciences Publication. No. 250.

Hibbert, A.-R. 1967. 'Forest Treatment Effects on Water Yield'. In: W.E. Sopper and H.W. Lull. 1976. *Proceedings of the International Symposium on Forest Hydrology*. New York, Pergamon Press.

Hoëg, K. 2000. 'Dams: Essential Infrastructure for Future Water Management'. Paper presented at the Second World Water Forum for the International Commission on Large Dams, 17–22 March 2000, The Hague.

Horne, R.-A. 1978. *The Chemistry of Our Environment*. New York, John Wiley and Sons.

IAH (International Association of Hydrogeologists). 2002. *The State of the Resource – Groundwater*. Kenilworth.

IAHS/UNESCO (International Association of Hydrological Sciences/United Nations Educational, Scientific and Cultural Organization). 1970. *Symposium on the Results of Research on Representative and Experimental* Basins. IAHS Pub. No. 96, Vols. I and II. Wallingford.

IJHD (International Journal of Hydropower and Dams). 2000. 'World Atlas and Industry Guide 2000'. pp. 37–8.

Jones, J.-A.-A. 1997. *Global Hydrology: Processes, Resources and Environmental Management*. London, Longman.

Kasser, P. 1967. *Fluctuation of Glaciers 1959–1965*. Paris, International Association of Hydrological Sciences, United Nations Educational, Scientific and Cultural

Organization, Series on Snow and Ice, Vol. 1.

Kirby, C.; Newson, M.-D.; Gilman, K. (eds.). 1991. *Plynlimon Research: The First Two Decades.* Wallingford, Centre for Ecology and Hydrology, Report No. 109.

Korzun, V.-I. 1978. *World Water Balance and Water Resources of the Earth.* Vol. 25 of Studies and Reports in Hydrology. Paris, United Nations Educational, Scientific and Cultural Organization.

Korzun, V.-I. et al. 1974. *Atlas of the World Water Balance.* USSR National Committee for the IHD (International Hydrological Decade). Gidromet. English translation: 1977. Paris, United Nations Educational, Scientific and Cultural Organization.

Laurans, Y.; Cattan, A.; Dubien, I. 1996. *Les Services Rendus par les Zones Humides à la Gestion des Eaux. Evaluation Economique pour le Bassin Seine-Normandie.* Paris, Seine-Normandy Water Agency-ASCA.

Llamas, M.-R. and Custodio, E. In press. *Intensive Use of Groundwater: Challenges and Opportunities.* Rotterdam, Balkema.

Lvovitch, M.-I. 1986. *Water and life.* Moscow, Mysl.

——. 1970. *World Water Balance* (General Report). Symposium on the World Water Balance. Wallingford, International Association of Hydrological Sciences, Pub. No. 93, Vol. II, pp. 401–15.

Maidment, D. and Reed, S. 1996. 'Water Balance in Africa'. Exercises prepared for a course on GIS in Water Resources for Africa, 12–14 November 1996, Rabat, Direction Générale de l'Hydraulique. Food and Agricultural Organization and United Nations Educational, Scientific and Cultural Organization.

Maksimovic, C. (ed.). 1996. *Rain and Floods in Our Cities: Gauging the Problem.* Geneva, WMO (World Meteorological Organization) Technical Reports in Hydrology and Water Resources. No. 53.

Margat, J. 1990a. *Les Eaux Souterraines dans le Monde.* Orléans, Bureau de recherches géologiques et minières [BRGM], Département eau.

Margat, J. 1990b. 'Les Gisement d'Eau Souterraine'. *La Recherche*, No. 221.

McKinney, D.-C; Maidment, D.-R.; Reed, S.-M.; Akmansoy, S.; Olivera, F.; Ye, Z. 1998. 'Digital Atlas of the World Water Balance'. In: H. Zebedi (ed.), *Water: A Looming Crisis?* Proceedings of the International Conference on World Water Resources at the Beginning of the 21st Century. Paris, United Nations Educational, Scientific and Cultural Organization/International Hydrological Programme.

Meybeck, M. 1998. 'Surface Water Quality: Global Assessment and Perspectives'. In: H. Zebedi (ed.), *Water: A Looming Crisis?* Proceedings of the International Conference on World Water Resources at the Beginning of the 21st Century. Paris, United Nations Educational, Scientific and Cultural Organization/International Hydrological Programme.

——. 1979. 'Concentration des Eaux Fluviales en Eléments Majeurs et Apports en Solution aux Océans'. *Revue de Géologie Dynamique et de Géographie Physique*, Vol. 21, pp. 215–46.

Nace, R.-L. 1971. *Scientific Framework of the World Water Balance.* Paris, United Nations Educational, Scientific and Cultural Organization, Technical Papers in Hydrology, No. 7.

Naef, F. 1981. 'Can We Model the Rainfall-Runoff Process Today?' *Hydrological Sciences Bulletin*, Vol. 26, pp. 281–89.

Neal, C. (ed.) 1997. *Water Quality of the Plynlimon Catchments.* United Kingdom, Hydrology and Earth System Sciences Special Issue, Vol. 1, pp. 381–763.

Paul, F. 2002. 'Combined Technologies Allow Rapid Analysis of Glacier Changes'. *EOS, Transactions of the American Geophysical Union*, Vol. 83, pp. 253 and 261.

Pieyns, S.-A. and Kraemer, D. 1997. 'WHYCOS, a Programme Supporting Regional and Global Hydrology'. In: A. Gustard, S. Blazkova, M. Brilly, S. Demuth, J. Dixon, H. van Lanen, C. Llasat, S. Mkhandi and E. Servat (eds.), *Friend '97 –Regional Hydrology Concepts and Models for Sustainable Water Resources Management.* Wallingford, International Association of Hydrological Sciences, Pub. No. 247.

Postel, S. 1992. *Last Oasis: Facing Water Scarcity.* The Worldwatch Environmental Alert Series. New York, W.W. Norton & Co.

Robinson, M. 1993. *Methods of Hydrological Basin Comparison.* Wallingford, Centre for Ecology and Hydrology, Report No. 120.

Rodda, J.-C. 1998. 'Hydrological Networks Need Improving!' In: H. Zebedi (ed.), *Water: A Looming Crisis?* Proceedings of the International Conference on World Water Resources at the Beginning of the 21st Century. Paris, United Nations Educational, Scientific and Cultural Organization/International Hydrological Programme.

——. 1976. 'Basin Studies'. In: J.-C. Rodda (ed.), *Facets of Hydrology.* New York, John Wiley and Sons.

Rodda, J.-C.; Pieyns, S.-A.; Sehmi, N.-S.; Matthews, G. 1993. 'Towards a World Hydrological Cycle Observing System'. *Hydrological Sciences Journal*, Vol. 38, pp. 373–8

Schultz, G.-A. and Engman, E.-T. 2001. 'Present Use and Future Perspectives of Remote Sensing in Hydrology and Water Management'. In: M. Owe, K. Brubaker, J. Ritchie and A. Rango (eds.), *Remote Sensing and Hydrology 2000.* Wallingford, International Association of Hydrological Sciences, Pub. No. 267.

Servat, E.; Hughes, D.; Fritsch, J.-M.; Hulme, M. (eds.). 1998. *Water Resources Variability in Africa During the Twentieth Century.* Wallingford, International Association of Hydrological Sciences, Pub. No. 252.

Shiklomanov, I.-A. Forthcoming. *World Water Resources at the Beginning of the 21st Century.* Cambridge, Cambridge University Press.

——. 1998a. 'Global Renewable Water Resources'. In: H. Zebedi (ed.), *Water: A Looming Crisis?* Proceedings of the International Conference on World Water Resources at the Beginning of the 21st Century. Paris, United Nations Educational, Scientific and Cultural Organization/International Hydrological Programme.

——. 1998b. *World Water Resources – a New Appraisal and Assessment for the 21st Century.* Paris, United Nations Educational, Scientific and Cultural Organization/International Hydrological Programme.

——(ed.). 1997. *Comprehensive Assessment of the Freshwater Resources of the World.* Stockholm, Stockholm Environment Institute.

Shiklomanov, I.-A.; Lammers, R.-B.; Vörösmarty, C.-J. 2002. 'Widespread Decline in Hydrological Monitoring Threatens Pan-Arctic Research'. *EOS, Transactions of the American Geophysical Union*, Vol. 83, pp. 13–16.

Snoeyink, V.-L. and Jenkins, D. 1980. *Water Chemistry.* New York, John Wiley and Sons.

Sopper, W.-E. and Lull, H.-W. 1967. *Proceedings of the International Symposium on Forest Hydrology.* New York, Pergamon Press.

Struckmeier, W. and Margat, J. 1995. *Hydrogeological Maps – a Guide and a Standard Legend.* International Association of Hydrogeologists, International Contributions to Hydrology, Vol. 17.

Swanson, R.-H.; Bernier, P.Y.; Woodward P.-D. (eds.). 1987. *Forest Hydrology and Watershed Management.* Wallingford, International Association of Hydrological Sciences, Pub. No. 187.

Syvitski, J.-P. and Morehead, M.-D. 1999. 'Estimating River-Sediment Discharge to the Ocean: Application to the Eel Margin, Northern California'. *Marine Geology*, Vol. 154, pp. 13–28

Toebes, C. and Ouryvaev, V. 1970. *Representative and Experimental Basins: An International Guide for Research and Practice.* Paris, United Nations Educational, Scientific and Cultural Organization Studies and Reports in Hydrology 4.

UN (United Nations). 1992. Agenda 21. Programme of Action for Sustainable Development. Official outcome of the United Nations Conference on Environment and Development (UNCED), 3-14 June 1992, Rio de Janeiro.

UNEP (United Nations Environment Programme). 1991. *Freshwater Pollution.* Nairobi, United Nations Environment Programme/Global Environment Monitoring System Environment.

Unninayar, S. and Schiffer, R.-A. 1997. *In-situ Observations for the Global Observing Systems: A Compendium of Requirements and Systems.* Office of the Mission to Planet Earth. NASA.

UNESCO (United Nations Educational, Scientific and Cultural Organization). 1993.

Discharge of Selected Rivers in the World: Mean Monthly and Extreme Discharges (1980–1984). Vol. III. Paris, United Nations Educational, Scientific and Cultural Organization/International Hydrological Programme.

UNESCO/IHP (United Nations Educational, Scientific and Cultural Organization/International Hydrological Programme). 2001. *Balance hídrico superficial de América latina y el Caribe.* Montevideo.

——. 1996. *Mapa hidrogeológico de Sud America.* Montevideo.

Van Lanen H.-A.-J. and Demuth, S. 2002. *FRIEND 2002 – Regional Hydrology: Bridging the Gap between Research and Practice.* Wallingford, International Association of Hydrological Sciences, Pub. No. 274.

Vörösmarty, C.-J.; Meybeck, M.; Fekete, B.; Sharma, K. 1997. 'The Potential Impact of Neo-Catorisation on Sediment Transport by the Global Network of Rivers'. In: D.-E Walling and J.-L. Probst (eds.), *Human Impact on Erosion and Sedimentation.* Wallingford, International Association of Hydrological Sciences, Pub. No. 245.

Vrba, J. and Zaporozec, A. 1994. *Guidebook on Mapping Groundwater Vulnerability.* International Contributions to Hydrogeology. Wallingford, International Association of Hydrological Sciences.

Wallace, J. 1996. 'Hydrological Processes'. In: *Institute of Hydrology Annual Report 1994–95.* Wallingford, International Association of Hydrological Sciences.

Walling, D.-E. and Webb, B.-W. 1996. 'Erosion and Sediment Yield: A Global View'. In: D.-E. Walling and B.-W. Webb (eds.), *Erosion and Sediment Yield: Global and Regional Perspectives.* Wallingford, International Association of Hydrological Sciences, Pub. No. 236.

WHO (World Health Organization). 1991. *GEMS/Water 1990–2000. The Challenge Ahead.* UNEP/WHO/UNESCO/WMO Programme on Global Water Quality Monitoring and Assessment. Geneva.

WHO/UNEP (World Health Organization/United Nations Environment Programme). 1991. *Water Quality: Progress in the Implementation of the Mar del Plata Action Plan and a Strategy for the 1990s.* Nairobi, Earthwatch Global Environment Monitoring System, World Health Organization, United Nations Environment Programme.

WMO (World Meteorological Organization). 1995. 'Infohydro Manual'. *Operational Hydrology Report,* No. 28. Geneva.

——. 1994. *Guide to Hydrological Practices.* Fifth edition. Geneva.

WMO/UNESCO (World Meteorological Organization/United Nations Educational, Scientific and Cultural Organization). 1997. *Water Resources Assessment: Handbook for Review of National Capabilities.*

WCD (World Commission on Dams). 2000. *Dams and Development: A New Framework for Decision-Making.* London, Earthscan Publications Ltd.

Wright, J.-F.; Sutcliffe, D.-W.; Furse, M.-T. (eds.). 1997. *Assessing the Biological Quality of Fresh Waters.* Proceedings of the International Workshop. Ambleside, Cumbria, Freshwater Biological Association.

Young, G.-J.; Dooge, J.-C.-I.; Rodda, J.-C. 1994. *Global Water Resource Issues.* Cambridge, Cambridge University Press.

Zektser, I. 1999. *World Map of Hydrogeological Conditions and Groundwater Flow.* Paris, United Nations Educational, Scientific and Cultural Organization.

Zektser, I. and Margat, J. Forthcoming. *Groundwater Resources of the World and Their Use.* Paris, United Nations Educational, Scientific and Cultural Organization/International Hydrological Programme Monograph.

Some Useful Web Sites[3]

Global Precipitation Climatology Centre (GPCC)
http://www.dwd.de/research/gpcc/
Global precipitation analyses for investigation of the earth's climate.

United Nations Educational, Scientific and Cultural Organization (UNESCO): International Hydrological Programme (IHP)
http://www.unesco.org/water/ihp/
UNESCO intergovernmental scientific programme in water resources.

United Nations Environment Programme (UNEP): Freshwater Portal
http://freshwater.unep.net
Information on key issues of the water situation.

United Nations Environment Programme (UNEP): Global Environment Monitoring System (GEMS/WATER)
http://www.cciw.ca/gems/gems-e.html
A multifaceted water science programme oriented towards understanding freshwater quality issues throughout the world. Major activities include monitoring, assessment and capacity-building.

3. These sites were last accessed on 19 December 2002.

World Meteorological Organization (WMO): Global Runoff Data Centre (GRDC)

http://www.bafg.de/grdc.htm

Collection and dissemination of river discharge data on a global scale.

World Meteorological Organization (WMO): Hydrology and Water resources Programme

http://www.wmo.ch/web/homs/

Collection and analysis of hydrological data as a basis for assessing and managing freshwater resources.

World Meteorological Organization (WMO): World Climate research Programme (WCRP)

http://www.wmo.ch/web/wcrp/wcrp-home.html

Studies of the global atmosphere, oceans, sea and land ice, and the land surface which together constitute the earth's physical climate system.

World Meteorological Organization (WMO): World Hydrological Observing System (WHYCOS)

http://www.wmo.ch/web/homs/projects/whycos.html

Global network of national hydrological observatories.

World Water Assessment Programme / United Nations Educational, Scientific and Cultural Organization (WWAP/UNESCO): Water Portal

http://www.unesco.org/water/

A new initiative for accessing and sharing water data and information from all over the world.

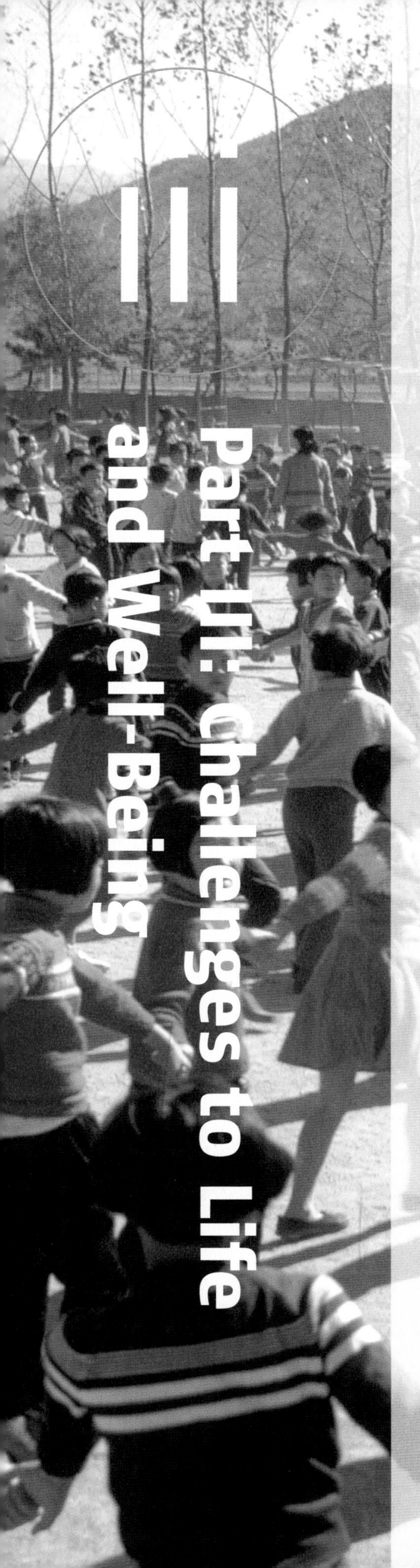

III

Part III: Challenges to Life and Well-Being

It is in the context of a global water crisis, informed by the intricate nature of the water cycle, that we begin our assessment of the challenge areas. Just as our definitions of what constitute basic needs have expanded to encompass a fuller understanding of human development and the natural environment, so must our evaluation reflect these changing expectations.

This section explores the ways in which we use water and the increasing demands we are placing on the resource. Signs of stress and strain are apparent across every sector: health, ecosystems, cities, food, industry and energy. As the following chapters show, with population growth and continuing pollution, these pressures are likely to increase. Our only hope is to learn to accommodate the competing uses and users in an equitable and responsible manner.

5 Basic Needs and the Right to Health

By: World Health Organization (WHO)
Collaborating agency: United Nations Children's Fund (UNICEF)

Table of contents

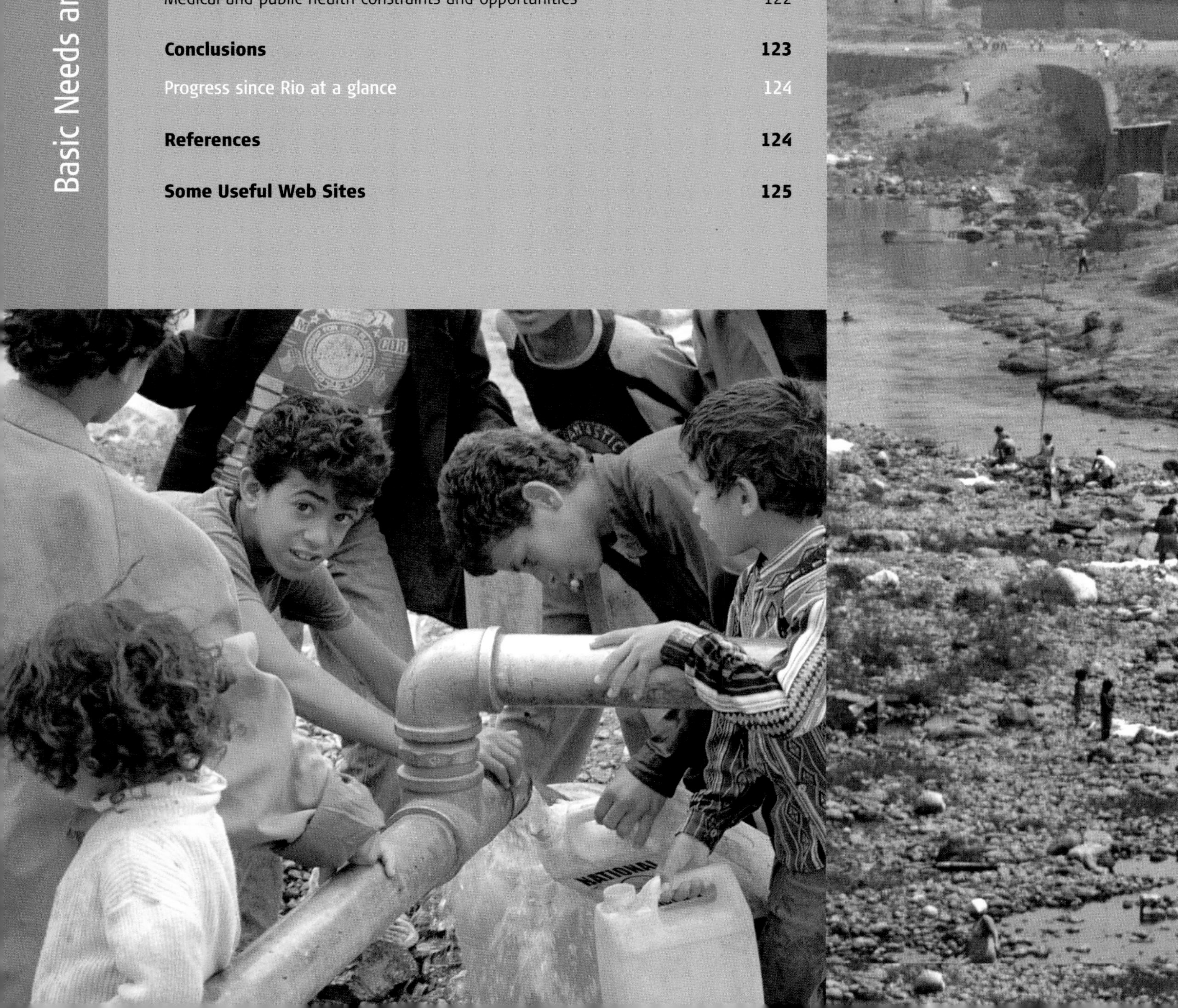
NATIONAL

He who has health has hope.
He who has hope has everything.

Arabic proverb

IN WEALTHIER PARTS OF THE WORLD, the connection between water, hygiene and health is taken for granted. But for the less fortunate majority, access to clean and adequate water is a daily struggle. This chapter describes why water supply and sanitation are essential for meeting basic needs, yet also underlines the terrible costs to society if broader conditions of physical, mental and social well-being are not factored into the health and human development equation as well. The importance of promoting and protecting health has risen to the top of the political agenda and much progress has been made. However, millions are still deprived of a basic human right and huge challenges remain in fulfilling the numerous promises made.

The Burden of Water-Associated Ill-Health

Every day, diarrhoeal diseases cause an estimated 5,483 deaths, mostly among children under five. The World Health Organization (WHO) global estimate of the number of deaths from infectious diarrhoeas in the year 2001 amounts to 2 million for all age groups, with a heavy toll among children under five: an estimated 1.4 million deaths.

Similar WHO estimates indicate that malaria kills over a million people every year, and a large percentage of them are under five as well, mainly in Africa south of the Sahara. Worldwide, over 2 billion people are infected with schistosomes and soil-transmitted helminths and 300 million of these suffer serious illness as a result. In Bangladesh alone, some 35 million people are exposed, on a daily basis, to elevated levels of arsenic in their drinking water, which will ultimately threaten their health and shorten their life expectancy.

These absolute numbers are dramatic in their own right. They stand out against a much larger cluster of water-associated ill-health; for example, certain malignant tumours with a suspected link to long-term exposure to pollutants in water, for which the fractional attribution to specific determinants still needs to be firmly established. The cluster of ill-health is even more insidious in its impact on the economics of countries and on livelihoods at the household level. Water-associated diseases hit the poor in a disproportionate way and this burden of ill-health maintains the vicious cycle in which poverty leads to more ill-health, and more ill-health implies further impoverishment.

Water, sanitation and hygiene are three intertwined determinants of the water/ill-health/poverty spectrum, with hygiene to be considered in its broadest sense, including environmental as well as personal hygiene. The associated burden of disease is not only felt in the world today, it also affects the potential of future generations. The most important category of personal hygiene-related diseases affecting school-age children is intestinal helminth infections. These parasites consume nutrients, aggravate malnutrition, retard children's physical development and result in poor school attendance and performance. The outlook is bleak. Helminth infections destroy the well-being and learning potential of millions of children. Each year 19.5 million people are infected with roundworm and whipworm alone, with the highest rate of infection among school-age children. Schistosomiasis (bilharzia) is also a young people's disease; an estimated 118.9 million children under fifteen years are infected.

The full picture of water-associated ill-health is complex for a number of reasons. As stated, the burden of several disease groups can only partly be attributed to water determinants. This holds true even for diseases whose link to water at first sight may seem obvious and exclusive. Infectious diarrhoeas for example, are in part food-borne. For the part that is water-borne, there may be numerous exposure routes, operating in combinations that are often location specific and may change over time. These multiple transmission pathways may be synergistically or antagonistically linked. Where water plays an essential role in the ecology of diseases, it may be hard to pinpoint the relative importance of aquatic components of the local ecosystem. And, when it comes to disease prevention and control, there are methodological obstacles to a cost-effectiveness analysis of health sector versus water management interventions that will satisfy all parties concerned.

The diseases and conditions of ill-health directly associated with water, sanitation and hygiene include infectious diarrhoea (which, in turn, includes cholera [see box 5.1], salmonellosis, shigellosis,

Box 5.1: Cholera in 2001

Cholera, caused by a variety of strains of the bacterium *Vibrio cholerae*, remains a global threat and a challenge to countries where access to safe drinking water and basic sanitation cannot be ensured for all. In 2001, fifty-eight countries from all regions of the world officially reported to the WHO a total of 184,311 cases and 27,728 deaths. Compared to the year 2000 the case fatality rate (CFR) dropped from 3.6 to 1.48 percent. This drop can be attributed almost entirely to the extremely low CFR of 0.22 percent in the cholera outbreak in South Africa which accounted for 58 percent of the global total of reported cases. Leaving aside the South African outbreak, the CFR diminished only slightly to 3.21 percent from 3.9 percent in 2000. With a total of 173,359 cases, Africa accounted for 94 percent of the global total of cholera cases. Case numbers reported from Asia remained stable (around 10,000), while an important decline was observed in the Americas.

Despite efforts by many countries to contain the spread of cholera, the disease is once more on the rise worldwide. Officially notified cases do not reflect the overall burden of the disease, because of underreporting, trade-related sanctions and other limitations in the surveillance and reporting system.

Source: WHO, 2002c.

amoebiasis and a number of other protozoal and viral infections), typhoid and paratyphoid fevers, acute hepatitis A, acute hepatitis E and F, fluorosis, arsenicosis, legionellosis, methaemoglobinaemia, schistosomiasis, trachoma, intestinal helminth infections (including ascariasis, trichuriasis and hookworm infection), dracunculiasis, scabies, dengue, the filariases (including lymphatic filariasis and onchocerciasis), malaria, Japanese encephalitis, West Nile virus infection, yellow fever and impetigo.

Yet there are also indications of water quality, quantity and/or hygiene links to conditions as diverse as ischemic heart disease or malignant bladder tumours. Unintentional drowning is a major cause of death in the category of accidents and injuries, and that same category also includes the permanent skeletal damage caused to women from carrying heavy loads of water over long distances day after day.

The burden of disease concept

An extensive debate took place in the 1980s on the issues surrounding the economic impact of ill-health, the ways in which the health sector should let economics influence decision-making and the use of its limited resources for interventions, and the transfer of hidden costs to the health sector resulting from development projects with adverse health impacts. This health-economics debate led to the development of a number of new indicators linking the costs of interventions to health outcomes in terms of quality of life and well-being, while striving for greater equity. The *1993 World Development Report*, with the theme *Investing in Health* (World Bank, 1993) formally introduced a new indicator of population health, the Disability-Adjusted Life Year (DALY).

The DALY is a summary measure of population health. One DALY represents a lost year of healthy life and is used to estimate the gap between the current health of a population and an ideal situation where everyone in that population would live into old age in full health. For each disease DALYs are calculated, on a population scale, as the sum of years lost due to premature mortality (YLL) and the healthy years lost due to disability (YLD) for incident cases of the ill-health condition (WHO, 2002e).

Within the context of the health sector, DALYs are used to estimate the burden caused by specific diseases, injuries and risk factors. Burden of disease estimates are used as a decision-making criterion to choose between the deployment of such disparate interventions as, for example, a measles vaccination campaign and indoor residual spraying for malaria control. Now that the effect of interventions can be expressed in fully comparable units it has become possible to establish the relative cost-effectiveness of each intervention. The WHO reports on the status of world health every year with global burden of disease estimates as the main statistics.

The mortality and burden of disease estimates for the year 2001 for some water-associated infectious diseases and for drowning are presented in table 5.1. The causes of death have been estimated based on data from national vital registration systems that capture about 17 million deaths annually. In addition, information from sample registration systems, population laboratories and epidemiological analyses of specific conditions has been used to improve the estimates. These figures illustrate how the extended debilitating nature of some diseases reflects in the burden they cause (e.g. lymphatic filariasis); they also indicate where major reductions in disease burden can be achieved beyond a strict reduction in mortality.

Water and health: an intricate relationship

The definition of health as contained in the 1948 Constitution of WHO has withstood the ravages of time: 'Health is a complete state of physical, mental and social well-being, and not merely the absence of disease and infirmity.' Certainly, having one's basic needs in terms of water met and being able to rely on a sustainable livelihood are crucial elements of social well-being, and contribute substantially to physical and mental well-being as well. From here on, however, the issue becomes more complex. Water and health are intricately linked. A workable public health perspective of all water issues requires a clear definition of the nature and magnitude of the links between the two.

There are basically two types of links, which facilitate the elucidation of cause-effect relationships between water management issues and impacts on health: water as the conveyance medium of pathogens (disease-causing organisms), and water providing the habitat for vectors and intermediate hosts of pathogens (for species that produce or maintain pathogens). Water plays a conveyance role for micro-organisms, chemical pollutants and sources of radiological risks. The importance of this role for health relates mainly to drinking water, but also indirectly to water applied to food crops and livestock, and through aerosols generated by air conditioning systems. This role alters to a health-promoting mechanism when water is used for purposes of hygiene: from this perspective, quantity is a more important determining factor than quality.

Aquatic ecosystems serve as breeding habitats for insect vectors of diseases and for snails that serve as intermediate hosts in the transmission cycle of certain parasitic diseases. These ecosystems can be permanent, with wetland areas as the most obvious example, or seasonal, linked to local weather patterns.

Lack of adequate sanitation is the most critical determinant of contamination of drinking water with micro-organisms. Pollution from urban and industrial waste and runoff of agrochemicals is by and large responsible for chemical contamination, although naturally occurring anorganic pollutants (fluor and arsenic) may also contribute substantially. Wastewater use in agricultural production

Table 5.1: Some water-associated diseases by cause and sex, estimates for 2001

	Deaths (in thousands)						Burden of disease (in thousands)					
	Both sexes		Males		Females		Both sexes		Males		Females	
	number	%	number	%	number	%	number	%	number	%	number	%
Total burden of disease (000s of DALYs)							1,467,183	100	768,064	100	699,119	100
Total deaths (000s)	56,552	100	29,626	100	26,926	100						
Communicable diseases, maternal and perinatal conditions and nutritional deficiencies overall	18,374	32.5	9,529	32.2	8,846	32.9	615,737	42.0	304,269	39.6	311.468	44.6
Infectious and parasitic diseases altogether	10,937	19.3	5,875	19.8	5,063	18.8	359,377	24.5	184,997	24.1	174,380	24.9
Diarrhoeal diseases	2,001	3.5	1,035	3.5	966	3.6	62,451	4.3	31,633	4.1	30,818	4.4
Malaria	1,123	2.0	532	1.8	591	2.2	42,280	2.9	20,024	2.6	22,256	3.2
Schistosomiasis	15	0.0	11	0.0	5	0.0	1,760	0.1	1,081	0.1	678	0.1
Lymphatic filariasis	0	0.0	0	0.0	0	0.0	5,644	0.4	4,316	0.6	1,327	0.2
Onchocerciasis	0	0.0	0	0.0	0	0.0	987	0.1	571	0.1	416	0.1
Dengue	21	0.0	10	0.0	11	0.0	653	0.0	287	0.0	365	0.0
Japanese encephalitis	15	0.0	8	0.0	8	0.0	767	0.1	367	0.0	400	0.1
Trachoma	0	0.0	0	0.0	0	0.0	3,997	0.3	1,082	0.1	2,915	0.4
Intestinal nematode infections	12	0.0	6	0.0	5	0.0	4,706	0.3	2,410	0.3	2,296	0.3
Ascariasis	4	0.0	2	0.0	2	0.0	1,181	0.1	604	0.1	577	0.1
Trichuriasis	2	0.0	1	0.0	1	0.0	1,649	0.1	849	0.1	800	0.1
Hookworm infection	4	0.0	2	0.0	2	0.0	1,825	0.1	932	0.1	893	0.1
Unintentional injuries overall	3,508	6.2	2,251	7.6	1,256	4.7	129,853	8.9	82,378	10.7	47,475	6.8
Drowning	402	0.7	276	1.0	126	0.5	11,778	0.8	8,150	1.1	3,628	0.5

The burden of disease is calculated through an indicator of population health, the DALY: a DALY represents a lost year of healthy life and is the unit used to estimate the gap between the current health of a population and an ideal situation in which everyone in that population would live into old age in full health. This table shows the total deaths and burden of disease caused by communicable diseases, maternal and perinatal conditions and nutritional deficiencies, non-communicable diseases and injuries related to water.

Source: WHO, 2002e.

systems carries specific risks of contamination, both with pathogenic organisms (e.g. intestinal helminths) and chemicals (e.g. heavy metals) through concentration and amplification. More sophisticated water conveyance and treatment systems may become the source of pathogens that are released into the environment as aerosols, as is the case for *Legionella ssp* (Legionnaire's disease) in association with air conditioning.

A number of diseases caused by bacteria or parasites, for example, trachoma and intestinal helminth infections, will proliferate because of lack of sufficient quantities of water for basic hygiene. These are traditionally referred to as the water-washed (as opposed to the water-borne) diseases. Behavioural changes, such as hand-washing and regular bathing, will only be effective if the required minimum amounts of water are available, but evidence-based standards and norms are still lacking.

A range of aquatic ecosystems supports the breeding of a great number of species that play a role in the transmission of diseases; these species are known as vectors. Biological diversity is a main feature of vectors and the diseases they transmit, making it necessary to consider epidemiological situations in the specific local context in which they occur. In different parts of the world the accepted knowledge bases related to the diversity of all these water-related organisms vary a great deal. Frequently, our potential to deal with these health problems in a sustainable way is limited by the lack of sufficient knowledge of local pathogen and vector ecologies. New knowledge continues to be generated, however, particularly at the molecular genetics level, together with new opportunities for improving public health.

The significance of this collective knowledge base linking water and health lies in the options provided on effective ways to prevent ill-health and disease, and to promote the health status of communities. Access to a supply of safe drinking water, combined with sanitation that prevents contaminants from reaching sources of drinking water, and hygiene behaviour such as hand-washing and proper food handling, supported by sufficient quantities of water, are the main tools in the fight against gastrointestinal infections. Water management practices that reduce the environment's receptivity to the propagation of disease vectors and intermediate hosts can, in specific settings, be the main contributor to reducing transmission risks of such diseases as malaria and schistosomiasis. In places where transmission levels are intense, such practices at least add sustainability and resilience to medical approaches delivered by the health services. All measures when combined, i.e. the supply of safe drinking water and adequate sanitation, improved hygiene behaviour and environmental management aimed at disease vectors, translate into a considerable reduction of the costs of the delivery of health services incurred to governments, and the costs incurred to households, directly and indirectly, as a result of ill-health or, worse, death of family members.

Diseases: status and trends

Diseases related to the lack of safe drinking water, sanitation and/or hygiene

A recent analysis based on health statistics for the year 2000 (Prüss et al., 2002) shows that globally between 1,085,000 and 2,187,000 deaths due to diarrhoeal diseases can be attributed to the 'water, sanitation and hygiene' risk factor, 90 percent of them among children under five. In terms of burden of disease, the attributable fraction ranges between 37,923,000 and 76,340,000 DALYs, a burden that can be up to 240 times higher in developing countries compared to industrialized countries.

This analysis tackles the various complexities in a consistent way, starting by mapping out the transmission pathways of faecal-oral diseases (see figure 5.1). It then continues by offsetting the fourteen regions distinguished in the WHO burden of disease statistics against six globally important exposure scenarios, indicating population percentages for each combination. The fourteen regions consist of the six WHO regions (Africa, the Americas, Eastern Mediterranean, Europe, South-East Asia, Western Pacific), subdivided on the basis of child and adult mortality levels as criteria. The six exposure scenarios were constructed using the WHO/UNICEF Joint Monitoring Programme's (JMP) database as shown in table 5.2.

Rather than following an attribution procedure based on experts, the analysis assessed exposure risks as derived from literature-based intervention reports, to arrive at relative risks for the different scenarios. The standard burden of disease methodology was applied, but to account for uncertainty a sensitivity analysis was included based on the assumptions for the most critical transition, namely from scenario IV to scenario II. By applying two different values to the relative risks comparing these scenarios, so-called minimal and realistic estimates were obtained (as reflected in the overall ranges presented earlier for the global figures).

To the global estimates for the number of deaths and burden of disease due to water/sanitation/hygiene-attributed infectious diarrhoeas were added the estimates of a number of other water-associated diseases (schistosomiasis, trachoma and intestinal helminth infections). The grand total of this first and incomplete estimate of water-, sanitation- and hygiene-associated ill-health amounted to 2,213,000 deaths and 82,196,000 DALYs per year.

In addition to the major methodological improvements achieved through this analysis, it also:

- confirms, with a stronger evidence base than before, that water, sanitation and hygiene are key determinants of health, with substantial mortality and morbidity occurring as a result of the lack of access to water and sanitation and of inadequate hygiene behaviour;
- underlines how diseases related to water, sanitation and hygiene disproportionately affect the poor; and

Figure 5.1: Transmission pathways of faecal-oral diseases

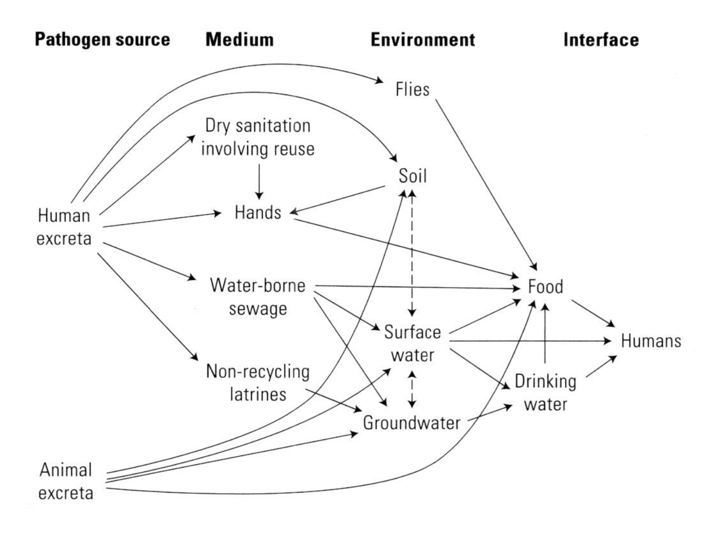

Most of the transmission pathways of faecal-oral diseases are related to water.

Source: Prüss et al., 2002.

Table 5.2: Six scenarios of exposure to environmental faecal-oral pathogens

Scenario	Description	Environmental faecal-oral pathogen load
VI	No improved water supply and no basic sanitation in a country that is not extensively covered by those services, and where water supply is not routinely controlled	Very high
Vb	Improved water supply and no basic sanitation in a country that is not extensively covered by those services, and where water supply is not routinely controlled	Very high
Va	Basic sanitation but no improved water supply in a country that is not extensively covered by those services, and where water supply is not routinely controlled	High
IV	Improved water supply and basic sanitation in a country that is not extensively covered by those services, and where water supply is not routinely controlled	High
IIIc	IV and improved access to drinking water (generally piped to household)	High
IIIb	IV and improved personal hygiene	High
IIIa	IV and drinking water disinfected at point of use	High
II	Regulated water supply and full sanitation coverage, with partial treatment for sewage, corresponding to a situation typically occurring in developed countries	Medium to low
I	Ideal situation, corresponding to the absence of transmission of diarrhoeal diseases through water, sanitation and hygiene	Low

This figure shows how the risks of human contamination from faecal-oral diseases vary according to different levels of provision for water, sanitation and hygiene. An obvious link is to be observed between health gains and improved service levels.

Source: Based on table 2 and figure 2 in Prüss et al., 2002, pp. 539–40.

- illustrates the high potential for disease reduction by simple methods such as safe drinking water storage and disinfection at the household level.

The paucity of data from the field continues to be a constraint, as illustrated in the box on cholera earlier.

WHO is currently embarking on analyses to clarify the attributable fraction of the burden of disease to ecosystem components of water resource development projects, for malaria, schistosomiasis, filariasis and Japanese encephalitis.

Chemical water quality parameters are partly responsible for the ill-health related to drinking water supply. These chemicals include the inorganic compounds fluor, arsenic, cadmium and uranium which may occur naturally in aquifers used for drinking water withdrawals. Because of the arsenic crisis in Bangladesh (35 million people exposed through drinking water supply boreholes) attention has recently been focused on this compound, but it is estimated that the health impact of naturally occurring fluor is more widespread. The health condition resulting from exposure to fluor over and above a certain threshold is known as fluorosis and affects the skeletal tissues. The heavier elements cadmium and uranium cause kidney damage with resorption of minerals from the bone as a secondary effect.

Industrial waste and agricultural runoff are the major sources of organic pollution, and the impact of exposure to these compounds and their residues may be characterized as one or more of the following: mutagenic, carcinogenic, teratogenic, toxic to the embryo or causing reproductive toxicity; details are provided in the Guidelines for Drinking Water Quality (WHO, 1996). Of growing concern are the endocrine disrupting chemicals (EDCs), exposure to which may modulate the endocrine system. While drinking water is a potential source of human exposure to EDCs, it is not considered a major exposure pathway unless unusual contamination has occurred (WHO, 2002d).

Vector-borne diseases

Two characteristics distinguish the water-related vector-borne diseases from the water-borne diseases: their transmission mode and the nature of their association with water. Insects, essentially bloodsucking species, play a key role in the transmission of diseases such as malaria, filariasis and various infections of a viral etiology (e.g. yellow fever, West Nile virus, Japanese encephalitis and dengue). This role is not simply a mechanical one of taking pathogens from an infected to a not-infected person. Usually, the vector itself provides conditions required for the development of part of a parasite lifecycle. Aquatic and amphibious snails that serve as intermediate hosts for the *Schistosoma* parasites equally provide a crucial habitat for the development of their infective larvae, but they do not actually play an active role in the transmission from one person to another. In the case of schistosomiasis (bilharzia), transmission is a passive process initiated by the contamination of water with excreta containing parasite eggs, and ending with the penetration by the infective larvae (*cercariae*) of the skin of people in contact with that water. From a disease control perspective, however, the term vector is used generously to include all species (true vectors, intermediate hosts and rodent reservoirs of pathogens) whose elimination will interrupt disease transmission.

These diseases are linked to water through the ecological requirements of the vector species. Mosquito vectors all have an aquatic larval stage. The distribution of the diseases reflects the differences between genera and individual species. Lymphatic filariasis and malaria provide an example. The *Culex* mosquitoes transmitting filariasis breed in organically polluted water and distribution of the disease is therefore linked to the urban environment, particularly where open sewers and treatment ponds provide the required breeding habitat. On the other hand, Anopheline species of mosquitoes transmitting malaria generally require standing or slowly running clean freshwater, a habitat often not available in cities. In fact, in many African cities a downward gradient in malaria transmission intensity can be seen when moving from the outskirts to the city centre, considered to reflect the increasing pollution of water collections.

Malaria

Malaria is a life-threatening parasitic disease transmitted from one person to another through the bite of female *Anopheles* mosquitoes. The disease exerts its heaviest toll in Africa south of the Sahara, where about 90 percent of the annual global rate of over 1 million deaths from malaria occur. It is the leading cause of deaths in young children.

In 2001, the estimated global burden of malaria amounted to 42.3 million DALYs out of 359.4 million DALYs for all infectious and parasitic diseases, and 1,467.2 million DALYs for the overall global burden of disease. The malaria burden constitutes 10 percent of Africa's overall disease burden. Malaria causes at least 396.8 million cases of acute illness each year. Pregnant women are the main adult risk group in most endemic areas of the world. As one of the major public health problems in tropical countries, malaria contributes substantially to the erosion of development achievements and hampers poverty alleviation efforts in the world's poorest countries. It has been claimed that the disease has reduced economic growth in African countries by 1.3 percent each year over the past thirty years (Sachs and Malaney, 2002).

There are some 422 species of *Anopheles* mosquitoes in the world, but only seventy are vectors of malaria under natural conditions and of these some forty are of major public health importance. A wide range of aquatic ecologies provides the context for malaria transmission, including shaded and sunny pools and streams, rooftop drinking water tanks (in south Asia), rainwater pools of a certain duration, mangroves and brackish water lagoons, and the canals, ditches, night storage tanks and other components of irrigation schemes and hydraulic infrastructure.

There is no vaccine available to protect against infection with the *Plasmodium* parasites causing malaria. Parasite resistance against prophylactic and curative drugs is a permanent concern, as is poor people's access to drugs and health services constrained by many factors. Insecticide-impregnated mosquito nets are effective as a means of personal protection, particularly for children under five, but only where mosquito biting patterns overlap with the use of nets. Chemical vector control is faced with insecticide resistance as well as increasing regulations concerning the use of insecticides. The 2001 Stockholm Convention on Persistent Organic Pollutants restricts production and use of DDT, with an exemption for the reportable application of this pesticide in disease vector control, while it aims to promote the development and testing of alternatives to manage vector populations (UNEP, 2001).

Depending on the species, malaria vector mosquitoes can rely on a range of aquatic ecologies to propagate. The example below (box 5.2) illustrates the impact of one type of water resource development in the Ethiopian highlands.

Schistosomiasis

Schistosomiasis, also known as bilharzia, is a disease caused by parasitic flatworms called trematodes of the genus *Schistosoma* for which aquatic (Africa and the Americas) or amphibious (western Pacific) snails serve as intermediate hosts. An estimated 246.7 million people worldwide are infected, and of these, 20 million suffer severe consequences of the infection, while 120 million suffer milder symptoms. An estimated 80 percent of transmission takes place in Africa south of the Sahara.

The estimated global burden of schistosomiasis was established as 1.8 million DALYs in 2001, but the parameters and assumptions used to arrive at this estimate are currently under review.

Box 5.2: The compounded malaria impact of microdams in Ethiopia

Recent studies in Ethiopia using community-based incidence surveys revealed a 7.3-fold increase of malaria incidence associated with the presence of microdams. The study sites were all at altitudes where malaria transmission is seasonal (in association with the rains). The increase was more pronounced for dams below 1,900 metres of altitude, and less above that altitude. In addition, observed trends in incidence suggest that dams increase the established pattern of transmission throughout the year, which leads to greatly increased levels of malaria at the end of the transmission season.

Source: Ghebreyesus et al., 1999.

There is no vaccine against schistosomes, which infect mainly children and adolescents, as well as those in frequent and close touch with contaminated water for occupational reasons, such as irrigation farmers. There is, however, an effective drug, Praziquantl, whose price has come down considerably in recent years. Nevertheless use of the drug has not succeeded in breaking through the cycle of water contamination and reinfection. The remaining prevalence of infection following various rounds of mass drug treatment reflects the state of sanitation facilities, hygiene behaviour and environmental receptivity. Without addressing these determinants through improved irrigation water management, for example, the economic feasibility of maintaining results of mass drug treatment campaigns is low.

Lymphatic filariasis

Lymphatic filariasis is a mosquito-borne parasitic worm infection, which in its most dramatic form is expressed in the symptoms of elephantiasis, the accumulation of lymph, usually in the legs. It is not a killer disease, but causes severe debilitation and social stigma. So while the number of deaths from filariasis is close to zero, the burden of disease value is relatively high, globally estimated at 5.6 million DALYs in 2001. Urban populations in Africa and south and South-East Asia are most affected, although some rural cottage industries also provide conditions that favour the *Culex* vectors: in Sri Lanka, coconut husk pits, where coconut shells are left to rot to harvest the fibre, are notorious breeding places and communities engaging in this practice tend to have high infection rates.

International efforts are underway to eliminate filariasis as a global public health problem. These have become possible thanks to greatly improved diagnostic techniques and dramatic advances in treatment methods – both for controlling the spread of this infection and for alleviating its symptoms. Partnerships with the pharmaceutical industry ensure that drugs are available wherever they are needed.

Improvements in urban water management will provide a solid basis for the progress that is being made through these health sector efforts. The improvements should focus on the infrastructure for sewage and stormwater drainage and for collection of wastewater before its treatment and possible reuse. Two parasite species cause lymphatic filariasis and the one of lesser public health importance (*Brughia malayi*) is transmitted by mosquitoes of the genus *Mansonia* whose breeding is associated with aquatic weeds. Proper management of reservoirs, i.e. the harvesting or elimination of aquatic weeds, meets dual objectives: reduced evapotranspiration from the reservoir and an effective interruption of transmission of this parasite.

Arboviral infections

Arboviral infections are acute and have high mortality rates; infection outbreaks occur in a cyclical way, and as a result, burden of disease estimates vary greatly from one year to another. In 2001, the estimate of the burden of two arboviral diseases (dengue and Japanese encephalitis) combined amounted to about 1.4 million DALYs.

The association with water is disease-specific. The distribution of Japanese encephalitis is from east to south Asia, with a strong link to flooded rice ecosystems, where the main vectors of the *Culex tritaeniorrhynchus/gelidus* group breed. The *Aedes* vectors of dengue breed in small, domestic collections of water (flower vases, dumped car tyres and standing water in solid waste dumps). The water management practices to prevent arbovirus transmission and outbreaks of arboviral diseases are similarly disease- and site-specific. In some countries, for example India, the provision of drinking water supplies in rural areas has contributed to increased dengue risks. In instances where there is an unreliable supply, people tend to collect water when it is available and store it in their house. Such domestic water storage may become a prime source of *Aedes* mosquitoes.

Water Management for Health

Drinking water supply and sanitation

Water supply, sanitation and hygiene improvements have a long history as public health interventions, and the rationale for their promotion has mainly relied on the substantial reductions in morbidity and mortality they can achieve, especially in the developing world. Yet the *Global Water Supply and Sanitation 2000 Assessment* (WHO/UNICEF, 2000) shows that 1.1 billion people lack access to improved water supply and 2.4 billion to improved sanitation. In the vicious poverty-ill-health cycle, inadequate water supply and sanitation are both the underlying cause and the outcome: invariably, those who lack adequate and affordable water supplies are the poorest in society. If improved water supply and basic sanitation were extended to the present-day 'unserved', it is estimated that the burden of infectious diarrhoeas would be reduced by some 17 percent; if universal piped, well-regulated water supply and full sanitation were achieved, this would reduce the burden by some 70 percent. Improvements such as these would also lead to reductions in other water-, sanitation- and hygiene-related diseases, such as schistosomiasis, trachoma and infectious hepatitis.

The *2000 Assessment* established that most of the unserved populations are in Asia and Africa. In absolute terms, Asia has the highest number of underserved, but proportionally this group is bigger

Figure 5.2: Water supply, distribution of unserved populations

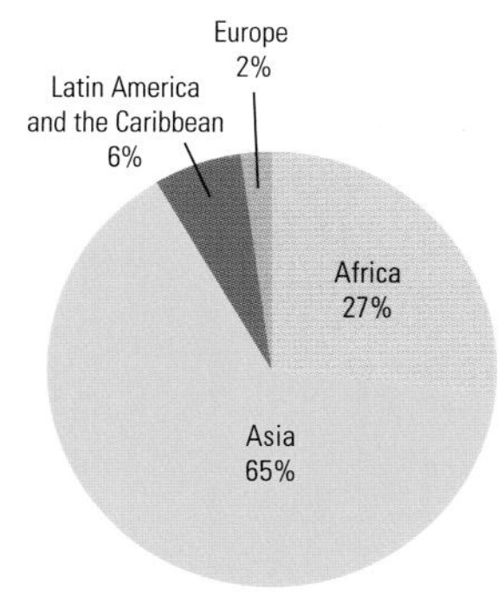

Total unserved: 1.1 billion

Figure 5.3: Sanitation, distribution of unserved populations

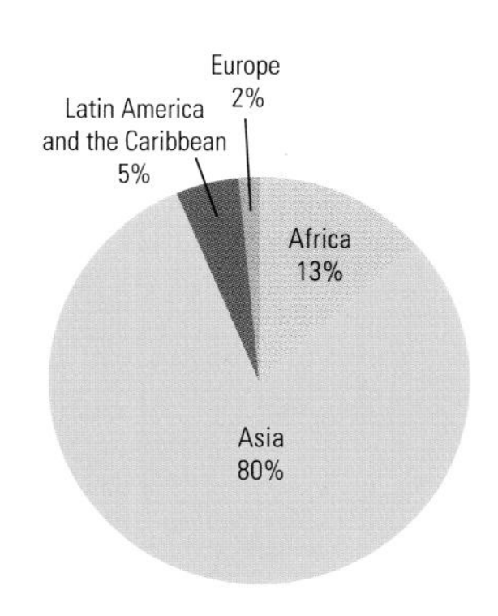

Total unserved: 2.4 billion

Asia shows the highest number of people unserved by either water supply or sanitation; yet it is important to note that proportionally, this group is bigger in Africa because of the difference of population size between the two continents.

Source: WHO/UNICEF Joint Monitoring Programme, 2002. Updated in September 2002.

Table 5.3: Africa, Asia, Latin America and the Caribbean: access to water supply and sanitation services by level of services – evolution during the last decade

		Water supply			Sanitation		
		Access to improved water supply facilities (%)	Access through household connections (%)	Not served (%)	Access to improved sanitation facilities (%)	Access through house connections to sewer systems (%)	Not served (%)
Africa	1990	59	17	41	59	11	41
	2000	64	24	36	60	13	40
Asia	1990	73	43	27	29	13	71
	2000	81	49	19	47	18	53
Latin America and the Caribbean	1990	82	60	18	72	42	28
	2000	87	66	13	78	49	22
Total	1990	72	41	28	38	16	62
	2000	79	47	21	52	20	48

This table shows a global improving trend of people having access to water supply and sanitation during the last decade. Asia has made the most visible progress, yet 53 percent of its population still does not have access to water and sanitation. The African trend is more worrying, since it shows no clear progress, especially concerning water and sanitation.

Source: WHO/UNICEF Joint Monitoring Programme, 2002. Updated in September 2002.

in Africa. The regional breakdown is presented in figures 5.2 and 5.3.

Levels of service are an important parameter to consider, because of their relevance for health. In Latin America and the Caribbean, an estimated 66 percent of the population has access to piped water through household connections, while these figures are only 49 percent and 24 percent for Asia and Africa, respectively. With regard to sanitation linked to a sewage system, these proportions are 66 percent for Latin America and the Caribbean, 18 percent for Asia and 13 percent for Africa. Table 5.3 reviews the trend over the last decade in these coverage figures and compares them with figures for access to improved water sources and sanitation and no access at all.

The United Nations Millennium Development Goals

One of the goals set by the United Nations (UN) Millennium Declaration, adopted by all UN member states in September 2000, is to ensure environmental sustainability. Significantly, one of three key targets defined to achieve this goal is to halve, by 2015, the proportion of people without sustainable access to safe drinking water. This target was originally a Vision 21 recommendation and was as such adopted by the Ministerial Conference of the Second World Water Forum in The Hague (WSSCC, 2000). The accompanying indicator (proportion of population with sustainable access to an improved water source) is one of the indicators that will continue to be covered by JMP. Moreover, the next target in the Millennium Declaration ('by 2020 to have achieved a significant improvement in the lives of at least 100 million slum dwellers') uses as one of its indicators the proportion of people with access to improved sanitation, another JMP-covered statistic.

Taking into account the projected growth of the world population, these targets imply that an additional 1.5 billion people will require access to some form of improved water supply by 2015 and, adapting the sanitation target to similar terms for rural and urban dwellers combined, about 1.9 billion people are required to gain access to improved sanitation. The information on actual and target global water supply and sanitation coverage, overall and broken down for rural and urban settings, is presented in figures 5.4 and 5.5.

While there has been massive investment in the extension of drinking water supplies since 1980, poor progress in the management of human excreta has become the limiting factor when it comes to realizing the full potential of the health benefits. Attitudes, cultural beliefs and taboos with respect to human excreta are among the challenges to be faced when proposing possible sanitation solutions to a community. Education to successfully improve the sanitation conditions should be participatory and should link the value of excreta (faeces and urine) to ecology and health protection.

At the 2002 World Summit for Sustainable Development (WSSD), the UN Millennium Development Goal for access to drinking water was reconfirmed. In addition, the Summit set the target for access to sanitation, i.e. halving, by 2015, the proportion of people who do not have access to basic sanitation, with the following elaboration on action needed:

- develop and implement efficient household sanitation systems;
- improve sanitation in public institutions, especially schools;
- promote safe hygiene practices;
- promote education and outreach focused on children, as agents of behavioural changes;
- promote affordable and socially and culturally acceptable technologies and practices;
- develop innovative financing and partnership mechanisms; and
- integrate sanitation into water resources management strategies.

The coverage target most likely to be achieved by 2015 is that of rural water supply. Rural populations are projected to decline in number, and existing levels of rural water supply coverage are relatively high compared to rural sanitation coverage. Urban services, on the other hand, face the greatest challenge: more than a billion additional people will need access to both water supply and sanitation over the next fifteen years in order to meet the targets. Even to maintain the year 2000 proportional level of coverage in urban areas until 2015 will require an estimated 953 million people to gain access to water supply and an additional 838 million to sanitation; an effort equivalent to water supply and sanitation infrastructure development for a population three times the size of that of North America. All these projections are made on the assumption that existing services will be sustained. This may be overly optimistic as water supply and sanitation services continue to face major constraints, including limited financial resources, insufficient cost recovery of services provided and inadequate operation and maintenance capacity.

Extrapolating past experience into future trends, offset against the 2015 targets, resulted in the following conclusions:

- Meeting the 2015 target of halving the fraction of the population without sustainable access to drinking water globally means providing services for an additional 100 million people each year (274,000/day) from 2000 to 2015. By comparison, during the decade of the 1990s an estimated total of only 901 million people gained access to water supply. With the exception of Africa south of Sahara, all regions will have achieved or will be close to achieving this target, provided the level of investment of the 1990s is maintained over the period 2000–2015.

- The challenge for sanitation is more daunting: the pace that allowed about 1 billion people to gain access to improved sanitation during the 1990s, will need to be accelerated to allow for sanitation provision of 125 million people each year

Figure 5.4: Actual and target supply coverage

Global water supply coverage

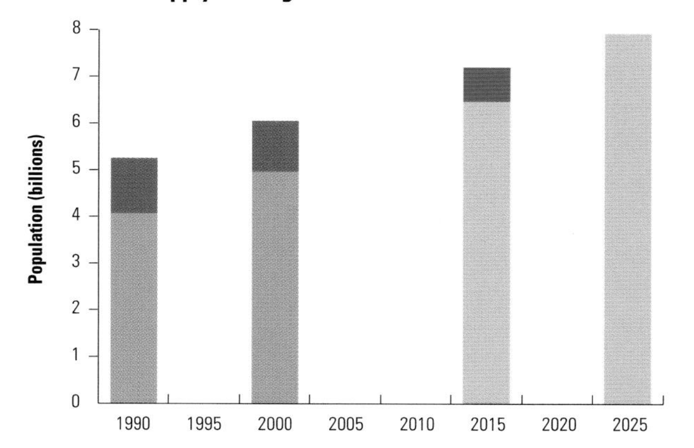

Urban water supply coverage

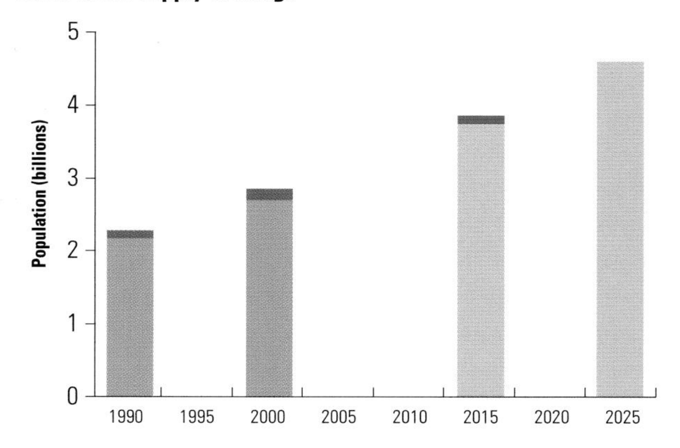

Rural water supply coverage

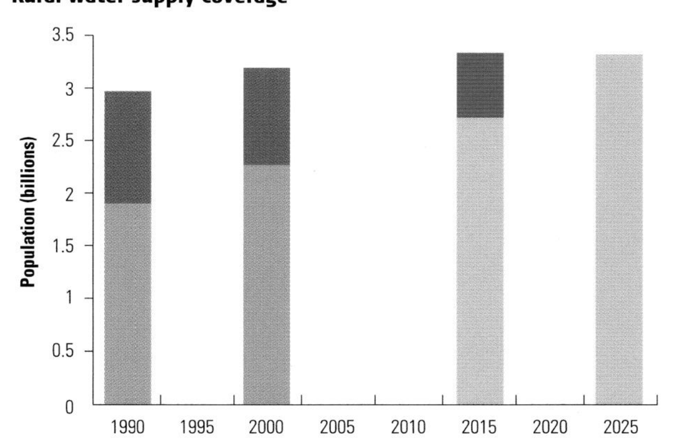

Taking into account the projected growth of the world population, Millennium targets imply that an additional 1.5 billion people will require access to some form of improved water supply by 2015, in other words 100 million people each year (or 274,000 people each day).

Source: WHO/UNICEF Joint Monitoring Programme, 2002. Updated in September 2002.

Figure 5.5: Actual and target sanitation coverage

Global sanitation coverage

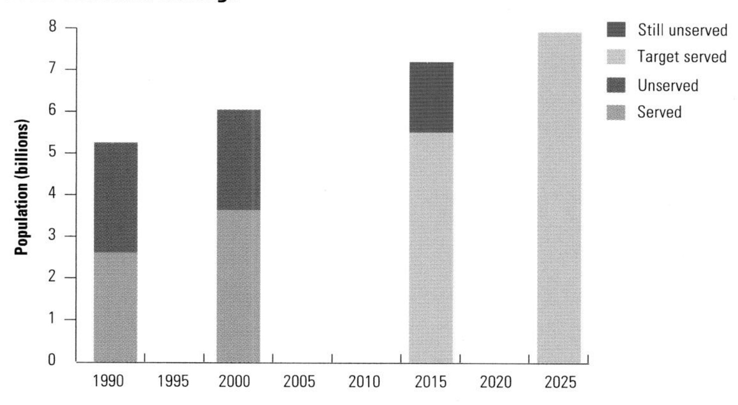

Urban sanitation coverage

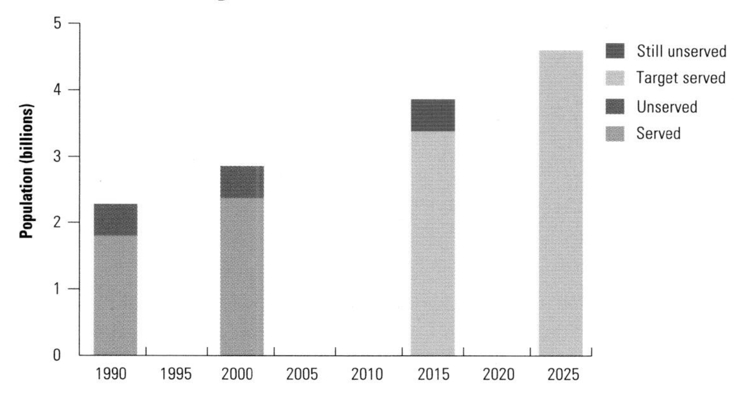

Rural sanitation coverage

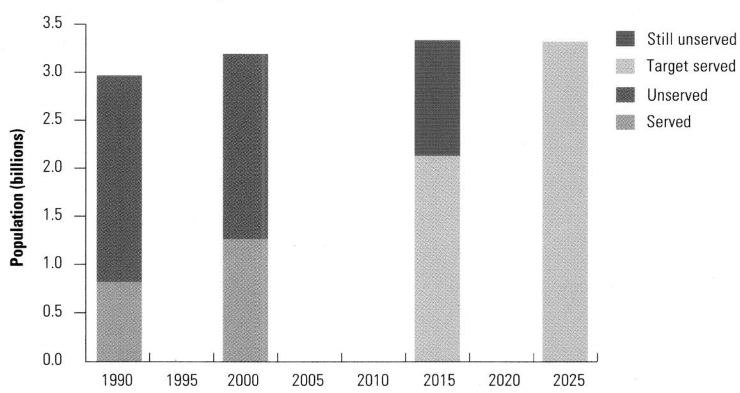

The 2002 World Summit for Sustainable Development (WSSD) set the target of halving, by 2015, the proportion of people who do not have access to basic sanitation. Given the projected growth of the world population, this target implies that an additional 1.9 billion people will require access to improved sanitation by 2015 (in other words 125 million people each year, or 342,000 people each day).

Source: WHO/UNICEF Joint Monitoring Programme, 2002. Updated in September 2002.

(342,000/day) until 2015. Again, it is expected that all regions with the exception of Africa south of the Sahara will have reached or will be close to reaching this target by 2015, provided the current level of investment is maintained.

- In absolute terms, the investment needs of Asia outstrip those of Africa, Latin America and the Caribbean combined.

Putting aside the overriding constraints imposed by the lack of sufficient resources, the challenges faced in achieving the stated goals circle around the issues of institutional strengthening and financial and economic arrangements for drinking water supply, and those of demand management and marketing for sanitation extension. Institution-building, either with a traditional public sector orientation or through more innovative public-private partnerships, is critical both to attract initial investments as well as to ensure the long-term sustainability of infrastructure and services once they have been established.

The extension of basic sanitation is mainly a household concern, as sophisticated piped sewage systems require an unrealistic level of investment to meet the needs of the poor. Unlike water supply, sanitation suffers from a lack of natural demand, and to overcome this, a marketing approach has been proposed, backed up by effective health and hygiene education, geared to the capabilities of often illiterate community groups. WHO has developed and promoted a methodology to change community hygiene behaviour and to improve water supply and sanitation facilities, the so-called PHAST (Participatory Hygiene and Sanitation Transformation) methodology (Sawyer et al., 1998). It aims to enable community workers to assist communities in improving hygiene behaviours, preventing diarrhoeal diseases and encourages community management of water and sanitation facilities. Critical steps in this process include demonstrating the relationship between sanitation and health, increasing the self-esteem of community members and empowering the community to plan environmental improvements and to own and operate water and sanitation facilities.

Recently, hygiene has made a comeback based on the rapidly expanding evidence that relatively small changes in hygiene behaviour will have large impacts in protecting individuals at the household level. While access to improved water sources and basic sanitation are crucial preconditions to hygiene behavioural change, access alone will not bring about these changes automatically. As such, the water-sanitation-hygiene conglomerate may be a public good when it comes to enabling infrastructure, but in its implementation it operates at the household level. The household focus is critical in the extension of sanitation coverage and this extension will often be achieved through community action without direct involvement of the formal service providers.

The sustainability of established water supply and sanitation systems can be broken down into two aspects: functional sustainability relates to the conditions upon which systems can continue to operate, with resources and capacities as the key constraints. Environmental sustainability takes into account, from a cross-generational perspective, both the environmental and health impacts of systems' operation and the impact of outside changes on the long-term viability of the system. In connection to the latter, there are worrying trends both in terms of water quantity (receding groundwater levels in different parts of the world) and quality (increased pollution levels and the further discovery of natural contaminants such as fluor and arsenic).

Water supply and sanitation monitoring

For the International Decade for Drinking Water Supply and Sanitation (IDWSSD), WHO was mandated by the UN system to monitor progress in water supply and sanitation coverage. The WHO monitoring relied exclusively on data and information provided by governments of its member states. These data were based on criteria that varied from one country to another and within individual countries over time. The data were frequently inaccurate and the information derived from them inconsistent and not representative of the situation on the ground. The statistics reflected the biases and sometimes interests of the agencies responsible for the supply of drinking water, not the real or perceived needs of the users.

Following the Decade, WHO and UNICEF decided to combine their experience and resources in the JMP. Since then, four assessments have been carried out. The objective of this endeavour has been broadened from a simple monitoring task to a capacity-building effort, with countries improving their institutional and human-resource capacities to plan and manage monitoring, through their active participation in the programme. The first three reports of the JMP (published in 1991, 1993 and 1996) still followed the conventional approach: coverage data linked to reporting on progress in national monitoring capacities. The *Global Water Supply and Sanitation 2000 Assessment*, presented in JMP's fourth report (WHO/UNICEF, 2000), marked a departure from the earlier methodology in several ways.

The *2000 Assessment* used broader and better verifiable data sources, including information from national surveys, linked to a more comprehensive analysis beyond strict coverage. Specifically, it differed from previous assessments in three important ways.

- The assessment covers the entire world through presentation of data from six regions: Africa, Asia, Europe, Latin America and the Caribbean, North America and Oceania as defined by the United Nations Population Division; previous assessments were limited to developing countries.

- Household survey data have been used extensively to estimate coverage figures. Assessment questionnaires were sent to all WHO representations in member states, and local WHO and UNICEF staff worked jointly with relevant national agencies to complete the questionnaires, following detailed instructions. A first step in this task was the preparation of an inventory of existing population-based datasets on access to water supply and sanitation, which could include national census reports, Demographic Health Surveys (DHS) and UNICEF's Multiple Indicator Cluster Surveys (MICS).

- The JMP report presents a more comprehensive package of information beyond strict coverage. This includes general planning and management, with a focus on target setting, investment patterns and trends, financial aspects, including urban water services' tariffs and costs, quality of services and constraints.

Any assessment will need to apply clearly defined criteria in order to achieve consistency and comparability over time and between locations. The definitions below applied to the criteria used in the *2000 Assessment*.

- 'Access to water supply and sanitation' was defined in terms of types of technology and service afforded. For water supply, access included house connections, public standpipes, boreholes with hand pumps, protected dug wells, protected springs and rainwater collections. Allowance was made for other, locally defined technologies.

- 'Reasonable access' was defined as the availability of at least 20 litres per person per day from a source within 1 kilometre of the user's dwelling. Tanker trucks, bottled water and other source types not giving reasonable access to water for domestic hygiene purposes were not included.

- 'Sanitation' was defined to include connection to a sewer or septic tank system, pour-flush latrine, simple pit or ventilated improved pit latrine, again with allowance for acceptable local technologies. The excreta disposal system was considered adequate if it was private or shared (but not public) and if it hygienically separated human excreta from human contact.

By introducing the concept of 'improved' technologies, an attempt is made to establish a simple classification reflecting a high likelihood that water supply and safe sanitation are adequate. Ongoing studies aim to strengthen the evidence base for the designation criteria. The designation 'improved' versus 'non-improved' was applied to water supply and sanitation technologies as shown in table 5.4.

Table 5.4: Improved versus non-improved water supply and sanitation

	Improved	Non-improved
Water supply	Household connection Public standpipe Borehole Protected dug well Protected spring water Rainwater collection	Unprotected well Unprotected spring Vendor-provided water Bottled water[1] Tanker-truck provided
Sanitation	Connection to a public sewer Connection to a septic system Pour-flush latrine Simple pit latrine Ventilated improved pit latrine	Service or bucket latrines[2] Public latrines Latrines with an open pit

[1] Considered as 'non-improved' because of quantity rather than quality of supplied water.
[2] Latrines from where excreta are manually removed.

Source: WHO, 2002. Prepared for the World Water Assessment Programme (WWAP).

A number of caveats need attention when considering the JMP assessment results. Access to improved water and sanitation does not imply that the level of service or quality is 'safe' or 'adequate', terminology previously used. It merely indicates the likelihood that it is safe or adequate. Coverage figures have not been discounted for intermittence or poor quality of the water supplies. Instructions stated, however, that piped systems should not be considered as functioning unless they were operating at over 50 percent of capacity on a daily basis; hand pumps should not be considered as functioning unless they were operating at least 70 percent of the time, with a lag between breakdown and repair not exceeding two weeks. These aspects were taken into consideration only when estimating coverage for countries where national surveys had not been conducted. In most cases, however, national survey data were available.

Monitoring the state of water supply, sanitation and hygiene behaviour remains critical to ensure progress and an accelerated coverage, together with greater and more active household participation. Several new developments in the concepts of monitoring have been noted (Shordt, 2000) and the four most important are:

- More groups and stakeholders have been brought into the process of data collection, analysis, interpretation and use;

- There is an increased emphasis on monitoring behavioural change;

- With the development of participatory appraisal and qualitative research techniques, a wider range of strategies and measurement tools to support the strategies, has emerged;

- There is increased emphasis on the timely use of monitoring and evaluation results.

The relative importance of water in adequate quantities, as compared to water quality, sanitation and hygiene, has been the subject of debate for many years, yet so far, no international norms for minimum domestic water quantities have been formally proposed. International targets, including the Millennium Development Goals, tend to omit this aspect. Domestic water is defined as water used for all usual household purposes including consumption, bathing and cooking, yet it must be borne in mind when interpreting and applying minimum values for water quantity, that some of these uses occur at home, while others (for example, laundry and bathing) occur away from the home. Some household-level use may also go beyond the conventional concept of what is domestic, and enter in the productive sphere: horticulture, livestock watering, construction and wholesale preparation of food and beverages. These home products may be essential for the livelihood of poor households.

It has been suggested that poor hygiene may, in part, be traced back to lack of water in sufficient quantity (Cairncross and Feachem, 1993), but reviews of numerous studies on various different single and multiple water and sanitation interventions have been inconclusive on the relative contribution of water quantity, and have sometimes detected contradictory results. Early studies seem to suggest that increased water quantity has an edge over water quality in its effectiveness to reduce diarrhoeal disease incidence, but later studies contradict this and in fact seem to indicate that neither quality nor quantity improvements result in significant health gains without concomitant sanitation improvement. From the variability and inconsistencies observed, the most tenable conclusion is that all interventions potentially have a significant impact and that the scale and relative impact of a single intervention depends strongly on the exposure route that is predominant under local conditions at a specific point in time. The complex exposure routes of most faecal-oral infections (see figure 5.1) make it hard to predict which intervention, whether single or a combination of measures, will be most effective. For some infections, however, the exposure pathway is simple. This is the case of the Guinea worm infection (see box 5.3) for which the exposure pathway is singular (although biologically complex) and has permitted the success of the Guinea worm eradication programme. In this case, however, interrupting transmission is not a matter of quantity but of quality.

There is evidence that the public health gains derived from use of increased volumes of water typically occur in two major increments. The first increment occurs when the total lack of basic access is overcome, leading to the availability of volumes adequate to support basic personal hygiene. A further significant health gain occurs when water supply becomes available at the household level (Howard and Bartram, in prep.). Some studies indicate that the health gains of increased water quantities are greater for some age groups than for others. Studies from India suggest that water quality is more critical for the health of children under three, while water quantity becomes a crucial health determinant above the age of

Box 5.3: The role of improved water supply in the eradication of Guinea worm infection

Guinea worm (*Dracunculus medinensis*) is a parasitic worm that causes an infection called dracunculiasis or Guinea worm disease. The disease eradication efforts of the past decade or so dramatically demonstrate the power of water supply interventions.

Water is fundamental in the dracunculiasis transmission cycle. People get the infection from drinking water with infected intermediate hosts of the genus Cyclops. The larvae develop into long parasitic worms that lodge in the joints, particularly the knees, where they cause a blister. From the blister, eggs are released, develop into larvae and complete the cycle in the Cyclops intermediate host.

Dracunculiasis is a disabling disease with a seasonal pattern, often peaking at times in the agricultural year when labour is in maximum demand. For this reason dracunculiasis is also known as the 'empty granary' disease.

The unique ecology of the *Drancunculus* parasite make the provision of an improved water supply a singular, critical intervention to interrupt transmission. The evidence for the impact of improved water supplies on dracunculiasis is clear in, for example, India, where it was responsible for a 80 to 98 percent reduction in annual incidence (adapted from Cairncross et al., 2002).

The Guinea worm eradication campaign started in 1989 and it is lead primarily by WHO, UNICEF and the Carter Center. Before the beginning of the campaign more than three million cases were estimated worldwide (Watts, 1987). Since then, WHO has certified 151 countries as Guinea worm disease-free and five more are in pre-certification phase. Among the certified countries, India and Pakistan achieved interruption of transmission after the beginning of the global eradication campaign in the 1980s. Nowadays only thirteen nations still have the disease and reported a total of 60,000 cases in 2001 (WHO, 2002a).

Sources: Cairncross et al., 2002; Watts, 1987; WHO, 2002a.

Box 5.4: Effect of improved water supply and sanitation on global problem of blindness

Trachoma is an eye infection caused by *Chlamydia trachomatis* which can lead to blindness after repeated reinfections. It spreads easily from one family member to another by ocular and respiratory secretions. Flies can also transmit the infection. WHO estimates that roughly 146 million people presently suffer from trachoma and associated infections, primarily among the poorest rural communities in developing countries. Roughly 6 million people are blind or severely visually disabled because of trachoma, making it one of the leading causes of preventable blindness worldwide. Central to controlling trachoma is easy access to sufficient quantities of water, facilitating the frequent washing of children's faces and improved environmental hygiene.

The WHO Global Alliance for the Elimination of Trachoma by 2020 has adopted the 'SAFE' strategy, consisting of four components: Surgery, Antibiotic treatment, promotion of Facial cleanliness and the initiation of Environmental changes. Recent reviews have emphasized the importance of the F and E components of the 'SAFE' strategy, concluding that improved personal and community hygiene has great potential for a sustainable reduction of trachoma transmission. They also concluded that there is likely to be a long-term beneficial effect of a combination of improved water supplies, provision of latrines, facial hygiene promotion and control of eye-seeking flies. Trachoma is a just one example of a number of human eye and skin infections that can be reduced through improvements in water supply, sanitation and hygiene promotion.

Sources: Prüss and Mariotti, 2000 and Emerson et al., 2000.

three. A reduction in the time required to collect water may translate not only into greater net water availability but also into more time for mothers to engage in childcare activities, including feeding and hygiene. Studies by Prost and Négrel (1989) point out that a twenty-fold reduction in time used for water collection resulted in thirty times more water used for child hygiene. The impact of such a considerable increase on diarrhoeal disease is hard to quantify, but it certainly plays a key role in the reduction of trachoma (see box 5.4).

Water management practices

Management decisions on water resources and water services will all potentially have an impact on human health, yet they are made by actors in many different public sectors as well as in the private sector, often with little awareness of the nature and magnitude of these implications. In many countries, institutional arrangements and other mechanisms for intersectoral coordination are rudimentary at best. As a result, considerable opportunities are lost to protect and promote health in the management process. As long as water policies and programmes remain fragmented, it will continue to be a challenge to address water-associated health issues in a cross-cutting way.

The World Water Assessment Programme (WWAP) Secretariat took note of these concerns; as a result, irrigation water management for health, for example, is covered in chapter 8, while ample coverage of urban water supply and sanitation can be found in chapter 7.

For drinking water supply services, water quality management is critical. A wide range of both chemical and microbial constituents of drinking water can cause adverse human health effects. The detection of these constituents in both raw water and water delivered to consumers is often slow, complex and costly, which limits early warning capability and affordability. Reliance on water quality determination alone is insufficient to protect public health.

The most effective and protective means of consistently assuring a supply of acceptably safe drinking water is the application of a preventive 'quality assurance' framework. A preventive framework, developed to manage drinking water quality, works in an iterative cycle encompassing assessment of public health concerns, risk assessment, establishing health-based water quality targets and risk management. Feeding into this cycle are the determination of environmental exposure levels and the estimation of what constitutes a tolerable risk (Davison et al., 2002).

Drinking water quality management may be established through a combination of protection of water sources, control of treatment processes and management of the distribution and handling of water. It has five key components:

1. water quality targets based on critical evaluation of health concerns;

2. system assessment to determine whether the water supply chain (up to the point of consumption) as a whole can deliver water of a quality that meets the above targets;

3. monitoring of the control points in the supply chain that are of particular importance in securing drinking water safety;

4. management plans documenting the system assessment and monitoring; and describing actions to be taken under normal and incident conditions. This includes documentation and communication; and

5. a system of independent surveillance that verifies that the above are operating properly.

It is important that water quality targets, defined by the relevant national health authority, be realistic under local operating conditions and set to protect and improve public health. Formal water supply agencies have a basic responsibility to provide safe water and would be expected to develop and implement management plans to address points 2 through 4 above.

The management plans, or Water Safety Plans (WSPs), developed by water suppliers, should address all aspects of the water supply and focus on the control of water production, treatment and delivery of drinking water. The control of microbial and chemical quality of drinking water requires the development of WSPs which, when implemented, provide the basis for process control to ensure that pathogen and chemical loads are acceptable. Implicit within this process is the acceptance that a tolerable disease burden has been defined at national and local levels and that water quality targets, which have been established to improve public health, are achievable.

Water resource development projects offer a range of options for water management practices that will contribute to a reduction of risks to human health. The earliest possible opportunity in the planning of water resource development needs to be seized to initiate the process of Health Impact Assessment (HIA).

HIA is a combination of procedures, methods and tools by which a policy, programme or project may be judged as to its potential effects on the health of a population, and the distribution of those effects within the population. It presents evidence, infers changes and recommends actions to safeguard, mitigate and enhance human health with the ultimate goal of providing decision-makers with sound information on the health implications of any given policy or project. By their nature, water resource development projects imply changes in landscape and in demography, which will in turn have an impact on both environmental and social health determinants.

There is an imperative need to integrate HIA into the formulation and planning of water resource development policies, and into the planning of water resource development programmes and projects, because it:

- improves the quality of development outcomes through decision-making and recommendations that help minimize negative health impacts and make maximum use of health opportunities;

- addresses the determinants of health in an integrated way rather than concentrating on single risk factors. While vertical disease control programmes tend to ignore environment and development links, HIA provides a logical platform for the process of decentralization and integration of health services;

- is conducive to intersectoral cooperation: systematic assessments of health effects are needed to inform decision-makers in the various sectors responsible for water resource management of their responsibility for the impact of their actions on health; and

- keeps additional, and usually more costly, burdens on the health sector limited to a minimum. Therefore taking human health into consideration at the planning stage makes good economic sense, by eliminating the transfer of hidden costs of water resource development to the health sector.

The key principles of HIA (equity, democracy, sustainability and ethical use of evidence) ensure full compatibility with more established Environmental Impact Assessment (EIA) practices. Routinely, EIAs tend to assess health issues only superficially and recommend health-sector-confined measures. When they do consider the health impacts of water resource development, EIAs usually focus on physical aspects of the environment and on pollution-related health hazards or communicable diseases, but often fail to address social determinants of health, including poverty, cultural alienation and community disruption.

The creation of a separate and parallel procedure for HIA is desirable. There should be effective links to EIA procedures and the assessments should draw on information collected for the feasibility study on development of water resources. Local conditions will dictate a specific balance between integration and maintaining a separate profile for health in the overall impact-assessment picture. As an example, a proposed small-scale irrigation and medium-size dam project in the northern part of Zimbabwe was submitted to a HIA. The results clearly showed potential threats to health such as increased incidence of malaria and schistosomiasis and of increasing sexually transmitted infections due to an influx of migrant workers, but it also indicated improved nutrition through greater agricultural output. The cost of enhancing the positive and mitigating the negative was estimated at about 1.8 percent of the total project cost – not too high a price for health.

National authorities cannot use such instruments as HIA to their full potential until there is a body of trained personnel, and this is clearly lacking throughout the world at the present time. A favourable policy climate is essential for this body of trained personnel to function properly, with special attention to policies in all water-associated sectors, conducive to intersectoral cooperation. Health sector personnel will benefit from training in impact assessment procedures and methods, and will be better placed to appreciate the concerns of the other sectors responsible for water projects. The health sector structure also needs renewal to allow for a better response to the needs incurred by HIA, i.e. contributing to terms of reference, HIA appraisal, monitoring health determinants and of compliance with recommended measures. In turn, water resources and water user sectors should work towards developing an understanding of the association between their decisions and human health. Where lacking, all groups should develop skills in intersectoral communication, collaboration and community participation. A course manual on developing such skills in support of HIA will be published in 2003 (Bos et al., 2003).

Water management practice options to protect or promote health range from canal lining in irrigation schemes to specially designed reservoir management programmes in dam projects. Dams often illustrate these issues in a clear, admittedly sometimes extreme way. The following best practice list was compiled as part of WHO's input into the work of the World Commission on Dams.

There are a number of fully or partially validated options that can mitigate the adverse effects of dam construction on human health. Examples of such health safeguards relating to operational water management include (WHO, 2000):

1. Design options
   - Multiple depth off-takes that allow the release of first flush inflows that may contain high levels of contaminants and nutrients, and allow a high level of control of fluctuation in reservoir water level (which can be advantageous in the control of disease vectors such as snails and mosquitoes).
   - Double spillways in areas where onchocerciasis (river blindness) is endemic. Spillways have been shown to provide an appropriate habitat for the breeding of blackflies (*Simulium ssp.*), the vectors of the *Onchocerca* parasite; their alternate use will prevent such breeding.
   - At all potential sites, ensuring careful examination of reservoir bathymetry so as to avoid dam sites that have extensive shallow areas conducive to insect and snail breeding. While shallow margins can never be totally avoided, catchment topographies that give rise to large reservoirs of low average depth (and therefore large wetted perimeters) should be avoided. Such reservoirs will also be undesirable from an evaporative-loss point of view.
   - Provision of simple infrastructure (jetties, for example) at critical places along the reservoir shore to reduce water contact for specific target groups (fishermen, women, children).
   - A greater than standard diameter of off-takes will allow the rapid draw-down of reservoirs, allowing a rapid drop in shoreline water levels, stranding and killing mosquito vectors (provided no pool formation occurs) and intermediate host snails of schistosomiasis. This will also allow an artificial flood downstream that will flush out any vector breeding places in rock pools.
   - Careful settlement planning that ensures that, wherever possible, and in balance with other planning and social needs, population settlement occurs away from areas of impounded and slow-flowing water. This will minimize human exposure to disease-carrying vectors.
   - Adequate planning for, and design of, community water supply and sanitation, including careful management of sewage and waste. This will reduce the rate of reservoir eutrophication and the occurrence and severity of toxic cyanobacterial blooms, as well as reducing water pollution generally.
2. Reservoir options
   - In-reservoir management to prevent eutrophication and excessive growth of problematic organisms such as toxic cyanobacteria and aquatic weeds. The development of massive blooms of toxic cyanobacteria is an area of increasing concern, especially in poorer countries where drinking water treatment may be less common or absent, and where exposure to toxic blooms may go unmanaged or unreported (see box 5.5).
   - Well-formulated dam environmental management plans that will support sustainable fisheries practices, sustain populations of natural predators of disease vectors and minimize excessive growth of aquatic weeds.
   - Catchment management to minimize negative impacts on the impoundment, including those from population growth and agricultural development in the upper catchment.
   - Adequate in-flow forecasting for disaster prevention in case of increased settlement on the downstream floodplain and heavy livelihood dependence on new agricultural production system.
   - Water release regimes that minimize impacts on downstream ecology and productivity especially in regions where there is

Box 5.5: Freshwater cyanobacterial toxins – an emerging dam-related health issue

In tropical, subtropical and arid regions of the world it is inevitable that new dams will become eutrophied (nutrient-enriched) rather quickly, often within the first few years of filling and operation. Eutrophication brings with it problems of excessive aquatic weed growth or 'blooms' of toxic cyanobacteria (a type of microscopic algae). Arid zones of the world are particularly at risk, where the artificial impoundment of water in hot climate creates the perfect ecological environment for the growth of toxic cyanobacteria. Added to this natural climatic effect is the enhanced rate of nutrient input that accompanies the growth of towns and the development of agriculture in the catchment around a dam, often with inadequate effluent collection and treatment facilities.

Blooms of freshwater algae and cyanobacteria have always occurred in eutrophied waterways, but the toxicity of these organisms has only been elucidated in recent years. There are several types of cyanobacterial toxins found throughout the world, all of which are potentially lethal to humans and animals if consumed in sufficient quantities. Additionally, some cyanobacterial toxins can promote liver cancer during chronic low-level exposure, and most cyanobacteria can cause a range of gastrointestinal and allergenic illnesses in humans exposed to toxins in drinking water, food or during swimming. A norm for drinking water concentrations for the common cyanobacterial toxin microcystin has recently been developed by WHO.

The most severe and well-documented case of human poisoning due to cyanobacterial toxins occurred in the Brazilian city of Curaru in 1996. Inadequately treated water from a local reservoir was used for patients in a local kidney dialysis clinic. As a consequence, more than fifty people died due to direct exposure of the cyanobacterial toxin to their blood stream during dialysis. Elsewhere in South America, in 1988, more than eighty deaths and 2,000 illnesses due to severe gastro-enteritis have also been directly linked with toxic cyanobacteria in a newly constructed dam. In China, a high incidence of primary liver cancer has been linked to the presence of cyanobacterial toxins in drinking water.

Source: Chorus and Bartram, 1999.

a significant nutritional reliance on traditional production methods such as recession agriculture.

- Sensitive management of flood plains and water resources to ensure wetland conservation, while at the same time minimizing excessive propagation of water-borne and water-related vector-borne diseases. As with irrigated agricultural production systems, natural seasonal wetting and drying cycles will be an important management tool. Conventional irrigation and drainage practices often lead to permanent inundation and wetting of previously ephemeral wetlands. The outcome of this is both degradation of the wetland and increased health risks.

3. Design and management options in irrigation schemes
   - Minimizing low flow zones in artificial channel networks to eliminate habitats for the propagation of disease vectors.
   - Concrete lining of irrigation canals to reduce seepage and prevent pools of standing water where mosquito vectors propagate. This also has water-saving benefits for irrigation.
   - Management of irrigated cropping systems to maintain wetting and drying cycles (while ensuring efficiency in water use), crop diversification and synchronization of cropping patterns. Regular wetting and drying of flooded rice fields provides an important tool to control water-associated vector-borne diseases such as malaria and Japanese encephalitis. In particular, there should be no advocacy of excessive multiple cropping within a single production year, and synchronization of cropping cycles is recommended.
   - Management plans for irrigated areas that minimize long-term salination and water logging and therefore impact on food security and scheme viability.

The cost-effectiveness of water interventions

In order to allow informed decision-making on interventions aimed at disease prevention and control, it is crucial to carry out a sound economic evaluation of the various options available in specific settings. This will permit the selection of an option or combination of options ensuring maximum health benefits within the constraints of a limited budget, or the achievement of defined goals at the lowest possible costs. The evaluation method of choice is cost-effectiveness analysis, with the reduction in burden of disease as the effectiveness indicator. Even within the confines of the health sector, analysing the cost-effectiveness of various interventions, such as case detection and treatment, vaccination when a vaccine is available, disease-vector control and health education, is not an easy matter. Expanding the scope of the economic evaluation to include water supply, water management and sanitation measures adds further complexities.

- The effectiveness indicator for health outcomes (reduction of the burden of disease expressed in DALYs) is not only fairly new and therefore relatively unknown outside of the health sector, but it is an indicator that, ultimately, does not express the health gains in monetary terms. This is a weak point when integrating health economics into the economic evaluation of water resource development and management, which operate at the cost-benefit level with outcomes expressed in monetary units.

- In accordance with the main purpose of water supply, water management and sanitation measures, different types of benefits will be considered as 'externalities'. A straightforward example is the lining of irrigation canals. If canal-lining is done with the purpose of saving water for agricultural production the health benefits (reduced malaria transmission risks because of a reduction in mosquito breeding potential) will be externalities in the economic evaluation of water-saving options. If the purpose is reducing malaria transmission risks, then the water-saving benefits to the agricultural production system will be considered externalities. Since these benefits can be expressed in monetary terms, one possible way of including them into the calculation is by deducting them from the cost of the measures.

- Many health sector interventions are of the recurrent type, while a substantial part of water and sanitation measures require a capital investment in infrastructure development. As a result, the economic evaluation of water and sanitation measures strictly for public health purposes will be disadvantaged in comparison to recurrent medical and other public health interventions whenever the discount rate is elevated.

WHO is currently supporting modelling studies that apply these concepts and principles in a cost-effectiveness analysis of drinking water supply and sanitation options in relation to diarrhoeal diseases control, estimating the costs, the health benefits and the non-health benefits of selected interventions for each of the fourteen subregions considered in WHO's burden of disease estimates.

The definition of types of water supply and sanitation follows that established for the WHO/UNICEF *2000 Assessment* for improved water sources and improved sanitation. The interventions reflect steps between the exposure scenarios described in the beginning of this section and are linked to achieving specific targets. The six interventions included in the model are:

- achieving the Millennium target of halving the proportion of people who do not have sustainable access to safe drinking water, with priority given to those who already have access to improved sanitation (i.e. moving from scenario VI to Vb and from Va to IV in table 5.2);

- halving the proportion of people without access to both improved water sources and basic sanitation facilities, essentially giving priority to populations living in scenarios Va and Vb over those in scenario VI;

- achieving disinfection at point-of-use through chlorine treatment and safe storage vessels combined with limited hygiene education to people currently without access to improved water sources;

- providing improved water supply and basic sanitation to people currently without access, to reach a total coverage of 98 percent (moving from scenarios VI, Va and Vb to IV);

- providing improved water supply and sanitation plus household treatment, safe storage and limited hygiene education to people currently without access, to reach a total coverage of 98 percent (moving from scenarios VI, Va and Vb to III);

- providing piped water to houses, with treatment to remove pathogens, quality monitoring and pollution control as well as sewage connection with partial treatment of wastewater, to reach a total coverage of 98 percent (moving from scenarios VI, Va and Vb, IV to II);

- judging the effectiveness of different interventions depending on the amount of exposure reduction in the population and the existing levels of overall morbidity and mortality within a given WHO subregion; and

- judging interventions with the assumption being that they will run for a period of ten years.

For the cost aspects of the equation, information was collected from member states through the mechanisms of the WHO/UNICEF JMP. The analysis of the attributable fraction of burden of disease reduction associated with different interventions was presented in the paragraph on 'the burden disease concept'. The incremental analysis of non-health benefits addressed three issues: the avoided direct expenditures due to a reduced diarrhoeal disease burden, both for the collective health sector and at the household level, the avoided days lost, whether affecting formal or informal employment, other productive activities or school attendance, and reduced opportunity costs related to the location of water supply and sanitation facilities.

Preliminary results of this modelling study present the following picture.

- In absolute terms, the first interventions at a global cost of US$12.6 billion, with a global gain of 30.2 million DALYs avoided.
- Disinfection at point-of-use through chlorine treatment and safe storage vessels combined with limited hygiene education would result in 122.2 million DALYs avoided at a relatively low incremental cost (total cost US$11.4 billion).
- Disinfection at point-of-use proved consistently to be the most cost-effective intervention across all subregions and would be classified as very cost-effective in all areas where it was evaluated.
- Interventions targeted at key behaviours such as improving hand washing would also provide a highly cost-effective way of achieving substantial health gains.
- In many developing countries, these cost-effectiveness data warrant a policy shift, towards better household water quality management (together with improved individual hygiene) to complement the continued expansion of coverage and upgrading of services, with a greater emphasis on achieving health gains associated with access to drinking water at the household level. As a reticulated water supply piped into individual homes remains a long-term aim for most developing countries, a focus on low-cost solutions with a great impact on health per unit of investment is desirable.

Health Sector Issues Associated with Water

From basic needs to human rights

Since the 1970s, the basic needs concept has been a dominating element in the development debate. It comes as no surprise that water was first among the issues addressed in the efforts to meet basic needs and to support a decent minimum standard of living and an acceptable level of human livelihood. The Mar del Plata Conference in 1977 focused almost exclusively on the drinking water supply and sanitation needs of the poor and vulnerable. This resulted in the designation of the period from 1981 to 1990 as the International Drinking Water Supply and Sanitation Decade (IDWSSD).

The focus of this period certainly resulted in the mobilization of considerable additional resources. It also facilitated the accelerated development of a more functional policy framework for drinking water supply and sanitation in many countries, and supported institutional strengthening and the establishment of arrangements between institutions formerly working without proper coordination. Great progress was made, certainly more than would ever have been achieved had the 1980s not been designated as IDWSSD. In the case of provision of safe drinking water, substantial progress was made towards the goal of universal coverage. Provision of access to sanitation, however, only managed to keep pace with population growth and started trailing access to safe drinking water in an increasingly disproportionate way.

At the beginning of the 1990s, the concept of water as a basic need became more differentiated. Sustainable development had appeared on the scene (WCED, 1987) and water was among the natural resources that needed to be used wisely to serve the needs of the present generation without jeopardizing the needs of future generations. Poverty and excessive consumption patterns were identified as the key driving forces behind an unsustainable use of natural resources. Poverty was also increasingly recognized as the key driving force for ill-health, in a vicious circle where ill-health led to a further deepening of poverty at the household and community level. The concept of sustainable livelihoods became complementary to that of basic needs. It translated into issues such as the judicious stewardship of water resources as well as personal hygiene and water security at the household level.

In the basic needs approach it had always been taken for granted that the promotion and protection of human health was the implicit goal of the provision of safe water and adequate sanitation. The results of the IDWSSD and the new concept of sustainable livelihoods raised new questions about the nature and scope of the links between water and health. As Integrated Water Resources Management (IWRM) entered the development arena after the UN

Conference on Environment and Development (UNCED) in Rio, the multiple health dimensions of water, for people, for food, for the environment required a new logical framework. These changes also took drinking water supply and sanitation out of their subsectoral confinement and placed them firmly on the broader agenda of human development, with a key role of the provision of these services in the fight against poverty.

From a strictly physiological perspective, the basic water need for each individual human being amounts to an intake of about 5 litres a day. Unlike food for proper nutrition, the other basic need for survival, mortality due to an acute lack of water has a low profile, except in natural disaster events of extreme drought. While an estimated 25,000 people die each day of hunger, records on the number of people dying from thirst are not apparent. So basic is the need for water, that, as long as they are physically able, people will leave areas of drought in search for water to survive, and emergency and humanitarian aid programmes will always give the provision of drinking water first priority. The true nature of water as a basic need therefore lies in the safety of the water available for drinking and other domestic purposes, and the adequacy of sanitation, considering that human excreta are the major source of contamination of water intended for domestic purposes. Beyond that, our knowledge of the health risks associated with aquatic ecosystems provides a handle on water management practices to reduce the burden of disease that results from them.

The enjoyment of the highest attainable standard of health is a fundamental right of every human being, enshrined in the Constitution of WHO of 1948. The reduction of vulnerability and the impact of ill-health sit among the several complex linkages between health and other human rights and steps must be taken to respect, protect and fulfil human rights (WHO, 2002b). In May 2000, the Committee on Economic, Social and Cultural Rights, which monitors the covenant of the same name, adopted a General Comment on the right to health. The latter states *inter alia*: 'Water is fundamental for life in human dignity. It is a pre-requisite to the realization of all other human rights.'

Box 5.6: Human rights to water

The General Comment on the right to water, adopted by the Covenant on Economic, Social and Cultural Rights (CESCR) in November 2002, is a milestone in the history of human rights. For the first time water is explicitly recognized as a fundamental human right and the 145 countries which have ratified the International CESCR will now be compelled to progressively ensure that everyone has access to safe and secure drinking water, equitably and without discrimination.

The General Comment states that 'the human right to water entitles everyone to sufficient, affordable, physically accessible, safe and acceptable water for personal and domestic uses'. It requires governments to adopt national strategies and plans of action which will allow them to 'move expeditiously and effectively towards the full realization of the right to water'. These strategies should be based on human rights law and principles, cover all aspects of the right to water and the corresponding obligations of countries, define clear objectives, set targets or goals to be achieved and the time-frame for their achievement, and formulate adequate policies and corresponding indicators.

Generally, governmental obligations towards the right to drinking water under human rights law broadly fall under the principles: respect, protect and fulfil. The obligation to respect the right requires Parties to the Covenant to refrain from engaging in any conduct that interferes with the enjoyment of the right, such as practices which, for example, deny equal access to adequate drinking water or unlawfully pollute water through waste from state-owned facilities. Parties are obligated to protect human rights by preventing third parties from interfering in any way with the enjoyment of the right to drinking water. The obligation to fulfil requires Parties to adopt the necessary measures directed towards the full realization of the right to drinking water.

The General Comment is important because it provides a tool for civil society to hold governments accountable for ensuring equitable access to water. It also provides a framework to assist governments in establishing effective policies and strategies that yield real benefits for health and society. An important aspect of the value it provides is in focusing attention and activities on those most adversely affected including the poor and vulnerable.

Before the adoption of the General Comment, the right to water had been more or less implicitly recognized in the General Comment on the right to health, 2000, in the Convention on the Rights of the Child (CRC), 1989, and in the Convention on the Elimination of All Forms of Discrimination Against Women (CEDAW), 1979.

The WHO Programme on Water, Sanitation and Health fully supports the right to water, which, as indicated above, is inextricably linked to the right of all humans to attain the highest possible standard of health. To achieve the goal of ensuring access for all to an adequate supply of safe drinking water, WHO proposes standards and regulations for drinking water quality, through its Guidelines for Drinking Water Quality (WHO, 1997, 1996, 1993). Among the substantive elements of a rights-based approach to health, paying attention to those population groups considered most vulnerable in society is a critical one. In this context, the principle of equity, i.e. the fact that the distribution of opportunities for well-being is guided by people's needs rather than through their social privileges, is increasingly serving as an important, non-legal generic policy concept.

At the core of WHO's work is the estimation of the burden of water-related ill-health, which reflects the link between the right to water and the right to health, and the promotion of safe water supply and safe water management practices to affirm these rights. Basic needs, sustainable livelihoods and the human rights approach will continue to be the guiding principles in water supply and management. An update on the status of the human right to water issue is presented in box 5.6.

Decentralization

Integration of health services and decentralization of their operations has been going on for the past ten years as part of the ongoing public sector restructuring. The goal is a more efficient health service that is more responsive to health issues as they arise and better targeted at the needs of vulnerable groups in society. The integration aims to reduce the disproportionate costs of so-called vertical programmes (malaria control programmes are among the examples) that were once established as time-limited operations, but became entrenched as singular routine operations within the health sector.

Decentralization also foresees a devolution of planning and decision-making at the local level, with standard setting, quality control and expert technical cooperation remaining functions at higher levels. In many countries, the period of transition is characterized by problems related to lack of adequate capacities at the local level and resistance within the system against the change imposed. In some instances, the actual decentralization is hampered by the fact that major resource decisions continue to be made at the national level, leaving little room for local health centres to adjust their programmes to local needs.

This process has a number of implications that are specific to water-associated diseases. On the one hand, a number of these diseases were covered by vertical programmes in the past, which used to have strong epidemiological surveillance components (though usually at a high price). With integration, it has been observed that part of this surveillance capacity disappears, and what at first sight would seem to be a decline in disease incidence often turns out to be the result of a reduced surveillance effort that captures fewer cases. With decentralization, the often specialized knowledge related to the links between water parameters and disease situations may lose its 'home'. This hampers the development and design of water management solutions and interventions, and creates a bias towards more strictly medical interventions. On the positive side, provided capacity is present or has been built at local level, decentralization allows for a more detailed epidemiological analysis of local disease situations and favours the design of local solutions to replace universal or blanket interventions, such as, for example, spraying of residual insecticides for interrupting the transmission of malaria. There will, therefore, be greater opportunities as well for water management solutions that address local health problems and fit with the local state of water resources and aquatic environment.

Devolution of operations to the local level may imply discrepancies between the administrative boundaries and the natural boundaries of watersheds. At the surveillance stage, such discrepancies may create wrong impressions about the links between water-related determinants of health; at the time of implementing interventions, they may obstruct optimal water management solutions, because the site where action is needed lies outside of the jurisdiction of authorities that have to deal with the health problem. Such problems can be overcome, either by dealing with them at a higher level of government, such as a provincial government, or by establishing effective institutional arrangements between the health authorities and, for instance, river basin authorities.

Medical and public health constraints and opportunities

The health sector is under pressure to control many of the water-associated diseases. For a number of diseases, prevention through vaccination campaigns is not an option, simply because a vaccine does not (yet) exist. This is the case for malaria, dengue and the gastrointestinal infections. Even the existing cholera vaccine is of too low an efficacy to contribute significantly to public health efforts. Insecticides for transmission interruption of vector-borne diseases become increasingly less effective because of the development of resistance in important vector species, while legally binding international instruments such as the Stockholm Convention on Persistent Organic Pollutants are also limiting use in some cases. As soon as a curative approach is required, resistance of disease-causing organisms against antibiotics and drugs becomes a phenomenon of growing importance, undermining the treatment of bacterial as well as of some parasitic infections. Even where effective tools are still available, they are often out of reach of the poor, who either can not afford them or are not adequately covered by resources-strapped health services.

It is against this backdrop of health sector constraints that the potential of access to improved water sources and best water management practice, basic sanitation and improved hygiene behaviour must be assessed. Major health gains can be achieved at the household level through personal protection as indicated earlier. Farming communities can be informed about the water management options that both benefit agricultural production and reduce health risks. Communities can also be mobilized to work towards improved drinking water facilities, as well as being taught about drinking water contamination risks at the household level and safe storage of drinking water from unreliable supplies. Health workers operating at the district level can verify the promotion of basic sanitation and hygiene behaviour. In many instances, these local health workers will liaise with the health sector's environmental health programme, through sanitary engineers or environmental health inspectors.

The health sector structure is made up of a well-defined core of health services delivery institutions with a more nebulous margin where many of the more prevention-oriented programmes reside. These include environmental health services, which tend to be characterized by a lack of functional programme structure, poor career opportunities and a general lack of resources. Yet the functions of environmental health services are of great health importance in relation to the regulation of environmental and social health risk factors. A number of these relate to water resources, water supply and water management.

Strengthening of this programmatic weakness in the health sector of most developing countries requires a number of important points to be addressed, including:

- the identification and definition of essential environmental health functions, combining some of the traditional functions, such as those related to drinking water supply and sanitation, with new functions, such as those related to health-impact assessment of water resources development;

- the readjustment of the balance between operational functions and regulatory functions, to ensure that sectors responsible for water resource development and management decisions are accountable, within existing public health legislation, for adverse health impacts of their actions;

- from its vantage point on the interface between the health sector and other sectors, maintaining intersectoral coordination and cooperation between the health sector core (epidemiological surveillance and health services delivery) and those responsible for water resource development and management in other sectors.

- regular economic evaluations of the hidden costs transferred to the health sector because of water resource development that does not consider health issues, and cost-effectiveness analyses of water supply and management interventions in comparison with conventional health sector ones.

Conclusions

We all need 20 to 50 litres of water free from harmful contaminants each and every day. However, the number of people who do not benefit from anywhere near this amount is staggeringly high. Deficiencies in water supply and sanitation coverage significantly hamper economic opportunities for every one of those people individually, and decrease the quality of their lives. Two critical problems face the sector: first, keeping pace with a net population growth of more than a billion people over the next fifteen years; and second, closing the coverage and service gap, with an emphasis on sanitation that lags considerably behind water supply.

Despite the fact that progress has been made during the past ten years, and despite the fact that the right to water has been internationally recognized as a human right, one sixth of the world's population is still without water, and two fifths are without sanitation.

Even more shocking is the number of deaths, mostly of children, that are largely preventable through water/hygiene-related measures, as well as deaths from water-related vector-borne diseases, such as malaria and schistosomiasis, both of which are rampant. They result in the loss of millions of life years every year, and affect the physical, social and economic well-being of populations. They reinforce the cycle of poverty and powerlessness that keeps people trapped and unfulfilled as well as slowing the ability of societies to develop. What can be done?

On one level the answer is simple. Water management solutions exist that can make significant inroads in combating both disease and poverty. We need only apply them. More difficult, however, is finding the will to do so. Taking an integrated approach can help, for it then becomes clear how water and sanitation and health all fit within a broader context of promoting human development. If we are to meet our goals, we must not rest until the privileges of the fortunate have been extended to those millions still deprived of water, sanitation and health.

Progress since Rio at a glance

Agreed action	Progress since Rio
Establish concrete targets on water supply and sanitation	
Provide better access to potable water supply and sanitation services to rural communities worldwide	
Ensure international cooperation to improve access to safe water in sufficient quantities and adequate sanitation	
Assist developing countries in their water treatment	
Reverse trends of resource degradation and depletion	

Unsatisfactory — Moderate — Satisfactory

References

Bos, R.; Birley, M.-H.; Engel, C.; Furu, P. 2003. *Health Opportunities in Development: A Course Manual on Developing Intersectoral Decision-Making Skills in Support of Health Impact Assessment.* Geneva, World Health Organization/Danish Bilharziasis Laboratory.

Cairncross, S. and Feachem, R. 1993. *Environmental Health Engineering in the Tropics: An Introductory Text.* Second edition. Chichester, John Wiley and Sons.

Cairncross, S.; Muller, R.; Zagaria, N. 2002. 'Dracunculiasis (Guinea Worm Disease) and the Eradication Initiative'. *Clinical Microbiology Reviews*, Vol. 15, No. 2, pp. 223–46.

Chorus, I. and Bartram, J. 1999. *Toxic Cyanobacteria in Water: A Guide to Their Public Health Consequences, Monitoring and Management.* World Health Organization. London, E & FN Spon.

Cifuentes, E.; Blumenthal, U.; Ruiz-Palacios, G.; Bennett, S.; Quigley, M. 2000. 'Health Risk in Agricultural Villages Practising Wastewater Irrigation in Central Mexico: Perspectives for Protection'. In: I. Chorus, U. Ringelband, G. Schlag and O. Schmoll (eds.), *Water Sanitation and Health.* London, International Water Assessment Publishing.

Davison, A.; Howard, G.; Stevens, M.; Callan, P.; Kirby, R.; Deere, D.; Bartram, J. 2002. *Water Safety Plans (working draft).* Document WHO/SDE/WSH/02.09. Geneva, World Health Organization.

Emerson, P.-M.; Cairncross, S.; Bailey R.-L.; Mabey, D.-C. 2000. 'Review of the Evidence Base for the "F" and "E" Components of the SAFE Strategy for Trachoma Control'. *Tropical Medicine and International Health,* Vol. 5, No. 8, pp. 515–27.

Ghebreyesus, T.-A.; Haile, M.; et al. 1999. 'Incidence of Malaria Among Children Living Near Dams in Northern Ethiopia: Community-Based Incidence Survey'. *British Medical Journal*, Vol. 319, pp. 663–6.

Howard, G. and Bartram, J. (in prep.). *Domestic Water Quantity, Service Level and Health: What Should Be the Goal for the Water and Health Sectors?*

Mara, D. and Cairncross, S. 1989. *Guidelines for the Safe Use of Wastewater and Excreta in Agriculture and Aquaculture.* Geneva, World Health Organization.

Prost, A. and Négrel, A.-D. 1989. 'Water, Trachoma and Conjunctivitis'. *Bulletin of the World Health Organization*, Vol. 67, No. 1, pp. 9–18.

Prüss, A. and Mariotti, S.-P. 2000. 'Preventing Trachoma through Environmental Sanitation: A Review of the Evidence Base'. *Bulletin of the World Health Organization*, Vol. 78, No. 2, pp. 258–66.

Prüss, A.; Kay, D.; Fewtrell, L.; Bartram, J. 2002. 'Estimating the Burden of Disease from Water, Sanitation and Hygiene at a Global Level'. *Environmental Health Perspectives,* Vol. 110, No. 5, pp. 537–42.

Sachs, J. and Malaney, P. 2002. 'The Economic and Social Burden of Malaria'. *Nature*, Vol. 415, pp. 680–85.

Sawyer, R.; Simpson-Hébert, M.; Wood, S. 1998. *PHAST Step-by-Step Guide: A Participatory Approach for the Control of Diarrhoeal Disease.* Geneva, World Health Organization.

Shordt, K. 2000. *Trainer's Manual: Action Monitoring for Effectiveness.* Delft, International Water and Sanitation Centre.

UNEP (United Nations Environment Programme). 2001. *Stockholm Convention on Persistent Organic Pollutants, Text and Annexes.* Geneva, UNEP Chemicals, Interim Secretariat for the Stockholm Convention, United Nations Environment Programme.

Watts, S.-J. 1987. 'Dracunculiasis in Africa: Its Geographic Extent, Incidence, and At-Risk Population'. *The American Journal of Tropical Medicine Hygiene*, Vol. 37, pp. 119–25.

WCED (World Commission on Environment and Development). 1987. *Our Common Future.* Report of the World Commission on Environment and Development. New York, Oxford University Press.

WHO (World Health Organization). 2002a. *Report on the Status of the Dracunculiasis Eradication Campaign in 2001.* Document WHO/CDS/CPE/CEE/2002.30. Geneva.

——. 2002b. *Questions and Answers on Health and Human Rights.* WHO Health and Human Rights publication series. Geneva.

——. 2002c. *Weekly Epidemiological Record*, Vol. 77, No. 31, pp. 257–64.

——. 2002d. *Global Assessment of the State-of-the-Science of Endocrine Disruptors.* Document WHO/PCS/EDC/02.2. Geneva.

——. 2002e. *World Health Report 2002.* Geneva.

——. 2000. *Human Health and Dams: The World Health Organization's Submission to the World Commission on Dams (WCD).* Document WHO/SDE/WSH/00.01. Geneva.

——. 1997. *Guidelines for Drinking-Water Quality.* Second edition. Vol. 3: *Surveillance and Control of Community Supplies.* Geneva.

——. 1996. *Guidelines for Drinking-Water Quality.* Second edition. Vol. 2: *Health Criteria and Other Supporting Information.* Geneva.

——. 1993. *Guidelines for Drinking-Water Quality.* Second edition. Vol. 1: *Recommendations.* Geneva.

——. 1992. *The International Drinking Water Supply and Sanitation Decade: End of Decade Review.* Document WHO/CWS/92.12. Geneva.

WHO/UNICEF (World Health Organization/United Nations Children's Fund). 2000. *Global Water Supply and Sanitation Assessment 2000 Report.* New York.

World Bank. 1993. *The World Development Report 1993: Investing in Health.* Washington DC.

WSSCC (World Supply and Sanitation Collaborative Council). 2000. 'Vision 21: A Shared Vision for Hygiene, Sanitation and Water Supply and a Framework for Action'. In: *Proceedings of the Second World Water Forum (The Hague, 17–22 March 2000).* Geneva.

Some Useful Web Sites[1]

United Children's Fund (UNICEF) Statistics: Water and Sanitation

http://www.childinfo.org/eddb/water.htm

UNICEF's key statistical database with detailed country-specific information that was used for the end-of-decade assessment. Among the main themes: Water and Sanitation, Child Survival and Health, Child Nutrition, Maternal Health, Education, Child Rights.

World Health Organization (WHO): Statistical Information System (WHOSIS)

http://www3.who.int/whosis/menu.cfm

A guide to health and health-related epidemiological and statistical information available from WHO, and elsewhere.

World Health Organization (WHO): Water, Sanitation and Health Programme

http://www.who.int/water_sanitation_health

Regularly updated information of the WHO's Programme on Water Sanitation and Health, including all publications and documents in PDF and HTML format files.

World Health Organization/Pan-American Health Organization/Pan-American Centre for Sanitary Engineering and Environmental Sciences (WHO/PAHO/CEPIS): Virtual library on health and the environment: Assessment of Drinking Water and Sanitation in the Americas

http://www.cepis.ops-oms.org/enwww/eva2000/infopais.html

Data and indicators on Urban and Rural Population, Water, Sanitation, General Health and Hygiene.

World Health Organization/United Children's Fund (WHO/UNICEF) Joint Monitoring Programme on Water Supply and Sanitation (JMP)

http://www.wssinfo.org

Data on access to water supply and sanitation services at the country, regional and global levels.

1. These sites were last accessed on 20 December 2002.

'Protection of ecosystems must remain central to sustainable development because environmental security, social well-being and economic security are intricately intertwined and fundamentally interdependent. Degradation of any one worsens the condition of all three.'

Protecting Ecosystems for People and Planet

By: UNEP (United Nations Environment Programme)
Collaborating agencies:
UNECE (United Nations Economic Commission for Europe)/WHO (World Health Organization)/UNCBD (United Nations Secretariat of the Convention on Biological Diversity)/UNESCO (United Nations Educational, Scientific and Cultural Organization)/UNDESA (United Nations Department of Economic and Social Affairs)/UNU (United Nations University)

Table of contents

**If you eat from the river, you must protect it,
and if you drink from the river you must conserve it.**

Saying of the Karen people

A HEALTHY AND UNPOLLUTED NATURAL ENVIRONMENT is essential to human well-being and sustainable development. Our rivers and wetlands – and the communities of plants, fish, birds, insects and wildlife they support – are an integral part of our lives and provide the resource base that helps us meet a multitude of needs. But as this chapter shows, the world's freshwater ecosystems are under great pressure. We use them as dumping grounds for waste, we alter their natural flow by building dams, diversions and canals, and we drain them for agriculture and other uses. How extensive is the damage? Is it reversible? What are the trends? The chapter reports on many local and national efforts to clean up and protect our water resources, but the global picture remains disturbing.

THROUGHOUT THE WORLD, people's use of water is exerting pressure on the environment. Many rivers, lakes and groundwater resources are already dried up and polluted. Drinkable water is increasingly scarce. By the year 2025, it is predicted that water abstraction will increase by 50 percent in developing countries and 18 percent in developed countries, as population growth and development drive up water demand. Effects on the world's ecosystems have the potential to dramatically worsen the present situation, and current assessments suggest that existing practices are not adequate to avert this.

The Dublin Statement, emanating from the International Conference on Water and the Environment of January 1992, stressed that a holistic approach was required if water resources were to be managed effectively. This was among the first attempts to articulate policy goals in which the specific needs identified by the water supply sector were integrated with respect to broader issues relating to sustainability and environmental conservation. Environmental issues were given prominence at the United Nations Conference on Environment and Development (UNCED) held in Rio in 1992 and a number of targets relating specifically to biodiversity and ecosystem protection were then incorporated into the Convention on Biological Diversity (CBD).

Details were laid out in the freshwater Chapter 18 of Agenda 21, which devotes a Programme Area to the Protection of Water Resources, Water Quality and Aquatic Ecosystems. The chapter outlines overall direction for holistic freshwater management, with supporting objectives in the protection of ecosystem integrity for environmental health, water quality and water source protection, and of biodiversity protection in its own right. The Programme Area on Water and Sustainable Food Production and Rural Development was underpinned by the sustainability of the environment to support rural livelihoods.

In 1998, the World Water Council set up a commission to produce a 'Vision' for the world's water, towards the end of a decade that had seen rapid growth in awareness of freshwater resources management as a global environmental issue. The primary contribution on environmental aspects was the *World Vision for Water and Nature* (IUCN, 2000), based on the understanding that the protection of ecosystems must remain central to sustainable development because environmental security, social well-being and economic security are intricately intertwined and fundamentally interdependent. Degradation of any one will worsen the condition of all three. This requires that two basic concepts be acknowledged:

- ecosystems have intrinsic values and provide essential goods and services;
- sustainability of water resources requires participatory ecosystem-based management.

The Vision is supported by targets defined at the Second World Water Forum (2000), including:

- setting national strategies for sustainable development in every country by 2005;
- setting national standards to ensure the health of ecosystems in all countries by 2005, and implementing programmes to improve the health of freshwater ecosystems by 2015; and
- implementing programmes to protect surface and groundwater resources by 2003 with defined standards achieved by 2010.

These targets and the new ethic of environmental stewardship provide a framework for assessing progress towards protecting ecosystems for environmental health, water quality, natural resources and biodiversity. This chapter offers an overview of the significance and use of freshwater ecosystems, the nature of the pressures to which they are subject and ways to measure ecosystem quality. It assesses the current status of a number of different ecosystem types and discusses progress towards effective management.

Significance of Freshwater Ecosystems

Water has the pivotal role in mediating global ecosystem processes, linking together the atmosphere, lithosphere and biosphere by moving substances between them and enabling chemical reactions to take place. Not only is it essential for maintaining living organisms, but its physical properties allow it to be used by humans for energy generation, transport and waste disposal, and in a variety of industrial processes. Access to a sufficient amount of good-quality water is essential to human health; productive freshwater ecosystems are crucial to the livelihood of fishers and other littoral communities; and healthy freshwater systems provide a range of services to people worldwide.

Characteristics of freshwater ecosystems

The term 'ecosystem' refers to the communities of plants, animals and other organisms, and the physical environment of any given place, these elements being linked by the flow of energy and the circulation of materials from producers to consumers and decomposers. Because the term refers primarily to system processes rather than a defined place, it can be used at a range of spatial scales, so that one may refer to aquatic – as opposed to terrestrial – ecosystems, or generic 'lake ecosystems' or to an individual lake as 'an ecosystem'.

The presence of liquid water is one of the defining features of planet Earth, and water is virtually ubiquitous in the global environment. This chapter is concerned only with freshwater ecosystems in the traditional sense, that is surface water including rivers, streams, lakes, ponds, marshes and other wetlands, together with the ground or soil water to which they are linked and deeper aquifers.[1] These include intrinsically valuable natural or semi-natural systems, and also increasingly a number of man-made or artificial aquatic environments. Artificial ecosystems may be very extensive and retain their own biological value, for example the flooded rice agro-ecosystems in Asia, or also be important in terms of the function they provide. In a broader sense, however, almost the entire terrestrial environment can be considered a 'freshwater ecosystem' because all known life forms depend on water. Almost all landscapes and microhabitats interact with water during the hydrological cycle and in fact, the evaporative part of the hydrological cycle would constitute by far the greatest 'use' of water.

While surface waters can include biological communities of considerable complexity, with representatives from many different phyla, or categories of organisms, species in soil water mainly comprise fungi and other micro-organisms, and deep aquifers are sterile or inhabited by a very few kinds of archaea and bacteria. Cave waters, notably in karst regions, often include specialized restricted-range species of invertebrates, fish and amphibians. At higher taxonomic levels, global diversity in freshwaters is lower than on land or at sea (with only one phylum, Gamophyta, entirely confined to freshwaters). At the species level, diversity appears to be high in relation to habitat extent; for example, globally the number of fish species per unit volume of water is more than 5,000 times greater in freshwaters than in the sea. Terrestrial or hydrological barriers to dispersal promote endemism in freshwater species, and many fish and other inland water species are restricted to individual water systems, such as river basins, lakes and caves, or to particular stretches of river or lake edge.

For many purposes, the catchment (watershed, basin), defined as the entire land surface from which water flows downhill to a given point, is the principal unit of management. While the catchment approach is desirable, it is not alone sufficient to address all issues related to inland water ecosystems. The quantity and quality of water reaching the coast affects estuarine and coastal waters; catchments can be interconnected by infrastructure; aquifers in which groundwater occurs may bear little relationship to the surface topography that determines the geometry of river catchments; and economic and social factors external to a single catchment can exert significant impacts.

Uses and benefits of freshwater ecosystems

The living and the abiotic components of an ecosystem (organisms, sediment, water) interact in many ways, engaging in processes that may be biological, physical, chemical or hydrological in nature (organic production, nutrient cycling, carbon storage, water retention, habitat maintenance). Some ecosystem components may be regarded in economic terms as 'goods' and the functional results of ecosystem processes as 'services'. Humans derive direct and indirect benefits from both aspects of ecosystems, and from properties such as 'biodiversity' that can be attributed to ecosystems. Several different classifications of these benefits have been produced as aids to discussion and analysis; one example is given in table 6.1. Some benefits depend on the presence of particular individual species. Elements of biodiversity thus underlie most of the benefits and functions cited.

1. In practice, the terms 'freshwater ecosystem' and 'inland water ecosystem' (or derivatives) tend to be used interchangeably, although the latter is more inclusive in covering enclosed saline waters and estuaries. In traditional usage, 'wetland' refers to areas of waterlogged soil and of land, such as river floodplain, that is seasonally or permanently covered by relatively shallow water. In the broader usage of the Ramsar Convention, wetlands are 'areas of marsh, fen, peatland or water, whether natural or artificial, permanent or temporary, with water that is static or flowing, fresh, brackish or salt, including areas of marine water the depth of which at low tide does not exceed six metres'.

Table 6.1: Simplified classification of services provided by water ecosystems

Production
Water (drinking, irrigation)
Food (fish, rice)*
Raw materials (reeds)*
Energy (hydroelectric)
Species habitat
Genetic resources*
Regulation
Buffering (storm protection, flood control, storage)*
Biogeochemical cycling (oxygen production, carbon storage, methane)*
Waste removal (filter-feeding invertebrates, sediment micro-organisms)*
Climate (local)
Biological control (pest control, pollination)*
Other
Recreation and tourism*
Cultural uses*
Transport

* Ecosystem services include all functional results of ecosystem processes. All items marked with an asterisk depend on elements of biodiversity, that is, on the presence of species and communities of organisms, and their ecological attributes.

Source: Modified from IUCN, 2000.

Maintenance of the entire range of goods and services derived from any ecosystem depends on the continued presence of key components (e.g. water, key fishery species, sediment communities, marginal vegetation) and continuation of ecosystem processes (e.g. water retention, removal of pollutants). In the case of natural freshwaters – unmodified systems with all components, processes and attributes essentially intact at all relevant scales are generally of the greatest intrinsic value and can be considered to have high ecological integrity. In different terms, this may be taken as broadly equivalent to good 'ecosystem health' status. In artificial systems, however, ecosystem integrity is best reflected by the purpose they are designed to serve and the quality of the specific goods or services derived. While pristine natural systems may be of greatest value, particularly in terms of biodiversity and hence their potential use for future human generations, it is important to note that systems with some degree of impact are often also of high significance. Highly modified systems may have their own value (for example the low-intensity grazing marshes of southern England). Degraded systems may be restorable where this can be identified as a shared societal goal.

Pressures on Freshwater Ecosystems

A wide range of human uses and transformations of freshwater or terrestrial environments has the potential to alter, sometimes irreversibly, the integrity of freshwater ecosystems. Activities and their potential impacts on ecosystem health are summarized in table 6.2.

It is a familiar observation that human numbers, biomass and technological capacity have had enormous impacts on the biosphere, mainly through land conversion, consumptive use of resources and disposal of waste. Persistence of the human species will, as a consequence of simple ecological principles, continue to exert such pressures, leading to a number of challenges:

- potential conflict between the interests of upstream users versus downstream users, at local, national or international levels; and
- the need to assess and prioritize the range of water uses, in particular those that can immediately support local human development (for example. drinking water, irrigated agriculture, hydroelectric power, fisheries) in relation to those with benefits that may be realized mainly at larger scales or over longer periods of time such as flood control, recreation, biodiversity maintenance. An interesting illustration is the historic approach of wetland drainage for malaria control (see box 6.1).

Water is a resource that is not permanently transformed by use in the same way that many other natural resources (e.g. timber, fisheries, rubber) are. At the global level, essentially the same amount remains after use as before; although it can be reused, its quality is likely to decline progressively without processing through industrial or natural biological systems. Regardless of the global stock of water, at a local (or catchment) level it is common for water to be either in very short supply, or in great excess, at any one time or place, and water uses can be regarded as consumptive (e.g. irrigation) or non-consumptive (e.g. hydropower). Some uses of water, for example for irrigation, power generation or industrial cooling, do not have narrow water quality requirements. Other uses, of which the single most important is human drinking water, require water of a certain quality (in terms of solutes, micro-organism populations, etc.). Potential conflicts of interest arise when any given stock of water is used for different purposes and at different times and places (see chapter 12 on sharing water).

From the water quality and human environmental health perspectives, control of chemical and biological pollution is of key significance in protecting ecosystems. Many human activities, from water supply and sanitation to transport, mining and the chemical industry, have the potential to pollute water. This pollution may

Table 6.2: Pressures on freshwater ecosystems

Human activity	Potential impact	Function at risk
Population and consumption growth	Increases water abstraction and acquisition of cultivated land through wetland drainage. Increases requirement for all other activities with consequent risks	Virtually all ecosystem functions including habitat, production and regulation functions
Infrastructure development (dams, dikes, levees, diversions)	Loss of integrity alters timing and quantity of river flows, water temperature, nutrient and sediment transport and thus delta replenishment, blocks fish migrations	Water quantity and quality, habitats, floodplain fertility, fisheries, delta economies
Land conversion	Eliminates key components of aquatic environment, loss of functions; integrity, habitat and biodiversity, alters runoff patterns, inhibits natural recharge, fills water bodies with silt	Natural flood control, habitats for fisheries and waterfowl, recreation, water supply, water quantity and quality
Overharvesting and exploitation	Depletes living resources, ecosystem functions and biodiversity (groundwater depletion, collapse of fisheries)	Food production, water supply, water quality and water quantity
Introduction of exotic species	Competition from introduced species alters production and nutrient cycling, and causes loss of biodiversity among native species	Food production, wildlife habitat, recreation
Release of pollutants to land, air or water	Pollution of water bodies alters chemistry and ecology of rivers, lakes and wetlands. Greenhouse gas emissions produce dramatic changes in runoff and rainfall patterns	Water supply, habitat, water quality, food production. Climate change may also impact hydropower, dilution capacity, transport, flood control

A wide range of human uses and transformations of freshwater or terrestrial environments has the potential to alter, sometimes irreversibly, the integrity of freshwater ecosystems.

Source: IUCN, 2000.

Box 6.1: Draining wetlands for malaria control: a conflict of interests?

With the 1898 discovery that anopheline mosquitoes alone transmitted the malaria parasite (and, about the same time, the discovery of other mosquito genera and vector-borne diseases) a period of disease control activities started focusing on 'source reduction'. By eliminating the aquatic breeding habitats of mosquitoes, their population densities were reduced and disease transmission interrupted. Anticipating later integrated rural development approaches, malariologists, engineers and agronomists collaborated to reduce the disease burden and promote agricultural development at the same time. The drainage of the Pontine Marshes near Rome in the 1930s remains the most well-known example, though it remains small compared to what colonial governments achieved in, for example, India, the Malaysian peninsula and Indonesia. The advent of residual insecticides put an end to this approach.

From today's perspective, with the high value given to wetlands, such an approach to disease control is inconceivable. Yet the conflict of interest between human health and environmental integrity needs to be assessed for each specific location. To start with, wetland drainage would not, generally, be the practice of choice in Africa south of the Sahara, where the world's main malaria burden (90 percent) persists. The local vector species is ecologically too versatile and transmission patterns too intense to be influenced by reducing the density of mosquito population alone. In other parts of the world, the trade-offs between impacts on the environment of limited drainage (or other interventions in the hydrology) and of the recurrent use of insecticides as mainstay interventions will need to be assessed locally. Wetland communities will need to be protected by specially targeted interventions for malaria control such as placing mesh screens in house windows, doors and eaves, and the intensified use of mosquito nets. Also, in an environment with an increased malaria risk, access to health services, often already more difficult in wetland areas, will need to be greatly facilitated to allow for early detection and treatment.

arise from point sources, such as discharge pipes, or be more diffuse in nature, arising for example from agricultural land use. Pollutants can be classified into a number of classes (table 6.3).

The most frequent sources of pollution are human waste (with 2 million tons a day disposed of in water courses), industrial wastes and other chemicals including agricultural pesticides and fertilizers. It has been calculated that humans currently use about 26 percent of total terrestrial evapotranspiration and 54 percent of accessible runoff (Postel et al., 1996). By some estimates (Tilman et al., 2001), the expansion of agricultural demand for food by a wealthier and 50 percent larger global population could drive the conversion of an additional billion hectares of unmodified ecosystems to agriculture by 2050. This could result in nitrogen- and phosphorus-driven eutrophication of terrestrial, freshwater and near-shore marine ecosystems being more than doubled, with comparable increase in pesticide use. An example of current and predicted emissions of nitrogen and organic pollutants from different sectors is shown in figure 6.1.

Impacts of water use on quality differ between sectors and are not symmetrical between uses. Thus, recreational use such as swimming and fishing upstream will have no impact on downstream waste disposal; inversely, waste disposal upstream can have an impact on recreational use downstream. Water conditions, including sediment load and volume of inflow, may also be affected by land use decisions applied to terrestrial parts of the catchment basin. Many uses of aquatic ecosystems (e.g. fisheries, recreation, water purification, biodiversity maintenance, some forms of flood reduction) depend on ecosystems that are at or close to natural conditions. Other uses (e.g. hydropower, irrigation, transport) are less dependent on ecosystem condition, and often lead to highly engineered environments.

Table 6.3: Types of pollutants affecting freshwater ecosystems

Types of pollutants
Nutrients such as nitrogen and phosphorus from fertilizers and manures
Faecal and other pathogens from livestock and human waste
Soil particles from farming, upland erosion, forestry, urban areas and construction and demolition sites
Pesticides, veterinary medicines and biocides from industrial, municipal and agricultural use
Organic wastes (slurries, silage liquor, surplus crops, sewage sludge and industrial wastes)
Oil and hydrocarbons from vehicle use and maintenance
Chlorinated solvents from industrial areas
Metals, including iron, acidifying pollutants and chemicals from atmospheric deposition, abandoned mines, industrial processes
Endocrine-disrupting substances (particularly oestrogenic steroids deriving from human contraceptive pills, linked to feminization of male fish)

Figure 6.1: Emissions of water pollutants by sector

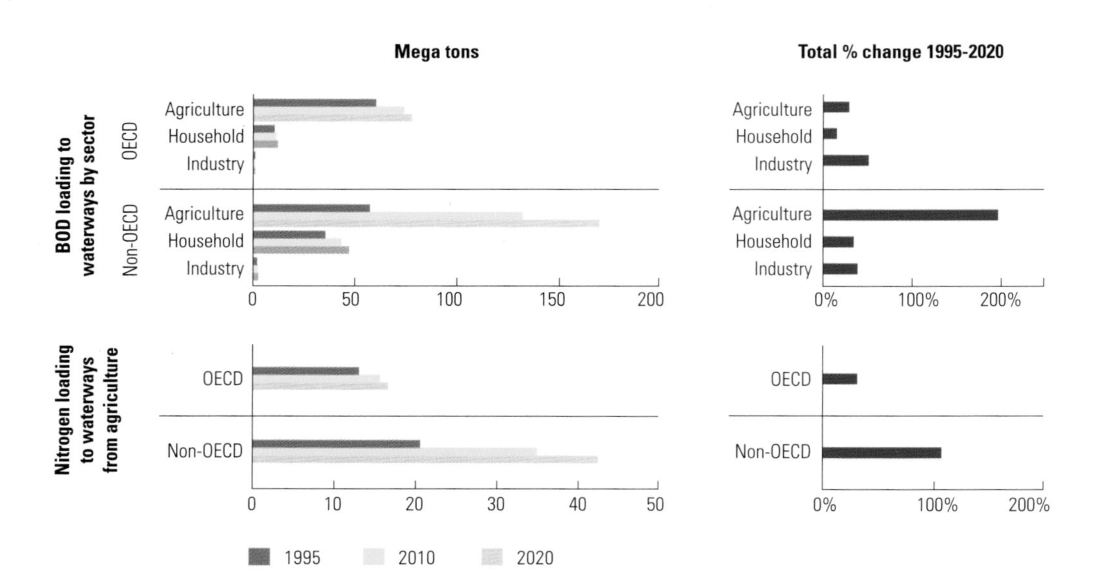

The figures give current and predicted pollutant percentages by sector. In 1995, agriculture was responsible for a more or less equivalent amount of biological oxygen demand (BOD) in both OECD (Organization for Economic Cooperation and Development) and non-OECD countries; yet this trend is reversing, and agriculture in non-OECD countries is expected to load two times more BOD in waterways than OECD countries by 2020.

Source: OECD, 2001.

In addition to considerations of water quality, the integrity and health of aquatic ecosystems depend on the maintenance of adequate quantities of water. Increasingly it is clear that the timing of water availability, for example flow regime and flooding events, is as important as absolute minimum quantities for most natural systems. While natural aquatic systems are typically able to withstand seasonal or annual variation in water supply, and hence have a degree of resilience to artificial perturbations, sustained reductions in water quantity can dramatically alter ecological balances and degrade the system. A major challenge in water resource management is therefore to identify critical ecosystem requirements in order to deliver appropriate water quantities for human social and economic needs, within the temporal and spatial constraints of environmental protection. This is the first step to restoring ecosystem health in such circumstances.

Taking the Measure of Ecosystem Health

Appropriate measures (or indicators) of ecosystem health, whether direct or indirect, are a prerequisite of holistic water management. With a key policy focus on environmental protection, measures must include tools for assessing status or monitoring trends in terms of public health, water quality, natural resource production and biodiversity.

To date, much discussion of terrestrial ecosystems has focused on changes in total area. At a gross level, loss of habitat (for example the loss of wetland through drainage) provides a useful general indicator of global trends in freshwater ecosystem condition, particularly in the context of natural resource provision. Rivers however are essentially linear systems and area is therefore inappropriate as an indicator. In addition, it is clear that more sophisticated measures of overall condition are needed, which are able to integrate extent and 'quality' of freshwater systems to allow tracking of changes over time. Ecosystem 'quality' as discussed earlier is reflected by the overall state of ecosystem processes together with the relative value of the individual components and/or the biodiversity of the system as a whole. A range of indicators and methodologies have been developed towards this goal. Broadly speaking, these can be divided into water quality indicators (both physico-chemical and biological), hydrological information and biological assessment, including measures of biodiversity. There are a few other indirect approaches that are of use in specific situations. These include evidence of change derived from observed changing patterns of human use of an aquatic ecosystem: for example, a decline in the number of fishers might indicate decline in the availability of target fish species.

Water quality indicators

Monitoring physico-chemical water quality has historically been a key means of assessing ecosystem health. In providing a direct measure of the concentration of substances known or believed to affect humans or other species, water quality provides an essential link between ecosystem health and environmental health (in its traditional sense of public health). Water quality monitoring standards have been effective in comparing sampled water for compliance, and regulating point and non-point source pollution. Other physico-chemical variables provide useful indicators of risk to human health, or water quality issues that might compromise other uses or the needs of a particular ecosystem component. These indicators include measuring the amount of faecal coliforms (as an indicator of potential pathogens from human or animal waste) and biological oxygen demand (BOD), which indicates organic loading and the capacity of a river system to purify industrial or other effluents.

Biological water quality indicators have been adopted, for example, by the United States Environmental Protection Agency (US EPA), by Environment Australia, by the United Kingdom Environment Agency and by many other similar national and subnational bodies (see figure 6.2 for an example of these indicators taken for United Kingdom rivers). These provide a complementary measure to chemical water quality and are useful in assessing intermittent pollution or impacts of unknown contaminants. The procedures used for analysis can vary significantly, and there is a growing technical literature on sampling and related statistical issues (Lillie et al., 2002; Wright et al., 2000; Barbour et al., 1999; TNC, 1999). Community structure indicators (such as the numbers and kinds of organisms dwelling adjacent to the bottom sediment) rather than individual-species indicators, may give greater accuracy and less uncertainty in detecting water quality change, but may also increase resources needed for data collection and analysis. Similarly, composite indices based on a numerical integration of multiple separate indicators of a range of ecological attributes can, in some circumstances, strengthen data interpretation and yield a more robust assessment of overall biological condition. In the United Kingdom, the River Invertebrate Prediction and Classification Scheme (RIVPACS) enables biological indices for benthic macro-invertebrates to be predicted, based on unimpacted sites in rivers of similar physical and chemical characteristics. Significant differences between observed and predicted scores can be used to highlight potential issues and biological classifications of rivers produced (see table 6.4). A similar approach is used in several catchment- and state-level programmes in the United States, for example, applying protocols developed by the US EPA (Barbour et al., 1999).

Figure 6.2: Biological quality of United Kingdom rivers, 1990–2000

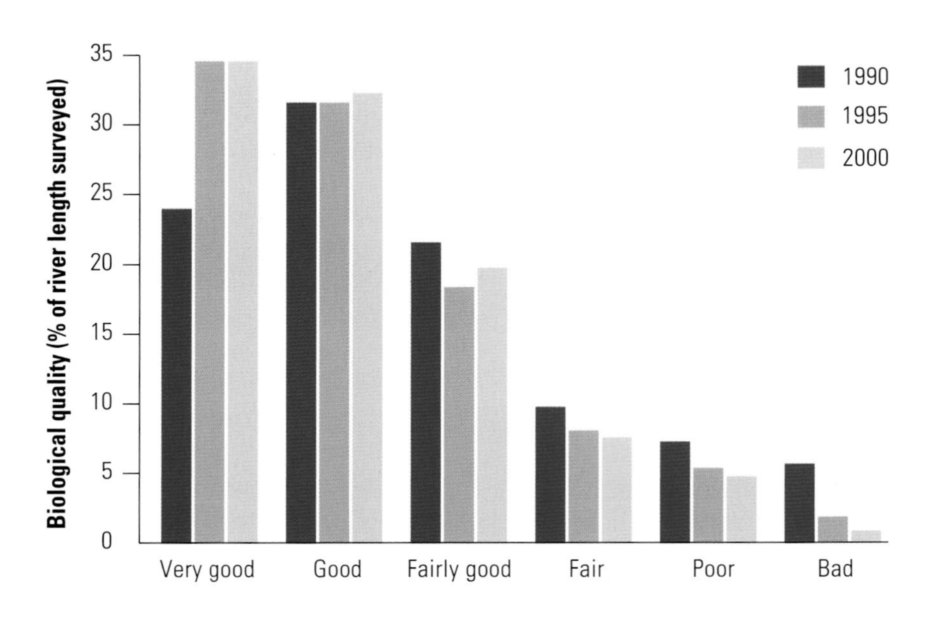

The biological water quality indicators here applied to United Kingdom rivers show that there are a far greater number of rivers classified as 'very good', and there is a decreasing number of those classified as 'poor' or 'bad'. The indicators therefore show the general improvement in water quality in the United Kingdom.

Source: Adapted from Environment Agency, UK, 2002.

Hydrological information

Hydrological information has also been widely used in determining the condition of ecosystems. River flows, water levels in wetlands, extent of flooding, storage capacity of aquifers and recharge rates, lake volumes and rainfall data are hydrological variables that provide a useful handle on the existing condition and changes in freshwater ecosystem condition. A long-standing challenge in hydrology, one which poses a constraint through its persistence, is that measurements of river flows (or other variables) contain the signal of humanity's changing impact within long data series. While this is actually of value in detecting climate change impacts, it hinders our ability to quantify the 'natural condition' of the hydrological cycle. Much effort is therefore devoted to the 'naturalization' of data by operational hydrological services – adjusting observed but 'unnatural' river-flow data series (over many years) to a consistent baseline that is representative of natural conditions. This is a precursor to many forms of impact assessment.

At the global level, there is a large amount of quantitative information on hydrological flows, thanks to initiatives such as the Global Runoff Data Centre (GRDC), and on several variables relevant to water quality (e.g. the Global Environment Monitoring Systems Freshwater Programme [GEMS/WATER]). Highly detailed regional and national data are available in most areas, albeit within the constraints of the purpose for which the monitoring was originally established, which was rarely ecological. Hydrological tools and datasets for water resources assessment are discussed more generally in chapter 4.

Biological assessment

Although no single direct measure of ecosystem condition is possible, and it is impractical to routinely assess its various components (resistance, resilience), biological assessments are generally held to be potentially indicative of ecosystem condition. Because organisms have a fundamental and integral role in ecosystem processes, tracking change in aspects of community organization is likely to be an efficient, if indirect way, to track causal changes in a wide range of variables that impact on ecosystem integrity. In contrast to physico-chemical indicators, these changes will be integrated temporally over a scale related to the lifespan of the organisms assessed. Biological criteria can be used to set the biological quality goal that is reached by implementing appropriate management of chemical and physical variables.

Many types of biological assessment aim to set a baseline corresponding to natural or relatively unmodified conditions, commonly designated as the target condition (e.g. Brink et al., 1991). The common first step then is to collect data on the occurrence of taxa of organisms at a range of reference sites. This information can be used to characterize the communities present in relatively unmodified natural water bodies of each kind, primarily in terms of the spectrum of taxa recorded in a sampling programme (often families, sometimes other taxonomic units and sometimes with additional data on abundance). This then defines the community expected to occur in other systems of the same kind in comparable regions that have not yet been surveyed. Communities present in these newly surveyed systems are liable to differ from what is expected if the ecosystem has been significantly modified in some way. They are usually less diverse taxonomically (often with an increased abundance of those species tolerant of changed conditions or of invasive elements), and this provides a measure of the degree of degradation in the system. Lowveld rivers of southern Zimbabwe, long-regulated by upstream reservoirs to supply the sugar industry, have suffered significant losses of species characteristic of ephemeral and seasonal rivers, and have been invaded by species that favour more constant, year-round flow.

A good example of the baseline approach is the Water Framework Directive (WFD) adopted across the European Union in 2000. The basis of the WFD is a requirement for member states to develop integrated management in all 'river basins' in order to achieve 'good' or 'high' ecological status in all water bodies by 2015. Ecological status is assessed by a series of measures that collectively may be taken as indicating a departure from natural or pristine conditions for any given category of water body (table 6.4). Data are evaluated against a set of normative definitions of five categories of ecological status. The categories range from 'high', representing absence of anthropogenic disturbance in all variables, or only very minor alteration, through to 'bad', reflecting extensive departure from natural conditions. Identifying the baseline reference conditions from which this can be defined has proved to be the key challenge, reflecting the virtual lack of truly 'pristine' sites in the developed world.

Table 6.4: Quality indicators for classifying the ecological status of rivers

Biological elements
composition and abundance of aquatic flora
composition and abundance of benthic invertebrate fauna
composition, abundance and age structure of fish fauna
Hydromorphological elements
hydrological regime
river continuity
morphological conditions
Chemical and physico-chemical elements supporting biological elements
thermal
oxygenation
salinity
acidification
nutrient status
specific pollutants

Source: Modified from EEC, 2000.

A constraint of biological assessment methodologies, particularly in terms of water quality assessment, is that while they have been shown to work well within the context of a single country, few attempts have yet been made to integrate the existing range of national measures into a global indicator system. Standard datasets do not readily lend themselves to harmonization between countries. It may be possible, however, to adopt a standard classification by which to interpret disparate national datasets, and so derive effective information on trends over time at a continental or even a global level. The development of the WFD is clearly a step in this direction.

A further constraint with many bioassessment approaches (particularly water quality indices) is that it is not yet clear how readily they can be applied to large and complex tropical freshwater ecosystems. Rather than developing a statistically robust profile of unmodified biodiversity, the focus in these systems has tended to be on performing a rapid inventory of biodiversity. A typical example is the AquaRAP joint initiative by Conservation International and the Field Museum of Chicago, assessing the biological and conservation value of tropical freshwater ecosystems.

In all systems, even with the relatively good availability of data on water quality, there remains a fundamental need for better information on biodiversity, in particular because biodiversity underlies so many of the services provided by freshwater ecosystems. Changes in biodiversity may occur in response to an enormous range of environmental factors including water quality, quantity and periodicity, the individual significance of which may be unclear. While not substituting for water quality information, which is essential for health-related management goals, biodiversity measures do have the potential to provide an integrated measure of overall ecosystem condition.

Such an approach may therefore be the most cost-effective option where the goal is to develop an integrated water management system that achieves ecosystem protection. The methods are relatively straightforward to implement, and can be repeated systematically in order to provide a strong indicator of trends in aquatic communities over time. They can also be supported where appropriate by narrower indicators, e.g. nitrate level, acidification, amenity value, or the presence or abundance of introduced species (box 6.2).

There is increasing interest in the links between ecosystem health and human health. Harvard University, in collaboration with the World Health Organization (WHO) and the United Nations Environment Programme (UNEP), is preparing a state-of-our-

Box 6.2: Non-native species

The 'National Census of River and Waterside' in Japan aims to provide information relevant to ecosystem condition, focusing on aquatic species (e.g. fish, molluscs, benthic organisms, plants) and human water-related activities. Alien species such as black bass Micropterus salmoides and other North American fish are included. The data show the number of reservoirs where the species were recorded during two survey periods. This evidence suggests that the distribution of introduced species is increasing in the country (seven additional reservoirs showed introduced species during the second survey), and the number of sites where they occur provides a useful indicator of ecosystem condition.

1st survey (1991–1995)	found	not found	found	not found
2nd survey (1996–2000)	found	found	not found	not found
Number of reservoirs	26	7	0	42

Source: Water Resources Environment Technology Centre (Japan), 2001.

knowledge publication on biodiversity and its implications for human health. Such links are highly complex with many confounding factors, but water is a common theme throughout. So far, it is clear that:

- high levels of biodiversity imply a great deal of diversity in pathogens and vectors, which in part explains the burdens of high infectious disease in the tropics; and

- where ecosystem degradation causes loss of biodiversity, more often than not this habitat simplification works in favour of species that play a role in the transmission of human disease.

In managing ecosystems, both ecosystem integrity and the environmental determinants of human health need to be taken into account. Local communities must be involved and their health (and the health of their children) will provide a strong incentive for doing so. The Primary Health Care approach developed by WHO and the United Nations Children's Fund (UNICEF) following the 1978 Alma Ata Conference (Declaration of Alma Ata, 1978) and the Primary Environmental Care approach championed by the World Conservation Union (IUCN) in the early 1990s should be integrated to serve as a basis for community action aimed at improving ecosystem and human health.

Lake Malawi (southern Africa) is an aquatic system that was originally endowed with a great deal of fish and also freshwater snail biodiversity. However, loss in fish biodiversity has resulted in the favouring of certain snail species that play a role in the transmission of schistosomiasis. The increased health risk has greatly affected the tourist industry of Malawi and the whole economy has declined. The Global Environment Facility (GEF) project for Lake Malawi is unique in that it is the only one combining biodiversity and human health.

Assessment of Current Ecosystem Condition and Trends

Global condition of freshwater ecosystems

There is a wealth of data on the condition of freshwater ecosystems around the world ranging from quantitative local and global data to anecdotal or unstructured qualitative information. A recent attempt to synthesize available information on freshwater ecosystems worldwide (Revenga et al., 2000) reviewed data on the extent of human modification, water quantity, water quality, fisheries and biodiversity (some twenty-two measures in all), and suggests that at a global scale the picture is not encouraging. Among the conclusions:

- 60 percent of the world's 227 largest rivers are strongly to moderately fragmented by dams, diversions and canals, and a high rate of dam construction in the developing world threatens the integrity of remaining free-flowing rivers;

- water quality appears to have declined worldwide in almost all regions with intensive agriculture and large urban/industrial areas;

- historical data for well-studied commercial fisheries show dramatic declines throughout the twentieth century, mainly from habitat degradation, invasive species and overharvesting.

Another study aimed at global assessment (Groombridge and Jenkins, 1998) used the terrestrial Wilderness Index (a measure of the spatial extent of roads, settlements and other human infrastructure) to estimate the probable degree of anthropogenic disturbance to major river basins. Because degradation of freshwater ecosystems is highly

Map 6.1: Relative naturalness of land in major world river basins

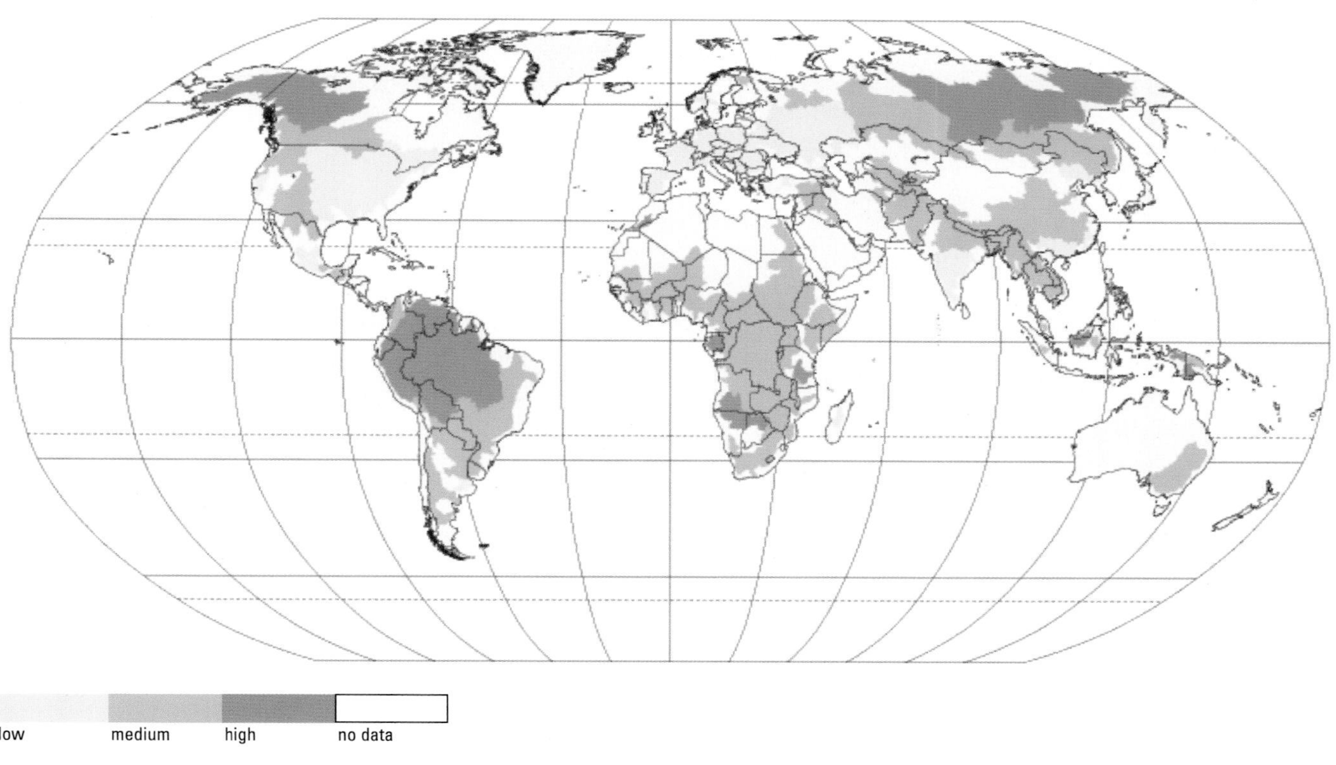

This map shows the area of spaces formed by the forces of nature, and where humankind's imprint is not yet significant. Vast regional and continental disparities can be noted, corresponding to a large degree to population density disparities.

Source: Map prepared for the World Water Assessment Programme (WWAP) by the Centre for Environmental Research, University of Kassel, Germany, based on data from UNEP-WCMC from 1998.

correlated with impacts on land, the results (see map 6.1) provide a guide to the possible condition of aquatic systems within each basin. This measure is indirect and therefore somewhat imprecise because, for example, pollutants from a point source may significantly impair water quality while the terrestrial catchment remains relatively undisturbed (as is the case for parts of the Amazon drainage basin in South America). Nonetheless, the impact of social and economic development on the natural state of the environment is clear, and lack of data for key areas is highlighted.

A more direct indication of current condition can be derived from the vast amount of global water quality data available. A global assessment (UNEP, 2002), developed with the participation of national and regional experts noted *inter alia* the widespread occurrence of poor water quality, the diversion of water from natural aquatic ecosystems, and the emerging problems with groundwater quality and recharge. Indications of relative water quality in 122 countries are given in table 6.5.

Pressures are particularly severe in developing countries where institutional and structural arrangements for the treatment of municipal, industrial and agricultural are often poor. This is reflected in increased pollution from industrial organic substances, acidifying substances from mining and atmospheric emissions, heavy metals from industry, ammonia, nitrate and phosphate pollution from agriculture, pesticide residues (again from agriculture), sediments from human-induced erosion to rivers, lakes and reservoirs and salinization.

In developed countries, waste-treatment facilities have reduced levels of bacteriological contamination and key issues with regard to water pollution are persistent substances, including pesticides, hydrocarbons and endocrine disruptors. In both developed and developing countries, loss of ecosystem integrity through poor water quality compromises the use of the resource for drinking water, food production and other aspects of human health. One of the most significant forms of river pollution is nutrients, although in recent years introduction of tertiary treatment of wastewater has resulted in improved levels of phosphorus and organic matter (but not in nitrate levels) in many more developed countries.

Table 6.5: Water quality indicator values in selected countries

Indicator value	Country	Rank	Indicator value	Country	Rank	Indicator value	Country	Rank
1.85	Finland	1	.11	Estonia	42	-.33	Kazakhstan	83
1.45	Canada	2	.11	Panama	43	-.33	China	84
1.53	New Zealand	3	.10	Slovakia	44	-.33	Libya	85
1.42	United Kingdom	4	.10	Turkey	45	-.35	Papua New Guinea	86
1.32	Japan	5	.10	Trinidad and Tobago	46	-.35	Malaysia	87
1.31	Norway	6	.09	South Africa	47	-.35	Israel	88
1.30	Russian Federation	7	.09	Croatia	48	-.36	Honduras	89
1.27	Republic of Korea	8	.08	El Salvador	49	-.37	Paraguay	90
1.19	Sweden	9	.06	Fiji	50	-.37	Uzbekistan	91
1.13	France	10	.04	Bulgaria	51	-.39	Azerbaijan	92
1.09	Portugal	11	.04	Botswana	52	-.40	Gabon	93
1.04	United States	12	-.01	Venezuela	53	-.42	Senegal	94
1.03	Argentina	13	-.02	Lithuania	54	-.47	Ukraine	95
.93	Hungary	14	-.04	Jamaica	55	-.49	Bhutan	96
.91	Philippines	15	-.06	Ecuador	56	-.49	Madagascar	97
.87	Switzerland	16	-.06	Germany	57	-.53	Togo	98
.86	Ireland	17	-.08	Zimbabwe	58	-.54	Tunisia	99
.85	Austria	18	-.08	Peru	59	-.59	Thailand	100
.74	Iceland	19	-.11	Lebanon	60	-.61	Haiti	101
.73	Australia	20	-.13	Romania	61	-.62	Nigeria	102
.70	Netherlands	21	-.14	Albania	62	-.64	Mozambique	103
.66	Mali	22	-.15	Egypt	63	-.64	Algeria	104
.64	Brazil	23	-.16	Sri Lanka	64	-.67	Zambia	105
.63	Slovenia	24	-.18	Saudi Arabia	65	-.69	Mexico	106
.62	Singapore	25	-.19	Armenia	66	-.70	Benin	107
.61	Greece	26	-.2	Bolivia	67	-.70	Uganda	108
.60	Cuba	27	-.2	Cameroon	68	-.74	Ethiopia	109
.58	Spain	28	-.22	Moldova	69	-.77	Indonesia	110
.55	Denmark	29	-.22	Tanzania, United Rep. of	70	-.77	Malawi	111
.52	Iran	30	-.22	Belarus	71	-.77	Mauritius	112
.47	Italy	31	-.23	Macedonia	72	-.78	Rwanda	113
.39	Uruguay	32	-.23	Viet Nam	73	-.81	Central African Republic	114
.39	Kuwait	33	-.24	Mongolia	74	-.95	Burundi	115
.37	Poland	34	-.26	Kenya	75	-1.0	Burkina Faso	116
.27	Colombia	35	-.28	Dominican Republic	76	-1.04	Niger	117
.27	Czech Republic	36	-.28	Kyrgyz Republic	77	-.06	Sudan	118
.23	Ghana	37	-.28	Nepal	78	-1.26	Jordan	119
.23	Costa Rica	38	-.29	Syria	79	-1.31	India	120
.19	Chile	39	-.30	Pakistan	80	-1.36	Morocco	121
.18	Bangladesh	40	-.30	Guatemala	81	-2.25	Belgium	122
.15	Latvia	41	-.32	Nicaragua	82			

Pressures on water quality are particularly severe in developing countries where institutional and structural arrangements for the treatment of municipal, industrial and agricultural are often poor.

Source: Esty and Cornelius, 2002.

Changes in freshwater biodiversity

Serious concern for inland water biodiversity at the global level was raised in the early 1990s (e.g. Moyle and Leidy, 1992), mainly referring to evidence on the conservation status of fish. Most of the relatively few reviews that have attempted a global perspective have appeared only during the past six years (Revenga et al., 2000; Groombridge and Jenkins, 1998; Revenga et al., 1998; McAllister et al., 1997; Abramovitz, 1996). These still rely heavily on information relating to fishes, but draw on many other case studies of groups where information is available, e.g. for molluscs in United States waters, and also deal in increasing detail with threat factors and their sources.

Table 6.6 shows data relating to a sample of the relatively few countries where the fish fauna is reasonably well-characterized and assessed. The data of interest relate to the number of species that have been evaluated under the IUCN categorization system and determined to be threatened with extinction, and the percent that this represents of the national freshwater fish fauna. In a geographically widespread sample of countries, the proportion is 20 percent or greater. At a global level, around 24 percent of mammals and 12 percent of birds (both of which groups have been comprehensively assessed) are in threatened categories (Hilton-Taylor, 2000). Only about 10 percent of the world's fish have been assessed, the great majority of these being from inland waters, but 30 percent of those are listed as threatened. More than 150 turtle species worldwide are restricted to, or occur in, freshwaters, and ninety-nine were categorized as threatened in 2000, equivalent to about 60 percent of all the freshwater forms. Excessive exploitation rather than habitat degradation alone is an important pressure on this group. Table 6.7 provides concise information on a small selection of the more than 3,500 vertebrate and invertebrate animals associated with freshwater habitats that were assessed as Critically Endangered (the category for populations at highest risk of extinction) in 2000.

Table 6.6: Numbers of threatened freshwater fish in selected countries

	Total species	Threatened species	% threatened
United States	822	120	15
Mexico	384	82	21
Australia	216	27	13
South Africa	94	24	26
Croatia	64	22	34
Turkey	174	22	13
Greece	98	19	19
Madagascar	41	13	32
Canada	177	12	7
Papua New Guinea	195	11	6
Romania	87	11	13
Italy	45	11	24
Bulgaria	72	11	15
Hungary	79	10	13
Spain	50	10	20
Moldova	82	9	11
Portugal	28	9	32
Sri Lanka	90	9	10
Slovakia	62	9	15
Japan	150	9	6

The countries listed here have the greatest number of globally threatened freshwater fish species, and are ordered by threatened species number. The fish faunas of these twenty countries have been evaluated completely, or nearly so.

Source: Groombridge and Jenkins, 2002; total species estimates (all approximate) from UNEP-WCMC database; threatened species data from online Red List http://www.redlist.org (4 March 2002).

Some of the most comprehensive national data are from the United States, where the Nature Conservancy and Natural Heritage Network released an analysis of the conservation status of more than 20,000 species in 1997. The four groups with the highest proportion of species extinct or at risk – freshwater mussels, crayfish, amphibians and freshwater fish – are all inhabitants of, or dependent on, inland water habitats (Master et al., 1998) (see figure 6.3). Similarly, in Australia, four (22 percent) of eighteen waterbird species assessed are listed as threatened, along with twenty-seven (13 percent) of the frog species and twenty-two (about 10 percent) of freshwater fishes.

Many extinctions have taken place in inland waters – at least thirty-four fish species (six since 1970) and possibly up to eighty, from the late nineteenth century onward – and inland water ecosystems have seen probably the largest known multispecies extinction episodes of the twentieth century. Lake Victoria, shared by Kenya, the United Republic of Tanzania and Uganda, was until recently the home of a species population of up to 500 haplochromine cichlid fish (not all yet formally described), as well as of a number of other fish species. Following introduction of the Nile perch *Lates niloticus* – and probably also as a result of heavy fishing pressure, increased sedimentation and oxygen depletion due to the increased organic and nutrient loading – about half of the native species are now believed extinct or nearly so, with little chance of recovery. In the Mobile Bay drainage basin in the United States, dam construction has had a catastrophic impact on what was probably the most diverse freshwater snail fauna in the world (Bogan et al., 1995). Nine families and about 120 species were known from the drainage basin. At least thirty-eight species are believed to have become extinct in the 1930s and 1940s following extensive dam construction in the basin: the system now has thirty-three major hydroelectric dams and many smaller impoundments, as well as locks and flood-control structures. These patterns are likely to have been repeated at a smaller scale in many other less well-documented parts of the world.

Table 6.7: Threatened inland water species: a selection of species assessed by IUCN as Critically Endangered in 2000

Common name	Scientific name	Description
Mammals		
Baiji	*Lipotes vexillifer*	A freshwater dolphin endemic to the Yangtze River, China. The fewer than 200 remaining individuals are threatened by entanglement in fishing gear, collisions with boats, pollution and hydroelectric schemes.
Birds		
Brazilian Merganser	*Mergus octosetaceus*	This little-known duck inhabits the shallow, fast-flowing rivers of eastern South America. Occurs in a few widely scattered populations, threatened by deforestation, hydroelectric development and hunting.
Lake Junin Grebe	*Podiceps taczanowskii*	A small grebe restricted to a single lake in west-central Peru. Abundant in the 1960s but down to about 200 birds by the 1990s. Nests in flooded reed beds, but water level change for hydroelectric plant can cause breeding to fail; also affected by mining sediments and possibly by El Niño events.
Reptiles		
Chinese Alligator	*Alligator sinensis*	Formerly widespread in the lower Changjiang (China) but declining at a rate of 5% annually because of loss of natural wetlands and persecution, and now restricted in the wild to a very small part of Anhui Province, where around 130 individuals persist. Large numbers exist in captivity.
Striped Narrow-Headed Softshell Turtle	*Chitra chitra*	Restricted to the Mae Klong and the Khwae Noi rivers in Thailand. Threatened by collection for food and the domestic pet trade.
Amphibians		
Lake Lerma axolotl	*Ambystoma lermaense*	Present in the remnants of Lake Lerma, Mexico. At risk because of the restricted range, within the Lerma-Chapala system, affected by drainage and declining water quality.
Mount Glorious Torrent Frog	*Taudactylus diurnus*	Known only from a few mountain rainforest streams in south-east Queensland, Australia. Not found in recent searches, possibly extinct. Reason for decline not known.
Fish		
Common Sturgeon	*Acipenser sturio*	A large anadromous fish previously widespread in large European river basins. Following habitat loss, pollution and overfishing the species now spawns only in the Gironde-Garonne-Dordogne basin of France and the Rioni basin of Georgia.
Cave Catfish	*Clarias cavernicola*	Endemic to Aigamas Cave lake, near Otavi, Namibia. The small population of catfish (<400 individuals) is threatened by a decrease in water level resulting from the depletion of local aquifers.
Smoky Madtom	*Notropis baileyi*	Known mainly from one small population of fish in Citico Greek, a tributary of the Little Tennessee River. At risk because of its small natural distribution. Restricted to riffle-edge habitats, and vulnerable to flow and quality changes.
Border barb	*Barbus trevelyani*	Restricted to Keiskamma and Buffalo systems in the Ciskei and eastern Cape Province, South Africa. Inhabits pools and riffles of clear rocky streams. Threatened by siltation, invading riparian plants and alien predators (notably trout).

Source: Earlier version in CBD, 2001; species from IUCN Red List of Threatened Species at http://www.redlist.org (March 2002); other information from miscellaneous sources.

Figure 6.3: Proportion of United States species at risk or extinct, by taxonomic grouping

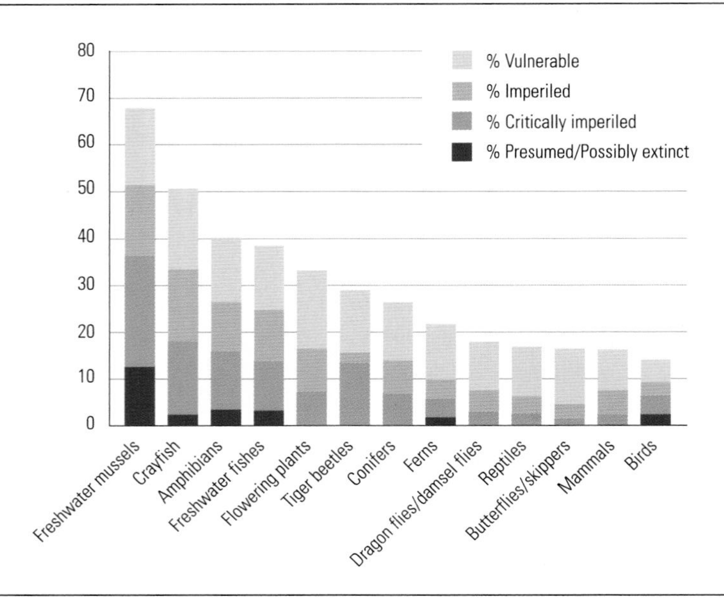

There is a large number of species from all taxa at risk. However, the quantity of critically imperiled species is alarming, and several taxa have been classified as extinct. There are also a great many species classified as vulnerable. All of these species are affected by freshwater ecosystems, in terms of their availability and their quality.

Source: Based on Master et al., 1998.

Figure 6.4: Living Planet Index 1999: inland waters

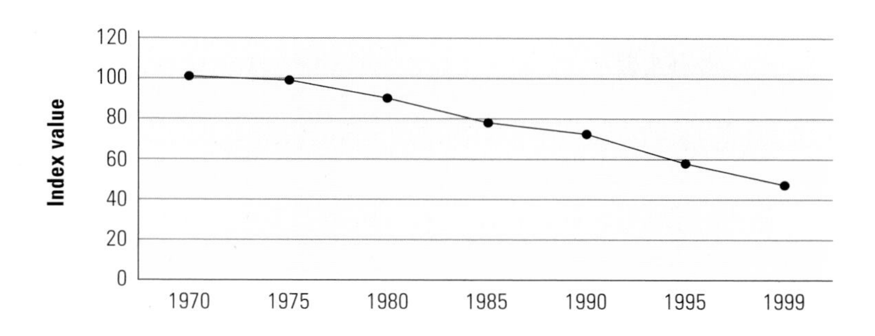

This figure shows a consistently declining number of inland freshwater species over the last thirty years. The system is based on estimates of population size of wild species available in the scientific literature. The index is calculated as a percentage of the population size estimated at 1970; the mean value of the index is calculated as an average of all the species included in the assessment at each time interval. Vertebrates other than fish (water birds, turtles, crocodilians, amphibians) are disproportionately represented.

Source: Loh et al., 1999.

The World Wide Fund for Nature (WWF) *Living Planet Report* (Loh et al., 2000, 1999, 1998) provides the overall trend in a large sample of inland water species for which indicators of population change are available. Some of the species represented had been assessed as globally threatened while several populations were actually increasing during the period, sometimes very steeply (many of these being waterbirds subject to management for hunting). The sample included a large number of wetland and water margin species in addition to truly aquatic forms. The method is designed to represent the average change in the size of sampled populations from one five-year interval to the next, starting in 1970. The sample in 1999 represented 194 species of mammals, birds, reptiles and fish associated with inland waters, and the index suggests a declining trend over the last three decades of the twentieth century (figure 6.4).

These global and national data on the status of species provide a strong indication that inland water biodiversity is widely in decline, and because in most cases the principal threats arise from habitat disturbance, rather than excess harvesting or other factors extrinsic to the ecosystem, this can be taken as evidence of declining ecosystem condition.

Lakes

The changing condition of freshwater lakes since 1970 has been assessed in a semi-quantitative global study using published information (Groombridge and Jenkins, 1998; Loh et al., 1998). A baseline was provided by Project Aqua, initiated by the Societas Internationalis Limnologiae in 1959, which collated and published information from local specialists on more than 600 water bodies (Luther and Rzóska, 1971). Many of these lakes were treated in later information sources relating to the 1980s and 1990s, and in some ninety-three cases it was possible to make an assessment that may be taken as indicative of changing conditions. Each lake was scored according to whether its condition appeared to have deteriorated (or impacts increased), or to have improved or whether no change was reported. Improvement was reported in a very small number of lakes, but the overwhelming trend was for deterioration in condition (see figure 6.5).

Wetlands

The most recent attempt to summarize information on wetland areas (Finlayson and Davidson, 1999) concluded that the information is patchy, inconsistent and not adequate to provide a precise picture of global change. Nevertheless it was reported that around 50 percent of the world's wetlands present in 1900 had been lost by the late 1990s, with conversion of land to agriculture being the main cause

Figure 6.5: Changes in lake condition, 1960s–1990s

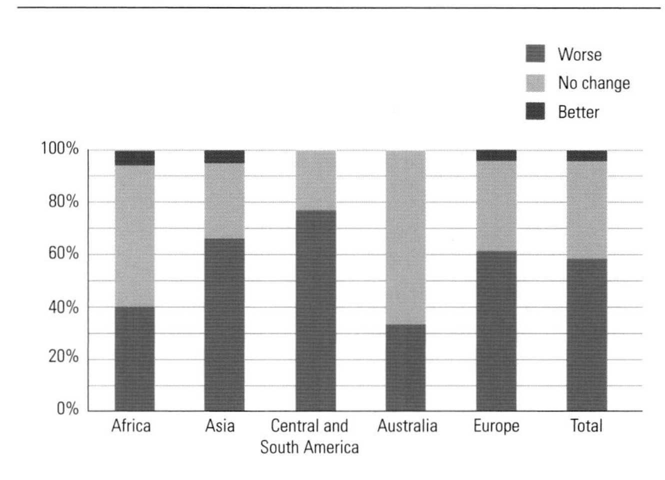

This figure is based on a sample of ninety-three lakes. Although there has been improvement of lake water condition in some areas of all regions, the overwhelming trend illustrated here is deterioration in quality, most notably in Central and South America where close to 80 percent of sampled lakes deteriorated in the studied period.

Source: Data collated for Loh et al., 1998.

Table 6.8: Examples of wetland loss in arid Eurasia

Wetland	Attributed causes	Immediate effects
Aral Sea		
Area reduced from more than 65,000 km^2 to about 28,500 km^2 in 1998. Total volume has fallen by 75 percent, and salinity has greatly increased	Excess water withdrawals from these rivers, primarily for cotton irrigation in latter half of twentieth century	Health status of the human population reduced; commercial fishery has collapsed; waterlogging and salinization have degraded agricultural land
Mesopotamian wetlands		
Loss of about 90 percent of the lakes and marshlands in the lower Mesopotamian wetlands in the last three decades	Primarily a result of major hydrological engineering works in southern Iraq (Major Outfall Drain diverts water to the head of the Gulf)	Marsh Arab populations have dispersed, and biodiversity is at great risk; there are likely to be a range of other hugely significant impacts
Azraq Oasis		
Much of the former marshland area lost; but significant resupply now underway within GEF-funded UNDP project	Increased groundwater extraction for urban needs in Jordan (about 2 million m^3 in 1979 to about 25 million m^3 in 1993) plus increased agricultural irrigation	Important wetland staging post for Eurasia-Africa migrant bird populations have been impacted

Source: Groombridge and Jenkins, 2002.

of loss. Table 6.8 provides an illustration of losses at three globally important wetland regions (in the broader sense of the definition). The wetland areas of Lake Peipsi/Chudskoe are a typical example of a vulnerable ecosystem under pressure from industrial discharge and pollution (see box 6.3 and chapter 17).

Rivers

Reviews of information around the world suggest that rivers also have become significantly degraded. In connection with the new EU Framework Directive (EEC, 2000), the WWF European Freshwater Programme collated and analysed data on fifty-five rivers (sixty-nine river stretches) in sixteen countries (WWF, 2001). Their key findings are the following:

- fifty out of sixty-nine river stretches in Europe are found to be in poor ecological condition due to the impacts of canalization, dams, pollution and altered flow regimes; and
- only five out of fifty-five rivers are considered almost pristine, and only the upper sections of the fourteen largest rivers in Europe retain 'good ecological status' as required by the EU WFD.

Similarly a recent review (UNESCAP, 2000) of the state of the environment in the Asia-Pacific region concluded, for example, that:

- the median count of *Escherichia coli* or thermo-tolerant coliforms in rivers of mainland Asia overall is fifty times higher than WHO guidelines, and higher still in the South-East Asia subregion;
- around half of the rivers have exceedingly high nutrient concentrations; many water bodies, particularly in South-East Asia, contain heavy metals in excess of WHO basic water quality standards; and
- sedimentation and rising salinity are widespread problems.

Levels of suspended solids in rivers in Asia have risen by a factor of four over the last three decades. Asian rivers also have a BOD some 1.4 times the global average, along with three times as many bacteria from human waste. They also include rivers that have twenty times more lead than that in surface waters of the Organization for Economic Cooperation and Development (OECD) countries. A report on the state of India's rivers concluded that

> *India's rivers, especially the smaller ones, have all turned into toxic streams. And even the big ones like the Ganga are far from pure. The assault on India's rivers – from population growth, agricultural modernisation, urbanisation and industrialisation – is enormous and growing by the day.... Most Indian cities get a large part of their drinking water from rivers. This entire life stands threatened* (CSE, 1999, p. 58).

Such a statement holds true for many other rivers in Asia and around the world.

In the United States, about 40 percent of streams, lakes and estuaries that were assessed in 1998 (that is about 32 percent of the country's waters) were not clean enough to support uses such

Box 6.3: Protecting ecosystems in the Lake Pskovskoe watershed

The Lake Peipsi/Chudskoe-Pskovskoe watershed is rich in wetland areas, which can be regarded as very vulnerable ecosystems – bogs and marshes occupy about 15 percent of the lake basin, and wet areas in general are spread across 35 percent of the territory. There are two main regions of international importance which are named as Ramsar sites: the Emajõe Suursoo Nature Reserve (Estonia) and the Remdovsky Reserve (Russian Federation) included in 'Lake Peipsi lowlands'. These wetlands serve as a resting place for numerous birds migrating between the breeding areas in the tundra and northern taiga, and the wintering places in western Europe. In addition several internationally Important Bird Areas (IBAs) have been designated along the lake shore. Moreover, these wetlands contain a number of birds and plants included in the Russian and Estonian Red Lists, which are rare and endangered in western Europe.

One of the main problems for this region is wastewater, which affects water quality and thereby threatens these Ramsar sites through the deteriorating quality of habitat for the endangered species.

The emerging need for an action plan has caused growing concern also on the international level. The Danish Environmental Protection Agency and Danish Cooperation for Environment in eastern Europe are at present financing the project aimed at developing and implementing a management plan for the designated Ramsar site, Lake Pskovskoe in the Russian Federation.[1]

1. You can refer to the map in chapter 17.

Source: Ministry of the Environment of Estonia and Ministry of Natural Resources of the Russian Federation, 2002. Prepared for the World Water Assessment Programme (WWAP).

as fishing and swimming. Leading pollutants in impaired waters include siltation, bacteria, nutrients and metals. Runoff from agricultural lands and urban areas are the primary sources of these pollutants (US EPA, 1998).

In New South Wales (Australia), rivers in highly urbanized catchments and in catchments where the predominant land use is cropping, showed the most signs of ecosystem stress – over half the sites are rated fair or good, but nearly half are rated poor or very poor (New South Wales EPA, 2000).

Progress in Managing Freshwater Ecosystems

The previous section paints a bleak picture of the current state of freshwater ecosystems and points to a general trend of degradation at the global level. Ideally, environmental protection should not only be coordinated with economic and social development, but should also be a tool for enhancing them. The absence of global monitoring of such actions limits the extent of conclusions that can be drawn on progress towards key objectives, or uptake of policy or practice-based measures targeted at environmental protection in all countries. However, at national, regional and local levels, there are also quite a number of positive actions in motion for protecting ecosystems.

This section describes some examples of management tools, which provide clear evidence that considerable progress is being made in many countries around the world. While these tools are not directly linked to any single objective arising from ecosystem protection, they each contribute through a multidimensional approach to the environment's protection and to its contribution to development. Ecosystem degradation is not an inevitable consequence of development, and consistent improvements in local ecosystem protection should eventually be reflected in global changes. The examples are taken mainly from among fifty-nine country reports recently submitted to the Commission for Sustainable Development (CSD). At this time, assessing progress in ecosystem protection relies on individual measures taken in national examples.

Policy, strategies and institutions

The primary group of management measures incorporates national environmental policies, legislation and strategies as a tool for environmental protection. These may set standards and overall targets, or invest in environmental institutions with regulatory and enforcement powers. Promotion of integrated land and water management is revealed as a key concept and has been furthered through legislation and the use of holistic (district/basin/national level) environmental action plans.

In Barbados, policy for integrated land and water management is contained in the Environmental Management and Land Use Planning for Sustainable Development project plans currently being finalized and a draft policy framework for Water Resources Development and

Management. Ghana's National Strategy incorporated in the Environment Action Plan adopted in 1991 aims to protect water and the general environment. This was supported by the Environmental Protection Agency Act passed in 1994, which conferred regulatory and enforcement powers on the EPA. In response, the EPA has provided guidelines for developments that affect the environment (a framework for Environmental Impact Assessment [EIA]). A wide regulatory framework has been developed in Hungary, including the Water Act of 1995 with other general provisions under the Act on Environmental Protection (1995). At the end of 1999, the National Assembly of Slovenia adopted the National Environmental Action Plan (NEAP), which defines objectives and basic guidelines in protection and use of waters as a public resource. The NEAP pays special attention to specific problems of the littoral, rural areas and karst regions, in accordance with principles of preserving biodiversity. Another main strategic goal identified in Slovenia's NEAP is the construction of water supply networks in water-deficient areas. Bringing action to a lower level, the government of Malawi has initiated a Programme of District Environmental Action Plans, emphasizing participatory planning and action. At a local level, the Local Health Inspection Boards of Iceland conduct on-site measurements of water supplies and enforce health regulations and standards.

International cooperation

International cooperation is an important requirement for countries that rely on water from shared groundwater or surface water systems. There are many examples of cooperative sharing, addressed in more detail in chapter 12 on sharing water. The Lithuanian example provided here illustrates the complexity of agreements required by some countries. Lithuania participates in the 1974 Helsinki Convention on the Protection of the Marine Environment of the Baltic Sea Area and the 1992 UN Economic Commission for Europe (UNECE) Convention on Protection and Use of Transboundary Watercourses and International Lakes. The Lithuanian Government has signed bilateral intergovernmental cooperation agreements in the field of environmental protection with the government of Flanders, Latvia, the Russian Federation and Sweden. The Ministry of the Environment has signed bilateral agreements with related ministries and other institutions of Austria, Belarus, Denmark, Finland, the Netherlands, Poland, Slovakia and United States. A trilateral agreement was also signed between the governments of the Republic of Estonia, Latvia and Lithuania.

Environmental education

Environmental education is essential in achieving goals such as reduction in water usage and also for the successful implementation of participatory approaches to environmental management. In Saudi Arabia, the Ministry of Agriculture and Water has undertaken a huge range of education activities aimed at all levels of society and sectors. These include:

- publication and distribution of booklets, bulletins and posters;
- education of farmers on optimal agricultural methods such as drip irrigation;
- rationalization of water consumption through television and radio programmes;
- contracts with advertising companies for awareness campaigns on water saving based on billboards along most major roads; and
- recruitment of a specialized work force.

The government of Poland has introduced a number of education campaigns geared towards sustainable water usage and minimization of waste, including the 'Save Water Campaign' promoting economical and rational water consumption, the 'Blue Thumb' – a nationwide programme on caring for resources, and 'Now Vistula Time' for the conservation of the natural course of the River Vistula.

Reporting

Reporting at a national level on environmental quality and changes is a prerequisite for monitoring trends at a more global level. Many countries now have the institutions and programmes in place to provide regular reports. In Austria, the Ministry of Agriculture, Forestry, Environment and Water Management is charged with delivering to Parliament at least triennial, published reports on the status of water protection in Austria. South Africa has begun publication of a regular national report on the state of the nation's water. The government of Barbados is testing the monitoring of indicators for sustainable development. Management and protection of aquatic resources in Peru is the responsibility of the General Directorate of Environmental Health (DIGESA), which is part of the Ministry of Health. DIGESA maintains databases/registers and reports on water quality throughout the country.

Flow maintenance

Mechanisms used to maintain required water quantities and periodicity include abstraction management, based on controlled abstraction in the interests of downstream abstractors, dilution of pollution, ecological river-flow objectives and water levels for river transport, floodplain/wetland needs and aesthetic values. Reservoir compensation flows are a further measure whereby downstream flows are maintained (but at lower flow volumes) by deliberate and scheduled release of water from storage.

In the United Kingdom, an urgent action programme was introduced to deal with over forty cases in which excessive abstraction has resulted in unacceptably low flows in rivers. Under the European Community Habitats Directive, all abstraction authorizations are being reviewed in the context of their impact on internationally important sites. A review of reservoir compensation flows has also been carried out as part of a shift from historically constant releases to varied releases that better match seasonal ecological requirements. The new National Water Act of South Africa embraces strong abstraction management regulations and requires authorizations for water use activities. The Act provides for coordination and joint decision-making across government departments in terms of water use activities. The Department of Water Affairs has focused on the wide set of streamflow reduction activities and has taken steps to require abstraction permits for the consumptive water use of commercial forestry.

Environmental Impact Assessment (EIA)

EIA has been widely adopted and frameworks are now in place in the majority of countries, with regard to major developments. There is a trend towards more Strategic Impact Assessments (at the policy level) and towards giving certain aspects (such as health) a higher profile within the EIA framework. In Thailand, all large dams and reservoirs will go through EIA studies and public hearing processes. A compulsory procedure is in place for EIA in Kazakhstan.

Site protection

Site-based measures typically involve designating protected areas and regulating access or use so as to maintain particular habitat elements, with or without steps to regulate water quality and flow characteristics. At the international level, various tools have been developed to protect ecosystems (see box 6.5). The 1971 Convention on Wetlands of International Importance is dedicated specifically to inland waters and coastal wetlands. More than half of all Ramsar sites occur in western Europe and these comprise around 20 percent of the total area of all Ramsar sites globally. Many wetlands, with their often abundant and highly conspicuous avifauna, have been designated as protected areas under this Convention: notable examples include the Moremi Game Reserve in the Okavango Delta (Botswana), Camargue National Reserve (France), Keoladeo (Bharatpur) National Park (India), Doñana National Park (Spain) and Everglades National Park (United States). The effectiveness of such measures is determined in part by the type of water bodies involved. Lake Titicaca (shared by Peru and Bolivia) provides an interesting example of a transboundary Ramsar site (see box 6.4 and chapter 21).

With awareness of the problems facing many aquatic ecosystems now far greater than a decade ago, there are increasing efforts to

Box 6.4: Lake Titicaca: a transboundary Ramsar site of vulnerable ecosystems and ancient cultures

The world's highest navigable freshwater lake, at 3,810 metres above sea level, straddles the frontier between Peru and Bolivia in the Andean mountains. The site includes a whole system of freshwater permanent lakes, brackish lakes, rivers, marshes and peatlands. The adjoining Lakes Poopó and Uru Uru are also designated as protected sites under the Ramsar Convention. The sites harbour many endemic and endangered fauna and flora, such as species of birds, vicuna and pumas, many kinds of rare flamingos, the cactus Opuntia and Trichocereus and the world's largest frog. Algae and floating vegetation are abundant and the dominant emergent species is the 'totora' *Schoenoplectus totora*, which can reach up to 7 metres. When the 'totora' drifts away from the shore, it forms islands that are used by some members of the local Uru community to live on. The Lake region is home to two pre-Hispanic cultures, the Aymaras and the Urus. The indigenous populations live on subsistence agriculture and fishing, and continue to practice a lifestyle that is true to traditional values and resistant to change. Poverty is endemic and like most wetland areas, the freshwater ecosystem is vulnerable to increasing pollution from nearby cities, untreated waste and mining activities.

There are several important initiatives to save this unique lake. Peru and Bolivia, through the 'Proyecto Especial Binacional Lago Titicaca', are implementing the 'Plan Director Global Binacional' for conservation, flood prevention and use of resources. They are carrying out research, environmental management and monitoring, public use, environmental education and fisheries development programmes, including the regulation of water use for agriculture and human consumption. There are also efforts on behalf of soil restoration and recovery of traditional agriculture techniques that are valuable for ecosystem conservation.

Source: Prepared for the World Water Assessment Programme (WWAP) by A. Clayson, 2002.

Box 6.5: International instruments for protecting ecosystems

The notion of a coordinated, international effort to protect the ecosystems of the world probably dates back to the 1960s with the beginnings of the environmental movement. UNESCO's Programme on Man and the Biosphere (MAB) in the 1970s specifically initiated a World Network of Biosphere Reserves with the aim of conserving representative examples of ecosystems based on community participation, rational use of natural resources and voluntary scientific cooperation. A zonation pattern is proposed to accommodate the multiple functions of biosphere reserves.

The Convention on Wetlands of International Importance (Ramsar Convention) was launched in 1971 in recognition that wetland ecosystems need special protective measures due to the multiplicity and magnitude of pressures bearing upon them to meet the needs of the expanding world population: this convention enshrined the idea of 'wise use' of wetlands areas, again underlining the need for a balanced approach to nature protection, combining rural development, the maintenance of wetland goods and services and also socio-economic and cultural features. The biosphere reserves of the MAB Programme and the sites designated under the Wetlands Convention thus have many features in common.

The protection of biodiversity in general, of which ecosystems are a part, was initially a primary motive for preparing a Convention on Biological Diversity (CBD) in the late 1980s. However, the convention negotiations expanded their scope to encompass the aims of the sustainable use of natural resources and the fair and equitable sharing of benefits arising out of the utilization of genetic resources. The CBD, first signed at the United Nations Conference on the Environment and Development (UNCED) in 1992 is thus a framework convention that does not set specific targets or obligations. To become effective it requires implementation at the national level. For this, the CBD countries have adopted the 'ecosystem approach' based on twelve principles, themselves founded on the application of scientific methodologies focused on levels of biological organization, and on the recognition that humans, with their cultural diversity, are an integral part of many ecosystems.

In practice, this means that in fact there is a healthy convergence of concepts underlying international efforts to protect ecosystems with special emphasis on water resources. This general agreement makes it easier for countries to contribute to these efforts.

Today, the UNESCO/MAB Biosphere Reserves, serve as 'living laboratories' for testing and demonstrating the ecosystem approach. This type of land and waterscape management calls for new forms of institutional cooperation and links between different levels of economic and political decision-making. Today, of the 409 biosphere reserves and the 1,107 Ramsar sites, fifty-nine sites in thirty-six countries carry both international designations, thus signalling their significance at the international level and conferring added protection. A joint work plan between the Convention on Wetlands and the UNESCO/MAB Programme makes it easier to pool resources and streamline national reporting. Such a streamlining of international effort is of mutual benefit to all parties, whether at the international, national or local level.

Source: UNESCO/MAB, 2002. Prepared for the World Water Assessment Programme (WWAP).

define priorities on the basis of biodiversity and other criteria (e.g. Duker and Borre, 2001; Groombridge and Jenkins, 1998). For example, at global level, WWF has selected fifty-three key freshwater eco-regions as a guide to prioritizing conservation investment and action (Olson and Dinerstein, 1998). These eco-regions are catchment-based and take into account biogeographic region, water body type, biological diversity and representativeness.

Since Iran adopted the Ramsar Convention in 1995, a number of wetlands have been designated under the Ramsar Convention and programmes are underway to prepare national reports on status. In the Philippines, the Ramsar Convention came into force in 1994.

Water quality standards

The concept of water quality standards as a tool for ecosystem protection encompasses a number of different principles as well as a wide variety of operational tools. In the United States, the Clean Water Action Plan (1998) is a collaborative intersectoral initiative to build a new framework for watershed protection. It focuses on achieving cleaner water by strengthening public health protection, targeting watershed protection efforts at high priority areas and providing communities with new resources to control polluted runoff and enhance natural resource stewardship. States also regulate water quality through federal programmes such as the National Pollutant Discharge Elimination System and the Non-point Source Programme. In the United Kingdom, implementation and compliance monitoring for European Community Directives provide a comprehensive framework of water quality standards. These include the Urban Waste Water Directive and Surface Water Abstraction Directive. In Syria, water pollution from unregulated discharges of industry, agriculture and public sewage are not yet fully controlled but basic water quality parameters are continuously monitored and a database/network for seven basins has been established. The Iranian Department of the Environment has standard limits for all toxic and hazardous substances from effluents from different sectors to maintain water quality standards.

Water source protection

The United States Department of Agriculture's Natural Resources Conservation Services programme focuses on upland watershed protection and on improving water management on farms, in rural areas and in small communities through voluntary efforts backed in large part by financial incentives. To safeguard water resources in Israel, the state of available resource is continuously monitored by the Hydrological Service Department of the Water Commission and reports are used to influence the planning process. A water resources conservation map restricting land use in critical areas to activities that are not harmful to water resources has been produced and is used in land use planning.

Species protection plans

Species level initiatives may include regulating persecution and hunting, or developing integrated recovery plans for species at risk of extinction. Such approaches are more often concerned with larger vertebrates, such as birds of water margins and other wetlands, than with other elements of inland water ecosystems. The Wetlands Working Group of Namibia was formed in 1997 and prioritizes taxa and wetlands for research attention, reflecting the importance of the Okavango wetlands. The government of Norway has expanded its National Protection Plan and has proposed a plan for protection of the most important salmon rivers and fjords.

Environmental economics

Economic factors tend to carry much weight in environmental decision-making. Where a costs/benefit approach to planning is taken, it is important that a comprehensive economic analysis also be undertaken, aiming in particular to attach value to environmental benefits that may not be routinely valued, and to take into account costs that would otherwise be externalized (see box 6.6). This approach appears of special importance in the case of freshwater ecosystems because these systems are assets that underpin such a wide range of human activities (e.g. Swanson et al., 1999).

Examples presented here include the use of environmental economics in the broader sense as a range of measures that are used in protecting ecosystems for specific aims. This includes pricing structures, incentives, fines, cost-benefit analysis and the polluter pays principle. The water-pricing policy of Singapore is based on cost recovery and support of water conservation objectives (see box 7.1). Measures include fiscal incentives and monetary penalties to discourage wasteful use. The water tariff structure comprises domestic and industrial/commercial categories and a water conservation tax (percentage of water charges) is imposed. The household tariff is a flat rate, up to 40 m^3 per month at which point a higher charge is imposed. Policy in Saudi Arabia treats water as an economic commodity and pricing has been updated so that unit price increases with consumption. In Thailand, the command and control approach, which is based on the polluter pays principle, has been complemented by economic incentives such as taxes and low interest loans from environmental funds. In Belgium, the Flemish Environment Holding Company was created (51 percent public, 49 percent private) to promote investments in the environment sector. Through its daughter AQUAFIN, supra-municipal infrastructure is being developed. In 1996 the government of Denmark adopted a tax on wastewater for discharges of nitrogen, phosphorous and organic substances and this has been complemented with additional investment in treatment plants.

Box 6.6: Taking hidden costs and benefits into account

An early partial analysis of a tropical wetland system (Barbier et al., 1991) calculated that agricultural, fuel-wood and fishery benefits of unmodified wetland amounted to about US$32.00 per 1,000 m^3 of water used, whereas the equivalent value of crops grown under irrigation in the region was about US$0.15. Similarly, a cost/benefit analysis was made of the economic consequences of maintaining four dams on the Snake River (north-west United States) or removing them and restoring the river and its salmon run. It was calculated that the restoration option would save a minimum of US$86.6 million annually compared with the costs of maintaining the dams.

Source: Oregon Natural Resources Council (ONRC, n.d.).

Restoration of degraded systems

Increasing recognition of deterioration in a range of ecosystems has led to a growth of interest in the science and practice of ecological restoration. The main aim is to re-establish the key characteristics of an ecosystem, such as composition, structure and function, which were present prior to degradation.

Many restoration projects have now been initiated in different parts of the world, focusing on a variety of ecosystem types, including freshwaters. Many projects are being undertaken by non-governmental organizations (NGOs), often as community-based initiatives. It is widely anticipated that restoration will become a central activity in environmental management in the future. Such efforts are being supported by development of national and international policies – for example, they are explicitly recommended in Article 8f of the UN Convention on Biological Diversity, which states that parties should 'rehabilitate and restore degraded ecosystems and promote the recovery of threatened species, through the development and implementation of plans or other management strategies'.

Depending on the nature of the disturbance, even heavily modified waters can in some circumstances recover if the source of impact is removed. This applies to river systems in particular, perhaps because of their dynamic nature, with inherent hydrological variability, and the consequent need for flexibility and colonization ability among their biological communities (see box 6.7).

The term 'restoration' may be applied to a range of measures that differ greatly in spatial scale and complexity. Most schemes to date have involved enhancement of water quality, often by controlling point source pollution in rivers or lakes, or enhancing geomorphic or habitat conditions along a river reach. In principle, a series of small-scale projects could be coordinated in order to meet restoration goals within an entire catchment, but this would require a wider scientific base and wider participation than anything yet achieved.

Some recent interest has focused on the physical removal of dams and related engineering structures. In some cases it has been shown that there is a greater long-term economic benefit to be gained from removing a dam and restoring fishery and amenity values, than from maintaining the dam, particularly where fish passage must be provided. However, several high-profile cases show that agreeing on and implementing a plan for dam removal can be contentious. A significant number of dams have recently been removed or are likely to be removed soon. Most such events have occurred in developed countries, mainly in North America and Europe (see box 6.8), and most of the dams have been small run-of-river impoundments (not hydroelectric). This phenomenon has become so important that the actual removals may have outpaced growth of the scientific understanding needed to manage the consequent ecological changes.

There is a large and increasingly accessible literature available on restoration of rivers (e.g. Nijland and Cals, 2001) and wetlands (Interagency Workgroup on Wetland Restoration, n.d.). Several works are in 'toolkit' form in order to assist locally based initiatives (Environment Agency, 2001).

Box 6.7: The 'Green Corridor': floodplain restoration along the Danube Valley

More than 80 percent of wetlands along the Danube River have been destroyed since the start of the twentieth century. Working with Romania, Bulgaria, Moldova and the Ukraine, WWF is developing a series of projects to maintain existing wetland protected areas, create new ones and reconnect the remnant marshlands and other wetlands. This is the largest international wetland restoration and protection initiative in Europe, and is intended to restore the Danube's natural capacities for pollution reduction, flood retention and nature conservation for the benefit of local people and ecosystems of the Danube River and Black Sea.

Source: WWF at http://www.panda.org/livingwaters/danube/index.cfm.

Box 6.8: The Edwards dam

The Edwards hydroelectric dam on the Kennebec River, Maine (United States), was removed in 1999 following a ruling by the Federal Energy Regulatory Commission that the environmental benefits of removal outweighed the economic benefits of continued operation and maintenance. It had been in place for more than 160 years. A year after removal, migratory fishes including Atlantic salmon *Salmo salar* and alewife *Alosa pseudoharengus* had returned to reaches above the dam site. In addition, water quality had improved, and both the biodiversity of aquatic invertebrates and recreational use of the river increased.

Source: Natural Resources Council of Maine.

Table 6.9: Review of National Reports submitted to the Convention on Biological Diversity (CBD)

Question	Response	Number of CBD Party countries
Has your country included inland water biological diversity considerations in its work with organizations, institutions and conventions affecting or working with inland water?	No	4
	Yes	66
Has your country reviewed the programme of work specified in annex 1 to the decision (Decision IV/4), and identified priorities for national action in implementing the programme?	No	20
	Under review	29
	Yes	11
Is your country supporting and/or participating in the River Basin Initiative?	No	35
	Yes	34
Is your country gathering information on the status of inland water biological diversity?	No	9
	Ongoing	56
	Completed	5
Has your country developed national and/or sectoral plans for the conservation and sustainable use of inland water ecosystems?	No	15
	Yes	54

Source: Seventy-two second National Reports submitted; data retrieved 24 June 2002 from second National Report Analyser at http://www.biodiv.org/reports/nr-02.asp (values have been rounded in two cases where 0.5 was assigned when a country made two responses to a question).

In Ukraine, a number of programmes are designed to restore and protect ecosystems, including the National Programme of Ecological Improvement of Quality of Drinking Water (1999), the Nationwide programme of Protection and Reproduction of the Azoz and Black Seas Environment, the Programme of Development of Water-supply and Sewer Facilities. Incentives for Private Land Stewardship in the United States have resulted in installation of about 1,159,000 km of protective buffers. Also US$1 billion has been allocated to remove sensitive lands from agricultural production and to encourage use of conservation practices. Technical guidance on restoration methods has been produced and demonstrated in twelve showcase watersheds. Against the background of a policy of 'no net loss of wetlands', a strategy has been implemented to achieve a net increase of 100,000 wetland acres per year by 2005. The Rawal Lake has been selected for special attention by the government of Pakistan, following identification of unsustainable practices such as construction of septic tanks/latrines in adjacent settlements. Three restoration programmes have been recommended including the development of a scheme to collect and recycle waste for use in irrigation.

Summary of progress

It is noted above that Chapter 18 of Agenda 21, arising from UNCED in 1992, set out key objectives for protection and integrated management of freshwater resources. Progress towards those objectives made after five years was formally reviewed by the UN CSD (1997). As illustrated, this review provided numerous examples of successes and promising changes in technical cooperation, participatory planning, and water quality monitoring, but noted *inter alia* the continuing global concern over deteriorating water quality, the incomplete knowledge of the pathways and impacts of pollutants, lack of appropriate legislation and financing, and the urgency of remedial action.

Reporting obligations under the 1992 CBD enable a partial assessment to be made of recent national initiatives. Although only a minority of CBD Parties (seventy-two out of a total 183) have submitted second National Reports to the CBD, a review of these suggests that significant progress has been made in certain aspects of biodiversity conservation and use in inland waters (see table 6.9).

Most of the seventy-two countries submitting reports indicate that they are gathering information on inland water biodiversity; similarly, most have reportedly developed national plans for conservation and sustainable use of inland water ecosystems, and have implemented relevant capacity-building measures. This suggests that a number of countries have already made some progress towards the strategic planning and target-setting discussed at The Hague. It is not yet possible to assess to what extent these measures have influenced the condition of ecosystems in these countries.

In addition to general measures for the conservation and sustainable use of genes, species and ecosystems, the CBD has established a thematic work programme on inland water biodiversity, and much effort has been devoted to developing this work, identifying obstacles and priorities, and establishing cooperation among relevant initiatives and organizations. The programme is being developed in cooperation with the Ramsar Convention, and now includes a catchment-focused River Basin Initiative.

However, while several developed countries are implementing some form of assessment and monitoring programme and undertaking remedial work, less-developed countries are less able to place these activities among their highest priorities. Protecting inland water ecosystems is therefore likely to remain problematic in areas where both population growth and the likelihood of agricultural expansion are highest.

In recent years, an extension of the holistic approach implied by Integrated Water Resources Management (IWRM) has emerged in which ecosystems themselves have become the subject of conservation efforts. Instead of restricting efforts to individual species and aquatic sites, planning has now to expand in scope:

- geographically, to a scale appropriate to the size of catchment basins; and
- conceptually, aiming to maintain ecosystem processes and the components of these systems, including human communities (the 'ecosystems approach', e.g. Reynolds, 1993).

In practice, reconciling the many different concerned interests and coordinating actions has proved difficult and is likely to remain so in many situations. Maintaining biodiversity and conserving inland water capture fisheries (other than some lucrative sports fisheries), for example, have not ranked highly among these competing interests, and it has thus proven difficult to impose catchment-wide regulations or remedial measures for their benefit.

Although it is difficult to assess the outcome of all these processes, in terms of beneficial effects on ecosystem condition (and results cannot yet be expected in many instances), the evidence reviewed above demonstrates the improving trend in certain variables that is apparent in some countries, and the determination of others to reverse recent decline in condition. However, such cases relate mainly to OECD and other developed countries, and the overall global situation remains very hard to evaluate.

Conclusions

Protection of ecosystems must remain central to sustainable development because environmental security, social well-being and economic security are intricately intertwined and fundamentally interdependent. Degradation of any one worsens the condition of all three. We protect our ecosystems and their creatures for their inherent values and because of the benefits they yield to waste disposal, environmental health, natural resource production, and as water sources. Pressures on ecosystems are reasonably well understood, and a number of practical tools exist that allow the condition of ecosystems to be assessed and monitored. These clearly show that freshwater ecosystems have been hit much harder than land or sea ecosystems. In effect, a majority of the world's largest rivers are substantially fragmented by dams, diversion and canals, and many fall far below accepted ecological quality guidelines. This is not only true for rivers, but also for lakes, the global condition of which is steadily deteriorating, and for wetlands, which are frequently converted to agricultural lands.

In many places, water quality has deteriorated dramatically, particularly in regions with intensive agriculture and large urban/industrial areas. Pollution from bacteria, human waste, high concentrations of nutrients, sedimentation and rising salinity, is an ever-growing problem throughout the world. Freshwater biodiversity has been severely affected by this decline in water quality, among other issues, and many species have become extinct.

Although there is evidence of local – and sometimes national – improvement as well as action plans that are being implemented, these are not yet stemming the global deterioration of ecosystem conditions. The future sum of the local gains will need to add up to much more if they are to have effect – and they must also counter the inevitable future population growth pressures from improvements in health, food, energy and industry. Thus, protecting ecosystems implies reconciling the many different water interests, through some kind of a global approach like that claimed by IWRM. If this is not done, the economic and social development targets to which the international community is committed will be jeopardized.

Progress since Rio at a glance

Agreed action | **Progress since Rio**

- Manage freshwater in a holistic manner, based on catchments and a balance of the needs of people and environment
- Adopt preventative approach to halt environmental degradation
- Evaluate the consequences that various users of water have on the environment, support measures controlling water-related diseases and protect ecosystems
- Develop outlines for protection, conservation and rational use
- Effectively prevent and control pollution
- Establish international water quality programmes and criteria
- Reduce prevalence of water-associated diseases
- Eradicate dracunculiasis (Guinea worm disease) and onchocerciasis by 2000
- Adopt an integrated approach to environmentally sustainable water and ecosystem management

Unsatisfactory | **Moderate** | **Satisfactory**

References

Abramovitz, J.-N. 1996. *Imperiled Waters, Impoverished Future: The Decline of Freshwater Ecosystems.* Washington DC, Worldwatch Paper No. 128.

Anon. 2000. Directive 2000/60/EC of the European Parliament and of the Council of 23 October 2000, establishing a framework for Community action in the field of water policy. Official Journal of the European Commission, L 327/1-72.

Barbier, E.-B.; Adams, W.-M.; Kimmage, E. 1991. *Economic Valuation of Wetland Benefits: The Hadejia-Jama'are Floodplain, Nigeria.* LEEC Discussion Paper, International Institute for Environment and Development.

Barbour, M.-T.; Gerritsen, J.; Snyder, B.-D.; Stribling, J.-B. 1999. *Rapid Bioassessment Protocols for Use in Streams and Wadeable Rivers: Periphyton, Benthic Macroinvertebrates, and Fish.* Washington DC, Environmental Protection Agency.

Bogan, A.-E.; Pierson, J.-M.; Hartfield, P. 1995. 'Decline in the Freshwater Gastropod Fauna in the Mobile Bay Basin'. In: E.T. LaRoe, G.S. Farris, C.E. Puckett, P.D. Doran and M.J. Mac (eds.), *Our Living Resources: A Report to the Nation on the Distribution, Abundance, and Health of U.S. Plants, Animals, and Ecosystems.* Washington, DC, U.S. Department of the Interior, National Biological Service.

Brink, B.-J.-E.; Hosper, H.; Colijn, F. 1991. 'A Quantitative Method for Description and Assessment of Ecosystems: The AMOEBA-Approach'. *Marine Pollution Bulletin*, Vol. 3, pp. 65–70.

CBD (Convention on Biological Diversity). 2001. *Global Biodiversity Outlook.* Montreal.

Cosgrove, B. and Rijsberman, F.-R. 2000. *World Water Vision: Making Water Everybody's Business.* London, World Water Council, Earthscan Publications Ltd.

CSD (Commission on Sustainable Development). 1997. *Overall Progress Achieved since the United Nations Conference on Environment and Development. Addendum. Protection of the Quality and Supply of Freshwater Resources: Application of Integrated Approaches to the Development, Management and Use of Water Resources* (Chapter 18 of Agenda 21). UN E/CN.17/1997/2/Add.17.

CSE (Centre for Science and Environment). 1999. *The Citizen's Fifth Report.* Delhi.

Declaration of Alma Ata. 1978. Official Outcome of the International Conference on Primary Health Care, 6–12 September 1978. Alma Ata.

Doyle, M.-W; Stanley, E.-H.; Luebke, M.-A.; Harbor, J.-M. 2000. *Dam Removal: Physical, Biological, and Societal Considerations.* American Society of Civil Engineers Joint Conference on Water Resources Engineering and Water Resources Planning and Management. Minneapolis.

Duker, L. and Borre, L. 2001. *Biodiversity Conservation of the World's Lakes: A Preliminary Framework for Identifying Priorities.* LakeNet Report Series No. 2. Annapolis, Monitor International.

EEC (European Economic Community). 2000. Framework Directive in the Field of Water Policy (Water Framework). Directive 2000/60/EC of the European Parliament and of the Council of 23 October 2000, establishing a framework for community action in the field of water policy [Official Journal L 327, 22.12.2001].

Environmental Protection Agency Act. 1994.

Environment Agency, UK. 2002. *River and Estuaries – a Decade of Improvements.*

——. 2001. 'Manual of Techniques for River Restoration – R and D Manual W5A-060/M'. In: *A Compilation of Case Studies Undertaken Mainly in the Context of Flood Risk Mitigation.*

Esty, D.-C. and Cornelius, P.-K. (eds.). 2002. *Environmental Performance Measurement: Global Report 2001–2002.* New York, Oxford University Press.

Finlayson, C-M. and Davidson, N.-C.. (collators). 1999. 'Global Review of Wetland Resources and Priorities for Wetland Inventory: Summary Report'. In: C.-M. Finlayson and A.-G. Spiers (eds.), *Global Review of Wetland Resources and Priorities for Wetland Inventory.* CD-ROM, Canberra, Australia Supervising Scientist Report 144.

Groombridge, B. and Jenkins, M.-D. 2002. *World Atlas of Biodiversity: Earth's Living Resources in the 21st Century.* Prepared at the United Nations Environment Programme-World Conservation Monitoring Centre, University of California Press.

Groombridge, B. and Jenkins, M. 1998. *Freshwater Biodiversity: A Preliminary Global Assessment.* Cambridge, United Nations Environment Programme-World Conservation Monitoring Centre, World Conservation Press.

Hilton-Taylor, C. (compiler). 2000. *2000 IUCN Red List of Threatened Species.* Gland and Cambridge, International Union for the Conservation of Nature and Natural Resources.

Interagency Workgroup on Wetland Restoration. N.d. *An Introduction and User's Guide to Wetland Restoration, Creation, and Enhancement.*

IUCN (International Union for the Conservation of Nature and Natural Resources). 2000. *Vision for Water and Nature. A World Strategy for Conservation and Sustainable Management of Water Resources in the 21st Century – Compilation of All Project Documents.* Cambridge.

Lillie, R.-A.; Garrison, P.; Dodson, S.-I., Bautz, R.-A.; LaLiberte, G. 2002. *Refinement and Expansion of Wetland Biological Indices for Wisconsin. Final Report to the United Nations Environment Protection Agency (US EPA) – Region 5.* Wisconsin Department of Natural Resources.

Loh, J.; Randers J.; MacGillivray, A.; Kapos, V.; Jenkins, M.; Groombridge, B.; Cox, N.; Warren, B. (ed.). 2000. *Living Planet Report 2000.* Gland, Switzerland, World Wide Fund International.

——. 1999. *Living Planet Report 1999.* Gland, World Wide Fund International.

——. 1998. *Living Planet Report 1998.* Gland, Switzerland, World Wide Fund International.

Luther, H. and Rzóska, J. 1971. *Project Aqua: A Source Book of Inland Waters Proposed for Conservation.* IBP Handbook No. 21. International Union for the Conservation of Nature and Natural Resources (IUCN) Occasional Paper No. 2. International Biological Programme. Oxford and Edinburgh, Blackwell Scientific Publications.

Master, L.-L.; Flack, S.-R. and Stein, B.-A. (eds.). 1998. *Rivers of Life: Critical Watersheds for Protecting Freshwater Biodiversity.* Arlington, The Nature Conservancy.

McAllister, D.-E.; Hamilton, A.-L.; Harvey, B.-H. 1997. 'Global Freshwater Biodiversity: Striving for the Integrity of Freshwater Ecosystems'. *Sea Wind*, Vol. 11, No. 3. Special issue (July–September 1997).

Moyle, P.-B. and Leidy, R.-A. 1992. 'Loss of Biodiversity in Aquatic Ecosystems: Evidence from Fish Faunas'. In: P.L. Fielder et al. (eds.), *Conservation Biology: The Theory and Practice of Nature Conservation Preservation and Management.* New York and London, Chapman and Hall.

New South Wales EPA (Environment Protection Agency). 2000. *State of the Environment 2000.* Sydney.

Nijland, H.-J. and Cals, M.-J.-R. (eds.). 2001. *River Restoration in Europe: Practical Approaches.* Proceedings from the Conference on River Restoration, Wageningen, The Netherlands, 2000. Report nr.2001.023. Lelystad, Institute for Inland Water Management and Waste Water Treatment.

OECD (Organization for Economic Cooperation and Development). 2001. *Environmental Outlook.*

Olson, D.-M. and Dinerstein, E. 1998. 'The Global 200: A Representation Approach to Conserving the Earth's Distinctive Ecoregions'. *Conservation Biology*, Vol. 12, pp. 502–15.

ONRC (Oregon Natural Resources Council). N.d. *Restoring the Lower Snake River: Saving Snake River Salmon and Saving Money.* Portland, Oregon.

Pirot, J-Y.; Meynell, P-J.; Elder, D. 2000. *Ecosystem Management: Lessons from Around the World. A Guide for Development and Conservation Practitioners.* Gland and Cambridge, International Union for the Conservation of Nature and Natural Resources.

Postel, S.-L.; Daily, G.-C.; Ehrlich, P.-R. 1996. 'Human Appropriation of Renewable Fresh Water'. *Science*, Vol. 271, pp. 785–8.

Revenga, C.; Brunner, J.; Henninger N., Kassem K.; Payne, R. 2000. *Pilot Analysis of Global Ecosystems: Freshwater Systems.* Washington DC, World Resources Institute.

Revenga, C.; Murray, S.; Abramovitz, J.; Hammond, A. 1998. *Watersheds of the World: Ecological Value and Vulnerability.* Washington DC, World Resources Institute and Worldwatch Institute.

Reynolds, C.-S. 1993. 'The Ecosystems Approach to Water Management: The Main Features of the Ecosystem Concept'. *Journal of Aquatic Ecosystem Health*, Vol. 2, pp. 3–8.

Richter, B.-D. et al. 1997. 'How Much Water Does a River Need?'.*Freshwater Biology*, Vol. 37, pp. 231–49.

RIKZ/RIZA (National Institute for Coastal and Marine Management /Inland Water Management and Waste Water Treatment). 1996. 'Future for Water'. In: *Report on the Dutch Aquatic Outlook.* The Hague, Ministry of Transport and Public Works.

Shiklomanov, I.-A. (ed.). 1997. *Comprehensive Assessment of the Freshwater Resources of the World.* Stockholm, Stockholm Environment Institute.

Soussan, J. and Harrison, R. 2000. Commitments on Water Security in the 21st Century: An Analysis of Pledges and Statements at the Ministerial Conference and World Water Forum, The Hague, March 2000.

Swanson, T.; Doble, C.; Olsen, N. 1999. *Freshwater Ecosystem Management and Economic Security.* Framework paper for IUCN workshop at World Water Council, Bangkok.

Tilman, D.; Fargione, J.; Wolff, B.; D'Antonio, C.; Dobson, A.; Howarth, R.; Schindler, D.; Schlesinger, W.-H.; Simberloff, D.; Swackhamer, D. 2001. 'Forecasting Agriculturally Driven Global Environmental Change'. *Science*, Vol. 292, pp. 281–4.

TNC (The Nature Conservancy). 1999. Evaluating Ecological Integrity at Freshwater Sites. Freshwater Initiative Workshop Proceedings.

UN (United Nations). 2000. United Nations Millennium Declaration. Resolution adopted by the General Assembly. A/RES/55/2.

UNEP (United Nations Environment Programme). 2002. *Global Environmental Outlook 3.* London and Stirling, Earthscan Publications Ltd.

UNEP/CBD (United Nations Environment Programme/Convention on Biological Diversity). 2000. *Progress Report on the Implementation of the Programmes of Work on the Biological Diversity of Inland Water Ecosystems, Marine and Coastal Biological Diversity, and Forest Biological Diversity (Implementation of Decisions iv/4, iv/5, iv/7).* Note by the executive secretary. UNEP/CBD/COP/5/10.

UNESCAP (United Nations Economic and Social Commission for Asia and the Pacific) and ADB (Asian Development Bank). 2000. *State of the Environment in Asia and the Pacific 2000.* New York, United Nations.

US EPA (United States Environment Protection Agency). 1998. *Water Quality Conditions in the United States: A Profile from the 1998 National Water Quality Inventory Report to Congress.* Washington DC.

Water Resources Environment Technology Centre. 2001. *Damu no Kankyo* (Dam Environment).

Wright, J.-F.; Sutcliffe, D.-W.; Furse, M.-T. (eds.). 2000. *Assessing the Biological Quality of Freshwaters – RIVPACS and Other Techniques.* Ambleside, Freshwater Biological Association.

WWF (World Wide Fund for Nature) European Freshwater Programme. 2001. *Water and Wetland Index: Assessment of 16 European Countries. Phase I Results.* Copenhagen.

Some Useful Web Sites[2]

United Nations Convention on Biological Diversity (UNCBD)

http://www.biodiv.org

Promotes nature and human well-being, focusing on the importance of biological diversity for the health of people and the planet.

United Nations Environment Programme (UNEP): Freshwater Portal

http://freshwater.unep.net/

Information on the key issues of the global water situation. A section dedicated to water and ecosystems.

United Nations Environment Programme (UNEP): Global Environment Monitoring (GEMS/WATER)

http://www.cciw.ca/gems/

A multifaceted water science programme oriented towards understanding freshwater quality issues throughout the world.

World Resources Institute (WRI): EarthTrends

http://earthtrends.wri.org/

Environmental Information Portal. Among the issues featured: coastal and marine ecosystems, water resources and freshwater ecosystems, human health and well-being, biodiversity and protected areas.

World Wide Fund for Nature (WWF): The Living Planet

http://www.panda.org/livingplanet/

The Living Planet Report is WWF's periodic update on the state of the world's ecosystems and the human pressures on them through the consumption of renewable natural resources.

2. The sites were last accessed on 19 December 2002.

‘A significant proportion of those living without proper water supply and sanitation are urban dwellers, mainly in the peri-urban areas. They are forced to draw on water sources that are unsafe, unreliable and often difficult to access. For sanitation, they have poor quality latrines – often shared with so many others that access and latrine cleanliness is difficult – or they have no provision for sanitation. Virtually all urban dwellers with inadequate provision live in the low- and middle-income nations of Africa, Asia and Latin America and the Caribbean.’

Cities: Competing Needs in an Urban Environment

By: UN-Habitat
Collaborative agencies:
WHO (World Health Organization)/
UNDESA (United Nations Department of Economic and Social Affairs)

Table of contents

This chapter draws from a comprehensive review of provision for water and sanitation in cities by UN-Habitat published by Earthscan Publications, London, in March 2003. UN-Habitat is particularly grateful to staff from WWDR, WHO and an external reviewer for their help in developing this chapter and in their comments and suggested additions to the first draft.

Cities: Competing Needs in an Urban Environment

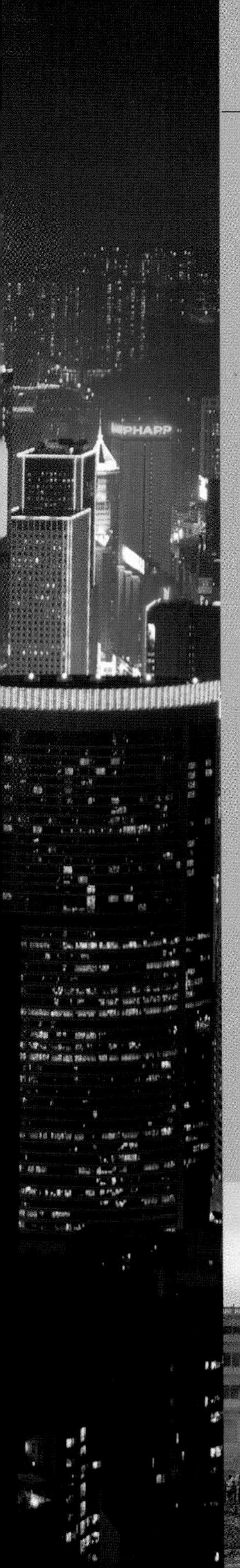

The cities of the world are concentric, isomorphic, synchronic. Only one exists and you are always in the same one. It's the effect of their permanent revolution, their intense circulation, their instantaneous magnetism.

J. Baudrillard, *Cool Memories*

CITIES SHARE MANY SIMILARITIES WITH NATURAL ECOSYSTEMS. They are dynamic, generating and recycling energy and waste, and responding to the interrelated cycles of needs, uses and demands of the institutions and people who live their lives within an urban environment. Cities are the future. The great migrations of people from rural to urban areas all over the world are proof of their power to attract and dazzle with their promise of a better life. Yet in this chapter we learn that half the urban population in Africa, Asia and Latin America suffers one or more diseases associated with inadequate water and sanitation. The World Health Organization (WHO) has recognized that when infrastructure and services are lacking, cities are among the planet's most threatening environments. Among the urban population of low-income countries, one child in six dies before the age of five. Gathering reliable data about water and sanitation in cities is extremely tricky, but indications are that targets are far from being met and enormous work remains to be done. Better management of resources in and around cities is therefore a major challenge the world over.

With over 60 percent of the world's population (nearly 5 billion people) expected to be living in urban areas by 2030 (compared with less than 15 percent in 1990 and 48 percent in 2002), cities are rising to the top of the policy agenda.

The Challenge of Water and Cities

One of the most significant urban changes has been the growth of cities to unprecedented sizes. The average size of the world's 100 largest cities grew from around 0.2 million in 1800 to 0.7 million in 1900 to 6.2 million in 2000 (Satterthwaite, 2002). By 2000 there were 388 cities with 1 million or more inhabitants (UN, 2002). Only in the late twentieth century did 'megacities' of 10 or more million inhabitants develop, with sixteen cities becoming 'megacities' in 2000, concentrating some 4 percent of the world's population (see table 7.2). Map 7.1 shows that the majority of these megacities lie within regions experiencing mild to severe water stress. Meeting the water needs of fast-growing cities can be extremely challenging.

Although rapid urban change is often viewed as an uncontrolled flood of people, there is an economic logic underpinning global urban trends. Most of the world's urban population and most of its largest cities are concentrated in the world's largest economies (Satterthwaite, 2002). In addition, the nations that urbanized most quickly over the last forty years are generally the countries with the greatest economic expansion (UN-Habitat, 1996). Table 7.1 highlights the fast urban growth in the less developed countries: while the most developed regions still have a much higher percentage of their population living in urban areas, the 2015 projection shows the beginning of a reversing trend, with half of the population of the less

Map 7.1: Water stress in regions around megacities

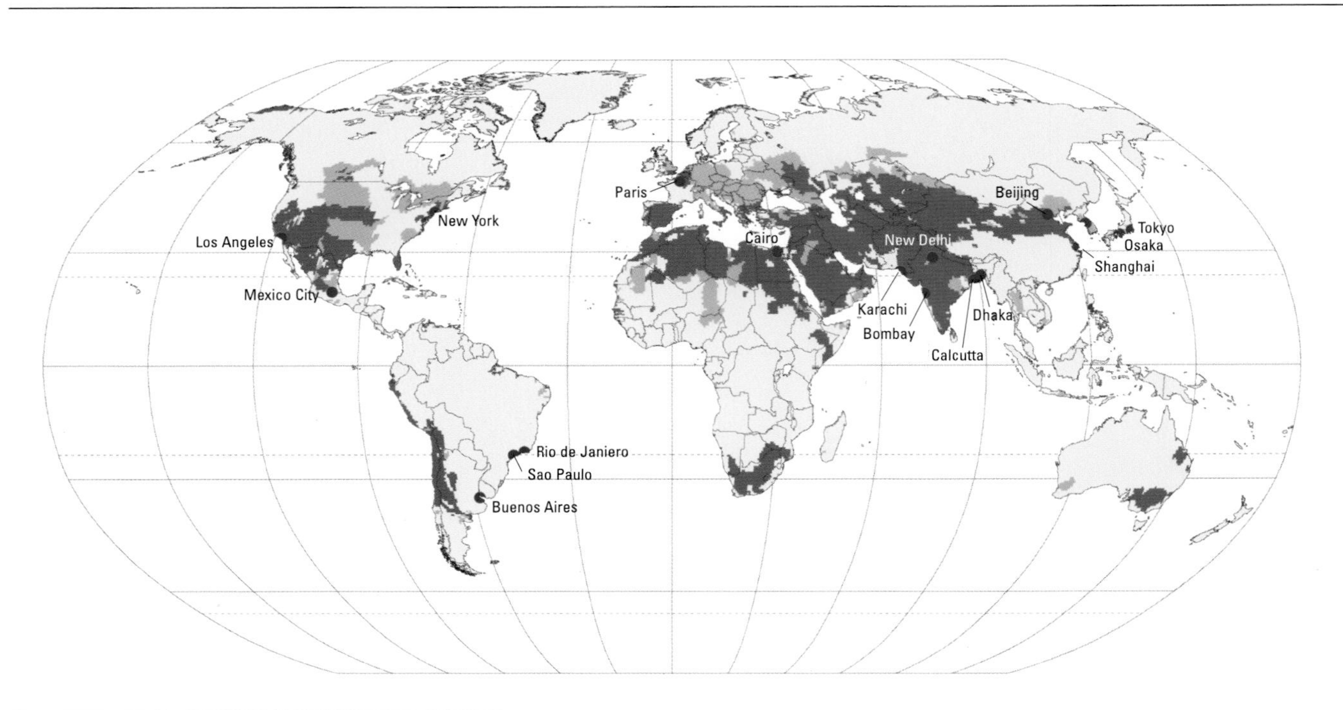

In 2000, the majority of the sixteen megacities were found along the coasts, within regions experiencing mild to severe water stress; this is particularly true for the cities located on the Asian continent. 'Water stress' is a measure of the amount of pressure put on water resources and aquatic ecosystems by the users of these resources, including the various municipalities, industries, power plants and agricultural users that line the world's rivers. The map uses a conventional measure of water stress, the ratio of total annual water withdrawals divided by the estimated total water availability. This map is based on estimated water withdrawals for 1995, and water availability during the 'climate normal' period (1961–1990).

Sources: Map prepared for the World Water Assessment Programme (WWAP) by the Centre for Environmental Research, University of Kassel, 2002. For the water stress calculation: data from WaterGAP Version 2.1.D; Cosgrove and Rijsberman, 2000; Raskin et al., 1997. For the megacities: UN, 2002.

Table 7.1: Distribution of urban population in more and less developed regions in 1975, 2000 and 2015

Development group	Area of residence and size class of urban settlement (number of inhabitants)	Population (millions)			Percentage distribution		
		1975	2000	2015	1975	2000	2015
World	Urban area	1,543	2,862	3,869	37.9	47.2	53.7
	Fewer than 500,000	844	1503	1950	20.8	24.8	27.1
	500,000 to 1 million	176	290	354	4.3	4.8	4.9
	1 million to 5 million	332	675	960	8.2	11.1	13.3
	5 million to 10 million	122	169	264	3.0	2.8	3.7
	10 million or more	68	225	340	1.7	3.7	4.7
	Rural area	2,523	3,195	3,338	62.1	52.8	46.3
	Total	**4,066**	**6,057**	**7,207**	**100.0**	**100.0**	**100.0**
More developed regions	Urban area	734	898	954	70.0	75.4	78.6
	Fewer than 500,000	422	498	522	40.3	41.8	43.0
	500,000 to 1 million	69	77	74	6.5	6.5	6.1
	1 million to 5 million	145	216	243	13.9	18.1	20.0
	5 million to 10 million	62	39	45	5.9	3.3	3.7
	10 million or more	36	67	71	3.4	5.7	5.8
	Rural area	314	294	259	30.0	24.6	21.4
	Total	**1,048**	**1,191**	**1,214**	**100.0**	**100.0**	**100.0**
Less developed regions	Urban area	809	1,964	2,915	26.8	40.4	48.6
	Fewer than 500,000	422	1,005	1,429	14.0	20.7	23.8
	500,000 to 1 million	108	213	280	3.6	4.4	4.7
	1 million to 5 million	186	458	718	6.2	9.4	12.0
	5 million to 10 million	60	130	218	2.0	2.7	3.6
	10 million or more	32	158	270	1.1	3.2	4.5
	Rural area	2,209	2,901	3,078	73.2	59.6	51.4
	Total	**3,017**	**4,865**	**5,994**	**100.0**	**100.0**	**100.0**

Fast urban growth is occurring in the less developed regions, reversing the trend that shows the more developed regions having the highest percentage of urban population. In this table, 75 percent of the urban population is foreseen to live in the less developed countries by 2015.

Source: UN, 2002.

developed regions living in urban areas, i.e. 75 percent of the world's urban population.

As centres of economic and social activity, cities provide a unique critical mass of highly productive skills and opportunities that drive development forward – but at a cost. Meeting competing demands from commercial, domestic and industrial users puts great pressures on freshwater resources. Many cities have to go ever deeper into groundwater sources and ever farther to distant surface water sources, at costs that are ultimately unsustainable in both economic and environmental terms. About 1.2 billion urban dwellers rely on groundwater and 1.8 billion on surface water sources. They are increasingly in competition with the rising demands for water from peri-urban agriculture and rural regions.

Many urban residents, and especially the poor, have only intermittent or no water supplies, and no sanitation. For the urban poor, this lack of access to safe water and basic sanitation causes widespread ill-health that further limits their productive capabilities. Ironically, the urban poor have often to buy their water from private vendors and pay far more per litre than their richer neighbours (see chapter 13 on valuing water).

Many urban water systems are poorly maintained, and it is not unusual for half the water to be lost in distribution. At the same time, revenue collection for much of the rest is poor, further restricting operation, maintenance and investment funds for expansion.

New ways of responding to rapid change and making the urban environment sustainable are being explored, especially through better management and service pricing, greater participation of community groups and women, and creative partnerships between public and private sector enterprises.

However, the success of these initiatives will be dependent on instituting better urban water governance, otherwise the degradation and depletion of freshwater resources will threaten the very livelihood of cities and the sustainability of economic and social development.

Table 7.2: The world's largest cities in 2000

Size	Urban centre	Country	Population (millions)
1	Tokyo	Japan	26.444
2	Mexico City	Mexico	18.066
3	Sao Paulo	Brazil	17.962
4	New York	United States of America	16.732
5	Mumbai (Bombay)	India	16.086
6	Los Angeles	United States of America	13.213
7	Calcutta	India	13.058
8	Shanghai	China	12.887
9	Dhaka	Bangladesh	12.519
10	Delhi	India	12.441
11	Buenos Aires	Argentina	12.024
12	Jakarta	Indonesia	11.018
13	Osaka	Japan	11.013
14	Beijing	China	10.839
15	Rio de Janeiro	Brazil	10.652
16	Karachi	Pakistan	10.032
17	Manila	Philippines	9.950
18	Seoul	Republic of Korea	9.888
19	Paris	France	9.630
20	Cairo	Egypt	9.462
21	Tianjin	China	9.156
22	Istanbul	Turkey	8.953
23	Lagos	Nigeria	8.665
24	Moscow	Russian Federation	8.367
25	London	United Kingdom	7.640
26	Lima	Peru	7.443
27	Bangkok	Thailand	7.372
28	Chicago	United States of America	6.989
29	Teheran	Iran (Islamic Republic of)	6.979
30	Hong Kong	China. Hong Kong SAR	6.860
31	Bogota	Colombia	6.771
32	Rhein-Ruhr North	Germany	6.531
33	Madras/Chennai	India	6.353
34	Bangalore	India	5.567
35	Santiago	Chile	5.467
36	Lahore	Pakistan	5.452
37	Hyderabad	India	5.445
38	Wuhan	China	5.169
39	Kinshasa	Dem. Rep of the Congo	5.054
40	Chongqing	China	4.900
41	Baghdad	Iraq	4.865
42	Shenyang	China	4.828
43	Toronto	Canada	4.752
44	Saint Petersburg	Russian Federation	4.635
45	Ho Chi Minh	Viet Nam	4.619
46	Riyadh	Saudi Arabia	4.549
47	Ahmedabad	India	4.427
48	Philadelphia	United States of America	4.427
49	Yangon	Myanmar	4.393
50	Milan	Italy	4.251
51	Belo Horizonte	Brazil	4.224
52	San Francisco-Oakland	United States of America	4.077
53	Singapore	Singapore	4.018
54	Madrid	Spain	3.976
55	Washington. D.C.	United States of America	3.952
56	Dallas-Fort Worth	United States of America	3.937
57	Sydney	Australia	3.907
58	Guangzhou	China	3.893
59	Lisbon	Portugal	3.861
60	Pusan	Republic of Korea	3.830
61	Detroit	United States of America	3.809
62	Abidjan	Côte d'Ivoire	3.790
63	Porto Alegre	Brazil	3.757
64	Hanoi	Viet Nam	3.751
65	Guadalajara	Mexico	3.697
66	Rhein-Main	Germany	3.681
67	Pune	India	3.655
68	Chittagong	Bangladesh	3.651
69	Alexandria	Egypt	3.506
70	Katowice	Poland	3.494
71	Montreal	Canada	3.480
72	Bandung	Indonesia	3.409
73	Houston	United States of America	3.386
74	Casablanca	Morocco	3.357
75	Recife	Brazil	3.346
76	Berlin	Germany	3.319
77	Chengdu	China	3.294
78	Monterrey	Mexico	3.267
79	Guatemala City	Guatemala	3.242
80	Salvador	Brazil	3.238
81	Rhein-Ruhr Middle	Germany	3.233
82	Melbourne	Australia	3.232
83	Jidda	Saudi Arabia	3.192
84	Nagoya	Japan	3.157
85	Ankara	Turkey	3.155
86	Caracas	Venezuela	3.153
87	Pyongyang	Dem. People's Republic of Korea	3.124
88	Xian	China	3.123
89	Athens	Greece	3.116
90	Changchun	China	3.093
91	Fortaleza	Brazil	3.066
92	Rhein-Ruhr South	Germany	3.050
93	Naples	Italy	3.012
94	San Diego	United States of America	3.002
95	Johannesburg	South Africa	2.950
96	Boston	United States of America	2.934
97	Capetown	South Africa	2.930
98	Harbin	China	2.928
99	Inch'on	Republic of Korea	2.884
100	Medellin	Colombia	2.866
101	Algiers	Algeria	2.761
102	Kitakyushu	Japan	2.750
103	Khartoum	Sudan	2.742
104	Nanjing	China	2.740
105	Barcelona	Spain	2.729
106	Atlanta	United States of America	2.706

Table 7.2: *continued*

Size	Urban centre	Country	Population (millions)
107	Surat	India	2.699
108	Luanda	Angola	2.697
109	Taegu	Republic of Korea	2.675
110	Zibo	China	2.675
111	Stuttgart	Germany	2.672
112	Hamburg	Germany	2.664
113	Rome	Italy	2.649
114	Addis Ababa	Ethiopia	2.645
115	Kanpur	India	2.641
116	Dalian	China	2.628
117	Phoenix	United States of America	2.623
118	Kabul	Afghanistan	2.602
119	Jinan	China	2.568
120	Santo Domingo	Dominican Republic	2.563
121	Curitiba	Brazil	2.562
122	Taipei	China	2.550
123	Guiyang	China	2.533
124	Kiev	Ukraine	2.499
125	Linyi	China	2.498
126	Surabaja	Indonesia	2.461
127	Taiyuan	China	2.415
128	Durban	South Africa	2.391
129	Minneapolis-St Paul	United States of America	2.378
130	Qingdao	China	2.316
131	Munich	Germany	2.291
132	Warsaw	Poland	2.274
133	Birmingham	United Kingdom	2.272
134	Jaipur	India	2.259
135	Havana	Cuba	2.256
136	Manchester	United Kingdom	2.252
137	Cali	Colombia	2.233
138	Nairobi	Kenya	2.233
139	Aleppo	Syrian Arab Republic	2.229
140	Miami-Hialeah	United States of America	2.224
141	Lucknow	India	2.221
142	Izmir	Turkey	2.214
143	Tashkent	Uzbekistan	2.148
144	Damascus	Syrian Arab Republic	2.144
145	Faisalabad	Pakistan	2.142
146	Guayaquil	Ecuador	2.118
147	Dar es Salaam	United Rep. of Tanzania	2.115
148	Seattle	United States of America	2.097
149	Nagpur	India	2.089
150	St Louis	United States of America	2.084
151	Dakar	Senegal	2.078
152	Beirut	Lebanon	2.070
153	Zhengzhou	China	2.070
154	Vienna	Austria	2.065
155	Tampa-St. Petersburg-Clearwater	United States of America	2.064
156	Baltimore	United States of America	2.053
157	Vancouver	Canada	2.049
158	Zaozhuang	China	2.048
159	Liupanshui	China	2.023
160	Brasilia	Brazil	2.016
161	Bucharest	Romania	2.001
162	Tel-Aviv-Yafo	Israel	2.001
163	Handan	China	1.996
164	Mashhad	Iran (Islamic Republic of)	1.990
165	Norfolk-Virginia Beach-Newport News	United States of America	1.963
166	Baku	Azerbaijan	1.948
167	Oporto	Portugal	1.940
168	Maracaibo	Venezuela	1.901
169	Campinas	Brazil	1.895
170	Valencia	Venezuela	1.893
171	Tunis	Tunisia	1.892
172	Puebla de Zaragoza	Mexico	1.888
173	Medan	Indonesia	1.879
174	Accra	Ghana	1.868
175	Kyoto	Japan	1.849
176	Jinxi	China	1.821
177	Budapest	Hungary	1.819
178	Liuan	China	1.818
179	Sapporo	Japan	1.813
180	Harare	Zimbabwe	1.791
181	Hangzhou	China	1.780
182	Tianmen	China	1.779
183	Changsha	China	1.775
184	Port-au-Prince	Haiti	1.769
185	Wanxian	China	1.759
186	Cleveland	United States of America	1.735
187	Pittsburgh	United States of America	1.735
188	Tripoli	Libyan Arab Jamahiriya	1.733
189	Lanzhou	China	1.730
190	Nanchang	China	1.722
191	Kunming	China	1.701
192	Riverside-San Bernardino	United States of America	1.699
193	Denver	United States of America	1.698
194	Barranquilla	Colombia	1.683
195	Yantai	China	1.681
196	Hai Phong	Viet Nam	1.676
197	Belgrade	Yugoslavia	1.673
198	Tangshan	China	1.671
199	Minsk	Belarus	1.667
200	Belem	Brazil	1.658
201	Patna	India	1.658
202	Lusaka	Zambia	1.653
203	Douala	Cameroon	1.642
204	Xuzhou	China	1.636
205	San Jose	United States of America	1.635
206	Brisbane	Australia	1.622
207	Quito	Ecuador	1.616
208	Rabat	Morocco	1.616
209	Xiantao	China	1.614

Table 7.2: *continued*

Size	Urban centre	Country	Population (millions)
210	Stockholm	Sweden	1.612
211	Rhein-Neckar	Germany	1.605
212	Antananarivo	Madagascar	1.603
213	Shijiazhuang	China	1.603
214	Heze	China	1.600
215	Indore	India	1.597
216	Pretoria	South Africa	1.590
217	Yancheng	China	1.562
218	Yulin	China	1.558
219	Xinghua	China	1.556
220	East Rand	South Africa	1.552
221	Ibadan	Nigeria	1.549
222	Taejon	Republic of Korea	1.522
223	Rawalpindi	Pakistan	1.521
224	Taian	China	1.503
225	Pingxiang	China	1.502
226	Fort Lauderdale-Hollywood-Pampano Beach	United States of America	1.471
227	Manaus	Brazil	1.467
228	Vadodara	India	1.465
229	Kaohsiung	China	1.463
230	Kansas City	United States of America	1.460
231	La Paz	Bolivia	1.460
232	Toluca	Mexico	1.455
233	Anshan	China	1.453
234	Luoyang	China	1.451
235	Khulna	Bangladesh	1.442
236	Jilin	China	1.435
237	Qiqihar	China	1.435
238	Leeds	United Kingdom	1.433
239	Suining	China	1.428
240	Bhopal	India	1.425
241	Palembang	Indonesia	1.422
242	Coimbatore	India	1.420
243	Yaounde	Cameroon	1.420
245	Kharkov	Ukraine	1.416
246	Wulumuqi	China	1.415
247	Fushun	China	1.413
248	Sacramento	United States of America	1.408
249	Yerevan	Armenia	1.407
250	Tbilisi	Georgia	1.406
251	Fuzhou	China	1.397
252	Neijiang	China	1.393
253	San Juan	Puerto Rico	1.388
254	Esfahan	Iran (Islamic Republic of)	1.381
255	Kuala Lumpur	Malaysia	1.379
256	Kwangju	Republic of Korea	1.379
257	Changde	China	1.374
258	Cordoba	Argentina	1.368
259	Ludhiana	India	1.368
260	Zhanjiang	China	1.368
261	Huainan	China	1.354
262	Lyon	France	1.353
263	Songnam	Republic of Korea	1.353
264	Yiyang	China	1.343
265	San Salvador	El Salvador	1.341
266	Kochi (Cochin)	India	1.340
267	Ulsan	Republic of Korea	1.340
268	Mecca	Saudi Arabia	1.335
269	Copenhagen	Denmark	1.332
270	Nizhni Novgorod	Russian Federation	1.332
271	Perth	Australia	1.329
272	Portland-Vancouver	United States of America	1.328
273	Sana	Yemen	1.327
274	Gujranwala	Pakistan	1.325
275	Xintai	China	1.325
276	Montevideo	Uruguay	1.324
277	Cincinnati	United States of America	1.323
278	Novosibirsk	Russian Federation	1.321
279	Baotou	China	1.319
280	Dongguan	China	1.319
281	San Antonio	United States of America	1.318
282	Nanning	China	1.311
283	Visakhapatnam	India	1.309
284	Brazzaville	Congo	1.306
285	Tijuana	Mexico	1.297
286	Bielefeld	Germany	1.294
287	Turin	Italy	1.294
288	Agra	India	1.293
289	Leon de los Aldamas	Mexico	1.293
290	Marseille	France	1.290
291	Weifang	China	1.287
292	Milwaukee	United States of America	1.285
293	Hanover	Germany	1.283
294	Rosario	Argentina	1.279
295	Tabriz	Iran (Islamic Republic of)	1.274
296	Santos	Brazil	1.270
297	Wenzhou	China	1.269
298	Multan	Pakistan	1.263
299	Asuncion	Paraguay	1.262
300	Hefei	China	1.242
301	Ciudad Juarez	Mexico	1.239
302	Conakry	Guinea	1.232
303	Huaian	China	1.232
304	Orlando	United States of America	1.226
305	Hyderabad	Pakistan	1.221
306	Ekaterinburg	Russian Federation	1.218
307	Kampala	Uganda	1.213
308	Yueyang	China	1.213
309	Prague	Czech Republic	1.203
310	Varanasi	India	1.199
311	Nuremberg	Germany	1.189
312	Suqian	China	1.189
313	Madurai	India	1.187
314	Sofia	Bulgaria	1.187

Table 7.2: *continued*

Size	Urban centre	Country	Population (millions)
315	Tianshui	China	1.187
316	Suzhou	China	1.183
317	Shantou	China	1.176
318	Omsk	Russian Federation	1.174
319	Ningbo	China	1.173
320	Panama City	Panama	1.173
321	Yuzhou	China	1.173
322	Bursa	Turkey	1.166
323	Datong	China	1.165
324	Mogadishu	Somalia	1.157
325	Jingmen	China	1.153
326	Amman	Jordan	1.148
327	Davao	Philippines	1.146
328	Meerut	India	1.143
329	West Palm Beach-Boca Raton-Delray Beach	United States of America	1.143
330	Leshan	China	1.137
331	Brussels	Belgium	1.135
332	Samara	Russian Federation	1.132
333	Mosul	Iraq	1.131
334	Shenzhen	China	1.131
335	Almaty	Kazakhstan	1.130
336	Wuxi	China	1.127
337	Shiraz	Iran (Islamic Republic of)	1.124
338	Xiaoshan	China	1.124
339	Zaoyang	China	1.121
340	Goiania	Brazil	1.117
341	Nashik	India	1.117
342	Bamako	Mali	1.114
343	Yixing	China	1.108
344	Amsterdam	Netherlands	1.105
345	Auckland	New Zealand	1.102
346	Ufa	Russian Federation	1.102
347	Jabalpur	India	1.100
348	Maracay	Venezuela	1.100
349	Yongzhou	China	1.097
350	Maputo	Mozambique	1.094
351	Adana	Turkey	1.091
352	Chifeng	China	1.087
353	Jamshedpur	India	1.081
354	Ottawa	Canada	1.081
355	New Orleans	United States of America	1.079
356	Rotterdam	Netherlands	1.078
357	Huzhou	China	1.077
358	Daqing	China	1.076
359	Zigong	China	1.072
360	Phnom-Penh	Cambodia	1.070
361	Dnepropetrovsk	Ukraine	1.069
362	Columbus	United States of America	1.067
363	Zagreb	Croatia	1.067
364	Peshawar	Pakistan	1.066
365	Asansol	India	1.065
366	Mianyang	China	1.065
367	Adelaide	Australia	1.064
368	Kazan	Russian Federation	1.063
369	Santa Cruz	Bolivia	1.062
370	Aachen	Germany	1.060
371	Nanchong	China	1.055
372	Lodz	Poland	1.053
373	Ujung Pandang	Indonesia	1.051
374	Dhanbad	India	1.046
375	Chelyabinsk	Russian Federation	1.045
376	Karaj	Iran (Islamic Republic of)	1.044
377	Allahabad	India	1.035
378	Rajshahi	Bangladesh	1.035
379	Fuyu	China	1.025
380	Nampo	Dem. People's Republic of Korea	1.022
381	Jining	China	1.019
382	Faridabad	India	1.018
383	Rostov-On-Don	Russian Federation	1.012
384	Torreon	Mexico	1.012
385	Managua	Nicaragua	1.009
386	Indianapolis	United States of America	1.008
387	Donetsk	Ukraine	1.007
388	Port Elizabeth	South Africa	1.006
389	Volgograd	Russian Federation	1.000

Source: UN, 2002.

Cities rarely develop with careful water and wastewater management and the governance structures these require. Most cities that today enjoy good water systems have histories of mismanagement and crisis, and only developed their water supply and wastewater management systems in response to the water problems generated by growth and accelerated by industrial development.

The availability of water has always been an important influence on where cities grow. Although many other factors have or once had importance for underpinning urban development in what are today large urban centres, cities grow because enterprises choose to concentrate there and people respond to the resulting concentration of jobs or other economic opportunities by moving there. City populations also grow through natural increase.

Most cities initially expanded with little government attention given to ensuring continued water availability and wastewater management. The process by which water was obtained and wastewater disposed of was (and often still is) a chaotic and poorly managed process, with each user seeking the cheapest source of water. For instance, it is common for users and even water utilities to draw on groundwater, with neither regulation nor coordination.

They also seek the cheapest and most convenient means to dispose of wastewater, which often results in their contaminating water sources for their 'downstream' neighbours or for other groundwater users. Even where reticulated water supply and sanitation systems are operational, maintenance may be the last item on the budget and therefore prone to resource cuts. Sewage leakage and pressure irregularities in the water supply system, resulting from insufficient maintenance, eventually create substantial contamination risks.

Good urban water management is complex, and requires not only water and wastewater infrastructure but also pollution control (especially from industries), sustainable use of water sources, wastewater management and flood prevention. In addition, it requires coordination across many sectors and, usually, between different local authorities as most cities' water supplies and wastewater are not limited to water catchments within their boundaries.

The importance of ensuring good water quality for water systems used for recreation and of limiting ecological damage to the water systems that receive wastewater, storm and surface runoff has added considerably to authorities' tasks. From a financial perspective, however, management decisions often have a cascading effect: the installation of a new or improved sewage system can imply investment in an increased capacity of treatment plants.

All these tasks require governance structures that provide a sound legal, institutional and financial basis. In cities with rapidly growing populations or those with weak economies and limited possibilities of raising funds for water management, these structures must also be adapted to the particular difficulties facing local authorities.

Water, Sanitation and Hygiene: 'Improved' vs. 'Adequate' Provision

Progress in providing water and sanitation services used to be expressed in coverage figures originating from national authorities and service providers. The *Global Water Supply and Sanitation 2000 Assessment Report* (WHO/UNICEF, 2000) made a significant step forward by defining access to improved water sources and sanitation in terms of probability of safety, and by basing itself on surveys and censuses among users. In line with the intention of the WHO/UNICEF (United Nations Children's Fund) Joint Monitoring Programme of Water Supply and Sanitation (JMP), it is argued here that the methodology should be further developed so that progress can be assessed by measuring the numbers of people with access to 'safe and sufficient' water and 'safe and convenient' sanitation that meets basic welfare and hygiene needs. In this context, adequate water provision would refer to the supply of water that meets drinking water quality standards (i.e. water that can be safely drunk and used for cooking), in sufficient quantity to allow for washing and other aspects of personal hygiene and domestic cleanliness.

All city dwellers have access to water in some way since no one can live without it. The issue is not whether they have access to water but whether the supplies are safe, sufficient for their needs and easily accessed at a price they can afford. Similarly, for sanitation, all city dwellers have to make some provision for defecation, even if this is defecating on wasteland. The issue is not whether they have provision but whether they have a quality of provision that eliminates their (and others') contact with human excreta and wastewater by making available toilets that are convenient, clean, easily accessed and affordable by all. Meeting these basic needs and thus reducing the burden of disease related to their insufficiency should be the driving force, and raising the health status of vulnerable groups the primary goal, of public action.

Water and sanitation provision standards

Over time, many cities have acquired governance structures that have greatly improved water management (including all the national or provincial laws, institutions and financial systems to support this). In cities of high-income nations, it is taken for granted that each home or business has a twenty-four-hours-a-day piped water supply that can be used for drinking, bathing and other domestic purposes, as well as hygienic, easily cleaned toilets available to all. Yet it was just over 100 years ago that provision of access to safe, piped water supplies, sanitation for all city dwellers and governance structures to ensure that these operations were carried out, began to be accepted, initially in Europe and North America, as a key part to any city's water management (Mumford, 1991).

This acceptance now seems universal. In 1976, at the United Nations (UN) Conference on Human Settlements (Habitat), 132 governments formally committed themselves to a recommendation stating that 'safe water supply and hygienic disposal should receive priority with a view to achieving measurable qualitative and quantitative targets serving all the population by a certain date' (UN, 1976). In 1977, at the UN Water Conference at Mar del Plata, governments agreed that national plans should aim to provide safe drinking water and basic sanitation to all by 1990 if possible, that is within the International Drinking Water Supply and Sanitation Decade (IDWSSD), (UN, 1977). In 1990, at the World Summit for Children, the many assembled governments made a commitment to achieving universal access to safe water and adequate sanitation by the year 2000. However, these targets were not met, and hundreds of millions of urban dwellers still suffer from very poor or non-existent provision for water and sanitation. There is also the worry that the targets set within the Millennium Development Goals, which imply improved provision for water, sanitation and drainage – i.e. to have achieved, by 2020, a significant improvement in the lives of at least 100 million slum dwellers – will achieve no more than previous commitments made by governments and international agencies.

There is, however, an add-on value of setting such goals on a relatively short-term basis and within a minimalist view of development: in retrospect, without the IDWSSD goals, the status of water and sanitation in the world would have been even worse.

In high-income nations, the need for all urban households to have safe, regular piped water supplies to their home, internal plumbing and their own sanitary toilet is unquestioned. These can be set as the standard for which universal provision is possible. But in lower-income nations, where universal provision to such a standard is not possible, other standards have to be set. From a public health perspective, it is better to provide a whole city's population with safe water supplies to taps within 50 metres of their home than to provide only the richest 20 percent of households with water piped to their home. The *Global Water Supply and Sanitation 2000 Assessment Report* suggests that 'reasonable access' for water should be broadly defined as 'the availability of at least 20 litres per person per day from a source within one kilometre of the user's dwelling'. For most urban settings, distance alone does not provide an appropriate standard; population density is a critical modulator. While in a sparsely populated rural area, a one-kilometre distance may indeed provide reasonable access, this is most unlikely to be the case for a high-density squatter settlement with a population of 100,000. Furthermore, standards for water provision should also consider the regularity of the supply alongside such issues as water quality and price. In recognition of the fact that assessments at all levels of access to water and sanitation need to account for multiple criteria, the WHO/UNICEF JMP is constantly refining existing criteria and developing and testing new ones. Future assessments should be increasingly sensitive and specific, and yield results that will facilitate informed decision-making.

Unfortunately, there are few detailed datasets available on the quality of provision for water and sanitation in most of the world's cities. Most of the data upon which national or global surveys of provision rely are drawn from censuses or household surveys that do not ask most of the critical questions regarding the adequacy of provision. In addition, the criteria used in most such censuses or household surveys for assessing adequacy do not address important differences between rural and urban contexts. Indeed, health risks from using toilets in many urban settings may be more linked to the number of people sharing each toilet than to the kind of toilet used (Benneh et al., 1993), but data are often not collected on the extent of toilet sharing. This explains why global assessments of water and sanitation provision for the world's urban (and rural) populations are not able to measure the proportion of people with access to 'safe and sufficient' water and 'safe and convenient' sanitation, but only the proportion with access to 'improved' sources of water and 'improved' sanitation. Reviewing the quality of water and sanitation provision in the world's cities shows a continuum between those with very high quality to those with little or none. Basically, there are three levels of provision.

- In high-income countries, there is more or less universal provision of advanced water and sanitation facilities and other city water services, most of which are provided by public sector utilities, although increasing use is being made of private sector provision. The main challenges revolve around prevention of microbial and chemical contamination of water distribution systems, optimizing the efficiency of utility operation both in economic and in ecological terms, dealing with issues of asset renewal and management of residuals from water treatment, and ensuring that the impact of effluents from wastewater discharges on receiving waters remains within acceptable levels.

- In middle-income countries, a great deal of water and sanitation infrastructure exists but it is often in poor condition. The service delivery systems are frequently underfunded, poorly managed and in a poor state of maintenance with high levels of water leakage, and inadequate wastewater treatment. Here the most pressing issues are usually related to improving efficiency, infrastructure maintenance, renewal and extension, pricing and revenue collection, and more effective supervision and enforcement of regulations on industrial pollutants. Governments are beginning to take steps to address these issues, especially in larger cities (see box 7.1).

- Lower-income countries have particularly difficult problems. They have less water and sanitation infrastructure than high- and middle-income countries. Likewise, their institutions and management systems are generally underdeveloped, and their overall capacity to deliver a reasonable water and sanitation service is very low. Big cities generally have some water and sanitation infrastructure in their central areas, and this is, in many cases, being improved and expanded by the introduction of private concessionaires or improved public utility operations. However, in many peri-urban areas of large cities and in most smaller urban centres, water and sanitation infrastructure is very limited, and there are problems with industrial pollution – often difficult to control as they stem from many small-scale operations. The overall result is widespread microbial and chemical pollution of water sources in and around the cities.

In each of these three situations, cities face the challenges of developing integrated urban water management systems appropriate to their needs and capacities, but nested within the system of

Box 7.1: Singapore Public Utilities Board: reducing Unaccounted-for-Water

With no rivers or lakes to tap for freshwater, Singapore relies on rainfall collected in its fourteen reservoirs and on imports from Malaysia; in the future it will increasingly depend on desalination. Because of this dependence on outside sources, the Public Utilities Board has made an effort in recent years to improve efficiency and reduce waste, especially in the area of Unaccounted-for-Water (UfW).

UfW is the difference between the water delivered to the distribution system and the water sold. It has two basic components: physical losses, such as water lost from pipes and overflows from tanks, and commercial losses, which include water used but not paid for.

Through a consistent monitoring programme, Singapore has achieved a UfW rate of an impressive 6 percent. Singapore's UfW reduction programme is based on metering, audits of commercial water use and leak detection. The metering programme aims to achieve universal metering with high accuracy on all meters by replacing domestic meters every seven years and industrial meters every four years, and by measuring and billing for water used for firefighting. Singapore has implemented programmes to identify consumption patterns and notify customers of excessive consumption, identify inconsistent meter readings, replace faulty meters and set average water rates close to the marginal cost of water to encourage careful inspection and evaluation of unaccounted-for and inefficient use of water. There are also substantial efforts to detect and stop leaks: the systems are tested annually and the surface pipes quarterly, and these are replaced if more than three breaks a year are reported. As a result of this programme, the number of pipe breaks has decreased from twelve per 100 km/year in 1985 to less than four per 100 km/year in 1992 (Yepes, 1995).

Although the annual cost of these investments over a thirty-year period works out to US$5.8 million, the programme has proven to be cost-effective when considering the current and long-term marginal cost of water to Singapore.

Source: Prepared for the World Water Assessment Programme (WWAP) by V. Srinivasan, P.-H. Gleick and C. Hunt at the Pacific Institute, 2002.

Integrated Water Resources Management (IWRM) of the basin in which each city lies.

The health impacts of inadequate provision

Urban areas – life-threatening environments

There are several common-sense reasons why urban areas should be better served than rural areas. The first is that urban areas provide significant economies of scale and proximity for the delivery of piped water and provision for good-quality sanitation and drainage, so unit costs are lower. The second is that many cities have a more prosperous economic base than rural areas, providing higher average incomes for large sections of the population and greater possibilities for governments or private utilities to raise revenues for such provision.

The third reason is that urban areas concentrate not only people and enterprises but also their wastes, and as WHO has recognized, when infrastructure and services are lacking, urban areas are among the world's most life-threatening environments (WHO, 1999a). The deficiencies in water and sanitation mean high disease burdens for much of the population. In addition, in many cities, insufficiencies in provision for drainage and flood protection expose large sections of the population to high levels of risk from flooding and the consequent spread of disease.

Human contamination from faecal-oral pathogens

The risks of human contamination from faecal-oral diseases vary with different levels of provision for water, sanitation and hygiene (see figure 5.1 in chapter 5). But one difficulty consists in assessing the provision for water and sanitation, since even ignoring the variation at each level, it is unclear where to draw the line between 'adequate' and 'inadequate' provision. What is clear, however, is that there are quantum health gains from improved service levels of water supply and sanitation. The transition from a situation of no services at all to one with the most basic of services allows the most substantial health gains, followed by those gained from extending the services to the individual household. The best health system in cities is water piped into everyone's home with internal plumbing feeding into bathrooms and kitchens, and with toilets, baths, showers and sinks connected to sewers. This is an accepted standard for cities in high-income nations and may be achieved or achievable in some cities in middle-income nations (see for instance levels of provision in Porto Alegre in Brazil, described in Menegat, 2002). However, in any city in which large sections of the population have very inadequate provision for water and sanitation and limited resources, it is inappropriate and not feasible. In this case, the priority should be to ensure that everyone has 'improved' provision, although higher standards can be provided to areas of the city where the inhabitants are willing and able to pay the full cost of so doing. This suggests the need for assessments of who has access to different levels of provision in each settlement – i.e. detailed data for each household within each settlement. This information would then allow the authorities to focus on ensuring 'improved' provision to everyone and supporting better quality provision wherever possible.

Child mortality rates

In cities served by piped water, sanitation, drainage, waste removal and a good healthcare system, child mortality rates are generally around 10 per 1,000 live births and few child deaths are the result of water-related diseases (or other environmental hazards). In cities or neighbourhoods with inadequate provision, it is common for the rate to be ten to twenty times this. This is the case in many low-income countries, some of whose child mortality rates have increased in recent years. Many middle-income nations still have urban child mortality rates of 50 to 100 per 1,000 live births (Montgomery, 2002).

These are average figures for entire urban populations and as such, they obscure the higher rates within the lower-income settlements. In a well managed city, the difference in mortality rates for children between the lowest and highest income areas is not very large; in a badly managed city they can vary by a factor of 10, 20 or more. Surveys in seven settlements in Karachi (Pakistan) found that infant mortality rates varied from 33 to 209 per 1,000 live births (Hasan, 1999).

Diarrhoeal diseases

Diarrhoeal diseases are still a primary cause of infant and child death for large sections of the world's urban population. When provision for water and sanitation is poor, diarrhoeal diseases and other diseases linked to contaminated water (such as typhoid) or contaminated food and water (such as cholera and hepatitis A) are among the most serious health problems within urban populations. Case studies in low-income settlements have shown the high proportion of children who have debilitating intestinal worms (Bradley et al., 1991). The prevalence of various skin and eye infections such as scabies and trachoma associated with a lack of water supplies for washing is particularly high among children living in poor-quality homes and neighbourhoods (see for instance Landwehr et al., 1998). Chapter 5 on health gives more details on water-related diseases.

Unsanitary environments contribute also to malnutrition by challenging children's immune systems. Long-term impacts for children are not restricted to health; research in poor urban settlements in Brazil has related early childhood diarrhoeas to impaired cognitive functioning several years later (Guerrant et al., 1999).

Factors contributing to the inadequacies in provision for water and sanitation exist at every level, from the most local to the international, as illustrated by figure 7.1.

Figure 7.1: Examples of causes for the prevalence of diarrhoeal diseases in a squatter settlement

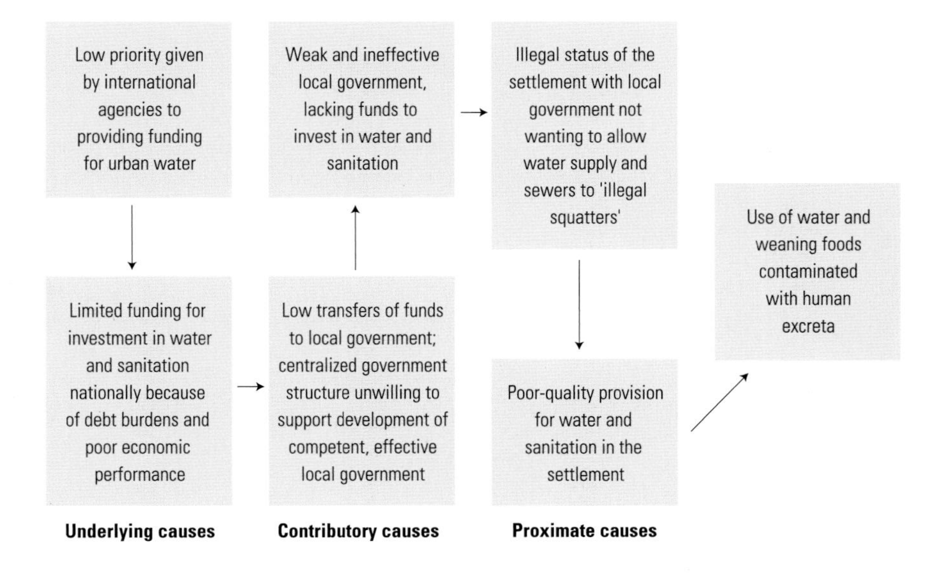

Drawing on the particular example of the causes for the prevalence of diarrhoeal diseases, this figure shows that many factors contribute to the inadequacies for water and sanitation, from the international to the local level.

Source: UNICEF, 2003.

Other health problems

Poor water management in cities contributes to many other health problems. Malaria, often considered a rural disease, is now among the main causes of illness and death in many urban areas. In south Asia it is related to drinking water storage on rooftops (so-called overhead tanks), to which the *Anopheles stephensi* mosquito has adapted its breeding habits; in Africa and Latin America it is more often associated with poorly drained locations where the mosquitoes breed in clear standing water (WHO, 1999a). In some cities, a gradient has been observed where malaria transmission declines towards the city centre, with pollution of open water being the key determinant. *Aedes* mosquitoes that transmit a number of viral diseases, including dengue fever, dengue haemorrhagic fever and yellow fever, breed in small water collections and containers. These are related to poor drainage or to solid waste (car tyre dumps are notorious) and also include small domestic water collections. The latter may be the result of inadequate or intermittent water supplies, which force people to keep drinking water containers in their homes (Cairncross and Feachem, 1993).

The vectors of lymphatic filariasis breed in organically polluted waters, including open sewage canals. Many other disease vectors thrive where there is poor drainage and inadequate provision for rubbish collection, sanitation and piped water – these include house flies, fleas, lice and cockroaches (Satterthwaite et al., 1996).

Water provision

Table 7.3 shows the proportion of the urban population in each of the world's regions with access to 'improved' water sources and improved sanitation in 2000. For this assessment, certain types of technology were designated as 'improved' based on the likelihood that the water and sanitation provided were safe. WHO and UNICEF are carrying out pilot studies in a number of countries in an attempt to move from a qualitative designation of access to improved sources, to a quantitative method of assessment.

Individual city studies and the data collected by the *2000 Assessment* for large cities suggest that if it were possible to widen the assessment to measure the proportion with access to safe, sufficient supplies, the number of urban dwellers inadequately served would be much higher.

Safe and sufficient provision

The quality and level of detail in the *2000 Assessment* was a considerable advance on earlier assessments since it recognized that

Table 7.3: The proportion of urban populations with access to 'improved' water supply and sanitation

Region	Urban population (millions)	% with 'improved' provision	Number of people unserved (millions)	% of people unserved
Global				
Urban water supply	2,862	95	157	5.5
Urban sanitation		83	481	16.8
Africa				
Urban water supply	295	86	40	13.6
Urban sanitation		80	59	20
Asia				
Urban water supply	1,376	93	90	6.5
Urban sanitation		74	360	26
Latin America and the Caribbean				
Urban water supply	391	94	24	6.1
Urban sanitation		86	55	14
Oceania				
Urban water supply	23	98	0.4	1.7
Urban sanitation		99	0.2	0.9
Europe				
Urban water supply	534	100	2.4	0.4
Urban sanitation		99	6.1	1.15
North America				
Urban water supply	243	100	0	0
Urban sanitation		100	0	0

According to the *Global Water Supply and Sanitation 2000 Assessment*, 'improved' water supply and sanitation designate safe water and sanitation, and mostly rely on quality. This table explicitly shows that most of the unserved populations are in Asia and Africa.

Source: WHO and UNICEF Joint Monitoring Programme, 2002. Updated in September 2002.

'the definition of safe or improved water supply and sanitation facilities sometimes differed not only from one country to another, but also for a given country over time.' It also noted that some of the data from individual countries 'often showed rapid and implausible changes in level of coverage from one assessment to the next'. Details of coverage were inevitably limited by the restricted range of questions asked about water and sanitation within these surveys (see chapter 5 on water and health).

Studies of provision for water and sanitation in particular cities reveal a gap between the proportion of people with 'improved' supplies and the proportion that could be considered to have 'safe, sufficient provision'. For instance, 99 percent of the urban population of Bangladesh had access to 'improved' water supply in 2000 (WHO/UNICEF, 2000). Studies drawn from individual cities within Bangladesh, however, show that the population proportion with 'safe and sufficient' provision is much less (see box 7.2). Similarly, in India, 92 percent of the urban population has 'improved' water supply, yet descriptions of water provision in many city case studies suggest that a much smaller proportion has access to safe, sufficient provision (see, for instance, Hardoy et al., 2001; Schenck, 2001; Anand, 1999; TARU Leading Edge, 1998; Ghosh et al., 1994). In Pakistan, 96 percent of the urban population may have 'improved' water supply but descriptions of conditions in Karachi and Faisalabad suggest that a much lower proportion has safe, sufficient provision (Hasan, 1999). These and many other case studies of large cities show that a high proportion of all inhabitants have very inadequate provision. Analyses drawn from the demographic and health surveys also suggest that provision for water in these same nations is worse in smaller urban centres than in the larger cities (Hewett and Montgomery, 2001).

This same gap between the proportion of urban dwellers with 'improved' provision and 'safe, sufficient provision' is evident in many African nations (see box 7.3). For Kenya, 87 percent of its urban population had improved water supplies by 2000, yet detailed studies in Kenya's two largest cities, Nairobi and Mombasa, show a much smaller proportion with safe, sufficient provision (Hardoy et al., 2001). This is also evident in many other sub-Saharan African nations, including the United Republic of Tanzania and Nigeria (UN-Habitat, forthcoming).

The *2000 Assessment* also collected statistics on water supplies for large cities. For Africa, this covered forty-three cities and showed that 31 percent of the population was unserved, with only

Box 7.2: Deficiencies in provision for water and sanitation in cities in Bangladesh and Pakistan

In Dhaka, Bangladesh, if 'access to water' is defined as the availability of water within 100 metres, the proportion counted as having access would be much lower than the 99 percent given in official statistics (Islam et al., 1997). In 2002, the head of Dhaka's water and sewerage authority estimated that there were 2.5 million people in Dhaka's 'slums', most with very inadequate provision for water and sanitation. 70 percent of the population have no sewers. Tens of thousands of children die each year because of water-borne diseases and polluted water (Vidal, 2002). A survey by the International Centre for Diarrhoeal Disease Research found that for half the population in 'slum' areas, it takes more than thirty minutes each time they collect water (UNICEF, 1997). In addition, 42 percent of those living in 'slums' used a pit or open latrine, while 2 percent had no fixed arrangement and 2.7 percent used an open field (Islam et al., 1997). In Chittagong, about a quarter of the population of 1.6 million is served by individual house connections, 200,000 by 588 street hydrants, and the rest collect water from other sources such as natural springs, canals, ponds and rainwater catchments (Kaneez Hasna, 1995). A 1993 survey found that nearly three quarters of the metropolitan 'slum' population relied on bucket or pit latrines (Bangladesh Bureau of Statistics, 1996).

More than half of Karachi's (Pakistan) 12 million inhabitants lives in katchi abadis, squatter camps made of mud or timber huts. The most recent figures available (for 1989) show that only half the katchi abadis have piped water and only 12 percent have provision for sanitation, compared to the planned areas where more than four fifths have piped water and sanitation. For the whole city, only 40 percent of the population is connected to the official sewer system (Hasan, 1999). In Faisalabad, some two thirds of the city's 2 million inhabitants live in largely unserviced areas; over half have no piped water supply and less than a third have sewers (Alimuddin et al., 2000).

Box 7.3: Inadequacies in provision for water and sanitation in cities in Kenya and the United Republic of Tanzania

A report in 1994 stated that 55 percent of Nairobi's (Kenya) population lived in informal settlements that are squeezed onto less than 6 percent of the city's land area. Only 12 percent of plots in these settlements have piped supplies. Most people have to obtain water from kiosks. A survey found that 80 percent of households complained of water shortages and pipes running dry. This same survey suggested that 94 percent of the inhabitants of informal settlements do not have access to adequate sanitation. Only a minority of dwellings have toilets. Significant proportions of the total population have no access to showers and baths, and in most areas drainage is inadequate (Alder, 1995). Kibera is the largest low-income urban area in Nairobi, covering an area of 225 hectares with an estimated population of 470,000. Traditional pit latrines are the only excreta-disposal system available, and a high proportion of households have no toilet within or close to their home. There are often up to 200 persons per pit latrine. Pits fill up quickly and emptying is a problem due to difficult access. Space to dig new pits is often not available (WHO, 1996). For the whole of Nairobi, only 40 percent of households are connected to sewers (Mosha, 2000).

In the United Republic of Tanzania, more than 60 percent of communities in Dar es Salaam live in areas with minimal or no infrastructure such as water supply, sanitation and drainage (Mazwile, 2000). Only a small proportion of the population of the country's largest cities have sewage connections (Shayo Temu, 2000). 83 percent of households in Dar es Salaam use pit latrines among which 10 percent have septic tanks and 6 to 7 percent have sewers; the sewage network covers only the central part of the city and a small section outside the centre. The system is old and unreliable, owing to deferred maintenance (Mosha, 2000). Many Tanzanian cities had water available for only a few hours a day on average, ranging from Dodoma (7 hours) to Singida (2 hours), (Njau, 2000).

43 percent having a house connection or yard tap and 21 percent relying on public taps (see figure 7.2). If the definition of 'adequate provision for water' were to be set as a house connection or yard tap, then more than half the population in these cities has inadequate provision (Hewett and Montgomery, 2001).

Sanitation provision

Safe and convenient sanitation

Detailed city studies show that a large proportion of the population that has 'improved' sanitation does not have safe, convenient sanitation. The cities listed in boxes 7.2 and 7.3 are among the largest in each of their countries. However, provision for sanitation is worse in smaller urban centres than in the larger cities so the gap between those with 'improved' sanitation and 'safe, convenient' sanitation may be even larger here. Many city studies in other African nations also suggest that the proportion of people with safe, convenient sanitation is much lower than the proportion with 'improved' sanitation (Hardoy et al., 2001).

The *2000 Assessment*'s statistics for sanitation for forty-three of Africa's large cities showed that 19 percent of the population remains unserved (see figure 7.2). Among these populations, only 18 percent have toilets connected to sewers, a very low proportion as confirmed by an analysis of the Demographic and Health Surveys suggesting that a mere 25 percent of Africa's urban population has access to toilets connected to sewers (Hewett and Montgomery, 2001). This conclusion is also supported by statistics on the proportion of households with sewer connections in the largest city in each African nation. In most of these cities, less than 10 percent of the population has sewer connections while in many, including Abidjan (Côte d'Ivoire), Addis Ababa (Ethiopia), Asmara (Eritrea), Brazzaville (Congo), Cotonou (Benin), Kinshasa (Congo), Libreville (Gabon), Moroni (Comoros), N'Djamena (Republic Of Chad), Ouagadougou (Burkina Faso), and less than 2 percent have connections (WHO/UNICEF, 2000).

Toilet use

In most urban settings, shared toilets and pit latrines are not adequate to fulfil the primary health function of a toilet – to ensure the safe disposal of human excreta so it does not contaminate hands, clothes, water or food and is inaccessible to flies and other disease vectors. If the only available toilets are shared, they will not

Figure 7.2: The proportion of households in major cities connected to piped water and sewers

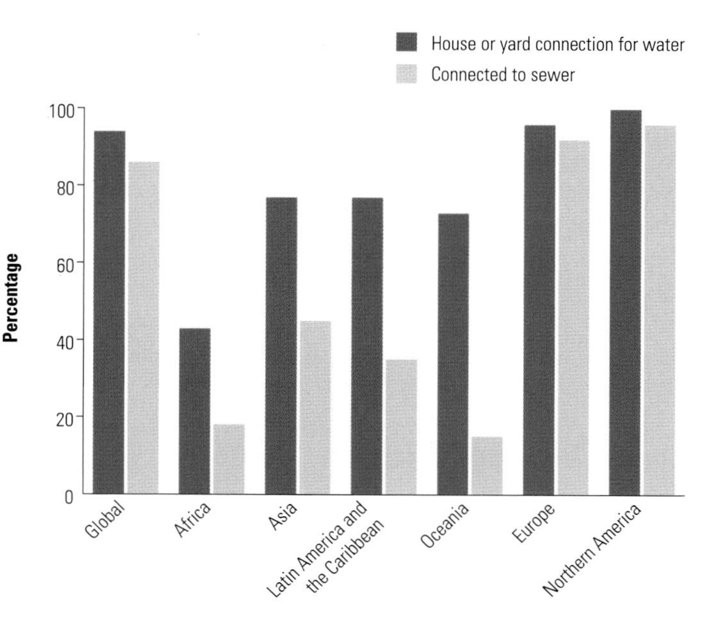

These figures are based on information provided by 116 cities. In no region was there a representative sample of large cities, although the figures for each region are likely to be indicative of average levels of provision for major cities in that region. If adequate provision for sanitation in large cities is taken to mean a toilet connected to a sewer, then these figures indicate there is a significant lack of adequate provision in cities throughout Africa, Asia, Latin America, the Caribbean and Oceania.

Source: WHO/UNICEF, 2000.

be used if they are too distant or not easily available; it is also difficult to ensure that shared toilets are kept clean.

In urban areas with a high concentration of low-income households it is common for each toilet to be shared by dozens of people or for the only toilets available to be public toilets with hundreds of people to each toilet. Tens of millions of households in informal settlements in Africa and Asia have access only to overused and poorly maintained communal or public toilets. It is also rare for official statistics on communal or public toilets to assess whether the toilets are in use or maintained. In Mumbai (India), as in many other places, official estimates of toilet availability overlook the fact that many toilets are in fact unusable; in one community, 80 percent of municipally provided seats were found to be unusable (NSDF et al., 1997).

The use of public toilets is particularly problematic for young children. Taking a small child any distance to a toilet or having to queue to use it is impractical. Women and adolescent girls may be reluctant to use public toilets because these facilities offer little privacy and the lack of security increases the risk of sexual abuse or violence. Public latrines can also be expensive: in Kumasi (Ghana) the use of public toilets once a day by each family member can use up 10 to 15 percent of the main income-earner's wages (Korboe et al., 2000). In many Indian cities as well, low-income households cannot afford to use public toilets.

Provision for sanitation is so poor in many cities that significant proportions of their population resort to open defecation or to defecation in some waste material (such as waste paper or a plastic bag) – what is termed 'wrap and throw' in Cebu (Philippines) or 'flying toilets' in Accra (Ghana), (Etemadi, 2000). Studies of many individual cities, including Addis Ababa, Bangalore (India), Colombo (Sri Lanka), Dhaka (Bangladesh), Kingston (Jamaica), and Ouagadougou, have found open defecation to be a serious problem (Hardoy et al., 2001).

Public health issues

The absence of provision for drainage and for the collection of household wastes within low-income settlements also contributes to the likelihood of faecal contamination and the large disease burden this brings. Most informal settlements have no service to collect solid waste. In many African cities, only 10 to 30 percent of all urban households' solid wastes are collected, and services are inevitably most deficient for informal settlements (Hardoy et al., 2001). Uncollected garbage, along with excreta, is often disposed of in drainage ditches, which can become quickly clogged. Also, in overcrowded city settings where space is particularly problematic, it is difficult to ensure regular and safe sludge removal from latrines or septic tanks. When wastewater and stormwater cannot be easily drained, flooding spreads waste and excreta through the surrounding area. Standing water can also be contaminated by blocked sewers and overflowing septic tanks, and pathogens are then quickly disseminated. Organically polluted water also becomes a breeding place for certain disease vectors, i.e. the *Culex quinquefasciatus* mosquito. Drainage is an especially serious concern for the many urban communities on steep or swampy land (Cairncross and Ouano, 1990).

If adequate provision for sanitation in large cities is taken to mean a toilet connected to a sewer, then figure 7.2 indicates that there is a significant lack of adequate provision in cities throughout Africa, Asia, Latin America and the Caribbean. While this overstates the case, since a large proportion of the population of many large cities is adequately served, it does highlight how the scale of the 'sanitation gap' is very dependent on the criteria used with respect to what is said to be 'adequate'.

Hygiene behaviour

There is an apparent contradiction between the increased stress given to the role of hygiene behaviour in reducing the incidence of many water-related diseases, and the criteria used to define what is adequate for water and sanitation. Although it is recognized that the role of good hygiene really increases significantly once water is piped to the home or the yard, many governments still regard water available within 100 or 200 metres as adequate.

Similarly, good hygiene depends on young children not defecating in the open, yet pit latrines are considered as 'improved sanitation' even when most children will not use them. A survey conducted by UNICEF's India office found that only 1 percent of children under six use latrines, that the stools of an additional 5 percent are thrown into latrines, and that the remainder end up in drains, streets or yards, furthering the likelihood of contamination (UNICEF, 2000).

Hygiene improvements for young children depend heavily on the quality of childcare. Care providers face great difficulties where provision for water and sanitation is inadequate since managing water supplies, keeping children clean and safe, dealing with waste and excreta in the absence of adequate services and handling food and utensils hygienically are very difficult and often compete with many other tasks (Bartlett, 2002).

Urban Development and Water Management

City-region issues[1]

Environmental impacts

Cities transform the environments and landscapes around them at considerable distances, and can have environmental impacts into large areas usually defined as rural. Environmental impacts on this wider region, both positive and negative, arise from:

- the expansion of the built-up area and the transformations this brings – for instance as land surfaces are reshaped, valleys and swamps fill, large volumes of mineral resources are extracted and moved, water sources are tapped and rivers and streams channelled;
- the demand from city-based enterprises, households and institutions for the products of the region around the city; and
- the solid, liquid and air-borne wastes generated within the city and transferred to the region around it.

1. This section draws on chapter 5 of Hardoy et al., 2001.

Water supplies, wastewater and drainage

Cities require a very large input of freshwater (and other natural resources), and the more populous the city and the richer its inhabitants, the greater the demand on resources and, in general, the larger the area from which these are drawn. Water demand is also critically influenced by the nature of the city's economic base.

Wastewater is then returned to rivers, lakes or the sea at a lower quality than that originally supplied. Storm and surface runoff also collects large pollution loads as it flows through cities, especially where there is inadequate provision for solid waste collection as much of the uncollected waste generally finds its way into water bodies, adding to the pollution. There are also the problems of pollution surges as urban stormwater can be up to ten times more polluted than dry-weather wastewater owing to wash-off from roads and open spaces, pollution gathered during percolation through the soil and resuspension of deposits accumulated in sewers and drains.

Uncontrolled expansion

In most low- and middle-income nations, in the absence of any effective land use plan or other means to guide and control new developments, cities generally expand haphazardly. Uncontrolled physical growth impacts most on the immediate surrounding area. Within this area, agriculture may disappear or decline as land is bought up by people or companies in anticipation of its urbanization and its resulting increases in land value, as the city's built-up area and transport system expands. There is usually a lack of effective public control of such changes or on the profits that can be made from them, even when public investment (for instance the expansion of road networks) creates much of the increment in land value. Around prosperous cities, expansion is also encouraged by the scale of profits that can be made – and it is difficult to develop governance structures that prevent powerful vested interests from being the prime beneficiaries (Kelly, 1998).

There are usually illegal squatter communities, too, who originally located to the surrounding area because the inaccessibility, lack of infrastructure and poor quality of the site reduced the chance of eviction. In many cities, including Buenos Aires (Argentina), Delhi (India), Manila (Philippines), Santiago (Chile) and Seoul (Republic of Korea), the surrounding area also contains settlements formed when their inhabitants were being evicted from their homes by 'slum' or squatter clearance. Low-income groups are often segregated in the worst located and often the most dangerous areas for which it is difficult and expensive to provide water, sanitation and drainage. The haphazard expansion of settlements generally results in greatly increased costs for providing basic infrastructure to developments that need connection to existing water, sewage and drainage networks some distance away. These costs are also augmented by the layouts of many illegal settlements and other site characteristics such as unclear plot boundaries and unstable soils (Roche et al., 2001). However, this is not always the case and some illegal settlements or illegal subdivisions develop with careful attention to site layouts that allow for future infrastructure installation. In addition, the agencies or companies whose responsibility it is to provide piped water and provision for sanitation and drainage may be reluctant to install the needed infrastructure in illegal settlements or may be prohibited from doing so.

Other water problems related to uncontrolled city expansion include:

- difficulties in protecting water sources as new urban developments occur within watersheds and surface water sources become polluted;
- damage to drainage systems, such as land clearance and deforestation, which greatly increases silt loads that clog drainage channels, thus increasing the risks of flooding (Vlachos and Braga, 2001);
- increases in wastewater, storm and surface runoff flows with the expansion of impermeable surfaces and extraction, use and disposal of available water sources, which often bring increased risks of flooding and reduced infiltration and recharge of aquifers (Ellis, 1999);
- new possibilities for disease vectors, as the expansion of the built-up area, water reservoirs and drains, together with land clearance and deforestation can drastically change the local ecology (natural foci for disease vectors may become entrapped within the suburban extension, and new ecological niches for the animal reservoirs may be created; within urban conurbations, disease vectors may adapt to new habitats and introduce new infections among the urban population [see chapter 5 for more details]); and
- the impact on peri-urban agriculture, as most major cities grew within fertile agricultural areas, with the result that expanding cities often cover agricultural land while urban water demand can pre-empt water previously used by farmers.

Water inputs and outputs

Urban water and sanitation benefits and costs

It is difficult to draw up any 'balance sheet' regarding the benefits and costs of urban-based demand for water and other rural resources. This section concentrates on the environmental costs but urban-based demands for rural resources is also an important (and in fact often the most important) basis for rural incomes and livelihoods, as the rural incomes produced by urban demand form the basis for prosperous, well-managed farms, fisheries and forests. In discussing cities' environmental impacts there is a tendency to forget the key role of cities in providing lower-income nations with stronger, more robust economies.

Water resource availability

Many urban centres face difficulties in obtaining sufficient freshwater and this is even the case in cities where half or more of the population is not adequately served with safe, sufficient supplies. Many cities have outgrown their capacity to provide adequate water supplies – as all nearby surface water sources have been tapped and/or as groundwater resources are being drawn on much faster than the natural rate of recharge. Acute problems may occur during periods of low rainfall. Problems of water scarcity are particularly serious in the many urban centres of relatively arid areas.

Surface water sources on which cities draw are often of poor quality – for instance, saline because of return water from irrigation, contaminated with agricultural chemical and human and livestock wastes, or heavily polluted by urban areas, industries or other upstream users. Groundwater resources are also often contaminated from wastewater discharges or from overuse resulting in saltwater intrusion (coastal aquifers). Consequently, cities face problems in financing the expansion of supplies to keep up with demand, as the cheapest and most easily tapped water sources have been polluted and drawing on newer sources implies much higher costs per unit volume of water (Bartone et al., 1994).

Historically, Latin American cities have relied on their closest sources of water which in most cases were freshwater streams, lakes or springs, but many large cities such as Bogota (Colombia), Lima (Peru) and Mexico City (Mexico) have had to invest heavily in the construction of dams and pipelines to supplement local water supplies with water from more remote sources (Anton, 1993). By the 1990s, among sixty-seven urban centres in twenty-nine nations in Africa (including most of the continent's largest cities), 58 percent were using rivers 25 or more kilometres away, and just over half of the urban centres that relied on rivers depended on interbasin transfers. Comparisons between the water sources used in thirty-eight of these urban centres in the 1970s and the 1990s found an increasing number relying on interbasin transfers (Showers, 2002), even though provision for water in most of them is still very inadequate.

Nations with very limited rainfall such as Algeria have had to make very large investments in dams and interbasin transfers. In 1962, Algeria had ten dams supplying 480 million cubic metres (Mm^3) a year, by 2002, it had forty-eight dams supplying nearly 2.8 billion m^3 a year. Despite this, most cities in Algeria suffer problems of inadequate and irregular supplies, in part linked to much-below-average rainfall in recent years.

Contamination from wastewater discharges

The contamination of rivers, lakes, seashores and coastal waters is an example both of the impact of city-generated wastes on the wider region and of governments' negligence in controlling pollution and managing surface and wastewater flows. This often leads to serious health problems for large numbers of people whose water supply is drawn from these water sources. In cities on or close to coasts, untreated sewage and industrial effluents often flow into the sea with little or no provision to pipe them far enough out to sea to protect the beaches and inshore waters, thereby posing a major health risk to bathers.

The possibilities for improvement vary greatly. In Europe and North America, great improvements have been made in reducing water pollution, mostly through stricter controls on industrial discharges and more sophisticated and comprehensive treatment of sewage and stormwater. In most cities in low- and middle-income nations, the problems are not so easily addressed as they have much more serious non-point sources of water pollution because of the lack of sewers and drains in many city districts and peripheral areas. Additionally, many uncollected wastes are washed into streams, rivers and lakes.

Liquid wastes from city activities often have far-reaching environmental impacts. It is common for river and coastal fisheries to be damaged or destroyed by liquid effluents from city-based industries with significant numbers of people losing their livelihood as a result. River pollution of this sort can lead to serious health problems in settlements downstream (see for instance CSE, 1999), and the source may become unusable for agriculture downstream.

Cities and water-related natural disasters

The most common form of disaster associated with water is flooding, which is generally linked to storms or periods of unusually high rainfall and, for coastal cities, exceptionally high tides and storm surges (for more information on risk management, see chapter 11).

However, while most floods arise from natural causes, most of the deaths, injuries and loss of property they cause in cities are the result of inadequate attention given to flood warning, flood preparation and post-disaster response. The ways in which cities develop can exacerbate risk from floods – or lessen them. In well-governed cities, flooding risks can be minimized through not only provision and maintenance of drains but also good watershed management (to limit the volume of flood waters and the speed with which they build up) and land use control (for instance to limit the exposure of soils during construction of new developments). Many cities also use recreational water bodies, parks and other open spaces to provide flood storage. Among high-income cities at risk of flooding, it is now common for detailed, sophisticated flood-risk maps to have been developed. These maps make each local authority aware of the risk within their jurisdiction and help identify optimum investment patterns to limit risk (see the Greater Tokyo case study, chapter 22). However, as noted earlier, in poorly governed cities, large population concentrations frequently develop in locations that are known to be at high risk of flooding. To the hazards inherent in the site are added those linked to a lack of investment in infrastructure and services, especially drainage, and little or no watershed management. It is also common for such settlements to combine high densities, shelters built with flammable materials and widespread use of open fires and kerosene stoves/lamps which mean high risks of accidental fires with, usually, little provision for effective firefighting and emergency treatment for those who are burnt.

However, even with careful management, urbanization can increase the consequences of floods. For instance, in Japan, rapid urbanization over the last fifty years has concentrated more people in flood-prone areas; today close to half the population and three quarters of the property value are located within the flood-prone areas of rivers. While flood mitigation measures have decreased the occurrence of major rivers overflowing their banks and of embankment failures, reducing both the severity and the area of flood damage, the increased concentration of people and high-value property in floodplains have increased the costs of flood damage. Meanwhile, river basins that are being rapidly urbanized are losing their natural water-retaining and water-retarding functions – as can be seen again in the Tokyo case study.

Cities' ecological footprints

The extent of the environmental changes caused by any city on its surroundings and the size of the affected area is much influenced by the city's size and wealth, as well as the nature of its production base and of the resource endowments of the region around it. It is also significantly influenced by the quality of environmental management both within the city and in the surrounding region.

Although much of the literature on the generation and transfer of environmental costs from cities concentrates on the surrounding region, the demands made by larger and wealthier cities for food, fuel and raw materials may be increasingly met by imports from distant ecosystems. This makes it easier to maintain high environmental standards in the region and also to preserve forests and water bodies. In addition, the goods whose fabrication involves high levels of water use and dirty industrial processes and discharges can be imported. This capacity of wealthy cities to draw on the productivity of 'distant elsewheres' led to the concept of cities' ecological footprints, which seek to calculate the land area on whose production the inhabitants of any city depend for food, other renewable resources and the absorption of carbon to compensate for the carbon dioxide emitted from fossil-fuel use (Rees, 1992).

A more effective approach to water management

It is difficult to generalize about more effective approaches to water management when considering all the world's cities. Clearly, there is a very urgent need to improve and extend provision for water, sanitation and hygiene in cities in low- and middle-income nations – but in small towns and rural areas, as well. This includes the need for financially sound, operationally efficient, consumer-oriented water and sanitation agencies, regardless of whether these are public, private (commercial or non-profit) or community-based. For most cities, there is a need for water governance systems that improve watershed management, lessen the ecological disruption caused by water withdrawals and wastewater returns and that make better use of existing water resources. One interesting example is provided by New York City (see box 7.4). Most cities also need to invest in disaster avoidance, as well as disaster-preparedness. Achieving all of this generally includes a need for coordinated action across different administrative boundaries.

Good governance in urban areas

Perhaps the only generalization concerning improved water management that is valid across all the world's cities is the need for good local governance, as the most appropriate solutions are always site-specific. The root of the problem in low- and most middle-income nations is that, generally, governance structures have not developed to efficiently and equitably address these problems and resolve the inevitable trade-offs. Good water governance implies not only frameworks to ensure provision but also regulations (to protect water sources and to protect and promote health) and revenue-raising (to pay for the system's functioning, maintenance and expansion). This involves more than just effective government institutions but also more efficient relationships between government and civil society (McCarney, 1996). Better water governance means that all stakeholders' water needs are considered and that the institutions responsible for water and wastewater management are accountable to them.

Good water management in cities also means setting limits on where industries can locate and developers can build, as well as on what local water sources they can tap and what wastes they can dispose of. A basin-wide perspective is required. This is difficult to achieve since political or administrative boundaries were never set to serve water basin management. In most large cities, there are many different political divisions within the water basin, with local governments controlled by different political parties and politicians refusing to collaborate with their neighbours to ensure an ecologically sound and fair regional water management system. Another contributing factor in some nations is the retention of powers and revenues that are needed for local governance.

In low- and many middle-income nations, there are serious difficulties in raising the funds for the major investments needed. Large and fast-growing cities face particularly serious problems given a large backlog of households and businesses in need of better provision and the continuing rapid growth of the population and economic base. But even here, there are many examples of local innovation showing how good-quality water provision and management is financially feasible in low-income cities.

Inadequate city governance generally has two aspects: local government institutions that are weak, unaccountable to citizens and underfunded (including water and sanitation utilities with little or no investment capacity), and higher levels of government that are unwilling to allow local institutions sufficient resources and revenue-raising powers. Of course, the problem is compounded by the low incomes of hundreds of millions of urban dwellers – yet this is not in and of itself sufficient explanation for the inadequacies. It is common for low-income groups to be paying two to fifty times more per litre of water than higher-income groups because low-income groups have to purchase from vendors while higher-income groups are being undercharged for water piped to their homes (Hardoy et al., 2001; World Bank, 1988). In addition, there are examples of low-income settlements with good-quality provision for water and sanitation and full cost recovery from user charges – or with users paying enough for the overall costs to be affordable to existing local authorities. For the underserved urban population, there may be insufficient funding for conventional utilities to extend piped water and sewer connections to each home but it is possible to greatly improve provision through community-municipal partnerships or support for community provision. Similarly, where provision for water and sanitation has been privatized, effective demand in lower-income areas of cities is rarely sufficient to motivate profit-seeking companies to extend good quality provision for water and sanitation (although it can often support water supply alone). Nonetheless, there are often intermediate solutions that combine public, private and community provision in ways that greatly improve provision and can recover most or all costs.

The inadequacies of urban governance in low- and middle-income nations have also meant that many issues other than those directly

Box 7.4: New York City – maximizing public participation while protecting water quality

New York City (NYC) has one of the largest public water supply systems in the world, providing water to over 8 million city residents. The water is under the public jurisdiction of the city's Department of Environment. However, as 73 percent of the watershed was privately owned, the NYC Department of Environmental Protection (DEP) developed a strategy to integrate private landowners into the planning process. As part of this strategy, the city works with farmers to foster understanding of how their behaviour affects downstream water quality and to provide incentives for implementing protection programmes (DEP, 2001). DEP's current success in implementing watershed control has relied extensively on this kind of public and private involvement in management decisions.

NYC has developed an innovative set of economic alternatives for protecting water quality for one of the world's largest public water systems. The DEP's programme includes the following six measures.

- Watershed Agricultural Programme: NYC funds the development and implementation of best management practices for the watersheds' farmland, including the design, implementation and subsidizing of Whole Farm Plans that address environmental issues on a farm while upholding farm business.

- Land Acquisition: of the total 480,000 hectares of watershed land, NYC has identified approximately 101,250 as especially vital to protecting future water quality. As of 2001, NYC has purchased more than 3,650 hectares of this land.

- Watershed Regulations: a set of stricter standards passed in 1997 regulates development and projects in the watershed.

- Environmental and Economic Partnership Programmes: NYC has funded partnership programmes to encourage cooperation among watershed stakeholders.

- Wastewater Treatment Plant (WWTP) Upgrades: thirty-four of the watershed's total fifty-seven WWTPs are privately owned and operated. All have agreed to DEP-approved upgrades. NYC has also upgraded the city-owned plants.

- Protection of Kensico Reservoir: both the Catskill and Delaware aqueducts drain into the Kensico Reservoir, which funnels 90 percent of NYC's water supply. The DEP is aiming to reduce the local non-point source pollution in this reservoir.

Source: Prepared for the World Water Assessment Programme (WWAP) by V. Srinivasan, P.-H. Gleick and C. Hunt at the Pacific Institute, 2002.

related to water and sanitation provision have been poorly managed. Existing freshwater resources remain unprotected and are often continuously degraded or depleted. Surface water sources are often polluted; very few cities in Africa, Asia and Latin America have rivers flowing through them that are not heavily contaminated, and much the same applies to nearby lakes, estuaries and seas. Watersheds are often degraded because of ineffective controls on industrial and urban developments. Although most nations have the environmental legislation in place to limit water pollution, it is rarely enforced (Hardoy et al., 2001). It is also common for urban expansion to take place over ecologically important areas such as wetlands and mangroves. Meanwhile, for many large cities, powerful industrial and commercial interests, allied to the higher-income groups that have piped water, can appropriate freshwater resources from other watersheds, often drawing on them from large distances with negative consequences for the ecology and the water users in these areas.

Although the need for additional international funding is often stressed, without improved local governance, additional resources may bring few benefits to low-income groups and little improvement in overall water management.

Effective monitoring is critical to support an ongoing process of improving governance and management. The WHO/UNICEF JMP has made important quality advances, but is limited by what a global assessment can achieve. Part of its merit is to assist member states in the building of local monitoring capacity. There is a need for complementary, locally-driven assessments that serve local service providers and reveal the equity issues at the local level.

There are many innovations that show more effective approaches, from sophisticated basin-wide water governance systems that incorporate all stakeholders (see the Seine-Normandy case study, chapter 19) to simple innovations in a particular squatter settlement that cheapened water costs while greatly improving access. Some are described below. Their relevance is more in the application of 'good local governance' than in their ability to be transferred to other cities. The community toilets built in Pune (India), (see box 7.5) are obviously not the most appropriate solution for cities where it is possible to provide good-quality sanitation to each house or apartment. But what does have wider relevance is the way in which representative organizations of the urban poor and the local non-governmental organizations (NGOs) that work with them were able to work with local authorities in developing major improvements in sanitation that both the users and the government could afford.

Various institutional or financial innovations can lessen the gap between what low-income households can afford and good-quality provision. For instance, the work of the Pakistan NGO Orangi Pilot Project has shown how its technical support to community organizations can produce sewers where unit costs are so reduced that even low-income households can afford them. This is no longer a 'pilot project' but a twenty-year programme through which hundreds of thousands of low-income groups have obtained good-quality sanitation in Karachi and in other cities in Pakistan (Hasan, 2001), (see box 7.6). This example has also demonstrated an alternative way of expanding provision – as the agencies responsible for water and sanitation provide the citywide water mains, trunk sewers and storm drains into which community organizations manage the installation of the piped service to each household.

For cities and smaller urban centres within low-income nations with less possibility of good governance, perhaps the most that can be hoped for in the more deprived neighbourhoods is regular, reasonably priced, protected water supplies available to each home or yard where possible, through enough well-maintained public standpipes to ensure that there are not excessive queues and long distances to carry water. In addition, one can hope for either household or community investment in good-quality toilets or (where densities are high and many people rent) good-quality, accessible public toilets (as in box 7.5). If the most appropriate local solution for improved sanitation is pit latrines, this in turn requires better provision for emptying them; box 7.7 gives an example of a technology to support this. It is also possible to install one of the lower-cost technologies initially and upgrade it as the households' capacity to pay for and desire more convenient systems increases (Kalbermatten et al., 1980).

Box 7.5: Community toilets in Pune and other Indian cities

Two fifths of Pune's 2.8 million inhabitants live in over 500 slums. Although various local government bodies are meant to provide and maintain public toilets in these settlements, provision is insufficient. The quality of toilet construction was often poor and the design inappropriate, with limited water supplies and no access to drainage. The toilets frequently went uncleaned and fell into disuse, the space around them used for open defecation and garbage dumping.

In 1999, Pune's Municipal Commissioner sought to improve the situation by inviting NGOs to make bids for toilet construction and maintenance. One NGO, the Society for the Promotion of Area Resource Centres (SPARC), had a long partnership with the National Slum Dwellers Federation and Mahila Milan and became a principal contractor. This alliance designed and costed the project, the city provided the capital costs and the communities developed the capacity for management and maintenance. A total of 114 toilet blocks were built, including 2,000 adult and 500 children's seats.

In many places, the inhabitants were involved in the design and construction of these toilets. Some women community leaders took on contracts and managed the whole construction process, supported by engineers and architects from SPARC. Unlike the previous models, they were bright and well-ventilated, with better-quality construction (which also made cleaning and maintenance easier). They had large storage tanks to ensure there was enough water for users to wash after defecation and to keep the toilets clean. Each toilet block had separate entrances and facilities for men and women. A block of specially designed children's toilets was included and in many blocks there were also toilets designed for easier use by the elderly and the disabled. Running costs were lower thanks to the inclusion of a room where the caretakers and their families could live. Even with these innovations, the cost of the toilet blocks was 5 percent less than the municipal corporation's costing.

This programme was also noteworthy in its transparency and accountability, with constant communication between government and community, weekly stakeholder meetings, and all aspects of costing and financing made public, thus curbing petty corruption. Similar programmes are now being developed in other cities.

Source: Information draws on Patel and Mitlin, 2001; Burra, 2000; SPARC, 2001.

Given the wide array of technologies, and the community-wide impacts of unhealthy water and sanitation systems, community involvement is important in deciding on the most appropriate technologies when truly 'safe' options are unaffordable. Given weak and underfinanced utilities and local governments, community involvement can also be important in determining how the chosen systems are to be financed, developed and maintained.

Community-supported upgrading programmes

For most cities, the upgrading of programmes has particular importance for water and sanitation for two reasons:

- it has been the primary means by which governments and international agencies have improved provision for water, sanitation and drainage within low-income urban settlements over the last thirty years; and

Box 7.6: Karachi, Pakistan – Orangi Pilot Project: when the community takes charge

The Orangi Pilot Project (OPP) was initiated in 1980, by a renowned social scientist. At that time, the squatter settlement of Orangi had no access to sanitation. A majority of the residents used bucket latrines emptied by a scavenger every four or five days. Wealthier households constructed soakpits; but these filled up after a few years and did not solve the wastewater problem. In the absence of an underground sewer system, open sewers crisscrossed the lanes. The infant mortality rate was high at 137 deaths per 1,000 live births.

Research by the OPP team revealed that the connection fee charged by the Karachi Water and Sewerage Board was extravagantly overpriced and that this cost could be much reduced, all the while providing technical assistance to the residents.

For each lane, the process involved overcoming the psychological and economic concerns of the residents, organizing them into lane organizations, providing technical surveys and labour and cost estimates. The residents would select a lane manager who would formally ask the OPP for assistance, collect the money and hold meetings. The OPP supervised the process but at no time handled the residents' money.

By the year 2000, about 98 percent of Orangi homes had in-house latrines. The entire construction was financed, supervised and constructed by the local population with no external subsidies.

The OPP programmes have also strengthened the position of women in Orangi society and their participation in community affairs. Infant mortality has dropped to 37 per 1,000 live births – far faster that the rest of Karachi. Because less money is spent on medical treatment, income is more available for other uses.

The residents of Orangi were legal landowners with an incentive to invest their own time and money on building a sewer system, a situation not commonly found in squatter settlements. Also, the natural slope of Orangi, which made it possible to dispense with secondary sewer lanes to a large extent, would not have been possible in flat terrain. However, the successful replication to over 35,000 homes elsewhere in Pakistan where there was no natural slope has shown that these issues are not insurmountable.

Because the community organizations and lane associations often do not survive once the construction is complete, long-term maintenance of the system can be a problem. Recently there has been a trend towards the formation of 'community-based organizations' to address this issue.

The one-lane-at-a-time, open-ended, exploratory philosophy adopted by OPP has been criticized by some development agencies as lacking a master plan. However, the Orangi model has actually achieved better results in slum areas than conventional target-oriented approaches. Also, the overall project was developed according to a master plan. As early as 1983, the OPP had prepared Circle Handbooks, identifying the secondary sewers required, the disposal points and the slope of the land.

To avoid reliance on outside entities, the OPP has preferred component-sharing rather than cost-sharing, and donor funding or government subsidies have been accepted only for construction of trunk sewers.

Source: Prepared for the World Water Assessment Programme (WWAP) by V. Srinivasan; P.-H. Gleick; C. Hunt at the Pacific Institute, 2002.

- it recognizes the rights of the inhabitants of the area being upgraded to have basic infrastructure and services, even though they may have occupied or developed the land illegally.

While many upgrading programmes have had limitations, e.g. inadequacies in the improvements to water and sanitation and the failure to ensure adequate provision for maintenance, a new generation of upgrading programmes has sought to address these (see for instance, Stein, 2001; Fiori et al., 2000). Perhaps more to the point, there is a recognition of the need to shift from support for upgrading projects to developing the institutional capacity of city and municipal authorities This would allow them to work continuously with the inhabitants of low-income settlements to help upgrade the quality and extent of infrastructure and service provision. Community-supported upgrading programmes provide perhaps the most important means to meet the Millennium Development Goals.

Self-built low-income housing

Many nations have developed innovative ways of increasing the possibilities for low-income households to buy or build their own new homes, which in turn provide for better water, sanitation and drainage. Support for low-income groups' savings schemes and for their acquisition of land with infrastructure on which they can construct their own homes is an important part of improving provision for water and sanitation. This is demonstrated by the large number of low-income households that have acquired better-quality housing through community-managed schemes in India (within the National Slum Dwellers Federation and Mahila Milan) (Patel and Mitlin, 2001), South Africa (within the Homeless People's Federation), (Baumann et al., 2001) and Thailand (with the support of the former Urban Community Development Office, now the Community Organizations Development Institute), (UCDO, 2000).

This is an approach that has to be 'demand-driven' for low-income households; many government schemes to provide low-income households with 'sites and services' on which they can build have been 'supply-driven' with the result that new sites were in the wrong location or were too expensive for low-income households.

Demand-Side Water Management and the Urban Poor

Demand-side water management can be defined as the implementation of policies or measures that serve to control or influence the consumption or waste of water. The benefits from demand-side management include:

- reduced water consumption;
- a more cost-effective means of meeting demand, wholly or in part, compared with investment in new resources;

Box 7.7: Urban sanitation micro-enterprises: the UN-Habitat Vacutug development project

In recent years, the problems associated with the disposal of human waste have escalated with the growth of unplanned settlements in low- and middle-income countries. In these settlements, there are often more than 100 persons to each pit latrine. Although the latrines are now generally constructed of modern materials, the problem of renewing them when they become full has proved a difficult challenge. Often the settlements have no road access for disposal tankers to use so a novel solution is needed. UN-Habitat has been developing the prototype technology that can empty latrines. This technology, 'Vacutug', is being constructed in association with a private-sector engineering company and a Kenyan water NGO. The micro-enterprise earned over US$10,000 over a two-year period and employed four people.

UN-Habitat has now launched the second phase of the project to expand operations and assess the technology in a number of different conditions in different countries. The Vacutug is seen as a simple but very effective solution and is currently supported by the governments of the United Kingdom, Ireland and Denmark. The programme is shortly to be extended to many other cities on a partnership basis and presents an ideal opportunity to greatly improve health aspects related to excreta disposal while at the same time generating much-needed income for the urban poor.

Source: Prepared for the World Water Assessment Programme (WWAP) by G.P. Alabaster, UN-Habitat (for more details, contact graham.alabaster@unhabitat.org).

- protection of the environment by making the best use of existing water resources and minimizing wastewater discharges; and

- the responsibility for implementing demand management can be spread between the utility and its users, as benefits will be felt by both.

Demand-side management options for influencing water demand are the following:

- reducing and controlling leakage from the utility's mains networks;

- encouraging industrial and commercial users to reduce their dependency on potable water supplies by increasing the level of recycling and implementing waste minimization strategies;

- encouraging domestic users to reduce their usage;

- recycling of rainwater by users;

- recycling of domestic washwater (grey water systems) by users; and

- volumetric charging by revenue metering of users.

It should be noted that simple and economic demand-side management techniques that can be implemented by domestic users and that do not drastically change their normal water habits, are more likely to be implemented and maintained than schemes that would require a significant capital or time expenditure. Several of the more common measures are:

- inserting a brick or similar object in a toilet cistern as a simple way of reducing the volume of water used to flush a toilet;

- recycling rainwater to reduce dependence on treated water supplies, and to use for watering gardens or other non-potable domestic purposes. This can be achieved at little or no expense to the user through the use of appropriate water storage facilities; and

- installing grey water systems, which involve recycling of bath and shower water for toilet flushing, systems which are starting to be considered in western Europe and North America. Although this is an ideal solution, it may take some time to filter into common use due to the cost and inconvenience of having a fitted system.

Revenue metering, particularly when combined with a punitive tariff structure that penalizes consumption above the basic needs level can have a large impact on consumption, although the reading and maintenance of the meters places a burden on the utility provider.

Educating all users in the benefits of demand-side management will increase the likelihood of success for any demand management scheme. The utility may find greater support by recognizing that careful water usage will reduce consumption and personal water costs, and make available water to meet demand elsewhere.

One of the critical shifts in better water governance is that from supply-side alone to a mix of supply-side and demand-side approaches. For example, supply-side investments in new water supplies can be undertaken in parallel with more cost-effective investment to achieve demand-side water savings, thus offering a 'least cost' approach to meeting water demand. The form of this shift will vary considerably, depending on local context, as illustrated by table 7.4, and depends on the existing service level.

Water utilities have been criticized for taking a 'supply-fix' approach by assuming that increasing demands have to be met by increasing supplies. Advocates of demand-side management argue that by increasing end-use efficiency and reducing waste, these demands could be met, wholly or in part, from the water saved.

Demand-side management in low-income cities should not only focus on water conservation but should also give attention to two issues: securing better access to water for the urban poor; and promoting hygiene.

In informal settlements and other areas with low average per-capita incomes, many households do not consume sufficient water to meet their basic needs for health. It is important to prevent conservation-oriented measures from further reducing the water consumption of deprived households. This could mean that a punitive tariff structure should consider setting the 'basic needs' tariff at a level affordable by very poor households with significantly higher tariffs imposed for consumption above the basic needs level.

In conditions of poverty, it is important that demand-side management recognizes that improved health is one of the major benefits water can provide, but that the health outcome depends upon how the water is used. Users often lack a relevant knowledge of hygiene, and it is therefore important for demand-side management to focus on reducing *unnecessary* water consumption in low-income settlements.

It is also important to consider the role of utilities. Although the initial successes of demand-side management relied on certain regulatory or economic restraints, this is no longer the case. In the absence of restraints, commercially-oriented utilities that get their revenue from selling water will favour higher prices: there is, in effect, no reason to assume that utilities have the economic incentive to engage in demand-side management.

Table 7.4: Comparison of different approaches to demand-side water management in the household sector

	The conservation argument	The hygiene argument	The marginal cost pricing argument[1]	The community action argument
Guiding concern	Water stress is a growing problem in most parts of the world, due to excessive water consumption.	Water-related diseases still constitute a large share of the global burden of disease.	Water is a scarce commodity, with an economic value in numerous alternative uses.	Adequate water is a basic need, without which people cannot live healthy and fulfilling lives.
Key insight	There are numerous unexploited opportunities for saving water without reducing the services water provides.	Achieving health depends on how water is used as well as how much water (of adequate quality) is provided.	Piped water is typically priced well below its (marginal) economic value.	Disorganized (poor) communities are at a disadvantage in both addressing their own water needs and negotiating with outsiders.
Contributory factors	Householders using piped water often cannot tell how much of their water is going to which purposes, are not aware when they are wasting water, and do not have the means of judging water-conserving technologies.	Householders cannot discern the health consequences of their water use practices, and often rely on social norms which, especially in crowded and generally hazardous living environments, may be unhealthy.	Water is often treated as a social good, with provision organized as a non-commercial enterprise. Even commercial providers rarely bear the full (marginal) costs of water withdrawal and in any case do not operate in a competitive market.	Water utilities are not responsive to the needs and demands of low-income communities, especially if they are located in informal settlements. Local organization is often suppressed for political reasons.
Demand-side consequences	Users are unaware and unconcerned about water conservation, and there is unnecessary water wastage.	Users often fail to adopt safe water practices, and do not achieve the potential health benefits even when they receive piped water.	Consumers overuse water, either leading to resource problems and/or depriving others of valuable water.	Residents receive inappropriate or inadequate water services, or must rely on informal and often costly and inadequate water sources.
Recommendation	Conservation education and promotion should become an integral part of piped water provision.	Hygiene education and promotion should become an integral part of water provision.	Piped water pricing should be based on long-run marginal costs, giving users the incentive to manage their own demand efficiently.	Poor communities should mobilize (or be mobilized) around local water issues, and providers should be responsive to community as well as individual demands.

1. This column concentrates on the economic arguments for marginal cost pricing, and ignores the economic arguments more specific to low-income communities.

The limited water management capacity in many cities may mean that adding new management burdens is counterproductive. As such, forms of demand-side management that also ease overall management burdens are far more likely to be successful than initiatives that give greater responsibilities to already struggling utilities and government agencies.

There is a danger that, rather than creating a more integrated form of water management, as most proponents hope, demand-side management will accentuate conflicts between ecological and human health and welfare goals. It is one thing to recognize that water is often wasted even in poor areas (leakage in particular is often a serious problem); it is quite another to treat water resource abuse as the defining environmental problem in areas where water-related health problems are pervasive.

Thus, there is a case for a form of demand-side water management that actively serves both conservation and environmental health/welfare goals, recognizing that the relative importance of these different goals, and the appropriate strategies for pursuing them, is very context dependent.

Conclusions

A significant proportion of those living without proper water supply and sanitation are urban dwellers, mainly in the peri-urban areas. They are forced to draw on water sources that are unsafe, unreliable and often difficult to access. For sanitation, they have poor quality latrines – often shared with so many others that access and latrine cleanliness is difficult – or they have no provision for sanitation. Virtually all urban dwellers with inadequate provision live in the low- and middle-income nations of Africa, Asia and Latin America and the Caribbean. Inadequacies in the provision of water, sanitation and hygiene bring an enormous health burden: half the urban population in Africa, Asia and Latin America suffer one or more of the diseases associated with inadequate water and sanitation, and among the urban population of low-income countries, one child in six dies before the age of five.

The numbers alone are striking, but when added to current projections of a world urban population set to reach 80 percent by 2030, they become thoroughly frightening. Efforts are being made to provide better water and sanitation services to the world's cities, but much remains to be done. There is a clear need to broaden and deepen the coverage of global assessments of the quality of provision for water, sanitation and hygiene in urban areas so they show the proportion of people with 'safe', 'sufficient' and 'convenient' as well as 'improved' provision. If the ultimate goal remains to improve such provision, what is needed are not only surveys based on representative samples of rural and urban populations, but also site-specific information, which documents the deficiencies and can be used by water and sanitation providers to plan improvements or by local residents to articulate their demands.

There is an equally urgent need to limit the extent to which uncontrolled urban development impacts on the surrounding areas, and to implement more effective infrastructure and water governance systems. A shift from supply-side only to a mix of supply- and demand-side initiatives could provide a more cost-effective solution in many settlements.

The situation is becoming increasingly serious, as urban populations continue to grow throughout the world: action must be taken now in order to provide the world's urban populations, especially the poor, with the safe, clean, accessible water and sanitation that is their right.

Progress since Rio at a glance

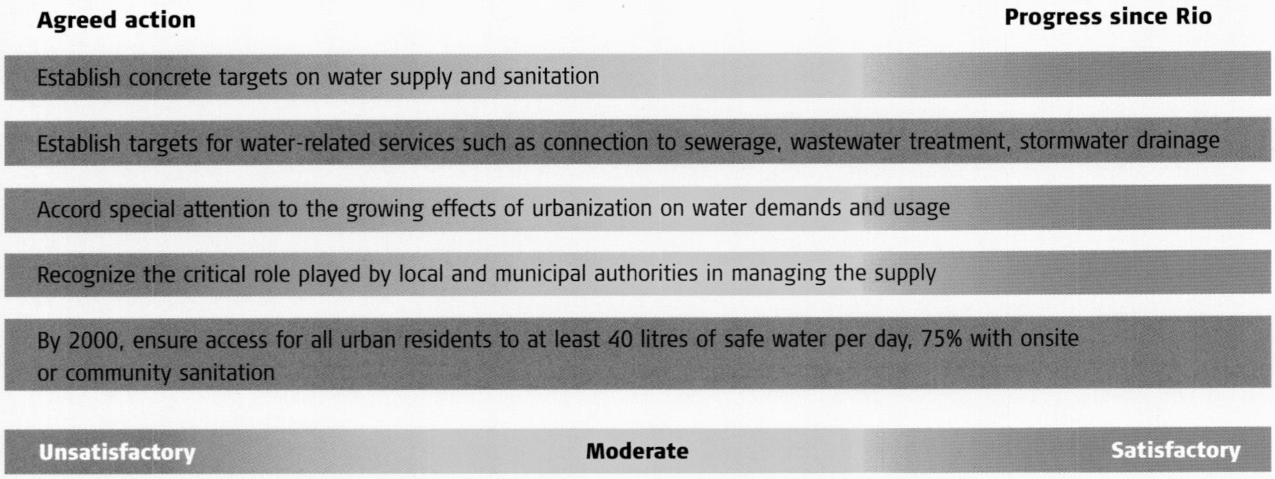

Agreed action	Progress since Rio
Establish concrete targets on water supply and sanitation	
Establish targets for water-related services such as connection to sewerage, wastewater treatment, stormwater drainage	
Accord special attention to the growing effects of urbanization on water demands and usage	
Recognize the critical role played by local and municipal authorities in managing the supply	
By 2000, ensure access for all urban residents to at least 40 litres of safe water per day, 75% with onsite or community sanitation	

Unsatisfactory — **Moderate** — **Satisfactory**

References

Acho-Chi. 1998. 'Human Interference and Environmental Instability: Addressing the Environmental Consequences of Rapid Urban Growth'. In: *Environment and Urbanization*, Vol. 10, No. 2, pp. 161–74.

Alcamo, J.; Döll, P.; Henrichs, T.; Lehner, B.; Kaspar, F.; Rösch, T.; Siebert, T. Forthcoming. WaterGAP: Development and Application of a Global Model for Water Withdrawals and Availability'. *Hydrological Sciences Journal*.

Alder, G. 1995. 'Tackling Poverty in Nairobi's Informal Settlements: Developing an Institutional Strategy'. *Environment and Urbanization*, Vol. 7, No. 2, pp. 85–107.

Alimuddin, S.; Hasan, A.; Sadiq, A. 2000. *The Work of the Anjuman Samaji Behbood and the Larger Faisalabad Context*. Working Paper. London, International Institute for Environment and Development.

Anand, P.-B. 1999. 'Waste Management in Madras Revisited'. *Environment and Urbanization*, Vol. 11, No. 2, pp. 161–76.

Anton, D.-J. 1993. *Thirsty Cities: Urban Environments and Water Supply in Latin America*. Ottawa, International Development Research Centre.

Bangladesh Bureau of Statistics, Ministry of Planning, Government of the People's Republic of Bangladesh (with assistance from the United Nations Children's Fund). 1996. *Progotir Pathey Achieving the Mid-Decade Goals for Children in Bangladesh*.

Bartlett, S. 2002. *Water, Sanitation and Children*. Background paper for the UN-Habitat Report on the State of the World's Water and Sanitation in Cities.

Bartone, C.; Bernstein. J.; Leitmann, J.; Eigen, J. 1994. *Towards Environmental Strategies for Cities: Policy Considerations for Urban Environmental Management in Developing Countries*. UNDP/UNCHS/World Bank Urban Management Program, No. 18. Washington DC, World Bank.

Baumann, T.; Bolnick, J.; Mitlin, D. 2001. *The Age of Cities and Organizations of the Urban Poor: The Work of the South African Homeless People's Federation and the People's Dialogue on Land and Shelter*. IIED Working Paper 2 on Poverty Reduction in Urban Areas. London, International Institute of Environment and Development.

Benneh, G.; Songsore, J.; Nabila, J.-S.; Amuzu, A.-T.; Tutu, K.-A.; Yangyuoro, Y.; McGranahan, G. 1993. *Environmental Problems and the Urban Household in the Greater Accra Metropolitan Area (GAMA) Ghana*. Stockholm, Stockholm Environment Institute.

Bradley, D.; Stephens, C.; Cairncross, S.; Harpham, T. 1991. *A Review of Environmental Health Impacts in Developing-Country Cities*. Urban Management Program Discussion Paper No. 6. Washington DC, The World Bank, United Nations Development Programme and UN-Habitat.

Burra, S. and Patel, S. 2002. 'Community Toilets in Pune and Other Indian Cities'. In: *PLA Notes*. Special Issue on Participatory Governance. London, International Institute of Environment and Development.

Cairncross, S. and Ouano, E.-A.-R. 1990. *Surface Water Drainage in Low-Income Communities*. Geneva, World Health Organization.

Cairncross, S. and Feachem, R.-G. 1993. *Environmental Health Engineering in the Tropics: An Introductory Text*. Second edition. Chichester, John Wiley and Sons.

CSE (Centre for Science and Environment). 1999. *State of India's Environment: The Citizens' Fifth Report*. New Delhi.

The City of New York Independent Budget Office. 2002. *Letter to Cathleen Breen, Watershed Protection Coordinator, New York Public Interest Research Group, 18 March*.

Committee to Review the New York City Watershed Management Strategy, Water Science and Technology Board, Commission on Geosciences, Environment and Resources, National Research Council. 1999. 'Watershed Management for Potable Water Supply: Assessing the New York City Strategy'. Washington DC, National Academy Press.

Cosgrove, B. and Rijsberman, F.-R. 2000. *World Water Vision: Making Water Everybody's Business*. London, World Water Council, Earthscan Publications Ltd.

DEP (Department of Environmental Protection). 2001. 'New York City's 2001 Watershed Protection Program Summary, Assessment and Long-Term Plan'. New York.

Döll, P.; Kaspar, F.; Lehner, B. Forthcoming. 'A Global Hydrological Model for Deriving Water Availability Indicators: Model Tuning and Validation'. *Journal of Hydrology*.

Döll, P. and Siebert, S. 2002. Global Modelling of Irrigation Water Requirements'. *Water Resources Research*, Vol. 38, No. 4, pp. 8.1–8.10. DOI 10.1029/2001WR000355.

——. 1986. 'Urban Geomorphology'. In: P.-G. Fookes and P.-R. Vaughan (eds.), *A Handbook of Engineering Geomorphology*. Glasgow, Surrey University Press (Blackie and Son).

——. 1983. *The Urban Environment*. London, Edward Arnold.

Ellis, J.-B. 1999. 'Preface'. In: J.-B. Ellis (ed.), *Impacts of Urban Growth on Surface Water: Groundwater Quality*, International Association of Hydrological Sciences Publication No. 259.

Etemadi, F.-U. 2000. 'Civil Society Participation in City Governance in Cebu City'. *Environment and Urbanization*, Vol. 12, No. 1, pp. 57–72.

Fiori, J.; Riley, L.; Ramirez, R. 2000. *Urban Poverty Alleviation through Environmental Upgrading in Rio de Janeiro: Favela Bairro*. London, Development Planning Unit, University College London.

Ghosh, A.; Ahmad, S.-S.; Maitra, S. 1994. *Basic Services for Urban Poor: A Study of Baroda, Bhilwara, Sambalpur and Siliguri*. Urban Studies Series No. 3. New Delhi, Institute of Social Sciences and Concept Publishing Company.

Guerrant, D.-I.; Moore, S.-R., Lima, A.-A.-M.; Patrick, P.; Schorling, J.-B.; Guerrant, R.-L. 1999. 'Association of Early Childhood Diarrhoea and Cryptosporidiosis with Impaired Physical Fitness and Cognitive Function 4 to 7 Years Later in a Poor Urban Community in Northeast Brazil'. *American Journal of Tropical Medicine and Hygiene*, Vol. 61, No. 5, pp. 707–13.

Hardoy, J.-E.; Mitlin, D.; Satterthwaite, D. 2001. *Environmental Problems in an Urbanizing World: Finding Solutions for Cities in Africa, Asia and Latin America*. London, Earthscan Publications Ltd.

Hasan, A. 2001. *Working with Communities*. Karachi, City Press.

——. 1999. *Understanding Karachi: Planning and Reform for the Future*. Karachi, City Press.

Hewett, P.-C. and Montgomery, M. 2001. 'Poverty and Public Services in Developing-Country Cities'. Policy Research Division Working Paper No. 154. New York, Population Council.

Islam, N.; Huda, N.; Narayan F.-B.; Rana, P.-B. (eds.). 1997. *Addressing the Urban Poverty Agenda in Bangladesh, Critical Issues and the 1995 Survey Findings*. Dhaka, The University Press Limited.

Kalbermatten, J.-M.; Julius, D.-S.; Gunnerson, C.-G. 1980. *Appropriate Technology for Water Supply and Sanitation: A Review of the Technical and Economic Options*. Washington DC, The World Bank.

Kaneez Hasna, M. 1995. 'Street Hydrant Project in Chittagong Low-Income Settlement'. *Environment and Urbanization*, Vol. 7, No. 2, pp. 207–18.

Kelly, P.F. 1998. 'The Politics of Urban-Rural Relationships: Land Conversion in the Philippines'. *Environment and Urbanization*, Vol. 10, No. 1, pp. 35–54.

Korboe, D.; Diaw, K.; Devas, N. 2000. 'Urban Governance, Partnership and Poverty: Kumasi'. In: *Urban Governance, Partnership and Poverty Working Paper 10*. Birmingham, International Development Department, University of Birmingham.

Landwehr, D.; Keita, S.-M.; Ponnighaus, J.-M.; Tounkara, C. 1998. 'Epidemiological Aspects of

Scabies in Mali, Malawi, and Cambodia'. *International Journal of Dermatology*, Vol. 37, No. 8, pp. 588–90.

Maksimovic, C. and Tejada-Guibert. J.-A. (eds.). 2002. *Frontiers in Urban Water Management: Deadlock or Hope.* London, United Nations Educational, Scientific and Cultural Organization, International Water Association Publishing.

Mazwile, M. 2000. 'Involvement of Women and Community in "Watsan" Activities'. Paper read at the 19th Annual Water Experts Conference, Arusha.

McCarney, P.-L. 1996. 'Considerations on the Notion of "Governance" – New Directions for Cities in the Developing World'. In: P.-L. McCarney (ed.), *Cities and Governance: New Directions in Latin America, Asia and Africa.* Toronto, Centre for Urban and Community Studies, University of Toronto.

Menegat, R. 2002. 'Environmental Management in Porto Alegre'. *Environment and Urbanization*, Vol. 14, No. 2.

Montgomery, M. 2002. *Analysis of 86 Demographic and Health Surveys Held in 53 Different Nations between 1986 and 1998.* New York, Population Council.

Mosha, J.-P.-N. 2000. 'Small-Scale Independent Providers in Provision of Water and Sanitation Services'. Paper read at the 19th Annual Water Experts Conference, Arusha.

Mumford, L. 1991. *The City in History.* London, Penguin Books.

Ndezi, T.-P. 2000. 'Willingness and Ability to Pay for Water and Sanitation within Low-Income Communities in Dar es Salaam'. Paper read at the 19th Annual Water Experts Conference, Arusha.

Njau, B.-E. 2000. 'Demand-Management Consideration in Preparation of Urban Water Supply Programmes'. Paper read at the 19th Annual Water Experts Conference, Arusha.

NSDF (National Slum Dwellers Federation); Mahila Milan; SPARC (Society for the Promotion of Area Resource Centres). 1997. *Toilet Talk.* Issue No. 1. Bombay.

Patel, S. and Burra, S. circa 2000. *A Note on Nirmal Bharat Abhiyan (NBA).* Bombay, Society for Promotion of Area Resource Centres.

Patel, S. and Mitlin, S. 2001. *The Work of SPARC and Its Partners Mahila Milan and the National Slum Dwellers Federation in India.* IIED Working Paper 5 on Urban Poverty Reduction. London, International Institute for Environment and Development.

Prüss, A.; Kay. D.; Fewtrell, L.; Bartram, J. 2002. 'Estimating the Burden of Disease from Water, Sanitation and Hygiene at a Global Level'. *Environmental Health Perspectives*, Vol. 110, No. 5, pp. 537–42.

Raskin, P.; Gleick, P.; Kirshen, P.; Pontius, G.; Strzepek, K. 1997. *Comprehensive Assessment of the Freshwater Resources of the World. Water Futures: Assessment of Long-Range Patterns and Problems.* Stockholm, Stockholm Environment Institute, Box 2142. S-103 14.

Rees, W.-E. 1992. 'Ecological Footprints and Appropriated Carrying Capacity'. *Environment and Urbanization*, Vol. 4, No. 2, pp. 121–30.

Rice, A.-L.; Sacco, L.; Hyder, A.; Black, R.-E. 2000. 'Malnutrition as an Underlying Cause of Childhood Deaths Associated with Infectious Diseases in Developing Countries'. *Bulletin of the World Health Organization*, Vol. 78, No. 10, pp. 1207–21.

Roche, P.-A.; Valiron, F.; Coulomb, R.; Villessot, D. 2001. 'Infrastructure Integration Issues'. In: C. Maksimovic and J.-A. Tejada-Guibert (eds.), *Frontiers in Urban Water Management: Deadlock or Hope.* London, International Water Association Publishers.

Satterthwaite, D. 2002. *Coping with Rapid Urban Growth.* London, RICS International Paper Series, Royal Institution of Chartered Surveyors.

Satterthwaite, D.; Hart, R.; Levy, C.; Mitlin, D.; Ross, D.-A., Smit, J.; Stephens, C. 1996. *The Environment for Children: Understanding and Acting on the Environmental Hazards That Threaten Children and Their Parents.* London, United Nations Children's Fund, Earthscan Publications Ltd.

Schenk, H. (ed.). 2001. *Living in India's Slums: A Case Study of Bangalore.* Delhi, Indo-Dutch Programme on Alternatives in Development.

Shayo Temu, S. 2000. 'Cost Recovery in Urban Water Supply and Sewerage Services'. Paper read at the 19th Annual Water Experts Conference, Arusha.

Showers, K. 2002. 'Water Scarcity and Urban Africa: An Overview of Urban-Rural Water Linkages'. *World Development*, Vol. 30, No. 4, pp. 621–48.

Slum Rehabilitation Authority. 1997. *Guidelines for the Implementation of Slum Rehabilitation Schemes in Greater Mumbai.* Bombay.

SPARC (Society for Promotion of Area Resource Centres). 2001. Video on Pune Toilets.

Stein, A. 2001. *Participation and Sustainability in Social Projects: The Experience of the Local Development Programme (PRODEL) in Nicaragua.* IIED Working Paper 3 on Poverty Reduction in Urban Areas, International Institute for Environment and Development.

TARU Leading Edge. 1998. *Bangalore Water Supply and Sewerage Master Plan: A Situation Analysis.* Prepared for AUS AID. New Delhi.

UCDO (Urban Community Development Office). 2000. *UCDO Update No 2.* Bangkok.

UN 2002. *World Urbanization Prospects: The 2001 Revision. Data Tables and Highlights.* New York, Population Division, UN Secretariat, Department of Economic and Social Affairs.

——. 2001. *World Urbanization Prospects: The 1999 Revision.* New York, Population Division, UN Secretariat, Department of Economic and Social Affairs.

——. 2000. *United Nations Millennium Declaration.* Resolution adopted by the General Assembly. A/RES/55/2.

——. 1977. Proceedings of the United Nations Water Conference, 14–25 March, Mar del Plata, Argentina.

——. 1976. Conference on Human Settlements (Habitat). Recommendation C12 from The Recommendations for National Action endorsed at the UN Conference on Human Settlements in 1976.

UN-Habitat. Forthcoming. *The State of the World's Cities for Water and Sanitation.* London, Earthscan Publications Ltd.

——. 1996. *The Habitat Agenda: Istanbul Declaration on Human Settlements.* Nairobi.

UNICEF (United Nations Children's Fund). 2003. *Poverty and Exclusion Among Urban Children.* Florence, Innocenti Centre.

——. 2000. *Multiple Indicator Survey.* Delhi.

——. 1997. *The Dancing Horizon Human Development Prospects for Bangladesh.* Dhaka, quoting Survey by the International Centre for Diarrhoeal Disease Research.

Vassolo, S. and Döll, P. 2002. 'Development of a Global Data Set for Industrial Water Use'. Unpublished manuscript. Germany, University of Kassel, Center for Environmental Systems Research.

Vidal, J. 2002. 'Water of Strife'. *The Guardian Society.* 27 March, pp. 8–9.

Vlachos E. and Braga, B. 2001. 'The Challenge of Urban Water Management'. In: C. Maksimovic and J.-A. Tejada-Guibert (eds.), *Frontiers in Urban Water Management: Deadlock or Hope.* London, International Water Association Publishing.

WHO (World Health Organization). 2001. *Water Supply and Sanitation Sector Assessment 2000.* Regional Office for Africa, Harare.

——. 1999a. 'Creating Healthy Cities in the 21st Century'. In: D. Satterthwaite (ed.), *The Earthscan Reader on Sustainable Cities.* London, Earthscan Publications Ltd.

——. 1999b. *World Health Report: 1999 Database.* Geneva.

——. 1996. 'Developments in Water, Sanitation and Environment'. *Water Newsletter.* No. 245. International Water and Sanitation Centre, World Health Organization Collaborating Center.

——. 1992. *Our Planet, Our Health.* Report of the WHO Commission on Health and Environment. Geneva, World Health Organization.

WHO/UNICEF (World Health Organization/United Children's Fund). 2000. *Global Water Supply and Sanitation Assessment, 2000 Report.* Geneva.

World Bank. 1988. *World Development Report 1988.* New York, Oxford University Press.

Yepes, G. 1995. *Reduction of Unaccounted for Water – the Job Can Be Done.* Washington DC, World Bank.

Some Useful Web Sites*

European Commission Database on Good Practice in Urban Management and Sustainability

http://europa.eu.int/comm/urban/

This database, part of the European Commission's site, is designed to help authorities work towards sustainable urban management through the dissemination of good practice and policy.

Society for the Promotion of Area Resource Centres (SPARC)

http://www.sparcindia.org/

A non-governmental organization devoted to working with the urban poor. The site provides information on projects, news events, publications and related links, including information on the National Slum Dwellers Federation (NSDF).

United Nations Development Programme (UNDP), Public-Private Partnerships for the Urban Environment (PPUE)

http://www.undp.org/ppp/

The PPUE supports the development of public-private partnerships at the local level in order to ensure more sustainable urban management practices. The aim is to facilitate meeting the challenges faced by cities in providing basic services to populations.

UN-Habitat: Global Urban Observatory

http://www.unchs.org/programmes/guo

Part of the UN-Habitat site, this section provides policy-oriented urban indicators, statistics and other information on global urban conditions and trends.

World Bank, Urban Development

http://www.worldbank.org/urban/

A site committed to promoting sustainable cities by improving the lives of the poor. This department of the World Bank provides *inter alia* information on a variety of aspects of urban management.

World Health Organization (WHO), Healthy Cities and Urban Governance

http://www.who.dk/eprise/main/WHO/Progs/HCP/Home

This site provides information on health issues in urban areas, as well as ideas on making cities healthier, news events from around the world and links to related sites.

* These sites were last accessed on 3 January 2003.

'Agriculture will remain the dominant user of water at the global level. In many countries, in particular those situated in the arid and semi-arid regions of the world, this dependency can be expected to intensify.'

8 Securing Food for a Growing World Population

By: FAO (Food and Agriculture Organization)
Collaborative agencies: WHO (World Health Organization)/ UNEP (United Nations Environment Programme)/ IAEA (International Atomic Energy Agency)

Table of contents

Securing Food for a Growing World Population

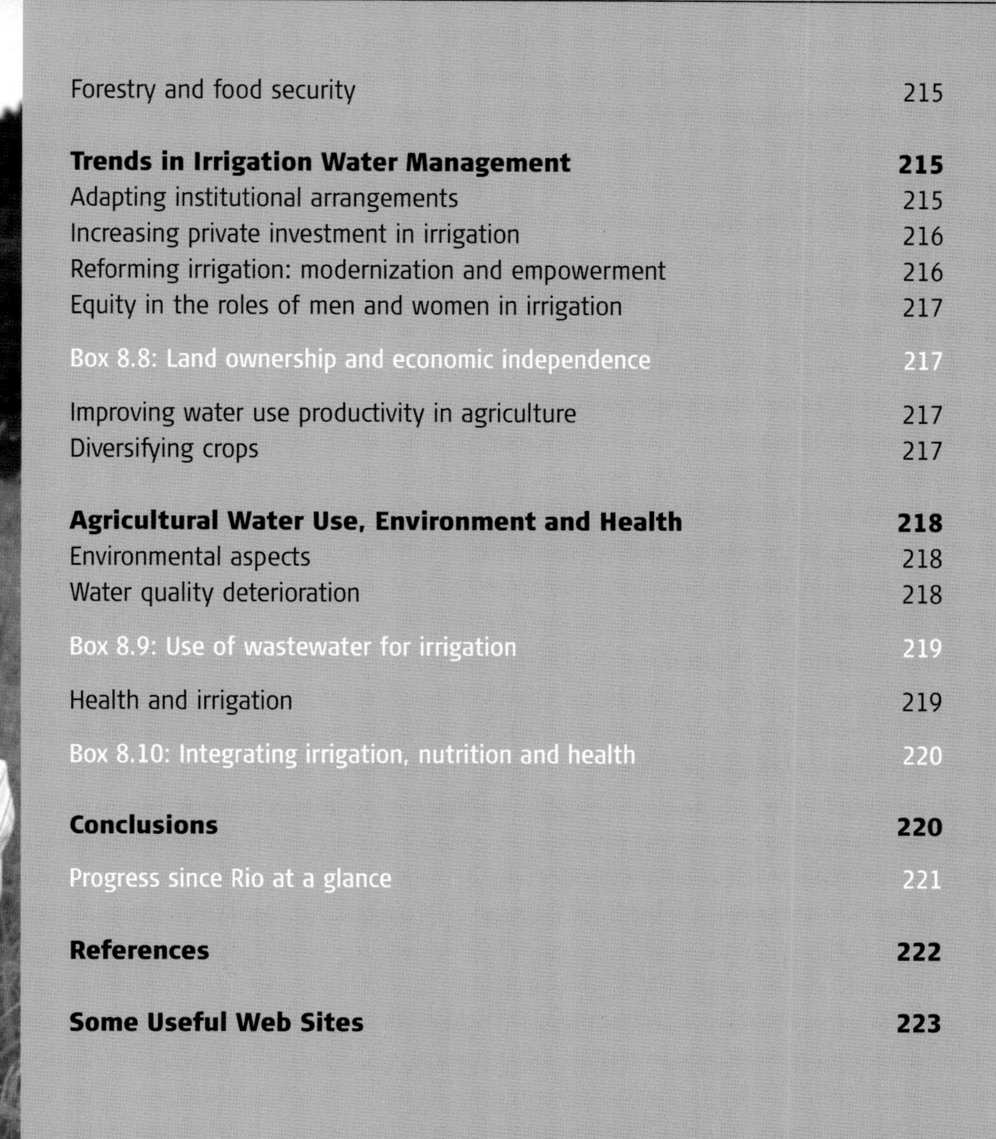

Securing Food for a Growing World Population

There are people in the world so hungry that God cannot appear to them except in the form of bread.

Mahatma Gandhi

FOOD AND AGRICULTURE are by far the largest consumers of water. They require perhaps one hundred times more than we use to meet basic personal needs, with up to 70 percent of the water we take from rivers and groundwater going into irrigation. Global food production has kept pace with population growth in recent decades; yet nearly 800 million people remain undernourished, and the population shift from rural to urban environments will certainly increase the pressures and problems associated with food security. A growing population will need more food and thus more water. What is the status of food production in the world? How can it be made more efficient without compromising the environment? What are the contributions from rainfed and irrigated agriculture and from fisheries? What role does the market play? What is the connection between food security and poverty? These questions have answers. Less certain, however, is whether we shall rise to the challenge of feeding the world's hungry by being more efficient and productive in our use of water while still respecting the resource base and demands from competing sectors.

SINCE THE 1960S, THE WORLD FOOD SYSTEM has responded to a doubling of the world population, providing more food per capita at progressively lower prices. Global nutrition has consistently improved. This performance was possible through a combination of high-yielding seeds, irrigation, plant nutrition and pest control. In the process, large quantities of water were appropriated for agriculture. As population keeps increasing, albeit at a slower rate, more food and livestock feed will need to be produced in the future and more water applied for this purpose. Water withdrawals for irrigation in developing countries are expected to increase by an aggregated 14 percent until 2030, while irrigation water use efficiency is expected to improve by an average 4 percent. Water scarcity stress is foreseen to grow locally and in some cases regionally and a number of countries will have to rely more on trade for their food security.

While food production is satisfying market demand at historically low prices, an estimated 777 million people in developing countries do not have access to sufficient and adequate food because they do not have the resources to buy it or, in the case of subsistence farmers, to produce it. In spite of the overall improvement in the nutritional situation, the absolute number of undernourished people is reducing at a much slower rate than has been anticipated. The 1996 World Food Summit (see box 8.1) set a target of reducing the number of chronically hungry people to about 400 million, but current projections indicate that this figure may not be achieved until fifteen years later than targeted, i.e. by 2030, unless decisive policy and financial action is taken (see figure 8.7 further on).

Irrigated agriculture will by necessity claim large quantities of water to produce the food required to feed the world. Irrigation water management has a long way to go to adapt to the new production requirements and reconcile competing claims from other economic sectors and calls for environmental protection. However, water-saving technologies are available and can significantly reduce the waste of water. In addition, the political, legal and institutional framework to support improved water productivity in irrigated agriculture also shows signs of adaptation. Water management trends point to empowering stakeholders, with a priority for the poor and the marginalized. At the same time, the water needs for human health and for the aquatic environment call for closer attention. Overall, the message from agriculture, which will remain globally the largest water user, is cautiously optimistic.

At the start of the twenty-first century, agriculture is using a global average of 70 percent of all water withdrawals from rivers, lakes and aquifers. The Food and Agriculture Organization (FAO) anticipates a net expansion of irrigated land of some 45 million hectares in ninety-three developing countries (to a total of 242 million hectares in 2030) and projects that agricultural water withdrawals will increase by some 14 percent from 2000 to 2030 to meet future food production needs. The analysis indicates a projected annual growth rate of 0.6 percent, compared with the 1.9 percent observed in the period from 1963 to 1999.

Only a part of agricultural water withdrawals are effectively used in the production of food or other agricultural commodities; a large

Box 8.1: The right to food

There is a trend towards recognizing access to food as a human right, as highlighted in the 1948 Universal Declaration of Human Rights, the 1966 International Covenant on Economic, Social and Cultural Rights and, with regard to armed conflicts, the 1977 Protocols to the 1949 Geneva Convention. While the right to food as humanitarian aid in the case of disaster or war is generally accepted, there is growing agreement that no nation has the right to starve its people. However, few governments are prepared, or have the capacity, to secure access to food for poor and undernourished citizens. As stated in 1999 by Mary Robinson, former United Nations High Commissioner for Human Rights, 'few economic rights are violated on such a scale as food and nutrition rights'.

At the 1996 World Food Summit, some argued that increased trade, achieved by cutting the global market free from government interventions, would result in greater food security, while others reasoned that food is a basic human right which national governments, not the global market, have a primary responsibility to make accessible. The following statement was included in the Rome Declaration on World Food Security:

> *We pledge our political will and our common and national commitment to achieving food security for all and to an ongoing effort to eradicate hunger in all countries, with an immediate view to reducing the number of undernourished people to half their present level no later than 2015.*

proportion of water may not reach the crop plants because it evaporates or infiltrates during conduction, evaporates from the soil in the field, or is used by non-productive growth such as weeds. Irrespective of the actual outcomes, it is important to highlight the fact that water allocations for agriculture will face increasing competition from other higher utility uses – municipal, industrial uses and calls for water to be left in the environment. Under these circumstances it is crucial that the role of water in securing food be understood and the potential for improving overall agricultural productivity with respect to water be fully realized.

In this section, the facts about past, present and future water demand in food production and food security are discussed. Information has been collected at the national level – much of this obtained through the Internet. For the purpose of discussion, three groups of countries are identified: developing countries, industrialized countries and countries in transition. Developing countries call for special attention because demographic growth rates are high and the potential demand for food is not yet satisfied. The countries are organized in regional groupings, that is: sub-Saharan Africa, Near East/North Africa, Latin America and the Caribbean, south Asia and east Asia. It should always be kept in mind that aggregate and average figures tend to hide as much as or more than they reveal. Water problems are always local or, at most, regional in nature, and may vary over time. Countries with large territories also have a large diversity of situations, including arid and humid regions and plains as well as mountains.

This section is largely based on FAO's technical report *World Agriculture: Towards 2015/2030*, the most recent edition of FAO's periodic assessments of likely future developments in world food, nutrition and agriculture. The report provides information on a global basis, with more detailed emphasis on ninety-three developing countries. The section also relies extensively on the data, information and knowledge provided by FAOSTAT, the FAO statistical database, and AQUASTAT, FAO's information system on water and agriculture. The contribution of the International Water Management Institute (IWMI) in the preparation of this section is acknowledged with thanks. National values of key indicators in 251 countries are presented in table 8.1. The significance of each indicator is highlighted in the relevant part of the discussion by reference to this table.

How the World Is Fed

The world food system: sustained improvement in food availability

Between the early 1960s and the late 1990s, while world population almost doubled, the productive potential of global agriculture met the growth of effective demand. Figure 8.1 shows that total investment in irrigation and drainage tended to correspond to food prices. But even with the observed decline in food prices, the nutritional status of the world's population continued to improve. Clearly, some of the early investment in agriculture paid off and productivity gains were being made. Irrigation played an important role in ensuring the needed growth in food production. Today, as the food production issue becomes less critical, concern arises over future large-scale irrigation in terms of its overall performance and the political and institutional viability of transferring the management of public irrigation schemes to users. The significance of non-structural irrigation and water management reform will grow as world agriculture in general becomes more responsive to demand. These issues are taken up later in the chapter.

Figure 8.1: Food prices and investment in irrigation and drainage

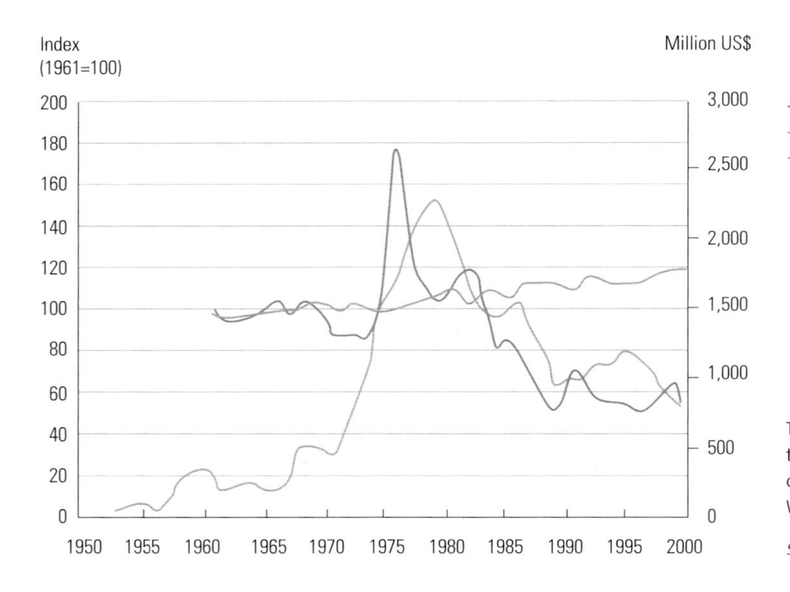

Food price (indexed)
Food per capita (indexed)
Lending for irrigation and drainage (million US$)

This figure shows that total investment in irrigation and drainage tends to correspond with food prices. Lending does not include lending by commercial banks to private farmers; it only includes lending by the World Bank.

Source: Thompson, 2001.

Table 8.1: National values of key indicators in securing the food supply

Country	Number of people undernourished 1990–92 (millions)	Number of people undernourished 1997–99 (millions)	Proportion of undernourished in total population 1990–92 (%)[1]	Proportion of undernourished in total population (%)[1]	Cultivated land area in 1998 (1,000 ha)[2]	Irrigated land area in 1998 (1, 000 ha)[2]	Irrigated land as % of cultivated land in 1998	Agricultural water withdrawal in 1998 (km^3/year)	Total renewable water resources (km^3/year)[3]	Agricultural water withdrawal as % of total renewable water resources in 1998
Afghanistan	9.3	12.1	64	58	8,054	2,386	30	22.84	65.00	35
Albania	0.5	0.3	14	10	699	340	49	1.00	41.70	2
Algeria	1.3	1.7	5	6	8,174	560	7	3.94	14.49	27
Angola	6.0	6.3	61	51	3,500	75	2	0.21	184.00	0.1
Antigua and Barbuda	–	–	–	–	8	0	–	0.001	0.05	2
Argentina	0.7	0.4	0	0	27,200	1,561	6	21.52	814.00	3
Armenia	–	1.3	–	35	560	287	51	1.94	10.53	18
Aruba	–	–	–	–	2	0	–	–	–	–
Australia	–	–	0	0	53,786	2,365	4	6.70	492.00	1
Austria	–	–	0	0	1,479	4	0.3	0.02	77.70	0.03
Azerbaijan	–	2.9	–	37	1,972	1,455	74	11.65	30.28	38
Bahamas	–	–	–	–	10	0	–	–	0.02	–
Bahrain	–	–	–	–	6	5	83	0.17	0.12	147
Bangladesh	39.2	44.1	35	33	8,332	3,850	46	70.20	1,210.64	6
Barbados	–	–	–	–	17	1	6	0.02	0.08	23
Belarus	–	0.1	–	0	6,311	115	2	0.84	58.00	1
Belgium – Luxembourg	–	–	0	0	832	40	5	0.11	21.40	0.5
Belize	–	–	–	–	89	3	3	0.0002	18.56	0.001
Benin	0.9	0.9	19	15	1,850	12	1	0.19	24.80	1
Bhutan	–	–	–	–	160	40	25	0.40	95.00	0.4
Bolivia	1.7	1.7	25	22	2,203	128	6	1.12	622.53	0.2
Bosnia and Herzegovina	–	0.2	–	4	650	2	0.3	–	37.50	–
Botswana	0.2	0.3	17	23	346	1	0.3	0.06	14.40	0.4
Brazil	19.3	15.9	13	10	65,200	2,870	4	36.12	8,233.00	0.4
Brunei Darussalam	–	–	0	0	7	1	14	–	8.50	–
Bulgaria	0.2	0.9	3	11	4,511	800	18	1.97	21.30	9
Burkina Faso	2.8	2.6	31	24	3,450	25	1	0.69	12.50	5
Burundi	2.8	4.1	48	66	1100	74	7	0.19	3.60	5
Cambodia	4.3	4.6	43	37	3,807	270	7	4.00	476.11	1
Cameroon	3.4	3.6	29	25	7160	33	0.5	0.73	285.50	0.3
Canada	–	–	0	0	45,700	720	2	5.41	2,902.00	0.2
Cape Verde	–	–	–	–	41	3	7	0.02	0.30	8
Central African Republic	1.4	1.5	46	43	2,020	0	–	0.001	144.40	0.001
Chad	3.5	2.5	58	34	3,550	20	1	0.19	43.00	0.4
Chile	1.1	0.6	8	4	2,294	1,800	78	7.97	922.00	1
China	192.6	116.3	16	9	135,365	52,878	39	414.76	2,896.00	14

Table 8.1: *continued*

Country	Number of people undernourished 1990–92 (millions)	Number of people undernourished 1997–99 (millions)	Proportion of undernourished in total population 1990–92 (%) [1]	Proportion of undernourished in total population (%) [1]	Cultivated land area in 1998 (1,000 ha)[2]	Irrigated land area in 1998 (1,000 ha)[2]	Irrigated land as % of cultivated land in 1998	Agricultural water withdrawal in 1998 (km³/year)	Total renewable water resources (km³/year)[3]	Agricultural water withdrawal as % of total renewable water resources in 1998
Colombia	6.1	5.3	17	13	4,115	850	21	4.92	2,132.00	0.2
Comoros	–	–	–	–	118	0	–	–	1.20	–
Congo, Dem. Republic of	13.7	31.0	35	64	7,880	11	0.1	0.11	1,283.00	0.01
Congo, Republic of	0.8	0.9	35	32	218	1	0.5	0.004	832.00	0.0005
Costa Rica	0.2	0.2	6	5	505	105	21	1.39	112.40	1
Côte d'Ivoire	2.5	2.4	19	16	7,350	73	1	0.60	81.00	1
Croatia	–	0.7	–	15	1,587	3	0.2	–	105.50	–
Cuba	0.5	1.9	5	17	4,465	870	19	5.64	38.12	15
Cyprus	–	–	0	0	144	40	28	0.17	0.78	22
Czech Republic	–	0.1	–	0	3,337	24	1	0.06	13.15	0.4
Denmark	–	–	0	0	2,374	460	19	0.55	6.00	9
Djibouti	–	–	–	–	1	1	100	0.007	0.30	2
Dominica	–	–	–	–	15	0	–	0	–	–
Dominican Republic	1.9	2.0	27	25	1,550	265	17	2.16	21.00	10
East Timor	–	–	–	–	80	0	–	–	–	–
Ecuador	0.9	0.6	8	5	3,001	865	29	13.96	432.00	3
Egypt	2.6	2.4	5	4	3,300	3,300	100	54.00	58.30	93
El Salvador	0.6	0.7	12	12	810	38	5	0.72	25.25	3
Equatorial Guinea	–	–	–	–	230	0	–	0.001	26.00	0.004
Eritrea	–	2.0	–	57	500	22	4	0.30	6.30	5
Estonia	–	0.1	–	4	1,135	4	0.4	0.008	12.81	0.1
Ethiopia	–	29.6	–	49	10,650	190	2	2.47	110.00	2
Fiji Islands	–	–	–	–	285	3	1	0.05	28.55	0.2
Finland	–	–	0	0	2,170	64	3	0.066	110.00	0.1
France	–	–	0	0	19,517	2,000	10	3.56	203.70	2
French Guiana	–	–	–	–	13	2	15	–	–	–
French Polynesia	–	–	–	–	21	0	–	–	–	–
Gabon	0.1	0.1	11	9	495	15	3	0.05	164.00	0.03
Gambia	0.2	0.2	19	15	200	2	1	0.02	8.00	0.3
Gaza Strip (Palestine)	–	–	–	–	25	12	48	–	0.06	–
Georgia	–	1.0	–	18	1,062	470	44	2.13	63.33	3
Germany	–	–	0	0	12,107	485	4	9.31	154.00	6
Ghana	5.4	2.7	35	15	5,300	11	0.2	0.25	53.20	0.5
Greece	–	–	0	0	3,882	1,422	37	6.12	74.25	8
Greenland	–	–	–	–	0	0	–	–	–	–
Grenada	–	–	–	–	11	0	–	–	–	–

Table 8.1: *continued*

Guadeloupe	–	–	–	–	26	3	12	–	–	–
Guam	–	–	–	–	12	0	–	–	–	–
Guatemala	1.3	2.3	14	22	1,905	130	7	1.61	111.27	1
Guinea	2.6	2.7	40	34	1,485	95	6	1.36	226.00	1
Guinea–Bissau	–	–	–	–	350	17	5	0.10	31.00	0.3
Guyana	0.1	0.1	19	14	496	150	30	1.60	241.00	1
Haiti	4.5	4.5	63	56	910	75	8	0.93	14.03	7
Honduras	1.1	1.3	23	21	1,875	76	4	0.66	95.93	1
Hungary	0.0	0.1	0	0	5,045	210	4	2.45	104.00	2
Iceland	–	–	0	0	7	0	–	0.0002	170.00	0.0001
India	214.6	225.3	25	23	169,650	57,000	34	580.81	1,896.66	31
Indonesia	16.7	12.0	9	6	30,987	4,815	16	75.60	2,838.00	3
Iran, Islamic Rep. of	2.7	3.5	4	5	18,803	7,562	40	66.78	137.51	49
Iraq	1.2	3.0	7	14	5,540	3,525	64	39.38	75.42	52
Ireland	–	–	0	0	1,088	0	–	0.0002	52.00	0.0004
Israel	–	–	0	0	440	199	45	1.31	1.67	78
Italy	–	–	0	0	11,137	2,698	24	20.00	191.30	10
Jamaica	0.3	0.2	12	8	274	25	9	0.20	9.40	2
Japan	–	–	0	0	4,905	2,679	55	56.03	430.00	13
Jordan	0.1	0.2	3	5	384	75	20	0.76	0.88	86
Kazakhstan	–	1.7	–	11	30,135	2,332	8	28.41	109.61	26
Kenya	11.5	13.4	47	46	4,520	67	1	1.01	30.20	3
Korea, Dem. People's Rep.	3.4	8.8	17	40	2,000	1,460	73	4.96	77.14	6
Korea, Republic of	0.8	0.7	0	0	1,910	1,159	61	8.99	69.70	13
Kuwait	0.5	0.1	23	4	7	6	86	0.20	0.02	1,000
Kyrgyzstan	–	0.5	–	10	1,428	1,072	75	9.45	20.58	46
Lao, PDR	1.2	1.4	29	28	940	168	18	2.59	333.55	1
Latvia	–	0.1	–	4	1,871	20	1	0.04	35.45	0.1
Lebanon	0.1	0.1	0	0	308	120	39	1.06	4.41	24
Lesotho	0.5	0.5	28	25	325	1	0.3	0.01	3.02	0.3
Liberia	0.8	1.0	37	42	390	3	1	0.06	232.00	0.03
Libyan Arab Jamahiriya	0.0	0.0	0	0	2,150	470	22	5.13	0.60	854
Lithuania	–	0.1	–	3	3,004	8	0.3	0.02	24.90	0.1
Macedonia, The Fmr. Yug. Rep.	–	0.1	–	5	635	55	9	–	6.40	–
Madagascar	4.3	6.1	35	40	3,108	1,090	35	14.31	337.00	4
Malawi	4.8	3.8	49	35	2,000	28	1	0.81	17.28	5
Malaysia	0.6	0.4	3	0	7,605	365	5	5.60	580.00	1
Maldives	–	–	–	–	3	0	–	0	0.03	0
Mali	2.2	3.0	25	28	4,650	138	3	6.87	100.00	7
Malta	–	–	–	–	9	2	22	0.01	0.05	28
Martinique	–	–	–	–	23	3	13	–	–	–
Mauritania	0.3	0.3	14	11	500	49	10	1.50	11.40	13
Mauritius	0.1	0.1	6	6	106	20	19	0.37	2.21	17
Mexico	4.3	5.0	5	5	27,300	6,500	24	60.34	457.22	13

Table 8.1: *continued*

Country	Number of people undernourished 1990–92 (millions)	Number of people undernourished 1997–99 (millions)	Proportion of undernourished in total population 1990–92 (%) [1]	Proportion of undernourished in total population (%) [1]	Cultivated land area in 1998 (1,000 ha) [2]	Irrigated land area in 1998 (1,000 ha) [2]	Irrigated land as % of cultivated land in 1998	Agricultural water withdrawal in 1998 (km^3/year)	Total renewable water resources (km^3/year) [3]	Agricultural water withdrawal as % of total renewable water resources in 1998
Moldova, Republic of	–	0.4	–	10	2,182	307	14	0.76	11.65	7
Mongolia	0.8	1.0	34	42	1,322	84	6	0.23	34.80	1
Morocco	1.4	1.8	5	6	9,976	1,291	13	11.36	29.00	39
Mozambique	9.6	9.5	69	54	3,350	107	3	0.55	216.11	0.3
Myanmar	3.9	3.2	9	7	10,143	1,692	17	27.86	1045.60	3
Namibia	0.4	0.6	30	33	820	7	1	0.17	17.94	1
Nepal	3.5	5.0	19	23	2,968	1,135	38	9.82	210.20	5
Netherlands	–	–	0	0	941	565	60	2.69	91.00	3
New Caledonia	–	–	–	–	13	0	–	–	–	–
New Zealand	–	–	0	0	3,280	285	9	0.89	327.00	0.3
Nicaragua	1.2	1.4	30	29	2,746	88	3	1.08	196.69	1
Niger	3.3	4.2	42	41	5,000	66	1	2.08	33.65	6
Nigeria 12.0	7.6	14	7	30,738	233	1	5.51	286.20	2	
Norway	–	–	–	–	903	127	14	0.23	382.00	0.1
Oman	–	–	–	–	77	62	81	1.23	0.99	125
Pakistan	26.5	24.4	24	18	21,970	18,000	82	161.84	418.27	39
Panama	0.5	0.4	19	16	655	35	5	0.23	147.98	0.2
Papua New Guinea	0.9	1.2	24	26	670	0	–	0.001	801.00	0.0001
Paraguay	0.8	0.7	18	13	2,285	67	3	0.35	336.00	0.1
Peru	8.9	3.1	41	13	4,170	1,195	29	16.42	1,913.00	1
Philippines	16.0	17.2	26	24	10,000	1,550	16	21.10	479.00	4
Poland	0.3	0.3	0	0	14,379	100	1	1.35	61.60	2
Portugal	–	–	0	0	2,620	650	25	3.60	68.70	5
Puerto Rico	–	–	–	–	81	40	49	–	3.40	–
Qatar	–	–	–	–	21	13	62	0.21	0.05	398
Reunion	–	–	–	–	38	12	32	–	5.00	–
Romania	0.7	0.3	3	0	9,843	2,880	29	14.23	211.85	7
Russian Federation	–	8.1	–	6	127,959	4,663	4	13.83	4,507.25	0.3
Rwanda	2.2	2.6	34	40	1070	4	0.4	0.02	5.20	0.4
Samoa	–	–	–	–	122	0	–	–	–	–
Sao Tome and Principe	–	–	–	–	41	10	24	–	2.18	–
Saudi Arabia	0.3	0.4	0	0	3,785	1,620	43	15.42	2.40	643
Senegal	1.7	2.1	23	24	2,266	71	3	1.43	39.40	4
Seychelles	–	–	–	–	7	0	–	–	–	–
Sierra Leone	1.9	1.7	46	41	540	29	5	0.34	160.00	0.2
Singapore	–	–	–	–	1	0	–	–	–	–

Table 8.1: *continued*

Slovakia	–	0.1	–	0	1,604	174	11	–	50.10	–
Slovenia	–	0.0	–	0	203	2	1	–	31.87	–
Solomon Islands	–	–	–	–	60	0	–	–	44.70	–
Somalia	4.8	6.0	67	75	1,065	200	19	3.28	13.50	24
South Africa	2.7	3.5	4	5	15,750	1,350	9	10.03	50.00	20
Spain	–	–	0	0	18,516	3,652	20	24.22	111.50	22
Sri Lanka	5.0	4.3	29	23	1,889	651	34	11.74	50.00	23
Saint Helena	–	–	–	–	4	0	–	–	–	–
Saint Kitts Nevis	–	–	–	–	7	0	–	–	0.02	–
Saint Lucia	–	–	–	–	17	3	18	–	–	–
Saint Vincent/Grenadines	–	–	–	–	11	1	9	0	–	–
Sudan	7.9	6.3	31	21	16,900	1,950	12	36.07	64.50	56
Suriname	0.0	0.0	12	11	67	51	76	0.62	122.00	1
Swaziland	0.1	0.1	10	12	180	69	38	0.75	2.64	28
Sweden	–	–	0	0	2,784	115	4	0.26	174.00	0.2
Switzerland	–	–	0	0	439	25	6	0.05	53.50	0.1
Syrian Arab Republic	0.2	0.2	0	0	5,484	1,213	22	18.96	26.26	72
Tajikistan	–	2.8	–	47	864	719	83	10.96	15.98	69
Tanzania, United Republic of	9.1	15.5	34	46	4,650	155	3	1.79	91.00	2
Thailand	16.9	12.9	30	21	18,297	4,749	26	79.29	409.94	19
Togo	0.9	0.7	27	17	2,300	7	0.3	0.08	14.70	1
Tonga	–	–	–	–	48	0	–	–	–	–
Trinidad and Tobago	0.1	0.2	12	13	122	3	2	0.02	3.84	0.4
Tunisia	0.1	0.0	0	0	5,100	380	7	2.23	4.46	50
Turkey	0.9	1.2	0	0	26,968	4,380	16	27.11	229.30	12
Turkmenistan	–	0.4	–	9	1,800	1,800	100	24.04	24.72	97
Uganda	4.2	6.2	24	28	6810	9	0.1	0.12	66.00	0.2
Ukraine	–	2.6	–	5	33,821	2,446	7	20.00	139.55	14
United Arab Emirates	0.1	0.1	3	0	132	74	56	1.53	0.15	1,021
United Kingdom	–	–	0	0	6,306	108	2	0.28	147.00	0.2
United States of America	–	–	0	0	179,000	22,300	12	209.43	3,069.00	7
Uruguay	0.2	0.1	6	3	1,307	180	14	3.03	139.00	2
Uzbekistan	–	0.9	–	4	4,850	4,281	88	54.37	50.41	108
Venezuela	2.3	4.8	11	21	3,490	570	16	3.94	1,233.17	0.3
Viet Nam	18.0	14.2	27	19	7,250	3,000	41	48.62	891.21	5
West Bank	–	–	–	–	209	12	6	–	0.75	–
Yemen	4.4	5.7	36	34	1,680	490	29	6.19	4.10	151
Yugoslavia, Fed. Rep. of	–	0.5	–	5	4,047	57	1	–	208.50	–
Zambia	3.6	4.7	43	47	5,279	46	1	1.32	105.20	1
Zimbabwe	4.6	4.8	43	39	3,350	117	3	2.24	20.00	11

– No data available. (1) Values marked as 0 are < 2.5%. (2) Values marked as 0 are < 1,000 ha. (3) Aggregation of data can not be carried out as it would result in double counting of shared water resources.

Source: FAO estimates.

Table 8.2: Per capita food consumption from 1965 to 2030 (kcal/person/day)

	1965	1975	1985	1998	2015	2030
World	2,358	2,435	2,655	2,803	2,940	3,050
Developing countries	2,054	2,152	2,450	2,681	2,850	2,980
Sub-Saharan Africa	2,058	2,079	2,057	2,195	2,360	2,540
Near East/North Africa	2,290	2,591	2,953	3,006	3,090	3,170
Latin America/Caribbean	2,393	2,546	2,689	2,826	2,980	3,140
South Asia	2,017	1,986	2,205	2,403	2,700	2,900
East Asia	1,957	2,105	2,559	2,921	3,060	3,190
Industrial countries	2,947	3,065	3,206	3,380	3,440	3,500
Transition countries	3,222	3,385	3,379	2,906	3,060	3,180

There is a global food security situation that is steadily improving, with a consistently increasing global level of food consumption per capita.

Source: FAO, 2002.

Figure 8.2: Per capita food consumption from 1965 to 2030

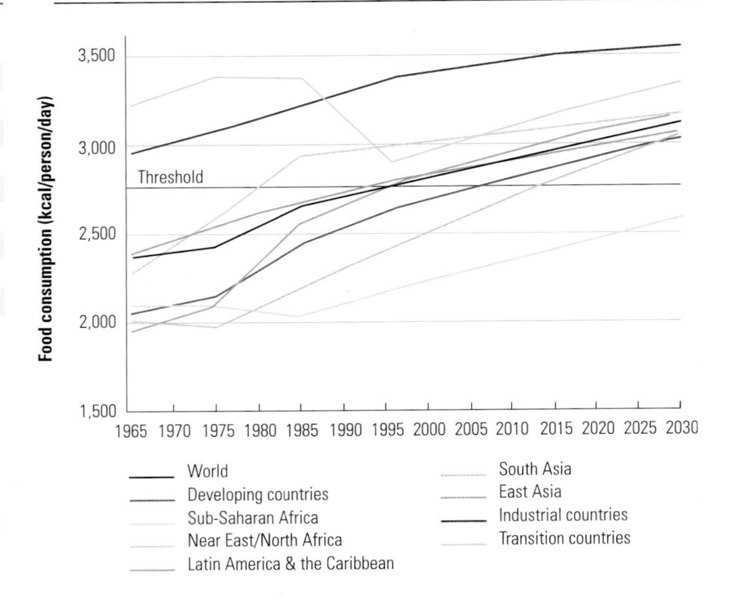

This figure shows a global food security situation that is consistently improving, at both global and developing country levels. The threshold of 2,700 kcal is taken as an indicator of the large proportion of people still affected by malnutrition.

Source: FAO, 2002.

Per capita food consumption, expressed in kcal/person/day, is used as the indicator of food intake. The evolution of per capita food consumption in 1965 and 2030 is given in table 8.2, based on historical data and on FAO projections for the years 2015 and 2030.

Table 8.2 and figure 8.2 show a global food security situation that is consistently improving, with a steady increase in per capita food consumption at the global level as well as at the level of developing countries. Demand for food tends to saturate at the level of 3,500 kcal/person/day. Figure 8.2 and table 8.2 also clearly show that per capita food consumption in sub-Saharan Africa remained disappointingly low over the last forty-five years although recent improvement trends are expected to continue. It should be noted that gains in overall food consumption are not necessarily translated into commensurate declines in the absolute numbers of undernourished people, in particular when there is high population growth.

The main sources of food supply

The main source of food for the population of the world is agriculture. The term agriculture, as broadly used here, also includes livestock husbandry, managed fisheries (aquaculture) and forestry. The composition of meals changes gradually as demand for food strengthens and lifestyles change. For those that can afford it, many products that are grown out of season or are exotic now appear on their local market. What agriculture produces is driven by consumer demand, and changes in consumer preferences have an influence on the water needed for food production.

It has been estimated that unmanaged natural systems could provide food for 600 million people, one tenth of the current world population (Mazoyer and Roudart, 1998). Thus, about 90 percent of the present world population could not be sustained without agriculture. Yet while few people live from only fishing, gathering and hunting, these unmanaged, or only loosely managed, natural food systems provide a strategically important contribution to the nutrition of indigenous people as well as to the existence and survival of many displaced, poor and marginal people. Except for marine fisheries, which are monitored, the diffuse reality of food resources directly obtained from natural ecosystems escapes most data collection and is usually not reflected in agricultural and economic statistics.

Therefore, the bulk of global food production (cereals, oils, livestock and fish) is dependent upon a whole range of agricultural systems in which water is a critical factor of production (FAO/World Bank, 2001).

Major crops

The prominent role of cereals and oil crops

Cereals are by far the most important source of total food consumption as measured in calories. In developing countries, consumption of cereals thirty years ago was 141 kilograms (kg)/person/year, representing 61 percent of total calories. At present it is 173 kg/person/year and provides 56 percent of calories. Thus, cereal use has increased, but less than other components of food intake. The fact that the growth of global demand for cereals is declining reflects diet diversification, as more countries achieve higher levels of nutrition. However, it is expected that cereals will continue to supply more than 50 percent of the food consumed in the foreseeable future.

To satisfy the cereal demand for a growing population using more cereals per capita, the annual world production of cereals grew by almost a billion tons from 0.94 billion tons in the mid-1960s to 1.89 billion tons in 1998. In the late 1990s, a slow-down in the growth of world consumption of cereals was recorded. It was, however, not caused by production constraints but by slowing demand. The annual world production of cereals is projected by FAO to increase by another billion tons from the 1998 level, to 2.8 billion tons. Within the cereal group, the relative importance of rice is expected to decline slightly, while consumption of wheat will continue to grow in per capita terms.

A large proportion of cereals are not produced for direct human consumption. Thus, of the future increment in cereal production projected by FAO, just under half will be for food, about 44 percent for animal feed, with the balance going to other uses, such as seed, industrial non-food and waste.

One out of every five calories added to food consumption in developing countries in the last two decades originated in the group of oil crops, which includes palm oil, soybean, sunflower, groundnut, sesame and coconut. In projections towards the future, it is expected that 44 percent of additional calories may come from these products. This projection reflects the prospect, in the majority of developing countries, of only modest further growth in the direct consumption of staples such as cereals, roots and tubers, in favour of non-staples such as vegetable oils. The major driving force of the world oil crops economy has been the growth of food demand in the developing countries, but additional demand growth has been experienced in the non-food industrial uses of oil and also in feed for the livestock sector. The future growth of aggregate world demand for, and production of, oil crops is expected to remain well above that of total agriculture. In terms of trade, developing countries have so far been net exporters of oil crops, but this position is likely to change as local consumption increases.

Sources of growth in crop production

There are three sources of growth in crop production:

- arable land expansion;
- increase in cropping intensities (multiple cropping and shorter fallow periods); and
- yield growth.

Since the early 1960s, land in agricultural use (arable land and land under permanent crops) in the world has increased by 12 percent to about 1.5 billion hectares. This amounts to 11 percent of the globe's land surface. During the same period, the world population nearly doubled from 3.1 billion to over 5.9 billion in 1998. By implication, arable land per person declined by 40 percent, from 0.43 hectares in 1962 to 0.26 hectares in 1998. As the world food system kept providing enough food for a growing population, a secular decline in the real price of food took place and the global situation of nutrition improved, both in relative terms and in absolute numbers. In the context of low food prices and consequent relatively low value of agricultural land, prime agricultural land is being converted to higher value urban and industrial uses. Also, irrigated land with inadequate or non-existing drainage infrastructure is being gradually lost to salinization which results in lowering yields. Yield increase and intensification have more than compensated for the reduction in per capita land availability.

As an example of growing crop yields, the world average grain yields doubled from 1.4 tons/hectares/crop in 1962 to 2.8 tons/hectares/crop in 1996. The average cropping intensity probably increased by some 5 percentage points, so that the arable land required to produce a given amount of grain declined by some 56 percent. It is expected that in the future, 80 percent of increased crop production in developing countries will come from intensification through higher yields, increased multiple cropping and shorter fallow periods. The remaining 20 percent would come from expansion of agricultural land in those developing countries and regions where the potential for expansion exists and where the prevailing farming systems and general demographic and socio-economic conditions favour it.

From 1998 to 2030, arable land in developing countries is projected to increase by 13 percent (120 million hectares). The bulk of the projected expansion is expected to take place in sub-Saharan Africa and Latin America, with a smaller part in east Asia. The slow-down in the expansion of arable land is mainly a consequence of the projected decrease in the growth of crop production.

Livestock: an increasing role

Food production from the livestock sector includes meat (beef, pork, poultry), dairy products and eggs. In the last few decades, consumption of meat in developing countries has been growing at a rate of about 5 to 6 percent per year, that of milk and dairy products at 3 to 4 percent per year. Much of the growth has occurred in a small number of countries, including such populous countries as Brazil and China. Many developing nations and whole regions, including sub-Saharan Africa and parts of the Near East/North Africa, where the need to increase protein consumption is the greatest, have not been participating in the buoyancy of the world meat sector. Worldwide, the poultry sector has been expanding fastest, and its share in the total meat output went from 13 percent in the mid-1960s to 28 percent today. The share of poultry in meat production is expected to continue in the future. The forces that made for the rapid growth of the meat sector in the past are, however, expected to weaken in the future owing to lower population growth and the deceleration of growth that follows the attainment of a fairly high level of consumption. Intensive forms of livestock production have led to a strong demand for cereals used as animal feed and production is rising steadily to meet this demand.

Fisheries and the growing importance of aquaculture

Average world per capita consumption of fish reached about 16.3 kg per year in 1999, with large differences ranging from countries with virtually no fish consumption to countries that reach over 100 kg per year. Per capita consumption could grow to 19 to 20 kg by 2030, raising total fish use to 150–160 million tons. Of the total world fisheries and aquaculture production in 2000 (130 million tons), some 74 percent (97 million tons) was available for direct human consumption as food fish. The remainder was reduced into fishmeal and fish oil for use in animal feeding (livestock and aquaculture) or for industrial purposes. Marine capture fisheries production, excluding aquaculture was located in the range of 80 to 85 million tons per year in the 1990s. The long-term yearly sustainable yield of marine capture fisheries is estimated at no more than around 100 million tons per year; overfishing of some species in certain parts of the worlds threatens the resource base. Achieving and sustaining these levels assume more efficient utilization of stocks, healthier ecosystems and better conservation of critical habitats. Inland catches (excluding aquaculture) were recorded at about 7 to 8 million tons per year. However, a recent in-depth study of inland fisheries revealed that actual catches may be double this amount. It is important to note that fishery resources in many inland water bodies such as rivers and lakes are under increasing environmental threats resulting from continued trends of growing aquatic pollution, habitat degradation, water abstraction and other human-made pressures.

The bulk of the future increase of fish supply will have to come from aquaculture, which has been growing at a rate of 10 percent per year during the 1990s and has increased its share in world fish supplies to about 27 percent. Most aquaculture development took place in Asia (some 70 percent of world aquaculture production is in China). At present, aquaculture production amounts to 35 million tons, of which 21 million is inland and 14 million marine. Over 90 percent of total aquaculture food fish production in 1995 came from developing countries, compared with 51 percent of terrestrial animal meat production. Fish exports from developing countries have been growing rapidly and now far exceed earnings from commodities such as coffee, cocoa, bananas or rubber. Strong growth may continue for some time, but constraints such as lack of feedstuffs and suitable sites, diseases and environmental constraints are becoming more binding. Major factors affecting both the sustainability of capture fisheries and the expansion of aquaculture will include improved management in the sector and a better understanding of aquatic ecosystems, as well as prevention and better management of environmental impacts affecting fishery resources and aquatic biodiversity.

Food trade

Developing countries increasing their imports

At the global level, food production equals consumption. For individual countries and clusters of countries, however, production and consumption differ depending on agricultural trade. In general, the growth rates of food production in the developing countries have been below growth rates of demand, and food imports of these countries have been growing faster than their agricultural exports. For example, the net cereal imports of developing countries increased from 39 million tons in the mid-1970s to 103 million tons in 1998. Notwithstanding lower growth in the demand for cereals in the future, the dependence of developing countries on cereal imports is expected to continue to grow owing to limited potential in these countries to increase production. One production constraint is scarcity of water resources for irrigation, but inadequate access to credit and markets, and poor agricultural policy and management have also hampered production increases. The course towards a widening net trade deficit of the developing countries is projected to continue: net food imports are expected to rise fairly rapidly to 198 million tons in 2015 and 265 million tons in 2030. This compares to a projected cereal production in developing countries to the order of 1,650 million tons in 2030.

Few countries pursue a policy of 100 percent food self-sufficiency, and likewise, few countries depend on imports for more than 20 percent of their food demand. A number of countries with a chronic trade balance deficit and high population growth already have difficulty in raising the foreign exchange needed to satisfy the growing demand for food imports. While in the past such a foreign exchange situation would have called for an increase in import taxes and encouragement for local food production to supply the local market, the structural adjustment programmes and market liberalization policies implemented in the 1980s and 1990s have precluded the adoption of national policies leading in the direction of food self-sufficiency (Stiglitz, 2002). Yet farmers in many developing countries with weak infrastructure and no access to capital and technology cannot face competition from the international market. This is particularly the case when their production competes with that of the heavily subsidized agriculture of industrial countries where the productivity of labour can be 1,000 times higher than theirs (Mazoyer and Roudart, 1997).

The concept of virtual water

The term 'virtual water' was coined in the 1990s in support of a trade and water policy point: for the food security of arid countries, where water is needed for domestic use and in support of the services and industrial sector, it is not necessary to use water for local food production, because the easier and economically more attractive alternative is to import food, in particular the inexpensive cereal base of the national diet. Thus, using a hydrologic perspective, trade in food was called trade in virtual water, that is the water consumed to produce an agricultural commodity. For example, a crop such as wheat consumes about 1 to 5 cubic metres (m^3) of water to produce 1 kg of cereal. For poultry with a feed/meat conversion factor of 4:1, the virtual water content would be 6 m^3/kg of poultry meat. For cattle, with a conversion factor of 10:1, the virtual water content of 1 kg of beef would be 15 m^3. The amount of virtual water imported by a country is a measure of the degree to which the country depends on the international market for its food supply.

Manipulation of the virtual water concept is subject to some caveats, one of which is that the water actually used by a crop may have stemmed partially or totally from rain, which is free of cost, whereas piped water has a cost. In the case of meat, one has to keep in mind that free-roaming animals are efficient collectors of virtual water: in arid areas, the rainfed pasture they consume usually would have no other use.

The Use of Water in Agriculture

Water for food production

For vegetative growth and development, plants require, within reach of their roots, water of adequate quality, in appropriate quantity and at the right time. Most of the water a plant absorbs performs the function of raising dissolved nutrients from the soil to the aerial organs; from there it is released to the atmosphere by transpiration: agricultural water use is intrinsically consumptive. Crops have specific water requirements and these vary depending on local climatic conditions. Whereas an indicative figure for producing 1 kg of wheat is about 1 m^3 of water which in turn is returned to the atmosphere, paddy rice may require twice this amount. The production of meat requires between six and twenty times more water than for cereals, depending on the feed/meat conversion factor. Specific values for the water equivalent of a selection of food products are given in table 8.3. Water required for human food intake can be derived from these specific values in a grossly approximate way, depending on the size and composition of the meals (see box 8.2).

Food production: the dominant role of rainfed agriculture

Non-irrigated (rainfed) agriculture depends entirely on rainfall stored in the soil profile. This form of agriculture is possible only in regions where rainfall distribution ensures continuing availability of soil moisture during the critical growing periods for crops. Non-irrigated agriculture accounts for some 60 percent of production in the developing countries. In rainfed agriculture, land management can have a significant impact on crop yields: proper land preparation

Box 8.2: Assessing freshwater needs for global food production

The amount of water involved in food production is significant, and most of it is provided directly by rainfall. A rough calculation of global water needs for food production can be based on the specific water requirements to produce food for one person. Depending on the composition of meals and allowing for post-harvest losses, the present average food ingest of 2,800 kcal/person/day may require roughly 1,000 m^3 per year to be produced. Thus, with a world population of 6 billion, water needed to produce the necessary food is 6,000 km^3 (excluding any conveyance losses associated with irrigation systems). Most water used by agriculture stems from rainfall stored in the soil profile and only about 15 percent of water for crops is provided through irrigation. Irrigation therefore needs 900 km^3 of water per year for food crops (to which some water must be added for non-food crops). On average, about 40 percent of water withdrawn from rivers, lakes and aquifers for agriculture effectively contribute to crop production, the remainder being lost to evaporation, deep infiltration or the growth of weeds. Consequently, the current global water withdrawals for irrigation are estimated to be about 2,000 to 2,500 km^3 per year.

Table 8.3: Water requirement equivalent of main food products

Product	Unit	Equivalent water in m^3 per unit
Cattle	head	4,000
Sheep and goats	head	500
Fresh beef	kg	15
Fresh lamb	kg	10
Fresh poultry	kg	6
Cereals	kg	1.5
Citrus fruits	kg	1
Palm oil	kg	2
Pulses, roots and tubers	kg	1

This table gives examples of water required per unit of major food products, including livestock, which consume the most water per unit. Cereals, oil crops, and pulses, roots and tubers consume far less water.

Source: FAO, 1997a.

leading surface runoff to infiltrate close to the roots improves the conservation of moisture in the soil. Various forms of rainwater harvesting can help to retain water *in situ*. Rainwater harvesting not only provides more water for the crop but can also add to groundwater recharge and help to reduce soil erosion. Other methods are based on collecting water from the local catchment and rely either on storage within the soil profile or else local storage behind bunds or ponds and other structures for use during dry periods. Recently, practices such as conservation tillage have proven to be effective in improving soil moisture conservation.

The potential to improve non-irrigated yields is restricted where rainfall is subject to large seasonal and interannual variations. With a high risk of yield reductions or complete loss of crop from dry spells and droughts, farmers are reluctant to invest in inputs such as plant nutrients, high-yielding seeds and pest management. For resource-poor farmers in semi-arid regions, the overriding requirement is to harvest sufficient foodstuff to ensure nutrition of the household through to the next harvest. This objective may be reached with robust, drought-resistant varieties associated with low yields. Genetic engineering has not yet delivered high-yield drought-resistant varieties, a difficult task to achieve because, for most crop plants, drought resistance is associated with low yields.

Role of irrigation in food production

In irrigated agriculture, water taken up by crops is partly or totally provided through human intervention. Irrigation water is withdrawn from a water source (river, lake or aquifer) and led to the field through an appropriate conveyance infrastructure. To satisfy their water requirements, irrigated crops benefit both from more or less unreliable natural rainfall and from irrigation water. Irrigation provides a powerful management tool against the vagaries of rainfall, and makes it economically attractive to grow high-yield seed varieties and to apply adequate plant nutrition as well as pest control and other inputs, thus giving room for a boost in yields. Figure 8.3 illustrates the typical yield response and water requirements in irrigated and non-irrigated agriculture. Irrigation is crucial to the world's food supplies. In 1998, irrigated land made up about one fifth of the total arable area in developing countries but produced two fifths of all crops and close to three fifths of cereal production.

The developed countries account for a quarter of the world's irrigated area (67 million hectares). Their annual growth of irrigated area reached a peak of 3 percent in the 1970s and dropped to only 0.2 percent in the 1990s. The population of this group of countries is growing only slowly and therefore a very slow growth in their demand and production of agricultural commodities is foreseen. The focus of irrigation development is consequently expected to concentrate on the group of developing countries where demographic growth is strongest. Increasing competition from the higher valued industrial and domestic sector results in a decrease in the amount of overall water used for irrigation. Figure 8.4 illustrates the case for the Zhanghe irrigation system in China.

Map 8.1 shows irrigated land as a percentage of arable land in developing countries. A high proportion of irrigated land is usually found in countries and regions with an arid or semi-arid climate. However, low proportions of irrigated land in sub-Saharan Africa point also to underdeveloped irrigation infrastructure. Data and projections of irrigated land compared to irrigation potential in developing countries are shown in figure 8.5. The irrigation potential figure already takes into account the availability of water. The graph shows that a sizeable part of irrigation potential is already used in the Near East/North Africa region (where water is the limiting factor) and in Asia (where land is often the limiting factor), whereas a large potential is still unused in sub-Saharan Africa and in Latin America.

According to FAO forecasts, the share of irrigation in world crop production is expected to increase in the next decades. Especially in developing countries, the area equipped for irrigation is expected to have expanded by 20 percent (40 million hectares) by 2030. This means that 20 percent of total land with irrigation potential but not yet equipped will be brought under irrigation, and that 60 percent of all land with irrigation potential (402 million hectares) will be in use by 2030. The net increase in irrigated land (40 million hectares, 0.6 percent per year) projected to 2030 is less than half the increase over the preceding thirty-six years (99 million hectares, 1.9 percent per year). The projected slowdown in irrigation development reflects the projected lower growth rate of food demand, combined with the increasing scarcity of suitable areas for irrigation and of water resources in some countries, as well as the rising cost of irrigation investment. The first selection of economically attractive irrigation projects has already been implemented, and prices for agricultural commodities have not risen to encourage investment in a second selection of more expensive irrigation projects.

Most of the expansion in irrigated land is achieved by converting land used for rainfed agriculture or land with rainfed production potential but not yet in use into irrigated land. The expansion of irrigation is projected to be strongest in south Asia, east Asia and Near East/North Africa. These regions have limited or no potential for expansion of non-irrigated agriculture. Arable land expansion will nevertheless remain an important factor in crop production growth in many countries in sub-Saharan Africa, Latin America and some countries in east Asia, although to a much smaller extent than in the past. The growth in wheat and rice production in the developing countries will increasingly come from gains in yield, while expansion of harvested land will continue to be a major contributor to the growth in production of maize.

Figure 8.3: Yields and water requirements of irrigated and non-irrigated agriculture

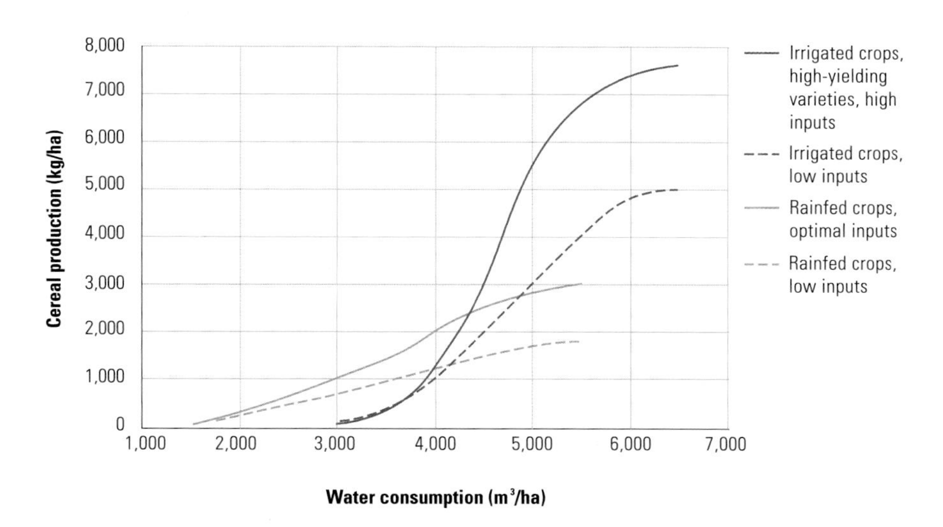

The graph shows the yield response and water requirements in irrigated and non-irrigated agriculture. Irrigated crops produce better than rainfed crops with high water consumption, even if those rainfed crops get optimal inputs. The graphs for rainfed agriculture stop at a certain point (5,500 m^3/ha) as it is impossible for 'typical' rainfed crops to consume more water without irrigation.

Source: Smith et al., 2001.

Figure 8.4: Competing uses of water in the Zhanghe irrigation district, China

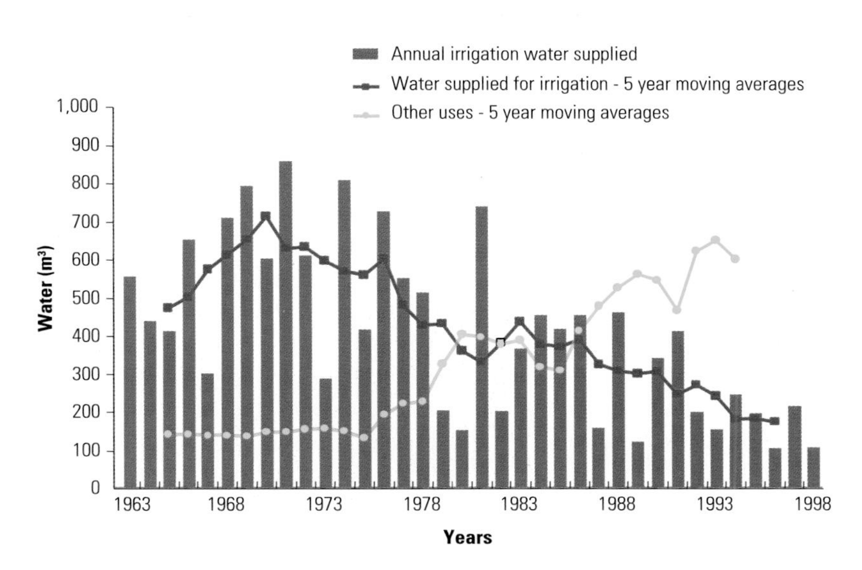

This figure illustrates that increasing competition and demand from the industrial and domestic sectors result in the decrease of irrigation's share of water use.

Source: Molden, unpublished.

Figure 8.5: Irrigated area as proportion of irrigation potential in developing countries

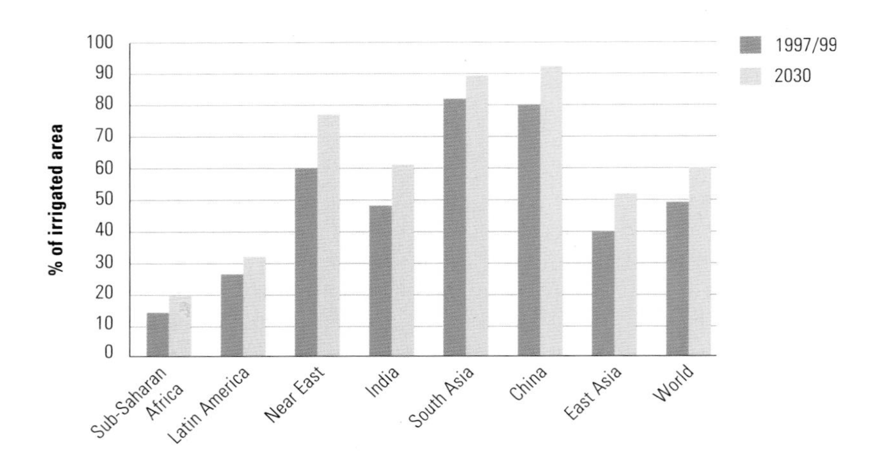

This figure shows that a vast share of the irrigation potential is already being used in sub-Saharan Africa and Latin America, as well as in Asia, but there remains a large potential still untapped in the Near East and India.

Source: FAO, 2002.

Future investments in irrigation

In many developing countries, investments in irrigated infrastructures have represented a significant share of the overall agricultural budget during the second half of the twentieth century. The unit cost of irrigation development varies with countries and types of irrigated infrastructures, ranging typically from US$1,000 to US$10,000 per hectare, with extreme cases reaching US$25,000 per hectare (these costs do not include the cost of water storage as the cost of dam construction varies on a case-by-case basis). The lowest investment costs in irrigation are in Asia, which has the bulk of irrigation and where scale economies are possible. The most expensive irrigation schemes are found in sub-Saharan Africa, where irrigation systems are usually smaller and the development of land and water resources is costly.

In the future, the estimates of expansion in land under irrigation will represent an annual investment of about US$5 billion, but most investment in irrigation – between US$10 and 12 billion per year – will certainly come from the needed rehabilitation and modernization of ageing irrigated schemes built during the period 1960–80. In the 1990s, annual investment in storage for irrigation was estimated at about US$12 billion (WCD, 2000). In the future, the contrasting effects of reduced demand for irrigation expansion and increased unit cost of water storage will result in an annual investment estimated between US$4 and 7 billion over the next thirty years.

Typically, investment figures in irrigation do not include the portion of investment provided by the farmer as land improvement and on-farm irrigation; this can represent up to 50 percent of the overall investment. In total, it is estimated that annual investment in irrigated agriculture will therefore range between US$25 and 30 billion, about 15 percent of annual expected investments in the water sector.

Water use efficiency

Assessing the impact of irrigation on available water resources requires an estimate of total abstraction for the purpose of irrigation from rivers, lakes and aquifers. The volume extracted is considerably greater than the consumptive use for irrigation because of conveyance losses from the withdrawal site to the plant root zone. Water use efficiency is an indicator often used to express the level of performance of irrigation systems from the source to the crop: it is the ratio between estimated plant requirements and the actual water withdrawal.

On average, it is estimated that overall water use efficiency of irrigation in developing countries is about 38 percent. Map 8.2 shows the importance of agriculture in the countries' water balance, and figure 8.6 shows the expected growth in water abstraction for irrigation from 1999 to 2030. The predictions are based on assumptions about possible improvements in irrigation efficiency in each region. These assumptions take into account that, from the farmer's perspective, wherever water is abundant and its cost low, the incentives to save water are limited. If the cost of water is irrelevant in the financial equation, it may be left flowing unless it is causing erosion or waterlogging. Conversely, if farmers can profitably irrigate more land using their allocation in an optimum way, irrigation efficiency may reach high levels.

Improving irrigation efficiency is a slow and difficult process that depends in large part on the local water scarcity situation. It may

Map 8.1: Area equipped for irrigation as percentage of cultivated land by country (1998)

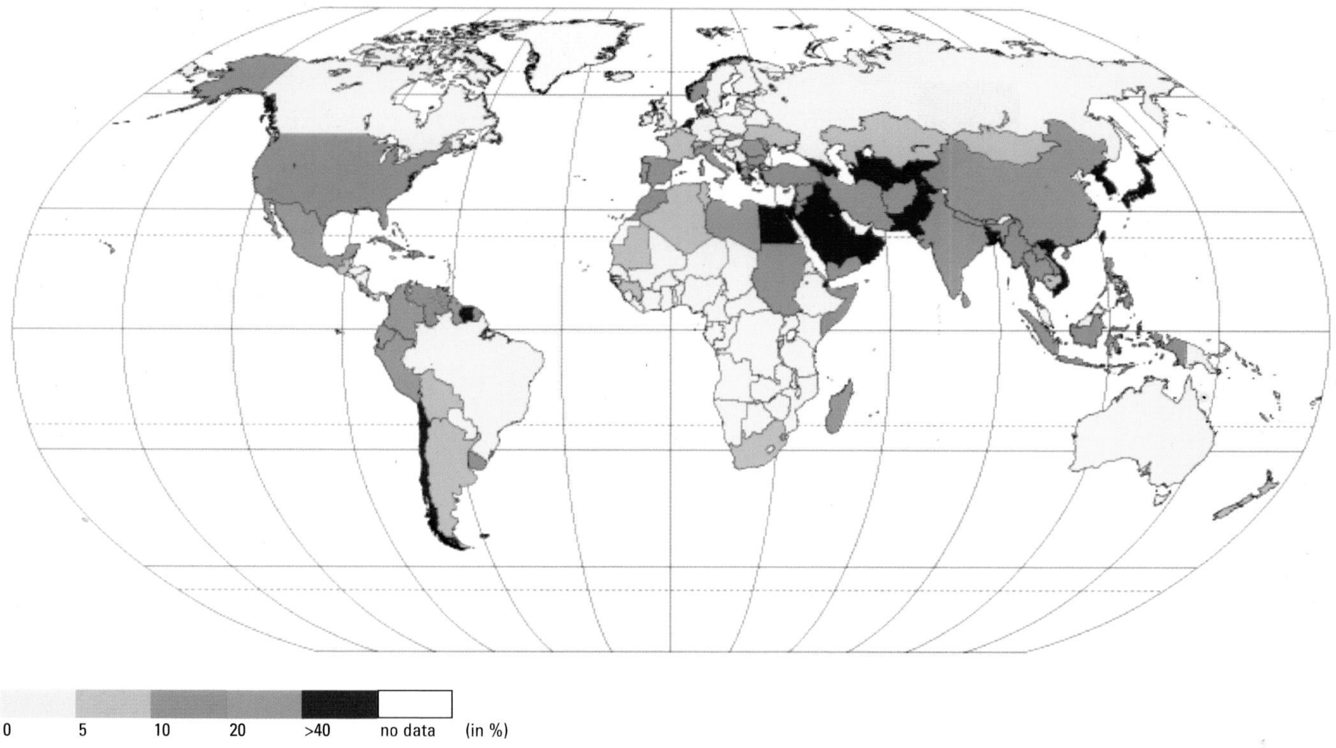

0 5 10 20 >40 no data (in %)

Expansive irrigated areas can be seen in the Middle East and Asia compared to the significantly under-irrigated arid and semi-arid regions such as sub-Saharan Africa. Developed countries worldwide show a consistently heavy irrigated area.

Source: Map produced for the World Water Assessment Programme (WWAP) by the Centre for Environmental Research, University of Kassel. Data source: FAOSTAT, 2002.

be expensive and requires willingness, know-how and action at various levels. Table 8.4 shows current and expected water use efficiency for developing countries in 1998 and 2030, as estimated by FAO. The investment and management decisions leading to higher irrigation efficiency involve irrigation system management and the system-dependent farmers. National water policy may encourage water savings in water-scarce areas by providing incentives and effectively enforcing penalties. When upstream managers cannot ensure conveyance efficiency, there may be no incentives for downstream water users to make efficiency gains. With groundwater, this caveat may not apply since the incentive is generally internalized by the users, and in many cases groundwater users show much greater efficiency than those depending on surface resources. Box 8.3 provides an overview of different aspects of potential improvements in agricultural water use efficiency.

Future water withdrawals for irrigation

Irrigation water withdrawal in developing countries is expected to grow by about 14 percent from the current 2,130 km^3 per year to

Figure 8.6: Irrigation and water resources: actual (1999) and predicted (2030) withdrawals

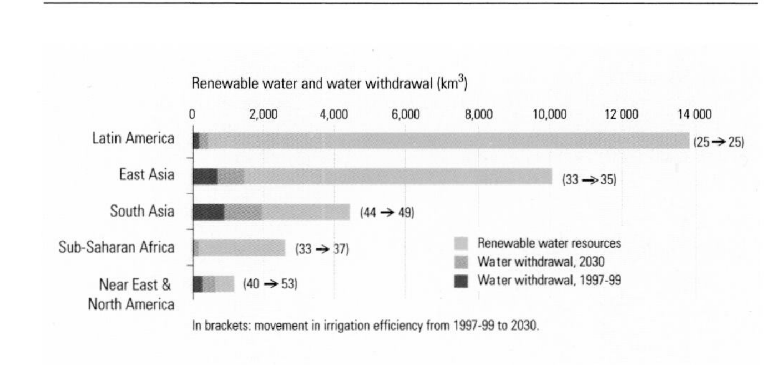

This figure shows the expected growth in water abstraction for irrigation for the period 1999 to 2030. There is a potential increase in all regions, most notably in south Asia, whereas the sub-Saharan Africa region is predicted to maintain its very low level of irrigation water withdrawals.

Source: FAO data and projections.

Map 8.2: Agricultural water withdrawals as percentage of renewable water resources (1998)

(%)
0 5 10 20 >40 no data

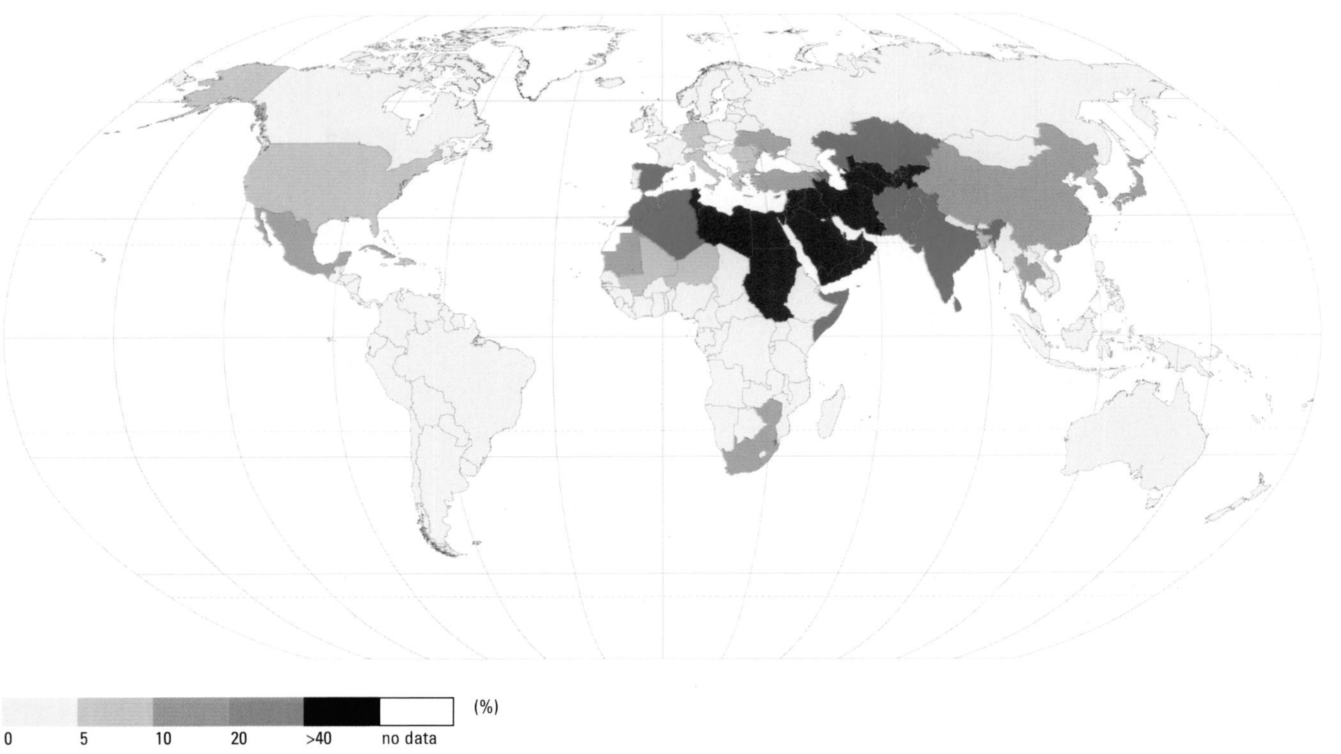

The importance of agriculture in countries' water balance is shown here. Whereas agricultural water withdrawals account for little of the total renewable water resources in the majority of countries, certain regions, such as north-east Africa and western Asia, are notable in that their agricultural withdrawals add up to more than 40 percent of their total water resources – and in the Near and Middle East, they represent a staggering 1,000 percent.

Source: Map produced for the World Water Assessment Programme (WWAP) by the Centre for Environmental Research, University of Kassel. Data source: AQUASTAT, 2002.

Table 8.4: Water use efficiency in 1998 and 2030 (predicted) in ninety-three developing countries

	Sub-Saharan Africa	Latin America	Near East – North Africa	South Asia	East Asia	All countries
Water use efficiency in irrigation (%)						
1998	33	25	40	44	33	38
2030	37	25	53	49	34	42
Irrigation water withdrawal as a percentage of renewable water resources (%)						
1998	2	1	53	36	8	8
2030	3	2	58	41	8	9

The projected rates of water use efficiency show a global potential, although some regions that have already largely tapped into their potential water use efficiency should remain stable, for example the Near East. The percentage of renewable water resources used for irrigation is set to increase in almost all regions as agriculture will continue to be the major user of water in the foreseeable future.

Source: FAO, 2002.

2,420 km^3 in 2030. This finding is consistent with the one given in box 8.2 earlier but it is based specifically on individual assessments for each developing country. Harvested irrigated area (the cumulated area of all crops during a year) is expected to increase by 33 percent from 257 million hectares in 1998 to 341 million hectares in 2030. The disproportionate increase in harvested area is explained by expected improvements in irrigation efficiency, which will result in a reduction in gross irrigation water abstraction per hectare of crop. A small part of the reduction is due to changes in cropping patterns in China, where consumer preference is causing a shift from rice to wheat production.

While some countries have reached extreme levels of water use for agriculture, irrigation still represents a relatively small part of the total water resources of developing countries. The projected increase in water withdrawal will not significantly alter the overall picture. At the local level, however, there are already severe water shortages, in particular in the Near East/North Africa region.

Of the ninety-three developing countries surveyed by FAO, ten are already using more than 40 percent of their renewable water resources for irrigation, a threshold used to flag the level at which countries are usually forced to make difficult choices between their agricultural and urban water supply sectors. Another eight countries were using more than 20 percent, a threshold that can indicate impending water scarcity. By 2030, south Asia will have reached the 40 percent level, and the Near East and North Africa not less than 58 percent. However, the proportion of renewable water resources allocated to irrigation in sub-Saharan Africa, Latin America and east Asia in 2030 is likely to remain far below the critical threshold.

Box 8.3: Potential for improvements in agricultural water use efficiency

Global water strategies tend to focus on the need to increase agricultural water use efficiency, reduce wastage and free large amounts of water for other, more productive uses as well as sustaining the environmental services of rivers and lakes. While there is scope for improved use of water in agriculture, these improvements can only be made slowly and are limited by several considerations. First, there are large areas of irrigated agriculture located in humid tropics where water is not scarce and where improved efficiency would not result in any gain in water productivity. Second, water use efficiency is usually computed at the level of the farm or irrigation scheme, but most of the water that is not used by the crops returns to the hydrological system and can be used further downstream. In these conditions, any improvement in water use efficiency at field level translates into limited improvement in overall efficiency at the level of the river basin. Finally, different cropping systems have different potential for improvement in water use efficiency. Typically, tree crops and vegetables are well adapted to the use of localized, highly efficient irrigation technologies, while such equipments are not adapted to cereal or other crops.

The special role of groundwater

Water contained in shallow underground aquifers has played a significant role in developing and diversifying agricultural production. This is understandable from a resource management perspective: when groundwater is accessible it offers a primary buffer against the vagaries of climate and surface water delivery. But its advantages are also quite subtle. Access to groundwater can occasion a large degree of distributive equity, and for many farmers groundwater has proved to be a perfect delivery system. Because groundwater is on demand and just-in-time, farmers have sometimes made private investments in groundwater technology as a substitute for unreliable or inequitable surface irrigation services. In many senses, groundwater has been used by farmers to break out of conventional command and control irrigation administration. Some of the management challenges posed by large surface irrigation schemes are avoided, but the aggregate impact of a large number of individual users can be damaging, and moderating the 'race to the pump-house' has proved difficult. However, as groundwater pumping involves a direct cost to the farmer, the incentives to use groundwater efficiently are high. These incentives do not apply so effectively where energy costs are subsidized; such distortion has arguably accelerated groundwater depletion in parts of India and Pakistan.

The technical principles involved in sustainable groundwater and aquifer management are well known but practical implementation of groundwater management has encountered serious difficulties. This is largely due to groundwater's traditional legal status as part of land property and the competing interests of farmers withdrawing water from common-property aquifers (Burke and Moench, 2000). Abstraction can result in water levels declining beyond the economic reach of pumping technology; this may penalize poorer farmers and result in areas being taken out of agricultural production. When near the sea, or in proximity to saline groundwater, overpumped aquifers are prone to saline intrusion. Groundwater quality is also threatened by the application of fertilizers, herbicides and pesticides that percolate into aquifers. These 'non-point' sources of pollution from agricultural activity often take time to become apparent, but their effects can be long-lasting, particularly in the case of persistent organic pollutants. For an example of this, see the Seine-Normandy case study in chapter 19.

Fossil groundwater, that is, groundwater contained in aquifers that are not actively recharged, represents a valuable but exhaustible resource. Thus, for example, the large sedimentary aquifers of North Africa and the Middle East, decoupled from contemporary recharge, have already been exploited for large-scale agricultural development in a process of planned depletion. The degree to which further abstractions occur will be limited in some cases by the economic limits to pumping, but promoted where strong economic demand from agriculture or urban water supply becomes effective (Schiffle, 1998). Two countries, Libyan Arab Jamahiriya (see box 8.4) and Saudi Arabia, are already using considerably more water for irrigation than their annual renewable resources, by drawing on fossil groundwater reserves. Several other countries rely to a limited extent on fossil groundwater for irrigation. Where groundwater reserves have a high strategic value in terms of water security, the wisdom of depleting such reserves for irrigation is questionable.

Ensuring Access to Food for All

The markets fail to provide food for all

Since the 1960s, market food prices have been low while food production satisfied market demand. However, FAO's estimates indicate that in 1998 there were 815 million undernourished people in the world: 777 million in the developing countries, 27 million in countries in transition and 11 million in the industrialized countries. The world is capable of producing sufficient food to feed its population until 2030 and beyond (actually, a growing part of cereal production is already dedicated to animal feed). The 1996 World Food Summit set a target of reducing the number of undernourished people to 400 million by 2015. FAO projections indicate that this target may not be achieved before 2030. The normative target and the projection of the current course of events are illustrated in figure 8.7.

The plight of undernourished people needs to be addressed through pro-active implementation of food security programmes. Necessary policy adjustment should be tailored to ensure that people can apply their initiative and ingenuity to access food and establish a livelihood. Food security programmes should identify the most vulnerable categories of population and consider their assets and constraints in order to emerge from poverty. FAO has developed specific indicators for this purpose (see box 8.5). A first level of support is emergency assistance to households that have been hit by natural or man-made disasters, or by conflicts. Households weakened by hunger and disease need to be restored to the levels of strength necessary to apply themselves to the construction of a viable livelihood. At this point, people may need timely support to realize their plans. External support may take a variety of forms, from provision of seeds and tools to capacity-building and infrastructure development. Many poverty alleviation activities bear some relation with water. The role of irrigation is discussed further on.

The undernourished: where, who and why?

Figure 8.8 and map 8.3 identify the countries with the highest prevalence of undernourished people. Many of these countries have been stricken by war and natural disasters, including extended periods of drought. Within the countries, large numbers of undernourished people live in environmentally degraded rural areas and in urban slums. During the 1990s, the number of undernourished people fell steeply in east Asia. In south Asia, although the proportion fell, the total number remained almost constant. In sub-Saharan Africa, the proportion remained virtually unchanged, which meant that the number of undernourished people rose steeply. Food security action has therefore a special focus on sub-Saharan Africa.

Many undernourished people are refugees who have lost their physical and social assets in displacement caused by war or natural disaster. The cause of displacement can also be unmitigated externalities stemming, for example, from urban development and consequent water pollution, as well as construction of dams and

Box 8.4: Libya – the Great Man-Made River Project

Libya is situated in the northern part of the African continent. Despite having an area exceeding 1,750,000 km^2, only a small part of the country has escaped the grasp of the desert. However, cave paintings in the south of the country indicate that the area once enjoyed considerable rainfall, which raised the question, is it possible that huge quantities of high-quality water might still exist in the depths of the Libyan Sahara Desert?

Geological surveys revealed that more than 120,000 km^3 of pure freshwater had been lying undisturbed beneath the desert for between 14,000 and 38,000 years. In 1984, work began on laying a pipeline that would convey 6 million m^3 of water per day on a journey of 3,500 km from the groundwater source in the Sahara to the Mediterranean coast in the north. The pipeline extends over an area roughly the size of western Europe. The total depth of wells drilled in the desert for the Great Man-Made River Project is over seventy times the height of Mt. Everest.

The pipelines re-emerge at the end and feed into huge coastal reservoirs. Today there is enough freshwater to supply each citizen in the Great Jamahiriya with over 1,000 litres per day and provide irrigation for 135,000 hectares of arid land. The water brings new life to the desert and offers a brilliant showcase of human inventiveness and engineering skill.

Source: Prepared for the World Water Assessment Programme (WWAP) by UNESCO/IHP, 2002.

Figure 8.7: Progress towards the World Food Summit target

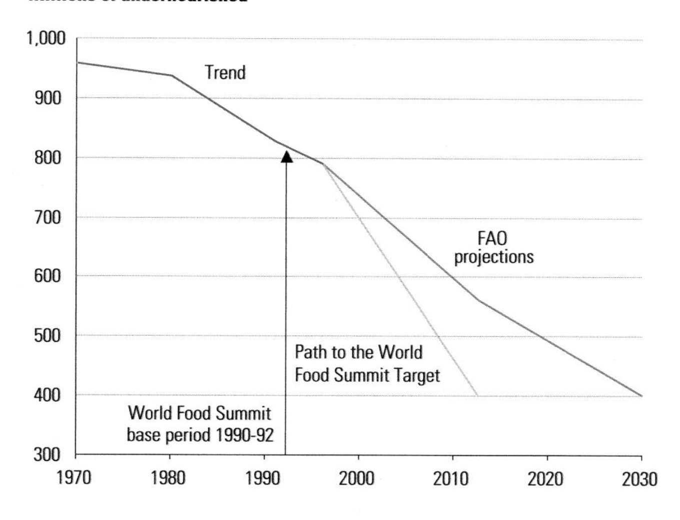

FAO projections of attainment of the World Summit food security goal are clearly at odds with the targeted goal. According to FAO estimates, the World Summit goal would not be achieved before 2030, fifteen years behind schedule.

Source: FAO, 2002.

consequent flooding. Some national macropolicies have failed to recognize the importance of agriculture and have also contributed to the forces pushing rural people into poverty. In rural areas, the people most affected include smallholders, landless labourers, traditional herders, fishermen and generally marginalized groups such as refugees, indigenous peoples and female-headed households. Children are particularly vulnerable to the full impact of hunger, which can lead to permanent impairment of physical and mental development.

Undernourishment is a characteristic feature of poverty. Poverty includes deprivation of health, education, nutrition, safety, and legal and political rights. Hunger is a symptom of poverty and also one of its causes, these dimensions of deprivation interacting with and reinforcing each other. Hunger is a condition produced either by human action, or a lack of human action aimed at correcting it. For example, in the early 1990s, nearly 80 percent of all malnourished children lived in developing countries that produced food surpluses. Lack of access to water to provide basic health services and support reliable food production is often a primary cause of undernourishment. To eradicate hunger, abundant food production is a requirement, but in addition, existing food needs to be accessible to all.

Box 8.5: Food security and its indicators

Food security is defined by FAO as physical, social and economic access for all people to sufficient, safe and nutritious food that meets their dietary needs and food preferences for an active and healthy life. Its converse, food insecurity, applies when people live with hunger and fear of starvation. Food security requires that:

- sufficient quantities of food of appropriate quality be available – a production issue;
- individuals and households have access to appropriate foods – a poverty issue; and
- nourishment is taken under good conditions, including regular meals, safe food, clean water and adequate sanitation – a public health issue.

The individual state of health is also relevant for food security as disease-stricken people are hampered or unable to contribute to their own and their household's food security. By the same token, undernourished people are much more prone to disease.

For regional and global assessments, per capita food intake per day in kilocalories is used as the indicator for food security. This indicator is derived from agricultural production and trade statistics. At the national level, a per capita food intake of less than 2,200 kcal/day is taken as indicative of a very poor level of food security, with a large proportion of the population affected by malnutrition. A level of more than 2,700 kcal/day indicates that only a small proportion of people will be affected by undernourishment. As people are enabled to access food, per capita food intake increases rapidly but levels off in the mid-3,000s. It must be stressed that per capita food intake in terms of kilocalories is only an indicator of food security: adequate nutrition requires, in addition to calories, a balanced diversity of food including all necessary nutrients.

Figure 8.8: Proportion of undernourished people in developing countries, 1990–1992 and 1997–1999

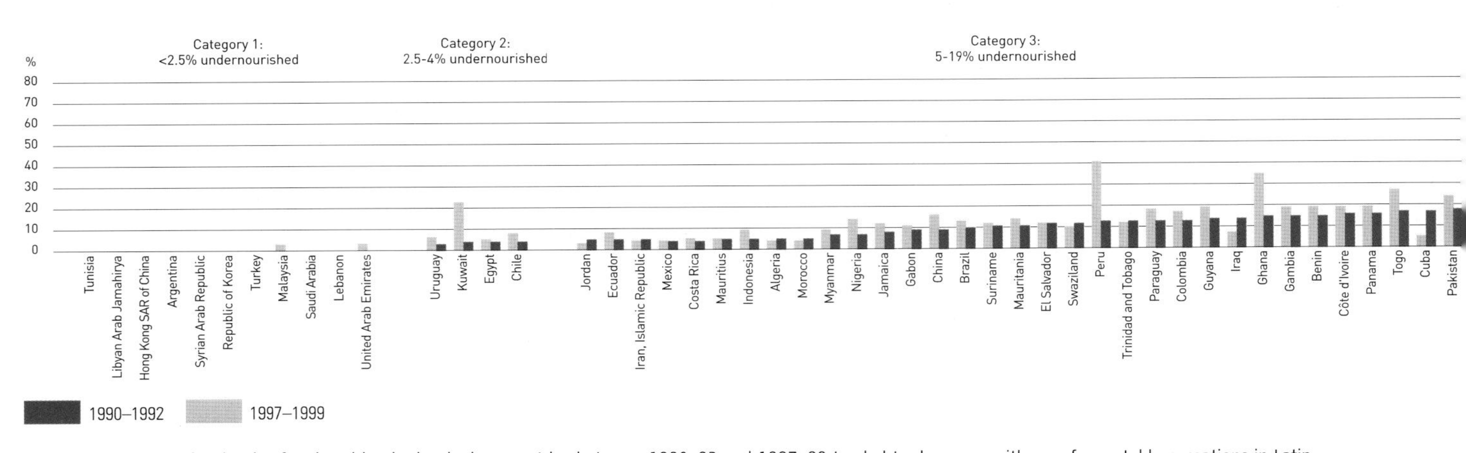

This graph indicates that levels of malnutrition in developing countries between 1990–92 and 1997–99 tended to decrease, with very few notable exceptions in Latin America and Africa.

Source: FAO, 2001b.

Map 8.3: Percentage of undernourished people by country (1998)

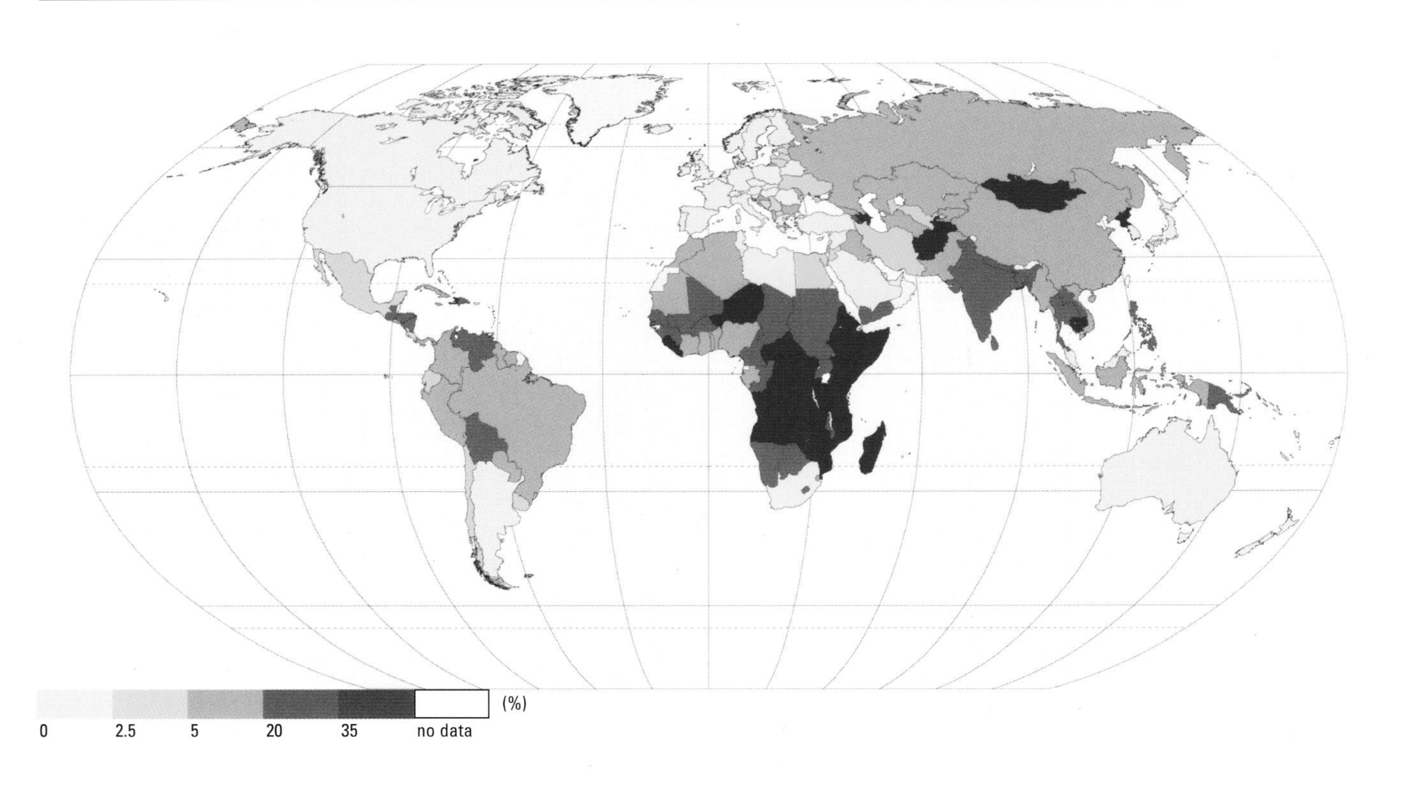

This map shows large regional disparities in the proportion of undernourished people throughout the world, and illustrates the typical division between developed and underdeveloped countries. Whereas western Europe and North America appear to have reasonable food security levels, much of Africa, Latin America and eastern Europe does not enjoy the same luxury.

Source: Map produced for the World Water Assessment Programme (WWAP) by the Centre for Environmental Research, University of Kassel, based on FAO, 2001b.

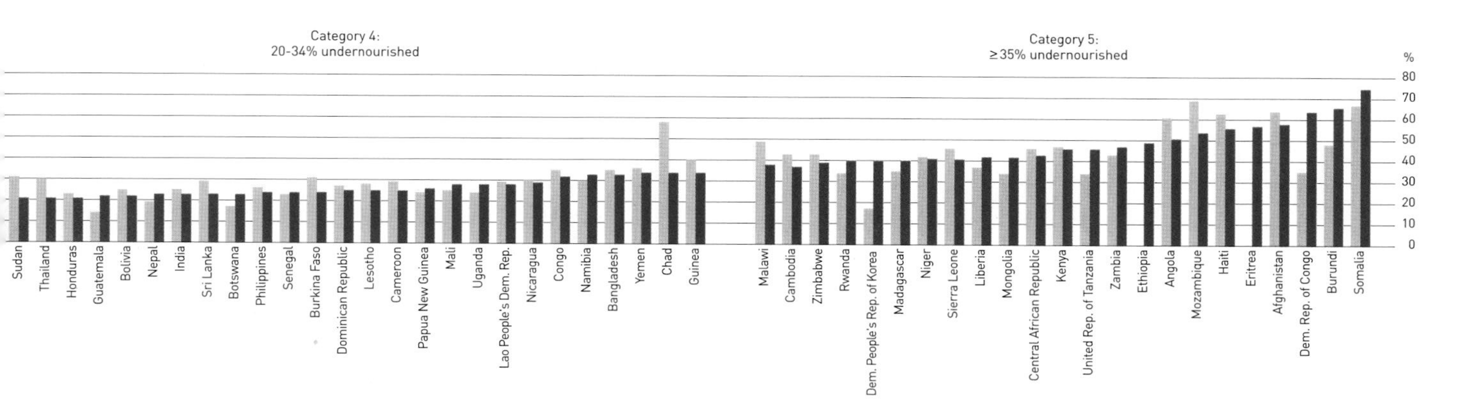

The role of irrigation in alleviating poverty and improving food security

There is a positive, albeit complex, link between water services for irrigation and other farm use, poverty alleviation and food security (FAO, 2001a; FAO/World Bank, 2001; IFAD, 2001). Many of the rural poor work directly in agriculture, as smallholders, farm labourers or herders. The overall impact can be remarkable: in India, for example, in non-irrigated districts, 69 percent of people are poor, while in irrigated districts, only 26 percent are poor (World Bank, 1991). In rural areas, income can be boosted by pro-poor measures, such as ensuring fair access to land, water and other assets and inputs, and to services, including education and health. Relevant reforms of agricultural policy and practices can strengthen these measures.

The availability of water confers opportunities to individuals and communities to boost food production, both in quantity and diversity, to satisfy their own needs and also to generate income from surpluses. Irrigation has a land-augmenting effect and can therefore mean the difference between extreme poverty and the satisfaction of the household's basic needs. It is generally recognized that in order to have an impact on food security, irrigation projects need to be integrated with an entire range of complementary measures, ranging from credit, marketing and agricultural extension advice to improvement of communications, health and education infrastructure (see box 8.6 for an example from Senegal). Land tenure may also represent a major constraint: irrigation schemes controlled by absentee landowners and serving distant markets, even when highly efficient, may fail to improve local food security when both commodities and benefits are exported.

Irrigation projects are as diverse as the local situations in which they are implemented. Generally, small-scale irrigation projects based on shallow groundwater pumping provide a manageable framework that can give control to the local poor and avoid leaking resources to the non-poor. Large-scale irrigation, as may be determined by the need to carry out large-scale engineering works to harness water and convey it to the fields, can also be made to work for the poor provided that the benefits can be shared equitably, and investment, operation and maintenance costs are efficiently covered.

Managing agricultural risk for sustainable livelihoods

Community-managed small-scale irrigation systems, by improving yields and cropping intensities, have proved effective in alleviating rural poverty and eradicating food insecurity. Marketing of agricultural produce, both locally and at more distant places when adequate transport and communication infrastructure is available, can make a significant contribution to the income of farmers, and especially of women. Bank deposits and credits, as well as crop insurance, can be used to finance agricultural operations and buffer against climatic risk. However, banking services are not usually accessible to people who have no collateral assets, or who are of poor social status. Many rural credit systems may also not accommodate pay-back over a period of years – the time needed to realize the benefits from investment in irrigation technology. However, non-conventional credit systems based on trust and social solidarity can support poor farmers. Improvement in poor people's food commodity storage facilities reduces post-harvest losses and can save significant amounts of food, thus contributing to food security. Similarly, where technically and financially possible, storing water in surface reservoirs and aquifers is a strategic means for managing agricultural risk. Water in the reservoir or the aquifer is, in a way, equivalent to money in the bank.

Irrigation and off-farm employment

Irrigation, supported through inputs such as high-yielding varieties, nutrients and pest management, together with a more extended agricultural season, higher cropping intensity and a more diverse

Box 8.6: Water and food security in the Senegal River basin

Parts of the Senegal River basin in Senegal and Mauritania are entirely situated in the arid Sahelian zone with, in the lower valley and the delta, a total annual rainfall that rarely exceeds 400 mm. Thus, when used in the low-lying areas, the rain-dependent crops of the plateau and flood-recession crops barely meet the food needs of farming families. During periods of drought like those of the 1970s, local populations are drastically affected; for this reason one of the four fundamental tasks for the Organization for the Development of the Senegal River (OMVS) at its founding in 1972 was 'to create food self-sufficiency for populations in the Senegal River basin and, by extension, for the subregion.'

To this end the goal set was to develop 375,000 hectares out of a potential irrigable area of 823,000 hectares, by combined operation of the Manantali and Diama dams. The crops targeted for irrigation were rice and wheat, to be added to sorghum, maize and market gardening – the traditional rain-dependent and recession crops. With respect to this goal, a total of some 100,000 hectares has so far been developed. However, a 1996 study by the International Institute for Environment and Development (IIED) indicated that no single type of farming could guarantee the survival of the family unit and that crop diversification was essential. The study also pointed out the major issues in optimizing flood plain development as part of the fight for food security. These geographical areas are vital for agriculture, fishing, livestock pasture and forest regeneration.

This is why, after the dams were filled in 1986 and 1987, OMVS decided to simultaneously expand the irrigated areas by delivering artificial flooding that would guarantee between 50,000 hectares and 100,000 hectares of recession crops and ensure 63,000 ha of pasture and wooded areas for 2.7 million cattle and 4.5 million sheep and goats. Fishing is also an economically and socially important activity in the Senegal River basin. With annual catches estimated at between 26,000 and 47,000 tons, it represents a substantial income for the populations concerned. The storage lakes of the Diama and Manantali dams – 11.5 million m^3 spread over 500 km^2 – have attracted large fishing communities since the dams became operational. The programmes begun by OMVS thus contribute to achieving food security in the area. For this aim to be met as quickly as possible, the OMVS High Commission sees a need, on the one hand, for increased technical, institutional and financial means so as to accelerate development and ensure sound management, and on the other hand, for technical improvements permitting more intensive farming, higher yields and a close association between farming, fishing, livestock-raising, forestry and the water economy.

Source: Prepared for the World Water Assessment Programme (WWAP), by the Organization for the Development of the Senegal River (OMVS), 2002.

assortment of crops, can generate rural employment in other non-agricultural services. The productivity boost provided by irrigated agriculture results in increased and sustained rural employment, thereby reducing the hardships experienced by rural populations that might otherwise drift to urban areas under economic pressure. Growth in the incomes of farmers and farm labourers creates increased demand for basic non-farm products and services in rural areas. These goods and services are often difficult to trade over long distances. They tend to be produced and provided locally, usually with labour-intensive methods, and so have great potential to create employment and alleviate poverty. Studies in many countries have shown multipliers ranging from two (in Malaysia, India and the United States) to six (in Australia), (World Bank, 2002).

The contribution of fisheries and aquaculture to food security

Fish has a very good nutrient profile and is an excellent source of high-quality animal protein and of highly digestible energy. Refugees and displaced people facing food insecurity may turn to fishing for survival, where this possibility exists. Staple-food fish, often comprising low-valued fish species, is in high demand in most developing countries, because of its affordability.

Inland fish production provides significant contributions to animal protein supplies in many rural areas. In some regions, freshwater fish represent an essential, often irreplaceable source of cheap high-quality animal protein, crucial to a well-balanced diet in marginally food-secure communities. Most inland fish produce is consumed locally and marketed domestically, and often contributes to the subsistence and livelihood of poor people. The degree of participation in fishing and fish farming is high in many rural communities. Fish production is often undertaken in addition to

Box 8.7: Rice-fish farming in Lao People's Democratic Republic

Lao People's Democratic Republic has extensive water resources in the form of rivers, lakes and wetlands. Fisheries and the collection of aquatic animals during the rainy season are important activities in the country and fish forms an important part of the national diet. Rice cultivation is widespread in rainfed, irrigated and terraced fields. Rice is mostly cultivated on a one-crop-per-year basis, but in irrigated areas two crops per year are possible.

In upland rainfed fields, bunds are often raised to increase water depth for fish cultivation. In some cases, a small channel is constructed to facilitate fish capture. In the Mekong River plain, rice-fish farming is practiced in rainfed rice fields where soils are relatively impermeable as well as in irrigated rice fields, which offer ideal conditions for fish cultivation. As elsewhere, there is little reliable data available concerning production levels from rice-fish farming but productions of 125 to 240 kg/hectare/year have been reported for upland rice-fish production systems. Carp, tilapia and other species cultured in this system are mostly for consumption in the farmer's home.

While rice-fish farming is popular with farmers, there are some constraints that need to be addressed through proper support. Integrated pest management practices should be applied to reduce pesticide use. Furthermore, fingerlings need to be made available more easily and farmers' access to credit should be improved.

Source: Dixon et al., 2001.

agricultural or other activities. Yields from inland capture fisheries, especially subsistence fisheries, can be very significant, even though they are often greatly underreported. Yields from inland capture fisheries are highest in Asia in terms of total volumes, but are also important in sub-Saharan Africa. Fishery enhancement techniques, especially stocking of natural and artificial water bodies, are making a major contribution to the total catch (FAO, 2000a).

Rural aquaculture contributes to the alleviation of poverty directly through small-scale household farming of aquatic organisms for domestic consumption and income. It also contributes indirectly through employment of the poor as service providers to aquaculture or as workers on aquatic farms. Poor rural and urban consumers can greatly benefit from low-cost fish provided by aquaculture. For aquaculture to be effective in alleviating poverty, it should focus on low-cost products favoured by the poor, and emphasis should be placed on aquatic species feeding low in the food chain. The potential exists for aquaculture production for local markets and consumers. Combined rice-fish systems are possible and carry high benefits because they provide cereals and protein at the same time. These systems have also been shown to have beneficial effects on the malaria situation, where mosquito vectors breed in rice fields and where the selected fish species feed on the mosquito larvae. This is the case in parts of South-East Asia (see box 8.7).

Forestry and food security

A large number of forest products contribute to food security: FAO estimates that about 1.6 billion people in the world rely to a certain extent on forest resources for their livelihoods. For most rural people in the world, firewood provides the fuel for cooking food, so its availability is an integral part of food security. Furthermore, the bioenergy sector generates employment and income in developing countries.

Most forests and tree plantations subsist on rainwater, or may develop around irrigation schemes. Some tree species can use large amounts of water drawn from water storage in the soil profile and in shallow aquifers. Trees in forests and outside of forests provide significant benefits to poor people and contribute to food security. The benefits from water used by forests are largely seen in terms of wood and non-wood products, as well as in protecting the environment, reducing soil degradation and preserving biodiversity.

Trends in Irrigation Water Management

Adapting institutional arrangements

The institutional arrangements governing irrigation water appropriation and use have been established over the centuries in various countries under different environmental and social circumstances. Adapting to new pressures that call for higher water productivity and increased user participation and liability in cost recovery has proved difficult. With increasing competition for water, both within the irrigated subsector (between farmers) and with other economic sectors – principally municipalities, industry and hydropower – it is often the case that irrigation institutions are not adequately equipped to adapt to changing circumstances and expectations. The competing demands from municipalities, industry and the energy sector are forcing transfers from agriculture to higher value uses. Figure 8.4 earlier showed an example of this process at work in one district in China.

Irrigated agriculture has played a pivotal role in keeping up with global food demands in the twentieth century, but as we enter the twenty-first century, there is continued evidence of lacklustre performance in the public sector alongside a frustrated private sector. In many cases, distorted markets, ill-conceived incentives and institutional rigidity plague the irrigation subsector. Producers are seeking financial returns against tight margins for food commodities, subsidies for rainfed agriculture and varying degrees of competition for raw water from other sectors, and are also expected to maintain environmental integrity.

One troubling aspect is the continued expectation that the physical endowments of land and water equate to a 'potential' that needs to be realized, without parallel analysis of economic, financial, institutional and environmental constraints and without a realistic market analysis. This has conditioned the approach that many governments continue to take in implementing irrigation and water policies, establishing irrigation institutions and in allocating public sector budgets. There has been a dominance of supply-led approaches predicated upon large-scale irrigation infrastructure, and governments have shown determination in continuing to play a role in operating public infrastructure for the supply of irrigation services. In many instances, there is a sharp discontinuity in policy, institutional capacity, investment between the provision of irrigation services and the promotion of agricultural systems. This applies to many of the key developing economies. It could be argued that if irrigated agriculture has its failures, it is in a large part because it has focused mainly on providing water and not enough on the productivity of the agricultural systems and their responsiveness to agricultural markets.

Increasing private investment in irrigation

Investment in large irrigation projects increased in the 1970s and then fell by more than 50 percent in the 1980s. It further declined in the 1990s. Most of the water-related infrastructure projects over the past forty to fifty years were financed by governments with significant intervention by the international development banks. Development costs for new irrigation lands have increased markedly in recent years; for example, costs have been increasing by more than 50 percent in the Philippines, 40 percent in Thailand and have nearly tripled in Sri Lanka. With declining crop prices, it is difficult to justify new irrigation development. Financial capability is lacking for new infrastructure as well as for modernizing present structure and ensuring system sustainability.

Recently, there has been an increase in the private sector financing of large water sector infrastructure and small-scale irrigation systems. According to the World Bank, 15 percent of the infrastructure is currently being financed through private funds, and this is part of a growing trend. Groundwater development has proved a particularly attractive target for private investment because of the private level of control it offers.

Reforming irrigation: modernization and empowerment

During the 'green revolution' of the 1960s and 1970s, high priority was given to making irrigation water available to farmers. Governments had surface irrigation systems established and managed through public sector agencies. Some large-scale systems were poorly designed, with insufficient provision for drainage and consequent soil degradation. System management often failed to respond to the needs of users, in particular of smallholders and sectors carrying low social and political weight. Water use fees were not collected or not applied to proper system operation and maintenance. Large rehabilitation needs emerged and as governments and international lending institutions found it difficult to raise funds for this purpose, it became clear that the economic and social context of large-scale irrigation needed to be reformed.

Reforming efforts focused on transferring responsibility for operating and maintaining irrigation systems to the farmers, organized in water user associations. This highlighted the need for building and developing managerial capacity among stakeholders, while confining irrigation system administration to the role of water service provider. Empowerment alone may not be sufficient unless deficiencies in design and operation and/or upgrade of infrastructures are addressed. Irrigation modernization is a process of change from supply-oriented to service-oriented irrigation. It involves institutional, organizational and technological changes and transforms a traditional irrigation scheme from protective to productive irrigation. The modernization and transfer of some management responsibilities of government-held irrigation systems to water user associations and service-providing companies have been carried out in several countries such as Mexico, China and Turkey and have proved beneficial in certain cases. However, governments have tended to be only half-hearted when enacting complementary policies and institutional reforms that are necessary to provide the appropriate environment for effective operation of the new irrigation management entities. The process of empowering marginalized stakeholders, including smallholders, and removing political influence from irrigation management is as yet not completed. Irrigation management transfer is complicated because of a number of factors. Firstly, there is a need to help poor stakeholders gain equity with non-poor stakeholders and reconcile conflict between upstream and downstream users. Secondly, the transaction costs of water user associations may be higher than under properly working 'command and control' management establishments. Finally, the apportionment of financial and operation risk and liability is difficult to make when a large-scale infrastructure is transferred to user associations or service companies not prepared for such responsibilities.

Equity in the roles of men and women in irrigation

Gender equity is a case in point. Women are major actors in poverty alleviation, in food production, and in ensuring and dispensing nutrition at the household level. A majority of the poor, estimated at 70 percent, are rural people, and rural poverty has become feminized as men of productive age migrate away from impoverished rural areas to the more promising urban environment, or are forcibly recruited by warring factions, leaving behind the women, the elderly, the sick and the children. In rural areas subject to recurrent conflicts, whatever little infrastructure exists collapses or is wantonly destroyed, thus increasing both insecurity and the burden on women. Households headed by women are recognized among the poorest of the poor. Against a biased view that 'women do not irrigate', women are now recognized as being actively involved in irrigation, often demonstrating high levels of skill (see box 8.8).

Box 8.8: Land ownership and economic independence

A recent study in Dakiri, Burkina Faso, shows that allocating smaller plots to men and women separately instead of allocating larger plots to household heads has produced both higher yields and social benefits. When both men and women have irrigated plots, the productivity of irrigated land and labour is higher than in households where only men have plots. Women are equally good or even better irrigation farmers than men, and those who have obtained irrigated plots are proud of their increased ability to contribute to the needs of their households. Women prefer to contribute to their households by working on their own plots rather than providing additional labour to their spouse's or to the collective plots. As they become economically less dependent upon their husbands, they can help support their relatives and increase their own opportunities for individual accumulation of wealth in the form of livestock. The effects of having an individual plot significantly improve the bargaining position of a woman within a household and are a source of pride in the household and the community.

Source: OECD/DAC, 1998.

Improving water use productivity in agriculture

Improving water use productivity is often understood in terms of obtaining as many kilograms of crop as possible per cubic metre of water – 'more crop for the drop'. Financially astute farmers may prefer to target a maximum income per cubic metre – 'more dollars for the drop', while community leaders and politicians could be looking for maximum employment and widespread income generated through the crop and its derivatives – 'more jobs for the drop'. In a broad sense, increasing productivity in agriculture involves deriving more benefit, or achieving more welfare, for every unit of water withdrawn from natural water bodies.

Technology permits accurate water application in the optimum quantity and timing for crop development. Drip irrigation for example can respond to existing soil moisture conditions by leading the required quantity of water to the root zone of the plant. Laser-supported land-levelling devices allow accurate bounded field irrigation. When such techniques are applied in water-scarce regions with high-value outputs, their application results in profit for the farmer. Further research now reaching the application phase has crops grown in greenhouses in such a way that water transpired by the plants is not released to the atmosphere but condensed and reused. The application of advanced technology is dependent on a level of investment and capacity as well as on an economic incentive to make it worthwhile. Most irrigation in the world was initially established to take advantage of otherwise unused water. It should not be surprising that water use efficiency makes only slow progress wherever water is cheap, either because it has no other uses and therefore a low opportunity value, or because it is subsidized. Actually improving efficiency can be a slow and laborious process that requires system modernization, therefore upgrading the technological environment and the knowledge and capacity of irrigation operators.

From the perspective of a national economy, a key goal for water use productivity is to improve net economic returns per dollar invested in water use, favouring investment in the urban and industrial sectors. However, such a view may not adequately recognize the need for food and the social and environmental benefits derived from agriculture.

Diversifying crops

Crop diversification, made possible by irrigation, has a beneficial influence on local food security in remote rural areas by supporting a longer growing season and providing a healthier, more diversified diet including fresh produce. At a medium scale, crop diversification strengthens the rural economy and reduces the uncertainty associated with the market vagaries that affect monocultures. In 1990 in Asia for example, cereals, pulses and other crops accounted for 66 percent, 8 percent and 26 percent of the total cropped area.

By 1997, this had changed to 56 percent, 7 percent and 37 percent. Per capita cereal production nevertheless increased owing to higher yields. Irrigation systems designed for cereal cultivation often do not have sufficient regulation and the effective water control structures required for crop diversification. Diversification also calls for higher levels of management capacity, as it is not enough to produce a variety of crops: in addition to being produced, these have to be marketed. Policy and economic factors such as market incentives and availability of labour influence the crop choice of farmers. Availability of low-cost pumping technology has supported the expansion, under private initiative and finance, of diversified cropping systems.

Agricultural Water Use, Environment and Health

Environmental aspects

Agriculture may have distinct negative externalities in terms of water quantity and quality. Pasture and crops take up 37 percent of the earth's land area. Agriculture is the largest water user and the main source of nitrate pollution of ground and surface waters, as well as the principal source of ammonia pollution. It is also a major contributor to the phosphate pollution of waterways and of release to the atmosphere of the greenhouse gases methane and nitrous oxide. Land degradation, salinization, overabstraction of groundwater and the reduction of genetic diversity in crop and livestock affect the basis of agriculture's own future. The vanishing Aral Sea is a clear example of the irreversible impacts of excessive withdrawals. The irrigation sector is coming under increasing public scrutiny as amenity and ecosystem values are lost while the expected economic and social benefits of irrigation systems are not fully realized. Competition between urban dwellers and agriculture is also a growing issue and may worsen environmental pressure. In developed countries, environmental concerns have been a key driver for modernizing irrigation systems.

Historically, reclamation of wetlands has made a major contribution to agricultural growth. Because of the presence of water during a large part of the year and in view of the relative fertility of their soils, many wetlands have a good potential for agricultural use. However, this use leads to serious environmental damage, and the importance of protecting them has been recognized by the adoption of the Convention on Wetlands (Ramsar, 1971). The developing countries still have some 300 million hectares of wetlands that may be suitable for crop production but only a relatively small percentage is currently used to this end. Where no alternative additional land resources are available to exploit, wetlands will inevitably be converted to crop production. This is the case in many parts of sub-Saharan Africa, where the nutritional situation is bad and wetlands represent an attractive opportunity for agricultural development.

However, unwise use of wetlands may result in environmental degradation. Draining of wetlands has often been carried out under the wrong assumption that wetlands are useless and worthless. Sustainable use of wetlands can be achieved by selecting crops adapted to wetland conditions, using appropriate water and soil management technologies and planning wetland development and management carefully within the framework of the watershed area. Wetlands of particular international or national importance on account of their significance in terms of ecology, botany, zoology or biodiversity should be protected from any agricultural use and from the influences of agricultural activities in upstream areas.

Water pollution, habitat degradation and massive water withdrawals can deprive fishing communities of their livelihood and push them into food insecurity. The resulting environmental impacts affecting fishery resources in inland waters can be devastating. Even in estuarine and coastal zones at the lower end of river basins, fishery resources are impacted by pollution, habitat degradation and upstream water withdrawal and use. It is increasingly recognized that agriculture also has positive externalities, including environmental services and products. The multifunctional nature of agriculture is increasingly acknowledged and encouraged, so that farmers are seen not only as commodity-producers but also appreciated as self-employed citizens, stewards of the landscape and stakeholders in vibrant communities. Trade-offs between food security and the environment can be further reduced through already available or emerging technologies and land management practices. By using more sustainable production methods, the negative impacts of agriculture on the environment can be attenuated. Agriculture can play an important role in reversing negative impacts by, for example, environmentally sound water use, biological treatment of waste, enhancing the infiltration of water to reduce flood runoff, preserving agricultural and natural biodiversity, and storing carbon in soils.

Water quality deterioration

With rising demands for water, concerns over water quality have risen rapidly. Pollutant loads have increased enormously, and at the same time, the amounts of water available for dilution are decreasing. The situation is particularly alarming in developing nations while, in developed countries, the enforcement of water quality measures has resulted in improved conditions for most rivers. Deterioration of water quality poses a serious threat to the sustainability and the safety of food produced by intensive farming systems upon which global food security has become increasingly dependent. Security and stability in food supplies in this century will be closely linked to success in water quality control. Organic matter, if free of pathogens, can actually be beneficial to irrigated agriculture (see box 8.9), but water contamination with hazardous chemicals makes it unusable for food production.

It is estimated that poor drainage and irrigation practices have led to waterlogging and salinization of about 10 percent of the world's irrigated lands, thereby reducing productivity. In particular, mobilization of resident salts is a widely occurring phenomenon in irrigated river basins in arid regions. Waterlogging and salinization in large-scale irrigation projects are often the result of unavailable drainage infrastructure which was not included in the engineering design in order to make projects look economically more attractive. These problems are generally associated with large-scale irrigation development under arid and semi-arid conditions, as in the Indus (Pakistan), the Tigris-Euphrates (Middle East) and the Nile (eastern Africa) river basins. The solutions to these problems are known but their implementation is costly.

Health and irrigation

Water-related diseases are described earlier in this book, together with figures on the burden of disease. The key irrigation-related vector-borne diseases are malaria, schistosomiasis and Japanese encephalitis.

Irrigation has in the past sometimes been accompanied by adverse impacts on the health of local communities. The principal causes are rooted in ecosystem changes that can create conditions conducive to the transmission of vector-borne diseases, as well as drinking water supply and sanitation conditions that lead to gastrointestinal conditions. The attribution of the burden of each of these diseases to irrigation or components thereof in specific settings is complex. Only where irrigation is introduced in an arid region where the diseases previously did not occur, is the association between the resulting dramatic landscape changes and the explosive rise of disease incidence and prevalence clear-cut. In most cases, there is a complex mixture of contextual determinants of the diseases, combined with a number of confounding factors. For example, in parts of Africa south of the Sahara, the transmission of malaria is so intense throughout the year that the additional risk factors from irrigation development will not add to the disease burden. Schistosomiasis, rightly equated with irrigation in Africa, is also determined by human behaviour and by the state of sanitation.

Many vector-borne disease problems in irrigated areas can be traced to absent or inadequate drainage. The various forms of surface irrigation all impose increased vector-borne disease hazards, while overhead irrigation and drip irrigation are virtually free of such hazards. Crop selection can be important. In that sense, flooded rice and sugarcane are crops that carry increased vector-borne disease risk. Irrigated agriculture often requires additional chemical inputs for crop protection, and the application of pesticides can disrupt the ecosystem balance favouring certain disease vectors; it can also contribute to an accelerated development of resistance to insecticide in disease vector species.

Box 8.9: Use of wastewater for irrigation

The cost of disposing of urban wastewater is all too often externalized against the aquatic environment and downstream users in rivers, estuaries and coastal zones and hardly, if ever, appears in the benefit and cost accounts. However, wastewater is recognized as a resource, particularly in water-scarce regions. If the polluter actually pays, wastewater is free or has only a low cost, is reliable in time and close to urban markets. In addition to direct benefits to farmers who would otherwise have little or no water for irrigation, wastewater improves soil fertility and reduces water contamination downstream. The total land irrigated with raw or partially diluted wastewater is estimated at 20 million hectares in fifty countries, somewhat below 10 percent of total irrigated land in developing countries. For irrigation use, wastewater should be subject to primary and secondary treatment, but in poor countries that is often not the case and raw sewage is applied.

Disadvantages and risks related to use of insufficiently treated wastewater concern the exposure of irrigation workers and food consumers to bacterial, amoebic, viral and nematode parasites as well as organic, chemical and heavy metal contaminants. In a context of prevailing poverty, such water is used in the informal, unregulated sector, but sanitary concerns preclude the export of the products and, at least partially, the access to local food markets. Governments and the development community promote efforts to lead wastewater reuse into sustainable channels, but countries and municipalities short of resources are slow in facing the cost of water treatment. Given water scarcity and the relatively high cost of obtaining potable freshwater for municipal uses, the use of treated wastewater in the urban context is projected to increase in the future, mostly for irrigation of trees in the urban and peri-urban landscape, including parks and golf courses.

There are many opportunities in the planning, design and operation of irrigation schemes to incorporate health safeguards: hydraulic structures, for instance, can be designed so they provide less or no habitat for vector breeding. Improved water management practices such as alternate wetting and drying of irrigated rice fields, rotational drying of parallel irrigation canals, flushing of canals with pools of standing water and clearing canals of aquatic weeds can reduce vector breeding. Moreover, the very infrastructural development that usually accompanies irrigation development and the economic development that follows in its wake imply improvements in access to health services and increased buying power to purchase drugs, mosquito nets and other preventative and protective tools and products.

Until the 1980s, a drinking water supply component was often overlooked in the context of irrigation development. While this situation has improved, the two types of water use are occasionally at odds. Highly increased chemical inputs may pollute groundwater, and local communities have been known to revert to canal water because the quality of water from their pumps had deteriorated. Easy access to large quantities of water in irrigation canals for domestic needs other than drinking will contribute positively to overall hygiene. There are also important overlaps between operation and maintenance tasks for irrigation and drainage schemes and for drinking water supply and sanitation services that would allow important economies of scale to be achieved. A study carried out in three African countries (see box 8.10) for instance, showed that small dams and wells acted as catalysts for change, initiating actions that generated income and allowed people to diversify diets, afford health services and better cope with hungry periods of the year.

Conclusions

Agriculture will remain the dominant user of water at the global level. In many countries, in particular those situated in the arid and semi-arid regions of the world, this dependency can be expected to intensify. Water appropriation by agriculture experienced strong growth with the 'green revolution'. The contribution of irrigated agriculture to food production is substantial but the rate of growth will be lower than in the past. However, for various reasons that include limited water resources, only about one third of arable land bears irrigation potential. Both irrigated and non-irrigated agriculture still have scope for increasing productivity, including water productivity. Arguably, the expansion of irrigated agriculture protected people on the nutritional fringe from premature death, and preserved tracts of land under forest and wetlands from encroachment by hard-pressed farmers. However, pressures to encroach on such lands persist.

Within the current demographic context, the global food security outlook is reasonably good and towards 2050, the increased world population could enjoy access to food for all. The fact that close to 800 million people are at present ravaged by chronic undernourishment in developing countries is not due to a lack of capacity of the world to produce the required food, but to global and national social, economic and political contexts that permit, and sometimes cause, unacceptable levels of poverty to perpetuate. Poverty is being alleviated, albeit slowly, and as the share of food slips lower in household budgets, the prospects for

Box 8.10: Integrating irrigation, nutrition and health

FAO assessed the impact of three small-scale irrigation projects on the health and welfare of villagers in Burkina Faso, Mali and the United Republic of Tanzania. The assessment showed that small dams and wells acted as catalysts for change, initiating actions that generated income and allowed people to better cope with the hungry periods of the year, diversify diets and afford health services. These projects encouraged production, processing and preparation of a variety of indigenous foods, nutrition education and the participation of women's groups. In all three cases, irrigation increased food production or income by enough to provide one additional meal per day, even during the 'hungry season' before the harvest (FAO, 2001b).

In the arid countries of Asia, there are often large areas where groundwater is brackish and people have to obtain water from irrigation canals for all uses, including domestic. A study by the International Water Management Institute in Pakistan showed that safe use of canal irrigation water is possible if households have a large water storage tank in their house and have a continuous water supply for sanitation and hygiene. The results also showed that children from households having a large storage capacity for water in the house had much lower prevalence of stunted growth than children from families lacking such a storage facility. Increasing the quantity of irrigation water available for domestic use and providing toilet facilities are the most important interventions to reduce the burden of diarrhoeal disease and malnutrition.

agriculture to internalize its costs will improve. Water still has a large unmet potential to help alleviate poverty and undernourishment. In taking this direction, agricultural water management will continue to require better integration with rural household water uses, and to make more positive contributions in environmental management. Food for all could be achieved much earlier than present projections indicate, provided that the necessary policies are backed with the necessary resources. The economic, social and environmental cost of continued food insecurity for hundreds of millions of people is high.

Agriculture can use water more efficiently than present practices indicate. Technology for efficient transport of water from the site of abstraction to the field, and for delivering it to the crop plants with a minimum of losses, is available and is actually applied wherever water is scarce. Irrigation water use efficiency increases when the right policy and market incentives are in place. As competition for limited water resources and pressure to internalize environmental impacts intensify in a number of countries, agriculture and, in particular, irrigation comes under growing pressure to review and adapt its policies and institutions, including the water rights and allocation system. Under such circumstances as in the Near East/North Africa region, current water efficiency is relatively high and projected to further increase. Data show an aggregate low water use efficiency in the water-rich Latin America region, and it is not predicted to increase significantly in the future as there are no other large-scale users in competition with agriculture; locally, however, wherever water is scarce, high efficiency is also more frequent throughout the region. Agriculture can also increase the use of recycled water and water stemming from non-conventional sources.

Progress since Rio at a glance

Agreed action	Progress since Rio
Agree that population growth has caused great strains on food production	
Agree that food security not only involves food production, but also access to food, post-production handling and storage, and nutritional value	
Suitably expand irrigated area	
Provide associated inputs such as credit, input supplies, markets, appropriate pricing and transportation to maximize returns from agriculture	
Implement water-saving technology and management methods in both rainfed and irrigated agriculture	
Enable communities to introduce institutions and provide incentives for rural populations to adopt new approaches	
Provide water supplies of suitable quality for livestock production	
Maximize yield of aquatic food organisms in an environmentally sound manner	

Unsatisfactory — **Moderate** — **Satisfactory**

References

Alcamo, J.; Döll, P.; Henrichs, T.; Lehner, B.; Kaspar, F.; Rösch, T.; Siebert, T. Forthcoming. 'WaterGAP: Development and Application of a Global Model for Water Withdrawals and Availability'. *Hydrological Sciences Journal.*

Burke, J.-J. and Moench, M. 2000. *Groundwater and Society, Resources, Tensions and Opportunities: Themes in Groundwater Management for the 21st Century.* New York, United Nations.

Dixon, J., Gulliver, A.; Gibbon, D. 2001. *Farming Systems and Poverty: Improving Farmer Livelihoods in a Changing World.* Washington DC, Food and Agriculture Organization / World Bank.

Döll, P.; Kaspar, F.; Lehner, B. Forthcoming. 'A Global Hydrological Model for Deriving Water Availability Indicators: Model Tuning and Validation'. *Journal of Hydrology.*

Döll, P. and Siebert, S. 2002. 'Global Modelling of Irrigation Water Requirements'. *Water Resources Research*, Vol. 38, No. 4, 8.1–8.10, DOI 10.1029/2001WR000355.

FAO (Food and Agriculture Organization). 2002. *World Agriculture: Towards 2015/2030, an FAO Study.* Rome.

——. 2001a. *Crops and Drops: Making the Best Use of Water for Agriculture.* Rome.

——. 2001b. *The State of Food Insecurity in the World.* (Annual report issued also in 1999 and 2000; see web link for the ongoing series of annual reports.) Rome.

——. 2000a. *The State of World Fisheries and Aquaculture 2000.* Rome.

——. 2000b. *World Agriculture towards 2015/2030, Summary Report.* Rome.

——. 1999. 'Integrating Fisheries and Agriculture for Enhanced Food Security'. In: *The State of Food and Agriculture in 1998.* Rome.

——. 1997a. *Water Resources of the Near East Region: A Review.* Rome.

——. 1997b. 'Irrigation Potential in Africa: A Basin Approach'. In: *FAO Land and Water Bulletin*, Vol. 4. Rome.

——. 1995. 'Irrigation in Africa in Figures'. In: *FAO Water Report No 7.* Rome.

FAO/World Bank. 2001. *Farming System and Poverty: Improving Farmers' Livelihoods in a Changing World.* Rome.

IFAD (International Fund for Agricultural Development). 2001. *Rural Poverty Report 2001: The Challenge of Ending Rural Poverty.* Oxford, Oxford University Press.

IHE (Institute for Infrastructural, Hydraulic and Environmental Engineering). 2000. *A Vision of Water for Food and Rural Development.* Publication for the Second World Water Forum, The Hague.

IUCN (World Conservation Union). 2000. *A Vision for Water and Nature.* Publication for the Second World Water Forum, The Hague.

Mazoyer, M. and Roudart, L. 1998. *Histoire des agricultures du Monde, du Néolithique à la crise contemporaine.* Paris, Editions du Seuil.

——. 1997. 'Développement des inégalités agricoles dans le monde et crise des paysanneries comparativement désavantagées'. In: *Land Reform, 1997/1.* Food and Agriculture Organization, Rome.

Molden, D. Unpublished. 'Competing Uses of Water in the Zhanghe Irrigation District, China'. Colombo, International Water Management Institute.

OECD/DAC (Organization for Economic Cooperation and Development/Development Assistance Committee). 1998. *Guidelines for Gender Equality and Women's Empowerment in Development Cooperation.* Paris.

Pitman, G.K. 2002. *Bridging Troubled Waters: Assessing the World Bank Water Resources Strategy.* A World Bank Operations Evaluation Study. Washington DC, World Bank.

Ramsar Convention on Wetlands. 1971. Ramsar, Iran.

Schiffle, M. 1998. *Economics of Groundwater Management in Arid Countries.* London, Frank Cass.

Smith, M.; Fereres, E.; Kassam, A. 2001. 'Crop Water Productivity Under Deficient Water Supply'. Paper presented on the occasion of the Expert Meeting on Crop Water Productivity Under Deficient Water Supply, 3–5 December 2001, Rome, Italy.

Stiglitz, J. 2002. *Globalization and its Discontents.* New York, W. W. Norton.

Thompson, R.-L. 2001. 'The World Bank Strategy to Rural Development with Special Reference to the Role of Irrigation and Drainage'. Keynote address on the occasion of the 52nd IEC meeting of the International Commission on Irrigation and Drainage, 16–21 September 2001, Seoul.

Vassolo, S. and Döll, P. 2002. 'Development of a Global Data Set for Industrial Water Use'. Unpublished manuscript. University of Kassel, Centre for Environmental Systems Research.

WCD (World Commission on Dams). 2000. *Dams and Development: A New Framework for Decision-Making.* London, Earthscan Publications.

World Bank. 2002. *Water Resources Sector Strategy* (draft). Washington DC.

——. 1991. *India Irrigation Sector Review*, Vols. 1 and 2. Washington DC.

Some Useful Web Sites*

Food and Agriculture Organization (FAO), AQUASTAT

http://www.fao.org/ag/agl/aglw/aquastat/main/

Provides data on the state of water resources across the world, including an online database on water and agriculture, GIS, maps, etc.

Food and Agriculture Organization (FAO), FAOSTAT

http://apps.fao.org/

Time series records covering production, trade, food balance sheets, fertilizer and pesticides, land use and irrigation, forest and fishery products, population, agricultural machinery, etc.

Food and Agriculture Organization (FAO), Fisheries Global Information System (FIGIS)

http://www.fao.org/fi/figis/tseries/index.jsp

Global fishery statistics on production, capture production, aquaculture production, fishery commodity production and trade, and fishing fleets.

Food and Agriculture Organization (FAO), Food Insecurity

http://www.fao.org/SOF/sofi/

Provides information on the state of food insecurity in the world and on global and national efforts.

International Water Management Institute (IWMI), Water for Agriculture

http://www.cgiar.org/iwmi/agriculture/

Provides information on issues related to water for agriculture: research activities, list of publications and links. This site is part of a larger site that houses information on a plethora of water management-related topics, such as the environment, health, etc.

Food and Agriculture Organization (FAO), WAICENT

http://www.fao.org/Waicent/

FAO's information portal: this is a programme for improving access to documents, statistics, maps and multimedia resources.

* These sites were last accessed on 6 January 2003.

'Some 20 percent of the world's freshwater abstraction is currently used by industry, corresponding to about 45 litres per day per person. Globalization, with its accompanying move of labour industries from high-income to low-income countries, is creating high water demand outside of its abundant sources, often in urban areas.'

Promoting Cleaner Industry for Everyone's Benefit

By: UNIDO (United Nations Industrial Development Organization)
Collaborating agencies: WHO (World Health Organization)/ UNDESA (United Nations Department of Economic and Social Affairs)

Table of contents

Promoting Cleaner Industry for Everyone's Benefit

I'm not sure what solutions we'll find to deal with all our environmental problems, but I'm sure of this: they will be provided by industry; they will be products of technology. Where else can they come from?

G.-M. Keller, *Nation's Business*, 12 June 1988

IT IS DIFFICULT TO IMAGINE ANY TYPE OF INDUSTRY in which water is not used – as an ingredient of the product itself, for heating or cooling, or as part of the manufacturing and cleaning process. Bulk processing industries need bulk supplies of water, while specialized firms such as pharmaceutical enterprises may require smaller amounts of higher quality water. All require water to be available on a regular basis. While the supply of water is certainly an important issue, this chapter also draws attention to water and pollution as outputs of industrial activity. Both affect the environment and the lives of downstream communities. The chapter provides examples of various economic and legislative instruments available for encouraging industries to exercise responsible citizenship. It suggests that both 'the carrot and the stick' can play a role in minimizing waste and encouraging good practice.

INDUSTRY IS AN ESSENTIAL ENGINE OF ECONOMIC GROWTH. As such, it is key to economic and social progress and thus contributes positively to two of the three components that must develop in harmony if sustainable development is to be achieved. All too often, however, the need to maximize economic output, particularly in developing countries and countries with economies in transition, has excluded the careful and balanced consideration of the third component – environmental protection – from the planning process. Adequate water resources of good quality, for example, are not only important for sustaining human communities and natural ecosystems but also represent a critical raw material for industry. By this approach, short- and medium-term economic gains have been mortgaged against long-term environmental harm and may ultimately be rendered unsustainable.

In recent decades, the large-scale transfer of manufacturing industry from developed to developing communities has exacerbated this imbalance. Water-intensive industries, such as textiles, originally located to take advantage of abundant and well-managed water supplies, may now find themselves relocated to communities where they compete for scarce or underdeveloped water supplies. In this way, the economic benefits derived from lower manufacturing costs are achieved in part by placing additional burdens on local supply-side water management or are offset, at least in part, by additional, and unplanned charges. Changes arise from the need to overcome inadequate supply and interrupted working, impaired water quality and increased product spoilage, and, in many cases, to avoid additional capital expenditure as enterprises take direct control of their water supply management.

Water and Sustainable Industrial Development

The international community has recognized the important role played by water in the framework of sustainable industrial development for many years. The Dublin International Conference on Water and the Environment stated in 1992 that 'human health and welfare, food security, industrial development and the ecosystems on which they depend, are all at risk, unless water and land resources are managed more effectively'.

Agenda 21, which was released the same year, gave considerable attention to water and industrial development while setting out the necessary framework for sustainable development. Chapter 18 implicitly highlights the need to promote cleaner production methodologies and 'innovative technologies ... to fully utilize limited water resources and to safeguard those resources against pollution'. Chapter 30 is completely dedicated to strengthening the role of business and industry as crucial drivers of social and economic development, but at the same time it recognizes that, all too often, industry uses resources inefficiently and is responsible for avoidable spoilage of those resources.

Industry impacts on water may be considered two-fold.

- **Quantity:** Water, often in large volumes, is required as a raw material in many industrial processes. In some cases it may be a direct raw material, bound into the manufactured product and thus 'exported' and lost from the local water system when these products are sent to market. In other cases, and perhaps more commonly, water is an indirect raw material, used in washing and cooling, raising steam for energy, cooking and processing and so on. In the latter case, the wastewater may be returned to the local water system through the sewerage system or directly to watercourses.

- **Quality:** Although industry requires water of good quality for manufacturing, the water it discharges may not meet the same quality standards. At best, this represents a burden on treatment plants responsible for restoring water quality to appropriate standards and suitable for recycling. At worst, industrial wastewater is discharged without treatment to open watercourses reducing the quality of larger water volumes and, in some cases, infiltrating aquifers and contaminating important groundwater resources. This endangers downstream communities that rely on those resources for their primary water supply.

In many developing countries, industry is effectively taking advantage of weak local water governance; passing liability for demand-side considerations either to already overburdened local utilities or to local communities and water users. Typically, the additional financial and environmental costs borne by the local water systems, or directly by other water users, are not taken into account in the preparation of statistics to demonstrate national economic development. Indeed, governments may show the capital costs of water supply and wastewater treatment as development advances rather than as costs passed to government by industry investors.

Although both precautionary and the polluter pays principles are widely adopted by governments, lack of resources within water governance means that they are not yet fully implemented. So the principles are not providing the protection and benefits originally envisaged and the water industry systems are unsustainable, being based on the exploitation of one by the other. In many countries this lack of sustainability is becoming increasingly evident. Projected growth in demand for water cannot be met from existing finite resources by supply-side considerations alone.

It follows that integrating improved supply-side considerations with enhanced demand-side management must be invoked both at government and enterprise levels to restore the balance between economic and environmental objectives.

Figure 9.1: Competing water uses for main income groups of countries

	Agricultural use (%)	Industrial use (%)	Domestic use (%)
World	70	22	8
Low income	87	8	5
Middle income	74	13	12
Lower middle income	75	15	10
Upper middle income	73	10	17
Low & middle income	82	10	8
East Asia & Pacific	80	14	6
Europe & central Asia	63	26	11
Latin America & Caribbean	74	9	18
Middle East & North Africa	89	4	6
South Asia	93	2	4
Sub-Saharan Africa	87	4	9
High income	30	59	11
Europe Economic and Monetary Union (EMU)	21	63	16

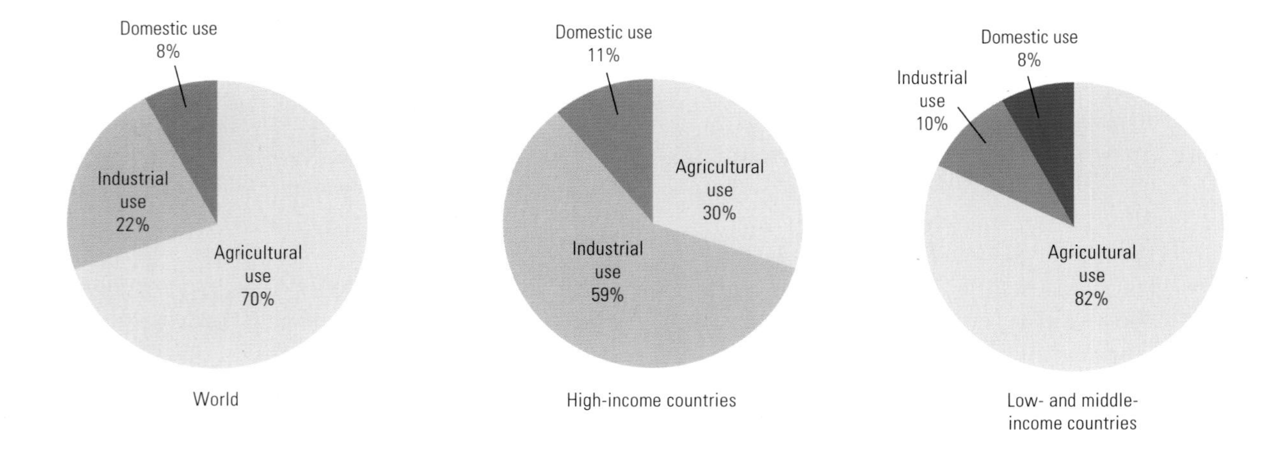

Industrial use of water increases with country income, going from 10 percent for low- and middle-income countries to 59 percent for high-income countries.

Source: World Bank, 2001.

Demand-side initiatives can play an important role in:

- increasing the efficiency of those industrial processes that place the greatest demands on water supply through the adoption of best available techniques; and

- lowering the pollutant loads of water discharged by industry through the recognition that much of this pollutant load represents excess raw materials that should not be discarded by an enterprise but rather captured for reuse.

These initiatives offer opportunities to break the prevailing paradigm whereby industrial growth and environmental protection are seen as incompatible alternatives. In existing industry, these demand-side initiatives can be driven, at least in part, by economic considerations at enterprise level. Thus, industry may be attracted to take up such work for reasons of enhanced competitiveness rather than for reasons of compliance with the negative drivers of regulation and enforcement. For new industrial investment, ensuring the incorporation of resource-efficient technologies and best-operating practices should be a key element of industrial planning by national investment promotion agencies.

Figure 9.2: Contribution of main industrial sectors to the production of organic water pollutants

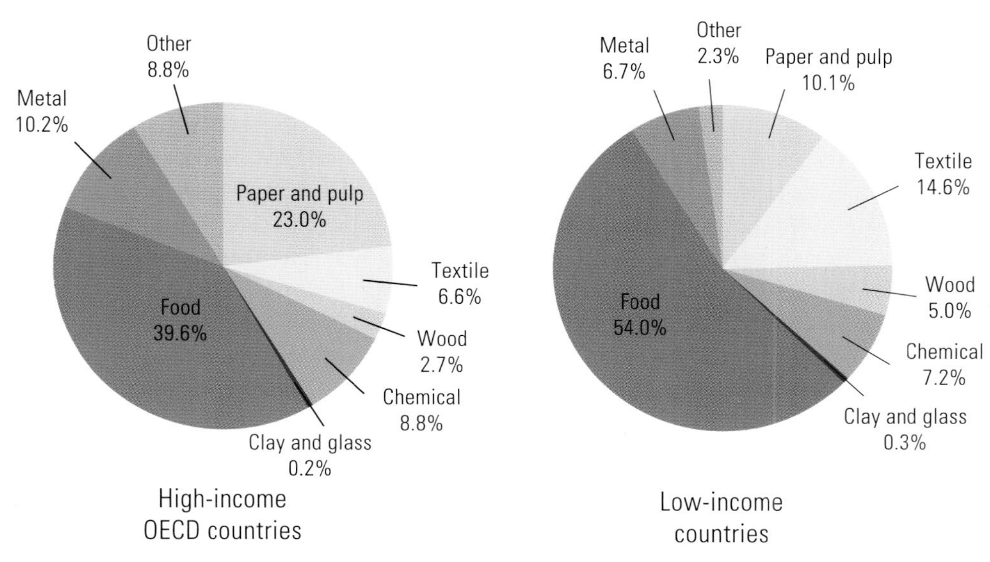

Industries based on organic raw materials are consistently the most significant contributors to the organic water pollutant load, with the food and beverages sector being the most important polluter.

Source: World Bank, 2001.

Water demand and industrial development

Global impacts on water by industry

Freshwater data set out in the *World Development Indicators Report* (World Bank, 2001) show that water for industrial use represents approximately 22 percent of total global freshwater use. In general, industrial use of water increases with country income, representing 59 percent of total water use in high-income countries but only 8 percent for low-income countries (see figure 9.1). The *World Water Resources and Their Use* database (Shiklomanov, 1999) forecasts that the annual water volume used by industry will rise from 752 cubic kilometres (km^3)/year in 1995 to an estimated 1,170 km^3/year in 2025, at which time the industrial component is expected to represent about 24 percent of total freshwater withdrawal.

One consequence of trade liberalization and the globalization of industry has been the migration of manufacturing industries from high-income countries to lower-income countries, sometimes by the simple relocation of production plants. In this way, industrial technologies developed in relatively water-rich regions are inherited by communities in areas where water may be a more scarce commodity or where governments are less able to match infrastructure growth to increasing demand. In these ways, both water stress and conflicts between users are likely to increase. The poorest groups in society, who typically have greatest difficulty in negotiating fair access, may be increasingly marginalized as conflicts increase. It is necessary now to consider precautionary and innovative approaches to prevent irreparable loss or damage to water resources.

On a global scale, however, industry may not be the most significant source of pollutants responsible for reductions in water quality. The runoff of agricultural inputs and untreated sewage from human communities create more widespread degradation of water resources (Kroeze and Seitzinger, 1998). In addition, the direct discharge of contaminants into water bodies is not the only vector by which industry degrades water quality at this scale.

Many of the chemicals and compounds discharged by industry as gaseous emissions have the potential for long-range transport, dispersal and deposition. This mechanism is recognized as an important factor in the degradation of fresh and marine waters in non-industrial regions and has stimulated a variety of multinational environmental agreements such as the Convention on Long-Range Transboundary Air Pollution and the Stockholm Convention on Persistent Organic Pollutants.

Global estimates of emissions of organic water pollutants by different industry sector are shown in figure 9.2. Inevitably, those industries based on organic raw materials are consistently the most significant contributors to the organic pollutant load with the food and beverages sector representing the most important polluter across the income range of the countries surveyed. Wood-based industries, including pulp and paper, and textiles are also important contributors, their respective values being determined by the relative importance of different industry sectors in the different country income groups.

Nevertheless, the location of industry in human settlements means that the industrial contamination of resources used for water supply is a common feature across the globe. It follows that considerable advances may be possible where successful local and regional initiatives can be shared and replicated. Land-based activities have a significant impact on coastal zone and, through long-range marine transport, may represent a global problem (see box 9.1).

In the high-income country group, it has been estimated that more than half of total freshwater abstractions are used for cooling in thermoelectric plants (Vassolo and Döll, 2002). Much of this water is returned without significant quality impacts at global and regional scales but with a persistent temperature increment that impacts on local ecosystems (see chapter 10 on energy for more details).

Regional impacts on water by industry

In high-income countries, industry has grown up to take advantage of local raw materials, including surface and groundwater resources of good quality, so that the regional distribution of industry reflects their geographical distribution. Although this pattern has become blurred by socio-economic factors, many examples remain. In these countries, water management is increasingly based on the holistic assessment of resources, supply and demand at the basin scale (EEC, 2000). This incorporates both consideration of potentially conflicting demands and encouragement of proper valuation of water through strict control of abstractions and rewards for efficient use. In general, a high proportion of the population has reliable and non-contentious access to good-quality water. In recent years, the real and transparent valuation of water resources has been encouraged as the management of the water industry has passed progressively to private sector corporations keen to establish 'bankable' assessments of resources and appropriate charging regimes. In these cases, the licensing of water enterprises needs to emphasize sustainable supply and protection of resources as principal objectives rather than short-term profit; this is particularly important for groundwater where robust resource estimates may be difficult to determine.

The development of integrated approaches to water management has coincided with the progressive transfer of manufacturing industry to developing countries and has resulted in the reduction in industrial water abstraction in high-income countries. Water management in formerly heavily industrialized regions in these countries now faces the challenge of rising groundwater levels and flooding. In many cases, this rising groundwater is recharging shallow aquifers last saturated many decades ago. Here the quality of this 'new' water resource may be reduced as it encounters contaminants that have infiltrated over the years from industrial sources above.

In lower-income countries, while primary industries such as mining are located so as to exploit natural resources, manufacturing industries may not be so dependent. Relatively small and impoverished local markets mean that much of this industry is export-led and located to take advantage of low production costs – particularly low labour rates and advantageous taxation regimes – and ease of shipment of products. In these circumstances, responsibility for water resource development may be divorced from investment or industrial development planning or becomes a 'bargaining chip' during investment negotiations. In many cases, perceived needs for economic and social development drive local governments to subsidize or assume supply-side responsibilities, placing additional burdens on local resources and infrastructure.

This problem may be exacerbated by the rapid, and often informal, growth of urban centres as people move to take up new employment opportunities. Local water resources and existing supply and wastewater treatment infrastructure may be overwhelmed by these changes. In this way, the well-intentioned sustainable development objectives of economic and industrial policies are undermined, with the most vulnerable sectors of society suffering further loss of access to good-quality water resources.

The concentration of industry on major transboundary rivers is an important factor in the degradation of regional water quality. Two forms of quality impairment of surface waters can be discerned:

- chronic quality impairment, whereby industry continuously discharges poorly treated or untreated contaminants so that pollutant loadings increase or high loadings are maintained over a long period of time; and

- acute quality impairment, whereby a catastrophic failure in industry generates extreme pollutant loadings in an uncontrolled but relatively short-lived event.

Chronic quality impairment is usually the result of a lack of dedicated treatment plants or because the industrial discharges are treated communally in municipal facilities unsuited for the purpose. In addition, the overall effectiveness of municipal facilities may be significantly impaired by the inclusion of uncontrolled mixed industrial pollutant loads. Different legal instruments have been formulated to deal with chronic quality issues (e.g. the Helsinki Convention on the Protection and Use of Transboundary Water Courses and International Lakes) and many projects developed by regional governmental groupings and their development partners.

Acute quality impairment typically results from inadequate safety management either of production processes or of production wastes. The failure of a waste impoundment structure at the Baia Mare mine in Romania provides an example (see box 9.2).

Box 9.1: Industrial water pollution control in the Gulf of Guinea basin (western Africa)

The Gulf of Guinea is one of the world's most productive marine areas, rich in fishery resources, oil and gas resources, and precious minerals, and an important global reservoir of marine biological diversity. It is bordered by countries in West and Central Africa adjacent to the Atlantic Ocean where pollution from residential and industrial sources has affected its waters and resulted in habitat degradation, loss of biological diversity and productivity, and degenerating human health.

To reverse this trend the countries of the region have adopted an integrated and holistic approach, applying the large marine ecosystem concept to the sustainable management of the regional environment and its living resources. Their communal project, 'Water Pollution Control and Biodiversity Conservation in the Gulf of Guinea Large Marine Ecosystem', financed by the Global Environment Facility (GEF), recognized that pollution from land-based sources contributes most of the pollution flux to the Gulf and placed high priority on the assessment, prevention and control of this pollution. An initial environmental survey used fish, benthic invertebrates and other biological indicator species to measure pollution effects on the marine and coastal ecosystems.

A major focus of the project was on industrial-pollution control. A rapid, semi-quantitative assessment of the land-based sources of pollution in the region was undertaken in each participating country. Industries situated within the 30–50 kilometres strip of the shoreline in the countries were assessed in terms of the manufacturing processes employed, the types and quantities of waste generated, and the nature of waste treatment and discharge practices.

The results of the survey demonstrated:

- an absence of pollution abatement infrastructure in the region leading to uncontrolled discharge of untreated wastes and effluents;
- an absence of common effluent discharge standards;
- an absence of environmental impact assessment, or environmental auditing during operation;
- insufficient human and material resources assigned to monitoring and enforcement;
- inadequate financial resources for implementation and compliance enforcement;
- an absence of reliable data and information in topographical maps with Global Positioning System (GPS) coordinates of the selected industries; and
- insufficient public awareness of pollution issues.

These assessments have provided the basis for:

- elaborating suggestions to improve industrial performance through the adoption of cleaner production methodologies and improved production process technology;
- the establishment of national cleaner production centres;
- the development of strategies and policies to encourage reduction, recycling, recovery and reuse of industrial wastes;
- the formulation of a draft version of the Regional Effluent and Discharge Standard; and
- collaborative Integrated Coastal Zone Management (ICZM) planning to regulate development in the region.

A pilot initiative in Ghana, the Waste Stock Exchange Management System, incorporating reuse and recycling philosophies to reduce waste input to the coastal and freshwater environments, was enthusiastically embraced by manufacturing industries with the slogan 'one person's waste, another person's raw material'. Cleaner production methodologies are being transferred to industries based on coastal lagoons in the Accra and Tema areas via a demonstration project led by the United Nations Industrial Development Organization (UNIDO).

Box 9.2: Environmental management and pollution control in the Tisza River basin (eastern Europe)

The Baia Mare (Romania) accident that took place in January 2000 drew international attention to the Tisza River when an earth dam impounding gold-mine wastes failed, releasing tons of cyanide-contaminated slurry to the Szamos River and then through the Tisza River to the Danube and the Black Sea.

The transboundary nature of the problem required an integrated approach to water resource management, based upon consideration of the river basin as an entity, combined with capacity-building, which can serve the needs of all stakeholders. Continuing protection and enhancement of the environment has to be ensured through the encouragement of a precautionary approach that requires the consideration of environmental risks in industrial planning and operation.

UNIDO is currently implementing a pilot project to promote an integrated approach to risk management in the Tisza River basin within the international legal framework of the European Union Water Framework Directive (EEC, 2000), the Seveso II Directive on the control of major-accident hazards involving dangerous substances (EEC, 1996b) and the OECD recommendation on the prevention of, and response to, accidents involving hazardous substance (OECD, 1988).

The project aims to support Tisza basin countries in:

- applying precautionary principles to industrial water pollution;
- improving emergency preparedness and response to accidental releases of toxic substances into the environment;
- improving communication between industry, government and community stakeholders with regard to risks and emergency systems;
- developing, with industry, practical preventative measures that can be implemented quickly; and
- transferring safe technology with hands-on experience.

Immediate project objectives are to:

- perform quantitative risk assessment of water contamination at selected industrial sites;
- assess gaps of the existing monitoring and Early Warning System (EWS) with respect to EU regulations;
- identify risk mitigation measures reducing both the frequency of occurrence of a water-polluting accident and the magnitude of the associated consequences;
- develop recommendations for external emergency plans and communication; and
- train identified actors and local communities.

Groundwater resources may be irreversibly reduced by both of these mechanisms as opportunities to remediate aquifers are more restricted. Additionally, the draw-down of freshwater levels in shallow aquifers by non-sustainable abstraction for coastal centres of population and industry may lead to saltwater intrusion, rendering these resources unsuitable for production.

Local impacts on water by industry

At the local level, inefficient water use and wastages by enterprises are the product of:

- the lack of technical capacity within the management of both government departments and enterprises;
- an understandable unwillingness to hinder industrial and economic performance; and
- the use of obsolete, inefficient or inappropriate technologies.

In many cases, industrial managers in middle- and lower-income countries are unaware of where and why water is being used in their enterprises. Its consumption may not be measured beyond its initial entry point so that management of its use at individual stages in the production process is not possible. For these reasons, water consumption is taken as an 'inevitable' cost, rather than as one of the array of manufacturing inputs that can, and should, be managed to maximize efficiency and minimize waste.

Encouraging change within industry cannot be done effectively by government-imposed regulation and licensing alone, particularly in developing countries where there are limited resources to monitor industrial performance. Rather, such policies need to be reinforced by schemes to improve the skills of managers and production staff so they are aware of both the environmental and economic advantages of using water resources carefully and efficiently. Such education, combined with the introduction of simple systems to determine water use and distribution, can lead to dramatic reductions in volumes consumed, often with little capital investment and initially without technological changes.

Of course, in many middle- and low-income countries, a large proportion of total employment and industrial effort may be concentrated in small- and medium-scale enterprises drawing water from domestic supplies. It is likely that this water demand is largely uncontrolled and unmeasured. Schemes to improve the water management skills within enterprises operating at these scales need to be integrated in general entrepreneurship development and may be particularly pertinent to women's groups as these are well represented amongst small-scale entrepreneurs and stand to gain most by release from water-gathering chores or from reductions in water pollution.

In a similar way, pollutant loadings of water discharged by enterprises can be significantly reduced by raising awareness of the value of raw material inputs being discharged as waste; capturing and recycling dyes from textile rinsing waters will, for example, reduce the biological oxygen demand (BOD) of discharged waters and increase the efficiency of dye use within the enterprise – thereby providing both environmental and economic gains.

Inevitably, pollutant loads cannot be eliminated by such methods. End-of-pipe treatment will remain a requirement. The capital investment for such plants can be derived, at least in part, from efficiencies made through the introduction of cleaner production methodologies and the progressive introduction of environmentally sound technologies and management practices. In this manner, industry is positively engaged in approaches to the sustainable use of water and other natural resources.

Monitoring industrial development and industrial impacts on water resources

Monitoring and developing indicators can be powerful tools to review and benchmark current environmental and economic performance at global, regional and local scales. They need to be able to evaluate trends and to indicate areas of concern where appropriate policies, assistance and investment strategies need to be developed. Data availability and reliability are necessary precursors to the derivation of robust indicators of current patterns of industrial water use.

At the global scale, present data relating directly to industry impacts on water resources may not be adequate for this purpose as they:

- are available for too few parameters;
- have been collected at different times and by different methods;
- represent estimates derived from a range of indirect data sources;
- do not adequately discriminate between industrial and other uses for water; and
- do not adequately discriminate between gross and net water consumption, which is particularly important with regard to water used in cooling thermoelectric plants, the bulk of which may be quickly available for reuse.

World Development Indicators 2001 tables of freshwater resources and industrial water pollution probably represent the most complete set of industry- and water-related data at the global level. The *World Water Resources and Their Use* database provides valuable data on renewable water resources and water use by region. Within these datasets, many of the individual data points require qualification, reducing the value of overall assessments; definitions of industrial use are inconsistent and vary from country to country while water quality data, for example, may refer to any year from 1993 to 1998 and are estimated as the product of estimated sectoral emissions per unit employment and sectoral employment numbers.

Other data sources, such as AQUASTAT and data published in The World's Water web site,[1] provide information on freshwater and industrial withdrawals share, but these sources do not significantly change the global picture of water demand. Industry impacts on water are not yet adequately discriminated within other systems, such as that operated by the World Meteorological Organization (WMO), the Global International Water Assessment (GIWA) and the Global Environment Monitoring System, Freshwater Quality Programme (GEMS/WATER), both managed by the United Nations Environment Programme (UNEP).

1. http://www.worldwater.org/

Nevertheless, the available data can be considered as a starting point for the development of indicators of demand-side performance. A crude index representing the economic value (in US$) obtained by industry per cubic metre (m^3) of water is obtained by comparing the quantity of freshwater consumed by industry each year and the Industrial Value Added (IVA) – at constant 1995 US$ (World Bank, 2001).[2]

The transformation of water quality data into indicators presents many of the same difficulties described above. In addition:

- estimation methodologies relate to particular industrial technologies and employee productivity rates that may be inappropriate, particularly in developing countries;
- BOD is not a total measure of industrial impacts on water quality as some contaminants do not affect BOD;
- industrial impacts are not only created from direct discharges to local water courses; and
- local and regional variations in water chemistry play a role in determining the ability of an ecosystem to 'cope' at different BOD levels.

Indicators developed for the regional or basin scale are mainly focused on the identification and evaluation of 'hot spots', the preparation of risk assessments and basin management plans. Indicators are needed to raise awareness and develop consensus among the different stakeholders and identify priorities for action. Many regional indicators are developed from dedicated datasets obtained by water monitoring networks. To obtain good statistics and reliable data for analysis and planning, it is not only necessary to establish such networks, but also to maintain them so that long-term changes in water availability and quality can be detected. This implies that ownership of monitoring initiatives established within the context of development agency programmes must be successfully passed to national or basin water managers.

Although planning may be focused at the regional scale, it is at the local level that the performance of individual enterprises can be influenced. Hot spots of water stress identified by the regional monitoring described above, may be addressed by reference either to international benchmarks, set by consideration of best available techniques, or, perhaps more likely for developing countries, by establishing local 'relative' benchmarks based upon existing performance. This approach is adopted, for example, by the cleaner production methodologies used by a number of international development agencies.[3]

The State of the Resource and Industry

Water quantity and water quality at the global scale

Initial estimates of the industrial water productivity (industrial value added per cubic metre of water used) gained by countries in the different incomes classes of the *World Development Indicators 2001* are given in table 9.1 and shown graphically in figure 9.3. Although there is considerable variation within each class, not least created by data difficulties described in the previous section, the different income groups fall into overlapping – but distinct – fields, particularly when revalorized on a per capita basis. The following broad conclusions may be drawn:

- for any given volume of water used by industry, the high-income users derive more value per cubic metre of water used than lower-income states;
- lower-income states can achieve similar water productivities as developed countries but do so only at significantly smaller total volumes of water used by industry;
- as total water consumption by industry increases, water productivity appears to fall in each income class; and
- economic growth from 'low-income' through 'lower-middle-income' to 'upper-middle-income' countries appears to have been achieved largely by additional consumption without significantly increased water productivity. It may, therefore, be limited by the availability of the resource.

Clearly, these conclusions must be treated with caution for the following reasons:

- There is considerable variation within each country income class with, in some classes, a small number of countries providing a considerable proportion of the total economic value.
- There are significantly different industry profiles across the countries sampled.
- Industry sectors based upon organic raw materials and addressing local markets may be heavily dependent on water but may have only limited opportunities for highly geared economic value addition.

2. Tables 3.5 and 4.2 in *World Development Indicators, 2001.*
3. UNIDO-UNEP National Cleaner Production Centres Programme.

Table 9.1: Industrial water efficiency

Country	Total annual freshwater withdrawal in billion m³ (1)	% for industry(2)	Industrial Value Added (IVA), in million US$ (3)	Population, in millions(4)	IVA/Industrial annual withdrawal (US$ /m³/capita)(5)
Algeria	4.5	15	22,618	30	1.11
Angola	0.5	10	4,182	12	7.26
Argentina	28.6	9	77,171	37	0.84
Armenia	2.9	4	1,029	4	2.14
Austria	2.2	60	76,386	8	7.14
Azerbaijan	16.5	25	1,213	8	0.04
Bangladesh	14.6	2	11,507	128	0.31
Belarus	2.7	43	9,543	10	0.81
Benin	0.2	10	333	6	3.59
Bolivia	1.4	20	1,529	8	0.68
Botswana	0.1	20	2,593	2	58.94
Brazil	54.9	18	231,442	168	0.14
Cameroon	0.4	19	2,360	15	2.07
Central African Republic	0.1	6	211	4	13.46
Chad	0.2	2	233	7	8.75
Chile	21.4	11	24,385	15	0.70
China	525.5	18	498,292	1,254	0.00
Colombia	8.9	4	23,120	42	1.40
Congo, Dem. Rep.	0.0	27	852	3	26.29
Costa Rica	5.8	7	4,456	4	2.88
Côte d'Ivoire	0.7	11	3,039	16	2.47
Croatia	0.1	50	4,995	4	31.22
Czech Republic	2.5	57	20,512	10	1.42
Denmark	0.9	9	40,142	5	100.23
Dominican Republic	8.3	1	5,530	8	16.58
Ecuador	17.0	6	6,535	12	0.57
Egypt, Arab Rep.	55.1	8	22,221	63	0.08
El Salvador	0.7	20	3,158	6	3.57
Estonia	0.2	39	1,494	1	23.88
Ethiopia	2.2	3	726	63	0.17
Finland	2.4	82	48,807	5	4.89
Gabon	0.1	22	2,752	1	208.45
Gambia	0.0	2	50	1	83.74
Georgia	3.5	20	378	5	0.11
Germany	46.3	86	760,536	82	0.23
Ghana	0.3	13	1,927	19	2.60
Guatemala	1.2	17	3,468	11	1.60
Guinea	0.7	3	1,431	7	9.21
Guinea-Bissau	0.0	4	46	1	63.33
Haiti	1.0	1	641	8	13.62
Honduras	1.5	5	1,234	6	2.71
India	500.0	3	113,041	998	0.01
Indonesia	74.3	1	85,633	207	0.56
Iran, Islamic Rep.	70.0	2	34,204	63	0.37
Italy	57.5	37	323,494	58	0.27
Jamaica	0.9	7	1,619	3	8.33
Jordan	1.0	3	1,738	5	10.43
Kenya	2.0	4	1,325	29	0.57
Korea, Rep.	23.7	11	249,268	23	4.16

Table 9.1: *continued*

Country	Total annual freshwater withdrawal in billion m³ (1)	% for industry(2)	Industrial Value Added (IVA), in million US$ (3)	Population, in millions(4)	IVA/Industrial annual withdrawal (US$ /m³/capita)(5)
Kyrgyzstan	10.1	3	699	5	0.46
Latvia	0.3	32	1,627	2	8.71
Lithuania	0.3	16	2,156	4	13.56
Malawi	0.9	3	288	11	0.82
Malaysia	12.7	13	43,503	23	1.14
Mali	1.4	1	580	11	3.88
Mauritania	16.3	2	284	3	0.32
Mauritius	0.4	7	1,419	1	57.13
Mexico	77.8	5	96,949	97	0.25
Moldova	3.0	65	508	4	0.07
Mongolia	0.4	27	362	2	1.56
Morocco	11.1	3	12,558	28	1.40
Mozambique	0.6	2	1,020	17	4.92
Namibia	0.3	3	971	2	57.12
Netherlands	7.8	68	116,700	16	1.37
New Zealand	2.0	13	15,683	4	15.08
Nicaragua	1.3	2	538	5	3.97
Niger	0.5	2	376	10	3.76
Nigeria	4.0	15	14,918	124	0.20
Norway	2.0	68	47,599	4	8.61
Pakistan	155.6	2	14,685	135	0.04
Panama	1.6	2	1,561	3	19.84
Papua New Guinea	0.1	22	1,779	5	16.17
Paraguay	0.4	7	2,334	5	15.51
Peru	19.0	7	20,714	25	0.61
Philippines	55.4	4	26,364	74	0.16
Poland	12.1	67	47,846	39	0.15
Russian Federation	77.1	62	97,800	146	0.01
Rwanda	0.8	1	356	8	4.13
Senegal	1.5	3	1,235	9	3.05
Sierra Leone	0.4	4	170	5	2.30
Slovak Republic	1.4	50	7,036	5	2.01
Slovenia	0.5	50	7,337	2	14.67
South Africa	13.3	11	49,363	42	0.81
Sri Lanka	9.8	2	3,862	19	1.04
Sweden	2.7	30	74,703	9	10.07
Tanzania, United Republic of	1.2	2	928	33	1.15
Thailand	33.1	4	64,800	60	0.81
Togo	0.1	13	309	5	5.21
Tunisia	2.8	2	6,297	9	13.01
Turkey	35.5	11	51,575	64	0.20
Turkmenistan	23.8	1	2,957	5	2.49
Uganda	0.2	8	1,191	21	3.55
Ukraine	26.0	52	17,854	50	0.03
United Kingdom	9.3	8	330,097	60	7.10
Uruguay	4.2	3	5,703	3	15.61
Uzbekistan	58.0	2	4,340	24	0.16
Venezuela	4.1	10	30,083	24	3.12
Viet Nam	54.3	10	9,052	78	0.02

Table 9.1: *continued*

Country	Total annual freshwater withdrawal in billion m³ (1)	% for industry(2)	Industrial Value Added (IVA), in million US$ (3)	Population, in millions(4)	IVA/Industrial annual withdrawal (US$ /m³/capita)(5)
Yemen	2.9	1	1,683	17	3.07
Zambia	1.7	7	996	10	0.84
Zimbabwe	1.2	7	2,005	12	1.96

(1) Data refer to any year from 1980 to 1999.
(2) Withdrawal shares are mostly estimated for 1987.
(3) US constant dollar 1995, data for 1999 .
(4) Data estimated for 1999.
(5) Population is expressed in millions, US constant dollar 1995.

The industrial water productivity shows the economic value (in US$) obtained annually by industry per cubic metre of water used. Very high differences can be noted, between high-income countries such as the United Kingdom, showing a per capita industrial water efficiency of US$ 7.10/m³, and many low-income countries, such as Moldova, with only US$ 0.07/m³. Observe, however, that countries having small populations or highly specialized industries (high-value gems, tourism) – such as Gabon, Namibia or Mauritius – have also achieved high productivity.

Source: World Bank, 2001.

Figure 9.3: Industrial Value Added from water use for main income classes of countries

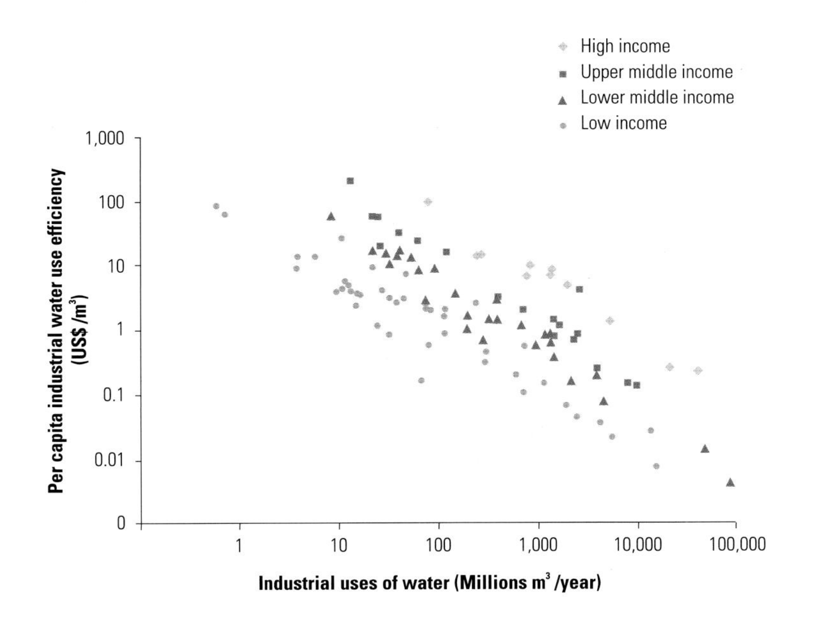

At any given volume of water withdrawn by industry, the per capita industrial water productivity increases with income class, but in any income class, water efficiency seems to fall with increasing industrial water withdrawal. Per capita industrial water efficiency is calculated as the ratio between country Industrial Value Added (IVA) and the volume of water withdrawn by industry and country total population (in millions). While Industrial Value Added data refer to the year 1999, the total annual freshwater withdrawal refers to any year from 1980 to 1999 and the published industrial shares are mostly estimated for 1987. Population data are estimated for 1999.

Source: World Bank, 2001.

- Consumption of water for cooling thermoelectric plants represents a considerable proportion of water abstracted in the high-income group of countries.

- The data have been measured in different ways, in different years or have been estimated from other economic statistics.

Information on the industrial degradation of water quality through emissions of organic water pollutants is given in section 3.6 of *World Development Indicators 2001*. This provides BOD data, the widest and most reliably measured indicator, for a range of countries together with the estimated shares contributed by various industry sectors.

The data also indicate the change in BOD loadings with time. Comparisons of data for 1980 and 1996 indicate that while BOD loadings for high-income countries have been reduced, those in middle- and low-income countries have risen substantially. The data also indicate that the contributions of two developing countries – India and China – are statistically significant within the overall data, with China contributing 32 percent and India 8 percent of estimated global emissions of organic water pollutants in 1996.

Many actions to restrict *inter alia* industry impacts on water have been undertaken by the international community and have led to multinational environmental agreements such as the Global Plan of Action for the Protection of the Marine Environment from Land Based Activities (GPA), the Mediterranean Action Plan (MAP), the Basel Convention and the Stockholm Convention.

Reducing industry impacts at a basin scale

Map 9.1 shows the distribution, by river basin, of water withdrawals for manufacturing industry. Assessing demand-side concerns in this way, rather than subdivided by political boundaries, enables transboundary risks and conflicts to be identified and managed on the basis of natural hydrological units. The map demonstrates the correlation between levels of industry withdrawals and areas of high population density; in particular, parts of India, much of eastern China, the eastern seaboard of the United States and Canada, much of Europe and central Russia, the Nile basin in Africa, and the Middle East. The water bodies in many of these areas suffer from water stress.

Map 9.1: Water withdrawals for manufacturing industries according to drainage basins

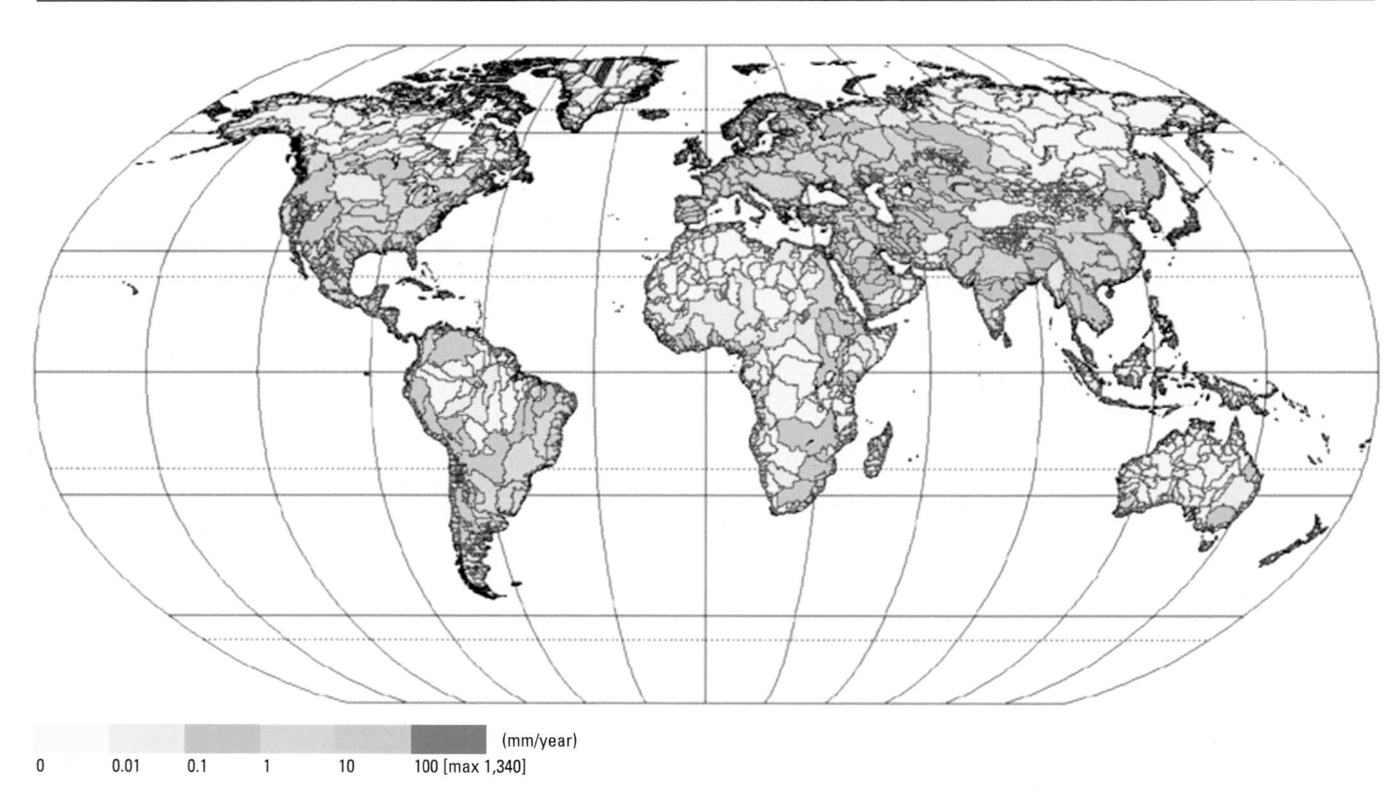

This map demonstrates the correlation between levels of industry withdrawals and areas of high population density, such as India, eastern China, the eastern seaboard of the United States and Canada.

Source: Map prepared for the World Water Assessment Programme (WWAP) by the Centre for Environmental Research, University of Kassel, based on data from WaterGAP, Version 2.1.D.

Box 9.3: Convention on cooperation for the protection and sustainable use of the Danube River (central-eastern Europe)

In the Danube River basin, regional agreements in 1991 and 1994 have resulted in the Convention on Cooperation for the Protection and Sustainable Use of the Danube River, or Danube River Protection Convention (DRPC). This took effect in 1998, securing the legal basis for protecting water resources. It has been ratified by eleven parties: Austria, Bulgaria, Croatia, Czech Republic, Germany, Hungary, Moldova, Romania, Slovakia, Slovenia and the European Union (EU).

The main objective of the convention is cooperation of the parties concerned in taking appropriate legal, administrative and technical measures to maintain and improve the environmental and water quality conditions of the Danube River and its catchment. This includes, among others:

- the improvement and rational use of surface water and groundwater;
- pollution reduction from point and non-point sources;
- the reduction of pollution loads entering the Black Sea; and
- the development of accident prevention and response measures.

The middle and downstream Danube countries with economies in transition (Bulgaria, Croatia, Hungary, Romania and Slovakia) are facing serious economic and financial problems in responding to the objectives of the convention and in implementing pollution reduction and environmental protection measures, such as the for European Union Integrated Pollution Prevention and Control (IPPC) Directive (EEC, 1996a), required for accession to the EU.

To assist them, the GEF-funded Pollution Reduction Programme for the Danube River basin identified those major manufacturing enterprises contributing the bulk of transboundary pollution, predominantly in the form of nutrients and/or persistent organic pollutants. One hundred and thirty such 'hot spots' were identified. The Transfer of Environmentally Sound Technology (TEST) project, begun in 2001, is taking up the challenge of demonstrating to selected industries in these countries that it is possible to respect environmental standards while maintaining or enhancing their competitive position.

Eco-efficiency indicators have been developed as:

- a benchmarking tool to help industry to monitor, assess and improve financial and environmental performance;
- a step towards the introduction of environmental accountability and a wider diffusion of environmental corporate responsibility; and
- an encouragement towards the development, endorsement and implementation of an Environmental Management System (EMS).

The endorsement of such an EMS and, ultimately, ISO14001 registration demonstrates the commitment of an enterprise to take the actions necessary both to comply with legal requirements and to make industry environmentally friendly by continued improvements in environmental performance through the responsible use of energy, water and raw materials.

The sustainability of the TEST programme is assured by two capacity-building objectives; transferring environmental management skills to industry both demonstrates the economic advantages of environmental compliance and builds demand for environmental services. This demand is satisfied by the locally available skills.

A successful basin-wide strategy to alleviate industry impacts on the water environment has been implemented in the Seine-Normandy basin in which 40 percent of France's national industrial production is concentrated (see chapter 19). The integrated management strategy complies with the European Water Framework Directive (WFD) that requires member states to prepare a water management plan by river basin to protect aquatic ecosystems, drinking water resources and bathing water on the basis of a combined approach that requires both source pollution control and the setting of water quality targets for the receiving environment.

International action to address water stress and chronic water quality issues at the regional scale have resulted in the development of multinational agreements supporting standing bodies for the planning and management of a number of transboundary basins and river systems. The ratification of the Danube River Protection Convention (DRPC) represents an example of this approach (see box 9.3).

An example of actions to prevent future acute water quality concerns arising from industrial activity is given by work being undertaken in the catchment of the largest tributary of the Danube, the Tisza River (see box 9.2).

Developing countries have the challenge and the opportunity to take advantage of such experiences and instigate integrated approaches to the management of water resources at the regional scale. A relevant example of this can be seen in Sri Lanka (see chapter 18), where most water abstraction is used for agricultural purposes, but where the industrial development proposed to alleviate poverty will result in rapid socio-economic transformation and significantly higher demand for water. Water scarcity is currently a major concern and wastewater effluents have already contaminated water bodies and affected domestic water supply. Industrialization and increasing population pressure are expected to worsen current conditions and threaten ecosystem well-being unless integrated management strategies, that include land planning, infrastructures enhancements, the development of legal and regulatory frameworks, and capacity-building are planned and put into action.

Regional actions to address the impacts of industry on coastal zones

The concentration of industry and population in the coastal zones of many developing and transitional economy countries, in particular tropical developing countries, has given rise to an alarming rate of destruction of critical coastal habitats. Toxic industrial discharges, solid and liquid urban wastes, destructive fishing, sediment inputs from land construction activities and dam development, mangrove conversion for aquaculture and agriculture development, coral mining, sand filling and canalization in wetlands, groundwater draw-down and aquifer salination, and so on, are producing long-term changes, especially in coastal water quality. These affect the ecological efficiency, sustainability, biological productivity and health of the environment and threaten the ability of coastal ecosystems to sustain their primary functions.

Coastal zones are especially vulnerable as they represent a receiving point for the pollution flux transported through the river system from land-based activities within the catchment. The particular physico-chemical conditions operating at the interface of fresh and marine waters serve to concentrate much of this pollutant load. Nevertheless, coastal ecosystems form a continuum with river basins so that integrated management of the latter provides important and tangible benefits to coastal systems and the livelihoods of those dependent on their natural riches (UNEP, MAP, PAP, 1999).

UNIDO has recognized the importance of reducing pollutant loads arriving in coastal zones from industrial sources within river basins, and has adopted the strategy of preventing or reducing pollution at source by facilitating the introduction of best environmental practices to key developing countries industries. UNIDO has also provided technical assistance to some developing countries in the Gulf of Guinea basin of West and Central Africa in the introduction and adoption of policies and strategies focusing on Integrated River Basin Management (IRBM) and Integrated Coastal Zone Management (ICZM) for the protection and management of coastal and freshwater resources (see box 9.1).

Local improvements of industrial practices with global/regional benefits

Many countries have moved to incorporate precautionary and polluter pays principles within water governance. However, a large number of developing countries lack the resources necessary for preventive planning or, indeed, for regular monitoring and enforcement. As a result, the application of these principles is at best only responsive in character, often based on concerns and complaints raised by local communities. This situation is inadequate because:

- it does not prevent excessive water use or the impairment of water resources;
- only 'obvious' pollution is addressed and important but 'invisible' pollution may be missed;
- considerable delays may occur between pollution events and remediation;
- water authorities may not have the technical capacity to identify individual polluters or, in some countries, liability may belong to, or have passed to, government; and
- some communities do not have access to industry, due to political boundaries, for example.

For these reasons, actions to use water efficiently and to eliminate contaminating discharges must be based on precautionary actions positively engaging industry in the sustainable development agenda. This requires the consideration of water governance issues, and implies the development of consensus between community, industry and government actors, early on in planning and investment processes. Within industry, a progressive package of environmentally sensitive improvements needs to be incorporated into production management and combined with the raising of technical capabilities at all levels.

In high-income countries preventive methodologies have long been incorporated into the 'toolkit' of the production manager as all manufacturing inputs and emissions have economic implications. This is less true of developing countries and countries with

Box 9.4: Regional African leather and footwear industry scheme

Leather-making is a significant source of income in most African countries but it is also a major cause of industrial pollution. Processes in tanneries are very water intensive and tannery wastewater carries a large amount of spent chemicals as well as organic matter. Cleaner technologies and better housekeeping and production practices can help reduce water use and chemical consumption as well as wastewater contamination. Nonetheless, end-of-pipe water treatment is essential in containing the adverse environmental impact of the leather industry.

UNIDO has been assisting Africa's leather industries over the last three decades and, since 1988, has provided assistance in pollution control to some thirty tanneries in Ethiopia, Kenya, Malawi, Namibia, Sudan, Uganda, the United Republic of Tanzania, Zambia and Zimbabwe. These activities confirm that a mix of waste management and of cleaner technologies (such as high-exhaustion chrome tanning, low sulphide dehairing, carbon dioxide deliming, wet-white processing) makes the tanning industry more environmentally friendly; increases productivity; reduces costs; cuts water, energy and chemical consumption; and strengthens the manufacturer's image amongst consumers. For example, using conventional technologies, up to one-third of the chrome used in tanning ends up in the effluent; with high-exhaustion chrome tanning technology these effluent loads are reduced as 90 percent of the chrome is taken up in the leather. In consequence, less chrome is required.

The overall benefits from these projects are expected to be reduced water consumption and a significant reduction in the main components of the effluent load. Evidence from work to date indicates that chemical oxygen demand (COD) and BOD values can be reduced by up to 60 percent while suspended solids, chromium and sulphide can be reduced by more than 90 percent. The introduction of improved 'house-keeping' during production, including better process controls, reduced overall water consumption by more than 14 percent in an Ethiopian tannery.

End-of-pipe water treatment represents a last but important mitigation strategy. Improving or installing treatment facilities and capacity-building for monitoring of the effluent treatment process are important components of most projects.

New ways of treating effluents and reducing solid waste volumes have also been developed. At the Zimbabwe Bata Shoe Company, a small-scale anaerobic digester of tannery sludge has been successfully tested. The results confirmed the feasibility of installing a facility capable of handling 150 m^3 of sludge per day resulting in zero solid waste but generating biogas that could be used as an energy source. The tannery wastewater is collected in a small pond and from there gradually discharged into a larger pond where Spirulina algae thrive on what is left of the effluent pollution load and make the pond a natural environment hospitable to fish, frogs and other aquatic life.

In Kenya, increasing human and industrial pollution to Lake Nakuru over the past two decades resulted in a significant deterioration in water quality and a sharp decline in the flamingo population. Nakuru Tanners, situated in the immediate vicinity of Lake Nakuru National Park and a major exporter of wet-blue leather products, joined the pollution control efforts in the area and cleaner production and effluent treatments led to substantial reductions in polluter indicators. In 1998, flamingos began to return to the lake in large numbers and most scientists agree that some of the credit is due to the pollution control efforts made by the tannery.

economies in transition. Subsidiaries of multinationals may benefit from the in-house transfer of skills while local industries linked to transnational supply chains may be contractually obliged to take up such improvements. In many countries, these represent a relatively small proportion of the national industrial effort. For this reason, considerable efforts are made by countries and their international development partners to transfer cleaner production, environmental management systems, and the application of best available practices to governments and industry at all scales.

Cleaner production methodologies seek to generate improvements within the manufacturing cycles of industry leading to:

- significant reductions in industry emissions;
- more efficient production providing considerable reductions in the raw materials consumption; and
- improved product quality.

In this way, cleaner production provides a number of key commercial incentives for companies to participate in the proactive consideration of environmental benefits even where regulatory drivers are weak.

Capacity-building in cleaner production methodologies is delivered through demonstration, or sector-related projects, often as part of integrated programmes of industry support to member countries. A global network of National Cleaner Production Centres (NCPC) has been established jointly by UNIDO and UNEP. Over twenty such centres are in operation worldwide, with more being established. The centres provide practical technical assistance and training to industry managers, supporting service providers and the staff or regulatory authorities. In addition, trainees benefit from access to information and experience gained by other centres of the global network and related institutions around the world.

An example of the successful promotion and diffusion of cleaner technologies is proved by the results achieved in Africa by projects with the leather and footwear industry (see box 9.4). The improvements in terms of reduced water demand and of better effluent quality confirm that a mix of cleaner technologies and waste management makes tanning industry more environmentally

Box 9.5: Impressive gains from cleaner food production in Viet Nam

A tenfold return on an initial investment of US$62,000 has put the Thien Huong Food Company on a steady, cleaner production course that has already yielded impressive environmental and business benefits.

The company, one of the largest food manufacturers in Ho Chi Minh City, faced a daunting double challenge in 1998. The management was under growing pressure to improve the economic performance of the company. At the same time it had been singled out as a major pollution culprit in a large residential area and been entered into the 'Black Book' of the city's environmental authorities.

The same year, Thien Huong joined a UNIDO cleaner production project carried out in cooperation with the local department of science, technology and environment. A task team led by the production manager was established. Assisted by local and international experts, the team conducted a thorough analysis of the waste streams generated by the various operations entailed by the production of instant noodles – the company's most important product. The purpose was to find ways of reducing the pollution load by changing or fine-tuning manufacturing processes.

The scrutiny resulted in sixty-two cleaner production options. Twenty-four of them, mostly low-cost or no-cost measures, were selected for immediate implementation and supported by:

- closely monitoring key production inputs, including materials, water and energy; and
- the introduction of a shop-floor incentive system for meeting efficient resource utilization targets.

By December 1999 the factory wastewater volume had been cut by 68 percent with a concomitant 35 percent reduction in organic pollution as well as a significant drop in gas emissions.

At least equally compelling were the business benefits. While the application of the twenty-four options cost the company a total of US$62,000 it saved a total of US$663,700. In addition to cost savings, the cleaner production measures improved product quality and consistency, lengthened the shelf life of the product, and helped increase production capacity by 25 percent.

Box 9.6: Removal of barriers for cleaner artisanal gold mining

Over the last few decades, artisanal gold mining activities have increased steadily and account now for approximately one-quarter of total world gold output. Despite the current low gold price, this gold rush in the artisanal sector continues. In many developing countries, artisanal gold mining has become a major safety valve, cushioning the worst effects of structural adjustment, recession and drought by providing people in the rural areas with an alternative way of securing a livelihood.

However, water siltation resulting from small-scale gold mining along rivers has resulted in a decrease of the fish population and has made water unfit for human consumption in regions where this resource was already scarce.

Most small-scale miners use mercury amalgamation to prepare final gold concentrates. Mercury is one of the most toxic substances in the world with long-term and far-reaching effects causing significant damage to the environment and to the health of people who handle it. The mercury released into watercourses travels long distances and can be transformed by micro-organisms into more toxic forms (methyl-mercury), which then enter the food chain.

It is estimated that 2 to 5 grams of mercury are necessary to produce 1 gram of gold and that, with the methods currently used, all this mercury is lost to the environment. In addition to mercury discarded or spilled directly into streams and rivers during the amalgamation process, a considerable volume of mercury vapour is released each year to the atmosphere. Much of this quickly returns with rain to the river ecosystem.

The environmental impacts resulting from mercury use by the artisanal mining sector require concerted and coordinated global responses. In recent years, UNIDO has developed projects to address the problem in Ghana, the Philippines, the United Republic of Tanzania and Zimbabwe. Through training campaigns, miners are informed of the dangers of mercury, are trained in improved mining and processing practices, and are made aware of the need to protect water resources, both for their own use and for the use of communities living downstream. Alternative cleaner techniques to amalgamation are demonstrated on-site and the technology then transferred to local manufacturers. Retorts, which allow the recycling of mercury during the burning process are introduced, thereby reducing environmental releases and reducing overall mercury consumption. Local laboratory capacities are also enhanced to monitor the environmental and human health impacts of the mining activity on the rivers. UNIDO also assists governments in the development of monitoring and enforcement programmes.

UNIDO has recently won approval for GEF funding for a global project that plans to address mercury pollution caused by small-scale mining activities undertaken in Brazil, Indonesia, Lao PDR, Sudan, the United Republic of Tanzania and Zimbabwe as well as seeking to reduce mining-related risks to international water bodies such as Lake Victoria, the Amazon, Mekong, Nile and Zambezi Rivers and the Java Sea. The strength of this global project lies in the exchange of experience between the countries and the facilitation of technology transfer. The project will reduce the environmental burden of artisanal gold mining on watercourses and public health while at the same time improving the skills and income of small-scale miners.

friendly, increases productivity, reduces costs by cutting water, energy and chemical consumption, and strengthens the manufacturer's image. Water savings made in this way may be allocated to other uses while improved water quality restores ecosystem functioning and potable water resources.

A similar experience in food-processing enterprises in Viet Nam indicates that significant reductions in water consumption in emissions can be achieved by introducing cleaner production methodologies, often with little or no capital investment in early stages (see box 9.5).

The introduction of cleaner technologies for the artisanal gold mining activities (see box 9.6) provides an example of actions to address local water contamination issues where the hazard is less obvious to industry and affected communities. Improving the management of mercury within the gold amalgamation process reduces consumption and significantly reduces mercury releases to the environment, reducing risks to both human and ecosystem health.

Recommendations for Future Development Strategies

Requirements and recommendations for future actions and initiatives include, at the global level:

- developing more robust water consumption indicators based on verifiable statistics to aid the identification of key problem issues and areas of concern;
- including industrial water efficiency, water reuse and recycling indicators with those for water stress;
- developing more robust water quality indicators including parameters beyond BOD;
- focusing global action on the improvement of industrial performance with regard to water in those countries most at risk from water scarcity and use conflicts; and
- identifying existing technologies and developing new technologies of use in improving the water-related performance of key industry sectors such as food processing, wood-based industries and textiles.

At the regional level, future actions and initiatives should focus on:

- developing collaborative and consensual governance schemes and multinational agreements to protect river basins, transboundary waters and water bodies isolated from centres of industry;
- mainstreaming water demand management into industry and investment planning and regulation;
- focusing industrialization strategies on sectors appropriate to regional water resource availability;
- establishing sustainable water governance and monitoring to enable the application of precautionary approaches and the recognition of the significant incremental cost penalties of remediation; and
- diffusing and promoting the precautionary approach to industry as an economic driver for change.

At the local level, it is recommended that future actions and initiatives focus on:

- identifying positive drivers and economic benefits to engage industry in proactive environmental management;
- developing voluntary and consensus agreements between industry, its regulatory authorities and surrounding communities in order to develop balanced dialogues concerning water use;
- identifying data collection needs and developing sensitive indicators useful to industry and its regulators; and
- developing cleaner production initiatives by adopting new and water-efficient technologies.

Conclusions

Some 20 percent of the world's freshwater abstraction is currently used by industry, corresponding to about 45 litres per day per person. Globalization, with its accompanying move of labour industries from high-income to low-income countries, is creating high water demand outside of its abundant sources, often in urban areas. Furthermore, lower-income states derive less value per cubic metre of water used than high-income states, and economic growth from low-income through lower-middle-income to upper-middle-income countries appears to have been achieved largely by additional consumption without significantly increased water efficiency.

Information on the industrial degradation of water quality is given by the emissions of organic water pollutants; comparison of data for 1980 and 1996 indicating that while BOD loadings for high-income countries have been reduced, those in middle- and low-income have risen substantially.

Progress since Rio at a glance

Agreed action	Progress since Rio
Increase concern about, and awareness of, the need and effect of industries on water resources	
Promote treatment, recycling and safe reuse of industrial wastewater	
Develop clean technologies, including control of industrial waste discharges	

Unsatisfactory — Moderate — Satisfactory

References

Alcamo, J.; Döll, P.; Henrichs, T.; Lehner, B.; Kaspar, F.; Rösch, T.; Siebert, T. Forthcoming. 'WaterGAP: Development and Application of a Global Model for Water Withdrawals and Availability'. *Hydrological Sciences Journal.*

Döll, P.; Kaspar, F.; Lehner, B. Forthcoming. 'A Global Hydrological Model for Deriving Water Availability Indicators: Model Tuning and Validation'. *Journal of Hydrology.*

Döll, P. and Siebert, S. 2002. 'Global Modeling of Irrigation Water Requirements'. *Water Resources Research*, Vol. 38, No. 4, pp. 8.1–8.10, DOI 10.1029/2001WR000355.

EEC (European Economic Community). 2000. Framework Directive in the Field of Water Policy (Water Framework). Directive 2000/60/EC of the European Parliament and of the Council of 23 October 2000, establishing a framework for EEC action in the field of water policy [Official Journal L 327, 22.12.2001].

——. 1996a. Council Directive 96/61/EC on the Integrated Pollution Prevention and Control (IPPC Directive).

——. 1996b. Seveso II Directive: Council Directive 96/82/EC on the Control of Major-Accident Hazards Involving Dangerous Substances.

Federal Ministry for the Environment, Nature Conservation and Nuclear Safety, and Federal Ministry for Economic Co-operation and Development. 2001. Ministerial Declaration, Bonn Keys, and Bonn Recommendation for Action. Outcomes of the International Conference on Freshwater held in Bonn, 3–7 December 2001.

Kroetze, C. and Seitzinger, S.-P. 1998. 'Nitrogen Inputs to Rivers, Estuaries and Continental Shelves and Related Nitrous Oxide Emissions in 1990 and 2050: A Global Model'. *Nutrient Cycling in Agroecosystems,* Vol. 52, pp. 195–212.

OECD (Organization for Economic Cooperation and Development). 1988. Decision-Recommendation of the Council Concerning Provision of Information to the Public and Public Participation in Decision-Making Processes Related to the Prevention of, and Response to, Accidents Involving Hazardous Substances. C(88)85/Final.

Shiklomanov, I.-A. 1999. *World Water Resources and Their Use.* St. Petersburg, State Hydrogeological Institute, part of the International Hydrological Programme of the United Nations Educational, Scientific and Cultural Organization. St. Petersburg.

UN (United Nations). 1992. Agenda 21: Programme of Action for Sustainable Development. Official outcome of the United Nations Conference on Environment and Development (UNCED), 3–14 June 1992. Rio de Janeiro.

UNDESA (United Nations Department of Economic and Social Affairs). 1997. 'EarthSummit+5'. Document presented at the Special Session of the General Assembly to Review and Appraise the Implementation of Agenda 21, 23–27 June. New York.

UNDP (United Nations Development Programme). 1997. *Implementing the Rio Agreements: A Guide to UNDP's Sustainable Energy and Environment Division,* section 2.1. New York.

UNEP/MAP/PAP (United Nations Environment Programme/Mediterranean Action Plan/Priority Actions Programme/). 1999. *Conceptual Framework and Planning Guidelines for Integrated Coastal Area and River Basin Management.* Split, Priority Actions Programme. Nairobi.

UNIDO (United Nations Industrial Development Organization). 2002. *Developing Countries' Industrial Source Book.* First edition. Vienna.

——. 2001. *Integrated Assessment, Management and Governance in River Basins, Coastal Zones and Large Marine Ecosystems.* A UNIDO Strategy Paper. Vienna.

Vassolo, S. and Döll, P. Forthcoming. *Development of a Global Data Set for Industrial Water Use.* University of Kassel, Centre for Environmental Systems Research.

——. 2002. *Industrial Water Use: A New Global Data Set.* Poster presented at the European Geophysical Society Conference, April. Nice.

World Bank. 2001. *World Development Indicators* (WDI). Washington DC. Available in CD-ROM.

Some Useful Web Sites*

United Nations Environment Programme (UNEP): Cleaner Production Activities
http://www.uneptie.org/pc/cp/
Information on cleaner production and UNEP's related activities.

United Nations Industrial Development Organization (UNIDO): Industrial Development Abstracts
http://www.unido.org/IDA.htmls
Source of information on the activities of UNIDO to assist industrialization in developing countries. Indexed abstracts of UNIDO documentation and descriptions of major studies and reports.

United Nations Industrial Development Organization (UNIDO): Section on Cleaner Production
http://www.unido.org/en/doc/5151/
Presentation of documents and projects related to cleaner production.

United Nations Industrial Development Organization/Organization for Economic Cooperation and Development (UNIDO/OECD): Industrial Statistics
http://www.unido.org/en/doc/3474/
Databases, publications, industrial country statistics compiled by UNIDO with the help of OECD.

World Bank: New Ideas in Pollution Reduction (NIPR)
http://www.worldbank.org/nipr/
Primary source for materials produced by the World Bank's Economics of Industrial Pollution Control Research Project.

* These sites were last accessed on 6 January 2003.

Developing Energy to Meet Development Needs

By: UNIDO
(United Nations Industrial Development Organization)
Collaborating agencies:
WHO (World Health Organization)/ UNEP (United Nations Environment Programme)/Regional Commissions/ World Bank

Table of contents

Developing Energy to Meet Development Needs

It is evident that the fortunes of the world's human population, for better or for worse, are inextricably interrelated with the use that is made of energy resources.

M. King Hubbert, *Resources and Man*

ENERGY, ESPECIALLY IN ITS MOST COMMON FORM of electricity, is central to our daily lives and productive capacity. Yet many people live their lives without electrical power. As a result, they pay a heavy toll in health, time spent gathering alternative sources of fuel for cooking, for example, and often gruelling labour. Water is used in most means of generating power, and in many countries hydropower is the only really sustainable source. This chapter assesses the contribution that water makes to our lives and livelihoods through energy, and the impact that the production of energy has on people and the environment.

SINCE THE RIO EARTH SUMMIT, the crucial role of energy as a component in sustainable development has been widely recognized. Although Agenda 21 did not have a chapter specifically devoted to energy, its comprehensive programme for action to achieve sustainable patterns of production and consumption revealed how closely such aims are linked to the availability of affordable energy. As more energy for domestic, industrial and agricultural use is required, the demand for electricity in particular will be even greater.

Current and Future Energy and Water Demands

According to the United States Energy Information Administration, world electricity consumption is expected to rise by 73 percent between 1999 and 2020, making electricity the fastest-growing form of energy. This growth will be driven mainly by developing countries.

However, according to the United Nations Development Programme (UNDP), at present 2 billion people have no electricity at all, 1 billion people use electricity from uneconomic sources (dry cell batteries, candles, kerosene) and 2.5 billion people in developing countries, mainly in rural areas, have little access to commercial energy services (UNDP, 2002).

Developing countries with low material resources incur heavy debts in foreign currency by having to import fuels for power generation and transport. Capital invested in the energy sector is one reason for the high debts of developing countries. In these nations, people living in poverty often pay a higher price per unit of energy services than the rich.

When energy and fuel supplies are absent, limited or too expensive, overall development aims are thwarted: employment prospects are poor; women spend a disproportionate amount of time fetching water and gathering fuel and so cannot participate fully in the community; the poor remain trapped in their poverty; provision of education and health services is hindered and economic growth is slowed. Lack of energy supplies can be seen therefore as a major threat to economic and social sustainability. However, meeting greater energy requirements to allow for the development of poorer countries must not be done at the expense of the environment: clean, environmentally friendly solutions are needed.

While energy is crucial to many development aims, it is still only one part of the overall picture. In power generation, energy and water objectives, concerns and constraints converge.

Demographic changes

The need to increase access to energy and water applies to both urban and rural areas. Urban and peri-urban needs are of growing importance because of both the internal expansion of cities with their high birth rates, and the ever-increasing migration of rural people to towns in search of improved livelihood prospects. As reported in chapter 7 on water for cities, the urban population is expected to grow by 1 billion in the less developed countries between 2000 and 2020, thus overtaking rural population figures (UN, 2000). Rural electricity supply in many developing countries is often hampered by low population densities, the limited purchasing power of rural people and the absence of decentralized supply options.

The Role of Energy in Meeting Development Targets

Generating income to overcome poverty

Electricity plays an important part in lifting people out of poverty. It is an essential resource for the emergence of small-scale entrepreneurial and industrial initiatives. Off-farm manufacturing can generate more income if reliable energy is available for mechanization, lighting and heating. Rural electricity services help promote non-farming and non-timber-based enterprises, thus creating opportunities for cash-generating activities and diversifying rural economies. Urban, small-scale business and micro-enterprises, in areas where levels of unemployment are very high, are often urban dwellers' only way out of poverty and are dependent on access to electricity.

Yet sources of electricity are still often costly, both in economic terms and in the levels of environmental pollution they cause. Grid electricity has the best cost-benefit ratio for lighting: a study in Indonesia showed that it is not only less costly than kerosene (112 *rupiahs* a month versus 218) but it also provides ten times more light (Peskin and Barnes, 1994). It is also much cheaper than car batteries for powering a television or a radio set in a poor rural household. Given the projected scale of increasing urbanization by 2015, it is clear that grid electricity will be an essential factor in helping to raise income levels.

Universal primary education

Availability of modern energy services frees children's – and especially girls' – time from helping with survival activities (gathering firewood, fetching water). Girls are likely to be kept home from school to help with household chores when women are overburdened and this contributes to the perpetuation of female poverty. Electricity enables good-quality lighting for home study or evening classes. It also allows access to educational media and communications (Information and Communication Technologies [ICTs]) in schools and at home, thus increasing educational opportunities and allowing distance learning. All in all, energy can help to create a more child-friendly environment (access to clean water, sanitation, lighting and space heating/cooling) thus improving attendance at school and reducing drop-out rates.

Gender equality and women's empowerment

Lack of energy services creates particular hardships for women since they are generally responsible for gathering fuels and performing household duties involving energy use such as cooking. Supplies of traditional fuel sources such as wood are being rapidly diminished and degraded due to a combination of economic and environmental pressures. Women often have to spend a great deal of time and physical energy searching for fuel far from home and hauling it back over long distances. This added investment of women's time and effort is necessary for the survival of families, but is not generally taken into account in calculating national energy needs and expenditures. These chores often leave little time for productive employment, education, community involvement or other activities beyond what is necessary for survival. Therefore it is obvious that a lack of energy services and infrastructure (water supply systems) limits women's productive and community development activities. Only a functioning infrastructure including energy and water supply allows women to escape this vicious circle because it frees up their time; having access to lighting in the evening makes it possible for them to pursue educational and entrepreneurial opportunities.

Women in developing countries have a lot of knowledge about traditional fuels, and those who are educated about energy alternatives can also play important roles as educators and activists in energy efficiency, renewable energy sources and better uses of traditional fuels within energy projects.

Women's micro-enterprises (an important factor in household income as well as in women's welfare and empowerment) tend to be either heat-intensive (food processing) or light-intensive (labour-intensive home industries with work in the evenings). Lack of adequate energy supplies and other coordinated support for these activities affects women's ability to operate these micro-enterprises profitably and safely (Cecelski, 2000).

Health

Some 3 billion people worldwide rely on biomass fuels and coal for their cooking and heating needs. Of these, about 800 million depend on agricultural residues and animal dung as sources of fuel due to severe wood-fuel shortages. Biomass accounts for 80 percent of all household fuel consumption in developing countries, most of it for cooking, which is done primarily by women.

Traditional low-efficiency cooking stoves produce a number of pollutants associated with incomplete combustion including fine particulates, carbon monoxide and carcinogenic compounds such as polycyclic aromatic hydrocarbons. Exposure to these pollutants can lead to acute respiratory infections, chronic obstructive lung diseases, lung cancer and eye problems. Exposure to these pollutants has also been linked with pregnancy-related problems such as stillbirths and low birth weight. One study in western India found a 50 percent increase in stillbirths associated with the exposure of pregnant women to indoor air pollution, which most likely contributes to excess heart disease as well. Other than these health problems, indoor air pollution is associated with blindness and changes in the immune system. A 1995 study in eastern India found the immune system of newborns to be depressed due to the indoor air pollution.

The official number of deaths among children under five is 10.8 million for the year 2000, 19.4 percent of which are due to acute respiratory infection. Estimates suggest that as much as 60 percent of the global burden of acute respiratory disease is associated with indoor air pollution and other environmental factors.

When fuel is scarce, the health of the whole family suffers as there is less cooked food prepared and less ability to boil water. The contributions of clean energy, particularly electricity, to health can be summarized as follows.

- It helps provide nutritious cooked food, space heating and boiled water, contributing to better health.
- It enables pumped clean water and purification.
- It can be safer (less burns, accidents and house fires).
- It helps provide access to better medical facilities for maternal care including medicine refrigeration, sterilization, electronic medical equipment and access to up-to-date medical information.
- In health centres, it enables night availability, helps retain qualified staff and allows equipment use (e.g. sterilization).
- It allows vaccine and medicine storage for preventing and treating diseases and infections.
- It enables access to health education media through ICT and broadcasting.

The Challenges for Electrification

While it is clear that access to electricity is a prerequisite for economic and social development, it will be difficult to achieve on a universal scale. At present in many developing countries, rural and urban electrification – whether hydro- or fossil-fuel-based – is often hampered by the economies of reticulated grid distribution, by inadequate management and billing practices, by illegal connections and by the limited purchasing power of the poorer customers. These problems are even more intense in sparsely populated, remote and isolated areas.

To foster sustainable development, it will be necessary to identify the most suitable means of generating electricity. The required service must be available at the lowest social, environmental and economic costs to the population. Yet this must be done rapidly so as not to unduly delay development. Furthermore, different systems will be required for different areas: densely populated urban centres with manufacturing activities might be well served by a centralized, compact electric system, whereas remote areas might be better served by decentralized, off-grid projects.

Electrification has to cope with all these economic, environmental, social, institutional and technical issues. Decision-makers necessarily have to make trade-offs between costs and benefits. One key issue is how to respect human rights and aim at social justice through a transparent, participative process so that all stakeholders are actively involved in the choices made.

Rural and urban energy needs

Table 10.1 shows the access to electricity of forty-three developing countries by region and income situation. It illustrates the strong correlation between electricity supply and income.

As a symbol of modernity and a basic aspiration for the poor, electricity is attractive, but the reality is that it is not always the most appropriate form of energy for the poorest households and villages. In fact, electric lighting has no cost-benefit in a poor rural household, while an efficient stove, fed with a reasonably clean fuel, will be cheaper and more readily usable for cooking and boiling water than an electric stove. While electricity is essential for conserving vaccines and medicines in village dispensaries, in the home, pumping water by hand may be more appropriate and cheaper than using an electric pump. These concerns all have to be weighed in the balance, as electricity, while being a key factor in the fight against poverty, must not be seen as providing the answer to all rural problems.

Access to electricity in urban and peri-urban areas, on the other hand, is essential for the emergence of entrepreneurial and industrial initiatives that can lead the way out of poverty. As with rural areas, however, electrification of shantytowns will not be a panacea. Reforms to world-trade barriers giving developing countries access to world markets are also needed if their increasing industrialization is to be a real source of development opportunity. In other words, looking at problems in a one-dimensional fashion will never provide all the solutions. While access to electricity remains an essential tool to improve manufacturing capabilities in urban areas, it needs to be part and parcel of integrated policies that tackle all facets of poverty.

A key element in electrifying both rural and urban areas is that the choices made must not aggravate unequal access to such resources, thus intensifying the displacement of rural poor to the cities. In addition, innovative billing techniques and foolproof connection procedures will be required to counter the massive abuse that occurs through illegal connections and default in payments.

Use of energy in rural areas

Energy use in rural areas can be broken down into household, agricultural and small-scale rural industry subsectors and services. The amount of energy use for services (health clinics, schools, street lighting, commerce, transport) is generally quite small in rural areas, and therefore is often included in the rural industries sector. Over 85 percent of energy is consumed by households mainly in the form of traditional energy sources used for cooking and heating. On the contrary, agricultural activities, depending on levels of mechanization, consume only about 2 to 8 percent of the total energy use, mainly in the form of commercial energy used to power mechanical equipment and irrigation pump-sets. Commercial energy, often kerosene and electricity where available, is mainly used for lighting, which on average constitutes about 2 to 10 percent of total rural consumption. Small amounts of electricity are used to operate radios, television sets and small appliances in electrified villages. The energy consumption of rural industries, including both cottage industries and village level enterprises, amounts to less than 10 percent of the rural aggregates in most countries. In a few cases in Asia and Africa, the share of traditional fuels in rural household energy rises to more than 95 percent.

Household energy consumption patterns are extremely varied in the rural areas of developing countries. How much is consumed and the types of energy sources used depend on a variety of factors, which include the availability and costs of energy sources. Among the poorest families in most developing countries, cooking and space heating (depending on the climate) accounts for between 90 and 100 percent of energy consumption. The remainder of the energy consumed is for lighting, provided either by the cooking fire, kerosene lamps, candles or electric torches. At higher income levels, better lighting is one of the first energy services sought to improve living standards and, frequently, to extend the working day. At still higher incomes, water heating, refrigeration and cooling begin to

Table 10.1: Distribution of households with access to electricity in forty-three developing countries (in percentages)

Country	Year	Poorest	Second	Wealth quintiles Middle	Fourth	Richest	Total
Africa							
Benin	1996	0.0	0.3	0.3	4.9	66.0	14.3
Burkina Faso	1992/93	0.0	0.0	0.0	40.0	29.9	6.0
Cameroon	1991	0.0	0.5	8.4	50.0	98.0	31.0
Central African Republic		0.0	0.0	0.0	0.2	25.0	5.0
Chad	1996/97	0.0	0.0	0.0	0.5	13.3	2.8
Comoros	1996	0.0	1.0	12.1	49.7	89.3	30.4
Côte d'Ivoire	1994	0.2	1.2	16.1	75.3	99.2	38.5
Ghana	1993	0.0	0.0	0.6	44.6	94.1	27.9
Kenya	1998	0.0	0.2	0.1	1.8	56.7	11.7
Madagascar	1997	0.0	0.0	0.1	0.4	55.1	11.1
Malawi	1992	0.0	0.0	0.0	0.0	18.5	3.7
Mali	1995/96	0.0	0.0	0.0	0.4	37.5	7.6
Mozambique	1997	0.0	0.0	0.0	2.2	47.5	9.9
Namibia	1992	0.0	0.0	0.0	9.2	92.0	20.2
Niger	1998	0.0	0.0	0.0	0.5	38.9	7.9
Nigeria	1990	0.0	0.2	0.9	31.6	94.9	25.5
Senegal	1997	0.0	0.0	3.0	59.2	98.4	32.2
Tanzania, United Rep. of	1996	0.0	0.0	0.0	0.8	44.0	8.9
Togo	1998	0.0	0.2	0.4	6.2	67.7	14.9
Uganda	1995	0.0	0.0	0.0	0.0	34.7	6.9
Zambia	1996	0.0	0.0	0.4	10.7	91.0	20.2
Zimbabwe	1994	0.0	0.2	0.4	19.3	96.6	23.2
Asia/Near East/North Africa							
Bangladesh	1996/97	0.0	0.0	6.8	37.8	79.8	24.9
Egypt	1995/96	80.7	99.0	99.6	99.9	100.0	95.8
India	1992/93	0.3	9.4	55.9	92.4	99.7	51.5
Indonesia	1997	35.0	76.3	91.9	98.9	99.9	80.4
Kazakhstan	1995	99.7	100.0	99.6	100.0	100.0	99.9
Kyrgyzstan	1997	98.7	100.0	100.0	100.0	100.0	99.7
Morocco	1993	0.3	10.2	33.2	91.2	99.9	46.9
Nepal	1996	0.0	0.0	0.6	19.1	70.4	18.0
Pakistan	1990/91	1.2	48.9	71.9	92.5	99.8	62.9
Philippines	1998	6.9	58.4	94.1	99.8	100.0	71.8
Uzbekistan	1996	97.6	100.0	100.0	100.0	100.0	99.5
Viet Nam	1997	20.5	78.9	87.9	96.6	99.9	76.7
Latin America and the Caribbean							
Bolivia	1998	3.6	59.5	96.2	99.8	100.0	71.8
Brazil	1996	65.3	99.6	99.7	99.8	100.0	92.9
Columbia	1995	54.6	99.2	99.8	100.0	100.0	90.7
Dominican Republic	1996	40.8	64.6	69.4	77.3	81.8	66.8
Guatemala	1995	1.6	22.6	71.8	97.5	99.8	58.7
Haiti	1994/95	0.0	0.1	5.3	51.7	98.1	31.0
Nicaragua	1997/98	2.4	48.7	92.4	98.5	98.6	68.1
Paraguay	1990	0.6	10.3	36.8	95.9	99.9	49.2
Peru	1996	2.6	43.7	90.3	99.1	99.9	67.1

This table illustrates the strong link between electricity supply and income: in general, higher electricity supply is supported by higher average incomes.

Source: Davis, 1995.

play an important role. In addition, at these higher income levels, the need for space heating may decline because houses may be better constructed.

Table 10.2 shows the use of energy for cooking and heating in rural Mexico. In each of the three regions, as incomes rise the percentage shares for cooking decline, while those for water heating increase sharply, and those for space heating first increase and then decline.

End-use of energy varies between regions. For eastern Africa, it has been estimated that 55 percent of the biomass fuel is used for cooking, 20 percent for water heating, 15 percent for space heating and 10 percent for ironing and other minor uses. In a survey of six low-income villages of southern India, where space heating needs are negligible, little variation in end-use shares was found, with cooking between 76 percent and 81 percent, water heating between 14 and 19 percent, and lighting by kerosene and some electricity between 2 and 3 percent. In contrast, in Chile's much cooler climate, a survey of eight rural villages found that cooking accounted for 42 to 55 percent and space heating for 23 to 52 percent, while water heating absorbed 14 to 22 percent, except one village with 6 percent. Clearly, understanding rural energy demands requires closer examination of the major household end-uses, cooking, lighting and space heating.

The provision of an adequate modern energy supply for water-related activities in rural areas of developing countries offers many advantages, including time saved not having to travel to collect water, thus increasing productivity; easier access to water through pumping of drinking water, irrigation water and water for animal husbandry; health benefits (ranging from water purification through filtration to reduced medical costs when boiling water for sterilization is unnecessary); and health and environmental benefits through the discharge of wastewater from canals, septic tanks and latrines. Energy also allows wastewater to be treated through aeration.

Hydropower Water and Energy Generation

Hydropower is available in a range of sizes from a few hundred watts to over 10 gigawatts (GW). At the low end of the spectrum, small hydropower can be divided into three categories. The definitions of the categories vary, but are broadly: micro (less than 100 kilowatts [kW]), mini (100 kW to 1 megawatt [MW]) and small (1 MW to 10 MW) hydro. Micro- and mini-hydro systems are generally stand-alone systems, that is, they are not connected to the electricity grid.

Worldwide electricity production by hydropower

Hydropower is already a major contributor to the world's energy balance, providing, as shown in figure 10.1, 19 percent of total electricity production (2,740 terawatts per hour [TWh] in 2001), (IHA, 2002). It is by far the most important and widely used renewable source of electricity. According to the IHA, there are 377 TWh either under construction or in the planning stage, and the equivalent of some 4,000 to 7,500 TWh of remaining potential.[1] Over the last ten years, the development of new hydropower capacity has kept pace with the overall generation increase in the electricity sector (in 1992 the total was 2,105 TWh). In fact, hydropower's relative contribution of about one fifth has remained constant.

A major benefit is that the continuing development of hydroelectric potential will reduce emissions of greenhouse gases and other air contaminants from thermal power plants. Each additional terawatt of hydropower per hour that replaces coal-generated electricity offsets 1 million tons a year of carbon dioxide equivalent. One third of the total carbon dioxide global emissions of 22,7 billion in 1995 (WRI et al., 1998) was produced by the energy

Table 10.2: End-use of energy for cooking and heating in rural Mexico (in percentages)

	Region 1 incomes			Region 2 incomes			Region 3 incomes		
	Low	Medium	High	Low	Medium	High	Low	Medium	High
Cooking	82.6	58.5	50.3	85.4	79.7	57.6	83.3	82.6	48.9
Water heating	2.0	9.1	34.0	10.5	36.7	–	4.3	4.3	–
Space heating	6.5	32.4	15.7	9.1	9.8	5.7	7.0	13.1	–
Total energy	11.5	10.2	8.3	9.1	7.9	5.9	9.1	7.9	5.9

In each of the three regions, there is a direct correlation between income and the end-uses: the higher the income, the lower the share of energy used for cooking, but the higher the share for water heating. Space heating energy use shows a tendency to rise with incomes, then to decrease.

Source: Guzman, 1982.

1. The World Energy Council estimates the remaining economically exploitable capability at 7,500 TWh/year: http://www.worldenergy.org/wec-geis/publications/reports/ser/hydro/hydro.asp/

Figure 10.1: World's electricity production

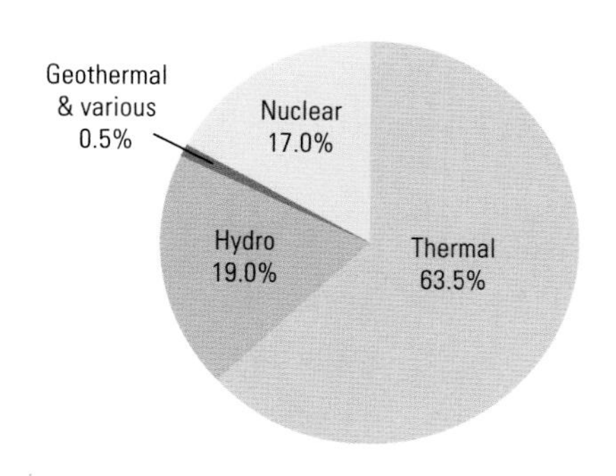

Thermal electricity generation accounts for two thirds of electricity production worldwide. The rest is almost equally split between nuclear and hydropower, the latter of which provides a full 19 percent of the world's electricity.

Source: IHA (International Hydropower Association) web site. Small-scale hydro section: http://europa.eu.int/comm/ energy_transport/atlas/htmlu/hydint.html. Data extracted in May 2002.

sector through the combustion of coal, oil and gas. Many of the countries producing a large percentage of their electricity with coal have underdeveloped hydropower potential. Hydropower in a mixed system enables the more efficient use of less flexible technologies, and reduces not only greenhouse gases but also particulate pollution, which causes respiratory disease, and compounds causing acid rain, thus neutralizing agricultural lands and forests.

Against this background of current and future demands for energy and water and the growing pressures of increased population and continuing urbanization, hydropower, dams and reservoirs, whether large- or small-scale, can make a significant and expanding contribution to satisfying these needs while meeting the criteria for sustainable development. Hydropower is an environmentally and economically efficient tool to permanently satisfy rural and urban needs, emitting virtually no greenhouse gases, air contaminants or wastes. With proper maintenance and renovations, the longevity and resilience of dams and reservoirs make them permanent infrastructures of development. Moreover, hydropower, by making use of local resources, can replace imported fossil fuels thus reducing foreign dependence and national trade deficits. Table 10.3 provides a view of the current and projected presence of hydropower around the world by region.

Hydroelectric power generation continues to be the most important and economic renewable source of commercial energy worldwide. In many developing countries, such as Afghanistan, Angola, Bhutan, Burundi, Cameroon, Congo, Ethiopia, Georgia, Lao People's Democratic Republic, Mozambique, Namibia, Nepal, Rwanda, Sri Lanka, Uganda and Zaire and in all of Central and South America, hydropower is already dominant in electricity production; even so, the potential that still remains economically exploitable is huge (see map 10.1).

Figure 10.2 shows the regional distribution of installed hydropower capacity at the end of 1996. Hydropower supplies at least 50 percent of electricity production in sixty-six countries, and at least 90 percent in twenty-four countries. About half of this capacity and generation is in Europe and North America. In Europe in particular, during the twentieth century, hydropower made a dramatic

Table 10.3: Deployment of hydropower

Location	Market area	Current deployment in 1995 (TWh/year)	Estimated deployment in 2010 (TWh/year)
World	Large hydro	2,265	3,990
	Small hydro	115	220
	Total hydro	2,380	4,210
EU + EFTA	Large hydro	401.5	443
	Small hydro	40	50
	Total hydro	441.5	493
CEE	Large hydro	57.5	83
	Small hydro	4.5	16
	Total hydro	62	99
CIS	Large hydro	160	388
	Small hydro	4	12
	Total hydro	164	400
NAFTA	Large hydro	635	685
	Small hydro	18	25
	Total hydro	653	710
OECD Pacific	Large hydro	131	138
	Small hydro	0.7	3
	Total hydro	131.7	141
Mediterranean	Large hydro	35.5	72
	Small hydro	0.5	0.7
	Total hydro	36	72.7
Africa	Large hydro	65.4	147
	Small hydro	1,6	3
	Total hydro	67	150
Middle East	Large hydro	24.8	49
	Small hydro	0.2	1
	Total hydro	25	50
Asia	Large hydro	291	1,000
	Small hydro	42	100
	Total hydro	333	1,100
Latin America	Large hydro	461.5	990
	Small hydro	3.5	10
	Total hydro	465	1,000

Key: EU + EFTA = European Union and European Free Trade Association; CEE = Central and Eastern Europe; Mediterranean = Turkey, Cyprus, Gibraltar, Malta; CIS = ex-USSR countries; OECD Pacific = Australia, Japan, New Zealand; NAFTA = United States, Canada, Mexico. Asia = all Asia excluding former USSR.

This table shows the current and projected deployment of hydropower throughout the world. It is set to expand in all regions, in particular in Africa, Asia and Latin America, where the potential for development is greatest.

Sources: Water Power and Dam Construction, 1995; International Journal on Hydropower and Dams, 1997.

Map 10.1: Proportion of hydropower electricity generation per country

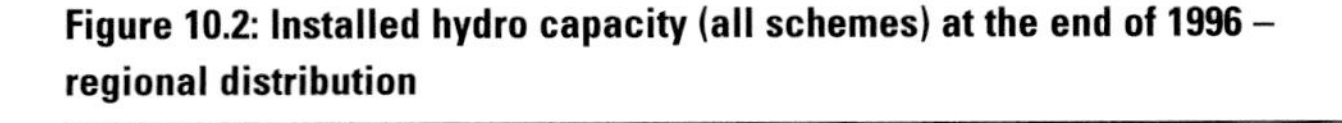

The map shows the percentage of hydropower in countries' electricity generation. Many areas, such as Canada, sub-Saharan Africa and Latin America, appear to depend mainly on hydropower for their electricity, to a much greater extent than Europe, North America and Asia. There remains, however, much hydro potential as yet untapped.

Source: Map produced for the World Water Assessment Programme (WWAP) by the Centre for Environmental Research, University of Kassel, based on data from the International Hydropower Association (IHA) and *International Journal on Hydropower and Dams*, 2002.

Figure 10.2: Installed hydro capacity (all schemes) at the end of 1996 – regional distribution

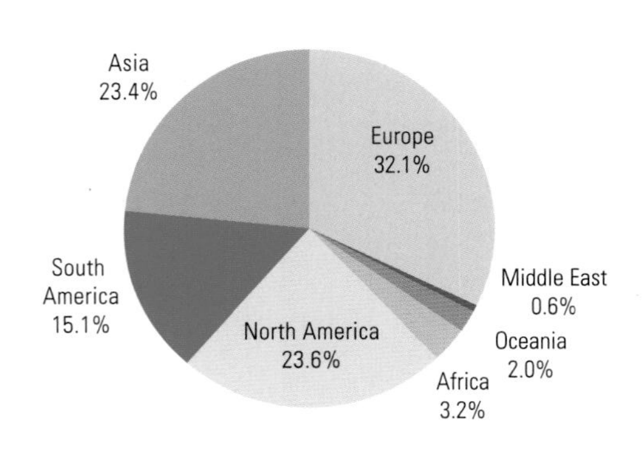

Hydropower schemes are a worldwide phenomenon, but their installation is not evenly spread. There is a noticeable disparity between areas with few hydropower schemes, such as the Middle East, Oceania and Africa, and the rest of the world, which has developed its hydropower potential to a much greater extent.

Source: WEC, 2001.

contribution to the development of the electricity sector and most of the best sites have been exploited for big plants. Not surprisingly, given the country's large size and population, the United States is the second largest producer of hydroelectricity in the world, with between 10 and 12 percent of the country's electricity supplied by hydropower. However, this proportion is changing as Asia and Latin America are building large amounts of new hydropower capacity. For poor, mountainous countries, such as Lao PDR or Nepal, hydropower for export offers one of the few new avenues for economic growth.

In rich and energy-intensive countries such as Canada, the United States or Norway, the opponents of new dam projects sometimes argue that what is needed is a reduction in demand rather than increases in supply. However, although more energy conservation is required, this does not negate the environmental benefits of building new hydropower schemes to replace the use of polluting fossil fuels.

In poor, developing countries, where average electricity use per household is less than a few hundred kilowatts per hour (kWh) a year, the energy conservation potential is much smaller in absolute terms than in the developed world. It is not an argument for depriving developing countries of additional generation capacity.

Moreover, in developing countries, the need is not only for electricity but also water supply to cities and for irrigation.

Shortcomings in large-scale and centrally controlled development efforts in poor and former socialist countries, have been equated with an alleged failure of the large-scale-dam option and the centralized distribution grid (Bello, 2001). However, problems with past large-scale projects should not be allowed to engender disillusionment about future developments. Given that it has proved difficult to remedy the management problems of large-scale projects, an alternative development path has emerged. Recently, there has been considerable progress on two fronts:

- the suggestions by the World Commission on Dams (WCD) for improved practice in the planning, construction, operation and de-commissioning of large dams provide useful lessons in evaluating the available options (see boxes 10.1 and 10.2); and
- in many circles, small-scale hydropower and off-grid autonomy are now considered capable of delivering 'green' or 'low-impact' hydro facilities and electrification.

In reality each scale of operations has its particular advantages and disadvantages. The type of project should be chosen according to the quantity of energy required and with consideration given to the appropriate context.

Principal physical requirements

Hydropower plants generate electricity or mechanical power by converting the power available in the flowing water of rivers, canals or streams. This requires a suitable rainfall catchment area, a hydraulic head, a means of transporting water from the intake to the turbine, such as a pipe or millrace, a power house containing the power generation equipment and valve gear needed to regulate the water supply as well as all primary mechanical and electrical equipment components, and, finally, a tailrace to return the water to its natural course.

Characteristic elements of hydropower

There is a general view that hydroelectricity is the renewable energy source par excellence, non-exhaustible, non-polluting and more economically attractive than other options. And although the number of hydropower plants that can be built is finite, only a third of the

Box 10.1: World Commission on Dams (WCD) – of risks, rights and negotiated agreements

The WCD is an independent, international, multistakeholder process which addresses the controversial issues associated with large dams. It provides a unique opportunity to bring into focus the many assumptions and paradigms that are at the centre of the search to reconcile economic growth, social equity, environmental conservation and political participation in the changing global context. The basic conclusions of the Commission are the following.

- Development choices made on the basis of such tangible trade-offs as cost and benefit neither capture the complexity of considerations involved, nor can they adequately reflect the values societies attach to different options in the broader context of sustainable development. To rectify this, the Commission proposed that an approach based on recognition of rights and assessment of risks be developed as a tool for future planning and decision-making.
- Considerable experience and good practice are encountered in implementing a rights-based approach. It requires a legal and procedural framework that provides for a free and informed negotiating process; the framework must provide for arbitration, recourse and appeal mechanisms ensuring equitable adjudication in cases where negotiated settlements are not achievable. Negotiated agreements become part of the project compliance framework.
- Seven strategic priorities are identified for an equitable and sustainable development of water and energy resources. All seven strategic priorities are supported by a key message and policy principles. They form the basis of the WCD Criteria and Guidelines. The criteria refer to critical decision-making points: for each point a detailed description of the process and its indicators is presented. This results in a twenty-six-point set of guidelines for good practice under the seven strategic priorities. Each of the twenty-six points is then elaborated in detail to serve as guidance to project proponents, planners and decision-makers.

Source: Based on WCD, 2000.

Box 10.2: Large-scale hydropower in Asia

In 1997, Asia had an installed hydroelectric capacity of about 100 GW. Asia is the continent with the fastest-growing hydroelectric industry, with many Asian countries stating that hydropower is the main focus for the development of their power sectors. China currently has the highest level of hydropower development activity in the world. The 18.2 GW Three Gorges dam, the 3.3 GW Ertan and the 1.8 GW Xiaolangdi hydroelectric projects are all under construction. Hydroelectric schemes with a total capacity of 50 GW are currently being built, and this will double the existing capacity in the country. The construction of an additional four large-scale projects will commence shortly: Xiluodo (14.4 GW), Xiangjiaba (6 GW), Longtan (4.2 GW) and Xiaowan (4.2 GW). Implementation of a further 80 GW of hydropower is planned, including thirteen stations along the upper reaches of the Yellow River, and ten stations along the Hongshui River.

A 280 MW hydropower station is being constructed in Paunglaung, Myanmar. In the Philippines, construction has started on the 70 MW Bakun AC scheme, which will be one of the first private hydropower projects in the country. Viet Nam has a large number of medium- to large-scale hydropower schemes planned for completion by the year 2010, including the 3.6 GW Son La scheme. India has 10 GW of hydropower under construction, with a further 28 GW planned. Indonesia has six large-scale hydro schemes planned, with a total capacity of 2 GW. In Malaysia however, due to environmental and economic pressures, the 2.4 GW Bakun hydroelectric project has been indefinitely delayed.

A number of Asian countries have major pumped storage development programmes, including Korea (2.3 GW under construction and 800 MW planned), Thailand (1 GW under construction and 1.46 GW planned) and Indonesia (1 GW planned).

Source: Based on The Australian Greenhouse Office web site, renewable energy section. http://www.greenhouse.gov.au/renewable/. Data extracted in May 2002.

sites quantified as economically feasible are tapped.

Hydropower plants emit much less greenhouse gas than thermal plants. Greenhouse gas emissions of hydropower are caused by the decay of vegetation in flooded areas and by the extensive use of cement in dam construction (not applicable for run-of-the-river type, mini-, micro- and small hydropower plants). Unfortunately, there are local impacts of the use of rivers, social as well as ecological, and they are gaining importance as people become aware of how those affect living standards. Most renewable sources of hydroelectricity generation are capital-intensive but have lower operational costs than thermal and nuclear options. The high initial cost is a serious barrier for its growth in developing countries where most of the untapped economic potential is located. It is a point of controversy whether hydropower is a clean, ecologically beneficial and sustainable form of energy generation, or whether the disadvantages associated with hydropower, such as the inundation of land, the associated disruption and displacement of communities, the adverse health effects for vulnerable groups as well as wildlife and fish habitats, are predominant. A rigorous environmental and health impact assessment is generally necessary for any project involving water storage.

Table 10.4 summarizes characteristic elements of hydropower in the form of its technical and environmental advantages and disadvantages.

Table 10.4: Advantages and disadvantages of hydropower

Advantages of hydropower	Disadvantages of hydropower
Renewable energy	High investment costs
Free fuel resource	Hydrology dependent (precipitation)
Fuel saver	In some cases, inundation of land and wildlife habitat
Flexible to meet load	In some cases, loss or modification of fish habitat
Efficient technology	Fish entrainment or passage restriction
Reliable, durable and proven technology	In some cases, changes in reservoir and stream water quality
Low operation and maintenance costs	In some cases, displacement of local populations
Few atmospheric pollutants	

Hydropower schemes present various costs and environmental and social advantages and disadvantages.

Proposed large-scale hydropower projects often face pressure from environment and human-rights groups as the resulting flooding of large areas of land often leads to the displacement of people living in this area, e.g. in the 18.2 GW Three Gorges dam project in China.

The south-eastern Anatolia project (see box 10.3) shows a new and integrated approach for large-scale hydropower projects.

Small-scale hydropower

Grid system/local supply

It can often be shown that decentralized systems such as hydropower are competitive with grid power at low consumption levels, but are not competitive at higher levels. This is because once the capital costs of grid extension have been met, any increases in consumption affect only the generation costs, while the costs of distribution per unit of consumption actually fall. In contrast, with a decentralized system each increment of energy use, or power, requires additional supply capacity.

Advantages/disadvantages of small-scale hydropower

Small hydropower plants and especially mini- and micro-hydropower plants normally do not imply huge dam constructions. Therefore the adverse health effects and the impact on the environment can be described as low (see box 10.4).

Environment

It has been estimated that a 1 MW small-scale hydro plant, producing 6,000 MWh in a typical year, avoids the emission of 4,000 tons of carbon dioxide and 275 tons of sulphur dioxide, compared with a coal-fired plant of equivalent output, while supplying the needs of 1,500 families. The present installed capacity of small hydro in the European Union (EU) avoids annual emissions of 36 million tons of carbon dioxide. The Italian Small Hydro Association, APEI, uses similar figures as a promotional slogan, which also states that, based on carbon dioxide absorption figures for woodlands, the non-construction of a 1 MW hydro plant is equivalent in terms of atmospheric carbon dioxide increase to the destruction of a 500-hectare wood. It can therefore be said that small hydropower technology involves low environmental impacts and is a key technology for sustainable development. This is demonstrated by the successful deployment and manufacture of small hydro equipment in such underdeveloped countries as Nepal and Peru.

There are several aspects of water flow that have to be taken into consideration.

- A minimum flow has to be maintained in the river to ensure the life and reproduction of fish, and the free passage of migratory fish.
- The determination of an acceptable minimum flow is a key issue in determining the economic viability of a scheme.
- Run-of-river hydropower schemes have minimal visual impact.
- The principal, permanent impacts are on the depleted stretch of the watercourse, where mitigating measures need to be taken to sustain the river ecology and fisheries.
- Some types of turbines also provide increased oxygenation of the tailrace water, leading to improved water quality.

Box 10.3: GAP – a paradigm shift in water resources development

The south-eastern Anatolia Project (GAP) has been conceived and implemented as a means of integrating water resources development (twenty-two storage dams and nineteen hydropower plants on the Euphrates and Tigris river systems) with overall human development in the poorest and most underdeveloped region of Turkey. Sustainability (social, spatial, environmental and economic), participation (of all stakeholders) and intersectoral coordination and integration are the hallmarks of this new approach. The strategy implements special programmes and projects concerned with basic social services (education, health, housing), gender equity, urban management, participatory irrigation management, agricultural extension, environmental protection, institutional and community capacity-building, and public empowerment, under the overall coordination umbrella of a special purpose vehicle known as the GAP Regional Development Agency. Initial results of instituting irrigation in the Harran plains are positive. The ratio of landless farmers has dropped from 40 percent to around 23 percent and the ratio of share croppers increased from 27 percent to 57 percent; migration from these areas dropped from 70 percent to 12 percent and in some areas in-migration started. Despite the existence of certain problems that need to be tackled, outcomes indicate that GAP shows promise, not only as a tool for achieving progress in hydraulic engineering, but also as an instrument for social change.

Source: Prepared for the World Water Assessment Programme (WWAP) by the south-eastern Anatolia Project (GAP), 2002.

Box 10.4: Advantages and disadvantages of small-scale hydropower projects

Micro-hydro systems are mostly 'run-of-the-river' systems, which allow the river flow to continue. This is preferable from an environmental point of view as seasonal river flow patterns downstream are not affected and there is no flooding of valleys upstream of the system. A further implication is that the power output of the system is not determined by controlling the flow of the river, but instead the turbine operates when there is water flow and the energy output is governed by the actual flow. Therefore, a complex mechanical governor system is not needed, which reduces costs and maintenance requirements.

Operational, design and construction experience is often available within the target countries. The systems can be built locally at low cost, and the simplicity gives rise to better long-term reliability. The technology is technically and commercially mature and small-scale hydro schemes can make a useful contribution to rural electrification strategies, presenting a suitable alternative to decentralized diesel generation, particularly where fuel supply is a problem.

There are, however, certain objections to or disadvantages of small-scale hydropower schemes. For example, the cumulative direct costs (purchase of land, development costs) and indirect costs (environmental pollution, population displacement) of adopting technology may sometimes overwhelm the reasons for implementing a scheme. The scheme may also fail to satisfy the merging energy needs of the area and constrain industrial development and/or expansion. Certainly, micro-hydropower often faces unfair competition from highly subsidized grid electricity, and from subsidized fossil fuels.

Hydropower implementation

Successful implementation of hydropower in developing countries does not finish with construction works: sustainable operation and maintenance of small-scale hydropower must also be taken into consideration. A small-scale hydropower plant works in a sustainable way when it is functioning and being used, when it is able to deliver an appropriate level of benefits (such as quality, quantity, continuity, affordability, efficiency), when it continues over a prolonged period of time and when its management is institutionalized. Its administrative and replacement costs, as well as its operation and maintenance, must be covered at a local level with limited but feasible external support. It also must not negatively impact the environment.

Therefore, before initiating a small-scale hydropower project the objectives have to be clearly defined. Such objectives could include social infrastructure (provision of energy for health services, schools, drinking water), physical infrastructure (provision of energy for irrigation systems, roads) and energy for the development of small profit-making businesses. Also, capacity-building processes such as teaching, training and awareness-raising are essential for a successful project although they take a lot of time. Local capacities to build, manage, operate and maintain micro-hydro plants reduce costs and are critical for sustainable success (see box 10.5 for an example involving social problems in Sri Lanka).

Regardless of ownership structure, the successful management of micro-hydro plants requires a corporate structure that minimizes political interference (e.g. from municipal authorities or powerful community members). The management is responsible for achieving clearly stated objectives related to financial situation, for coverage of the local grid system and quality (reliability) of the provided services.

There are various reasons for technical and economic problems of small-scale hydropower plants. A study on the functional status of existing micro-hydro plants in Nepal shows the following problems:

- poor site selection, inadequate/inaccurate surveys, wrong dimensioning, poor installations, faulty equipment;
- plants affected by floods and landslides;
- poor estimation of hydrological conditions, partly due to surveys conducted during rainy seasons;
- uneconomic canal length, bad canal design;
- neglect of civil works;
- owners' inability to replace generators after breakdown; and
- inaccurate estimation of raw materials, demand and end-use possibilities; oversized plants; overestimation of tariff collection; inappropriate rates; ignorance of competition with diesel.

Small hydropower development in industrialized countries: the EU case

An important role in achieving European renewable energy goals can still be played by small hydropower resources that are distributed on the continent and can offer all the benefits of dispersed renewable generation. Since about 1950, small hydropower has had a negative development in some EU member countries. Many small hydropower plants have been shut down because of age and competition from new, larger plants. The potential from reinstalling these plants and upgrading existing, underdeveloped small hydropower plants is estimated at an annual electricity production of approximately 4,500 GWh. Based on an analysis/study carried out in the years 1999–2000, the potential of new plants, reduced when economic and environmental constraints have been taken into account, is calculated to be about 19,600 GWh per year. According to the White Paper issued by the EU Commission in 1997 (EEC, 1997), the remaining estimated potential from small hydropower will be some 18 TWh in the year 2010. Based on the present annual production of 40 TWh, the EU White Paper foresees 55 TWh from 14,000 MW in 2010. If the economic situation for producers improves, and the environmental constraints decrease, the total contribution from small hydropower in the EU's fifteen member countries could probably reach 60 TWh in the decade 2020–30, and perhaps much more with the expected entry of ten or more countries in the EU in 2004.

Selected cases for small hydropower development in developing countries

Micro-hydro installations are widespread in Asia, where there is a significant resource potential for further development. China has a well-developed hydropower industry, with an estimated 60,000 small hydropower installations (less than 1 MW), and a combined capacity of about 17 GW.

The installed small hydropower (less than 10 MW) capacity in Viet Nam is 61.4 MW, with an estimated potential of about 1.8 GW. Some 3,000 sites have been identified for micro-hydro installations in the range of 1 to 10 kW. These sites will serve irrigation and drainage needs, in addition to generating electricity for 2 million households. Many areas in Viet Nam do not have access to the electricity grid, due to the high extension costs. In these areas, micro-hydro units are used by individual families for lighting and battery charging (for television and lighting use). It is estimated that over 3,000 family units of 1 kW or less are installed in Viet Nam.

Other countries in Asia and Oceania with micro-hydro resources include Bangladesh, Indonesia, Lao PDR, Papua New Guinea and the Philippines. In 1998, the Indonesian government announced its intention to electrify 18,600 villages using small and micro-hydro schemes.

Box 10.5: Hydropower and social problems in the Ruhuna basins (Sri Lanka)

Hydropower generation does not consume water. Energy generation can, however, result in location-specific water problems for other users. In the Ruhuna basins, water diverted for generations now returns the river downstream of an established irrigation system. Transfer of water from agriculture has socio-political repercussions as well as economic impacts. In this case, although the power generation authority is willing to compensate farmers for loss of water supplies, the agricultural community has so far preferred to continue irrigated cultivation. This may be due to the limited alternative to farming in the relatively remote area of the affected villages. In addition, payment of compensation in place of agricultural activities led to an increase in social and cultural problems. One farmer said: 'But sir, we do not want to live in a dead village. When we cultivate we work together and have ceremonies and parties. Vehicles come to bring inputs and take away our crops. Now no one comes, we just sit on the guard stones and wait.'

Intensive consultation amongst the concerned parties is transforming a developing conflict into a more harmonious agreement, resulting in better productivity in agriculture and regulation of river flow, while maximizing power generation. However, the importance and value of water to the rural community, beyond those derived from productive uses, has been recognized in arriving at these agreements.

Source: Prepared for the World Water Assessment Programme (WWAP) by the Ministry of Irrigation and Water Management of Sri Lanka, 2002.

Guinea has identified 150 mini- and micro-hydro sites; Nigeria plans to develop 700 MW of capacity in 236 different projects. In the sub-Saharan region of Africa, such as in Ethiopia, Malawi, Uganda or the United Republic of Tanzania, topographical and hydrological conditions would also allow the implementation of hydropower plants. Nepal is another example of a country with hydropower resources to be developed (see box 10.6).

Other Forms of Water/Energy Generation

Ocean energy

There are three known types of ocean energy generation. Tidal energy is transferred to oceans from the Earth's rotation through gravity of the sun and the moon and is the most advanced in terms of use, with a number of commercial plants. Wave energy is mechanical energy from wind retained by waves, and ocean thermal energy is stored in warm surface waters that can be made available using the temperature difference with water in ocean depths. However, this technology is still far from being mastered, and as such is an untapped source of energy.

Geothermal energy

Geothermal energy is generally defined as heat stored within the earth. The earth's temperature increases by about 3°C over 100 metres in depth. Two types of geothermal energy generation can be taken into consideration for practical application in developing countries: hydrothermal, which is hot water or steam at moderate depth (100 to 4,500 metres), and geothermal heat pumps. Table 10.5 presents the advantages and disadvantages of various forms of energy.

Box 10.6: Hydropower in Nepal

In Nepal, 85 percent of the population lives in rural areas. The national electrification rate is 13 percent, but above 80 percent in the cities. The electrification rate provides a large market for micro- and mini-hydro schemes. The power forecasts prepared by the Nepal Electricity Authority (NEA) foresee that 30 percent of the national households will be connected to the grid by the year 2020. Rural electrification develops mainly by the extension of the national grid in the southern lowlands, the Terai, where 40 percent of the national population will be living around 2005. In the inaccessible hill and mountain regions, electrification is achieved primarily via isolated grids.

Nepal has a feasible hydropower potential of around 80 GW. The political and expert opinion concerning the best strategy for the use of this potential and the promotion of national electrification is divided into two camps. One side argues for development of large-scale hydropower destined for exports to India with attached national grid electrification. Others argue that the primary focus of hydropower policy should be on the development of micro-hydropower. The latter technology is said to be relatively low-cost, to reply to a nationally manufactured technology, and would be well-suited to provide energy for the population in the isolated hills and mountain regions.

Sixty-three of Nepal's seventy-five districts have potential for hydropower application. With the support of international NGOs since the 1970s, Nepal has succeeded in building up an industrial sector capable of manufacturing or assembling all micro-hydropower components except the generators for micro-turbines up to 300 kW. The government supported the development of micro-hydro through various technical and financial support programmes. It is estimated that the economically viable micro-hydro potential in Nepal is about 42 MW. The cost of stand-alone plants is in the range of US$1,200 to 1,600 per kilo watt (in 1993 dollars).

The installed capacity of micro- and mini-hydropower is presented below, showing the actual use of micro-hydro generation.

Installed micro-and mini-hydropower in Nepal

Technology	Total number	Mechanical purposes	Add-on electricity	Electricity only	Installed capacity (MW)
Ghatta[1]	25,000	25,000	1	0	12
Peltric sets[2]	250			250	0.25
Micro-turbine[3]	950–1,000	800	150	30	9–10
Mini-turbine[4]	36			36	8.5

[1]Traditional water mill [2]0.5–3 kW integrated turbine and generator sets [3] >100 kW [4]100–1,000 kW.

Source: Based on Mostest, 1998

Thermal power generation requires substantial amounts of water at every stage of the energy cycle, from mineral extraction to delivery of power to the transmission/distribution system. But by far the greatest need for water comes from the cooling of turbines in the power plant. Power plant condenser coolant makes significant demands on water resources and water abstraction for this purpose is highest in regions with high installed capacities of thermal power generation: withdrawals in the eastern states of the United States and central Europe are in the range 10 to 1,500 mm/year, and 1 to 10 mm/year for much of eastern China, the Nile basin and northern India, compared with less than 1 mm/year for drainage basins in the rest of the world. The amount of water required depends on the type and size of power plant, and especially on the kind of cooling.

'Once-through cooling', in which condensed water is directly discharged, is responsible for thermal pollution of natural watercourses with attendant risk to downstream aquatic ecosystems and water users. 'Tower cooling' involves the turbines being cooled and the hot water being sent to a cooling tower, reused several times and eventually discharged from the plant.

Cooling towers enable the water to be recycled to the condensers, requiring less than 3 percent (per unit of energy generated) of the water withdrawals of once-through cooling. However, although tower cooling requires much lower withdrawals, it consumes twice as much water per unit of energy as once-through cooling since water is evaporated in the cooling process, thus requiring the addition of 'make-up water'. One of the advantages of tower cooling is that water is discharged back to waterways at much cooler temperatures, thereby protecting aquatic ecosystems and downstream water uses. Further, the heat stored in the cooling water can be extracted for utilization in district heating systems or for industrial heating purposes. Most water withdrawn by thermal power plants is returned to waterways without major degradation (except for thermal pollution caused by once-through-cooling).

Table 10.5: Advantages and disadvantages of various forms of energy

System	Technical advantages/disadvantages	Ecological advantages/disadvantages
Geothermal energy	**Advantages**	**Advantages**
	• Geothermal power stations are very reliable compared to conventional power plants. They have a high availability and capacity factor. Geothermal power plants are designed to run 24 hours a day, and operation is independent on the weather or fuel delivery. • Geothermal resources represent an indigenous supply of energy, providing energy supply security, reducing the need for fuel imports and improving the balance of payments. These issues are particularly important in developing countries, where geothermal resources can reduce the economic pressures of importing fuels, and can provide local technical infrastructure and employment.	• Geothermal energy is an abundant, secure and, if properly utilized, renewable source of energy. • Geothermal technologies, using modern emission controls, have minimal environmental impact. Modern geothermal plants emit less than 0.2% of the carbon dioxide of the cleanest fossil fuel plant, less than 1% of the sulphur dioxide and less than 0.1% of the particulates. Geothermal plants are therefore a viable alternative to conventional fossil fuel plants, particularly with respect to greenhouse gas emissions. • Geothermal power stations have a very low land area requirement.
	Disadvantages	**Disadvantages**
	• Geological uncertainties. • High initial investment. • High-tech technology.	• Geothermal energy produces non-condensable gaseous pollutants, mainly carbon dioxide, hydrogen sulphide, sulphur dioxide and methane. The condensed geothermal fluid also contains dissolved silica, heavy metals, sodium and potassium chlorides and sometimes carbonates. However, modern emission controls and reinjection techniques have reduced these impacts to a minimum. • Geothermal energy production has been associated with induced seismic activity. However, this is a debatable issue as most geothermal fields are located in regions that are already prone to earthquakes. In production plants where reinjection maintains reservoir pressures, seismic activity is not found to be greatly increased.

Table 10.5: *continued*

System	Technical advantages/disadvantages	Ecological advantages/disadvantages
Geothermal heat pumps	**Advantages**	**Advantages**
	• A slim hole of only 100–150 metres is necessary, with the heat being extracted from the borehole by means of a closed loop heat exchanger. • This is an electrically-based technology that allows high efficiency, reversible, water source heat pumps to be installed in buildings in most geographical and geological locations (worldwide). • Reduced need of fuel import. • Simple reliable technology.	• This technology can offer up to 40% reductions in carbon dioxide emissions against competing technologies. • If all of the electricity is supplied from non-fossil sources, there are no carbon dioxide emissions associated with heating and cooling a building.
	Disadvantages	**Disadvantages**
	• Technology needs electricity. • Geothermal heat pumps generate thermal energy and not electricity.	• Negative effects on groundwater (in case of incorrect installation).
Tidal energy	**Advantages**	**Advantages**
	• Improved transportation due to the development of traffic or rail bridges across estuaries. • Reduction of fuel imports.	• Reduced greenhouse gas emissions by utilizing non-polluting tidal power in place of fossil fuels.
	Disadvantages	**Disadvantages**
	• High construction costs.	• The construction of a tidal barrage in an estuary will change the tidal level in the basin. This change is difficult to predict, and can result in a lowering or raising of the tidal level. • This change will also have an effect on the sedimentation and turbidity of the water within the basin. • Navigation and recreation can be affected. • A raising of the tidal level could result in the flooding of the shoreline. • Effect of a tidal station on the plants and animals that live within the estuary.
Wave energy	**Advantages**	**Advantages**
	• Large theoretical potential.	• reduced greenhouse gas emissions by utilizing non-polluting tidal power in place of fossil fuels.
	Disadvantages	
	• High construction costs in relation to energy generation.	

Conclusions

The world is facing a situation in which 2 billion people have no electricity at all, and 2.5 billion people in developing countries, mainly in rural areas, have little access to commercial energy services. For many inhabitants of cities, losses of power supply during the day are a regular feature of life.

National energy plans need to focus more on rural and peri-urban electrification and to be better coordinated with other policies and women's development needs. From the perspective of technical management, river flows need to be managed according to needs of both upstream and downstream populations, to maintain and upgrade existing facilities, and to integrate ecological and technical knowledge into the project management. All of these different perspectives need to be taken together in order to enable and to ensure Integrated Water Resources Management (IWRM) in the area of hydropower. For many countries, implementing such sustainable energy development practices will be demanding and costly in terms of both project assessment and mitigation measures; it is therefore essential that improvement in public/private policies should include other energy options.

Hydropower has been the most significant source of energy from water, and it is likely that it will retain this position in the foreseeable future. Although over the last ten years the development of new hydropower capacity has kept pace with the overall increase in the energy sector, even conservative projections show that tremendous increases in energy from many sources will be required in the future. The current situation of irregular electricity supply is untenable, and gives an immediate impetus to the development of new capacity. However, while access to electricity remains an essential tool, understanding rural and other energy demands, and making intelligent choices with respect to it, is a key element to IWRM. Thus, there is also an urgent need to ensure that energy developments do not negatively impact on communities or the environment, and that all such aspects are taken into account for the future.

Progress since Rio at a glance

Agreed action	Progress since Rio
Develop multi-purpose hydropower schemes, taking into account environmental concerns	
Emphasize rural electrification	

Unsatisfactory — Moderate — Satisfactory

References

Alcamo, J.; Döll, P.; Henrichs, T.; Lehner, B.; Kaspar, F.; Rösch, T.; Siebert, T. Forthcoming. 'WaterGAP: Development and Application of a Global Model for Water Withdrawals and Availability'. *Hydrological Sciences Journal.*

Bello, W. 2001. 'The paradigm crisis behind the power crisis.' Article based on an author's talk at the International Forum on Globalization, 24–25 February, New York City.

Cecelski, E. 2000. *Enabling Equitable Access to Rural Electrification.* Briefing paper prepared for a brainstorming meeting, 26–27 January 2000. Washington DC, World Bank.

Davis, M. 1995. *Institutional Framework for Electricity Supply to Rural Communities: A Literature Review.* University of Cape Town, Energy and Development Research Centre.

Diaz, F. and del Valle, A. 1984. *Energy Sources and Uses in Poor Rural Sectors in Chile: Synthesis of Eight Case Studies.* Santiagio, Centro di Investigación y Planificación del Medio Ambiente.

Döll, P.; Kaspar, F.; Lehner, B. Forthcoming. 'A Global Hydrological Model for Deriving Water Availability Indicators: Model Tuning and Validation'. *Journal of Hydrology.*

Döll, P. and Siebert, S. 2002. *Global Modelling of Irrigation Water Requirements.* Water Resources Research, 38(4), 8.1–8.10, DOI 10.1029/2001WR000355.

EEC (European Economic Commission). 1997. 'Energy for the Future: Renewable Sources of Energy'. White Paper for a Community Strategy and Action Plan.

Floor, W. and Massé, R. 1999. *Peri-Urban Electricity Consumers – a Forgotten but Important Group: What Can We Do to Electrify Them?* Washington DC, Joint United Nations Development Programme/World Bank. Energy Sector Management Assistance Programme.

Guzman, O. 1982. 'Case Study on Mexico in MS Wionczek'. In: G. Foley and A. van Buren. *Energy in the Transition from Rural Subsistence.* Boulder, Westview Press.

International Journal on Hydropower and Dams. 2002. *2002 World Atlas of Hydropower and Dams.* Sutton, Aqua-Media International Ltd.

——. 1997. *1997 Atlas of Hydropower and Dams.* Sutton, Aqua-Media International Ltd.

Khennas, S. and Barnet, A. 2000. *Best Practices to the Sustainable Development of Microhydro Power in Developing Countries.* Department of International Development, United Kingdom.

Mostert, W. 1998. 'Scaling up Micro-Hydro, Lessons from Nepal and a Few Notes on Solar Home Systems'. Paper presented at the Conference on Village Power 98, Scaling Up Electricity Access for Sustainable Rural Development, 6–8 October 1997. Washington DC.

Peskin, H. and Barnes, D. 1994. *What is the Value of Electricity Access for Poor Urban Consumers?* Background paper for World Development Report. Washington DC, World Bank.

Reddy, A. 1982. 'Rural Energy Consumption Patterns: A Field Study'. *Biomass*, Vol. 2, pp. 255–80.

UN (United Nations). 2002. *World Urbanization Prospects: The 2001 Revision, Data Tables and Highlights.* New York, Population Division, UN Secretariat, Department of Economic and Social Affairs.

——. 2000. *World Urbanization Prospects: The 1999 Revision.* New York, Population Division, UN Secretariat, Department of Economic and Social Affairs.

UNDP (United Nations Development Programme). 1997. *A Guide to UNDP's Sustainable Energy and Environment Division.* New York.

Vassolo, S. and Döll, P. 2002. *Development of a Global Data Set for Industrial Water Use.* Unpublished manuscript. University of Kassel, Centre for Environmental Systems Research.

Water Power and Dam Construction. 1995. *International Water Power and Dam Construction Handbook.* Surrey, Sutton Publishing.

WCD (World Commission on Dams). 2000. *Dams and Development: A New Framework for Decision-Making. The Report of the World Commission on Dams.* London, Earthscan Publishing.

WEC (World Energy Council). 2001. *19th Edition Survey of Energy Resources* (CD-ROM). London.

WEC (World Energy Council) and FAO (Food and Agriculture Organization). 1999. *The Challenge of Rural Energy Poverty in Developing Countries.* London, World Energy Council.

WHO (World Health Organization). 1997. *Health and Environment in Sustainable Development: Five Years after the Earth Summit: Executive Summary.* Gland.

World Bank. 2002 (September). 'Energy and Health for the Poor'. *Indoor Air Pollution Newsletter*, No. 1.

WRI (Water Resources Institute); UNEP (United Nations Environment Programme); UNDP (United Nations Development Programme); World Bank. 1998. *World Resources 1998–1999.* Oxford, Oxford University Press.

Some Useful Web Sites*

ATLAS Project (small-scale hydro)

http://europa.eu.int/comm/energy_transport/atlas/

Part of the site of the European Commission. Comprehensive review of technologies relevant to the whole non-nuclear energy field.

World Energy Council

http://www.worldenergy.org

The World Energy Council web site includes energy statistics and information.

International Atomic Energy Agency

http://www.iaea.org/worldatom/

This agency provides techniques for planning and using nuclear science and technology for various peaceful purposes, including the generation of electricity.

International Network on Small Hydro Power: IN-SHP: Overview of Small Hydropower

http://www.inshp.org/small_hydro_power.htm

This site on Small Hydro Power provides information on small hydropower projects, services, partners, etc.

UNDP data by theme: Sustainable Energy and Environment Division

http://www.undp.org/seed/guide/intro.htm

Part of the United Nations Development Programme, this site provides access to the overall foci, themes and priorities of the Sustainable Energy and Environment Division.

World Bank Energy Programme

http://www.worldbank.org/energy/

The energy programme of the World Bank focuses on poverty alleviation, sustainability and selectivity, and promoting economic growth. Many topics are covered in this site, which also provides many links to related sites.

The World Commission on Dams

http://www.dams.org

Provides access to the Commission's mission statements, publications and dam-related events.

* These sites were last accessed on 6 January 2003.

iv

Part IV: Management Challenges: Stewardship and Governance

This section examines the ways in which the competing needs, uses and demands elucidated in the previous part might be met. It discusses a few of the many tools available to decision-makers and communities to help them mould policy and practice so as to encourage an efficient and equitable use of the resource.

As the examples show, a judicious mix of 'carrot and stick' may offer the best hope for achieving sustainable use and wise management. There are incentives available for promoting integrated approaches that reward efficiency, innovation, participatory consultations and the inclusion of measures for education, awareness-building, preparedness and prevention. There are also disincentives for discouraging waste and institutional inefficiency, sectoral blindness, inadequate cost-recovery mechanisms, and poor monitoring, maintenance and follow-up.

These tools, which are all interrelated aspects of governance, are discussed in separate chapters devoted to managing risks, sharing water, valuing water, ensuring the knowledge base and governing water wisely.

Mitigating Risk and Coping with Uncertainty

By: WMO (World Meteorological Organization)
Collaborating agencies: UNDESA (United Nations Department of Economic and Social Affairs)/ UNESCO (United Nations Educational, Scientific and Cultural Organization)/WHO (World Health Organization)/UNEP (United Nations Environment Programme)/ISDR (Secretariat of the International Strategy for Disaster Reduction)/ CCD (Secretariat of the Convention to Combat Desertification)/CBD (Secretariat of the Convention on Biological Diversity)/ Regional Commissions

Table of contents

You can't control the wind, but you can adjust your sails.

Yiddish proverb

THIS CHAPTER ASSESSES THE NATURE AND COSTS of water-related risks. Such disasters as floods and droughts take an enormous toll on human life, not to mention the social, economic and environmental losses they cause. Risk is growing, with human-induced emergencies now overtaking natural disasters. In 1999 alone, natural disasters accounted for at least 50,000 deaths. The burden of loss, of course, is greatest in poor countries, where thirteen times more people die from such events than in rich ones. Economic losses also tend to be larger as a proportion of the total economy in developing countries, and most are uninsured. Natural disasters may be unavoidable, but with better planning and prevention – risk management – their impact can be reduced.

THERE IS A GROWING REALIZATION that the scale and frequency of water-related disasters is increasing, and that effective management and mitigation of water-related risk is an important issue. This is reflected in the resolutions of international meetings and in the priorities and programmes of United Nations (UN) agencies, such as in the International Conference on Freshwater (Bonn, December 2001):

> *The world is experiencing a dramatic increase of suffering from the effects of disasters, ranging from extreme droughts to huge floods, caused by the poor management of water and land and possibly by climate change. Human society and particularly the poor are becoming more vulnerable to such disasters.*

Databases of the United States Office of Foreign Disaster Assistance (OFDA) and the Centre for Research on the Epidemiology of Disaster (CRED) reveal that more than 2,200 major and minor water-related disasters occurred in the world during the period 1990–2001. Of these, floods accounted for half of the total disasters, water-borne and vector disease outbreaks accounted for 28 percent and drought accounted for 11 percent of the total disasters. Thirty-five percent of these disasters occurred in Asia, 29 percent in Africa, 20 percent in the Americas, 13 percent in Europe and the rest in Oceania (see figure 11.1). These factors are restricting the potential for improving socio-economic development and in many cases causing a decline in real terms. The impacts from just one single disaster have, in some cases, lowered the Gross National Product (GNP) in poor economies by as much as 10 percent. It has been claimed that the economic losses from water disasters are currently equivalent to 20 percent of new investment needs in water.

The efficient and effective management of risk is fundamental to long-term prosperity. Risks arise from numerous human-made and natural phenomena, many of which relate in some way or other to various facets of water, including floods, droughts and pollution. The operation and management of water resources is not only exposed to extreme events generated as part of the natural weather cycle, but is also linked to economic and socio-political factors as well as human error.

Integrated Water Resources Management (IWRM), as the prevailing paradigm in water management, has been central to promoting the recognition that water meets a number of interrelated demands while forming an integral component of the economy and environment. Traditionally in the water sector, separate planning and operations strategies have been adopted for specific economic sectors, a practice that continues today. This narrow, sectoral approach, however, has probably limited the capacity for effective management of risk and uncertainty as well as

Figure 11.1: Types and distribution of water-related natural disasters, 1990–2001

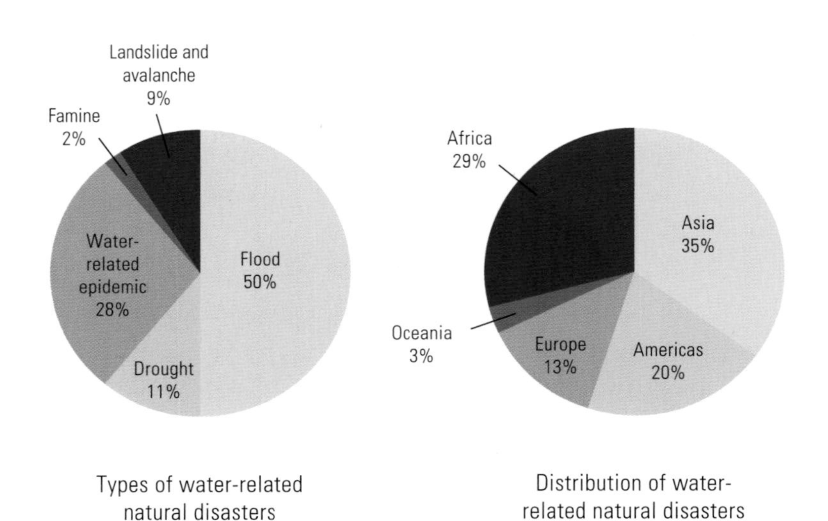

More than 2,200 major and minor water-related disasters occurred in the world between 1990 and 2001. Asia and Africa were the most affected continents, with floods accounting for half of these disasters.

Source: CRED, 2002.

obscuring certain types of risks inherent in the employed management measures (Delli Priscolli and Llamas, 2001).

The level of risk from sudden natural disasters is determined by the vulnerability of the society and the environment, combined with the probability that the hazard will occur. Climate change may exacerbate an existing vulnerability. For developing countries, the financial, human and ecological impacts are likely to be most severe, but their capacity to cope is weak. The floods in Bangladesh, Central America and Mozambique have illustrated the heavy human and environmental costs of extreme climate events. As such events become more frequent or more severe, the loss of life and livelihoods is likely to increase. The dimensions of social, economic and political vulnerability are equally related to inequalities, gender relations, economic patterns, and ethnical or racial divisions. It is also largely dependent on development practices that do not take into account the susceptibility to natural hazards.

Risk reduction refers to activities taken to reduce vulnerable conditions and, when possible, also the probability of occurrence. Vulnerability to disasters is a function of human action and behaviour. In the case of slow and unidentified disasters caused by environmental pollution, land degradation and climate change, risk reduction needs to focus on the source of the hazard.

Causes, Types and Effects of Disasters

General trends

A recent report by Munich Re (2001) notes that the number of major natural catastrophes has been increasing since 1950 (see figure 11.2). Along with this increase in the number of occurrences of floods, windstorms, earthquakes and volcanic eruptions, there has been a significant increase in overall economic losses and in uninsured losses in particular. According to Munich Re, natural disasters in 1998 caused at least 40,000 deaths around the world and 50,000 deaths in 1999. According to figures from the World Bank (2001), an estimated 97 percent of deaths related to natural disasters occur each year in developing countries.

Economic losses from the great natural catastrophes were estimated at around US$70 billion in 1999, compared with US$30 billion in 1990. In addition, more people were affected by disasters over the last decade – up from an average of 147 million per year (1981–1990) to 211 million per year (1991–2000). While the number of geophysical disasters has remained fairly steady, the number of hydrometeorological disasters since 1996 has more than doubled. During the past decade, over 90 percent of those killed by natural disasters lost their lives in hydrometeorological events such as droughts, windstorms and floods.

Figure 11.2: Trends in major natural catastrophes, 1950–2000

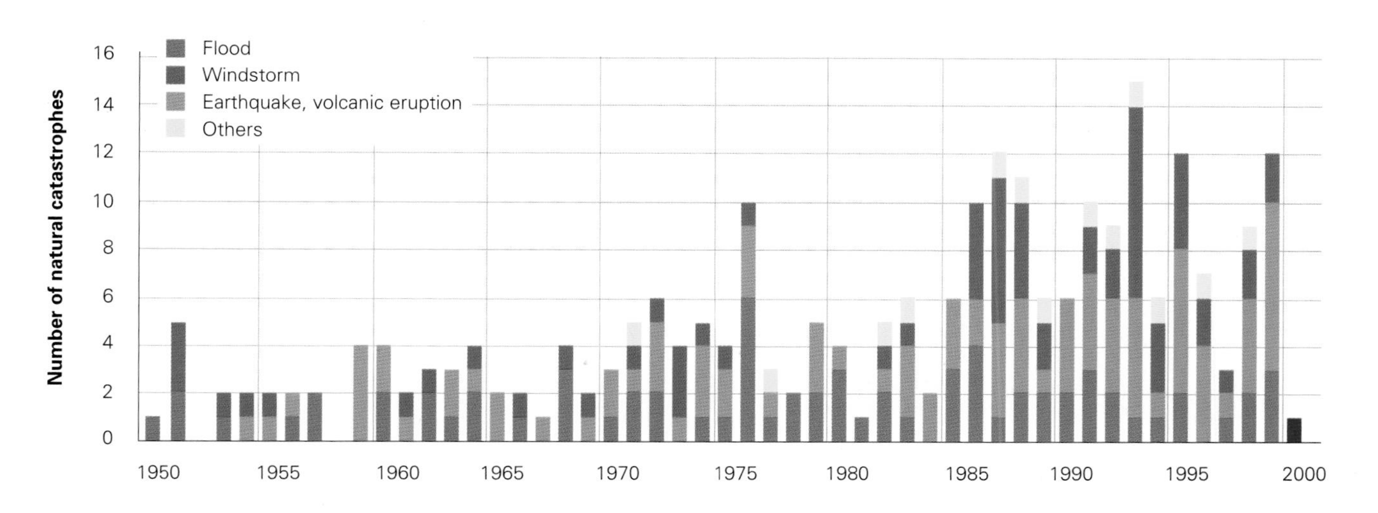

This figure clearly shows the increase of major natural catastrophes since 1950, leading to considerable human deaths (50,000 in 1999) and significant overall economic losses.

Source: Munich Re, 2001.

Floods

Floods accounted for over 65 percent of people affected by natural disasters, while famine affected nearly 20 percent. Between 1973 and 1997 an average of 66 million people a year suffered flood damage, making flooding the most damaging of all natural disasters (Cosgrove and Rijsberman, 2000). According to a study by the United Nations (UN), about 44 percent of the flood disasters that occurred in the world during the period 1987–1996 affected Asia. These disasters claimed some 228,000 lives – corresponding to 93 percent of the total number of flood-caused deaths in the world – and resulted in damage of US$136 billion to the Asian economy. In the 1990s alone, severe flooding devastated the Mississippi River basin (United States), and thousands lost their lives to flooding in Bangladesh, China, Guatemala, Honduras, Somalia, South Africa and Venezuela. Whether in loss of lives or property, the damage is enormous.

Despite these figures, floods proved less deadly, accounting for 15 percent of total deaths from natural disasters, compared to famine's 42 percent.

Several factors explain the increasing extent of disastrous flooding, among them the growing populations, denser occupancy of flood plains and other flood-prone areas, and the expansion of unwise forms of watershed land use. The hope of overcoming poverty drives many poor people to migrate. Frequently they move into places vulnerable to flooding and where effective flood protection is not assured. Informal settlements around megacities in developing countries are particularly at risk (see chapter 7 on cities).

Massive deforestation and urbanization reduce water storage capacity and amplify flood waves (Kundzewicz, 2001). In Asia, where the majority of recent large floods have occurred, rapid growth of industry and services in the 1990s resulted in considerable change of land use pattern. In Thailand, for example, it caused a reduction of natural retention and storage in the lower Chao Phraya River basin, contributing to the downstream flooding by about 3,000 cubic metres per second (m^3/s), (see chapter 16).

Even countries located in dry areas, such as Algeria, Egypt, Tunisia and Yemen have not been safe from floods. Although counterintuitive, it is a fact that in dry areas, more people die of floods than from lack of water, as the dryness is a normal state to which humans have adapted, while floods strike unprepared populations suddenly.

Floods can have widely different characteristics. In major rivers, they rise and fall relatively slowly while in urban areas they can rise and fall in a matter of hours, even minutes, offering little or no time for response. The freezing and thawing of rivers can result in ice dams that cause widespread flooding upstream and then collapse with

Box 11.1: The Rhine Action Plan

The global awareness of unanticipated high risk in international water courses was strongly influenced by the Sandoz Blaze in 1986, in the upstream section of the Rhine (western Europe). The accidental pollution incident had far-reaching consequences, triggering a successful and sustainable cooperation policy for protection of the Rhine by all countries bordering the river.

The 1993 and 1995 major floods in the shared Rhine and Meuse Rivers forced the evacuation of 250,000 people, when flood protection dikes came close to collapsing. These events provided the impetus for the adoption of policies focused on mitigating risk and reducing vulnerability to pollution and flooding. They also drew attention to the risks of deterioration of both the ecosystem and groundwater resources, as well as increasing concern about more frequent extreme flooding and drought events, as a result of climate change.

The Rhine countries have adopted the 1998 Rhine Action Plan on Flood Defence. It involves spending up to approximately US$12 billion to reduce the exposure to risk of assets with a possible value of about US$1.5 trillion. The Action Plan marks a change from planned and defensive action to management of risk, and has opened the door to a wider range of measures for reducing risk in shared rivers. It also represents a first example of an international commitment to reducing shared risk based on the value of the assets at risk, as well as an agreed shared financial commitment to manage these risks. The Action Plan for risk mitigation builds on integrated risk management at the local, regional, national and supranational level and includes water management, physical planning and urban development, nature conservation and alternative improved practices in agriculture and forestry.

Source: Based on Worm and de Villeneuve, 1999.

devastating consequences downstream. In coastal areas, floods may result from rising sea levels and tropical cyclone-driven storm surges.

It is important to remember that flooding also has positive benefits. Floods can play a positive role within our ecosystems and the environment at large, for floodwaters carry nutrients that allow for fertile flood plains and are important for various aquatic species. Integrated risk-based management provides the opportunity to take account of these benefits as well as mitigate the adverse impacts of floods (see box 11.1).

Droughts

Droughts are undoubtedly the most far-reaching of all natural disasters. From 1991 to 2000 alone, drought has been responsible for over 280,000 deaths and has cost tens of millions of US dollars in damage. For example, sub-Saharan Africa suffered its worst dry spell of the century in 1991/92 when drought covered a region of 6.7 million square kilometres (km^2) and affected about 110 million people. While droughts have always been a fact of life in Africa, the combination of drought with human activities such as overgrazing or deforestation may dramatically affect the desertification process and lead to a permanently or near-permanently degraded environment.

By the year 2025, the population projected to be living in water-scarce countries will rise to between 1 and 2.4 billion, representing roughly 13 to 20 percent of the projected global population. Africa and parts of western Asia appear to be particularly vulnerable.

Droughts have been categorized in three ways: as meteorological (due to a lack of precipitation), hydrological (lack of water in streams and aquifers) or agricultural (when conditions are unable to sustain agricultural and livestock production) (Hounam et al., 1975). The concept of what constitutes a drought varies between countries. In England, three weeks without rain is considered a problem; in many parts of the world much longer dry periods are normal.

Durations and extents of droughts vary greatly. Examples of severe, persistent droughts over large geographical areas include the Sahel covering 7.3 million km^2 from 1970 to 1988, continental Europe covering 9 million km^2 from 1988 to 1992, and India covering 3 million km^2 from 1965 to 1967. There are other examples of extreme droughts in North America and Australia. Table 11.1 provides a summary of major drought events and their associated loss of life and economic losses as established by Munich Re. Extreme drought can affect large tracts of land and large numbers of people anywhere in the world and can persist from a few months to several years. The impacts of a drought hazard can have far-reaching social, economic and environmental consequences.

Table 11.1: Major drought events and their consequences in the last forty years

Date	Country or continent	Fatalities	Economic losses (US$M)
1965–1967	India	1,500,000	100
1972–1975	Africa	250,000	500
1976	United Kingdom		1,000
1979–1980	Canada		3,000
April–June 1988	United States		13,000
June–July 1988	China	1,440	
1989–1990	Angola	10,000	
Summer 1989	France		1,600
Jan.–Oct. 1990	Greece		1,300
Summer 1990	Yugoslavia		1,000
Jan.–March 1992	Africa		1,000
May–Aug. 1998	United States	130	4,275
Jan.–Aug. 1999	Iran		3,300
Jan.–April 1999	Mauritius		175
June–Aug. 1999	United States	214	1,000

Source: Munich Re, 2001.

The effects of disasters

High stress results when many large disasters occur within a limited time span. The year 1999, for example, with several earthquakes, the Lothar storm in France, floods and mudslides in Venezuela and more than 50,000 deaths, still remains in political memory as the second most costly year in terms of global disaster and risk insurance indemnities. Stress is also high when one single region is hit by a sequence of several major catastrophes. One example is the Indian state of Orissa, which experienced massive flooding in 2000, followed in 2001 by the worst drought in a decade and new floods. Out of a population of 32 million, some 27 million people were affected.

In poor countries, natural disasters generally result in larger economic losses in proportion to the economies.

Depending on the robustness of the national economies, the negative consequences of disasters tend to be amplified as they erode the political and social stability of countries and upset the balance between the three necessary pillars for water resources management: economic development, environmental conservation and social stability (Appelgren et al., 2002). This is particularly the case when one disaster can virtually wipe out the investments made in infrastructure over the previous decade. Table 11.2 lists the severe natural disasters resulting in more than about 1,000 deaths in recent years. Most of these disasters occurred in developing countries, and the majority of the losses were uninsured.

The Zimbabwe drought of the early 1990s was associated with an 11 percent decline in Gross Domestic Product (GDP) and a 60 percent decline in the stock market; more recent floods in Mozambique led to a 23 percent reduction in GDP, and the 2000 drought in Brazil to a halving of projected economic growth. Even in developed countries, an extreme drought may cause considerable

Table 11.2: Severe natural disasters and their effects since 1994

Year	Date	Event	Area	Deaths	Economic loss US$M	Insured loss US$M	Remarks
1994	Summer	Flooding	China, whole country	1,700	>7,800		Landslides; broken dams; over 2 million dwellings destroyed; 50,000 km^2 of crops destroyed; 85 million affected
1995	7 Jan.	Earthquake	Japan: Kobe	6,348	100,000	3,000	RS 7.2; 150,000 buildings destroyed; 37,000 injured; 310,000 homeless
	May–July	Flooding	China: south	1,390	6,700	70	Over 1.1 million houses destroyed; 3.9 million damaged; severe damage to infrastructure; about 3,400 injured
	13 May	Earthquake	Russia: Sachalin	1,841	100		RS 7.6; numerous blocks of buildings collapsed; oil pipelines and production plants damaged; damage to infrastructure
1996	27 June–13 Aug.	Flooding, debris flows	China: central, south, west	2,700	24,000	445	Worst floods in 150 years; bridges, dams and over 5 million buildings destroyed; 8,000 factories affected (production stopped); damage to agriculture, infrastructure and supply facilities
1997	10 May	Earthquake	Iran: north-east; Afghanistan: west	1,573	500		RS 7.1; damage in 147 villages; over 100,000 homeless
	15 July–15 Sept.	Floods	Myanmar: central, south-east, south, esp. Pegu	1,000			Monsoon rain; over 6,000 houses destroyed; two million affected
	Oct.–Nov.	Flooding	Kenya: east; Somalia: central, south	1,850	2		Damage to over 9,000 buildings, roads and bridges; large areas cut off; food stores destroyed; water supply cut; over 250,000 people homeless
1998	4 Feb.	Earthquake	Afghanistan: north, Rostaq	c. 4,600			RS 6.1; affected region isolated; 28 villages destroyed
	15 May–16 June	Heatwave	India: north-west, esp. Rajasthan	3,028			Temperature up to 49°C; severest heatwave in fifty years
	30 May	Earthquake	Afghanistan: north, Rostaq	c. 4,500			RS 7.1; 90 villages destroyed or severely damaged
	9–11 June	Cyclone	India: west, Gujarat, Kandia	10,000	1,700	400	Gusts up to 185 km/h; waves up to 10 metres; storm surge; 170,000 houses damaged/destroyed; major losses to ports, storage buildings, oil tanks, salt fields, factories and wind parks; severe losses to power supply
	10 June–30 Sept.	Floods	Bangladesh: central, north, south; India: north, north-east, esp. Assam, west Bengal; Nepal: east, west	4,750	5,020		Worst floods in decades; 60,000 km^2 of land submerged; 1.2 million houses damaged; severe losses to agriculture, livestock and infrastructure; epidemics (hundreds of deaths); 66 million people affected
	15 Sept.–1 Oct.	Hurricane George	Caribbean: esp. Puerto Rico, Dominican Republic, Haiti; United States: Florida, Alabama, Missouri, Louisiana	>4,000	10,000	3,300	Gusts up to 260 km/h; hundreds of thousands of houses destroyed; losses to hotels, houseboats and yachts, and to agriculture and forestry; major losses to the infrastructure, esp. power supply
	22 Oct.–5 Nov.	Hurricane Mitch	Honduras, Nicaragua, Belize, El Salvador, Guatemala, Mexico, Costa Rica, Panama, United States	9,200	7,000	150	Wind speeds up to 340 km/h; fourth most powerful Atlantic hurricane of the twentieth century; 70 percent of Honduras's and Nicaragua's infrastructure severely damaged; five villages destroyed by mudslide from Casila volcano; major losses to agriculture; economic situation severely affected; 8,000 missing
1999	25 Jan.	Earthquake	Colombia: central, Quindo, Caterera, Armenia, Pereira	1,185	1,500	150	RS 6.2; 80,000 houses damaged or destroyed; severe damage to infrastructure
	17 Aug.	Earthquake	Turkey: north-west, Izmit, Kocaeli	>17,200	12,000	600	RS 7.4; 270,000 houses, businesses damaged or heavy damage in the industrial sector; thousands destroyed; missing; 44,000 injured; 600,000 homeless
	20 Sept.	Earthquake	Taiwan: central, Nantou, Chichi	2,474	14,000	850	RS 7.6; thousands of aftershocks; landslides; >50,000 buildings badly damaged/destroyed; 6 million households without electricity; severe damage to infrastructure; 310,000 homeless; 11,000 injured

Table 11.2: *continued*

Year	Date	Event	Area	Deaths	Economic loss US$M	Insured loss US$M	Remarks
	28–30 Oct.	Cyclone	India: east, Orissa	10,000–30,000	2,500	115	Worst storm in 100 years; torrential rain; 18,000 villages destroyed; Paradip Harbour badly damaged; 17,000 km^2 of paddy fields devastated
	13–16 Dec.	Floods, landslides	Venezuela: north, west	20,000	15,000	500	Devastating landslides and debris flows after nine days of rain; villages, towns destroyed; thousands injured or missing
2000	Feb.–March	Floods, tropical cyclone Eline	Mozambique, South Africa, Botswana, Swaziland, Malawi, Zambia	<1,000	660	50	Worst floods in fifty years; rivers burst their banks; dams breached; infrastructures destroyed; food and water supplies impaired; evacuations held up; 850,000 homeless; millions affected
	Aug.–Oct.	Floods	India: east, north, north-east; Nepal: central	1,550	1,200		Thousands of villages flooded; traffic routes blocked; severe losses in agriculture and livestock sectors; 3.5 million homeless/evacuated

(RS=earthquake magnitude on the Richter Scale)

Source: Munich Re, 2001.

disturbances in terms of environmental, economic and social losses. The 1988 drought in the United States may have caused direct agricultural losses totalling US$13 billion.

The general view is that the recorded losses are underestimated and could be at least doubled when the consequences of the many smaller and unrecorded disasters that cause significant losses at the community level are taken into account. Devastation in the aftermath of the floods that ravaged many countries in Africa, Asia and elsewhere, and droughts that plagued all of Afghanistan, Africa, Asia and Central America are major setbacks for communities

Table 11.3: Risk linkages and integration

Sector/ subsystem	Natural, man-made disaster risk	Flow systems production risk	Social risk	Industrial risk	International risk
Natural, man-made disaster risk[1]		Reduced capacity to mitigate natural disaster impact	Reduced capacity to mitigate natural disaster impact, cultural and ethical patterns	Chemical pollution disaster	Non-cooperation, conflict on water quantity/quality and regional instability
Flow systems production risk[2]	Affect strategic utilities and production		Malfunction, low production from social instability, collapse	Collapse in flow and production systems	Collapse of regional economic and environmental cooperation, food security, loss of economic resources
Social risk[3]	Produce human loss, health, inequity out-migration, instability	High costs, economic collapse, instability, inequity		Health hazards	Out-migration, civil conflict, health hazards
Industrial risk[4]	Insufficient investment for precaution, trigger chemical pollution	Production blackout, major industrial discharge and pollution	Labour instability and conflict		No access to technology, labour conflict
International risk[5]	International impact, consequential conflict	International pollution, food shortages	International, regional instability, conflict		

[1]Flood, drought, accidental pollution, health hazards.
[2]Water, waste, energy, transportation, communication, agricultural, industrial production.
[3]Population, migration, health, ethnical conflict, legal issues.
[4]Environment, labour force.
[5]Transboundary water courses and groundwater aquifers, upstream-downstream issues.

Source: Prepared for the World Water Assessment Programme (WWAP) by B. Applegren, 2002.

attempting to achieve sustainable development. With rapid urbanization and changed land use and settlement patterns the negative impacts continue to rise and the gap in matching measures for the management of water-related risks is widening.

The costs of risks to society are aggregated over the full range of events and different sectors (natural and man-made, flow-system and production, social, industrial and international risk). It is therefore important to develop tools for integrated management of natural and human-induced risk and uncertainty. The impacts of joint risk represent the total stress on the society, which is the result of the compounded risk of interlinked risk in the different systems and sectors and where water-related risk is just one of many. Table 11.3 presents examples of the principal interlinkages between risks or uncertainty in different sectors. The matrix demonstrates the importance of a joint assessment of risks that impact on society, based on an appreciation of the relative importance of both natural and human-induced risks in terms of the overall society and particular subsystems. The matrix proposes a model for risk-based integrated allocation for minimized social risk.

Responding to Disasters

The growing frequency of natural disasters led to the adoption by governments of the International Strategy for Disaster Reduction (ISDR) to succeed and promote implementation of the recommendations deriving from the International Decade for Natural Disaster Reduction (IDNDR, 1990–1999). The aim of the ISDR is to mobilize governments, UN agencies, regional bodies, civil society and the private sector to unite efforts in building resilient societies by developing a culture of prevention and preparedness. Different definitions and standard terminology are provided by the United Nations Department of Humanitarian Affairs (UNDHA, 1992) and have been adopted by the ISDR Secretariat after consultations with relevant agencies and sectors (see box 11.2).

Recognition of effective management and mitigation of water-related risk is growing, as reflected in the priorities and the programmes of UN agencies, national governments, international water lobbies, emergency groups and private enterprises (see box 11.3).

Management tools

Risk management traditionally possesses three major components as follows (Dilley, 2001).

- The first is knowing the risks. This means knowing the probability of the hazard to occur and establishing the vulnerability index and expected loss of life and possessions, injury and damage to the environment.
- The second is defining and implementing the measures to reduce these risks. Such approaches can include 'structural' and 'non-structural' measures such as early warning systems and preparedness. However, for uncertainties and unidentifiable risks, the only option is for society to share the risks.
- The third is in transference of risk, what is termed risk-sharing. It implies imposing risk on a wider group of people or in broadening the economic basis for support. In wealthy economies, this can be accomplished through insurance programmes[1] and other similar risk transfer mechanisms.

These instruments need to be used in combination with risk reduction, and are not themselves a solution to the cause of the risk.

Structural and non-structural measures

In attempts to mitigate the negative impacts of flooding, two general categories of measures or options exist: structural and non-structural. Structural measures include the building of physical works such as dams and dikes, but can also include channelization and dredging, creating additional artificial floodways, diversions and ponds, and armouring of channel walls to prevent erosion. Flood-proofing of existing structures has also witnessed rising popularity. Non-structural measures would include land use planning that promotes appropriate use of land within the floodplain and floodway and prohibits certain development and uses from occurring.

Acknowledging uncertainties

Risk identification and other risk management steps could provide invaluable indications and alert vulnerable groups, but could also establish perceptions of false security. As shown by the increasing number of unexpected large accidents, risk is often characterized by the uncertainty and incapacity of society to respond to early warnings. In this situation, there is an emergence of alternative approaches: the policy of fail-safe systems is now giving place to safe-fail ones. As stated during the International Conference on Water in Bonn: 'It is impossible to design a system that never fails (fail-safe). What is needed is to design a system that fails in a safe way (safe-fail)' (Kundzewicz, 2001). This acknowledges uncertainties and handles them through direct involvement of stakeholders as responsible managers and potential victims of the accidents, rather than through a technically refined analysis.

1. Insurance of water risk is marginal and not attractive to the private insurance industry and the coverage has come from governments that possess neither the authority nor the resources, especially in developing countries.

Box 11.2: Methodology and terminology adopted by the ISDR

Disaster: A serious disruption to the normal functioning of a community or a society, which causes widespread human, material, economic or environmental losses that exceed the ability of the affected community/society to cope using its own resources. Disasters are often classified according to their speed of onset (sudden or slow), or according to their cause (natural or human-induced).

Hazard: A potentially damaging physical event or phenomenon that can harm people and their welfare. Hazards can be latent conditions that may represent future threats, as well as being natural or induced by human processes.

Risk: The probability of harmful consequences or the expected loss (of lives, people injured, property or environmental damage, and livelihoods or economic activity disrupted) resulting from interactions between natural or human hazards and vulnerable conditions. Conventionally, risk is expressed by the equation:

Risk = Hazard x Vulnerability

The resulting risk is sometimes corrected and divided by factors reflecting actual managerial and operational capability to reduce the extent of the hazard or the degree of vulnerability.

For the purpose of economic assessments, risk is quantifiable and perceived in monetary terms. Taken from an economic perspective, risk is specified as the annual cost to the society from sudden accidental events and slow environmental degradation, determined from the product of the probability or frequency of occurrence and the vulnerability as social and economic losses in monetary terms:

Risk (economic cost per year) = Probability (once in n years) x Vulnerability (economic cost/event)

Risk assessment: Investigates the potential damage that could be caused by a specific natural or human-induced hazard to people, the environment and infrastructure. The assessment includes hazard or multi-hazard analysis, probability and scenario; the vulnerability analysis (physical, functional and socio-economic); and the analysis of coping capacities and mechanisms. Risk assessment forms the necessary basis for the development of disaster mitigation and preparedness measures.

Risk management: The systematic application of management policies, procedures and practices that seek to minimize disaster risks at all levels and locations in a given society. Risk management is normally based on a comprehensive strategy for increased awareness, assessment, analysis/evaluation, reduction and management measures. The risk management frameworks need to include legal provisions defining the responsibilities for disaster damage and longer-term social impacts and losses.

Uncertainty: Different from quantifiable risks, uncertainty is defined as unidentified and unexpected threats and catastrophic accidents. Uncertainty often manifests as large unexpected 'surprise' threats and catastrophes and requires management approaches different from traditional risk management. Recent uncertainties are exemplified by the major floods in Europe and the United States in the 1990s, but also by the chemical pollution of the Sandoz Blaze in 1986 in the Rhine (Europe) and the non-water-related industrial incident in Bhopal (India) in 1984 which resulted in more than 10,000 deaths and 200,000 people injured. The main management issues of uncertainty concern the society's incapacity to identify and react on weak signals; and as one consequence, the reluctance of decision-makers to assume related social responsibilities.

Vulnerability: A function of human actions and behaviour that describes the degree to which a socio-economic system is susceptible to the impact of hazards. Vulnerability relates to the physical characteristics of a community, structure or geographical area, which render it likely to be affected by, or protected from, the impact of a particular hazard on account of its nature, construction and proximity to hazardous terrain or a disaster-prone area. It also designates the combination of social and economic factors that determine the degree to which someone's life and livelihood are exposed to loss or damage by a specific identifiable threat or event in nature or society.

Box 11.3: Initiatives coping with water-related risks

Underlining the fact that risk forms a constraint on the willingness to invest and recognizing the large costs countries have to bear in adapting to the effects of water-induced shocks on their economies, the World Bank has emphasized the links between water resource variability and risk with the need for investments to mitigate the risks (World Bank, 2001).

The United Nations Development Programme (UNDP) has assumed the operational responsibilities for natural disaster mitigation, prevention and preparedness within the UN system. Disaster reduction and recovery comprise essential components within development priorities, including poverty eradication and sustainable livelihoods, gender equality and the advancement of women, environmental and natural resource sustainability, and sound governance.

The World Health Organization (WHO), with mandated responsibility for emergency-related health hazards, has emphasized the consequences of water disasters and the growing risks such as inadequate drinking water and sanitation and uncontrolled spreading of toxic waste.

The Food and Agriculture Organization (FAO), under the food security programmes, is concerned with the development of agricultural systems that can sustain production during drought and flood. FAO has a long tradition of forecasts and early warning systems for regional agricultural droughts.

The UN World Food Programme (WFP) is focused on emergency and post-disaster food relief and rehabilitation support during and after natural and human-induced water disasters.

The Global Water Partnership (GWP) has noted that risk management practices are important to the realization of IWRM goals, observing however that there has been 'relatively little attention ... paid to the systematic assessment of risk mitigation costs and benefits across water use sectors and to the consequent evaluation of various risk trade-off options' (GWP, 2000a).

National governments are exposed to water risk and recognize in general its critical importance. Several countries have recently made moves towards integrated risk-based approaches in water resources management.

- The Netherlands planned on turning its watercourses into a fully integrated technical and economic system but cancelled the scheme because of considerations of national and local risk. With intensive economic development and growing vulnerability, the social and economic risk from the emerging threat of sea level rise, aggravated by increasing subsoil subsidence and rainfall, is high. As a result, the country has adopted an integrated risk-based water management approach where water risks related to quality, flood control, ecosystems preservation and groundwater management form the guiding factors for the national spatial development policy.

- In the recent dialogue on the Swiss position at the World Summit for Sustainable Development (WSSD), the Swiss Government, together with Swiss Re, a private reinsurance company, emphasized risk mitigation as an important component of water resources management (Swiss Re and Swiss Government, 2002). The Swiss Agency for Development and Cooperation (SDC) 'fosters sustainable water management, with the focus on rural water supply and sanitation, integrated water resource management and disaster prevention and relief'. The Swiss WSSD document position will be applied to risk assessment, awareness-building and development of water-related risk-mitigating activities in the country.

- The Giulia-Venezia region in Italy is under high social stress from flood hazards and serious health hazards from long-term freshwater and coastal water pollution, together with other non-water-related social stress. The regional authorities are reviewing the management of the joint risk-based approach.

Hazard mapping

The World Meteorological Organization (WMO) and the Scientific and Technical Committee of the IDNDR (WMO, 1999), have initiated a joint assessment of risk from a variety of natural hazards – especially meteorological, hydrological, seismic and volcanic disasters. They found the need for standard methodology in different sectors where composite hazard maps are generally recognized as an important tool for joint assessment. Mitigation of risks from natural hazards depends on the implementation of risk management measures and political will. Risk management requires a long-term view, while governments operate more on short-term objectives and limited annual budgets, and are sometimes reluctant to commit expenditure for proactive risk management measures that are based on weak indications of possible future risks. Governments are, however, generally responsive to short-term emergency relief requirements.

An iterative process

Figure 11.3 shows that risk assessment and risk management are iterative processes wherein various options are assessed and an optimal solution with acceptable risk is attained, for which the possible measures and the cost are defined. It requires an assessment of what level of risk is acceptable, given that it is not generally possible to reduce risk to zero. The acceptability of certain risks may vary both in different countries and in time.

The economic grounds

Decisions on risk management may be taken on economic grounds. The costs and benefits associated with managing the risk from a particular hazard can be compared to the returns from other investments, both within the water sector and other public health and safety areas, to maximize the effectiveness of public funding. It is also important to note that social acceptability is a contributing factor in making good decisions. It should be recognized that decisions about allocation of risk are often made more as part of political economic processes than through public consultation or participation (Rees, 2001). Establishing effective mitigation of risk and uncertainty therefore depends on the level and the strength of the political economy (see box 11.4).

While the impact of water disasters (mainly from flooding, drought and pollution) is apparent, there is an even greater impact in terms of future social costs. Appropriate investment strategies and a redirection of resources into prevention offer the potential for significant

Figure 11.3: Framework for risk assessment

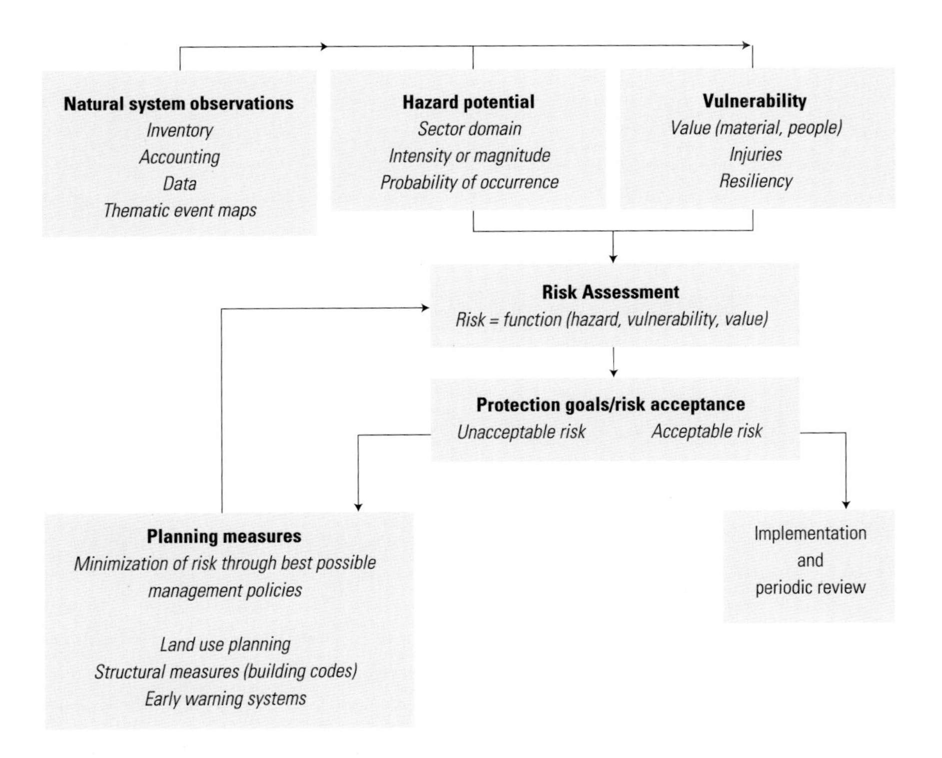

Risk assessment and risk management are iterative processes.

Source: Adapted from WMO, 1999.

Box 11.4: Political economy

Politicians have the incentive to balance allocation of budgets in a way that preserves political support. Political sustainability is important, and security and stability, together with distributional goals, make up important aspects of political agendas and are often given higher priority than efficiency. Efficiency losses to society and the microcosts imposed on the public could therefore be substantial, but are often disregarded. In the political environment, therefore, policy often responds more to shorter-term concerns than to the long-term consequences to society. The objective of policy selection clearly differs from maximizing efficiency. One practical consequence is that, to be effective, policy analysis and formulation need to be adapted to the preferences of policy-makers.

Source: Based on Just and Netanyahu, 1998.

economic benefits, as well as the reduction in the loss of human lives and improvements in welfare and social stability. This potential is particularly great in the fragile developing and transition economies.

Strategies for the management of risk can be broadly classified into two categories:

- private risk encompassing well-defined risks that can be covered by insurance policies; and
- common risk covering wide-ranging socio-economic factors that ultimately must be borne by governments.

Private individuals, communities and businesses must be encouraged to carry as much risk as possible within the constraints of their resources and abilities to manage the risks. However, in all societies there will always be categories of common risks that can only be effectively managed by national or international institutions. These include the impacts of events that lead to wide-scale homelessness, abandonment of property and loss of common environmental resources. Within the water sector, the challenge is to commit resources to the accurate identification of these risks, to prepare robust management strategies and to be able to respond in the event of natural or man-made disasters. The reduction in the exposure to risk can be recognized as a real socio-economic benefit that will influence the actions of people and promote sustainable growth and development. While supported by technically advanced methodology, programmes and investments, water sector risk management in terms of political economy is still based on reactive approaches and therefore has limited political currency. At the same time, however, the recognition of risk is sometimes already seen to be in contention with economic development and environmental conservation and could lessen economic prospects and competitiveness and repel investments.

Managing floods

Hazard potential in flooding is expressed as the relationship between the magnitude of the flood and its probability of occurrence. Vulnerability is a function of the land use at the location at risk of flooding.

Assessment of the probability of floods requires several years of high flow records and is also constrained by the uncertainty that the same high flow patterns might not persist into the future. There is evidence and growing concern that increased emissions of carbon dioxide and other greenhouse gases are producing changes in the world's climate. The Intergovernmental Panel on Climate Change (IPCC) stated that 'projections using the Special Report of Emission Scenarios of Future Climate indicate a tendency for increased flood and drought risks for many areas under most scenarios' (IPCC, 2002). This raises the concern of uncertainty over the challenge of predicting future extreme runoff conditions and meeting the need for alternative approaches for managing risk under uncertainty.

Both droughts and floods impact directly on human well-being and health, especially in urban and rural areas where water supply and drainage facilities are inadequate. The most vulnerable are always the poor – those living in crowded, poorly protected peri-urban settlements, and subsistence farmers in rural areas – who cannot afford to protect themselves, or are ignorant about how to cope with the impact of changes in their environment.

Apart from the major threat of climate change, alterations to land use within the basin can affect the magnitude and frequency of floods. Urbanization of the land increases runoff and flood peaks, as observed in the Greater Tokyo region for example, where urbanized areas expanded from 13 percent to 20 percent between 1974 and 2000 (see box 11.5).

Planning is an important non-structural measure for flood mitigation, covering the entire period from pre- to post-disaster. All parties, ranging from governmental and civil organizations to the local citizen, should know their roles and responsibilities. Everyone should also know exactly what is needed in terms of supplies and provisions, have them ready, and know how to respond when the hazard occurs. As part of this process, there should be a well-maintained contingency plan and vigilant flood forecast warning and alert programme. Experience has shown that local governments and citizens must be involved throughout the planning and implementation stages of such systems for them to be effective.

Losses from disasters do not cease once the flood has subsided. Post-disaster activities also decrease vulnerability and increase resilience. Relief during the first seventy-two hours of an event is critical in order to limit loss of life. Such relief requires the maximum possible lead time to enable pre-positioning and mobilizing of emergency assistance. Forecasts of non-significant events during periods of high alert are also important as they allow the re-assignment of human and other resources elsewhere. For example, according to the Chinese government, 90 percent of the 30,000 deaths from the floods in 1954 were a result of infectious diseases such as dysentery, typhoid and cholera, which struck in the period following the event (Worldwatch Institute, 2001). In contrast, after the 1998 flooding of the Yangtze, no such epidemics were reported. Additional post-disaster activities, beyond direct relief, include support to quickly restart the local economy and resume basic social services.

Flood control and mitigation is normally a high priority under International River Basin Commissions and in bilateral arrangements on transboundary rivers. Similar to the Rhine, flood management in the Lower Mekong basin will depend on being able to conserve the natural flood plain. Egypt and Sudan also have a long-established cooperation agreement for preparedness and operational management of the Nile summer floods.

Managing droughts

The onset of a drought is slow and very different from flooding with respect to the size of the affected area, the duration, the measures that can be taken to mitigate the impacts of the hazard and the ability to forecast the onset of the disaster. Nevertheless, many of the same principles put forth in the section on floods also apply to drought.

Drought is associated with significant human and social economic losses, especially in poor developing countries where livelihood and food security depend on vulnerable rainfed subsistence farming and livestock production. It is also often claimed that drought results from a lack of distribution, know-how, and human and capital resources in poor regions (Delli Priscolli and Llamas, 2001).

From a risk management perspective the questions that arise are: how frequently can one expect a certain type of drought to occur? What are the vulnerabilities and expected losses? And what mitigation efforts or options are plausible and at what cost? It is then necessary to balance the cost of the mitigation efforts with the potential costs of the risks, and identify measures that will provide an 'acceptable' risk to society for the specific drought at the lowest cost.

Drought mitigation strategies may aim to reduce the vulnerability factor by, for example, altering land use and agricultural practices, or may modify the severity of the drought by

Box 11.5: Comprehensive flood control measures in Japan

Flood mitigation measures in Japan during the post-war rehabilitation period have both decreased the occurrences of major rivers overflowing their banks and limited embankment failures, reducing the severity of flood damage and the total area affected by it. In recent years, however, due to the remarkable shift of population and social assets into urban areas since the beginning of the period of high economic growth in the 1960s, urbanization has progressed in areas with a high risk of disasters: lowland marshes, alluvial fans and cliffs. Today, 48.7 percent of the Japanese population and 75 percent of holdings are located within the flood-prone areas of rivers. The inflation of property values due to rapid economic growth and the continued concentration of urban property in floodplains have increased costs of flood damage in urban areas. Flood damage density (the ratio of damage to affected area) has risen sharply; and property damage due to river overflow and water collecting behind levees as a percentage of total damage has been on the increase as well.

River basins undergoing rapid urbanization are losing their natural water-retaining and retarding functions. At the same time, concentration of population and property in these urban river basins contributes to increasing 'damage potential' (the maximum amount of damage that could potentially result from a disaster). These problems are being dealt with through comprehensive flood control measures that consolidate the combined use of facilities to maintain the water-retaining and -retarding functions of river basins, the creation of incentives to use land safely and to build flood-resistant buildings, and the establishment of flood warning and evacuation systems.

Comprehensive flood control measures are implemented through the Council for Comprehensive Flood Control Measures, established for individual river basins and through the formulation of basin development plans that include improvement of the environment.

Source: Prepared for the World Water Assessment Programme (WWAP) by the National Institute for Land and Infrastructure Management (NILIM) and the Ministry of Land, Infrastructure and Transport (MLIT) of Japan, 2002.

providing irrigation from reservoirs, wells or water imports from areas unaffected by the drought. Other mitigation measures may include crop insurance and relief programmes to ensure water for basic needs and provide food supplements.

Mitigation efforts typically include contingency planning that might comprise arrangements for alternative supplies, putting into force water economy measures and protecting priority uses. The slow onset of drought combined with drought forecasting can enable such measures to be implemented in advance of the emergency occurring. The improvement in recent years in seasonal and long-term climate predictions, such as those issued by many national and regional institutes, including WMO's Drought Monitoring Centres in Africa, will assist effective implementation of contingency plans.

There are a variety of longer-term mitigation measures that can be taken. These include changing crop type, recognizing lands that are indeed marginal and appropriately changing agricultural practices, and constructing reservoirs. Populations will eventually have to build their security at a local and family level. An important requirement is, therefore, to identify and establish strategies that enable the community to cope with droughts, including revival of traditional customs for cultivation and livestock.

Relocation of populations is another possible long-term measure that is being debated. However, the social capacity to handle migration and resettlement are issues that warrant careful consideration.

Perspectives in the Management of Risks

Constraints to the achievement of effective risk management

There is a shortage of effective disaster-preparedness and mitigation measures, e.g. flood dikes, early warning systems, shelters, relief stockpiles and disaster response teams. Some of the barriers to effective risk reduction quoted by the International Federation of Red Cross (IFRC, 2002) are described below.

- Geopolitical conflicts of the 1990s dominated the humanitarian agenda, pushing aside the problem of vulnerability to natural hazards.
- Responsibilities for mitigating disasters are fragmented.
- Risk reduction is not an integral part of water resource management and development.
- Risk reduction is viewed as a technical problem, and often the underlying factors that compel people to live in insecure conditions are ignored.
- Donors dedicate far fewer resources to risk reduction than to relief.

While risk reduction technology and programmes are very important, the call is for enhanced responsibility for social water risk and recognition of a number of basic economic, institutional, legal and commercial constraints to the achievement of effective risk management.

- Economic constraint: one major constraint to successful management and reduction of risk is that of recovering the costs from the beneficiaries. The dilemma, expressed in economic terms, is that management and risk reduction measures respond to a public good. Different from quantity and quality, risk mitigation responds to a public good that is non-exclusive and non-rival and economic practice suggests that it will be underprovided by private markets or not provided at all due to the free-rider aspect of the good. For the same reason, risk mitigation has no efficient marginal price and cost recovery becomes complicated especially in a market-based economy. If risk reduction is the only service of an investment pricing at marginal cost, some lump sum measure to allocate and recover costs is necessary. These non-separable costs are however not often included or recovered as part of land taxes. Faced with budget constraints and the trend of transferring water management responsibilities to the private sector, national governments are experiencing difficulties supporting increasing risk management from public budgets.
- Institutional constraint: water constitutes one of several risks under the responsibility of Prevention and Civil Protection outside of the water sector. As a consequence, management of water-related risk is often relegated to a technical side issue at a subordinate level and not integrated into the political economic allocation process in the water sector or other strategic sectors.
- Legal constraint: for reasons already mentioned, risk is not always considered in project selection and resource allocation for economic development, notwithstanding that the responsibility of managing and mitigating natural and human-induced disasters is provided under the principles of precautionary and preventive action adopted in many national, regional and international water and environmental legislations.
- Commercial constraint: the accommodation of hydrological risks as a calculated insurance risk is hampered by the fact that the vulnerable objects are spread over extensive common areas where a multitude of other risks are added up to a very large covered and uncovered total insured risk. In this case not even reinsurance companies accept the risk, and governments might

not have the financial strength required to assume the ultimate financial responsibility.

Public and private responsibilities

In traditional systems, water users are accustomed to operating and sustaining production and social services under insecure and extreme climatic and economic conditions. As the systems grow more complex individual responsibilities to manage water-related risk are assumed by state administrations. Box 11.6 summarizes the evolution of responsibility for risk in the water sector over the last forty to fifty years as it has occurred in western Europe.

But deregulation, privatization and market liberalization, unless accompanied by a shift of responsibility for risk to the private sector, may leave the state with a responsibility that is not commensurate with its resources. For example, the community-based risk identification studies under the National Flood Insurance Programme (NFIP) in the United States amounted to about US$115 billion for 18,760 communities.

For a number of social, institutional and scientific reasons that range from climatic variability to economic globalization and liberalized market economies with their reduced roles of government, the management of risk and uncertainty has come to the forefront. Natural and human-induced hazards are becoming one of the major challenges to the management, protection and conservation of water. They risk undermining other efforts to attain development targets. While international organizations and in particular several national governments have adopted integrated risk-based management approaches, the institutional inertia and even resistance to full recognition of the wider scope of risk mitigation in the water sector persists. Risk and uncertainty continue to represent weak links in management of water systems. Parallel to the progress of traditional risk mitigation, there is a call to national governments and international organizations for alternative management approaches and adjusted governance responsibilities that need to be recognized, assessed and acted upon.

Citizens are becoming conscious of and affected by growing social costs of disasters such that the importance of risk management and uncertainty can be expected to emerge as major social and political challenges in water resources in the next decades. Figure 11.4 demonstrates the parallel and closely linked trend from natural to human-induced disasters in the causes of food emergencies. The evolution is towards wider social responsibility to control also the causes of disasters.

New challenges

New challenges have arisen in the control and mitigation of new and less well-known long-term risk and uncertainty, which are adding to the costs and problems of managing natural hazards. Traditional risks include unreliable or unsafe supplies for urban water supplies and irrigation, in particular for large urban settlements that depend on one source (hydrological risk) or one conveyance system (infrastructure risk). There are also localized risks

Box 11.6: Evolution of responsibility for risk-based water resource management

- **Up to 1945:** Single-structured economies; traditional management system based on individual responsibilities.
- **1945–1980:** Management responsibilities, including prevention and civil protection assumed by the State.
- **1970–1990:** Often in conflict with reduced government role and resources, the functions are expanded to provide for IWRM.
- **1980–2000:** In conflict with a trend of growing risk and crisis, no-risk-based and development-focused management is continued. Risks are dodged by integrated management and planning, with constructs for participation and far-reaching provisions for sustainable development, precaution and preventive action obscuring state responsibility.
- **Since 1990:** End of the no-risk era, with more frequent, severe human-induced and natural accidents including industry-, pollution- and health-related catastrophes, accompanied by civil accidents, violence, and civil and ethnic conflict at growing social costs. No-risk-based integrated management and planning approaches become insufficient to manage increasingly frequent, large and long-duration accidents;
- **Today:** No-risk to high-risk change renders current approaches non-operational, initiating a decline towards broader social crisis; and risk-based management approaches are needed that build on defined responsibility and timely, risk-based decision-making to pro-act on unanticipated and invisible threats.

Source: Appelgren et al., 2002.

Figure 11.4: Trends in food emergencies, 1981–2001

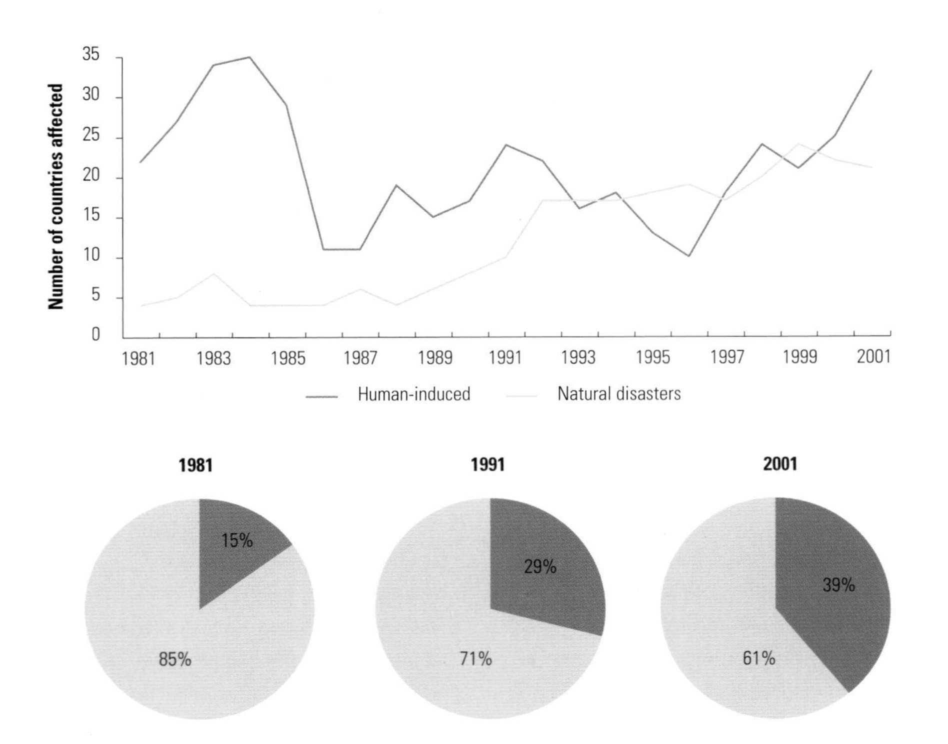

Human-induced disasters have taken an ever larger share in the causes of food emergencies in the last twenty years. This figure highlights the need for evolving towards wider social responsibility to control the causes of disasters.

Source: FAO web site, Global Information and Early Warning System on Food and Agriculture, 2001.

from improper sanitation, risks of chemical and toxic pollution from industrial sites, as wells as risks from accidental pollution of freshwater water bodies, coastal waters or aquifer systems. In addition to the sudden or relatively rapid-onset disasters, there are long-term environmental pollution risks such as accumulation of toxic agents in the water and the soil.

Data on the linkages between disasters and poverty are not always available, sometimes for political reasons. However, the statistics show that the victims of disasters in places where preparedness has been low are generally the poor and the marginalized, most of whom live in low-quality housing in flood-prone or drought-prone regions or along polluted watercourses. The poor are the most vulnerable to disasters, being exposed to the resulting health hazards but without the capacity to prepare for them or re-establish the life-supporting conditions after the catastrophes. Another tragic consequence is that flood and drought are also the main causes of poverty and the displacement and migration of poor populations.

Clearly, there is a close relationship between poverty eradication and the establishment of Comprehensive Disaster Risk Management (CDRM) strategies, which take account of economic, social and environmental dimensions in assessing risk and planning for preparedness measures in advance of water-related extremes. In making decisions about flood and drought management, it is important to broadly involve various stakeholders, even while devolving as much responsibility as is feasible to lower or community levels. There is an evident trend towards bringing together national and local governments, the private sector, non-governmental organizations (NGOs) and other representative groups of civil society in preparedness exercises. Such consultations build consensus about preparedness strategies and can help minimize risk. Encouraging public participation in self-protection strategies has been a successful strategy in some areas, as the Tokyo case study shows in chapter 22.

In many developing countries, women and female children are frequently the main providers of water for household use. Drought alleviation could reduce the annual expenditure of many million women years[2] of effort to carry water from distant sources. Women

2. A unit of measurement based on a standard number of woman-days in a year of work.

play a central part also in management and safeguarding of water, which makes it critical to involve them at all levels of the decision-making process. In some cultures, it is important to recognize that response to flood warning is largely gender-dependent: there are examples of married women ignoring flood warnings in the absence of their husbands.

Risk and uncertainty may exacerbate existing tensions in other ways too. In transition countries in particular, risk and uncertainty can worsen political differences and regional instability arising from disputes over shared water resources. In this case, sharing the risks becomes just as important as allocating of water resources and its associated benefits. Box 11.7 presents the example of common risk in water resources and the prospect for sharing the common risk between the non-stabilized post-Soviet economies in transition in Central Asia

Accountability, Monitoring and Indicators

In the present situation when the consequences of water-related disasters in all regions are severe, there is a need for practical indicators to support inter-regional comparison for the purpose of tracking trends and progress on management targets (policies) for reducing populations at risk from water-related hazards. In order to track progress and the sufficiency of policy, monitoring must be strengthened.

The indicators can be expected to foster country awareness and responsibility towards a coherent country-based approach to global and regional monitoring of water-related risk and uncertainty.

Table 11.4 gives examples of low-cost indicators focused on risk losses and progress of risk reduction. The country values of losses and benefits need to be adjusted for local income and fluctuations in local currency.

Box 11.7: Shared regional water resources in economies in transition

The risk for unexpected crises in transboundary water systems, where regulation is based on voluntary cooperation between sovereign states, is generally high and complicated to manage, even in stable and industrialized regions. In transitional and unstable regions and economies that are experiencing emerging and unanticipated risk, social instability and conflict, the need for a risk-based approach to the management of water resources is even more important.

In the post-Soviet central Asia subregion where most of the water resources are shared between the new states of Kazakhstan, Kyrgyzstan, Tajikistan, Turkmenistan and Uzbekistan, there is evidence of emerging undefined high risk and instability. This unstable and high-risk situation should allow for the development of joint management of common risk rather than engender further conflictive hydropolitical and development issues in the subregion. A joint risk-based approach for the management of the shared water resources will allow the central Asian countries to engage in joint preventive development and focus on identifying and managing emerging water risk to support and ensure regional stability.

Table 11.4: Examples of low-cost indicators focused on risk losses and progress of risk reduction

Risk losses **Past (10 years), present and future (25 years) losses (at different levels: basin, country, region, globally)**	**Risk reduction**
• in human life (number/year)	• legal and institutional provisions for risk-based management (established/not established)
• in real and relative social and economic values (total losses, percent of GNP, growth, investments and development benefits)	• budget allocation for water risk mitigation (total and percent of total budgets/year)
• population exposed to water-related risk (number of people/year, income groups)	• risk reduction in flood plains (percent of total flood plain populations)
• other than water-related risks (percent of losses from seismic, fire, industrial and civil stability risk)	• risk reduction and preparedness action plans formulated (percent of total number of countries)
	• risk-based resource allocation (country, international organizations, yes/no)

Source: Prepared for the World Water Assessment Programme (WWAP) by P.E. O'Connell; C.G. Kilsby; H.J. Fowler, 2001.

Conclusions

Global and local water-related risk is growing, resulting in an increasingly larger toll in human life and social, economic and environmental losses. Awareness of the enormous costs resulting from extreme events is encouraging decision-makers and water resource managers to give priority to the idea of integrating risk reduction and mitigation within normal management strategies.

The methodology and capacity to control natural disasters such as flood and drought have progressed in the last decades. However, as the trend is changing from natural to human-induced disasters and from known and manageable risk to higher uncertainty, the solutions based on engineering sciences are becoming increasingly *ad hoc*, fragmented and reactive and generally insufficient.

In the last decade, the debates on water governance and on management of water risk have increasingly been redirected to include ethical demands and management concepts based on political and economical realities. This change in governance responsibilities and attitudes has importance to the immediate victims of the disasters, who tend to be the poor and marginalized populations, but with special impact on women and children. The poor are exposed to the health hazards from natural and human-induced water disasters and lack individual capacity to prepare for or recover from them.

Agenda 21 emphasized the importance of mitigating floods and droughts. Eight years later, risk management, defined as 'managing risks to provide security from floods, droughts, pollution and water-related hazards', was given full recognition as one of the seven challenges listed at The Hague Ministerial Declaration in March 2000. The World Summit for Sustainable Development, held in Johannesburg in August/September 2002, reinforced this.

> *Combat desertification and mitigate the effects of drought and floods through such measures as improved use of climate and weather information and forecasts, early warning systems, land and natural resource management, agricultural practices and ecosystem conservation ... particularly in Africa, as one of the tools for poverty eradication* (*Plan of Implementation*, 2002).

The risk is likely to grow in the twenty-first century, heralded as the age of water scarcity, while flood losses also show a rising tendency. Increasing vulnerability to water-related disasters is due to growing exposure, which in many cases cannot be matched by an appropriate adaptive capacity.

As a consequence, there is a call for alternative and more sustainable risk management approaches. The steps forward need to be scheduled as a period of smooth transition from current integrated management to pragmatic and straightforward risk-based management. The initial steps should be focused on:

- adopting a required incremental improvement approach to handle water problems within existing legislation and institutional structures to not waste time and efforts in trying to achieve 'ideal' water management approaches, laws and institutions;

Progress since Rio at a glance

Agreed action	Progress since Rio
Encouraging countries to set feasible and quantifiable targets for reduction in water-related risks	
Facilitating the implementation of effective national countermeasures to climate change	
Recognizing the distinction between disasters resulting from natural hazards and those resulting from human action	
Improving collection of data and exchange information	
Improving forecasting systems, including early warning systems for the public	

Unsatisfactory — Moderate — Satisfactory

- changing educational and research approaches to pragmatic decision-making instead of searching for perfection based on unachievable data bases;
- inviting academics to build on the basis of the foundations they have constructed in IWRM, and allowing ethics to take a central role; and
- involving experts and officials with practical backgrounds in risk-based management from different sectors and disciplines (prevention and civil protection, legal and social medicine and public health, energy, financial and corporate sector, and the insurance and underwriting industry).

In order to achieve a more sustainable approach to managing risks, we must enhance recognition of water risks at political and economic level and entrench reduction and sharing of risk in development strategies. The management of risk and uncertainty to reduce social and stability uncertainty calls for integrated risk-based management approaches built on defined responsibilities at national and global levels.

References

Appelgren, B.; Burchi, S.; Garduno, H. 2002. 'No More Risk in Water Resources Management'. Discussion paper prepared for the Technical Consultation on Risk-Based Water Resources Management. Foggia, International Association on Water Law.

Cosgrove, B. and Rijsberman, F.-R. 2000. *World Water Vision: Making Water Everybody's Business.* London, World Water Council, Earthscan Publications Ltd.

CRED (Centre for Research on the Epidemiology of Disasters). 2002. The OFDA/CRED International Disaster Database. Brussels, Université Catholique de Louvain.

Delli Priscolli, J. and Llamas, R. 2001. 'Navigating Rough Waters: Ethical Issues in the Water Industry'. In: *International Perspectives on Ethical Dilemmas in the Water Industry.* Denver, American Water Works Association.

Dilley, M. 2001. 'The Use of Climate Information and Seasonal Prediction to Prevent Disasters'. Presented for the World Bank during the World Meteorological Organization Executive Council, Geneva.

Federal Ministry for the Environment, Nature Conservation and Nuclear Safety, and Federal Ministry for Economic Co-operation and Development. 2001. Ministerial Declaration, Bonn Keys, and Bonn Recommendation for Action. Official outcomes of the International Conference on Freshwater, 3–7 December 2001, Bonn.

GWP (Global Water Partnership). 2000a. *Integrated Water Resources Management.* Technical Advisory Committee background paper No. 4 (GWP-TAC4). Stockholm.

——. 2000b. *Towards Water Security: A Framework for Action.* Stockholm.

Guilhou, X. and Lagadec, P. 2002. *La fin du risque zéro* (The End of Zero Risk). Paris, Eyrolle.

Hounam, C.-E.; Burgosm, J.-J.; Kalik, M.-S.; Palmer, W.-C.; Rodda, J.-C. 1975. *Drought and Agriculture.* WMO technical note No. 138. Geneva, World Meteorological Organization.

IFRC (International Federation of Red Cross and Red Crescent Societies). 2002. *World Disasters Report.* Geneva.

IPCC (International Programme on Climate Change). 2002. *Climate Change 2001: Synthesis Report.* Geneva, International Programme on Climate Change Secretariat, World Meteorological Organization.

ISDR (International Strategy for Disaster Reduction). 2002. *Background Document on Natural Disasters and Sustainable Development.* Geneva.

Just, R.-E. and Netanyahu, S. 1998. *Conflict and Cooperation on Transboundary Water Resources.* Boston, Kluwer Academic Publishers.

Keith J. 1998. 'Economic Options for Managing Water Scarcity'. Paper presented during the FAO 2nd Virtual Conference on Water Scarcity, July 1998, Rome.

Kundzewicz, Z. 2001. 'Coping Capacity for Extreme Events'. Paper read at the International Conference on Freshwater, 3–7 December 2001, Bonn.

Melching, C. and Pilon, P. (eds.). 1999. Comprehensive Risk Assessment for Natural Hazards. WMO/TD No. 955. Geneva, World Meteorological Organization.

Ministerial Declaration of The Hague on Water Security in the 21st Century. 2000. Official outcome of the Second World Water Forum, 3–7 December 2001, The Hague.

Munich Re. 2001. *Topics, Annual Review: Natural Catastrophes 2000.* Munich.

Plan of Implementation. 2002 (Advance unedited). Official outcome of the World Summit for Sustainable Development (WSSD), 28 August–4 September 2002, Johannesburg.

Rees, J.-A. 2001. 'The Risk and Integrated Water Resources Management Paper'. Paper presented at the 9th International Conference on Conservation and Management of Lakes, 8–16 November 2001, Lake Biwa, Japan.

Swiss Re and Swiss Government. *Sustainable Development.* 2002. Document on risk mitigation for the World Summit on Sustainable Development, 28 August–4 September 2002, Johannesburg.

UN (United Nations). 1992. Agenda 21: Programme of Action for Sustainable Development. Official outcome of the United Nations Conference on Environment and Development (UNCED), 3–14 June 1992, Rio de Janeiro.

UNDHA (United Nations Department of Humanitarian Affairs). 1992. *Glossary: Internationally Agreed Glossary of Basic Terms Related to Disaster Management.* Geneva, United Nations Department of Humanitarian Affairs.

World Bank. 2001. *World Development Report 2000–1.* Washington DC.

WMO (World Meteorological Organization). 1999. *Final Report of the Scientific and Technical Committee of the International Decade for Natural Disaster Reduction.* Geneva.

Worldwatch Institute. 2001. *State of the World 2001.* New York, W.W. Norton & Co.

Worm, J. and de Villeneuve, C.-H.-V. 1999. 'Flood and Discharge Management: Rhine Action Plan on Flood Defence'. Selected contribution from the International Workshop on River Basin Management, The Hague, October 1999, International Hydrological Programme V. *Technical Documents in Hydrology.* No. 131. Paris, United Nations Educational, Scientific and Cultural Organization.

Some Useful Web Sites*

Secretariat of the United Nations Framework Convention on Climate Change (UNFCCC)

http://www.unfccc.int/

Part of the United Nations (UN) framework, this provides information relating to flood management and flood disaster reduction.

World Meteorological Organization (WMO)/Global Water Partnership (GWP), the Associated Programme on Flood Management (APFM)

http://www.wmo.ch/apfm/

Site promotes flood management within the context of integrated water resources management.

Secretariat of the United Nations Convention to Combat Desertification (UNCCD)

http://www.unccd.int/

Topics related to desertification, while staying in the context of sustainable development.

United Nations Environment Programme, Floods and Droughts

http://freshwater.unep.net/index.cfm?issue=water_flood_drought

Strategies, links, documents and other resources for coping with flood and drought.

United Nations International Strategy for Disaster Reduction (ISDR)

http://www.unisdr.org/

International site providing information, news events and training courses in order to increase awareness of the importance of disaster reduction.

* These sites were last accessed on 6 January 2003.

12

Sharing Water: Defining a Common Interest

By: UNESCO (United Nations Educational, Scientific and Cultural Organization)
Collaborating agencies: Regional Commissions

Table of contents

Sharing Water: Defining a Common Interest

'And you really live by the river? What a jolly life!'
'By it and with it and on it and in it,' said the Rat.
'It's brother and sister to me, and aunts, and company, and food and drink, and (naturally) washing. It's my world, and I don't want any other. What it hasn't got is not worth having, and what it doesn't know is not worth knowing.'

Kenneth Grahame, *The Wind in the Willows*

THERE WILL NEVER BE MORE WATER IN THE WORLD than there is right now. It is an endangered resource, essential to the everyday life of people and the planet in a plethora of ways. So many uses, so many demands. How can we accommodate everyone? The answer is simple – in theory, at least: we must share the resource. And in the best of all possible worlds, we would do so in a fair and equitable manner, ensuring that all needs were met. We would ensure that when water is withdrawn, it is also put back – in good condition – for others to use, even if they live in another country, or downstream. In practice, however, things are not so simple. This chapter analyses the issues involved and assesses whether we are making progress in sharing water: the balance sheet is mixed, but there are signs of hope.

WATER IS ESSENTIAL to national economic and social development, in the areas of health, food, industry and energy. As a resource that transcends most political and administrative boundaries, the world's available freshwater must be shared among and between individuals, economic sectors, intrastate jurisdictions and sovereign nations, while respecting the need for environmental sustainability. The challenges surrounding the equitable sharing of water resources are complex and have intensified in recent years due to population growth, development pressures and changing needs and values. There is already growing competition between different development sectors in countries, to varying degrees. This has placed increasing strain on freshwater supplies both in terms of quantity and quality, resulting in tensions and, indeed, conflict between uses, users and across political boundaries.

In any single country, the commitment to the Millennium Development Goals will increase water use in all key sectors, while the available water resource will remain the same. Thus, unless managed holistically with adequate sharing of water between sectors, each individual Millennium target involving water may not be achievable, to the detriment of those particular areas of economic or social development in most need of help and ecosystem integrity.

Furthermore, many countries rely on upstream member states for inflow, and therefore attainment of the Millennium Goals in any single country cannot depend solely on that water under the governance of national sovereignty alone. Attainment of targets and the associated people-centred outcomes by downstream countries will, to varying degrees, be dependent upon the actions of upstream countries. Conversely, in the case of longstanding arrangements for water sharing, later development of upstream countries may be constrained by precedents set for downstream use or existing agreements.

In recent years, as concern over the equitable allocation and sustainable development of scarce water stocks has heightened, efforts to improve the management of shared water resources have expanded. The Declaration of the Earth Summit in Rio in 1992, and the accompanying Agenda 21 Action Plan, for example, called upon the international community to recognize the multisector nature of water resources and to holistically manage the resource within and across national boundaries. Since then numerous policy and legal actions have been executed at all scales. These have included the passing of new national water laws, which have introduced or reinforced Integrated Water Resources Management (IWRM) techniques, the creation of national and international river basin organizations, and the adoption of international conventions, treaties and declarations concerning the management of freshwater supplies. Additionally, the concept of 'virtual water', which recognizes the sharing of water resources between water-rich and water-poor countries via trade in agricultural products and manufactured goods, has gained increasing attention in the post-Rio period.

Competition for Water within Countries

Analysis of total water abstraction as a proportion of river flow allows levels of water stress in a basin to be assessed. Map 12.1 shows the drainage basins that are considered to be under medium or high stress and the principal use of water in these regions.

There are many examples of how competition for scarce water within countries is increasing as a result of population growth and the need to satisfy social and economic development. These issues are of increasing importance in both the developed and developing world, and on a small and large scale. On a small scale, examples include Fiji, where indigenous peoples' water management, blended with ritual, is being threatened by the development of a government hydropower scheme, and the case of a small town in Colombia, where the community is experiencing the consequences of unfair allocation of water as illustrated in box 12.1.

On a larger scale, consider the case of China, which has achieved rapid economic and social development over the past twenty years, although the per capita resource is only a quarter of the global average. Through improvements in irrigated agriculture, China now provides food security for 22 percent of the world's population, using only 7 percent of the world's cultivated land. China has an annual average precipitation of 648 millimetres (mm), a runoff of 2,712 cubic kilometres (km^3), groundwater reserves of 829 km^3, and a total water resource volume that is sixth in the world at 2,812 km^3. Due to the huge population and area of farmland, the water volume per capita or per hectare is small. The resource is, in addition, unevenly distributed with the south benefiting from more water than the north, while there are large variations within and between years as well as in quality. In recent years, overall domestic and industrial water use in China has been steadily increasing, while irrigation water use is unchanged. The total volume of water use in the year 2000 reached 550 km^3, of which 58 km^3 was for domestic use, 114 km^3 for industry and 378 km^3 for irrigation. Today, industrialization is a main cause of the decrease in water availability and use for irrigation, as discussed in chapter 8. These factors have led to areas of intense competition between sectors, even within

Map 12.1: Drainage basins under high or medium stress and the sector that uses the most water

This map shows basins under medium or high water stress, spread throughout the world, but mainly concentrated in the northern hemisphere. Irrigation is clearly the main user of water withdrawals in these basins on a global scale, but in such areas as Europe and the East Coast of the United States, the industrial sector is the major cause of water stress.

Source: Map prepared for the World Water Assessment Programme (WWAP) by the Centre for Environmental Research, University of Kassel, based on Water Gap version 2.1. D.

'unstressed' basins with the result that the Chinese government has had to make great efforts to ensure the sustainable use of water resources. The Water Action Unit has now introduced the Nationwide Water Conservation Scheme. Other complementary measures include an integrated approach to flood control and disaster alleviation, water conservation and increased efficiency in all sectors, and better management practice in developing and allocating water resources.

Sri Lanka provides another example of competition for limited water resources. South-east Sri Lanka is less developed than much of the rest of the island and unemployment is a source of concern with regard to regional equity and political stability. The new Ruhunupura City and industrial area have been proposed to address this issue and will, if implemented, provide alternatives to the agriculture, fishing and tourism industries that currently dominate employment in the area. The development will, however, demand in excess of 100 million cubic metres (Mm^3) per year, which will have to be obtained by diversion from three main rivers, reducing water available to agriculture and affecting flows to wildlife resources and coastal fisheries. Balancing the demands of existing water users and the pressing needs surrounding the creation of new employment opportunities is expected to dominate water resource issues for several decades in the Ruhuna basins. For more details on this specific case, please refer to the Ruhuna basins case study in chapter 18.

Competition for water is not solely related to quantity, since adequate quantities are of little value if poor quality precludes the required use. In Europe, the development of statutory Water Quality Objectives and European Commission (EC) instruments such as the EC Surface Water Abstraction Directive have provided tools to protect different uses and to reflect increased intersectoral need for collaboration.

Water for ecosystems

Considerations of water consumption have thus far primarily dealt with human uses of water. How much water do the other species on this planet require? Our neglect of this need is discussed in

Box 12.1: Competition for water in Valle Province, Colombia

The small town of Felidia uses a mountain stream as the source of its piped gravity water supply system. The community has installed a multistage biological treatment system. The catchment area of the mountain stream is used for forestry and irrigated agriculture, to which fertilizers and pesticides are applied. The area is home to some 100 families who use the same stream for drinking water, tapping into it with either individual gravity systems or small group systems. They are not connected to the main gravity supply, as this would involve pumping. The catchment area is also becoming popular with people from the city of Cali (2 million inhabitants) for weekend recreation. Wealthy inhabitants have summer houses there.

The water and land use patterns of these different interest groups affect both the quality and the quantity of the water. The houses in the catchment area have latrines and pigsties draining directly into the river area. Soil erosion from land clearing for forestry and agriculture has increased the turbidity of the water to such an extent that the treatment system gets clogged. It is increasingly difficult to deal with the chemical and bacteriological pollution, and women in particular suffer from the poor quality of the water.

The rich summer-house owners use large volumes of water, primarily to fill swimming pools, and as the greatest beneficiaries of the flat water rate, they have so far resisted all attempts by the permanent population of the town to change the tariff system aided by their strong economic and political ties. The people most affected by the competition for resources are the women of the common town households who initiated and built the supply. Their water is inadequate in quantity and quality because, unlike weekend visitors, they require water seven days a week. There are no large reservoirs for storage and sedimentation and the low tariff prohibits enlarging the capacity of the scheme. Efforts are now being made to seek compromises through a more integrated watershed management in which all interest groups are involved.

Source: Van Wijk et al., 1996.

chapter 6 on the protection of ecosystems, but it is essential to bear in mind the needs of the environment when discussing the issues surrounding water sharing. One example of the cost of such neglect comes from the Florida Everglades in the United States. Channelling flow within the wetland to satisfy human needs for water supply and land development, has dried up the Everglades. Today values have changed and the costs of undoing these previous 'improvements' are being paid. A thirty-year agreement to ensure an adequate water supply for the restoration of the Everglades has now been signed by the federal and state governments. Water made available through the restoration plan will not be permitted for a consumptive use or otherwise made available by the state of Florida until such time as sufficient reservations of water for the restoration of the natural system are made. The US$7.8 billion Everglades restoration plan will restore about one million hectares of the Everglades ecosystem, providing the region with an additional 700 Mm^3 of freshwater per day. Estimates put the total cost of the restoration in excess of US$20 billion, but this value is similar in magnitude to the annual income from tourism in Florida, which would be significantly reduced if the environmental degradation continues.

Meanwhile, in South America it is proposed to convert the Paraguay-Paraná Hidrovia river system into an industrial shipping channel to expand agribusiness and mining activities. This could have irreversible impacts on the Brazilian Pantanal, the world's largest wetland area, and other valuable ecosystems in Argentina, Bolivia and Paraguay. The 300-member Ríos Vivos coalition of environmental, human rights and indigenous groups are opposing the project. Fortunately, at the same time a project funded by the Global Environment Facility (GEF) has resulted in the implementation of a detailed watershed management programme for the Pantanal and the Upper Paraguay River basin. Project activities will seek to enhance the environmental functioning of the system through the strengthening of basin institutions, capacity-building, public participation and integration of environmental concerns into economic development activities.

In Africa, the Okavango River Basin Commission (OKACOM), with GEF support, has conducted a transboundary diagnostic assessment that will lead to a strategic environmental programme to improve management of the basin's resources. The objectives of this project are firstly to develop methodologies to promote sustainable use of natural resources in critical areas, and secondly to develop methodologies to promote rehabilitation and conservation of ecosystem functions. At the same time, the project will strengthen capacity of the concerned central and local government bodies. Emphasis will be placed upon sustainability of the project by fully taking into account the socio-economic needs of local populations.

In Europe, the Danube River is in a watershed that includes eighteen countries – more than any transboundary basin on the planet. The long-term development objective is sustainable human development in the basin and the wider Black Sea area. This is to be done through reinforcing the capacities of the participating countries in developing effective mechanisms for regional cooperation and coordination, in order to ensure protection of international waters, sustainable management of natural resources and biodiversity. The overall objective of the Danube Regional Project supported by GEF is to complement the activities of the International Commission for the Protection of the Danube River (ICPDR). It will provide a regional approach to the development of national policies and legislation, as well as defining priority actions for nutrient reduction and pollution control. Particular attention will be paid to achieving sustainable transboundary ecological improvements within the Danube River basin and the Black Sea area.

Australians, too, are recognizing the need for better management of water. The Commonwealth Government's Natural Heritage Trust is funding Waterwatch – a national community water monitoring programme. Waterwatch enables Australians to become involved in the monitoring and management of waterways in their local catchment. It aims to build community understanding of water quality issues, and to encourage monitoring groups to undertake constructive actions to rectify water quality problems. Waterwatch is addressing issues for achieving healthy waterways through activating community groups and individuals in the protection and management of waterways: through increasing awareness, promoting community monitoring of local waterways and involvement in planning of waterway issues, thus realizing effective partnerships between all relevant sectors.

Activities such as those described above will eventually ensure that the environment is fully included in considerations of water sharing.

Mechanisms for Sharing Water within Countries

While competition for limited resources has increased rapidly in recent years, countries already have a large number of operational mechanisms for sharing. These measures address both sharing the routine bulk water and the limited resource available during times of shortage. Some examples include the following:

- National-level strategies and/or legislation on intersectoral allocation, which may be based on:
  - catchment socio-economic priorities (which are left vague so as to be determined at a local level, but the principle is established) as for example in Zimbabwe;
  - legislatively predefined priorities (including reserves for human consumption and environment), as for example in South Africa, or a hierarchy of uses (with for example industry and mining rated above agriculture) as in Zimbabwe; and
  - demand management (during shortages) targeted at certain sectors in priority.

- Harmonization of sectoral policies and laws. Examples where this approach has been implemented include Japan, where the Water Act was revised years ago to incorporate environmental issues through an intersectoral committee (see chapter 22 for more information). Also the Southern African Development Community (SADC) Protocol on Shared Watercourses involved harmonization of national laws with wider regional goals.

- Tariff disincentives and targeted subsidies as an economic influence on sharing through pricing differentials for different purposes. A recent example of this approach comes from China, where summer water prices have been tripled in Beijing in an attempt to reduce consumption.

- Abstraction management (issuance of permits/licences) to limit water use for particular purposes or during particular seasons, often with explicit conditions limiting abstraction when certain predefined conditions prevail (often to protect other downstream abstractions). This may also involve water management agencies recovering previously allocated but unused water.

- River water quality objectives (established on particular river systems), which through discharge/treatment/quality standards, ensure water of quality fit for downstream purposes.

- Prescribed flow points, whereby operations of a water management agency are set to maintain flow at particular river points in support of downstream uses, including the environment.

- Reservoir operating rules (often optimized to meet different purposes), multireservoir system management (optimized to meet different demands) and reservoir compensation flow releases.

Box 12.2: Drought conciliation in Japan

The volume of river flows in Japan varies significantly because of climate and topography, and water use from rivers was historically limited to a steady low level throughout the year. Massive population concentration and industrial development in major cities in the middle of the 1960s, however, promoted water resources development that focused on stabilizing river flow to meet new demands. Success in achieving a stable water supply accelerated the pace of economic growth and population drift to the cities, which in turn has created new demand and increased water use. Droughts thus pose an increasingly serious problem as the number of stakeholders increases.

When water resource development cannot keep up with demand, as is the case in some areas, new water abstractions are permitted only when extra river flow is sufficient to cover existing use. New abstractions are, however, generally for public drinking water supply, and although in principle should stop during severe drought, in reality must continue. This exacerbates water shortages during droughts.

Drought conciliation is therefore essential and drought conciliation councils have been established in many river basins. In theory, the conciliation process takes place among the water users themselves. However, conflicting interests among water users often give rise to difficulties with the procedure and a river administrator is usually required to facilitate the process, by providing information, drought conciliation proposals and advice as required.

Source: Prepared for the World Water Assessment Programme (WWAP) by the National Institute for Land and Infrastructure Management (NILIM) and Ministry of Land, Infrastructure and Transport (MLIT) of Japan, 2002.

Box 12.3: New technology helps share the resource in Saudi Arabia

Saudi Arabia is located in an arid region where the average annual rainfall ranges from 25 mm to 150 mm, and the average potential annual evaporation from 2,500 mm to about 4,500 mm. Faced with a growing population and large-scale industrial development, the country has begun to consider desalination as an increasingly viable option for meeting water needs.

In addition, the increase in oil revenues from 1975 enabled ambitious development in social, industrial and agricultural sectors, prompting a parallel increase in water demands. Domestic water demand rose from 6 percent of the country's total water use in 1990 to 10 percent in 2000, due to growth in the urban population. The larger revenues also allowed the government to use irrigated agriculture to support developments in rural areas and to settle nomads into prosperous agricultural communities. The cultivated area was thereby massively expanded, and irrigation water consumption still represents the large majority of national water use. Industrial water demands have also grown rapidly during the last two decades with significant industrial developments.

Even though the burden is shared between desalination plants and renewable groundwater resources, the dependence on non-renewable groundwater has increased. One of the alternatives being developed is the reuse of treated wastewater. Also, desalination technologies are being increasingly exploited and today show great promise as a partial solution to the country's struggle for water. Large plants have been constructed on the Gulf and Red Sea coasts to produce suitable drinking water, and pipelines constructed to transport the desalinated seawater to coastal and inland cities and towns. In 1997, a total of thirty-five desalination plants were in place, with capacities representing about 33 and 38 percent of the total domestic and industrial demands, respectively. By 2025, desalinated water is expected to supply about 54 percent of these demands. The new technology is, in effect, helping to transport the resource to those areas that need them, thus sharing it across the various sectors.

Source: Prepared for the World Water Assessment Programme (WWAP) by A. Abderrahman, Water Resources Management, King Fahd University of Petroleum and Minerals, 2002.

Such practical mechanisms are usually adopted by most, if not all countries, for intersectoral sharing. They represent a significant operational interface at a level below IWRM. The absence of assessment of the uptake of one or more of this suite of mechanisms precludes any authoritative conclusion on intersectoral sharing in practice. An example of a sharing mechanism used under drought conditions in Japan is described in box 12.2.

Not only are there institutional methods for facilitating the sharing of water resources, but some countries have implemented new technologies, such as desalination and vast pipelines, which enable the resource to be shared amongst regions and sectors alike. Box 12.3 provides details on how Saudi Arabia is attenuating its water scarcity problems, and denotes a commonly overlooked issue of sharing between sectors, as opposed to between countries.

Integrated Water Resources Management (IWRM)

Equitable and sustainable management of the world's shared water systems requires flexible, holistic institutions capable of responding to hydrological variation and changes in socio-economic needs, political regimes and societal values. Unfortunately, these and other key management components are noticeably absent from many national and international water institutions. Figure 12.1 provides a schema of the power and authority of different kinds of institutional mechanisms.

At the operational level, the challenge is to translate agreed principles into concrete action, while at a basic level the mechanisms discussed in the previous section provide a means of achieving this. The higher-level strategic response is often referred to as IWRM in contrast to 'traditional', fragmented water resources management. The concept of IWRM is widely debated and hence, regional or national institutions must develop their own IWRM practices using the collaborative framework that is emerging globally and regionally. The Global Water Partnership (GWP) has defined IWRM as 'a process which promotes the coordinated development and management of water, land and related resources to maximize the resultant economic and social welfare in an equitable manner without compromising the sustainability of vital ecosystems'.

The concept of IWRM at its most fundamental level is as concerned with the management of water demand as with its supply. Thus, integration can be considered in two basic categories, and must occur both within and between these, taking into account variability in time and space based on:

- the natural system, with its critical importance for resource availability and quality, and the wide range of environmental services that it provides; and
- the human system, which fundamentally determines resource use, waste production and pollution, and which must also set the development priorities.

While river basin organizations go some way towards meeting this goal, alone they are inadequate to deliver the full range of economic and social benefits associated with IWRM, which in reality has to be addressed by national or international laws. There is competition between demands for different uses of water, and geographic competition between upstream and downstream users in both of the above categories. Historically, water managers have tended to see themselves in a neutral role, managing the natural system to provide supplies to meet externally determined needs. IWRM approaches should assist them in recognizing that their behaviour

Figure 12.1: Power and authority of different institutional mechanisms

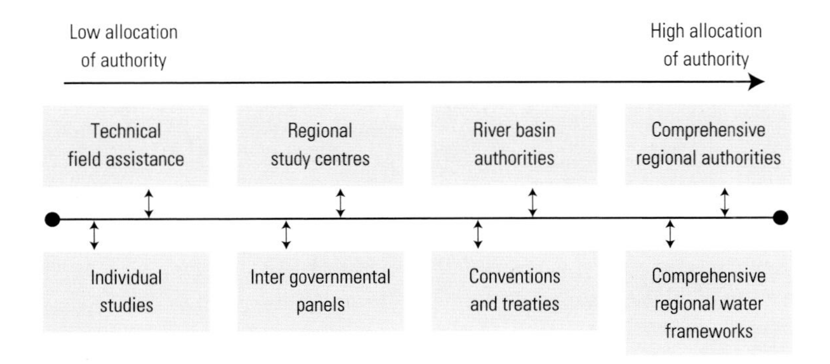

Source: Prepared for the World Water Assessment Programme (WWAP) by J. Delli Priscoli, 2001.

Box 12.4: Sharing water resources in the Seine-Normandy basin

The level of the Beauce water table, which is mainly used for irrigation, has dropped considerably since the beginning of the 1990s, provoking in particular the drought of its river outlets. To counter this problem, several solutions were attempted, resulting in an output that seems to be efficient and which has promoted stakeholder participation.

- **1993 - 1994:** irrigation was totally prohibited several days per week, during the summer. Irrigators did not accept these restrictions, in particular as the system was geographically inconsistent and they reacted by increasing abstraction on authorised days.

- **1995:** the Beauce Charter was passed between the administration and the irrigators' representative. This charter allowed different restrictions for irrigation, based on a comparison of the mean level of the water table (derived from representative piezometers) with three historically defined warning thresholds. Monitoring of groundwater levels allowed irrigation to be prohibited several days per week as required and permitted public awareness campaigns to be organized. This approach also proved unsuitable.

- **1999:** a smoother system was implemented which attributed a global volume to irrigation, negotiated with the irrigators' representative body, on the basis of the mean level of the water table. This volume was then shared between the administrative zones, each with its own rules for allocating to irrigators. A volumetric quota is attributed to each irrigator, depending on the location and the area of their irrigated land, and sometimes their crops and stockbreeding. The use of a hydrogeological model should improve the definition of the different parameters.

- **2002:** the water table is also soon to be classified as a 'water-sharing area', which will limit the volume withdrawn to a maximum of 8 m^3/h instead of the usual 80 m^3/h limit. In parallel, as proposed by the 1992 French Water Act, a local management plan (SAGE) has been in development for the water table since 1998 using a collaborative approach that will improve its suitability and acceptance.

Sources: Le Coz, 2000; Mouray and Vernoux, 2000.

also affects water demands. In effect, if IWRM were adopted as a standard approach everywhere, it would help to foresee and avoid issues that could create conflicts.

Chapter 15 on water governance discusses the issues of institutions at some length. Water managers generally understand and advocate the inherent powers of the concept of the watershed as a unit of management where surface and groundwater, quantity and quality, are all intrinsically connected and related to land use management. However, the institutions that have historically followed these tenets are the exception. One lesson that has been learnt about building institutions is that it is better to begin with those that cause the least disruption to existing power structures. Sharing information and cooperation at these levels will eventually lead to the desired institutions with the needed authority. This lesson is linked to a second: that the higher the degree of participation by all interested parties, the more sustainable the resultant institutional frameworks.

Adaptable management structures, clear and flexible water allocation, water quality management criteria and equitable benefit distribution further contribute to successful and sustainable institutions. An example of this process of increasing involvement of stakeholders and evolution of the policy framework is that of the management of the waters of the Beauce aquifer in the Seine-Normandy water basin in France (box 12.4).

Transboundary Water Management

The management of water resources across national boundaries strains the capabilities of institutions. Rarely do the boundaries of watersheds correspond to the existing administrative boundaries. Regional politics can exacerbate the already formidable task of understanding and managing complex natural systems, and disparities between riparian states – such as level of economic development, infrastructure capacity, political orientation or cultural values – can complicate the development of joint management structures.

Since the 1992 United Nations Conference on Environment and Development (UNCED) in Rio de Janeiro, the international community has expanded its involvement in the management of international freshwater resources. Actions taken have included the pronouncement of non-binding declarations, the creation of global water institutions and the codification of international water law principles. While more work is clearly required, these initiatives not only have raised awareness of the myriad issues related to international water resource management, but also have led to the creation of frameworks in which the issues can be addressed.

Box 12.5: Expanded freshwater programmes

One result of the Rio Conference and Agenda 21 has been an expansion of international freshwater resource institutions and programs. The World Water Council, a 'think tank' for world water resource issues was, for example, created in 1996 in response to recommendations from the Rio Conference. The World Water Council sponsored the First and Second World Water Forums that brought together government, non-government and private agency representatives to discuss and collectively determine a vision for the management of water resources over the next quarter century. The Council also sponsored an independent World Water Commission that guided the preparation of the vision.

At the Second Forum there was intensive discussion of the World Water Vision, a forward-looking declaration of philosophical and institutional water management needs. The Second World Water Forum also served as the venue for a Ministerial Conference in which the leaders of participating countries signed a declaration concerning water security in the twenty-first century. GWP, also established in 1996, has developed regional and national partnerships that facilitate actions 'on the ground'. The United Nations Millennium Assembly set targets for the provision of safe drinking water to the world population and called for improved management of water resources. These recent global water initiatives have, in addition, been supported by a number of interim appraisal meetings to review actions taken since the Rio Conference and the World Summit on Sustainable Development in Johannesburg. The Water Forum in Kyoto in 2003 is the most recent expression of this momentum.

Conventions and declarations

The Rio Declaration on Environment and Development and Agenda 21 outlined a set of principles, objectives and a related action plan for improving the state of the globe's natural resources in the twenty-first century. Water resources were given specific attention in Chapter 18 of Agenda 21, the overall goal of which is to ensure that the supply and quality of water is sufficient to meet both human and ecological needs worldwide. Measures to implement this objective are detailed in the Chapter's ambitious seven-part action plan for managing and protecting global freshwater resources.

The international community has reinforced its commitment to satisfy the water quality and quantity requirements of the global population and its surrounding environment and has identified attendant tasks and policy measures needed to fulfil its pledge (see box 12.5). While many of the strategies in Agenda 21 and subsequent statements are directed primarily at national water resources, their relevance extends to transboundary waters. In fact, the Ministerial Declaration at the Second World Water Forum included 'sharing water' (between different users and states) as one of its seven major challenges to achieving water security in the twenty-first century. Many of the other six challenges are also applicable to waters in international settings. In addition, policy measures prescribed by the international community to build greater institutional capacity, such as the creation of basin-level organizations applying the principles of IWRM, expanded stakeholder participation, and improved monitoring and evaluation schemes, are important examples of international water resource management.

None of these statements or declarations, however, focuses exclusively on international freshwater resources. And despite the efforts over the past decade to expand global institutional capacity over freshwater resources, no intergovernmental agency exists to facilitate management of transboundary resources. Thus, while many of the principles of national water management apply to the domestic component of international waters, the political, social and economic dynamics associated with waters shared between sovereign states require special consideration. National water courts exist in some countries, although it is difficult to see how these could serve as a model for management at an international level. In Spain there exists only one such court, el Tribunal de las Aguas de la Vega de Valencia, which has no overarching laws to guide it but judges on a case-by-case basis. Italy has a system of statutory water courts to hear and adjudicate certain water-related disputes. By contrast, the 'water courts' of Nordic European countries do not act in a judicial capacity but play the role of governmental institutions dealing with water.

Box 12.6: International public law related to the non-navigational uses of shared water resources

Article 38 of the statutes of the ICJ lists the sources of international law which govern the relations between sovereign states, including those related to the use of water resources. These are:

- International conventions:
  - The Helsinki Convention (1996) for the Protection and Use of Transboundary Watercourses and International Lakes obliges Parties to prevent, control and reduce water pollution from point and non-point sources. It includes provisions ensuring that transboundary waters are used in a reasonable and equitable way, provisions for monitoring, research and development, consultations, warning and alarm systems, mutual assistance, institutional arrangements, information exchange and public access to information.
  - The UN Convention on the Law of the Non-navigational Uses of International Watercourses (1997) is not yet in force, but represents the opinion of the leading experts and has an authoritative function. Above all it indicates broad agreement among states on the general principles relating to equitable and reasonable resource use, the duty not to cause significant harm, ecosystem protection, management obligations, information sharing, conflict resolution and the protection of the resource during armed conflicts.
- International custom: customary international water law provides a fairly 'global' set of rules, which highlight the following main principles:
  - the principle of restricted territorial sovereignty;
  - the prohibition of 'substantial injury in the territory of a co-basin state' and the obligation of the responsible state to compensate the suffering state;
  - the principle of the equitable and reasonable use and share; and
  - the pre-eminence of the last principle over the previous one.
- General principles of law recognized by civilized nations: the following general principles were adapted to the use of international water resources and became valid in this field as one source of international water law:
  - the principle of equity, meaning that international law does not operate in favour of any particular state or group of states;
  - the principle *Sic utero tuo ut alineum non laedas*, which means that a state's right to use shared waters is limited by the rights of the co-basin states to use the resources of the same watercourse without being harmed in a significant manner;
  - the principle of equitable apportionment, which entitles every basin state to an equitable and reasonable share of an international watercourse;
  - the principle of reciprocity according to which, when a state acts with its rights and obligations set by international law, it expects the same conduct from other states;
  - the obligation to settle disputes peacefully; and
  - the harmonious application of national laws in case of conflict between these.
- Judicial decisions and the teachings of the most highly qualified publicists: one of the most important rules of the judicial decisions related to the non-navigational uses of international watercourses reflects the principle of limited territorial sovereignty. This principle has also been recognized by highly qualified publicists as the most adequate rule applicable to the use of shared watercourses, among the other principles mentioned above.

Legal principles

The United Nations (UN) Convention on the Law of the Non-navigational Uses of International Watercourses, adopted in 1997 by the UN General Assembly following twenty-seven years of discussion and negotiations, is one post-Rio accomplishment that specifically focuses on transboundary water resources. It codifies many of the principles deemed essential by the international community for the management of shared water resources, such as equitable and reasonable utilization of waters with specific attention to vital human needs, protection of the aquatic environment and the promotion of cooperative management mechanisms. The document also incorporates provisions concerning data and information exchange and mechanisms for conflict resolution. If ratified, the UN Convention would provide a legally binding framework, at least upon its signatories, for managing international watercourses. Even without ratification, its guidelines are being increasingly invoked in international forums.

The UN's approval of the Convention, however, does not entirely resolve many legal questions concerning the management of internationally shared waters. First, the Convention would technically only be binding on those nations that sign and ratify it. Five years after its adoption by the UN General Assembly, the Convention has only been signed by sixteen countries[1] and ratified by nine, well below the requisite thirty-five instruments of ratification needed to bring the Convention into force, and it therefore has no legal status. Second, international law only guides conduct between sovereign nations, and cannot address the grievances of political or ethnic units within nations. Third, while the Convention offers general guidance to coriparian states, its vague and occasionally contradictory language can result in varied, and indeed conflicting, interpretations of the principles contained therein. For example, during the negotiations that preceded the adoption of the Convention, the wording of Article 7 devoted to the principle of 'non-significant harm' and its relation with Articles 5 and 6, both devoted to the principle of 'reasonable and equitable use', provoked one of the major confrontations between upstream and downstream countries. The final wording of Article 7 can be interpreted in favour of both upstream and downstream countries. Each group has considered that this new formulation was strong enough to support its allegations and accepted it, although the opposing group, which accepted it for the same reason, interpreted it in the exact opposite way. A fourth legal issue left unresolved by the Convention is that there is no practical enforcement mechanism to back up the Convention's guidance. The International Court of Justice (ICJ), for example, hears cases only with the consent of the parties involved and only on very specific legal points. Moreover, in its fifty-five-year history, the Court has decided only one case[2] pertinent to international waters – that of the Gabèikovo-Nagymaros System on the Danube between Hungary and Slovakia in 1997. Finally, the Convention addresses only those groundwater bodies that are connected to surface water systems – i.e. shallow, unconfined aquifers. Several nations are beginning to tap into deep and/or confined groundwater systems, many of which are shared across international boundaries. A more detailed summary of international public law related to shared water resources and the growing number of legal principles that are beginning to be accepted, is provided in box 12.6.

Developments in basin-level transboundary water management

A closer look at the world's international basins gives a greater sense of their significance in terms of area and conflict potential. There were 214 transboundary basins listed in 1978 (UN, 1978), the last time any official body attempted to delineate them, and there are 263 today (see table 12.1). The growth is largely the result of the 'internationalization' of national basins through political changes, such as the break-up of the Soviet Union and the Balkan States, as well as access to better mapping sources and technology.

Even more striking than the total number of basins is a breakdown of each nation's land surface that falls within these watersheds. A total of 145 nations include territory within transboundary basins. Twenty-one nations lie entirely within transboundary basins, and an additional twelve countries have greater than 95 percent of their territory within one or more transboundary basins. These nations are not limited to small countries, such as Andorra and Liechtenstein, but also include, for example, Bangladesh, Belarus, Hungary and Zambia.

Beyond their importance in terms of surface and political area consumed, a look at the number of countries that share individual watercourses highlights the precarious setting of many international basins. Approximately one third of the 263 transboundary basins are shared by more than two countries, and nineteen involve five or more sovereign states. Of these, one basin – the Danube – has eighteen riparian nations. Five basins – the Congo, Niger, Nile, Rhine and Zambezi – are shared by between nine and eleven countries. The remaining thirteen basins – the Amazon, Aral Sea, Ganges-Brahmaputra-Meghna, Jordan, Kura-Araks, Lake Chad, Mekong, Neman, La Plata, Tarim, Tigris-Euphrates and Vistula (Wista) – have between five and eight riparian countries.

1. As of 12 June 2002, the countries having signed the Convention were: Côte d'Ivoire, Finland, Germany, Hungary, Jordan, Luxembourg, Namibia, Netherlands, Norway, Paraguay, Portugal, South Africa, Syrian Arab Republic, Tunisia, Venezuela and Yemen.
2. The International Court of Justice was established in 1946 with the dissolution of its predecessor agency, the Permanent Court of International Justice. This earlier body did rule on four international water disputes during its existence from 1922–46.

Table 12.1: Transboundary river basins

Basin name	Total area of basin (km^2)	Countries	Area of country in basin (km^2)	Area of country in basin (%)
AFRICA				
Akpa	4,900	Cameroon	3,000	61.65
		Nigeria	1,900	38.17
Atui	32,600	Mauritania	20,500	62.91
		Western Sahara	11,200	34.24
Awash	154,900	Ethiopia	143,700	92.74
		Djibouti	11,000	7.09
		Somalia	300	0.16
Baraka	66,200	Eritrea	41,500	62.57
		Sudan	24,800	37.43
Benito/Ntem	45,100	Cameroon	18,900	41.87
		Equatorial Guinea	15,400	34.11
		Gabon	10,800	23.86
Bia	11,100	Ghana	6,400	40.28
		Côte d'Ivoire	4,500	57.58
Buzi	27,700	Mozambique	24,500	88.35
		Zimbabwe	3,200	11.65
Cavall	30,600	Côte d'Ivoire	16,600	54.12
		Liberia	12,700	41.66
		Guinea	1,300	4.22
Cestos	15,000	Côte d'Ivoire	2,200	14.91
		Liberia	12,800	84.99
		Guinea	20	0.11
Chiloango	11,600	Congo, Democratic Republic of (Kinshasa)	7,500	64.60
		Angola	3,800	32.71
		Congo, Republic of the (Brazzaville)	300	2.69
Congo/ Zaire	3,691,000	Congo, Democratic Republic of (Kinshasa)	2,302,800	62.39
		Central African Republic	400,800	10.86
		Angola	290,600	7.87
		Congo, Republic of the (Brazzaville)	248,100	6.72
		Zambia	176,000	4.77
		Tanzania, United Republic of	166,300	4.51
		Cameroon	85,200	2.31
		Burundi	14,400	0.39
		Rwanda	4,500	0.12
		Sudan	1,400	0.04
		Gabon	500	0.01
		Malawi	100	0.00
		Uganda	70	0.00
Corubal	24,000	Guinea	17,500	72.71
		Guinea-Bissau	6,500	27.02
Cross	52,800	Nigeria	40,300	76.34
		Cameroon	12,500	23.66
Cuvelai/ Etosha	167,400	Namibia	114,100	68.15
		Angola	53,300	31.85
Daoura	34,500	Morocco	18,200	52.72
		Algeria	16,300	47.28
AFRICA				
Dra	96,400	Morocco	75,800	78.65
		Algeria	20,600	21.33
Gambia	69,900	Senegal	50,700	72.48
		Guinea	13,200	18.92
		Gambia	5,900	8.51
Gash	40,000	Eritrea	21,400	53.39
		Sudan	9,600	24.09
		Ethiopia	9,000	22.52
Geba	12,800	Guinea-Bissau	8,700	67.69
		Senegal	4,100	31.88
		Guinea	50	0.42
Great Scarcies	12,100	Guinea	9,000	74.96
		Sierra Leone	3,000	25.04
Guir	78,900	Algeria	61,200	77.53
		Morocco	17,700	22.47
Incomati	46,700	South Africa	29,200	62.47
		Mozambique	14,600	31.20
		Swaziland	3,000	6.33
Juba-Shibeli	803,500	Ethiopia	367,400	45.72
		Somalia	220,900	27.49
		Kenya	215,300	26.79
Komoe	78,100	Côte d'Ivoire	58,300	74.67
		Burkina Faso	16,900	21.66
		Ghana	2,200	2.85
		Mali	600	0.82
Kunene	110,000	Angola	95,300	86.68
		Namibia	14,700	13.32
Lake Chad	2,388,700	Chad	1,079,200	45.18
		Niger	674,200	28.23
		Central African Republic	218,600	9.15
		Nigeria	180,200	7.54
		Algeria	90,000	3.77
		Sudan	82,800	3.47
		Cameroon	46,800	1.96
		Chad, claimed by Libya	12,300	0.51
		Libya	4,600	0.19
Lake Natron	55,400	Tanzania, United Republic of	37,100	67.00
		Kenya	18,300	33.00
Lake Turkana	206,900	Ethiopia	113,200	54.69
		Kenya	89,700	43.36
		Uganda	2,500	1.21
		Sudan	1,500	0.70
		Sudan, administered by Kenya	70	0.03
Limpopo	414,800	South Africa	183,500	44.25
		Mozambique	87,200	21.02
		Botswana	81,500	19.65
		Zimbabwe	62,600	15.08
Little Scarcies	18,900	Sierra Leone	13,000	69.12
		Guinea	5,800	30.88

Table 12.1: *continued*

Basin name	Total area of basin (km²)	Countries	Area of country in basin (km²)	Area of country in basin (%)
AFRICA				
Loffa	11,400	Liberia	10,100	88.56
		Guinea	1,300	11.38
Lotagipi Swamp	38,700	Kenya	20,300	52.33
		Sudan	9,900	25.54
		Sudan, administered by Kenya	3,300	8.52
		Ethiopia	3,200	8.32
		Uganda	2,100	5.29
Mana-Morro	6,800	Liberia	5,700	82.84
		Sierra Leone	1,200	17.16
Maputo	30,700	South Africa	18,500	60.31
		Swaziland	10,600	34.71
		Mozambique	1,500	4.98
Mbe	7,000	Gabon	6,500	92.97
		Equatorial Guinea	500	7.02
Medjerda	23,100	Tunisia	15,600	67.53
		Algeria	7,600	32.90
Moa	22,500	Sierra Leone	10,800	47.79
		Guinea	8,800	39.20
		Liberia	2,900	13.01
Mono	23,400	Togo	22,300	95.19
		Benin	1,100	4.81
Niger	2,113,200	Nigeria	561,900	26.59
		Mali	540,700	25.58
		Niger	497,900	23.56
		Algeria	161,300	7.63
		Guinea	95,900	4.54
		Cameroon	88,100	4.17
		Burkina Faso	82,900	3.93
		Benin	45,300	2.14
		Côte d'Ivoire	22,900	1.08
		Chad	16,400	0.78
		Sierra Leone	50	0.00
Nile	3,031,700	Sudan	1,927,300	63.57
		Ethiopia	356,000	11.74
		Egypt	272,600	8.99
		Uganda	238,500	7.87
		Tanzania, United Republic of	120,200	3.96
		Kenya	50,900	1.68
		Congo, Democratic Republic of (Kinshasa)	21,400	0.71
		Rwanda	20,700	0.68
		Burundi	12,900	0.43
		Egypt, administered by Sudan	4,400	0.15
		Eritrea	3,500	0.12
		Sudan, administered by Egypt	2,000	0.07
		Central African Republic	1,200	0.04
Nyanga	12,300	Gabon	11,500	93.56
		Congo, Republic of the (Brazzaville)	800	6.44
AFRICA				
Ogooue	223,000	Gabon	189,500	84.98
		Congo, Republic of the (Brazzaville)	26,300	11.79
		Cameroon	5,200	2.34
		Equatorial Guinea	2,000	0.89
Okavango	706,900	Botswana	358,000	50.65
		Namibia	176,200	24.93
		Angola	150,100	21.23
		Zimbabwe	22,600	3.19
Orange	945,500	South Africa	563,900	59.65
		Namibia	240,200	25.40
		Botswana	121,400	12.85
		Lesotho	19,900	2.10
Oued Bon Naima	500	Morocco	300	65.08
		Algeria	200	34.92
Oueme	59,500	Benin	49,400	82.98
		Nigeria	9,700	16.29
		Togo	400	0.73
Ruvuma	151,700	Mozambique	99,000	65.27
		Tanzania, United Republic of	52,200	34.43
		Malawi	400	0.30
Sabi	115,700	Zimbabwe	85,400	73.85
		Mozambique	30,300	26.15
Sassandra	68,200	Côte d'Ivoire	59,800	87.64
		Guinea	8,400	12.36
Senegal	436,000	Mauritania	219,100	50.25
		Mali	150,800	34.59
		Senegal	35,200	8.08
		Guinea	30,800	7.07
St. John (Africa)	15,600	Liberia	12,900	83.04
		Guinea	2,600	16.96
St. Paul	21,200	Liberia	11,800	55.75
		Guinea	9,400	44.25
Tafna	9,500	Algeria	7,000	74.39
		Morocco	2,400	25.60
Tano	15,600	Ghana	13,700	87.96
		Côte d'Ivoire	1,700	11.21
Umba	8,200	Tanzania, United Republic of	6,800	83.58
		Kenya	1,300	16.41
Umbeluzi	10,900	Mozambique	7,200	65.87
		Swaziland	3,500	32.44
		South Africa	30	0.27
Utamboni	7,700	Gabon	4,500	58.65
		Equatorial Guinea	3,100	40.40
Volta	412,800	Burkina Faso	173,500	42.04
		Ghana	166,000	40.21
		Togo	25,800	6.26
		Mali	18,800	4.56
		Benin	15,000	3.63
		Côte d'Ivoire	13,500	3.27

Table 12.1: *continued*

Basin name	Total area of basin (km^2)	Countries	Area of country in basin (km^2)	Area of country in basin (%)
AFRICA				
Zambezi	1,385,300	Zambia	576,900	41.64
		Angola	254,600	18.38
		Zimbabwe	215,500	15.55
		Mozambique	163,500	11.81
		Malawi	110,400	7.97
		Tanzania, United Republic of	27,200	1.97
		Botswana	18,900	1.37
		Namibia	17,200	1.24
		Congo, Democratic Republic of (Kinshasa)	1,100	0.08
ASIA				
Amur	2,085,900	Russian Federation	1,006,100	48.23
		China	889,100	42.62
		Mongolia	190,600	9.14
		Korea, Democratic People's Republic of (North)	100	0.01
An Nahr Al Kabir	1,300	Syria	900	67.60
		Lebanon	400	31.70
Aral Sea	1,231,400	Kazakhstan	424,400	34.46
		Uzbekistan	382,600	31.07
		Tajikistan	135,700	11.02
		Kyrgyzstan	111,700	9.07
		Afghanistan	104,900	8.52
		Turkmenistan	70,000	5.68
		China	1,900	0.15
		Pakistan	200	0.01
Asi/Orontes	37,900	Turkey	18,900	49.94
		Syria	16,800	44.32
		Lebanon	2,200	5.74
Astara Chay	600	Iran	500	81.64
		Azerbaijan	100	18.36
Atrak	34,200	Iran	23,600	68.86
		Turkmenistan	10,700	31.14
BahuKalat/ Rudkhanehye	18,000	Iran	18,000	99.83
		Pakistan	30	0.17
Bangau	60	Brunei	30	46.03
		Malaysia	30	49.21
Bei Jiang/Hsi	417,800	China	407,900	97.63
		Viet Nam	9,800	2.35
Beilun	900	China	800	84.92
		Viet Nam	100	15.08
Ca/Song Koi	31,000	Viet Nam	20,100	64.91
		Lao, People's Democratic Republic of	10,900	35.09
Coruh	22,100	Turkey	20,000	90.85
		Georgia	2,000	9.01
Dasht	33,400	Pakistan	26,200	78.42
		Iran	7,200	21.58

Basin name	Total area of basin (km^2)	Countries	Area of country in basin (km^2)	Area of country in basin (%)
ASIA				
Fenney	2,800	India	1,800	65.83
		Bangladesh	1,000	34.17
Fly	64,600	Papua New Guinea	60,400	93.40
		Indonesia	4,300	6.60
Ganges-Brahmaputra-Meghna	1,634,900	India	948,400	58.01
		China	321,300	19.65
		Nepal	147,400	9.01
		Bangladesh	107,100	6.55
		India, claimed by China	67,100	4.11
		Bhutan	39,900	2.44
		India control, claimed by China	1,200	0.07
		Myanmar (Burma)	80	0.00
Golok	1,800	Thailand	1,000	56.62
		Malaysia	800	43.38
Han	35,300	Korea, Republic of (South)	25,100	71.22
		Korea, Democratic People's Republic of (North)	10,100	28.67
Har Us Nur	185,300	Mongolia	179,300	96.81
		Russian Federation	5,600	3.04
		China	300	0.15
Hari/Harirud	92,600	Afghanistan	41,000	44.31
		Iran	35,400	38.27
		Turkmenistan	16,100	17.41
Helmand	353,500	Afghanistan	288,200	81.53
		Iran	54,900	15.52
		Pakistan	10,400	2.95
Ili/Kunes He	161,200	Kazakhstan	97,100	60.24
		China	55,300	34.32
		Kyrgyzstan	8,800	5.44
Indus	1,138,800	Pakistan	597,700	52.48
		India	381,600	33.51
		China	76,200	6.69
		Afghanistan	72,100	6.33
		Chinese control, claimed by India	9,600	0.84
		Indian control, claimed by China	1,600	0.14
		Nepal	10	0.00
Irrawaddy	404,200	Myanmar (Burma)	368,600	91.20
		China	18,500	4.58
		India	14,100	3.49
		India, claimed by China	1,200	0.30
Jordan	42,800	Jordan	20,600	48.13
		Israel	9,100	21.26
		Syria	4,900	11.45
		West Bank	3,200	7.48
		Egypt	2,700	6.31
		Golan Heights	1,500	3.50
		Lebanon	600	1.33

Table 12.1: *continued*

Basin name	Total area of basin (km^2)	Countries	Area of country in basin (km^2)	Area of country in basin (%)
ASIA				
Kaladan	30,500	Myanmar (Burma)	22,900	74.91
		India	7,300	23.84
Karnaphuli	12,500	Bangladesh	7,400	58.78
		India	5,100	41.14
		Myanmar (Burma)	10	0.09
Kowl E Namaksar	36,500	Iran	25,900	71.13
		Afghanistan	10,500	28.87
Kura-Araks	193,200	Azerbaijan	56,600	29.28
		Iran	39,700	20.55
		Armenia	34,800	18.03
		Georgia	34,300	17.77
		Turkey	27,700	14.32
		Russian Federation	60	0.03
Lake Ubsa-Nur	62,800	Mongolia	47,600	75.78
		Russia	15,200	24.22
Ma	30,300	Viet Nam	17,100	56.48
		Lao, People's Democratic Republic of	13,200	43.52
Mekong	787,800	Lao, People's Democratic Republic of	198,000	25.14
		Thailand	193,900	24.62
		China	171,700	21.79
		Cambodia (Kampuchea)	158,400	20.10
		Viet Nam	38,200	4.84
		Myanmar (Burma)	27,600	3.51
Murgab	60,900	Afghanistan	36,400	59.79
		Turkmenistan	24,500	40.21
Nahr El Kebir	1,500	Syria	1,300	85.61
		Turkey	200	13.87
Oral/Ural	311,000	Kazakhstan	175,500	56.43
		Russia	135,500	43.57
Pakchan	3,900	Myanmar (Burma)	1,900	49.11
		Thailand	1,800	47.24
Pandaruan	400	Brunei	200	60.65
		Malaysia	100	39.08
Pu Lun T'o	89,000	China	77,800	87.39
		Mongolia	11,100	12.48
		Russian Federation	80	0.09
		Kazakhstan	30	0.04
Red/Song Hong	157,100	China	84,500	53.75
		Viet Nam	71,500	45.50
		Lao, People's Democratic Republic of	1,200	0.74
Rezvaya	700	Turkey	500	74.66
		Bulgaria	200	25.34
Saigon	25,100	Viet Nam	24,800	98.67
		Cambodia (Kampuchea)	200	0.99
ASIA				
Salween	244,000	China	127,900	52.40
		Myanmar (Burma)	107,000	43.85
		Thailand	9,100	3.73
Sembakung	15,300	Indonesia	8,100	52.86
		Malaysia	7,200	47.14
Song Vam Co Dong	15,300	Viet Nam	7,800	50.68
		Cambodia (Kampuchea)	7,500	49.23
Sujfun	18,300	China	11,800	64.46
		Russian Federation	6,500	35.54
Tami	89,900	Indonesia	87,700	97.55
		Papua New Guinea	2,200	2.45
Tarim	1,051,600	China	1,000,300	95.12
		Chinese control, claimed by India	21,500	2.04
		Kyrgyzstan	21,100	2.00
		Tajikistan	6,600	0.63
		Pakistan	2,000	0.19
		Afghanistan	60	0.01
Tigris-Euphrates/ Shatt al Arab	789,000	Iraq	319,400	40.48
		Turkey	195,700	24.80
		Iran	155,400	19.70
		Syria	116,300	14.73
		Jordan	2,000	0.25
		Saudi Arabia	80	0.01
Tjeroaka-Wanggoe	6,600	Indonesia	4,000	61.57
		Papua New Guinea	2,500	38.43
Tumen	29,100	China	20,300	69.75
		Korea, Democratic People's Republic of (North)	8,300	28.59
		Russia	500	1.66
Wadi Al Izziyah	600	Lebanon	400	68.23
		Israel	200	31.60
Yalu	50,900	China	26,800	52.65
		Korea, Democratic People's Republic of (North)	23,800	46.82
EUROPE				
Bann	5,600	United Kingdom	5,400	97.14
		Ireland	200	2.86
Barta	1,800	Latvia	1,100	60.87
		Lithuania	700	37.71
Bidasoa	500	Spain	500	89.33
		France	60	10.67
Castletown	400	United Kingdom	300	76.12
		Ireland	90	23.88

Table 12.1: *continued*

Basin name	Total area of basin (km²)	Countries	Area of country in basin (km²)	Area of country in basin (%)
EUROPE				
Danube	790,100	Romania	228,500	28.93
		Hungary	92,800	11.74
		Austria	81,600	10.32
		Yugoslavia (Serbia and Montenegro)	81,500	10.31
		Germany	59,000	7.47
		Slovakia	45,600	5.77
		Bulgaria	40,900	5.17
		Bosnia and Herzegovina	38,200	4.83
		Croatia	35,900	4.54
		Ukraine	29,600	3.75
		Czech Republic	20,500	2.59
		Slovenia	17,200	2.18
		Moldova	13,900	1.76
		Switzerland	2,500	0.32
		Italy	1,200	0.15
		Poland	700	0.09
		Albania	200	0.03
Daugava	58,700	Belarus	28,300	48.14
		Latvia	20,200	34.38
		Russian Federation	9,500	16.11
		Lithuania	800	1.38
Dnieper	516,300	Ukraine	299,300	57.97
		Belarus	124,900	24.19
		Russian Federation	92,100	17.83
Dniester	62,000	Ukraine	46,800	75.44
		Moldova	15,200	24.52
		Poland	30	0.05
Don	425,600	Russian Federation	371,200	87.23
		Ukraine	54,300	12.76
Douro/Duero	98,900	Spain	80,700	81.63
		Portugal	18,200	18.37
Drin	17,900	Albania	8,100	45.39
		Yugoslavia (Serbia and Montenegro)	7,400	41.40
		Macedonia	2,200	12.18
Ebro	85,800	Spain	85,200	99.36
		Andorra	400	0.48
		France	100	0.16
Elancik	900	Russian Federation	700	71.32
		Ukraine	300	28.68
Elbe	132,200	Germany	83,100	62.86
		Czech Republic	47,600	36.02
		Austria	700	0.54
		Poland	700	0.56
Erne	4,800	Ireland	2,800	59.28
		United Kingdom	1,900	40.72
Fane	200	Ireland	200	96.46
		United Kingdom	10	3.54

Basin name	Total area of basin (km²)	Countries	Area of country in basin (km²)	Area of country in basin (%)
EUROPE				
Flurry	60	United Kingdom	50	73.77
		Ireland	20	26.23
Foyle	2,900	United Kingdom	2,000	67.30
		Ireland	1,000	32.70
Garonne	55,800	France	55,100	98.83
		Spain	600	1.07
		Andorra	40	0.08
Gauja	11,600	Latvia	10,400	90.42
		Estonia	1,100	9.58
Glama	43,000	Norway	42,600	99.00
		Sweden	400	0.99
Guadiana	67,900	Spain	54,900	80.82
		Portugal	13,000	19.18
Isonzo	3,000	Slovenia	1,800	59.48
		Italy	1,200	40.09
Jacobs	400	Norway	300	68.10
		Russian Federation	100	31.90
Jenisej/ Yenisey	2,557,800	Russian Federation	2,229,800	87.17
		Mongolia	327,900	12.82
Kemi	55,700	Finland	52,700	94.52
		Russian Federation	3,000	5.41
		Norway	10	0.01
Klaralven	51,000	Sweden	43,100	84.54
		Norway	7,900	15.46
Kogilnik	6,100	Moldova	3,600	57.82
		Ukraine	2,600	42.18
Krka	1,300	Croatia	1,100	89.55
		Bosnia and Herzegovina	100	8.93
		Yugoslavia (Serbia and Montenegro)	10	0.40
Lake Prespa	9,000	Albania	8,000	88.17
		Macedonia	800	8.50
		Greece	300	3.32
Lava/Pregel	8,600	Russian Federation	6,300	74.00
		Poland	2,000	23.84
Lielupe	14,400	Latvia	9,600	66.76
		Lithuania	4,800	33.22
Lima	2,300	Spain	1,200	50.88
		Portugal	1,100	49.04
Maritsa	49,600	Bulgaria	33,000	66.49
		Turkey	12,800	25.69
		Greece	3,700	7.55
Mino	15,100	Spain	14,500	96.18
		Portugal	600	3.70
Mius	2,800	Russian Federation	1,900	69.82
		Ukraine	800	30.07
Naatamo	1,000	Norway	600	57.73
		Finland	400	41.97

Table 12.1: *continued*

Basin name	Total area of basin (km²)	Countries	Area of country in basin (km²)	Area of country in basin (%)
EUROPE				
Narva	53,000	Russian Federation	28,200	53.20
		Estonia	18,100	34.09
		Latvia	5,900	11.13
		Belarus	800	1.57
Neman	90,300	Belarus	41,700	46.13
		Lithuania	39,700	43.97
		Russian Federation	4,800	5.30
		Poland	3,800	4.21
		Latvia	300	0.36
Neretva	5,500	Bosnia and Herzegovina	5,300	95.98
		Croatia	200	3.47
Nestos	10,200	Bulgaria	5,500	53.63
		Greece	4,700	46.36
Ob	2,950,800	Russian Federation	2,192,700	74.31
		Kazakhstan	743,800	25.21
		China	13,900	0.47
		Mongolia	200	0.01
Oder/Odra	122,400	Poland	103,100	84.20
		Czech Republic	10,300	8.38
		Germany	7,800	6.33
		Slovakia	1,300	1.09
Olanga	18,800	Russian Federation	16,800	89.37
		Finland	2,000	10.62
Oulu	28,700	Finland	26,700	93.20
		Russian Federation	1,900	6.78
Parnu	5,800	Estonia	5,800	99.85
		Latvia	10	0.15
Pasvik	16,000	Finland	12,400	77.46
		Russian Federation	2,600	16.15
		Norway	1,000	6.39
Po	87,100	Italy	82,200	94.44
		Switzerland	4,300	4.92
		France	500	0.54
		Austria	90	0.10
Prohladnaja	600	Russian Federation	500	76.90
		Poland	100	23.10
Rhine	172,900	Germany	97,700	56.49
		Switzerland	24,300	14.05
		France	23,100	13.34
		Belgium	13,900	8.03
		Netherlands	9,900	5.75
		Luxembourg	2,500	1.46
		Austria	1,300	0.76
		Liechtenstein	200	0.09
		Italy	70	0.04
Rhone	100,200	France	90,100	89.88
		Switzerland	10,100	10.05
		Italy	50	0.05
Roia	600	France	400	67.39
		Italy	200	30.45

Basin name	Total area of basin (km²)	Countries	Area of country in basin (km²)	Area of country in basin (%)
EUROPE				
Salaca	2,100	Latvia	1,600	78.52
		Estonia	100	5.70
Samur	6,800	Russian Federation	6,300	93.75
		Azerbaijan	400	6.22
Sarata	1,800	Ukraine	1,100	63.90
		Moldova	600	36.05
Schelde	17,100	France	8,600	50.03
		Belgium	8,400	49.28
		Netherlands	80	0.47
Seine	85,700	France	83,800	97.78
		Belgium	1,800	2.14
		Luxembourg	70	0.08
Struma	15,000	Bulgaria	8,600	57.66
		Greece	3,900	25.88
		Macedonia	1,800	12.22
		Yugoslavia (Serbia and Montenegro)	600	4.19
Sulak	15,100	Russian Federation	13,900	92.38
		Georgia	1,100	7.24
		Azerbaijan	60	0.38
Tagus/Tejo	77,900	Spain	51,400	66.06
		Portugal	26,100	33.50
Tana	15,600	Norway	9,300	59.71
		Finland	6,300	40.23
Terek	38,700	Russian Federation	37,000	95.39
		Georgia	1,800	4.61
Torne/Tornealven	37,300	Sweden	25,400	67.98
		Finland	10,400	28.00
		Norway	1,500	4.03
Tuloma	25,800	Russian Federation	23,700	91.85
		Finland	2,000	7.93
Vardar	32,400	Macedonia	20,300	62.83
		Yugoslavia (Serbia and Montenegro)	8,200	25.22
		Greece	3,900	11.94
Velaka	700	Bulgaria	700	95.25
		Turkey	30	3.74
Venta	9,500	Latvia	6,200	65.15
		Lithuania	3,300	34.72
Vijose	7,200	Albania	4,600	64.83
		Greece	2,500	34.66
Vistula/Wista	194,000	Poland	169,700	87.45
		Ukraine	12,700	6.55
		Belarus	9,800	5.03
		Slovakia	1,900	0.96
		Czech Republic	20	0.01
Volga	1,554,900	Russian Federation	1,551,300	99.77
		Kazakhstan	2,200	0.14
		Belarus	1,300	0.08

Table 12.1: *continued*

Basin name	Total area of basin (km²)	Countries	Area of country in basin (km²)	Area of country in basin (%)
EUROPE				
Vuoksa	62,700	Finland	54,300	86.48
		Russian Federation	8,500	13.52
Wiedau	1,100	Denmark	1,000	86.23
		Germany	200	13.32
Yser	900	France	500	53.63
		Belgium	400	46.37
LATIN AMERICA AND THE CARIBBEAN				
Amacuro	5,600	Venezuela	4,900	86.89
		Guyana	700	13.11
Amazon	5,883,400	Brazil	3,670,300	62.38
		Peru	956,500	16.26
		Bolivia	706,700	12.01
		Colombia	367,800	6.25
		Ecuador	123,800	2.10
		Venezuela	40,300	0.68
		Guyana	14,500	0.25
		Suriname	1,400	0.02
		French Guiana	30	0.00
Artibonite	8,800	Haiti	6,600	74.37
		Dominican Republic	2,300	25.55
Aviles	300	Argentina	200	88.72
		Chile	30	11.28
Aysen	13,600	Chile	13,100	96.07
		Argentina	500	3.93
Baker	30,800	Chile	21,000	68.15
		Argentina	9,800	31.83
Barima	2,100	Guyana	1,100	51.05
		Venezuela	1,000	47.84
Belize	11,500	Belize	7,000	60.86
		Guatemala	4,500	39.14
Cancoso/ Lauca	23,500	Bolivia	20,200	85.72
		Chile	3,400	14.28
Candelaria	12,800	Mexico	11,300	88.24
		Guatemala	1,500	11.7
Carmen Silva/ Chico	1,700	Argentina	1,000	59.70
		Chile	700	40.30
Catatumbo	31,000	Colombia	19,600	63.15
		Venezuela	11,400	36.75
Changuinola	3,200	Panama	2,900	91.29
		Costa Rica	300	8.33
Chira	15,700	Peru	9,800	62.23
		Ecuador	5,800	37.23
Chiriqui	1,700	Panama	1,500	86.17
		Costa Rica	200	13.83
Choluteca	7,400	Honduras	7,200	97.68
		Nicaragua	200	2.32
Chuy	200	Brazil	100	64.57
		Uruguay	60	32.57
LATIN AMERICA AND THE CARIBBEAN				
Coatan Achute	2,000	Mexico	1,700	86.27
		Guatemala	300	13.73
Coco/Segovia	25,400	Nicaragua	17,900	70.52
		Honduras	7,500	29.48
Comau	900	Chile	900	91.36
		Argentina	80	8.64
Corantijn/ Courantyne	41,800	Guyana	21,700	52.06
		Suriname	19,900	47.75
		Brazil	80	0.19
Cullen	600	Chile	500	83.00
		Argentina	100	17.00
Essequibo	239,500	Guyana	162,100	67.67
		Venezuela	52,400	21.87
		Suriname	24,300	10.13
		Brazil	200	0.07
Gallegos-Chico	11,600	Argentina	7,000	60.15
		Chile	4,600	39.85
Goascoran	2,800	Honduras	1,500	53.36
		El Salvador	1,300	46.64
Grijalva	126,800	Mexico	78,900	62.25
		Guatemala	47,800	37.72
		Belize	20	0.02
Hondo	14,600	Mexico	8,900	61.14
		Guatemala	4,200	28.50
		Belize	1,500	10.36
Jurado	700	Colombia	500	82.11
		Panama	100	17.89
La Plata	2,954,500	Brazil	1,379,300	46.69
		Argentina	817,900	27.68
		Paraguay	400,100	13.54
		Bolivia	245,100	8.30
		Uruguay	111,600	3.78
Lagoon Mirim	55,000	Uruguay	31,200	56.69
		Brazil	23,800	43.24
Lake Fagnano	3,200	Argentina	2,700	85.17
		Chile	500	14.83
Lake Titicaca-Poopó System	111,800	Bolivia	63,000	56.32
		Peru	48,000	42.94
		Chile	800	0.74
Lempa	18,000	El Salvador	9,500	52.45
		Honduras	5,800	32.01
		Guatemala	2,800	15.54
Maroni	65,000	Suriname	37,500	57.64
		French Guiana	27,200	41.90
		Brazil	200	0.27
Massacre	800	Haiti	500	62.03
		Dominican Republic	300	35.96
Mataje	700	Ecuador	500	73.98
		Colombia	200	26.02

Table 12.1: *continued*

Basin name	Total area of basin (km²)	Countries	Area of country in basin (km²)	Area of country in basin (%)
LATIN AMERICA AND THE CARIBBEAN				
Mira	12,100	Colombia	6,200	50.87
		Ecuador	5,800	47.97
Motaqua	16,100	Guatemala	14,600	90.85
		Honduras	1,500	9.11
Negro	5,800	Nicaragua	4,800	83.87
		Honduras	900	15.68
Oiapoque/ Oyupock	23,300	French Guiana	13,700	58.92
		Brazil	9,500	41.00
Orinoco	927,400	Venezuela	604,500	65.18
		Colombia	321,700	34.68
		Brazil	800	0.08
Palena	13,300	Chile	7,300	54.87
		Argentina	6,000	45.13
Pascua	13,700	Chile	7,300	53.51
		Argentina	6,400	46.46
Patia	21,300	Colombia	20,800	97.61
		Ecuador	500	2.38
Paz	2,200	Guatemala	1,400	64.47
		El Salvador	800	35.53
Pedernales	400	Haiti	269	67.32
		Dominican Republic	131	32.68
Puelo	8,400	Argentina	5,500	66.03
		Chile	2,900	33.97
Rio Grande	8,000	Argentina	4,000	49.74
		Chile	4,000	50.26
San Juan	42,200	Nicaragua	30,400	72.02
		Costa Rica	11,800	27.93
San Martin	700	Chile	600	87.44
		Argentina	80	12.56
Sarstun	2,100	Guatemala	1,800	87.63
		Belize	300	12.37
Seno Union/ Serrano	6,500	Chile	5,700	87.93
		Argentina	700	10.34
Sixaola	2,900	Costa Rica	2,500	88.65
		Panama	300	9.99
Suchiate	1,600	Guatemala	1,100	68.79
		Mexico	500	31.21
Tijuana	4,400	Mexico	3,100	70.57
		United States of America	1,300	29.43
Tumbes- Poyango	5,000	Ecuador	3,600	71.62
		Peru	1,400	28.38
Valdivia	15,000	Chile	14,700	98.39
		Argentina	100	0.69
Yaqui	74,700	Mexico	70,100	93.87
		United States of America	4,600	6.13
Yelcho	11,100	Argentina	6,900	62.14
		Chile	4,200	37.86
LATIN AMERICA AND THE CARIBBEAN				
Zapaleri	2,600	Chile	1,600	59.60
		Argentina	500	19.65
		Bolivia	500	20.75
Zarumilla	4,300	Ecuador	3,400	78.71
		Peru	900	20.51
NORTH AMERICA				
Alsek	28,400	Canada	26,500	93.50
		United States of America	1,800	6.50
Chilkat	3,800	United States of America	2,100	56.59
		Canada	1,600	43.35
Colorado	655,000	United States of America	644,600	98.41
		Mexico	10,400	1.59
Columbia	668,400	United States of America	566,500	84.75
		Canada	101,900	15.24
Firth	6,000	Canada	3,800	63.60
		United States of America	2,200	36.40
Fraser	239,700	Canada	239,100	99.74
		United States of America	600	0.26
Mississippi	3,226,300	United States of America	3,176,500	98.46
		Canada	49,800	1.54
Nelson- Saskatchewan	1,109,400	Canada	952,000	85.81
		United States of America	157,400	14.19
Rio Grande	656,100	United States of America	341,800	52.10
		Mexico	314,300	47.90
Skagit	8,000	United States of America	7,100	88.46
		Canada	900	11.54
St. Croix	4,600	United States of America	3,300	70.86
		Canada	1,400	29.14
St. John (North America)	47,700	Canada	30,300	63.50
		United States of America	17,300	36.22
St. Lawrence	1,055,200	Canada	559,000	52.98
		United States of America	496,100	47.02
Stikine	50,900	Canada	50,000	98.32
		United States of America	900	1.68
Taku	18,100	Canada	16,300	90.09
		United States of America	1,700	9.13
Whiting	2,600	Canada	2,000	80.06
		United States of America	500	19.94
Yukon	829,700	United States of America	496,400	59.83
		Canada	333,300	40.17
OCEANIA				
Sepik	73,400	Papua New Guinea	71,000	96.81
		Indonesia	2,300	3.19

Source: International River Basin register, updated August 2002.

Map 12.2: Country dependence on water resources inflow from neighbouring countries

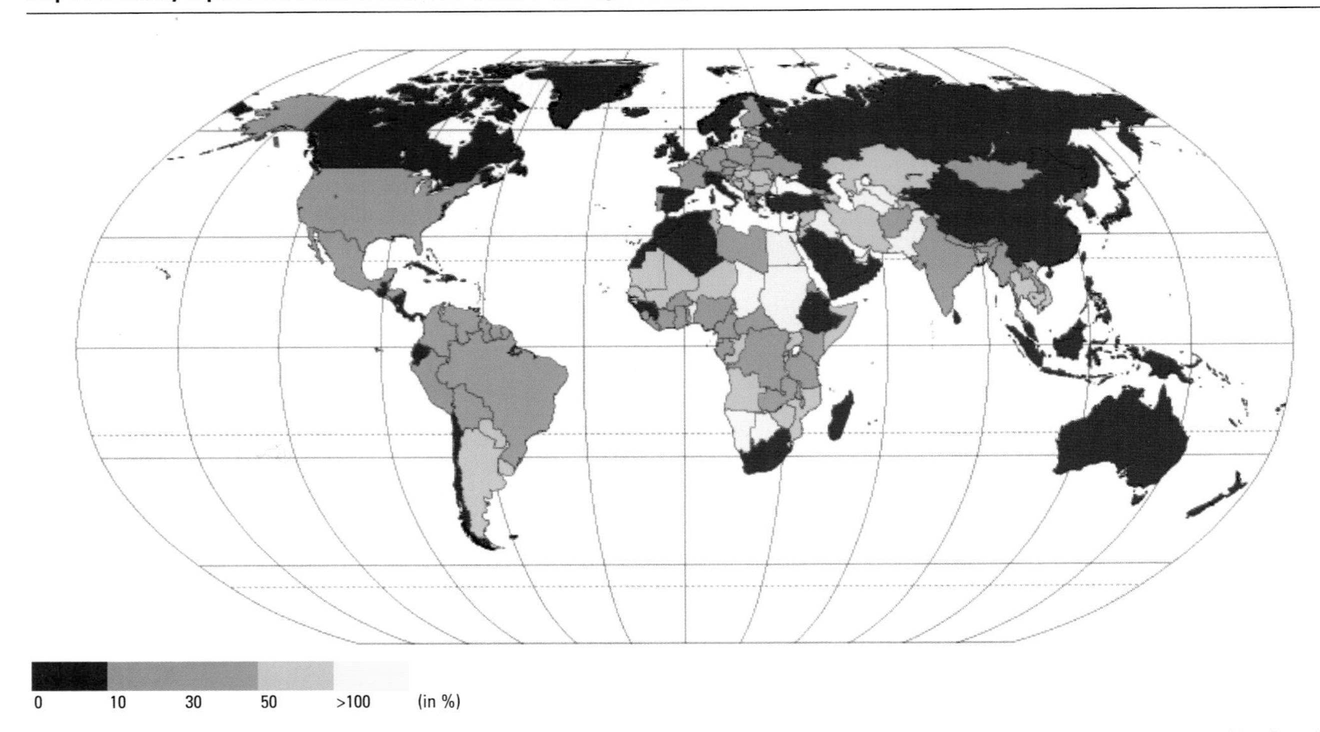

This map illustrates the current state of affairs with regards to countries' reliance upon water coming from neighbouring countries. This is an important consideration when assessing a given country's water supply and its quality. A good part of Africa and the Middle East depend upon foreign water resources for more than half of their own water, as does the southern tip of Latin America.

Source: Map prepared for the World Water Assessment Programme (WWAP) by the Centre for Environmental Research, University of Kassel, based on information from Water Gap version 2.1. D.

Another way to think of possible transboundary competition for water is to consider the extent to which some countries are dependent on neighbouring countries for their source of water. Map 12.2 illustrates the current state of affairs.

Conflict, Cooperation and the Importance of Resilient Institutions

The complex dynamics of managing transboundary waters

The largest empirical study of water conflict and cooperation, completed in 2001 at Oregon State University, documents a total of 1,831 interactions, both conflictive and cooperative, between two or more nations over water during the past fifty years.[3] The results are summarized graphically in figure 12.2. An analysis of the data yields the following general findings. First, despite the potential for dispute in transboundary basins, the record of cooperation historically overwhelms the record of acute conflict over international water resources. The last fifty years have seen only thirty-seven acute disputes (those involving violence) while, during the same period, approximately 200 treaties were negotiated and signed. The total number of water-related events between nations, of any magnitude, are likewise weighted towards cooperation: 507 conflict-related events, versus 1,228 cooperative ones, implying that violence over water is not strategically rational, effective or economically viable.

Secondly, nations find many more issues of cooperation than of conflict. Map 12.3 shows the location of transboundary catchments and the number of treaties associated with each. The distribution of these cooperative events indicates a broad spectrum of issue types, including quantity/quality of water, economic development, hydropower and joint management. In contrast, almost 90 percent of the conflictive events relate to quantity and infrastructure. Furthermore, if we look specifically at extensive military acts, of which there were only twenty-one (eighteen of them between Israel and its neighbours), almost all events fall within these two categories.

Thirdly, at the subacute level, water management issues act as both an irritant and as a unifier. As an irritant, issues can make good

3. Included in the study are interactions that involved water as a scarce and/or consumable resource or as a quantity to be managed as the *driver* of the event. Excluded are events where water is incidental to the dispute, such as those concerning fishing rights, access to ports, transportation or river boundaries, or where it is a tool, target or victim of armed conflict.

relations bad and bad relations worse. Threats and disputes have raged across boundaries with relations as diverse as those between Indians and Pakistanis and between Americans and Canadians. Water was the last and most contentious issue resolved in negotiations over a 1994 peace treaty between Israel and Jordan, and was relegated to 'final status' negotiations – along with other difficult issues such as Jerusalem and refugees – between Israel and the Palestinians.

Equally, transboundary waters, despite their complexities, can also act as a unifier in basins where relatively strong institutions are in place. The historical record shows that international water disputes do get resolved, even between bitter enemies, and even as conflicts erupt over other issues. The Senegal River in West Africa provides an example of a functioning institution created in 1972 to deal with issues surrounding competition for water (see box 12.7).

Some of the most vociferous enemies around the world have negotiated water agreements or are in the process of doing so, and the institutions they have created frequently prove to be resilient over time and during periods of otherwise strained relations. The Mekong Committee, for example, has functioned since 1957, and exchanged data throughout the Viet Nam War. Secret talks have been held between Israel and Jordan since the unsuccessful Johnston negotiations of 1953/55, even as these riparian states until only

Figure 12.2: Events related to transboundary water basins

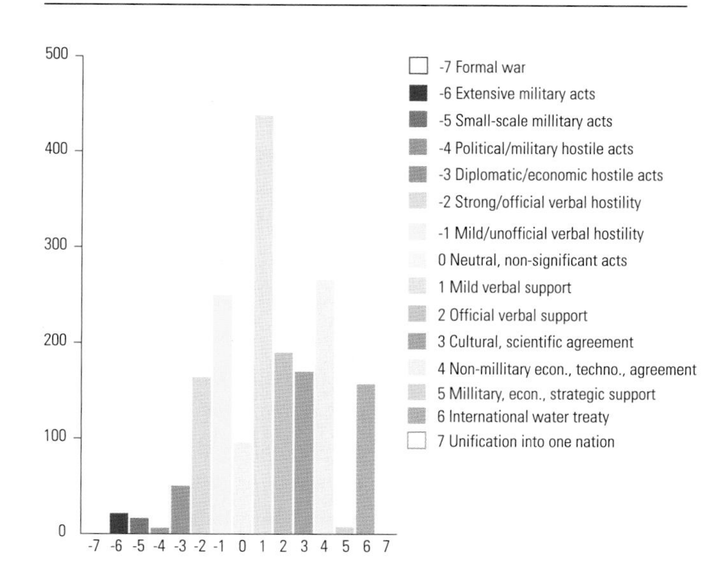

Although transboundary water resources can be fodder for hostility, the record of cooperation is vastly superior to that of acute conflict, that is to say, water is much more a vector of cooperation than a source of conflict.

Source: Wolf et al., forthcoming.

Map 12.3: Transboundary basins of the world and the number of associated treaties

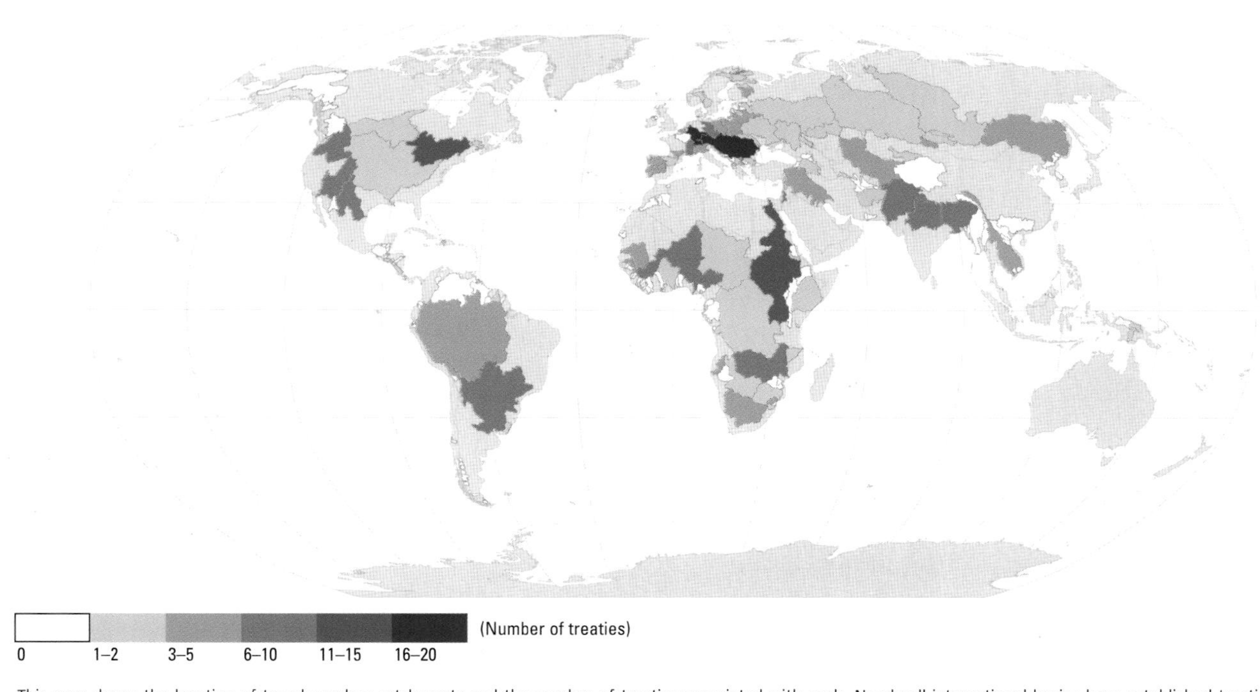

This map shows the location of transboundary catchments and the number of treaties associated with each. Nearly all international basins have established treaties to facilitate and legislate in some way the sharing of the resource. The distribution of these cooperative events indicates a broad spectrum of issue types, including quantity, quality, economic development, hydropower and joint management.

Source: Transboundary Freshwater Dispute Database, Oregon State University, 2002.

Box 12.7: Sharing water in the Senegal River basin

The Organization for the Development of the Senegal River (OMVS) legislative and regulatory framework clearly indicates, via the founding conventions of 1972 and the Senegal River Water Charter signed in May 2002, that sharing of the river's water should be agreed on by the different usage sectors. Sharing of the resource by the riparian states is not a matter of the volume of water to be withdrawn, but rather of optimal satisfaction of usage requirements. Usages to be taken into account include: agriculture, inland fishing, livestock and fish raising, forestry, flora and fauna, hydroelectric energy, provision of water to urban and rural populations, health, industry, navigation and the environment.

Principles, terms and conditions of water distribution for these usages have been drawn up and a Permanent Commission established as a consultative body for the OMVS Council of Ministers. The latter takes decisions and asks the High Commission to ensure their implementation.

The work of a permanent water commission and the ministers' decision-making criteria are founded on the following general principles:

- equitable and reasonable use of river water;
- the obligation to preserve the basin's environment;
- the obligation to negotiate in cases of disagreement/conflict over water use; and
- the obligation for each riparian country to inform the others before undertaking any measure or project that could affect water availability.

Source: Prepared for the World Water Assessment Programme (WWAP) by the Organization for the Development of the Senegal River (OMVS), 2002.

recently were in a legal state of war. The Indus River Commission survived through two wars between India and Pakistan. And all ten Nile riparian states are currently involved in negotiations over cooperative development of the basin (see box 12.8).

In the absence of institutions, however, changes within a basin can lead to conflict. To avoid the political intricacies of shared water resources, for example, a riparian stakeholder, generally the regional power,[4] may implement a project that impacts at least one of its neighbours. This might be to continue to meet existing uses in the face of decreasing relative water availability, as for example Egypt's plans for a high dam on the Nile or Indian diversions of the Ganges to protect the port of Kolkata (Calcutta). Or it might be to meet new needs and associated policies such as Turkey's Southeast Anatolia Project (GAP) on the Euphrates. When projects such as these proceed without regional collaboration, they can become a flash point, heightening tensions and regional instability, and requiring years or, more commonly decades, to resolve. Evidence of how institutions can diffuse tensions is seen in basins with large numbers of water infrastructure projects (e.g. in the Rhine and Danube basins). Coriparian relations have shown themselves to be significantly more cooperative in basins with treaties and high dam density than in similarly developed basins without treaties. Thus, institutional capacity together with shared interests and human creativity seem to ameliorate water's conflict-inducing characteristics, suggesting that an important lesson of international water is that as a resource it tends to induce cooperation, and to incite violence only as the exception.[5] A further example of this is provided between Turkey and Syria where an agreement for technical cooperation has been established.

The choice for the international community lies between a traditional chronology of events, where unilateral development is followed by a crisis and, possibly, a lengthy and expensive process of conflict resolution; and a process where riparian states are encouraged to get ahead of the crisis curve through crisis prevention, preventive diplomacy and institutional capacity-building (as is the case with the Nile Basin Initiative). It is alarming that the global community has often allowed water conflicts to drag on to the extent that they sometimes have. The Indus treaty took ten years of negotiations, the Ganges thirty, and the Jordan forty – meanwhile the water quality and quantity have degraded to a point where the health of dependent populations and ecosystems have been damaged or destroyed. A reread through the history of transboundary waters suggests that the simple fact that humans suffer and die in the absence of agreement apparently offers little

4. 'Power' in regional hydropolitics can include riparian position, with an upstream riparian having more relative strength vis-à-vis the water resources than its downstream riparian, in addition to the more conventional measures of military, political and economic strength.

5. It is important to understand there is history of water-related violence – but that it is a history of incidents at the subnational level, generally between tribes, water use sectors or states/provinces. In fact, what we seem to be finding is that geographic scale and intensity of conflict are inversely related.

Box 12.8: Water sharing as an instrument of regional integration – the Nile basin

The Nile River is the longest river in the world (nearly 6,700 km) and has historically been one of the world's greatest natural assets. It has nourished livelihoods, an array of ecosystems and a rich diversity of cultures since pharaonic times. It is a transboundary river shared among ten African countries (Burundi, Democratic Republic of Congo, Egypt, Eritrea, Ethiopia, Kenya, Rwanda, Sudan, Uganda and the United Republic of Tanzania). Its catchment covers one tenth of Africa's landmass and the population of its riparian states amounts to about 300 million, or 40 percent, of Africans.

Today the Nile basin faces the challenges of poverty (four riparians rank amongst the ten poorest counties in the world), instability (conflicts in the Great Lakes, Sudan and the Horn of Africa), rapid population growth, and severe environmental degradation (especially in the East African Highlands). The basic premise of the meeting to mobilize funds for joint regional development is that the Nile offers significant opportunities for cooperative management and development. When tapped, these will result in greater regional integration, which in turn will allow better socio-economic development to meet those challenges mentioned. Such socio-economic benefits are expected to exceed the direct benefits from the river alone.

Recognizing this, the Council of Ministers of Water Resources (NILE-COM) launched the Nile Basin Initiative (NBI) in February 1999. This initiative includes all riparian states and provides an agreed basin-wide framework to fight poverty and promote socio-economic development in the Nile basin. The NBI is guided by a shared vision 'to achieve sustainable socio-economic development through the equitable utilization of, and benefit from, the common Nile water resources'.

This vision is to be realized through a Strategic Action Programme comprising both basin-wide as well as sub-basin joint investment projects ranging from collaborative actions, experience and information sharing and capacity-building. A set of seven initial projects have been endorsed by the NILE-COM, and the first International Consultative Consortium on The Nile (ICCON) meeting was held in June 2000 to solicit funding for these projects and to support the NBI secretariat. These projects are:

- Nile Transboundary Environmental Action;
- Nile Basin Power Trade;
- Efficient Water Use for Agricultural Production;
- Water Resources Planning and Management;
- Confidence-Building and Stakeholder Involvement (Communication);
- Applied Training;
- Socio-economic Development and Benefit Sharing; and
- in addition to these Shared Vision Projects, groups of riparians – one in the Eastern (Blue) Nile and the other in the Nile Equatorial Lakes (White) – have identified joint and mutually beneficial investment opportunities at the sub-basin level called the Subsidiary Action Programs (the Eastern Nile Subsidiary Action Program [ENSAP], and the Nile Equatorial Lakes Subsidiary Action Program [NELSAP], respectively).

The Nile Basin Initiative is extremely promising as an example of water sharing as an instrument of regional integration.

Source: Based on a document prepared by UNECA for the *First Annual Report on Regional Integration in Africa*, 2002.

in the way of incentive to cooperate, even less so the health of aquatic ecosystems. This problem gets worse as the dispute gains in intensity. One rarely hears talk about the ecosystems of the lower Nile, the lower Jordan or the Aral Sea – they have effectively been written off to the vagaries of human intractability (although there are projects to stabilize the delta of the Aral Sea).

Basin-level water institutions: capacity-building opportunities

The cooperative water institutions referenced above are part of a larger history of basin-level treaty writing that has developed over the centuries. In contrast with the naturally vague and occasionally contradictory global declarations and principles, the institutions developed by coriparian nations have been able to focus on specific, regional conditions and concerns. An evaluation of these institutions over the past century, with particular attention to treaties signed since the Rio Conference, offers insights into how appropriately the emphasis areas highlighted in Agenda 21 and subsequent declarations on freshwater resource management in general address the specific needs of transboundary waters.

The Food and Agriculture Organization (FAO) has identified more than 3,600 treaties relating to international water resources dating from AD 805 to 1984, the majority of which relate to some aspect of navigation. In the last fifty years, approximately 200 treaties have been developed addressing non-navigational issues of water management, including flood control, hydropower projects or allocations for consumptive or non-consumptive uses in international basins.

Despite their growth in numbers and clear historic contribution to successful basin-wide management, a review of the treaties from the last half-century reveals a general lack of robustness. Water allocations, for example, the most conflictive issue between coriparian states, are seldom clearly delineated in water accords. Moreover, in the treaties that do specify quantities, allocations are often in fixed amounts to riparian nations, thus ignoring hydrological variation and changing values and needs. This was the case with the existing Nile basin accord between Egypt and Sudan. Likewise, review of such agreements shows that water quality provisions have played only a minor role in coriparian agreements historically. Monitoring, enforcement and conflict resolution mechanisms are also absent in a large percentage of the treaties. Finally, only exceptionally do transboundary water agreements include all riparian nations, which precludes the integrated basin management advocated by the international community.

One productive outgrowth from the transboundary treaty record has been a broadening in the definition and measurement of basin benefits. Traditionally, coriparian states have focused on water as a commodity to be divided – a zero-sum, rights-based approach. Precedents now exist for determining formulas that equitably allocate the benefits derived from water, not the water itself – a win-win, integrative approach. This is the approach now being followed in the Nile Basin Initiative (see box 12.7 earlier).

The treaties signed within the last ten years also reveal some encouraging developments. At least sixteen new bilateral and multilateral water agreements have been concluded since the signing of the Rio Declaration, covering basins in Asia, Africa, Europe and the Middle East. When compared with the treaties of the last half-century as a whole, a number of improvements can generally be seen in this more recent set of treaties. First, excluding agreements specifically focused on the environment, the majority of the treaties incorporate some type of water allocation measure. Second, provisions concerning water quality, monitoring and evaluation, data exchange and conflict resolution are included in most of the post-Rio treaties. Third, while not a majority, four of the agreements establish joint water commissions with decision-making and/or enforcement powers, a significant departure from the traditional advisory standing of basin commissions. Finally, country participation in basin-level accords appears to be expanding. Although few of the agreements incorporate all basin states, two thirds are multilateral, and in several of the treaties, references are made to the rights and interests of non-signatory nations.

Institutional vulnerabilities still exist, however, in a number of key areas. Few of the treaties, for example, possess the flexibility to handle changes in the hydrological regime or in regional values, and none specifically prioritize water usage in the basin. References to water quality, related groundwater systems, monitoring and evaluation, and conflict resolution mechanisms, while growing in numbers, are often weak in actual substance. Furthermore, public participation, an element that can greatly enhance the resiliency of institutions, is largely overlooked. It is likely that these weaknesses will be overcome as more experience is gained in the design of effective agreements and institutions.

Transboundary Aquifers: Groundwater Shared by Nations

While the debate about the equitable management of transboundary rivers basins has taken place for many years, the same cannot be stated about transboundary aquifers. As there are transboundary river basins, so there are also transboundary groundwater resources hidden below ground surface, in all parts of the world, which meet the basic needs of the rural and urban populations. Some transboundary aquifers contain huge freshwater resources, enough to provide safe and good-quality drinking water, as well as rural irrigation demands, thus securing food supplies. Though not visible as surface water, groundwater is widely occuring, if not ubiquitous in the global landmass. Chapter 4 of this book gives further details on groundwater availability throughout the world, in particular map 4.3.

Transboundary aquifers, due to their partial isolation from surface impacts, generally contain excellent quality water. Although such resources represent a substantial hidden global capital, they need prudent management. Competition for visible transboundary surface waters, based on available international law and hydraulic engineering, is evident in all continents. However, the hidden nature of transboundary groundwater and lack of legal frameworks invites misunderstandings by policy-makers. Not surprisingly therefore, transboundary aquifer management is still in its infancy, since its evaluation is difficult, suffering from a lack of institutional will and finance to collect the necessary information. Although there are

Map 12.4: Internationally shared aquifers in northern Africa

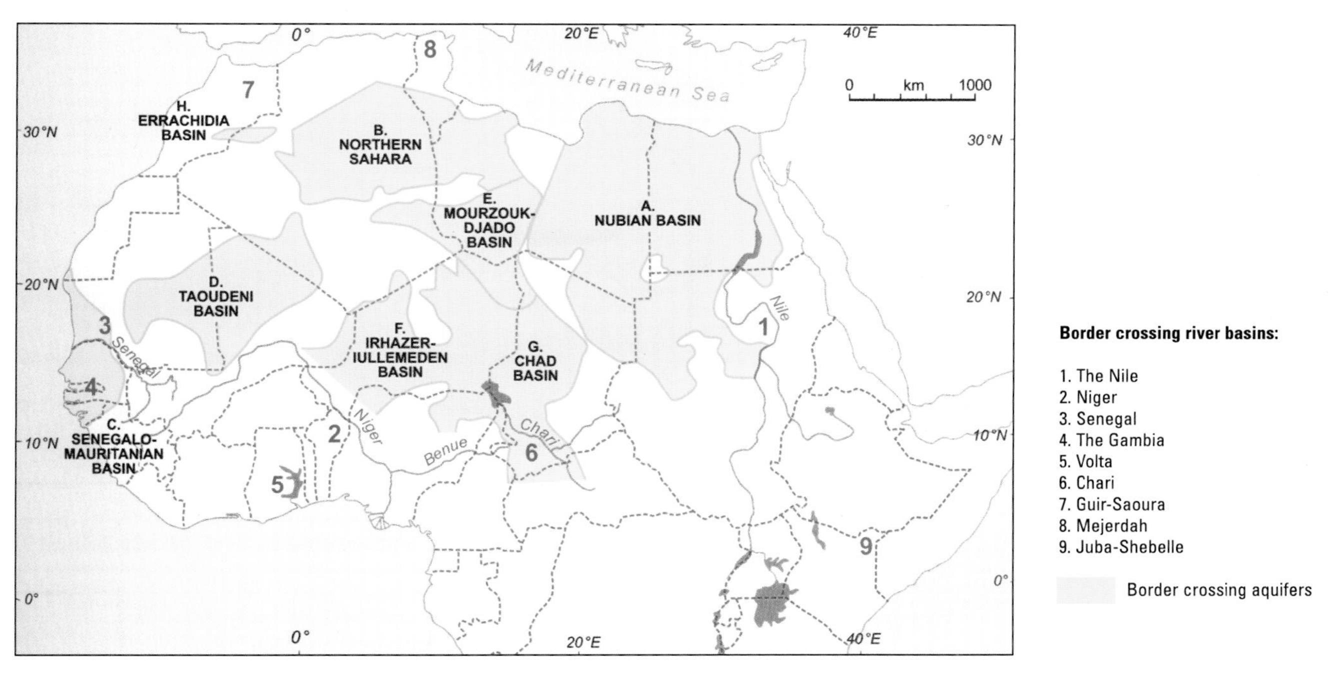

This map shows the distribution of several major transboundary aquifers underlying particularly water-stressed regions of northern Africa. Groundwater is an especially complex problem in terms of sharing the resource. The abundance of shared aquifers in this area underscores the importance of treaties and cooperative management.

Source: Based on OSS and UNESCO, 1997, in UNESCO, 2001.

fairly reliable estimates of the resources of the world's transboundary rivers, no such estimates exist for transboundary aquifers (World Bank, 1998).

While the groundwater component of the hydrological cycle is well understood, international water policy suffers from inadequate appreciation of its behaviour; for example the 1997 UN Convention on International Watercourses only refers to *some* groundwaters, but not all. The important role of groundwater in maintaining base flows in rivers and support to wetlands in the transboundary context has not yet been incorporated into most water-related conventions. Transboundary aquifers can also be the host for other human requirements that may be shared, notably geothermal heat (Roth et al., 2001). Consequently, integrated and holistic transboundary water resource management policies are constrained by this gap in international legislation. An important contribution to improved monitoring of transboundary aquifers has been made by the United Nations Economic Commission for Europe (UNECE), which has developed appropriate guidelines (UNECE, 2000). Pilot projects aimed at the application of these guidelines are just starting in Europe (Arnold and Uil, 2001).

As a resource essential for life, yet hidden from view, the sound national development of groundwater is sometimes constrained by contradicting socio-economic, institutional, legal, cultural and ethical policy frameworks. In a transboundary context, these can be even further amplified by contrasting levels of knowledge, capacities and institutional frameworks on either side of many international boundaries.

Whereas there are examples of how such issues have been dealt with in managing transboundary rivers, again there is no equivalent body of knowledge for the management of transboundary aquifers, the majority of which have not been inventoried. A recent UNECE survey of Europe indicated that there are over 100 transboundary aquifers in Europe (Convention on the Protection and Use of Transboundary Watercourses and International Lakes, 1999). Map 12.4 shows the distribution of several major North African aquifers underlying regions of acute water shortage. A worldwide survey of significant transboundary aquifers has recently been initiated through a collaboration of several international agencies (UNESCO [United Nations Educational, Scientific and Cultural Organization], FAO, IAH [International Association of Hydrogeologists], UNECE), under the ISARM (Internationally Shared Aquifer Resource Management) Initiative (Puri, 2001b). This survey will seek to fill the information gaps through a multidisciplinary assessment of internationally shared aquifers. For the sound management of transboundary aquifers the scientific, socio-economic and institutional and legal considerations need to be supplemented by such critical matters as capacity-building, participation, raising awareness, investment and appropriate technology (Puri, 2001a).

The key features of transboundary aquifers include a natural subsurface path of groundwater flow, intersected by an international boundary, such that water transfers from one side of the boundary to the other. In many cases, the aquifer might receive the majority of its recharge on one side, while the majority of its discharge would be on the other. It is this feature that requires wise governance and agreement on what constitutes equitable share. Activities such as withdrawals of the natural recharge on one side of the boundary could have subtle impact on base flows and wetlands on another side of a boundary, e.g. the aquifers in the delta areas of rivers flowing to the Aral Sea (Sydykov et al., 1998). In most transboundary aquifers these impacts can be widespread and delayed by decades. Many years may pass before the impacts are detected by monitoring. If the transboundary groundwater supports the biodiversity resources of a wetland, and a resident human population, valuing that aquifer remains a challenge to the policy-makers. Current international water law provides no guidance for these conditions.

It is therefore obvious that a thorough database is required to provide the information required to maintain sustainable groundwater resources for the needs of the planet.

Lessons in Hydrodiplomacy for the International Community

Conflict, cooperation and effective institutions

As demonstrated earlier in this chapter through comparisons of the number and nature of conflicts with the number of agreements, shared water is more often a catalyst for cooperation than a source of conflict. In an effort to explore this further, UNESCO launched a programme entitled From Potential for Conflict to Cooperation Potential (PCCP). While this programme is still ongoing, it has already identified certain critical lessons learnt from global experience in international water resource issues.

- Water crossing international boundaries can cause tensions between nations that share the basin. While the tension is not likely to lead to warfare, early coordination between riparian states can help prevent potential conflicts.

- Once international institutions are in place, they are tremendously resilient over time, even between otherwise hostile riparian nations, and even when conflict is waged over other issues.

- More likely than the occurrence of violent conflict, is the gradual decreasing of water quantity or quality, or both, which over time can affect the internal stability of a nation or region, and act as an irritant between ethnic groups, water sectors or states/ provinces. The resulting instability may have effects in the international arena.

- The greatest threat of the global water crisis to human security comes from the fact that millions of people lack access to sufficient quantities of water at sufficient quality for their well-being. This issue is receiving global attention that goes beyond individual basins.

Effective transboundary water resource management

The centrality of institutions both in effective transboundary water management and in preventive hydrodiplomacy cannot be overemphasized. Twentieth century water management offers lessons for the conception and implementation of transboundary water institutions. In combination with the existing efforts of the international community, the following lessons may help shape future policy and institution-building programmes directed specifically to the world's transboundary basins.

- Adaptable management structure: effective institutional management structures incorporate a certain level of flexibility, allowing for public input, changing basin priorities and new information and monitoring technologies. The adaptability of management structures must also extend to non-signatory riparian states, by incorporating provisions addressing their needs, rights and potential accession. The International Joint Commission (United States/Canada) has been particularly successful in dealing with such an evolving agenda of issues.

- Clear and flexible criteria for water allocations and quality: allocations, which are at the heart of most water disputes, are a function of water quantity and quality as well as political fiat. Thus, effective institutions must at least identify clear mechanisms for water allocation and water quality standards that simultaneously provide for extreme hydrological events, new understanding of basin dynamics and changing societal values. Additionally, riparian states may consider prioritizing uses throughout the basin. Establishing catchment-wide precedents based on the agreed principles may not only help to avert interriparian conflicts over water use, but also protect the environmental health of the basin as a whole.

- Equitable distribution of benefits: this concept, subtly yet powerfully different from equitable use or allocation, is at the root of some of the world's most successful institutions. The idea concerns the distribution of benefits from water use – whether from hydropower, agriculture, economic development, aesthetics or the preservation of healthy aquatic ecosystems – not the distribution of water itself. This forms the basis for the concept of 'virtual' water and distributing water use benefits allows for agreements where both parties benefit. Multi-resource linkages may offer more opportunities for creative solutions to be generated, allowing for greater economic efficiency through a 'basket' of benefits. The Colombia River Basin Treaty (United States/Canada) provides an example of such an approach. A similar approach was used in 1994 to form the Southern African Power Pool (SAPP) in which power generated from a hydropower scheme in one country feeds a regional distribution network supplying countries without sufficient hydropower generation capacity. This has been a positive driving force for regional cooperation and the efficient use of energy. The value of the 'virtual water' concept in relation to food benefits is discussed in greater detail in chapter 8.

- Detailed conflict resolution mechanisms: many basins continue to experience disputes even after a treaty is negotiated and signed. Thus, incorporating clear mechanisms for resolving conflicts is a prerequisite for effective, long-term basin management. The Rhine River basin is a good example of this.

As probably all of the examples of hydrodiplomacy involve support from the international community, one may have to conclude that their encouragement and participation is an essential ingredient for success.

Identifying Indicators for Transboundary Basins

Little exists in the environmental security literature regarding empirical identification of indicators of future water conflict. The most widely cited measure for water resources management is Malin Falkenmark's (1989) Water Stress Index, which divides the volume of available water resources for each country by its population. Though commonly used, Falkenmark's index has been critiqued on a number of grounds, mostly that it accounts neither for spatial variability in water resources within countries nor for the technological or economic adaptability of nations at different levels of development. To account for the latter critique, but not the former, Ohlsson (1999) developed a Social Water Stress Index, which incorporates 'adaptive capacity' into Falkenmark's measure, essentially weighting the index by a factor based on the United Nations Development Programme's (UNDP) Human Development Index. While Ohlsson's is a useful contribution, he also misses the spatial component. Similarly, neither Falkenmark nor Ohlsson suggest much about the geopolitical results of scarcity, focusing instead on implications for water management.

The only author to explicitly identify indices of vulnerability that might suggest 'regions at risk' for international water conflicts is Gleick (1993). He suggested four indicators:

- ratio of water demand to supply;
- water availability per person (Falkenmark's Water Stress Index);
- fraction of water supply originating outside a nation's borders; and
- dependence on hydroelectricity as a fraction of total electrical supply.

Gleick's indices, like Falkenmark's and Ohlsson's, focus on the nation as the unit of analysis and on physical components of water and energy. These indicators were neither empirically derived nor tested.

The need for an empirical methodology

Another more empirical approach[6] to identifying indicators of potential for water-related conflict or cooperation in international basins has been taken by researchers at Oregon State University in a recent three-year study of all[7] international river basins (as defined in Wolf et al., 1999)[8] over the period from 1948 to 1999.

Utilizing existing media and conflict databases, a dataset was compiled of every reported interaction, where water was the driver of the event,[9] between two or more nations, whether conflictive or cooperative. Each interaction involved water as a scarce and/or

6. For more information, see Wolf, et al. (forthcoming).
7. This includes a total of 265 basins for historical analysis. There are currently 263 transboundary basins: 261 were identified by Wolf et al. (1999), two from that list were merged as one watershed as new information came to light (the Benito and Ntem), and three additional basins were 'found' (the Glama, between Sweden and Norway; the Wiedau, between Denmark and Germany; and the Skagit, between the United States and Canada). In historical assessments, we also include two basins that were historically international, but whose status changed when countries unified (the Weser, between East and West Germany; and the Tiban, between North and South Yemen).
8. In that paper, 'river basin' is defined as being synonymous with what is referred to in the United States as a 'watershed' and in the United Kingdom as a 'catchment', or all waters, whether surface water or groundwater, which flow into a common terminus. Similarly, the 1997 UN Convention on Non-navigational Uses of International Watercourses defines a 'watercourse' as 'a system of surface and underground waters constituting by virtue of their physical relationship a unitary whole and flowing into a common terminus'. An 'international watercourse' is a watercourse, parts of which are situated in different states (nations).
9. Excluded are events where water is incidental to the dispute, such as those concerning fishing rights, access to ports, transportation or river boundaries. Also excluded are events where water is not the driver, such as those where water is a tool, target or victim of armed conflict.

consumable resource or as a quantity to be managed. A database for water conflict/cooperation was produced comprising 1,831 events – 507 conflictive, 1,228 cooperative and 96 neutral or non-significant (see figure 12.2). The information was then incorporated into a Geographic Information System (GIS) with approximately 100 layers of global and/or regional spatial data covering biophysical (e.g. topography, surface runoff, climate), socio-economic (e.g. Gross Domestic Product [GDP], dependence on hydropower) and geopolitical (e.g. style of government, present and historic boundaries) factors. Relevant parameters were backdated and formatted for historical consistency (e.g. 1964 boundaries coincide with 1964 GDPs and government types), such that the historical context of each event of conflict/cooperation could be fully assessed. By identifying sets of parameters that appear to be interrelated, and testing each set using single and multivariate statistical analyses, factors that seemed to be indicators of conflict/cooperation were identified.

In general, it was found that most parameters usually identified as indicators of water conflict are only weakly linked to disputes. Institutional capacity within a basin, whether defined by water management bodies or international treaties, or more generally by positive international relations, was as important, if not more so, than the physical aspects of a system. On the basis of this work, then, rapid changes either in institutional capacity or in the physical system would seem to have historically been at the root of most water conflicts.

These changes were reflected by the following three indicators:

- 'internationalized' basins, i.e. basins that include the management structures of newly independent states;
- basins that include unilateral development projects and the absence of cooperative regimes; and
- basins where basin states show general hostility over non-water issues.

By extrapolating these findings and using the above indicators in a predictive context, basins with current characteristics that suggest a potential for conflicting interests and/or a requirement for institutional strengthening over the coming five to ten years, were identified. These basins, shown in map 12.5, included the Aral, Ganges-Brahmaputra, Han, Incomati, Jordan, Kunene, Kura-Araks, Lake Chad, La Plata, Lempa, Limpopo, Mekong, Nile, Ob (Ertis), Okavango, Orange, Salween, Senegal, Tigris-Euphrates, Tumen and

Map 12.5: Status of cooperation in transboundary river basins

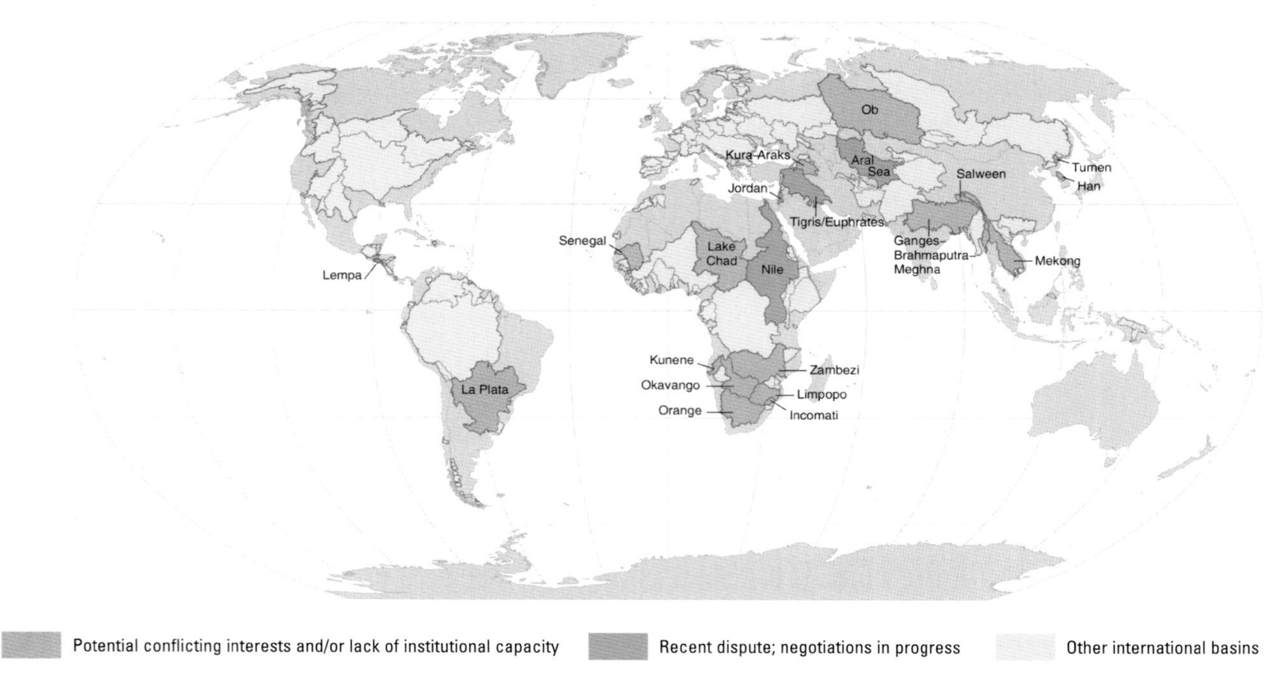

This map was extrapolated from indicators to pinpoint transboundary basins where current characteristics suggest a potential for conflict or a strengthening of institutions in the coming five to ten years.

Source: Wolf et al., forthcoming.

Zambezi. Of these basins, four (the Aral, the Jordan, the Nile and the Tigris-Euphrates) have been subject to recent dispute and are currently in various stages of negotiations.

Of possibly greater importance than this empirical analysis of the current and historical situation, is the potential for using these indicators to identify options for strategic institutional strengthening and diplomatic negotiations as basins become 'internationalized' or existing transboundary basins change in political or physical status. This ideal will require the adoption of suitable means of monitoring the indicators be so that changes can be identified at an early stage, therefore providing sufficient time to implement preventative measures. Information that is already available and that could provide a suitable handle on the indicators are listings of tenders for future water infrastructure projects (these are proposals for which funding is already available but which will usually have a lead time of three to five years), and details of increasingly active nationalist movements or unrepresented peoples. It is likely that other measurements will be identified with increasing experience.

Conclusions

As this chapter shows, despite the potential for dispute in international basins, the record of cooperation historically overwhelms the record of acute conflict over transboundary water resources. Indeed, there are a great many precedents, tools and instruments for cooperation, and a number of common mechanisms have been implemented within countries in order to protect the needs of all users, be they people, animals or ecosystems.

But the challenges have become increasingly important as population growth and development drive demand for the limited resource higher and higher. Intersectoral competition for water is growing, and water quality is more and more of an issue between upstream and downstream users as industrial, agricultural and domestic pollution takes its toll. Although progress has been made, the issue of sharing water has never been more timely, and there is an increasing urgency to develop sustainable and equitable means of sharing the resource.

Progress since Rio at a glance

Agreed action	Progress since Rio
Promote peaceful cooperation and develop synergies between different uses of water at all levels	
Manage river basins in a sustainable manner or find other appropriate approaches	
Sharing information for better sharing of resources	

Unsatisfactory — **Moderate** — **Satisfactory**

References

Almassy, E. and Buzas, Z. 1999. 'Inventory of Transboundary Ground Waters'. *UNECE Task Force on Monitoring and Assessment, under the Convention on the Protection and Use of Transboundary Watercourses and International Lakes.* Vol. 1. Lelystad, United Nations Economic Commission for Europe.

Arnold G. and Uil, H. 2001. 'International Initiatives on Monitoring and Assessment of Transboundary Groundwaters (the Implementation of the ECE Groundwater Guidelines in a Broader Perspective)'. Paper presented at the International Conference on Hydrological Challenges in Transboundary Water Resources Management, September 2001. International Hydrological Programme/United Nations Educational Scientific and Cultural Organization and Operational Hydrological Programme. Koblenz, Germany, Bundesanstalt fur Gewasserkunde (Federal Institute of Hydrography).

Dublin Statement. 1992. Official outcome of the International Conference on Water and the Environment: Development Issues for the 21st Century, 26–31 January 1992, Dublin. Geneva, World Meteorological Organization.

Falkenmark, M. 1989. 'The Massive Water Scarcity Now Threatening Africa—Why isn't it Being Addressed?' *Ambio*, Vol. 18, No. 2, pp. 112–18.

Federal Ministry for the Environment, Nature Conservation and Nuclear Safety, and Federal Ministry for Economic Co-operation and Development. 2001. Ministerial Declaration, Bonn Keys, and Bonn Recommendation for Action. Official outcomes of the International Conference on Freshwater, 3–7 December 2001, Bonn.

Gleick, P. 1993. 'Water and Conflict: Fresh Water Resources and International Security'. *International Security*, Vol. 18, No. 1, pp. 79–112.

IHP (International Hydrological Programme). 2001. *Internationally Shared (Transboundary) Aquifer Resources Management.* Non-serial Publications in Hydrology. Paris, United Nations Educational, Scientific and Cultural Organization.

Le Coz, D. 2000. 'Gestion durable d'une ressource en eaux souterraine – Cas de la Nappe de Beauce'. *La Houille Blanche*, Vol. 7/8, No. 66, p. 116.

Ministerial Declaration of The Hague on Water Security in the 21st Century. 2000. Official outcome of the Second World Water Forum, 3–7 December 2001, The Hague.

Mouray, V. and Vernoux, J.-F. 2000 (May). *Les risques pesant sur les nappes d'eau souterraine d'Ile-de-France*. Ile-de-France, Direction Régionale de l'Environnement d'Ile-de-France.

Ohlsson, L. 1999. *Environment, Scarcity and Conflict: A Study of Malthusian Concerns*. Department of Peace and Development Research, University of Göteborg.

Puri, S. 2001a (September). 'The Challenge of Managing Transboundary Aquifers – Multidisciplinary and Multifunctional Approaches'. Paper presented at the International Conference on Hydrological Challenges in Transboundary Water Resources Management, September 2001, International Hydrological Programme/United Nations Educational Scientific and Cultural Organization and Operational Hydrological Programme, Koblenz, Germany, Bundesanstalt fur Gewasserkunde (Federal Institute of Hydrography).

—— (ed.). 2001b. 'Internationally Shared Aquifer Resources Management – their significance and sustainable management'. A Framework Document: Non-serial Document in Hydrology, SC-2001/WS/40. Paris, United Nations Educational Scientific and Cultural Organization.

Roth, K.; Vollhofer, O.; Samek, K. 2001. 'German-Austrian Cooperation in Modelling and Managing a Transboundary Deep Groundwater Aquifer for Thermal Water Use'. Proceedings of the International Conference on Hydrological Challenges in Transboundary Water Resources Management, September 2001, Koblenz, Germany, German National Committee for International Hydrological Programme/United Nations Educational Scientific and Cultural Organization and Operational Hydrological Programme of the World Meteorological Organization.

Sydykov, Z.-S; Poryadin, V.-I; Vinniocova, G.-G; Oshlakov, V.-C; Dementiev; Dzakelov, A.-R. 1998. 'Estimation and Forecast of the State of Ecological-Hydrogeological Processes and Systems'. In: N. Aladin. *Ecological Research and Monitoring of the Aral Sea Deltas – A Basis for Restoration*. Aral Sea Project, 1992–1996 Final Scientific Research. Paris, United Nations Educational, Scientific and Cultural Organization.

UN (United Nations). 2000. *United Nations Millennium Declaration*. Resolution adopted by the General Assembly. A/RES/55/2.

——. 1992. Agenda 21. Programme of Action for Sustainable Development. Official outcome of the United Nations Conference on Environment and Development (UNCED), 3–14 June 1992, Rio de Janeiro.

——. 1978. *Register of International Rivers*. New York, Pergamon Press.

UNECE (United Nations Economic Commission for Europe). 2000. *Guidelines on Monitoring and Assessment of Transboundary Groundwaters*. Lelystad.

UNESCO (United Nations Educational, Scientific and Cultural Organization). 2001. *Internationally Shared (Transboundary) Aquifer Resources Management: A Framework Document*. Paris, IHP Non Serial Publications in Hydrology.

Van Dam, J.-C. and Wessel, J. (eds.). 1993. *Transboundary River Basin Management and Sustainable Development*. Vol. 2. Proceedings of the Lustrum Symposium Delft 1992. Paris, International Hydrological Programme/United Nations Educational, Scientific and Cultural Organization.

Van Wijk, C.; De Lange, E.; Saunders, D. 1996. 'Gender Aspects in the Management of Water'. *Natural Resources Forum*, Vol. 20, No. 2, pp. 91–103.

Wolf, A.; Natharius, J.; Danielson, J.; Ward, B.; Pender, J. 1999 (December). 'International River Basins of the World'. *International Journal of Water Resources Development*, Vol. 15, No. 4, pp. 387–427.

Wolf, A.; Yoffe, S.; Giordana, M. Forthcoming. 'International Waters: Identifying Basins at Risk'. *Water Policy*.

World Bank. 1998. International Watercourses – Enhancing Cooperation and Managing Conflict. Technical paper No. 414. Washington DC.

Some Useful Web Sites*

From Potential Conflict to Cooperation Potential (PCCP)

http://www.unesco.org/water/wwap/pccp/

In collaboration with Green Cross International, and part of the United Nations Educational, Scientific and Cultural Organization's (UNESCO) International Hydrological Programme (IHP), PCCP provides tools for aiding conflict resolution in transboundary water bodies. PCCP is also a UNESCO contribution to WWAP.

Green Cross International (GCI), Water Conflict Prevention

http://www.gci.ch/GreenCrossPrograms/waterres/waterresource.html

Aims to actively avoid and mitigate conflict in water-stressed regions. Provides news events, links, and bibliography.

International Water Law Project

http://www.internationalwaterlaw.org/

Joint United Nations initiative. Provides information, bibliography, and documents on water laws relating to transboundary water resources.

Transboundary Freshwater Dispute Database

http://www.transboundarywaters.orst.edu/

Searchable database of water-related treaties organized by basin, countries or states involved. Focuses on problems related to international waters.

*These sites were last assessed on 6 January 2003.

13

Recognizing and Valuing the Many Faces of Water

By: UNDESA (United Nations Department of Economic and Social Affairs)
Collaborating agencies: UNECE (United Nations Economic Commission for Europe)/World Bank

Table of contents

Recognizing and Valuing the Many Faces of Water

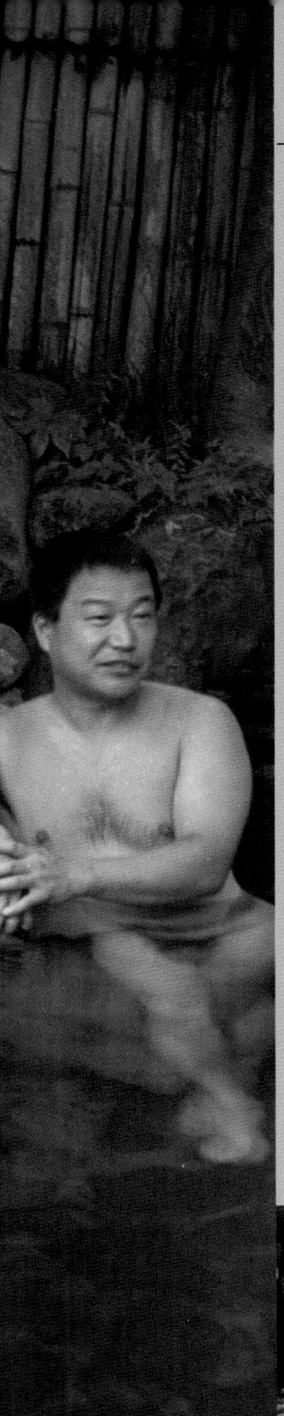

When the well's dry, we know the worth of water.

Benjamin Franklin

IN THIS CHAPTER WE DISCOVER THAT 'VALUE' is a multidimensional concept with many different meanings. The subject of valuing water is controversial and people get very emotional when they talk about it. Any discussion of value must therefore take into account people's perception of the world and their cultural and social traditions, as well as economic considerations and notions of full cost recovery. This chapter thus touches upon a number of issues important for policy-makers: investment strategies, public-private partnerships in providing water-related services, the polluter pays principle, resource allocation, gender considerations, community participation, accountability and governance. From the examples below, it seems that there exist as many approaches to valuing water as there are societies and cultural settings, but there is a growing consensus around general principles.

THE FOURTH GUIDING PRINCIPLE of the Dublin Statement[1], that 'water has an economic value in all its competing uses and should be recognized as an economic good', is immediately followed by an important qualifier: 'it is vital to recognize the basic right of all human beings to have access to clean water and sanitation at an affordable cost'. Although this principle is not directly quoted in Agenda 21 (UN, 1992), it is however detailed in Chapter 18 on Freshwater Resources (see box 13.1), with emphasis placed on the drinking water supply and sanitation subsector. Both the Dublin Statement and Agenda 21 wittingly or unwittingly tried to revise the conventional wisdom on the right of usage through 'prior appropriation' in order to take into account the social, economic and environmental values of water. The term 'economic value of water' was commonly referred to as the value imputed to its use in the productive process to emphasize that water should have a price. Because of this misunderstanding, several controversies have emerged in different parts of the world. This situation makes all the more clear the need for strategic approaches to freshwater management to be based on a set of well-defined principles so that they may progressively work towards the goals of equity and sustainability.

The equity concept in water use and management has been recognized as a central theme in the current debate on water issues discussed at the global level, notably in the sixth edition of the United Nations Commission for Sustainable Development (UNCSD, 1998), in the Ministerial Declarations of the Hague (2000) and Bonn (2001), and in the definition of the Millennium Development Goals (UN, 2000). This concept provides a direction towards maximizing the value of water among its various uses, while promoting 'equitable access and adequate supplies'. This chapter gives due consideration to these goals while reporting on progress made in the area of valuing water.

In recent international forums, it has been more clearly accepted that different users perceive the value of water differently. From a utility perspective, the same quality and quantity of water provide distinctly different values to consumers in different parts of the world. The value of water to people for domestic purposes is linked to their ability to pay, the use of the water (drinking, bathing, laundry, toilet flushing, garden watering), the availability of

Box 13.1: Water valuation and Agenda 21

Chapter 18 of Agenda 21 recommends the following economic measures for water management:

- Promoting schemes for rational water use through levying of water tariffs and other economic instruments, including the need for evaluation/testing of charging options that reflect true costs and ability to pay and for undertaking studies on willingness to pay.
- Charging mechanisms should reflect true cost and ability to pay.
- Developing transparent and participative planning efforts reflecting benefits, investment, protection, operation and maintenance (O&M) costs, and opportunity costs of the most valuable alternative use.
- Managing demand be based on conservation/reuse measures, resource assessment and financial instruments; changing perception and attitude so that 'some for all rather than more for some' be fully reflected in valuing water.
- Developing sound financial practices, achieved through better management of existing assets, and widespread use of appropriate technologies are necessary to improve access to safe water and sanitation for all.
- In urban areas, for efficient and equitable allocation of water resources, introducing water tariffs, taking into account different circumstances and, where affordable, reflecting the marginal and opportunity cost of water, especially for productive activities.
- In rural areas, providing access to water supply and sanitation for the unserved rural poor will require suitable cost recovery mechanisms, taking into account efficiency and equity through demand management.

1. The International Conference on Water and the Environment (ICWE), from which emerged the Dublin Statement, provided the major input for water to the United Nations Conference on Environment and Development held in Rio de Janeiro in June 1992.

alternative sources of inferior quality (river water may be a freely available source for bathing and laundry), and social factors. It is thus important to distinguish between the value of water, which is measured in terms of its benefit to the beneficiaries, the price of water (the charges to consumers) and the cost of supplying the water (the capital and operating costs of the works needed to abstract, treat and transfer the water to the point at which it is used). When aiming to meet the needs of the poor at affordable prices and to recover the costs of water supplies through tariffs, it is important to consider that water should not be sold at a price above the value placed on it by potential consumers. It will be necessary to adopt policies that provide an appropriate level of services to reconcile the need for equating water costs and prices with its value to the beneficiaries.

This chapter on water valuation has been prepared in the context of both developed and developing countries with the hope of helping to achieve internationally agreed targets such as those cited above. Examples have been drawn from as many different countries as possible to provide a global flavour to the effort. The concerns of individual countries have been recognized, with hopes that the allocation, demand management, water rights and pricing/subsidies being planned or implemented take due consideration of and help to achieve internationally agreed targets.

The Value of Water: Definitions and Perspectives

A controversial issue

A long-standing debate on how to value water has led to recognizing the need to have a clear analysis of what this means. Growth in population, increasing costs of water service delivery, changing consumption preferences, deterioration of water quality, dwindling supply and increasing realization of the opportunity cost of water are some of the 'eye-openers'. Most of these issues are discussed at greater length in other chapters of this report. However, a clear definition would advance the concept of valuing water both as an economic good and in terms of its social impact, which would then reflect better on the existing reality and promote resource equity. Controversy in valuing water using a pure market approach has originated from two sides: affordability by the poor and the marginalized, and externalities associated with the implementation of the fourth Dublin principle (cost recovery or other measures).

It is widely recognized that water has traditionally been regarded as a free resource of unlimited supply with zero cost at supply point and at best, water users have been charged only a proportion of the costs of extraction, transfer, treatment and

Box 13.2: Dreamtime and water in Aboriginal Australia

Australia is the driest inhabited continent on Earth. The bulk of the landmass is arid, most of it deserts, with extreme climatic conditions and little permanent surface water. The Aboriginal people occupy the whole territory, but the desert populations, in particular, have made water an intrinsic part of their culture.

The Aborigines have developed an intricate theory of their country's physical features, encompassing deep spiritual and mythical aspects. The foundation of this theory is the Tjukurrpa (the Dreamtime), which is a continuum wherein ancestors, animal or human, created the landscape through their actions.

The Dreamtime is part of an oral tradition, carried from generation to generation through stories and songs. In all of these, the relationship to water sources is fundamental. All water sources, whether permanent or intermittent, secret or public, were made by ancestors as they travelled and remain as living proof of their eternal presence. For that reason, water is very often the focus of sacred sites, some accessible to all group members, some reserved exclusively for youth, some for men and women's law and ceremonial business. These sites are primarily creeks, waterholes, springs and wells found throughout the desert, some of which are seasonal and others of which are 'living water', that is, permanent springs imbued with the life force of the ancestors.

Through the Dreamtime, the Aboriginal people identify places as part of their being with stories, and the mythological aspect of water is woven together with the many levels of understanding of the environment that the Aboriginal people possess. Water is the foundation of the Aboriginal people's beliefs. It is the physical manifestation of the process of creation itself, of the ancestors, and as such it is a sacred and protected element.

Source: Prepared for the World Water Assessment Programme (WWAP) by UNESCO, 2002.

disposal. All associated externality costs of water have been ignored and users are offered very little incentive to use water efficiently and not waste it. Major arguments for assigning price for the use of water have mostly originated from these concerns. Because costs of water supply delivery have escalated, it has become clear that economic measures such as pricing in general and demand management instruments have a distinct role to play in ensuring more efficient use of water.

The concept of water valuation is definitely not new. Communities and indigenous people have assigned religious and cultural values to water for generations (see box 13.2). The values of drinking water, domestic uses, irrigation and industrial uses have very often been socially established. Traditional management practices often reflect these socially determined norms for water allocation, demand management and sustainable practices.

The issue of water valuation was widely discussed during the Expert Group Meeting on Strategic Approaches to Freshwater Management held in Harare in 1998. The meeting considered valuing water within the broader context of Integrated Water Resources Management (IWRM) and came up with specific recommendations for discussion by the sixth session of the UNCSD 6, which specifically dealt with water resources issues. The meeting agreed on major guiding principles in valuing water, as outlined in box 13.3.

The Ministerial Declaration from the Second World Water Forum also identified the need for and the controversy surrounding the valuation of water and established valuing water as one of the seven challenge areas. Two specific targets were identified during the Forum: one is that the economic value of water should be recognized and fully reflected in national policies and strategies by 2005; the other that mechanisms should be established by 2015 to facilitate full cost pricing for water services while ensuring that the needs of the poor are guaranteed.

Box 13.3: Key recommendations for an integrated approach to freshwater resource management

Economics: Water planning and management needs to be integrated into the national economy, recognizing the vital role of water for the satisfaction of basic human needs, food security, poverty alleviation and ecosystem functioning, and taking into account the special conditions of non-monetary sectors of the economy.

Allocation: Water needs to be considered as a finite and vulnerable resource and a social and economic good, and the costs and benefits of different allocations to social, economic and environmental needs are to be assessed. The use of various economic instruments is important in guiding allocation decisions.

Accountability: It is essential to ensure efficiency, transparency and accountability in water resources management as a precondition to sustainable financial management.

Covering costs: All costs must be covered if the provision of water is to be viable. Subsidies for specific groups, usually the poorest, may be judged desirable within some countries. Wherever possible, the level of such subsidies and who benefits from them should be transparent. Information on performance indicators, procurement procedures, pricing, cost estimates, revenues and subsidies needs to be provided in order to ensure transparency and accountability, maintain confidence and improve investment and management capacities in the water sector.

Financial resources: Increased financial resources will need to be mobilized for the sustainable development of freshwater resources if the broader aims of sustainable economic and social development are to be realized, particularly in relation to poverty alleviation. Evidence that existing resources are being used efficiently will help to mobilize additional finance from national and international sources, both public and private.

Source: Extracts of Proceedings of the Expert Group Meeting for the United Nations Commission for Sustainable Development 6 (UNCSD 6), held in Harare, 1998. Published in UNDESA, 1998.

Conceptual premises and theoretical foundations

> *Water is needed in all aspects of life. For sustainable development, it is necessary to take into account water's social, environmental and economic dimensions and all of its varied uses. Water management therefore requires an integrated approach* (Ministerial Declaration, 2001).

A keynote paper on water valuation prepared by Peter Rogers (1997) was used as a framework for the Harare Meeting. This approach is shown in figure 13.1.

- Value to users of water: This value is calculated based on the marginal value of product, which is an estimate of the per unit output of industrial good or agricultural product for a unit of water use. These values are discussed in greater length in chapter 8 on food security and chapter 9 on industry.

- Net benefits from return flows: This is the value derived from the return flows, such as the value derived from aquifer recharge during irrigation, or downstream benefits of water diversion during hydropower generation. These values can be explained through the discussion of the hydrological cycle and consideration of evapotranspiration and sink losses. Chapter 4, on the natural cycle of water, makes an attempt to discuss such benefits.

Figure 13.1: Value of water

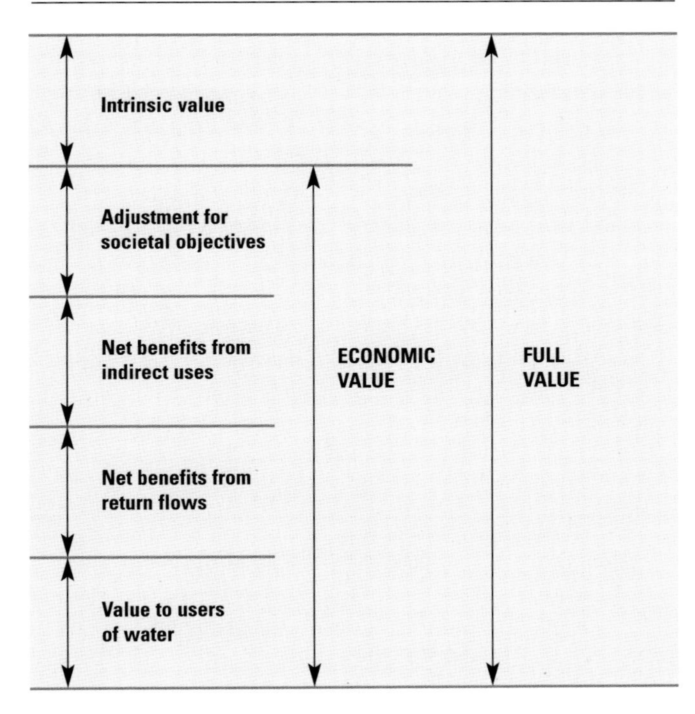

This figure shows the framework used by the Expert Group Meeting held in Harare in 1998, to provide conceptual clarity to valuing water.

Source: Rogers, 1997.

- Net benefit from indirect uses: Indirect benefits are derived when the water diverted for one purpose is used for another purpose. For example, in Bhutan, wastewater from small gravity water supply systems is also used for irrigating domestic gardens. Irrigation canals in India are also used for drinking water or other household chores. Consideration of such values is essential when calculating water value.

- Adjustment for societal objectives: Water supplied for irrigation and domestic use often also contributes to fulfilling societal objectives such as poverty alleviation, gender empowerment and food security. Adjustments that take into consideration such societal benefits are essential when valuing water, although they require a very careful method of estimation.

- Intrinsic value: In order to compute the full value of water, it is essential to take into consideration the bequest values and stewardship. It is widely recognized that the basic value derived from water often does not reflect environmental, social, cultural and other benefits gained from it. If this were the case, market economy instruments would suffice to analyse the value of water, and there would be no need to pay special attention to the issue. Environmental suitability, social and cultural worthiness and acceptance of resources management intervention are required. Such values are intrinsic to water and can be based on 'it is there' or 'it is not there' analyses.

In recent years, efforts have been made to estimate the intrinsic value of water using a national accounting system approach. This approach argues that the prices assigned are often notional, and the idea is to use them, rather than market prices, when elaborating public policies. However, the UNCSD 6 declaration called upon the member states,

> *when using economic instruments for guiding the allocation of water, to take into particular account the needs of vulnerable groups, children, local communities and people living in poverty, as well as environmental requirements, efficiency, transparency, equity and, in the light of the specific conditions of each country, at the national and local levels, the polluter pays principle* (UNDESA, 1998).

Value functions

> *The economics of water resources rarely influence water policy, even in water-short regions. As a result, the principal asset of the water resource base remains highly undervalued and readily used without much concern for its value to others, the structural role of water in the economy and its* in situ *value as an environmental asset* (UNDESA, 1998).

Water functions and values can be described from various points of view. Water contributes to a complex system of services and resources and each of these has an economic benefit, although it may not always be easy to estimate its value. Water benefits are not all the same: some are the result of economic activities; others have an indirect link with economic activities, while there are also benefits that do not come from economic activities. Defining water's different benefits or values is not only difficult but often also contentious. Protagonists of productive water use and those of water's nature values often engage in disputes over water use. However, valuing water is a means of providing information for participatory decision-making about water use. The value of water depends to a large extent on where it is available. Its value is site-specific and, because of this, it is also time-bound: water captured by a dam or in a lake can be used as and when required, while water in a river is only available when there is a flow.

As discussed above, water represents many values to society, and understanding the complex totality of these values is an important element in integrated water resources management. An assessment report completed by the Water Supply and Sanitation Collaborative Council (WSSCC), the World Health Organization (WHO) and the United Nations Children's Fund (UNICEF) in 2000 (WHO/UNICEF, 2000) found that water resources have traditionally been managed on the basis of water availability and historic demands, whereby priorities were set to serve one subsector before another. In about 50 percent of the countries, drinking water and sanitation have been given highest priority above other subsectors of water utilization.

Not surprisingly, the perceived value of water for domestic purposes is usually much higher than its value for irrigation. An important finding (similar to that emerging from the irrigation data) is that people, and notably poor people of developing countries, value a reliable supply much more than the intermittent, unpredictable supply commonly experienced (World Bank, 1993). In terms of opportunity cost, the short-term water value in hydropower in industrialized countries is typically quite low, often no higher than the value in irrigated agriculture (Gibbons, 1986). Long-term values are even lower. Whether or not hydropower is an economic proposition depends greatly on particulars – of the economy, of the power sector and of the water sector. In developing countries the demand for power is growing very rapidly. It has been put forth (Goodland, 1995) that the high environmental cost of alternative sources of power explains why hydropower is a particularly attractive option in many developing countries. It is generally stated that hydropower is a non-consumptive use and therefore does not impose costs on others. By modifying flow regimes and the timing of water to downstream users, however, hydropower installations can impose major costs on other users (see chapter 10 for more details on water and energy). The key issue is not consumptive or non-consumptive use, but the costs imposed on others by a particular use of a resource.

Surface water bodies and aquifers are generally hydraulically linked to aquatic ecosystems and they provide, usually on a seasonal basis, base flows that permit the good functioning of the ecosystems (e.g. the Lake Chad in West Africa or the Okavango Delta in Botswana). In return, water resources systems benefit from aquatic ecosystems that can play a role of buffer and filter (e.g. the Niger Delta in Mali, the Sud region on the Nile basin). Exchanges between water resources systems and aquatic ecosystems are usually intrinsically complex in terms of valuing water and insufficiently understood due to lack of monitoring. The multiple roles of the aquatic ecosystems confer a value to water and to humanity that may exceed the one derived from most other sources such as irrigation or hydropower. Costanza et al. (1997) valued the ecosystem services of different freshwater ecosystem types based on the Ramsar Convention on Wetlands and

Table 13.1: Value of aquatic ecosystem water services

Ecosystem types	Total value per hectare (US$ per year)	Total global flow value (US$ billion/ year)
Tidal marsh/mangroves	6,075	375
Swamps/floodplains	9,990	1,648
Lakes/rivers	19,580	3,231
Total		5,254

Global and per hectare values of ecosystems have been calculated based on the estimation of the indirect values of the aquatic ecosystems in flood control, groundwater recharge, shoreline stabilization and shore protection, nutrition cycling and retentions, water purification, preservation of biodiversity, and recreation and tourism.

Source: Costanza et al., 1997.

Box 13.4: Value of groundwater as a 'common property' resource

The value of groundwater – as distinct from surface water – is enhanced by virtue of its highly distributed occurrence. This occurrence reinforces groundwater's instrumentality (its value to others) in many complex and subtle ways. When attempting to manage groundwater, it is not a simple question of looking down a gradient (the upstream-downstream notion); instrumentality runs in all three spatial dimensions and also runs through a lagged temporal dimension. Therefore, the typical 'solutions of continuity' that would apply in the regulation of watercourses (catchments and basin boundaries) and the 'natural monopoly' conditions that apply in the provision of water services from single sources such as rivers or dams do not easily apply to aquifer systems, their linkages and their multiple users.

Groundwater is, in most situations, a common property resource with very high use value. It is also inherently vulnerable. Disposal of waste to aquifers and groundwater extraction affect neighbouring users in ways that are often difficult to predict and quantify. Conditions vary greatly between locations, and data, information and understanding are often lacking or in a form that cannot be readily understood by non-specialists who make up the bulk of the immediate users. As a result, there is little public awareness of groundwater, the benefits it confers and the limits under which it is available.

Lack of awareness and understanding of aquifer systems, combined with the common property nature of the resource, perpetuates chronic undervaluation of the resource base. Because few understand the complex nature of groundwater flow and contamination, the vulnerability of the resource base to irreversible damage through unconstrained use or waste disposal is rarely appreciated. At the same time, because the groundwater resource base common to large areas can easily be tapped through wells on small individual land holdings, its condition is dependent on the actions of many users. Unless each individual can be assured that others are behaving in ways that protect the resource base, the intrinsic value of the resource to individuals is low. Finally, because the full value of the resource is rarely reflected in the costs that individuals face when using services derived from groundwater, they are rarely aware of that value. As a result, individuals are often indifferent to the source of the water they are using and of the natural underground systems in which they dispose waste.

Very few formal attempts have been made to estimate both the direct and indirect uses and intrinsic values of groundwater. Economic or legal mechanisms through which intrinsic values could be communicated to groundwater users have yet to be systematically realized. One of the factors that complicate valuation issues is that many of the services provided by aquifer systems (such as base flow in rivers) depend on system characteristics. While there is scope for the direct use values to be documented through economic studies or communicated to users, documentation of the intrinsic values associated with aquifer systems is much more complicated.

Source: Burke and Moench, 2000.

estimated benefits in monetary terms based on energy and water interactions (see table 13.1). Per-hectare and total-flow values of ecosystems have been derived based on the estimation of indirect values of the aquatic ecosystems in flood control, groundwater recharge, shoreline stabilization and shore protection, nutrition cycling and retentions, water purification, preservation of biodiversity and recreation and tourism. Ecosystems are hugely diverse and share a complex interaction of their basic components – soil, animals and plants – that fulfil many functions and provide products for human use.

The value of water tends to change as society moves from rural-based small economies towards industrialization. Industry's share of total water withdrawals in the Organization for Economic Cooperation and Development (OECD) countries is increasing rapidly, although substantial variations have been reported between countries. The United Nations Industrial Development Organization (UNIDO) notes that although agriculture accounts for most water withdrawals worldwide (70 percent), industry currently accounts for 22 percent of all withdrawals. The current trend of water valuation in many countries of the world could change with the shift of water allocation away from agriculture, to other uses. Water valuation differs according to the water sources. In the last decade, progress has been made in valuing groundwater resources (see box 13.4).

To conclude, definite progress has been made in understanding the fact that the 'value of water' has economic, social, cultural and environmental dimensions, which are often interdependent. For example, the social value of water for 'health' has economic returns because a population in good health is more productive. Similarly, the environmental value of water has obvious economic and social implications. Such interactions can be characterized within three main clusters.

- Water is a vital common resource as it covers basic human needs and is required to sustain most life support systems.
- Water, in its productive capacity, helps to maintain economic activities and it has a fundamental role in managing other resources.
- Water provides both use and non-use benefits; it can generate taxes, derive products for consumption and help create employment of various kinds.

A recent survey by Buckley (1999) revealed that out of seventy-five water supply and sanitation projects financed by the World Bank and completed during 1990–97, only 33 percent were providing sustainable services. The remaining systems were reported to be unsustainable because of deficient operation and maintenance (O&M) and poor management. Country reports compiled by the United Nations Department of Economic and Social Affairs (UNDESA) as part of the World Summit on Sustainable Development (WSSD) further confirm that there is a growing cost of water services and only rarely is it considered as a value-free resource. For example, Japan's traditionally vested water rights became subject to a permit system in the 1960s, as illustrated in box 13.5. In other words, a resource unsustainably managed is a resource spent. The Heads of State in the Millennium Declaration (UN, 2000) stated their resolve to '... stop the

Box 13.5: Water rights – valuing water and historical background in Japan

Water resources in Japan are very tight as a result of meeting the copious demand of a large population. In order to cope with the limited resources, the water of Japan's rivers has, throughout history, been dominated by a large number of river users with vested water rights. These rights were generally founded and consolidated on customary practices, regional agreements and voluntary conciliation playing a primary role in the settlement of disputes so as to avoid drawing public authorities into the conflicts. Under these conditions, water rights evolved in a long historical process as substantive social rights. The agricultural interests with their customary right of water use accounted for almost the entire amount of water used from the major rivers around 1870.

The hallmark of this custom-based system of water use was that it tried to regulate disparate instances of water use. On the larger scale of the river basin as a whole, however, water used in the upper reaches was repeatedly used and reused in the lower reaches of the river. Irrigation ponds had a function of extending water use by collecting, storing and adjusting the water. In this way, a whole system in river basins maximized possible repetition of water use. The system was developed over the years to provide secured water availability when there was enough technology to increase the flow volume. The system was organized on the principle that old water rights had priority and that owners of established water rights would hold on to their rights with tenacity.

In the 1960s, the Japanese economy took off at a rapid pace. The explosive expansion in water demand could not be coped with. Since the demand for urban water, including drinking (household supply) water and industrial water, rose on a dramatic scale, water resource issues suddenly rose to prominence. Based on this consideration, the River Law, which controlled the river administration in Japan, was revised in 1964, and all water rights became subject to a permit system. However, the customary rights of the past were subject to the permit system authorizing the use of water under the same conditions as before. In order to create new potentials for water use through the construction of water resource development facilities such as dams, it has therefore been necessary to seek conciliation with these vested interests and file applications for development projects as constituting a new water right.

The river administrator has sought equitable, efficient and sustainable water use with respect to the historical background. To achieve sustainable water resource development and utilization, it is important to examine all possible measures, including water pricing, in order to identify the best strategy that fits local conditions or historical backgrounds.

Source: Prepared for the World Water Assessment Programme (WWAP) by the National Institute for Land and Infrastructure Management (NILIM) and Ministry of Land, Infrastructure and Transport (MLIT) of Japan, 2002.

unsustainable exploitation of water resources by developing water management strategies at regional, national and local levels'. Furthermore, efforts are underway to get this commitment to trickle to the lower levels. It has been increasingly realized that non-exploited water resources have many of the characteristics of a common good because no one can be excluded from using it in its natural state. Once water is mobilized and distributed, the resource is rarely properly utilized and the needs of others rarely adequately considered.

Complexities: theory versus practice

Although considerable progress has been made to better conceptualize the nature of water values and normative concerns, there are still many issues that hinder conversion of these theories into practice. A review of the country reports submitted to UNDESA provides the following insights into the complexities of such a conversion.

- The real value of water still tends to be vague because economic tools cannot fully assess social/cultural values or the intrinsic economic value of water.
- Most approaches to valuing water and analyses are complex and too 'academic' to be applied. For example, OECD (1999a) recommends the use of a hedonic tool to establish water prices for drinking water, a highly complex formula requiring special expertise for implementation.
- Many organizations have *ad hoc* definitions of sustainability and IWRM, some of which only consider the economic value of water as an important issue. A recent report based on a review of interventions concerning a water supply in Nepal (ICON/RWSSFDB, 2002) concluded that, apart from the World Bank-initiated water supply project, other projects do not recognize valuing water as one of the objectives to maximize while providing services to the beneficiaries.
- The strategic translation of normative understanding on valuing water into concrete actions has yet to take place and changes have to occur at the conceptual, local and operational levels.
- Water-related services are heavily subsidized, including in developed countries.
- Success stories are not uniform and do not easily lead to transferable practices, e.g. outcomes of irrigation management transfer efforts differ widely, and efforts to 'export' models of water management from France or from the Ruhr Valley in Germany have not been conclusive.
- A growing dichotomy exists among the social, environmental and financial objectives of IWRM: providing access to quality water and satisfaction of basic human needs, most notably for the poor, and at the same time, reducing national debts and being profitable to attract the private sector.
- IWRM can be seen as a purely integrating activity, stressing, for example, water resources assessments and development, multi-purpose resource planning or the integration of the interests of stakeholders within hydrological or administrative units. But a broader view sees IWRM as being concerned with the efficient use and enhancement of the value (in both direct and intrinsic senses) of water throughout the entire water service delivery chains, promoting more effective management within particular water use sectors or administrative areas. IWRM systems in different countries (or even within large countries) could exhibit large institutional variations, involving different types of organization (public, private, formal and informal), tools to allocate water, participatory mechanisms and policy implementation practices. System variations are essential to cope with the diversity of physical, political and socio-economic conditions within which the IWRM process must operate.

The Functions of Economic Water Valuation

Role in resource management and allocation

The most important role of water valuation relates to demand management and better allocation of water among its various uses. Supplying water requires a good knowledge of hydrology, hydrogeology and civil engineering, while demand is linked to various human and ecosystem needs, behaviours and processes, which are often time- and space-dependent and difficult to quantify and to forecast. Better water resource management requires decisions based on economic efficiency, social equity and ecological sustainability, and the value of water does not solely depend on its quantity but on at least four other factors – quality, location, reliability of access, time of availability. Box 13.6 provides an example from Sri Lanka, presenting how the collapse of an ancient hydraulic civilization, in existence for centuries, has affected demand management.

Although most countries recognize opportunity costs for water, the need for water supply for drinking and health as well as the need and attached costs of protecting aquatic ecosystems is beyond the strict focus of economics. Thus, valuing water is bridging the concern that water uses must be able to meet different social and environmental functions.

Box 13.6: Valuing water in Sri Lanka

Sri Lanka has a long tradition of developing hydraulic infrastructures and irrigation systems. Demand management has remained a basis for fair and equitable distribution of water among its various users. Traditions and the culture of the people have imposed sanctions against wastage of water for many years. In ancient royal decrees, fines for water wastage were specified to emphasize the need for optimum use of the limited resource.

These traditional water management systems were adversely affected by the collapse of the ancient hydraulic civilization. Where previously people could contribute to meeting the cost of development and maintenance by providing labour, more recent attempts to recover costs from farmers in cash for water and delivery services have had limited success. Where consumers have to pay for domestic and industrial water supply, charges are subsidized. Considering the socio-political situation in Sri Lanka, the options for greater contribution to O&M costs from users will probably come from greater participation by beneficiaries in the management of systems, thereby reducing the management cost for the state.

Source: Ministry of Irrigation and Water Management of Sri Lanka, 2002. Prepared for the World Water Assessment Programme (WWAP).

Role in financing

As the growing costs are adequately recognized, the concern for better valuation of water is also related to sector financing. There is significant ground still to be covered, in order to achieve the water-related Millennium Development Goals, particularly in the area of financing. A few issues become critical for analyses: what are the investment needs? What are the available sources? What is the actual investment and its gaps? The following sections present a brief assessment of these issues.

Financing the Millennium Development Goals is probably one of the most important challenges that the international community will have to face over the next fifteen years. It is unclear at the moment how much it will cost. In the case of water, wide-ranging estimates have emerged (see table 13.2). Not all of them are providing basis and hypothesis for estimating costs, which renders comparisons difficult. Different organizations have estimated that meeting the Millennium Development Goals on drinking water supply coverage would require between US$10 billion and US$30 billion a year on top of the amount already being spent.

Estimates for sanitation, for example, range from US$20 per person to US$500 per person per year. Further work is required to have a more accurate and better understanding of the global financial requirements to meet the water supply and sanitation Millennium Development Goals. One difficulty is the lack of knowledge in many developing countries of what can be rehabilitated and at what cost.

Total funding requirements for the whole water sector are estimated by three sources as ranging from approximately US$111 billion to US$180 billion a year (see table 13.3). The African

Table 13.2: Yearly funding requirements for water supply and sanitation

Source of estimation	Current funding (billion US$)	Required funding (billion US$)
WaterAid [a]	27–30	52–55
World Water Vision [b]	30	75
Vision 21 [c]	10–15	31–35
GWP [d]	14	30
IUCN (JpoA) [e]	10	20
J. Briscoe [f]	25	NA
Pricewaterhouse Coopers Report 2000 [g]	NA	30

Financing the Millennium Development Goals on water supply and sanitation is one of the most important challenges that the international community will have to face over the next fifteen years. Wide-ranging estimates have emerged, all stressing the lack of current funding.

Sources: (a) Narayan, 2002. (Director of WaterAid); (b) Cosgrove and Rijsberman, 2000; (c) WSSCC, 1998; (d) GWP, 2000; (e) IUCN, 2002; (f) Briscoe, 1999; (g) (Quoted from) UN, 2002.

Table 13.3: Yearly future funding requirements by the entire water and sanitation sector (in billion US$)

Water and sanitation sector component	Vision 21 [1]	PricewaterhouseCoopers Report [2]	IUCN [3]
Water supply and sanitation	75	30	20
Municipal wastes	–	70	–
Industrial effluents	–	30	–
Agriculture/food security	30	40	40
IWRM/protection	–	10	1
Hydropower	–	–	–
Environment, energy and industry	75	–	25
Water-borne diseases	–	–	25
Total	180	180	111

This table gives estimations of future requirements for the whole water sector. These global figures are only indicative of the magnitude of the challenge, although there is no consensus on exact requirement.

Sources: (1) Cosgrove and Rijsberman, 2000; (2) PricewaterhouseCoopers, 2001; (3) IUCN, 2002.

Ministerial Conference on Water recently announced that Africa requires US$10 billion a year to meet urgent water needs and an overall investment of US$20 billion a year for the development of water infrastructure in order to meet the Millennium Development Goals by 2015 (The Abuja Ministerial Declaration on Water, 2002). Global figures and ranges are only indicative of the magnitude of the challenge, although there is no consensus on exact requirements. In this regard, the most important step at this point would be to develop such estimates at the country level using a standard methodology with underlying assumptions adjusted to local conditions. Such an exercise is part of a sector investment plan and will help countries draw up strategies for resource mobilization and investment coordination.

It is quite clear that there is a massive investment gap and that the sources of financing are inadequate. The commitment by developed countries to provide 0.7 percent of Gross National Product (GNP) for official development assistance to developing countries is far from being met and this has caused a significant investment gap in development financing, including in the water sector. On this financing gap, the Ministerial Declaration (2001) reminded the developed countries to respect their commitments of 0.7 percent of GNP. The polluter pays and the user pays principles are difficult to apply because of poorly monitored legal or illegal water pollution and water abstractions.

Role in public-private partnerships

One of the options for financing urban water supply and sanitation infrastructure is to develop partnerships with the private sector. The concept of privatization (particularly when it involves divestiture of assets) is, however, controversial. It is argued that privatization can distort the notion of 'value' by replacing it with one of 'price' (GWP, 2000), and, in so doing, it can sideline all social objectives related to water. Maude Barlow, author of the book *Blue Gold: The Global Water Crisis and the Commodification of the World's Water Supply*, emphasized that

> *the privatization of municipal water services has a terrible record that is well documented. Customer rates are doubled or tripled; corporate profits rise as much as 700 percent; corruption and bribery are rampant; water quality standards drop, sometimes dramatically; overuse is promoted to make money; and customers who can't pay are cut off.... When privatisation hits the Third World, those who can't pay will die* (Barlow, 2000).

This diagnosis has not stopped many institutions and private companies from developed countries to continue promoting privatization of water and sanitation services as the only way to ensure efficiency and reliability of access. Long-term contracts continue to be executed between governments of developing countries and multinational water corporations with little or no external advice on their terms and conditions. It is claimed that protection of the poor will be insured if an adequate regulatory framework (mainly a regulating agency) is rapidly put in place. But developing such capacities cannot be achieved rapidly enough in most of the developing countries and the efficiency of new regulatory bodies remains questionable. Examples of these efforts to protect the poor can be found in Ghana, Indonesia, Nepal and Philippines.

The 'optimal' degree of privatization in water and sanitation services remains the subject of major debate. While significant gains in efficiency may be expected by privatizing some of the services, the 'natural monopoly' characteristics of water services argue in favour of a stronger public control in providing these services. In this context, the factual summary prepared by UNCSD (2001) notes that:

> *The effect of globalization is noticeable in the increasing interest of the private sector. However, globalization should not be seen as a panacea for sustainable water development and management. In order for globalization to take roots there needs to be sufficient funding, robust institutional structures, adequate human resources and a solid understanding and assessment of freshwater resources in relation to social, economic and environmental processes.*

In the preparatory process leading to the WSSD in Johannesburg (August/September 2002), all major groups (with the notable exception of business and industry groups) have raised concerns over privatization. It is essentially argued that control of the asset should remain in the hands of government and the users, and the needs of the poor should be duly considered.

There is an increasing involvement of private water companies in the urban water supply and sanitation of developed countries, as full or partial privatization is taking place in many OECD countries (OECD, 1999a). It is also notorious that world and regional development banks have been heavily criticized for pushing privatization as an investment agenda. In this context, the African Development Bank has been in the vanguard and has already adopted a policy statement on social issues (see box 13.7).

One valid criticism of public-private partnerships is that they may actually fail to increase investment. Where concessions for water utilities are won on the basis of the lowest tariff bid, the concession company may maximize its profit by minimizing investment. There is a risk that the existing assets will be run down, which may lead to a critical situation at the end of the concession.

Box 13.7: African Development Bank – policy statements on social issues

- Water is a social good and therefore a universal right and should be made available to all at an affordable cost.
- Gender issues should be taken into account in IWRM. The Bank will strongly support water resources development projects that show good prospects of reducing the time spent by women and girls in fetching and storing water.
- The Bank will bring up issues on Core Labour Standards in the process of discussions with Regional Member Countries and appraise programmes and projects in the water sector to ensure that they are in conformity with established criteria on labour.
- Where involved, the Bank will ensure that stakeholders are effective participants in all decision-making processes likely to affect them. Their willingness and capacity to pay for water resources development should be sought and not just assumed.
- The Bank will ensure that control and prevention measures on water-related diseases are integrated as part of the water management practices control and that prevention measures are based on proper health impact assessment and through effective stakeholder participation.

Source: AFDB, 2000.

Role in cost recovery

The need to move towards 'cost recovery' in providing water-related services is stated in the Ministerial Declaration of The Hague (2000). Most OECD countries have already adopted (or are in the process of adopting) 'full cost recovery' as an operating principle in the management of water resources. Definitions of exactly which costs should be included under this principle vary. The adoption of full cost recovery principles has been accompanied by significant reductions in total subsidies and in cross-subsidies between user groups.

The programme adopted by the Special Session of the General Assembly (UN, 1997) includes, with respect to water values, that

> *economic valuation of water should be seen within the context of its social and economic implications, reflecting the importance of meeting basic needs. Consideration should be given to the gradual implementation of pricing policies that are geared towards cost recovery.*

Many governments have taken action to put IWRM principles into action. While there is a consensus on broad principles, their implementation has been slow. A general conclusion on global progress is that water pricing has increasingly become an integral part of water sector management, but value of water is still insufficiently applied in the context of IWRM.

Agriculture is the largest water consumer worldwide. It is recognized that water pricing creates conditions for a more prudent use of the resource. Moreover, cost recovery has made water service providers more responsive to the demands of their users, resulting in improved management of irrigation systems. In Chile, a water market allows for water rights to be freely transferred within the irrigation subsector. Many countries have resisted liberalizing the water market, keeping a regulated form of allocation of water rights in order to secure the production of staple commodities. In such cases, water fees are established to recover costs. As in the drinking water supply subsector, many cases of cost recovery do not take into account investments in infrastructure, nor replacement costs. Table 13.4 shows

Table 13.4: Situation of cost recovery in irrigation for selected countries

Country	Investments in irrigation
Chile	Minimal development of large capital projects. Self-financing, supported by capital subsidies (to 45%) for improved technology.
China	In principle, beneficiaries pay. In practice, fee collection is erratic. Complex local financing and cross-subsidizing. Targeted state subsidies for major and special works.
India	Strong policy of construction in irrigation on equity grounds. Limited tariff-based cost recovery. Low recovery rates.
Indonesia	Continued high levels of expenditure in capital works (groundwater and rehabilitation). Cost recovery programme in rapid expansion phase.
Mexico	Continued development of large capital projects by state. Cost recovery of O&M.
Philippines	Major projects state-funded. Farmer funding for communal and small-scale projects. Cost recovery of O&M and some depreciation.

This table shows that many governments continue to invest in irrigation capital works, and that the payment received from water services is generally limited to the recovery of the costs for O&M.

Source: WWAP/Iwaco, 2002. Prepared for the World Water Assessment Programme (WWAP).

Box 13.8: Water pricing in Croatia

In Croatia, water user fees are determined by the decree of Hrvatske vode (Water Authority), based on the price level approved by the government. User fees are levied for all entities abstracting or drawing water directly from its natural sources. Companies supplying water to the consumers through public water supply systems collect the fee (part of the monthly water bills) and transfer the revenues to the Water Authority. Water user fees are calculated based on unit prices set by the government, the quantity of water used and a series of correction factors that reflect the intended use of water.

The water protection (pollution) fee is based on the price level approved by the government for a period of up to one year. In the case of direct discharge, the water protection fee is based on the direct measurements of pollution levels. Water supply companies collect the water protection fee directly from the polluters and transfer it to the Water Authority. The water protection fee for 1 cubic metre of discharged wastewater was €0.12 in 2000 (about US$0.12), this amount not having changed, in real terms, over the past ten years. The water protection fee is calculated based on the basic fee, the volume of water discharged and correction coefficients.

The penalties for non-payment of the water user fee, water protection fee or concession fee are defined by the Water Management Financing Act, which includes fines of €1,300–65,000 (about US$1,282–64,000).

The current price of water is sufficient only for financing essential maintenance and the most urgent investments. These expenses are fully recovered from the water tariffs, and it is often said that the price of water in fact defines the costs, and not the other way around. All the additional expenses that would be required to reduce the level of leakages, develop the water supply system and build necessary wastewater treatment facilities are not reflected in the existing water prices. Despite the provisions of the 1995 legislation (Water Act and Water Management Financing Act) aiming at full cost recovery by the respective groups of users, cross-subsidization between industrial and domestic sectors persists, mainly for social reasons. Except for the water use and water protection fees, economic instruments do not play an important role in water management. There are no waivers or tax exemptions for water abstraction and provision of wastewater services. Opportunities for trade in abstraction or effluent licences are not envisaged under the existing legislation.

Source: Adapted from Ostojic and Lukšic by UNDESA, 2001.

that many governments continue to invest in irrigation capital works, and that the payment received from water service is generally limited to the recovery of the costs for O&M. For many irrigation water users, the amount paid does not correspond to the costs of O&M.

An interesting development in the management of river basins comes under the Water Framework Directive (WFD) of the European Union (EEC, 2000). The Directive explicitly calls for decisions on river basin water management to be based on an understanding of the economic value of water. Since the adoption of the Framework Directive in September 2000, economic analysis of river basins has been declared a high-priority issue, for which member states need urgent guidance.

Another interesting approach is Australia's Strategic Water Reform Framework, which aims to restructure tariffs in line with the principles of: 'consumption-based' (i.e. volumetric) pricing and full cost recovery; reduction or complete elimination of subsidies; and transparency in essential subsidies and cross-subsidies. Educated and well-informed societies are a prerequisite for introducing these new options. According to the *Human Development Report 2001* (UNDP, 2001a), most developing countries do not have such well-informed societies.

Role in water pricing

Full cost recovery is linked to water pricing and water tariffs. Introduction of water tariffs infers that water has a recognized value. Growing unavailability of and increased competition for water resources require water demands to be well managed, so that water is efficiently used for valuable ends. Some forms of 'demand management must ultimately be applied and the use of pricing is an instrument which can be both effective and can be defended by rational and objective arguments' (Calder, 1998). Policies such as the 'No payment no project in Nepal' or the upfront cost contribution requirements in Pakistan may appear to be based on a stringent market-based approach, but in reality, they are treated quite similarly to other sectors policies. Dinar (2000) reported that *although water has several characteristics that make it different than other commodities, water-pricing reforms are affected by the same parameters as reforms in other sectors. A balance is sought between supply and demand, and the market mechanisms can play an important role in doing that, although governments may need to provide some form of check-and-control.*

For example, pricing for both use and pollution is levied in Croatia (see box 13.8).

Table 13.5 illustrates the wide range of variation in water pricing in developed countries of the world.

The practical implications of water pricing are contentious in many countries for the following main reasons:

- Water prices are expected to serve various and often conflicting purposes, including: cost recovery, economic efficiency and social equity.
- Water provides several services: drinking water, water for irrigation, hydropower, navigation, fisheries, tourism, pollution control, health of ecosystems. Each of these services are required to be valued differently, although in some cases services are provided simultaneously. This raises the question of how joint services should be valued and priced.
- Water resources are of critical importance for ecosystems and all efforts to embody this value in a price are fraught with conceptual difficulty and are highly debatable in practice.
- Social acceptance of paying for water is often low because of customary law and a long tradition of providing water at a minimal charge.
- Water prices are often made uniform across a country despite the fact that marginal costs of water mobilization and distribution vary between regions.
- There is often social strife and conflict between different water users within an area or between different regions within a country, lobbying for favourable terms of water supply.
- It is in many cases practically or economically not feasible to measure actual use or consumption of water. In such cases, water pricing becomes approximate and somewhat vague.
- The polluter pays principle can often not be applied because of uncontrollable (legal or illegal) water pollution.
- The user pays principle sometimes fails because of uncontrollable (legal or illegal) access to water resources.

As reflected in the above, the problem of water pricing is partly conceptual and partly institutional. At the Bonn Conference, the Ministerial Declaration acknowledged these problems, as did the major groups.

Table 13.5: Comparison of water pricing in developed countries

Country	US$/m^3
Germany	1.91
Denmark	1.64
Belgium	1.54
Netherlands	1.25
France	1.23
United Kingdom	1.18
Italy	0.76
Finland	0.69
Ireland	0.63
Sweden	0.58
Spain	0.57
United States	0.51
Australia	0.50
South Africa	0.47
Canada	0.40

These figures are based on supply for consumers in offices occupying 5,000 m^2 of city space and using 10,000 m^3/year. Developed countries show a wide range of variation in water pricing, from the lower cost in Canada to five times as much in Germany.

Source: Watertech Online, 2001.

Role in regulating water markets

Valuing water has an important role to play in regulating the water markets of the world. As mentioned before, in Chile, water rights can be freely traded within the irrigation subsector, like real estate property rights. About 30 percent of the households in Amman, Jordan, have decided to obtain additional water from the private market because accessible piped quantities are not sufficient. The choice to purchase water on the open market considers environmental and household characteristics and depends on the value prescribed to water by the households.

Economic valuation of water is at the core of water management: giving water a price helps to define how far we are from achieving the Millennium targets, and what economic efforts are needed. It helps to define a framework for sharing water in which all water users are fully responsible. It is an efficient tool, yet it has to be used with caution if water management is to promote the human right to water.

Box 13.9: The cost recovery approach in the Seine-Normandy basin

In the Seine-Normandy basin, and more generally in France, the bill paid by domestic and industrial users connected to the water system covers the cost of distribution and collection services: 'water pays for water'. This cost varies according to the local economic and technical configuration. In addition, a charge is levied by the Water Agency, which is made up of a charge on discharged pollution and a charge linked to resource withdrawal. The revenues drawn from these fees are redistributed by the Water Agency in the form of free loans or subsidies, which are intended to encourage the various end-users to improve the way they use the resource, via investments or better working techniques. The different categories of end-users are supported approximately in proportion to the fees they pay, although some transfers are made between both different types of end-users and different geographic zones in the basin.

The calculation of the pollution charge is in accordance with the polluter pays principle. The Water Agency revenues are entirely dedicated to supporting pollution reduction and clean-up actions. The nuisance caused by a specific water use over another one (negative externalities) still has to be measured and accounted for in the tariff. This is one of the WFD recommendations to be fulfilled by 2010.

In 1999, the price of water in the Seine-Normandy basin was, on average, 2.74€/m^3 (about US$2.70/m^3). On average, the amount spent on drinking water per household represented 1.03 percent of total income and 4 percent of housing costs. The water sector directly employs 18,700 people in the basin (in the drinking water and sanitation sectors) and represents an annual investment of about 60€/inhabitant per year (about US$60/inhabitant/year).

Source: Prepared for the World Water Assessment Programme (WWAP) by the Seine-Normandy Basin Agency (AESN), 2002.

The Changing Roles of Water Governance

Available options for better valuation of water and water infrastructure development depend very much on appropriate governance and institutional arrangements, whether it is a decision to get into contractual arrangements with operators for water and sanitation services or to set a tariff policy or to ensure better water allocation. With users becoming more involved in managing water resources, the concept of management transfer has been a central theme in valuing water. Programmes that transfer existing government-managed water systems to private firms, financially autonomous utilities and water user associations, are being implemented in many countries. In this context, governance modalities can be linked with water valuation and financing, as outlined below.

Everyone a part of the solution

Targets established through the Millennium Development Goals require adequate financing to be realized. As presented earlier, mobilizing the required financing is very difficult, and the existing absorption capacity of many governments is limited. From the point of view of water shortages, if consumers wish to avoid supply insecurities, they must 'buy' their security through contributions, which may, at times, be expensive. The role of consumers and other stakeholders is crucial in determining how to value water and introduce realistic prices. A preliminary review of fifty-nine country reports regarding progress in implementing Chapter 18 of Agenda 21, reveals that twenty-three countries have set up policies to involve stakeholders, including the private sector in water resources project development. For example, the Slovak Republic has formulated strong policies to value groundwater through people's participation, while the Ukraine's policies focus on dwindling supplies and encourage demand management through popular participation. Ghana has established a Water Resources Commission through an act of Parliament (1996) to ensure public participation in all aspects of water resource management. The Orangi project of Pakistan is one of the most lauded examples of the success of mobilizing resources through civic participation in improving the sanitation of a shantytown.

Decentralization and devolution

In the developing world, an approach that is increasingly being practiced is to decentralize water management responsibilities to the lowest possible levels of the administration. Examples exist in Sri Lanka and Thailand, where the municipal water supply responsibilities have been delegated to the municipal authorities. There has been a considerable effort in replicating the French Water Agency approach and its Associations Syndicales Autorisées (ASA).

The Associations recover investment costs and capital losses through water charges paid by farmers. Box 13.9 describes the system in the Seine-Normandy River basin.

It should be stressed however that unless adequate capacities exist at decentralized levels, such measures are not likely to be successful, in particular for protecting the interests of the poor. For example, price regulation by the municipalities or local authority may result in information asymmetry in developing countries. OECD (1999a) reports that lower-level authorities may not have the desired leverage to exercise control on service providers. The report also notes that the data collection and reporting procedures may also get complicated, as is the case in France or Italy. At the same time, the national reports to WSSD indicate that in many developing countries decentralization is regarded as a vehicle to achieve cost recovery and improved governance. For example in Ghana, the promulgation of a National Community Water and Sanitation Programme in 1994 first ensured community ownership and management with strong national and regional implementation. Since 2000, efforts have been made to further decentralize the development of water and sanitation projects through district assemblies. Similarly, Malawi's national Water Development Policy commits to decentralization and to bring on board communities, water boards, local authorities, the private sector, non-governmental organizations (NGOs) and government line agencies.

Polluter pays principle

As another tool for valuing water, the polluter pays principle has recently been adopted to reduce pollution by levying high charges on polluters. Apart from the OECD countries, actual implementation of this principle is rather slow in other countries of the world. However, a rapid assessment of country status reports to be delivered to the WSSD by fifty-nine countries of the world reveals that this policy has received wide recognition. The polluter pays principle has been duly included in water resources management and development policies and they are at different stages of implementation. However, it seems equally important to check whether policy and law are actually put into practice through decrees and regulations.

Technological options

Water and sanitation services can be better valued if appropriate technologies are selected. Although there is a small variation in the cost of developing different kinds of sanitation services, simple technologies such as the simple pit latrine, the ventilated pit latrine (VIP) and the small bore sewers are comparatively cheaper than other options. The International Research Centre for Water Supply and Sanitation (IRC) and the World Bank have shown that these technologies can be easily adapted to local conditions. Similarly a drilled borehole for drinking water supply in Asia can cost one fifth of a household connection to a water distribution network (WHO/UNICEF, 2000). Integrated Consultants Nepal (ICON/RWSSFDB, 2002) reports that small gravity water supply schemes managed by water users are one-third the cost of other available options and are also more sustainable. Thus, the challenge is to identify socially acceptable and relatively cheap solutions, and to avoid the errors committed in many countries with cheap technologies that have proved to be unsustainable.

Targeting Water Valuation

Although there is a clear need for proper valuation of water, it should be apparent that when the supply systems are deficient, the poor are the first to suffer. And as conditions of water stress develop, water becomes more expensive for those who are less privileged. A disturbing fact is that poor people with the most limited access to water supply have to pay significantly more for water (see table 13.6). The end result of these inequalities is that wealthier and more powerful persons can benefit from new opportunities at the expense of the poorer and less powerful. Therefore, if the attempt is serious about reducing inequality, it is a prerequisite that water valuation become more effectively 'targeted'.

It is estimated that more than 1.3 billion people in the developing world survive on less than a dollar a day and almost 3 billion survive on less than two dollars per day. Although the income of the upper and the middle classes in nearly every developing country has increased, the number of people living in poverty is rising at an above-average rate. In order to free people from the burden of disease and malnutrition, the need for secure access to water for the poor has been more strongly recognized. The Hague Ministerial Declaration also recognized that

> *combating poverty is the main challenge for achieving equitable and sustainable development, and water plays a vital role in relation to human health, livelihood, economic growth as well as sustaining ecosystems'* (Ministerial Declaration, 2000).

The legacy of publicly funded water services in excessive quantities to the few and at subsidized prices has created inefficient conditions resulting in severe environmental impacts on the resource itself. In many regions, the poor already subsidize those richest in society for their water use. According to NGOs lobbying for better interpretation of those challenges that were laid out during the Second World Water Forum:

Table 13.6: The poor pay more

City	Cost of water for domestic use (a) (house connections: 10 m³/month) US$/m³	Price charged by informal vendors (b) US$/m³	Ratio (b/a)
Vientiane (Lao PDR)	0.11	14.68	135.92
Male* (Maldives)	5.70	14.44	2.53
Mandalay (Myanmar)	0.81	11.33	14.00
Faisalabad (Pakistan)	0.11	7.38	68.33
Bandung (Indonesia)	0.12	6.05	50.00
Delhi (India)*	0.01	4.89	489.00
Manila (Philippines)	0.11	4.74	42.32
Cebu (Philippines)	0.33	4.17	12.75
Davao* (Philippines)	0.19	3.79	19.95
Chonburi* (Thailand)	0.25	2.43	9.57
Phnom Penh (Cambodia)	0.09	1.64	18.02
Bangkok* (Thailand)	0.16	1.62	10.00
Ulaanbaatar (Mongolia)	0.04	1.51	35.12
Hanoi (Viet Nam)	0.11	1.44	13.33
Mumbai* (India)	0.03	1.12	40.00
Ho Chi Minh City (Viet Nam)	0.12	1.08	9.23
Chiangmai* (Thailand)	0.15	1.01	6.64
Karachi (Pakistan)	0.14	0.81	5.74
Lae* (Papua New Guinea)	0.29	0.54	1.85
Chittagong* (India)	0.09	0.50	5.68
Dhaka (Bangladesh)	0.08	0.42	5.12
Jakarta (Indonesia)	0.16	0.31	1.97
Colombo* (Sri Lanka)	0.02	0.10	4.35

*Some water vending, but not common.

When the supply systems are deficient, the poor are the first to suffer. This table shows that in some countries, water from informal vendors is more than 100 times more expensive than water supplied by house connection.

Source: ADB, 1997.

Water is a public good, and all funds raised through full cost recovery must be reinvested in the provision of improved water services for people and the environment. Suitable mechanisms need to be developed for full cost transparency and classification of water price according to its quality.[2]

Although 'subsidies undermine the economic approach' and may obliterate the potential of eliminating wasteful practices and encouraging increased efficiency and conservation (Gleick, 1998), there is a growing consensus that if the subsidized water can generate positive social effects such as maintaining the rural landscape and traditions, supporting local economies or contributing to food security levels, subsidized water charges may be justified.

2. http://www.earthsummit2002.org/freshwater/ngo_major_group_discussion_paper.htm

Methodological Advances

Since the recognition of valuing water as one of the principal challenges in sustainable development and use of water resources, several important methodological developments have taken place. A few important steps are listed below.

- Major value domains are now better studied and elements of the total value for water have been prepared.
- Boyle and Bergstrom (1994) and National Research Council of the United States (NRC, 1997) have developed a cross-cutting analysis that lists the services derived from water and their effect on the value of water.
- There have been major methodological advances in terms of prices, productivity, hedonic pricing, travel cost, damage cost

avoided, replacement cost, substitute cost, contingent valuation and contingent choice in advancing the context of valuation.

- The development of valuation techniques has been greatly supported by the development in computing capabilities (see chapter 14 on ensuring the knowledge base for details).

- The quality of and access to databases have greatly improved, thus enabling the undertaking of water valuation studies.

Conclusions

The formulation of national policies that include an economic approach is the first step towards proper valuation of water. In this context, due consideration needs to be given to the opportunity costs of water as well as to environmental externalities to achieve water-related and internationally accepted goals, such as meeting basic needs and food security. In combination or independently, valuing water will require policies that can help realize normative reforms and well-prepared introduction of participatory and market-based instruments to meet the broad objective of sustainable water resources management.

The water sector interacts with almost all other sectors of the economy, and could potentially become a binding constraint on economic expansion and growth. This is especially true because while the amount of renewable water resources is practically fixed, water demands will continue to grow in the years ahead due to population growth, increased food demand and expansion and modernization of the industrial sector. Thus, the economic challenge is to maximize social and economic benefits from available water resources while ensuring that basic human needs are met and the environment is protected. This means implementing IWRM principles and mechanisms leading to the efficient allocation and use of water resources, rehabilitating and improving the performance of existing water supply systems and making future investments sustainable.

Identifying and mobilizing additional resources remains an important challenge. New exploitations of water resources – conventional or not – and minor adjustments in resource allocation patterns can have a significant impact in terms of extending the coverage. Establishment of an enabling environment that facilitates public-private partnerships could also contribute in meeting these challenges.

Allocation mechanisms should balance competing demands both within and between different water-using sectors as well as between countries and should incorporate the social, economic and environmental values of water. There is no magic formula because of the wide variability of country-specific conditions. Currently, water remains highly undervalued. The problem of cross-subsidization across sectors and different user groups makes it even more important to allocate water optimally, based on its value for different uses.

Progress since Rio at a glance

Agreed action	Progress since Rio
To manage water in a way that reflects its economic, social, environmental and cultural values for all its uses	
Move towards pricing water services to reflect cost of their provision	
Taking account of need for equity and the basic needs of the poor and vulnerable	

Unsatisfactory — Moderate — Satisfactory

References

The Abuja Ministerial Declaration on Water – a Key to Sustainable Development in Africa. 2002. African Ministerial Conference on Water (AMCOW), 29–30 April 2002, in Abuja.

ADB (Asian Development Bank). 2001. *Water for All: The Water Policy of the Asian Development Bank.* Manila.

——. 1997. *Second Water Utilities Data Book.* Manila.

AFDB (African Development Bank). 2000. *Policy for Integrated Water Resources Management.* OCOD (Organization for Cooperation in Overseas Development).

Albiac, J. 2002. 'Water Demand Management versus Water Supply Policy: the Ebro River Water Transfer'. Paper presented at the Water Forum 2002, World Bank, 6–8 May 2002, Washington.

Barlow, M. 1999. *Blue Gold: The Global Water Crisis and the Commodification of the World's Water Supply.* Council of Canadians, IFG Committee on the Globalization of Water.

——. 2000. 'Water Is a Basic Human Right – or Is It?'. *Toronto Globe and Mail,* Toronto (also quoted in *NEXUS New Times Magazine,* August-September 2000, Vol. 7, No. 5).

Boyle, K.-J. and Bergstrom, J.-C. 1994. *A Framework for Measuring the Economic Benefits of Ground Water.* Department of Agricultural and Resource Economics Staff Paper. Orono, University of Maine.

Briscoe, J. 1999. 'The Financing of Hydropower, Irrigation and Water Supply Infrastructure in Developing Countries'. *International Journal of Water Resources Development,* Vol. 15, No. 4, pp. 459–91.

——. 1998. 'The Changing Face of Water Infrastructure Financing in Developing Countries'. *International Journal of Water Resources Development,* Vol. 15, No. 3, pp. 301-8.

——. 1997. 'Managing water as an economic good'. In: M. Kay; T. Franks; L. Smith (eds.), *Water: Economics, Management and Demand.* London, E and FN Spon.

Bryce S. 2001. 'Hydrodollars: Water Privatisation'. *NEXUS New Times Magazine,* Vol. 8, No. 3.

Buckley, R. 1999. *1998 Annual Review of Development Effectiveness.* Washington, DC, World Bank, Operations Evaluation Department.

Burke, J. and Moench, M. 2000. *Groundwater and Society: Resources, Tensions and Opportunities.* New York, UNDESA (United Nations Department of Economic and Social Affairs), E.99.II.A.1.

Burke, J.; Sauveplane, C.; Moench, M. 1999. *Groundwater Management and Socioeconomic Responses. Natural Resources Forum,* Vol. 23, pp. 303–13.

Calder, I.-R. 1998. 'Water use by forests, limits and controls'. *Tree Physiology,* Vol. 18, pp. 625–31.

Cosgrove, W. and Rijsberman, F.-R. 2000. *World Water Vision: Making Water Everybody's Business.* London, UK, World Water Council and Earthscan Publications Ltd.

Costanza, R.; d'Arge, R.; de Groot, R.; Farber, S.; Grasso, M.; Hannon, B.; Limburg, K.; Naeem, S.; O'Neill, R.; Paruelo, J.; Raskin, R.; Sutton, P., van den Belt, M. 1997. 'The value of the world's ecosystem services and natural capital'. *Nature,* Vol. 387, pp. 253–60.

Dinar, A. (ed.). 2000. *The Political Economy of Water Pricing Reforms.* New York, World Bank and Oxford University Press.

Dublin Statement. 1992. Official outcome of the International Conference on Water and the Environment: Development Issues for the 21st century, 26–31 January 1992, Dublin. Geneva, World Meteorological Organization.

EEC (European Economic Community). 2000. Framework Directive in the Field of Water Policy (Water Framework). Directive 2000/60/EC of the European Parliament and of the Council of 23 October 2000, establishing a framework for Community action in the field of water policy [Official Journal L 327, 22.12.2001].

Ehrlich, P.-R., Ehrlich A.-H., Daily, G. 1995. *The Stork and the Plow: The Equity Answer to the Human Dilemma.* New York, G.P. Putnam's Sons.

FAO (Food and Agriculture Organization). 2001. *The State of Food and Agriculture.* Rome.

——. 1999. Irrigation and Drainage Paper # 58. Rome.

Federal Ministry for the Environment, Nature Conservation and Nuclear Safety, and Federal Ministry for Economic Co-operation and Development. 2001. Ministerial Declaration, Bonn Keys, and Bonn Recommendations for Action. International Freshwater Conference, 3–7 December 2001, Bonn.

Gibbons, D.-C. 1986. *The Economic Value of Water.* Washington DC, Resources for the Future.

Gleick, P.-H. 2001. 'Making Every Drop Count'. *Scientific American,* February, pp. 28–33.

——. 1998. *The World's Water 1998–1999: The Biennial Report on Freshwater Resources.* Washington DC, Island Press.

Goodland R. 1995. 'Environmental Sustainability Needs Renewable Energy: The Extent to Which Big Hydro is Part of the Transition'. Paper presented to the International Crane Foundation Workshop, 28 November – 2 December, 1995, Washington, DC.

Gutierrez, E. 1999. 'Boiling Point: Issues and Problems in Water Security and Sanitation'. A WaterAid Briefing Paper. London.

GWP (Global Water Partnership). 2000. *Toward Water Security: A Framework for Action to Achieve the Vision for Water in the 21st Century.* Stockholm.

ICON/RWSSFDB (Rural Water Supply and Sanitation Fund Development Board). 2002. *Detailed Demand Assessment Study.* Nepal.

Irrigation Newsletter. 2001. No. 54. 'Kathmandu, Irrigation Development and Management in Nepal'. Department of Irrigation, Irrigation Management Division.

IUCN (The World Conservation Union). 2002. *Johannesburg Programme of Action.* A document prepared for the World Summit on Sustainable Development (WSSD), 28 August-4 September, Johannesburg.

Ministerial Declaration of The Hague on Water Security in the 21st Century. 2000. Official outcome of the Second World Water Forum, 3–7 December 2001, The Hague.

Narayan, R. 2002. 'Sustainability: Water and the Choices We Face'. *Earth Times,* August.

NRC (National Research Council). 1997. *Valuing Ground Water: Economic Concepts and Approaches.* Washington, Committee on Valuing Ground Water, WSTB (Water Science and Technology Board), Commission on Geosciences, Environment, and Resources (CGER), NRC (National Research Council). Washington DC, National Academy Press.

NEPAD (New Partnership for Africa's Development). 2001. *The New Partnership for Africa's Development.*

OECD (Organization for Economic Cooperation and Development). 1999a. *Household Water Pricing in OECD Countries.* ENV/EPOC/GEEI(98)12/FINAL. Paris, Working Party on Economic and Environmental Policy Integration, Environment Directorate, Environment Policy Committee.

——. 1999b. *Agricultural Water Pricing in OECD Countries.* ENV/EPOC/GEEI(98)11/FNAL77608. Paris, Working Party on Economic and Environmental Policy Integration, Environment Directorate, Environment Policy Committee.

——. 1992. *The Polluter-Pays Principle, OECD Analyses and Recommendations.* OECD/GD(92)8. Paris, Organization for Economic Cooperation and Development.

Ostojic, Ž.; Lukšic, M. 2001. *Water Pricing in Croatia, Current Policies and Trends.* Zagreb, Regional Environmental Center for Central and Eastern Europe.

Pitman, G.-K. 2002. *Bridging Troubled Waters: Assessing the Water Resources Strategy Since 1993.* Washington, Operation Evaluation Division, World Bank.

'Polluter Pays Principle'. 2001. In: *Economic Issue of the Day*, Vol. II, No. 3.
PricewaterhouseCoopers. 2001. *Water: A World Financial Issue – a Major Challenge for Sustainable Development in the 21st Century*. Sustainable Development Series. Paris.
The Ramsar Convention on Wetlands. 2001. *Background papers on Wetland Values and Functions*. Gland.
Renwick, M.-E. 2001. 'Valuing Water in Irrigated Agriculture and Reservoir Fisheries: A Multiple Use Irrigation System in Sri Lanka'. IWMI Research Report 51. Colombo, International Water Management Institute.
Roger, P. 1997. 'Integrating Water Resources Management with Economic and Social Development'. Paper prepared for the United Nations Department of Economic and Social Affairs Expert Group Meeting, 1998, Harare.
Rogers, P.; Bhatia, R.; Huber, A. 1998. 'Water as a Social and Economic Good: How to Put the Principle into Practice Technical Advisory Committee', Background Paper No. 4. Stockholm, Global Water Partnership.
Tiwari D.; Dinar, A. 2001. *Role and Use of Economic Incentives in Irrigated Agriculture*. Washington DC, World Bank.
Tokyo Conclusion of the Workshop ICID. 2000. 'Toward Sustainable Development in Paddy Agriculture'. International Commission on Irrigation and Drainage.
UN (United Nations). 2002. A Framework for Action on Water and Sanitation. New York, WEHAB Working Group.
——. 2000. United Nations Millennium Declaration. Resolution 55/2 of the United Nations General Assembly. New York.
——. 1998. *Report of the Commission on Sustainable Development on its Sixth Session*. New York, E/CN.17/1998/20.
——. 1997. Programme for the Further Implementation of Agenda 21. Resolution Adopted by the General Assembly, S/19-2. New York.
——. 1992. Agenda 21: Programme of Action for Sustainable Development. Official outcome of the United Nations Conference on Environment and Development (UNCED), 3–14 June 1992, Rio de Janeiro.
UNCSD (United Nations Commission on Sustainable Development). 2001. *Water – a Key Resource in Sustainable Development*. Report of the Secretary General. New York, United Nations.
——. 2000. *Progress Made in Providing Safe Water Supply and Sanitation*. New York, United Nations.
——. 1998. *Report of the Expert Meeting on Strategic Approaches to Freshwater Management*. New York, United Nations.
UNDESA (United Nations Department of Economic and Social Affairs) 1998. Expert Group Meeting Report of Working Group 3, Economic and Financing Issues. New York.
UNDP (United Nations Development Programme). 2001a. *Human Development Report: Making New Technologies Work for Human Development*. New York.
——. 2001b. *Nepal Human Development Report, Poverty Reduction and Governance*. Kathmandu.
UNESCO (United Nations Educational, Scientific and Cultural Organization) MAB (Man and the Biosphere). 2002. *Cultural Value of Water in Australia*. Paris.
UNICEF (United Nations Children's Fund). 2002. *The State of the World's Children 2002*. New York.
WHO (World Health Organization) and UNICEF (United Nations Children's Fund). 2000. *Global Water Supply and Sanitation Assessment 2000 Report*. New York.
World Bank 1993. Policy Paper on Water Resources Management. Washington DC.
WSSCC (Water Supply and Sanitation Collaborative Council). 2000. *Vision 21: Water for People – A Shared Vision for Hygiene, Sanitation and Water Supply and A Framework for Action*. Geneva.

Some Useful Web Sites*

United Nations Statistics Division

http://unstats.un.org/unsd/

United Nations global statistics system on socio-economic and many other data fields.

World Bank: Data and Statistics

http://www.worldbank.int/data/

Data, statistics and indicators on national economic accounts (income, GDP, GNI, exchange rate), poverty, life expectancy, agriculture, education, etc.

World Bank: Water Economics

http://lnweb18.worldbank.org/ESSD/essdext.nsf/18ByDocName/SectorsandThemesWaterEconomics

Economic aspects of water including the economic analysis of water resources and water supply and sanitation projects and looks at recent advances in water economics methodology, tools and applications.

* These sites were last accessed on 06 January 2003.

14

Ensuring the Knowledge Base: A Collective Responsibility

By: UNESCO (United Nations Educational, Scientific and Cultural Organization)/WMO (World Meteorological Organization)
Collaborating agencies: UNDESA (United Nations Department of Economic and Social Affairs)/ IAEA (International Atomic Energy Agency)/World Bank/UNEP (United Nations Environment Programme)/ UNU (United Nations University)

Table of contents

Tell me and I forget. Show me and I remember.
Involve me and I learn.

Anonymous

KNOWLEDGE DOES NOT EXIST IN ISOLATION, and understanding even less so. This chapter investigates some of the subtle but powerful ways in which education, training, public information, cultural traditions, the media and modern telecommunications interact with each other to influence the way we behave and respond to change in the world around us. Our attitudes begin forming at a very early age, and those in a position to affect the way we perceive and manage our water resources must pay as much attention to non-formal communication as to formal education. Public awareness, community involvement and the inclusion of all actors – but especially women – in decision-making are keys to success. In this context, such high-tech tools as computers, geographic information systems (GIS) and electronic databases are almost irrelevant as long as millions of people are deprived of basic education, health and food.

KNOWLEDGE IS CRUCIAL to improved livelihoods, environmental conservation, broader participation and stronger democracies – in short, to development. The unprecedented revolution ushered in by the new information and communication technologies (ICTs) has placed an ever higher premium on knowledge as having a potential to not only generate wealth, but also contribute to sustainable development for the benefit of present and future generations. Generating and disseminating this knowledge requires political will, investment and international cooperation: towards expanding education, facilitating scientific research, building capacities at all levels and bridging the gap between rich and poor.

In the complex domain of water, a narrow definition would describe the knowledge base as all aspects of data gathering, information, experience and knowledge enabling countries and regions to produce a well-documented assessment of their water resources. However, because water affects every facet of life, from health to agriculture, to industry and the ecosystem at large, the knowledge and skills required to improve stewardship of this finite resource stretch across an extremely broad spectrum encompassing education, health, law, economics, communications, and science and technology. In particular, the constituencies required to govern water wisely cut across all socio-economic categories and generations to include grass roots communities, industry and business leaders, health specialists, educators, lawyers, scientists, engineers and government agencies.

Since the United Nations Conference on Environment and Development (UNCED) in 1992, the international community has made considerable efforts to raise awareness about water resources, building upon the goals of equity and sustainability elaborated during previous conferences, such as the 1990 New Delhi Global Consultation on Safe Water and Sanitation and the 1992 Dublin International Conference on Water. Following along the same lines, the 2000 World Water Vision reiterates that integrated management is the bottom line for addressing the current crisis. It relies *inter alia* on empowering women, men and communities, increasing public funding for research and innovation in the public interest and improving cooperation in international water basins. At the Second World Water Forum in The Hague in 2000, the Global Water Partnership (GWP) spelt out targets related to strengthening the knowledge base: water awareness initiatives instigated in all countries by August 2001, capacity for informed decision-making at all levels and across all stakeholders increased by 2005, investment in research on water issues increased by August 2001, hygiene education in 80 percent of all schools by 2010. These targets highlight the significance of awareness-raising, training and research, and sound data, and also underscore the evident links between better health and safer water practices.

Formidable barriers still stand in the way of progress. Driven by the revolution in information technology and increasing mobility of capital, globalization is generating new wealth and greater interconnectedness and interdependence. It has the potential to reduce inequality and poverty, but also holds the danger of creating a market place in knowledge that excludes the poor and the disadvantaged. Population growth, HIV/AIDS (Human Immunodeficiency Virus/Acquired Immune Deficiency Syndrome) and armed conflict have all contributed to increasing poverty and social inequalities in recent years. Climate change, environmental degradation and rapid urbanization make societies more vulnerable to disasters. The rise of market-oriented research risks undermining science as a public good, capable of responding to pressing environmental and social problems.

The media, both print and electronic, play a pivotal role in raising awareness of water's value, promoting safe health practices and facilitating dialogue between stakeholders. Furthermore, because water and climate know no national boundaries, initiatives in sharing knowledge require a high degree of international collaboration. The rise in information technology has made it easier to spread knowledge and pave new avenues for learning. However, the digital divide leaves large swathes of the globe isolated from accessing and contributing to this exchange.

Although no formal monitoring process has been established to track progress, trends in knowledge can be identified on several fronts, in more participatory approaches to managing water, in collaborative ventures to improve assessment of water resources, in the strengthening of global networks for sharing knowledge and advancing pro-poor policies to ensure water security. Adopting a broad approach that reflects the aspiration towards more integrated management of water resources, this chapter paints a picture of trends and progress by placing knowledge about water in the wider context of education, science and communication worldwide, while drawing more specific attention to community participation, data collection, basic and applied research and international cooperation.

Holistic Thinking: Making the Knowledge Base Everyone's Business

The notion of integrated management mirrors a progressive shift in our approach to development towards an all-inclusive endeavour that takes stock of the intricate connections between society, culture, science and the environment. It relies upon greater participation at all levels and profound changes in the relationship between state and society, with implications for the ways in which knowledge is created, acquired and shared.

In 1977, the Mar del Plata Conference underscored the importance of water resources management as a global issue; one that required the capacities of engineers, economists and environmentalists. Since then, the circle of stakeholders has grown steadily, spanning the natural and social sciences. Since UNCED, economic expansion, demographics and climate instability have raised water resource development and management to the status of a global social, economic and environmental issue. Rio was but one in a series of major international United Nations (UN) conferences spanning the 1990s which, working along the same lines, reaffirmed that people are at the centre of development, with education playing a determining role in alleviating poverty and building a more sustainable future. A clear understanding has emerged that knowledge is critical to achieving the goals of global security, poverty alleviation and water resource stewardship expressed in Agenda 21 (see box 14.1).

In the words of former World Bank Vice President Joseph Stiglitz:

> *Today, we recognize that knowledge is not only a public good, but a global or international good. We have also come to recognize that knowledge is central to successful development. The international community ... has a collective responsibility for the creation and dissemination of one global public good – knowledge for development* (World Bank, 1998).

Ensuring basic education, creating opportunities for lifelong learning and supporting tertiary education, especially in science and engineering, is one of the critical steps that must be taken to narrow knowledge gaps.

Education: The Base of the Knowledge Pyramid

Education – whether in the formal school setting, through a community-led initiative or in the context of an adult literacy class – is a fundamental pillar of human rights and a key to achieving sustainable development. It plays a critical role in empowering individuals with the knowledge and skills to reflect, make choices and enjoy a better life – in short, to become agents of change.

Every UN Conference in the 1990s, whether focusing on children, social development, women, cities, environment, science, human rights or population, made significant recommendations

Box 14.1: Pointers from the 1992 Earth Summit

UNCED's Agenda 21 called for national comprehensive policies for water resources management that were holistic, integrated and environmentally sound. It advised that improved water use policies and legal frameworks should embrace human health, service coverage, food, disaster mitigation and environmental protection. Chapter 18 of Agenda 21 dealing with freshwater resources pointed to several objectives and activities aiming to ensure the knowledge base. The importance of the following elements was particularly emphasized:

- inventory of water resources;
- dissemination of operational guidelines, and education for water users;
- development of interactive databases;
- public awareness-raising educational programmes;
- use of GIS and expert systems;
- training of water managers at all levels;
- strengthening training capacities in developing countries;
- training of professionals, and improving career structures;
- sharing of appropriate knowledge and technology, including knowledge needed to extract the best performance from the existing investment system; and
- establishing or strengthening research and development programmes.

Figure 14.1: Gross enrolment ratio in primary education

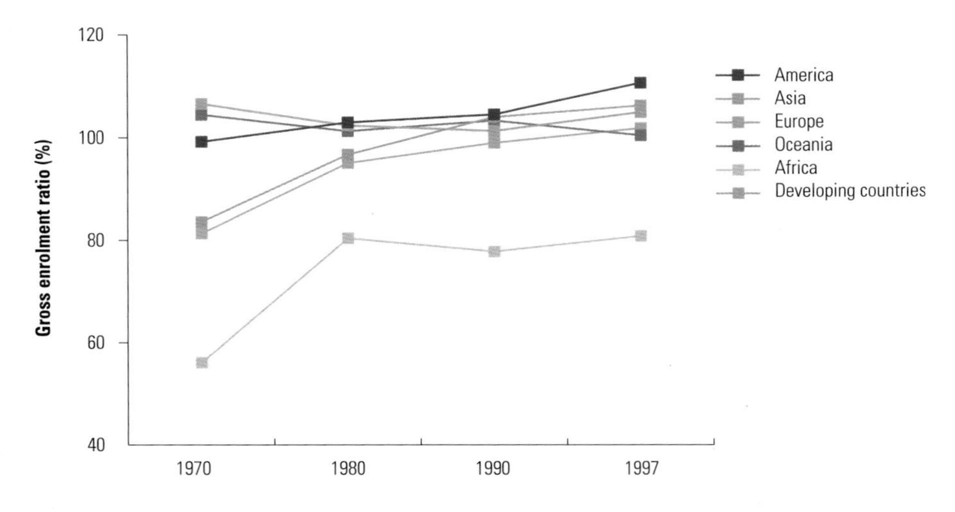

Gross enrolment ratio in primary education is defined as the total enrolment in primary education, regardless of age, expressed as a percentage of the population of the age group which officially corresponds to primary schooling. In the past thirty years, developing countries have made enormous strides in expanding enrolments at all levels. There has been a global increase in enrolment rates worldwide. However, the regional disparities are striking, with a much higher rate of enrolment in developed countries than in Africa, which lags far behind the rest of the world.

Source: UNESCO, 1999b.

concerning education. All reaffirmed that education was first and foremost a right, stipulated in Article 26 of the Universal Declaration of Human Rights. All underlined the positive correlations between better education – particularly of girls and women – and improved health practices. The Programme of the UNCED underscored that both 'formal and non-formal education are indispensable to changing people's attitudes so that they have the capacity to assess and address their sustainable development concerns'. Understanding why it is necessary to conserve our finite water resources is a starting point for protection and better management.

Basic education for all

As well as providing an understanding of the issues surrounding water resources, a good educational base is essential if suitable professionals capable of monitoring and managing water resources are to emerge. In the past thirty years, developing countries have made enormous strides in expanding enrolments at all levels: in 1960, fewer than half of the developing world's children aged six to eleven were enrolled in primary school, compared with 79 percent today (see figure 14.1). At the other end of the education spectrum, higher education witnessed a six-fold increase in enrolments worldwide, between 1960 and 1995. But as the United Nations Educational, Scientific and Cultural Organization (UNESCO) World Conference on Higher Education noted in 1998, the gap between industrially developed, developing and least developed countries with regards to access to and resources for learning and research has widened in this period. Africa's gross enrolment at the tertiary level stands at 5.2 percent, compared to 51.6 percent for developed countries (UNESCO, 1999c). Without adequate higher education and research institutions providing a critical mass of skilled and educated people, no country can ensure genuine, endogenous and sustainable development.

Despite the increase in enrolment rates, many countries, mostly concentrated in sub-Saharan Africa and south Asia, remain impoverished by poor educational achievement. Today, some 113 million children of primary school age, 60 percent of whom are girls, do not have access to education. Four out of every ten primary-age children in sub-Saharan Africa do not go to school (UNESCO, 2001b). Figure 14.2 shows the illiteracy rates, an indication of the general level of basic education. It is clear that while improvements have been made, there is still far to go.

Water factors, such as the need to collect domestic water, play a large part in school attendance. Many girls are prevented from attending school because of the lack of separate toilet facilities. In addition, many school days are lost due to illness as a result of water-related factors: improved environmental health is essential to allow more children to attend school. To this end, a UN interagency flagship launched in 2000 aims to assist governments in implementing school-based health programmes, including promotion of health skills and provision of safe water and sanitation. Basic

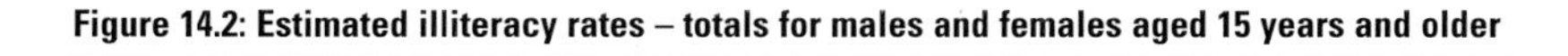

Figure 14.2: Estimated illiteracy rates – totals for males and females aged 15 years and older

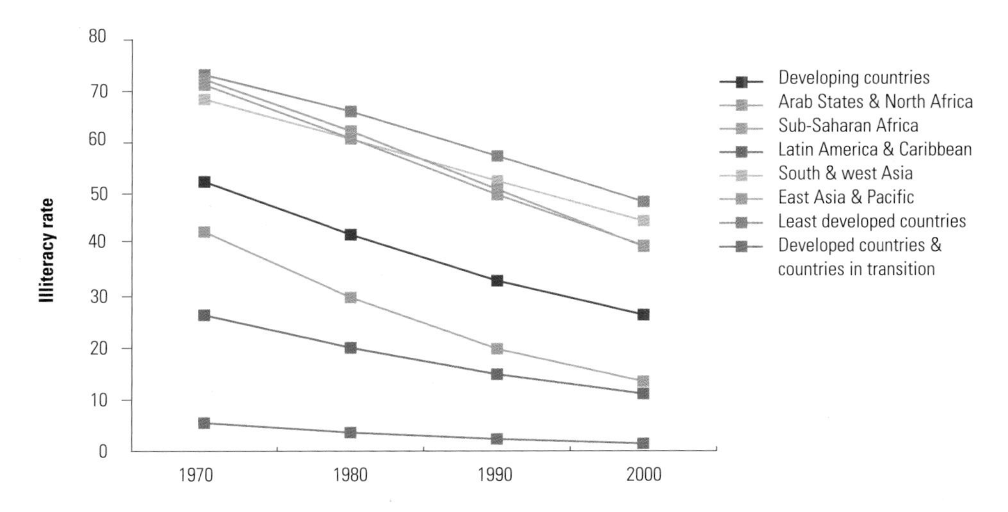

Illiteracy rates are decreasing rapidly throughout the world, but there is a vast difference between developed countries, which in 2000 had an illiteracy rate of close to 0, and the least developed countries, with a rate of nearly 50 percent. While improvements have been made, there is still far to go.

Source: UIS, 2002.

hygiene education is essential if children are to realize the benefits of water. Simple hand and face washing can prevent many water-related diseases. At the Second World Water Forum, a target was agreed to provide hygiene education in 80 percent of primary schools by 2040.

There is an emerging consensus that wise water management cannot be accomplished by technical or regulatory measures alone, but must encompass education and awareness-raising initiatives. This suggests the need for much more pronounced efforts to make curricula more practical, to train teachers in water education courses and to foster stronger linkages between schools, water companies, city managers, relevant government ministries and non-governmental organizations (NGOs). To monitor progress, an in-depth global assessment of water within formal school curricula worldwide would be a valuable tool for promoting broader individual and societal responsibility. A monitoring mechanism is also urgently required to track progress towards attaining hygiene education targets.

In light of the irrefutable links between education and poverty alleviation, the number of out-of-school children casts a pall over the ambition to foster better understanding of water's value and to teach basic health skills in the regions of the world where needs are greatest. The prospect is all the more disquieting given the disproportionate number of girls not receiving any education: as future mothers, they are more than likely to perpetuate unsound practices.

Education for sustainable development

Beyond the overarching priority of providing all children with universal primary education, attention has focused in recent years on fostering an environmental ethic. Agenda 21 noted that education is 'critical for achieving environmental and ethical awareness, values and attitudes, skills and behaviour consistent with sustainable development and for effective public participation in decision-making'. This suggests gearing curricula to address a range of health and environmental issues, while involving parents and communities at large in a wide range of awareness-raising and self-reliance initiatives.

Why is water valuable and how can it be managed in a sustainable manner? How is drinking water produced? How can we avoid conflicts over water? By answering these questions, water-related environmental education has gained ground in both industrialized and developing countries. An expert group meeting on Water Education in African Cities held in Johannesburg, South Africa (2001) provided valuable insight into initiatives underway on a continent faced with water scarcity. Water education was described as a 'strategic entry point to developing a new ethic for water governance in African cities ... an important agent for behavioural and attitude changes of key actors in the urban scene'. In particular, participants emphasized the richness of indigenous knowledge and experience in environmental education, both for enhancing cultural diversity and promoting ecological integrity.

In many countries, the broader concept of education for sustainable development is making headway. It aims to shape values, promote responsible behaviour and make children aware of their role in preserving the environment. In India, the Centre for Science and Environment produces quality resource materials and programmes to foster more ecologically conscious lifestyles. The Centre organizes workshops around the country to introduce students to the value and wise use of water as well as technologies such as water harvesting. The British NGO WaterAid has developed games and study materials aimed at children of all ages.

Introducing water education in schools is a complex, long-term endeavour: curricula are all too often overloaded and plagued by an academic, exam-oriented focus. Water-related contents are often scattered across several disciplines, failing to give a holistic view of water. The task of mainstreaming water education calls for revising curricula and textbooks, producing teachers' guides and providing adapted in-service training, particularly to promote a more hands-on, active pedagogy based on problem solving. A number of countries have introduced reforms in this spirit: in Mexico, the natural science curriculum for primary school was revised to broach topics such as the importance of water, different types of pollution and how technologies affect ecosystems. In Ethiopia as well, the school curriculum includes water-related topics in all grades, from understanding the sources and uses of water to conservation methods, hygiene and sanitation, and the effects of pollution.

Formal schooling is not the only channel for conveying such information. On the contrary: an expanded vision of education advocated since the World Conference on Education for All in 1990 and reinforced by the Delors Commission on Education for the twenty-first century, recognizes learning as a seamless process starting from early childhood and continuing through adult life. It necessarily entails a variety of learning methods and settings, both formal and non-formal: school equivalency programmes, adolescent and teenage literacy classes, skills training by local cooperatives and associations. Community-oriented environmental education has the potential to change behaviour and attitudes towards the environment and water management, provided it is geared to needs, for example, of the urban poor. Such initiatives are a stepping-stone for empowering users and promoting more effective participation. Water agencies also have their role to play: a good example is that of the Seine-Normandy Water Agency, which has been developing a programme for educating the general public by providing a range of teaching materials for the past fifteen years. The Agency's 'Water classes' can be adapted to all scholastic levels and to groups of adults, enabling the public to become more aware of water's importance and the active role that everyone can play in its preservation. These training courses also allow the public to discover various aspects of water management and how roles are shared out in this sector, as well as in other contexts worldwide. The water classes, of which there are presently more than 1,000 each year throughout the Seine-Normandy basin, are organized by teachers based on meetings with stakeholders of the water sector, visits to typical installations and workshops on water. They lead to a joint production such as an exhibition, a video or a magazine (for more details on the Seine-Normandy basin, see chapter 19).

Very often, these settings are testing grounds for educational innovation that adopt more creative approaches to learning. Programmes are often created and fine-tuned through a participatory approach so that materials are carefully targeted and understood. In Mexico, for example, a health and hygiene programme (PIACT – International Programme for the Improvement of Working Conditions and Environment) visited homes, learnt about healthcare habits and consulted experts about local customs to design materials on promoting sanitation and oral rehydration. Clearly, participation on all scales by all parties is essential in building a more sustained, equitable knowledge base.

Knowledge from the bottom up

Civil society mobilization in favour of fairer development has been a hallmark of the past decade. NGOs representing a vast range of interests have stepped into the global arena, determined to share their knowledge and influence the development agenda. International gatherings on key challenges of the twenty-first century have all reiterated that participation – making development everybody's business – is the *sine qua non* to reducing poverty and improving the welfare of people everywhere.

The Dublin principles (1992) stated that water development and management should be participatory, involving users, planners and policy-makers at all levels. The major groups involved include farmers, NGOs, local authorities, the scientific and technological community, trade unions, business and industry, indigenous people, children, youth and women. The principles reflect a shift in conventional water governance from a top-down towards a bottom-up approach. Participation opens up the way for more informed decision-making, and offers people opportunities to claim their rights as well as to meet their responsibilities. In principle, it gives a voice to relatively powerless groups, such as women – central to providing, managing and safeguarding water – and indigenous people, custodians of sound, ancestral water management practices.

Local people have a direct interest in improving the quality of life in their village. For projects to succeed, for a sense of ownership to exist, their input into decision-making is critical. Consulting with communities is important for determining the most appropriate intervention. In the Indian city of Pune, for example, a partnership between the municipal government, NGOs and community-based organizations has greatly improved sanitation for

Box 14.2: Guinea-Bissau – with training, women 'man' the pumps

Since 1987, the government of Guinea-Bissau has focused on developing a decentralized maintenance system and standardizing hand pumps. This has improved pump performance, supported the introduction of cost-recovery strategies and promoted the principle of user involvement. In 1993, these efforts began to yield results: users financed 5 to 10 percent of maintenance costs. Early in 1994, forty-six villages were surveyed to review the management performance of the water point committees. In almost all cases, the committees were functioning well. Some 53 percent of their members were female, with 20 percent of the women fulfilling management functions beyond their traditional task of cleaning pump surroundings.

Candidates for the position of area mechanic were selected at village meetings. The villagers preferred male mechanics because of the need to visit villages by bicycle and the physical labour involved in pump repair. However, though the job was popular, it did not pay enough to maintain the bicycles. The provincial promotion team encouraged the villagers to select women for this job, as they had a more direct interest in the maintenance of the pumps and were less likely to be absent or to leave the village in search of work. By mid-1993, a total of 177 village mechanics, including ninety-eight females, had been trained and were maintaining their hand pumps.

Source: Based on Visscher and Van de Werff, 1995.

more than half a million people (see box 7.5 in chapter 7 on water and cities). The city government recognized the capacity of community organizations to develop their own solutions, supported by local NGOs. In the process, water users generate their own knowledge as a springboard for action. Participatory rural appraisals now represent a key planning tool in rural development. They involve engaging with communities to determine outcomes and activities, replacing what was largely a central planning exercise.

In seeking to overcome their own water management problems, communities generate precious knowledge. The adoption of participatory approaches in water management, greater public consultation on proposed schemes and devolved responsibilities to water user groups have stimulated knowledge amongst wide numbers of people on specific issues. It has also contributed to challenging assumptions about gender division of labour, a first step towards giving women a greater say in planning water schemes. In Niger, research in several villages on the respective roles of men and women in handling water supply, sanitation and hygiene was a starting point for fostering discussion about gender divisions of work. In Guinea-Bissau (see box 14.2), women have successfully demonstrated their skills at maintaining pumps. The fact that thinking and experience have moved beyond women and development to gender and development is an important step forward. Capacity-building that pays attention to gender and opportunities for women in water resources management must be systematically reinforced, and indigenous knowledge more diligently explored.

Traditional local knowledge

Local communities are often custodians of knowledge with the potential to reverse water waste and shortages in hostile environments. Indigenous knowledge generally reflects a profound understanding of the water cycle. In Fiji for example, the indigenous population has long blended birth and death rituals with sound water management, allowing fish stocks to replenish after ceremonies. In southern Africa, the bushmen of the Kalahari desert had an inventive technology for extracting water (see box 14.3).

Today, through community initiatives, some of these ancient techniques are being revived. Water harvesting provides a rich example. An ancient technique, which comprises a range of methods of collecting and concentrating runoff from various precipitation sources (rain, dew, fog), has enabled villagers in several parts of India (western Rajasthan for example) to overcome perennial water shortages and ensure more reliable water supplies. These initiatives, now gaining ground across the country, highlight the benefits to be gained from a combination of public awareness, traditional knowledge and community-based approaches to self-reliance. More effort must be devoted to sharing and promoting these practices both at a national and regional level, and combining this knowledge with modern materials and techniques.

Community mobilization

The trend towards decentralization in many countries has placed more decision-making power into the hands of civil society and local government, particularly in countries where trimming the central

Box 14.3: The bushmen's loss is everyone's loss

Several centuries ago, the San, a semi-nomadic population of the Kalahari desert in southern Africa, invented an innovative technology for pumping water: at selected sip-wells (places where water was sucked up through a straw) holes were prepared, into which the San inserted a straw with grass filters at either end. They then compacted the sand around the straw, letting it accumulate moisture, after which they sucked the water out of the sand and stored it in ostrich eggs, sealed and buried for later use.

The San also knew of rare springs that provided freshwater throughout the year. Other water sources included pan-like areas that stored rainwater for long periods of time, but these were generally only used in cooler seasons when the threat of malaria decreased, and the San often discarded this water for drinking because of its brackish taste. In addition, certain tree species have holes out of which the San sucked collected rainwater and dew.

However, the San's methods of water management were drastically altered in the seventeenth century when their territories were invaded by Bantu and European settlers. New technologies such as boreholes were introduced, pumping vast quantities of groundwater that eventually dried up the San's sip-wells. The settlers' livestock ravaged the food supplies and the water-collecting trees. Now thoroughly dispossessed, the San are fighting to regain their territories, and their way of life and water management.

Source: UNESCO, 2002. Prepared for the World Water Assessment Programme (WWAP).

Civil Service was a main condition of the economic and structural adjustment programmes of recent decades. This has charged many local government agencies with new functions, for which they require adequate training. In many instances, responsibility for operating and maintaining systems has been transferred to local water user associations. While the local users have sound knowledge about the local context, all too often the associations are understaffed, rely on voluntary workers and experience difficulties in scaling up operations (for an example, see chapter 16, the Chao Phraya River basin case study).

New water laws (see box 14.4 on Brazil and Sri Lanka) have been enacted in many countries. They are changing the rules of water governance, empowering communities to manage their resources and develop sustainable water management policies. The new Water Law of Zimbabwe, for example, delegates catchment management responsibilities and day-to-day duties of water rights allocation to stakeholder-elected Catchment Councils.

If delegation is to be effective, it must address local human, financial and institutional capacities. Many governments maintain an overly instrumental view of local communities, and their active involvement is normally sought only for implementing water projects, whereas true participation would entail involvement throughout the whole policy or project cycle. New mechanisms are required to enhance links between government agencies, private sector and civil society organizations, particularly in many cities of Africa, Asia and Latin America, where governance structures are often inadequate to address issues of provision and revenue-raising. An open, transparent and continuous process of consultation and

Box 14.4: Power to society – water laws in Brazil and Sri Lanka

In 1997, Brazil adopted a new water law, based on the principle set out in the 1988 Constitution describing water as a public good. The law reflects a shift in thinking on water management towards a more democratic and decentralized exercise. It underlines the importance of giving society more decision-making power in water matters, specifically national state councils for water resources and river basin committees. The latter are seen as belonging to everyone, and become a conciliating element between the interests of the state and those of consumers.

In Sri Lanka, heated debate surrounds a draft water act aimed at fostering decentralized management of water through river basin organizations. The latter will become responsible for planning, implementing and regulating water allocations between water use sections in each basin. However, introduction of the water resource management concepts that underpin the act, including water rights, have proved contentious, largely due to fears over loss of rights for traditional water usage and apprehension over the possible introduction of new water charges. These fears are delaying presentation of the act to Parliament.

Source: Based on P. Alfonso Romano, former Secretary of Water Resources of Brazil, and the Ministry of Irrigation and Water Management of Sri Lanka.

Box 14.5: CapNet – the virtues of networking

CapNet works towards strengthening or establishing regional networks delivering education and training support for improved management of water resources. This objective is achieved through networking, awareness creation, training and education, and development of relevant materials and tools. Target groups include water professionals and decision-makers, with a strong emphasis on the inclusion of women. Partners include particular regional and national capacity-building institutions in both the north and the south (http://www.cap-net.org/).

participation is essential if national water resources are to be managed in an equitable and sustainable way.

Growing mobilization at the community level is also witnessed by the surge of river basin groups, usually staffed by volunteers interested in protecting the integrity of their river basin. The United States Environmental Protection Agency (US EPA) has identified more than 3,000 such groups, each composed primarily of individuals, identifying and using local knowledge relevant to their river basin. Some have discussion forums that encourage sharing of information and knowledge. South Africa and Australia have Internet sites with a wealth of knowledge and experience covering the full spectrum of river basin activities. Mirroring a worldwide trend, river basin organizations have established a worldwide network, the International Network of Basin Organizations (INBO), to exchange information and experience on a wide range of issues such as mechanisms for cooperative action, conflict mitigation, public-private partnerships and practices of water pricing and allocation of resources.

Buttressing this burgeoning of local initiatives, a number of national and international NGOs as well as civil society groups are working to consolidate this emerging knowledge base through building community awareness and local capacity. These organizations offer learning tools, collect cases of best practice and draw on a wealth of experiences through community networks. They are devoting significant resources to partnering and networking activities to more effectively carry out their mission. The rapid development of information technologies has further spurred the creation of planetary networks allowing communities of interest to share information and learn from each other. These knowledge bases range from simple bulletin-board-style communications systems to complex and information-rich compilations with well developed navigation and query systems. Some, such as CapNet, specifically aim to bolster capacities (see box 14.5).

Many networks are devoted to specific aspects of water, such as LakeNet, a global network of people and organizations dedicated to the conservation and sustainable management of lakes. The Oneworld Water and Sanitation Think Tank aims to promote sharing of experiences on a south-south basis and to enable water practitioners in developing countries to integrate these lessons into policy-making. Others adopt an integrated vision to improve water management. GWP works through a network of partners to identify critical knowledge needs, assists in designing programmes to meet them and serves as a broker between providers and donors. GWP stresses the need for innovative solutions and for all users of water to share information, understand data and work together to solve problems.

The Challenges

Shifts in the water industry

The participatory approach is also taking hold in the water industry as companies expand their operations and adopt more flexible management practices. Less than half a century ago, the knowledge base for a water supply company consisted primarily of design drawings, operation and maintenance manuals and inherent understanding of the system through the skills and experience of staff. Today, assessment of demand by customers, higher standards of treatment, improved concern for public health, metering and better operations and maintenance have all required a wider consultation with other professionals. Bridges are being built between consumers, water companies and government, requiring more accessible and comprehensive knowledge. They include Public Service Agreements detailing dates and terms of government delivery, Service Level Agreements by companies to meet consumer-consulted supply reliabilities and public consultations on government white papers or technical applications with a public impact. Further examples of these emerging connections between stakeholders include public consultations on particular water schemes and consumer association involvement in water.

Across the water sector, the number of generic guidelines on best practice relating to specific issues numbers into the tens of thousands. Against the backdrop of mergers and public-private partnerships, the water consultancy business has soared as organizations seek advice on ever-wider missions and responsibilities in the aim of integrating new knowledge into their operational practices. Development partners are increasingly stimulating growth in the local consultancy sector rather than turning to international expertise. Patterns in the supply of bilateral support are also beginning to change, with the untying of development assistance. As a result, national funds from a donor country are not necessarily tied to technical assistance from the same country.

As noted, the water industry relies on a very broad body of knowledge to operate: codes of practice, operating and training manuals for systems, plants and equipment, databases, government guidelines, research journals, reports of professional and trade associations, manufacturers and suppliers. Much of this operating knowledge, however, still tends to be based on the needs of advanced countries. All too often, low-income countries adopt laws, regulations and working practices from advanced countries, when in many cases they lack the capacity to enforce and apply them. While several low-income countries have developed local expertise to deal with challenges in their water sectors – examples include Mexican wastewater and irrigation reform practices and Brazilian expertise in water and sanitation in poor communities – these experiences are not systematically shared with other developing countries. Language, financial and cultural barriers impede the transfer of knowledge. Channels must be facilitated to boost this south-south collaboration and, in the process, place higher value on local expertise, which is more inclined to view water knowledge in context.

Overcoming poor data availability

Empowering stakeholders at all levels is one facet of creating a sound knowledge base. Another vital facet is the production of high-quality data. Agenda 21 warned that a lack of data is 'seriously impairing the capacities of countries to make informed decisions concerning environment and development'.

National water databases are the backbone of international data management. However, in many parts of the world hydrometric and water quality monitoring networks are deficient. Table 14.1 shows the regional variation in the total number and density of various monitoring stations. Data from many of the stations measuring river discharge are not available and only a relatively small number have data series of sufficient length and quality for use in analyses. Similarly, there is a dearth of quality data on groundwater in spite of its potential for future water supply. Studies have additionally shown that the density of these stations is much lower in Africa than elsewhere.

In addition to the problems in collecting data – often due to a lack of resources for maintaining observation stations – the ability to use available data to describe the status and trends of global water resources is hampered by divergent procedures for collecting data, different quality assurance procedures and unreliable telecommunications. The fragmentation of national organizations dealing with water resources assessment has meant that networks

Table 14.1. Increases and decreases in the number of hydrological observing stations in the world between 1974 and 1997

	World Meteorological Organization Regions						
Type of stations	**I Africa**	**II Asia**	**III South America**	**IV North and Central America**	**V South-west Pacific**	**VI Europe**	**Total (global)**
METEOROLOGICAL							
Precipitation	10,074 <>	9,445 ++	22,975 +++	20,174 <>	16,367 <>	35,091 -	114,126 <>
Evaporation (pans)	682 +++	1,011 +++	1,945 +++	871 <>	1,296 +++	1,129 +	6,934 ++
SURFACE WATER							
Discharge	1,748 +++	3,163 ++	7,568 +++	11,958 -	5,935 +	18,796 ++	49,168 +
Stage	1,798 +++	8,186 +++	7,022 +++	10,819 -	852 –	10,427 -	39,104 +
Suspended solids	560 +++	440 ++	1,187 +++	1,008 +	514 –	3,590 +++	7,299 ++
Bed load	6 +++	27 ++	339 +++	0 n.a.	0 —	1,423 +++	1,795 +++
Water quality	310 -	2,057 ++	3,076 +++	14,218 +++	1,415 —	14,974 +++	36,050 +++
GROUNDWATER							
Groundwater level	1,450 ++	3,776 <>	1,133 +++	4,344 ++	1,999 +++	45,782 –	58,484 -

Key: — = < -2% – = < -1% - = < -0.5% <> = no significant change + = > +0.5% ++ = > +1% +++ = > +2% n.a. = data not available

This table shows the regional variation in the total number and density of various monitoring stations. Data from many of the stations measuring river discharge are not available and only a relatively small number have data series of sufficient length and quality for use in analyses.

Source: WMO, the World's Hydrological Networks, 1997.

Box 14.6: New tools for the Mekong

The Lower Mekong suffered serious damages resulting in heavy loss of life during six floods in the last decade. Decision-making needs to be supported by scientific assessment of the causes of increased flood damage – change in climate or land use, population growth – but poor data availability has prevented this.

In response, the International Association of Hydrologic Sciences has developed its Prediction of Ungauged Basins (PUB), an international research initiative to assess water resources in basins with no records. One possible application is the Mekong River basin.

Using a downscaling technique that requires a sophisticated integrated hydrometeorological model, experts from the Public Works Research Institute of Japan conducted a blind test on Greater Tokyo to reconstruct historic rainfall over the area. The approach offers several advantages: it can be applied to any basin in the world including ungauged ones, in contrast to remote sensing that requires some ground rainfall data. This technique will be used to reconstruct rainfall over the Lower Mekong River basin in the last two decades and scientifically assess flood frequency.

Source: Prepared for the World Water Assessment Programme (WWAP) by the Ministry of Land, Infrastructure and Transport (MLIT) of Japan, 2002.

specialized in different hydrological observations (e.g. surface water, groundwater, management of reservoirs) are often poorly integrated. Coordination is similarly lacking between databases on water and similar ones on geology, land use, demographic data, health, economics and other fields that tie into a vision of integrated management. Climate change is also introducing elements of data uncertainty concerning the variability in the distribution of water resources.

New technologies such as remote sensing and GIS have improved water data collection and contributed to advancing scientific knowledge. Remote sensing enables a continuous monitoring of the globe at all scales. It has found important application in the mapping of snow and ice covers, glacier movement, currents in lakes and seas, biomass and algal growth, particles in water, water temperature and other variables. GIS technology has been instrumental in mapping Africa's water resources while the European environmental satellite, Envisat, launched in March 2002, will provide valuable input to the study of global climate change and the world's waters. New computing tools (see box 14.6 for an example from the Mekong River) promise to advance knowledge of flooding and strategies for risk mitigation.

Several international programmes are seeking to improve national capacity to assess water resources. The World Meteorological Organization (WMO), supported by the World Bank, has developed the World Hydrological Cycle Observing System (WHYCOS), establishing a global network of hydrological observatories, supplementing national networks. The International Atomic Energy Agency (IAEA), with WMO, contributes to building capacity for improved water resource understanding and management through the Global Network of Isotopes in Precipitation. The United Nations Environment Programme's (UNEP) Global Environmental Monitoring System gathers data from sixty-nine countries on variables such as organic matter, heavy metals, salinity and acidifying atmospheric emissions. An international groundwater resources assessment centre, under the auspices of UNESCO and WMO, is planned, and will focus on the development of procedures for collecting and processing data on the world's aquifers, promoting adequate monitoring systems and increasing awareness.

Sharing and harmonizing knowledge

In addition to the gathering of data, sharing knowledge and making knowledge and data available are aims that have been pursued by UN organizations and their partners, boosted by the opportunities offered by information technology. Many knowledge bases relating to agriculture, health and indigenous knowledge are now made freely available to decision-makers, stakeholders and the public at large. The Global Terrestrial Network provides an Internet-based mechanism to access data and meta-data information from independent centres on such variables as surface water discharge, groundwater fluxes, precipitation and soil moisture. The WMO's Global Runoff Data Centre (GRDC) collects and disseminates data on river discharge on a global scale, and provides data products and specialized services for the research community, water managers and water-related programmes.

Regional databases also exist but data characterization remains a major problem, rendering comparisons from different sources hazardous. Joint monitoring projects underway at Lake Peipsi and cooperation in drawing up a management plan for the Senegal River basin, shared between four countries (see boxes 14.7 and 14.8), are vital steps to enhancing the knowledge base. International cooperation is also urgently required to ensure a reliable, comparable global database.

Box 14.7: Senegal River basin – information for sustainable management

Reliable, comprehensive information is the foundation of better management. In Senegal, for example, OMVS (Organisation pour la Mise en Valeur du Fleuve du Sénégal, or Organization for the Development of the Senegal River) is preparing a development and management plan for the Senegal River basin, which is shared amongst Guinea, Mali, Mauritania and Senegal. For this task, authorities are relying on several tools developed over recent years, namely a flow rate monitoring network with statistics kept since 1904, a computer program designed to evaluate the effects of different management rules applying to the Manatali dam and two dam management manuals. The latter provide managers with guidelines on keeping the upstream storage lakes at high levels during periods of flooding, security guidelines and flow rates required to best meet the aims of regular output and electricity production.

Source: Based on a text by the Organization for the Development of the Senegal River (OMVS), 2002. Prepared for the World Water Assessment Programme (WWAP).

Box 14.8: Joint monitoring on Lake Peipsi

Sharing water is one of the major challenges to achieving water security in the twenty-first century. A joint project on Lake Peipsi, now divided between the Russian Federation and the Republic of Estonia, has resulted in coordinating sampling programmes and comparing water quality norms. The Narva Watershed Research Programme originated in 1998, supported by the Swedish Water Management Research Programme (VASTRA), to develop catchment-based strategies for sustainable water use. A separate three-year European Union project will result in extensive research papers and databases, including numerous GIS layers covering the whole basin. This project stands to benefit environmental authorities in both countries.

Source: Prepared for the World Water Assessment Programme (WWAP) by the Ministry of Natural Resources of the Russian Federation, and the Ministry of the Environment of Estonia, 2002.

A more holistic approach to water management can equally be gleaned from initiatives such as the UN Earthwatch system, which utilizes the combined resources of UN partners and others in producing major environmental assessments. A *World Water and Climate Atlas* was produced in 1999 by the International Water Management Institute (IWMI), while 'Our Fragile World', a forerunner to the *Encyclopaedia of Life Support Systems*, supported by UNESCO (2001c), presents an integrated vision of knowledge essential for global stability by connecting issues relating to water, energy, environment, food and agriculture.

While many organizations provide valuable data and information on trends in the arena of water management, the absence of a coordinated monitoring process is hampering appraisal of progress. The failure to measure progress against targets set out at major international meetings runs the risk, over time, of undermining the effectiveness of policies and of basing investments on poorly identified priorities.

Efforts are being made towards improving the coverage and effectiveness of the global statistical systems through establishment of large datasets. In certain instances, agencies within the UN are tasked by member states with particular monitoring and evaluation roles, such as that of the UN Economic and Social Council and the Commission on Sustainable Development (CSD) in tracking progress against Agenda 21.

The UN's data holdings permit the publication of authoritative reports, such as the recently introduced *World Development Indicators*, the World Bank's premier annual compilation of development data, with some 800 indicators. It is one of WWAP's major priorities to further harmonize data within the UN family. At present, there are at least three UN system-wide meta-data-gathering initiatives taking place:

- the UN Geographic Information Working Group (UNGIWG), which is looking at the coordinated development of GIS within the UN system;
- the UN statistics division initiative, which concentrates on the development of a standard set of statistics that are collected to support the UN as a whole; and
- Earthwatch, a mechanism for coordinating UN agency initiatives in the field of environmental management. Furthermore, the UN Department of Economic and Social Affairs (DESA) has compiled a survey of water-related databases.

However, UN efforts at creating these global datasets are only as good as a country's ability to assemble its own water information. The global statistical system is built up from the national level, where data are obtained from a number of different sources: central statistical offices, ministries and national banks. Efforts are being focused on improving collection of water-related data from censuses, surveys, consumer information and statutory data collection requirements. Reinforcing the capacity of countries to obtain and store data is an overriding priority for ensuring our ability to describe and assess the global water situation. Financial constraints have reduced the ability of public service institutions in charge of water resources to collect data at the field level in many developing countries. As noted earlier, there has been in many cases a decline in the quantity and quality of information on water resources and their uses. GIS and computerized databases of water resources and related socio-economic information still have to be made available on a much larger scale in developing countries, combined with capacity-building. Despite progress made in the Joint Monitoring Programme's 2000 Assessment (WHO/UNICEF, 2000), there is little detailed data available on the quality of provision for water and sanitation in most of the world's cities. More detailed data would mean significant changes in the questions asked within censuses – an evident strain on the resources of poor countries.

Meeting research and development needs: the production and use of scientific knowledge

Beyond its capacity to generate, analyse, store and share data, the scientific community requires observational rigour. Without an endogenous research base, how can countries adequately address issues from both a local and global perspective?

The World Conference on Science organized in Budapest in 1999 drew attention to the growing gap between science-rich and science-poor countries in the production and use of scientific knowledge. Today, North America, western Europe, Japan and the newly industrialized countries account for 85 percent of total research and development (R&D) expenditure worldwide. Figure 14.3 shows the number of scientists and engineers per 100,000 people, Gross Domestic Product (GDP) per capita (Purchasing Power Parity [PPP] dollars), Human Development Index (HDI) and the percentage of Gross National Product (GNP) spent on research in certain example countries, and illustrates clearly the difference between developed and developing countries.

Lack of investment in science and technology is maintaining the brain drain, depriving developing countries and those in transition of the high-level expertise required to accelerate their socio-economic progress and find solutions adapted to their needs. Up to 30 percent of African scientists are lost to the brain drain according to several

Figure 14.3: Number of scientists and engineers per 100,000 inhabitants, GDP per capita (PPP in US$), Human Development Index value and R&D expenditure

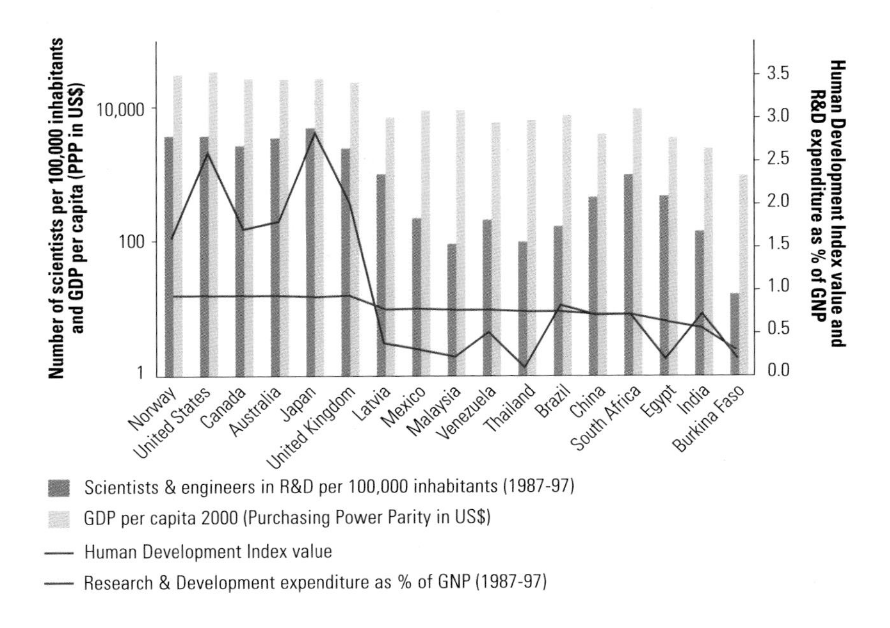

This figure shows the number of scientists per 100,000 people, GDP per capita, Human Development Index and the percentage of GNP spent on research and development in certain countries. The difference between developed and developing countries is clearly illustrated.

Source: UNDP, 2001 and 2002.

studies sponsored by the Research and Development Forum for Science-led Development in Africa.

The global economic context is ushering in a new framework that is transforming the social institution of science and research systems. According to UNESCO's regular *World Science Report*, science education at the post-secondary level is facing a severe crisis in many developing countries, marked by a growing perception that science is failing to tackle acute problems associated with water, sanitation, food security and the environment. Deteriorating working conditions in laboratories and universities and wide salary gaps favouring scientists in the private sector are accentuating the trend. This mostly holds true for non-water trades in which the trained water professional has valuable transferable skills. The impact of privatization is being felt in several countries through cutbacks in state funding, the closing of certain national research facilities and increasing dependence on foreign private and donor funding. Universities are knitting tighter relationships with the corporate world to become more responsive to the needs and demands of industry, but will this serve the public good and the needs of the poor? What are the consequences for research expenditure on achieving a finer understanding of basic water processes and developing more efficient techniques for water use and extraction, whether in the domestic, agricultural or industrial sector?

Assessments have pointed out that large numbers of technical and scientific personnel lack sufficient knowledge about overall water management and use. While important scientific and technological advances have been made – modelling capability stands out as an example – the specific needs of developing countries in monitoring and managing their water resources are not high on the research agenda. Many barriers to the effective management and supply of water lie in the institutional and managerial sphere and will not be solved by improved technologies alone. Research focused on effective institutional structures and management techniques is required.

A potential deadlock lies in the polarized standpoints within the water research community. At this stage, there remains ample scope for divergent policies and actions to evolve, all backed legitimately by the 'knowledge base'. The nature of the journal process – a key measure of academic performance which holds the influence to attract financing – makes it all the more difficult to find an objective middle ground. A vast array of academic papers exist. Two of the main bibliographic databases, Selected Water Resources Abstracts (1967–94) and Water Resources Abstracts (1994–present) identify 370,000 papers or abstracts. The figure is probably conservative given the lack of access to some bibliographies (particularly non-English ones). This leaves the knowledge base with an outstanding problem of consolidation. Conferences, thematic volumes, best practice documents and guidelines all attempt to streamline this fragmented body of information. The Toolbox for Integrated Water Resource Management, released in 2001 by GWP, provides a range of tools that users can select or modify according to their needs and local circumstances. Launched at the Freshwater Conference in Bonn in December 2001, the Tool Box draws together a wealth of experience and expertise in IWRM in one product. Divided into two main sections, Policy Guidance and Operational Tools, the tool box features a wide variety of options related to the enabling environment (what it takes to make policy changes on the ground), institutional roles and management instruments. To complement this information, case studies on IWRM practices are included. The Tool Box will continue to grow as users gain experience with it and start providing feedback on the success or failure of a particular set of actions in a given situation.[1]

This middle ground is reachable when a consensus, based on a shared vision, is sought by governments, the scientific community and society. As Michael Gibbons, secretary general of the Association of Commonwealth Universities wrote in the magazine *Nature* (Gibbons, 1999),

> *Under the prevailing contract between science and society, science has been expected to produce 'reliable' knowledge, provided merely that it communicates its discoveries to society. A new contract must now ensure that scientific knowledge is 'socially robust', and that its production is seen by society to be both transparent and participative.*

If this is to hold sway in the water field, there is an urgency to clearly demonstrate the links between new knowledge and socio-economic outcomes, and to generate stronger demand for access to clean water. The explicit linking of water issues to human development and economic productivity is generally lacking both in terms of national policy declarations and legislative and administrative support.

The traditional wisdom of applying general solutions to water issues is being replaced by the idea of developing locally specific and applicable ideas. However, new investigations into specific local issues require increased investment or knowledge generation. The countries where the need for focused knowledge to achieve results is highest are often the countries with the lowest investment.

1. For more details, see http://www.gwpforum.org.

The Road Ahead

New avenues for learning

A CD-ROM search of the World Higher Education Database (IAU, 2002) reveals that there are 3,873 institutes all over the world offering higher-education courses on water and water-related subjects (see figure 14.4).

Against this backdrop, a host of specific programmes and partnerships are seeking to offer new learning and training opportunities for water specialists. Development partners are increasingly recognizing the importance of building local educational capacity, in order to train people within their own countries and educational institutions. The past decade has witnessed an expansion of international freshwater resource institutions and programmes. International educational programmes have proliferated, particularly within the European Union, through such methods as university twinning.

Partnerships in water education (see box 14.9) are enhancing cooperation between academic centres and professional organizations. The UNESCO International Institute for Infrastructural, Hydraulic and Environmental Engineering (IHE) aims to build capacity in water education by linking educational organizations and regional networks.

Box 14.9: UNESCO-IHE – a partnership in water education

The UNESCO-IHE partnership is contributing to the post-graduate education and training of professionals and to building the capacity of knowledge centres and other organizations in the fields of water, the environment and infrastructure in developing and transition countries. This is done through a partnership network of academic centres and professional organizations offering demand responsive and duly accredited educational programmes at the local and regional level. UNESCO-IHE encourages all those involved in the water sector, including its 12,000 alumni from more than one hundred countries, and scientists and professionals in public, private and civil society organizations, to participate in this dynamic network. Through international projects, regional refresher seminars for alumni, symposia and other routes, UNESCO-IHE is able to fine-tune its education, training and research programmes to ensure that they continue to meet changing demands.

Figure 14.4: Institutes offering water-related subjects in higher education

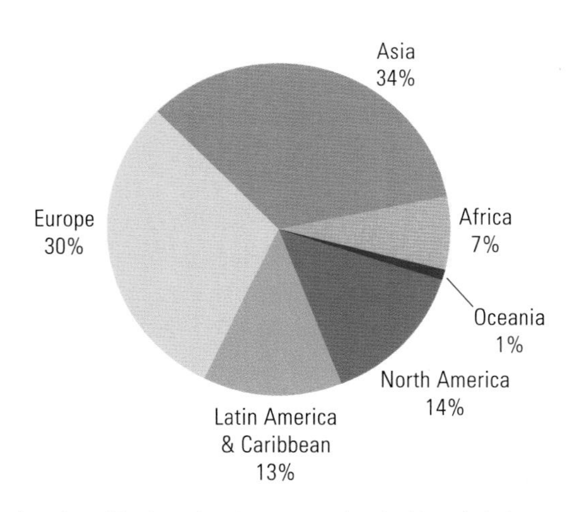

The total number of institutes is 3,873. Water-related subjects include Water Science in Agriculture, Water Resource Engineering, Civil Engineering, Environmental Engineering, Rural and Town Planning, Public Health and Sanitation, Meteorology, Arid and Arctic Studies, Ecology, Environmental Science, Wildlife, Waste Management and Natural Resource Management.

The maximum concentration of institutes offering water-related subjects in higher education is in Asia, with 34 percent of the world total. This is closely followed by Europe, with 30 percent. Northern America and Latin America and the Caribbean have very similar rates, with 14 and 13 percent respectively. Africa and Oceania lag behind, with only 7 and 1 percent.

Source: IAU, 2002.

ICTs are also creating new learning environments, ranging from distance-education facilities to complete virtual higher-education institutions and systems capable of bridging distances. Online education facilities have flourished over the past decade, provided by universities, private companies and individuals. According to the Canadian TeleCampus database, there are at present about 45,000 courses available internationally for online learning, including some ninety in environmental engineering and thirty in hydrology and water resources.

In addition, many operational organizations are granting more importance to the links between qualifications/certification, employee status and rewards. To nurture local expertise, training and education systems will require further expansion, through, for example, professional associations and networking. In addition to ensuring high scientific standards, curricula must be constantly adapted to concrete problems and graduates taught to operate in an integrated, multidisciplinary environment.

Media and public information: furthering the cause

For water to become a shared global concern, it has to be viewed as such by society. In this endeavour, the media and public information play a critical role. At the Second World Water Forum, an event attended by some 600 journalists, World Bank Vice President Ismail Serageldin stated that the role of the media was to aid in ensuring transparency and to help combat corruption. The need to expose corruption, he said, 'is why we want all social actors and the media to be involved in all the issues of water' (Roberts, 2002).

In many countries where a free press is still in its infancy, the task of exposing corruption has sent journalists to jail more than once in recent years. The fact points to the powerful influence of the press – in asking hard questions, in exposing malpractices, in exploring who benefits from privatization contracts, in holding companies accountable, but also in creating broad consumer awareness.

The end of the Cold War, the ensuing wave of democratization and the introduction of digital technology have led to an upheaval in the media landscape. We have witnessed the emergence of an independent press and audiovisual in countries where information had until recently been held under tight state control, such as the transition countries of the former Soviet Union and parts of Latin America and Africa. New forms of television have emerged: cable and satellite television, video on demand, theme channels and new interactive services. These offer untold potential for 'popular' and scientific programmes on water issues. In contrast to this burgeoning offer, a trend towards concentration can also be gauged: three major agencies process and disseminate 80 percent of international information broadcast each day around the world.

Access to the media remains highly uneven across the world. The spread of the written press is hampered by financial resources and high illiteracy rates, particularly in southern Asia and sub-Saharan Africa. Circulation of daily newspapers stands at 226 copies per 1,000 inhabitants in developed countries compared to under 33 per 1,000 inhabitants for the rest of the world. The access to television and radio, while increasing, is also unevenly spread across the world, as shown in figure 14.5.

The rapid development of community radio, an especially powerful medium in regions with low literacy rates, is a particularly distinctive factor with the potential to play a positive role in garnering attention towards water issues. Since 1989, these stations have flourished in all parts of the developing world thanks to technological innovations, lower FM transmitter costs and slacker controls on broadcasting by public monopolies. Through novel approaches, based on listener participation, programmes tackle such issues as health, hygiene and rural development. These stations enhance dialogue within communities and promote the free flow of information and public accountability. In Sri Lanka, for example, Radio Mahaweli was set up in 1979 with the assistance of UNESCO and the Danish Agency for Development Assistance (DANIDA) when national authorities launched a construction plan for a hydroelectric dam. Over the course of six years, the news programmes broadcast by the mobile radio station enabled the population to be efficiently moved to new home sites.

Boosting coverage of water issues in the media is a two-way street: journalists themselves must become more educated on the complexity of water issues. Towards this end, the Water Media Network, supported by the World Bank, is an initiative designed to help journalists examine the social, environmental, regulatory and financial issues relating to water. The network features workshops, field visits and distance-learning courses for the media.

While water issues appear to have garnered heightened press attention in recent years, there is an urgent need to monitor their coverage more comprehensively in order to assess how the tools of the media are being used to foster understanding and awareness, and to promote debate. Capacity-building programmes are required both to strengthen journalists' professional skills and to sensitize the media to sustainable development issues.

Public information sites and water portals are also vital tools for improving knowledge about water, providing specific information on water quality, risks of flooding or drought, tariffs and other matters. Access to specific and local information has been greatly facilitated through the Internet, flexible search engines and online computer services. Communication strategies are inherent components of public policy, with over 100 government departments for water maintaining official web sites. International and NGO web sites are contributing to building an improved knowledge base on global water resources. The World Water Portal (see box 14.10) is a gateway intended to enhance access to information related to freshwater available on the Internet, while the WWAP web site serves as a central communication network for this UN system-wide programme.

Hampering access: the digital divide

'Scan globally, Reinvent locally' runs a catch phrase urging stakeholders to take stock of this large circulating knowledge base in order to adapt it to local contexts in all domains. This recommendation, unfortunately, remains severely hampered.

The digital divide is the foremost hurdle. While ICT offers researchers and other communities in developing countries an unprecedented opportunity to overcome their economic and geographical isolation, the digital divide still prevents millions from taking advantage of this vast store of knowledge. The average numbers of fixed telephone lines – an accepted indicator of progress for knowledge in the Millennium Declaration – varies dramatically between regions, as shown in figure 14.6. OECD (Organization for Economic Cooperation and Development) countries account for 80 percent of people who use the Internet at work. In Asia, Latin America and Africa, more than 98 percent of the population is not connected to the Internet. The total international

Figure 14.5: Communications (television, radio, newspapers) per 1,000 inhabitants

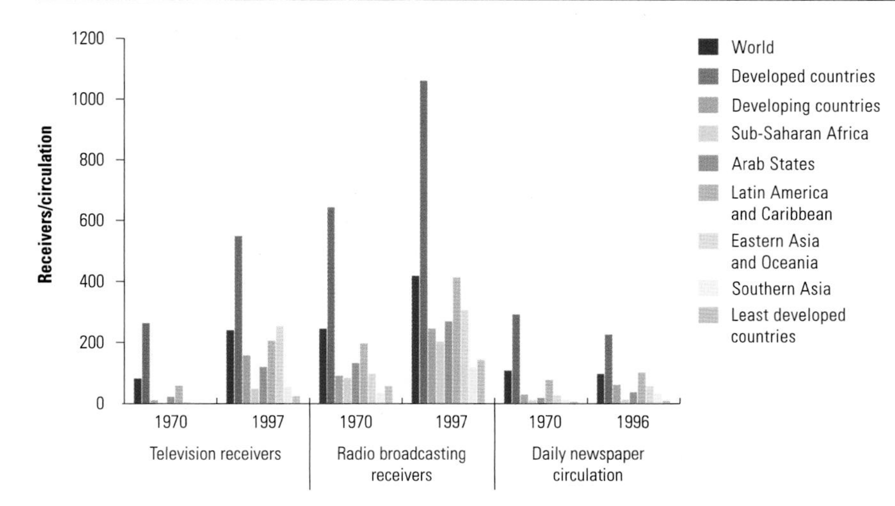

Access to the media remains highly uneven across the world. The spread of the written press is hampered by financial resources and high illiteracy rates, particularly in Southern Asia and sub-Saharan Africa. Circulation of daily newspapers stands at 226 copies per 1,000 inhabitants in developed countries compared to under 33 per 1,000 inhabitants for the rest of the world. The access to television and radio, while increasing, is also unevenly spread across the world.

Source: UNESCO, 1999b.

Box 14.10: The World Water Portal – a model for water information sharing and cooperation

WWAP, together with other water programmes and organizations, is developing the 'World Water Portal', a model for water information sharing and cooperation. This Internet portal will integrate various regional networks with the WWAP global water portal using common structures, protocols and standards to provide seamless access to a wide body of water information. Current priorities for the development of the World Water Portal include:

- development of a network of reliable water information providers;
- development of an organizational structure that will provide technical support (meta-data assistance/standards, 'good practice' guidance for database and web page development, search and database integration software, and development of processes for data acquisition) and ensure information quality through peer review processes (coordination/support of peer review process, discussion lists) and promote the adherence to sound information management standards;
- capacity-building in the area of information management and web site development for partners and contributing organizations, education and training for both managers and technicians enabling them to make more efficient use of the Internet; and
- the use of reliable information and the improvement of integrated water resource management decisions, with the goal of facilitating working partnerships. By accurately and consistently describing information resources, and linking to other information partners, the Portal aims to provide a valuable and self-perpetuating water information source for use by decision-makers, resource managers, researchers, students and the public at large.

In preparation for going global, a prototype water portal is now being developed for the Americas. If appropriate, its techniques for sharing and integrating information will provide a basis for the World Water Portal. This model will allow local, national and regional water organizations to develop the relationships and pursue the water information issues that are most important to them while contributing to the world's body of water knowledge. The prototype tools and technologies may then be easily implemented by other regions to rapidly expand the content and scope of the World Water Portal. (http://www.waterportal-americas.org).

Figure 14.6: Main telephone lines and Internet users per 1,000 inhabitants

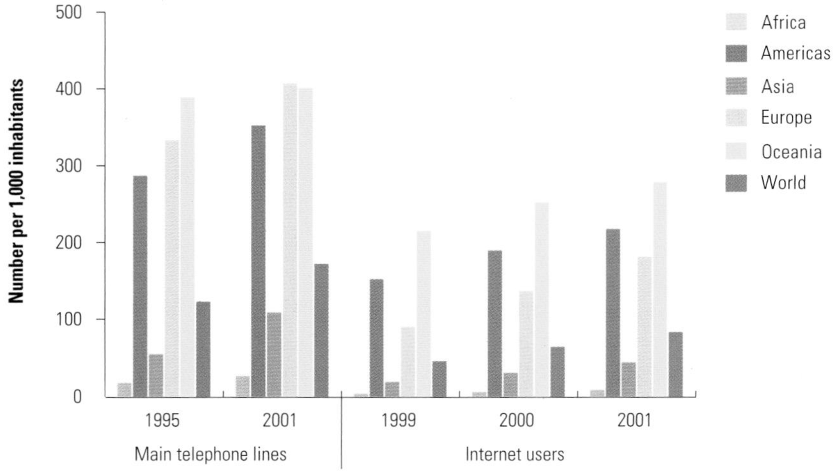

The average numbers of telephone lines varies dramatically between regions, as does the number of Internet users: in Africa, less than 10 people per thousand are connected to the Internet, compared to over 250 inhabitants per thousand in Oceania. Both the numbers of telephone lines and Internet users are increasing worldwide.

Source: ITU, 2002. Taken from the ITU web site: http://www.itu.int/ITU-D/ict/statistics/.

bandwidth available for Africa is less than for the city of Sao Paulo in Brazil. Until the digital divide has been narrowed, the emergence of a society founded on knowledge remains severely compromised, as expenditure on ICT is much higher in developed countries, as shown in figure 14.7.

Innovative strategies are making small dents in the digital divide: UNESCO's International Initiative for Community Multimedia Centres (CMC), for example, combines community broadcasting with Internet and related technologies. A typical centre features radio both by and for local people and telecentre facilities offering access to Internet, email, word processing and other services. During radio-browsing programmes, presenters search the Internet in response to listeners' queries and discuss the contents on air with studio guests. Centres can gradually build up their own database of materials that meet the community's needs in such areas as health, education and income generation.

The UN Special Initiative for Africa focused on harnessing information technology for development, while the United Nations Development Programme (UNDP) has initiated a programme aiming to reduce the knowledge gap. One of the targets of the Millennium Development Goal addressing a global partnership for development calls for cooperation with the private sector, to make available the benefits of new technologies, especially information and communications.

A separate trend towards the privatization of knowledge is another emerging obstacle. As a WMO resolution on the Exchange of Hydrological Data and Products underlines, water data should be unrestricted and freely accessible. This principle is crucial to ensuring that water data and derived knowledge do not become expensive merchandise in the hands of powerful interest groups. Private insurance companies, for example, are coveting the natural disaster market and collecting data and historical records for risk analysis. Such information must stay in the public domain if water is to remain everybody's business.

Conclusions

The lesson is straightforward: the wider the knowledge base, the more common ground there will be for negotiation and discussion, and advancing towards a shared vision of needs. This knowledge base must encompass the general populace as well as water sector professionals. Education, both formal and informal, is vital if the value of water is to be appreciated throughout the world and in ensuring that everyone has a knowledge of good water practice, in terms of efficient water use and safe hygienic practices. Water sector professionals require access to reliable, high-quality data as a prerequisite for understanding key changes, identifying significant trends and making informed policy decisions on water management.

To effectively utilize the existing knowledge base in the management and operation of the water sector, and to improve understanding of the issues involved, consultation and the participation of the various stakeholders are necessary. This is true at local, national and international levels, both for developing policy as well as implementation of individual water schemes.

Figure 14.7: ICT expenditure as percentage of GDP

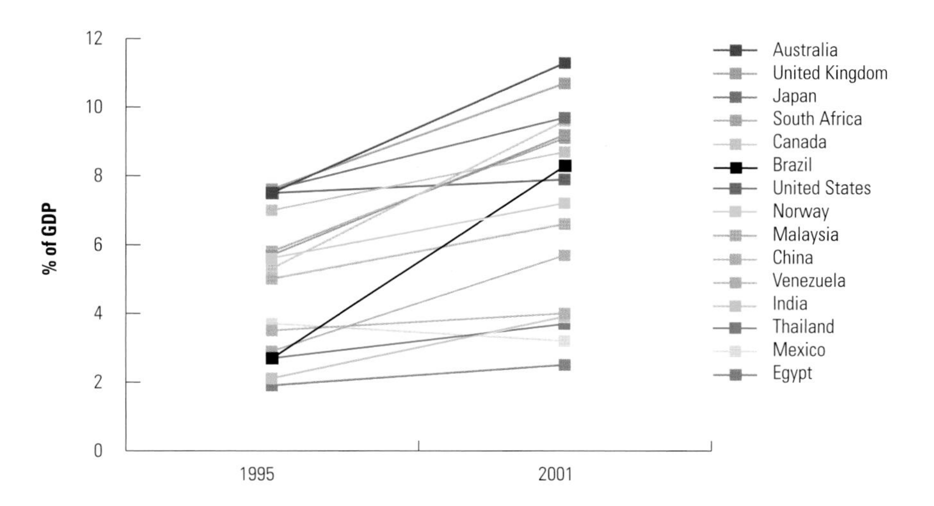

This figure shows the dramatic difference in expenditure on Information and Communication Technologies between developed and developing countries.

Source: Extracted from the World Bank Group web site, 2002. Data by Country – ICT at a Glance tables. Data and Statistics. http://www.worldbank.org/data/countrydata/countrydata.html

New technologies can improve the method and reliability of gathering data. Advances in ICTs can improve the ability of water professionals to gather, analyse and share data. They also provide a further avenue for spreading the word, enabling individuals and interested parties to access information more readily. But caution is needed – access to ICTs remains highly uneven across the world. Developing countries, which arguably have the greatest need for the benefits of ICT in overcoming their geographical and economic isolation, are hindered by the digital divide.

The challenges are clear. The overriding urgency is to enhance the capacities of low-income countries to develop their own expertise, while also ensuring they have full access to the global body of knowledge. The goal is fundamentally an ethical one, linking water to equity and social justice. It is a task that demands a sustained pursuit of international collaboration and investment to meet the UN Millennium Development Goal and the World Summit on Sustainable Development target of halving the proportion of people without access to safe drinking water and sanitation by 2015, and to pave the way for fairer, more sustainable development by integrating social, economic and environmental concerns.

Progress since Rio at a glance

Agreed action	Progress since Rio
Regard ensuring knowledge as a practical basis for sustainable water resource management	
Conduct feasibility of water resource assessment services by 2000	
Establish long-term target of fully operational services including hydrometric networks	
Ensure effective sharing of knowledge and technology	

Unsatisfactory — **Moderate** — **Satisfactory**

References

Colclough, C. Forthcoming. *Can the Millennium Development Goals for Education be Achieved?* Sussex, United Kingdom, Institute for Development Studies.

Cosgrove, B. and Rijsberman, F.-R. 2000. *World Water Vision: Making Water Everybody's Business.* London, World Water Council, Earthscan Publications Ltd.

ECOSOC (United Nations Economic and Social Council) and CSD (Commission on Sustainable Development). 2001. *Water: A Key Resource for Sustainable Development. Report of the Secretary General.* New York, United Nations.

Gibbons, M. 1999. 'Science's New Social Contract with Society'. *Nature,* Vol. 402, supplement, pp. C81–C84.

IAU (International Association of Universities). 2002. *World Higher Education Database 2001/2.* Paris, United Nations Educational, Scientific and Cultural Organization, CD-ROM.

IWMI (International Water Management Institute). 1999. *World Water and Climate Atlas. Colombo.*

Ministerial Declaration of The Hague on Water Security in the 21st Century. 2000. Official outcome of the Second World Water Forum, 3–7 December 2001, The Hague, The Netherlands.

Roberts, J. 2002. 'The Role of the Media in Reporting on Water Issues in the Middle East and North Africa'. Fourth Symposium on Water. Cannes, Journalists' Initiatives.

UIS (United Nations Educational Scientific and Cultural Organization Institute for Statistics). 2002. *UIS July 2002 Assessment.* Montreal.

UN (United Nations). 1992. Agenda 21. Programme of Action for Sustainable Development. Official outcome of the United Nations Conference on Environment and Development (UNCED), 3–14 June 1992, Rio de Janeiro.

UNDP (United Nations Development Programme). 2002. *Human Development Report 2002. Deepening Democracy in a Fragmented World.* New York.

——. 2001. *Human Development Report 2001. Making New Technologies Work for Human Development.* New York.

UNESCO (United Nations Educational, Scientific and Cultural Organization). 2001a. *Monitoring Report on Education for All.* Paris.

——. 2001b. *Regional Report on sub-Saharan Africa.* Paris, United Nations Educational, Scientific and Cultural Organization Institute for Statistics.

——. 2001c. 'Our Fragile World: Challenges and Opportunities for Sustainable Development.' In: *Encyclopaedia of Life Support Systems.* EOLSS Publishers.

——. 2000. 'The Right to Education: Towards Education for All Throughout Life' In: *World Education Report 2000.* Paris.

——. 1999a. *World Social Science Report 1999.* Paris.

——. 1999b. *Statistical Yearbook 1999.* Paris.

——. 1999c. *World Communication and Information Report 1999–2000.* Paris.

——. 1998a. *World Science Report 1998.* Paris.

——. 1998b. *Strategic approaches to freshwater management.* New York, United Nations.

UN-Habitat (United Nations Centre for Human Settlements). 2001. 'Water Education in African Cities'. Report of an Expert Group Meeting, 30 April–2 May 2001. Johannesburg.

Visscher, J.-T. and Van de Werff, K. 1995. *Towards Sustainable Water Supply: Eight Years of Experiences from Guinea-Bissau.* Delft, International Water and Sanitation Centre.

WHO/UNICEF (World Health Organization/United Nations Children's Fund). 2000. *Global Water Supply and Sanitation Assessment 2000 Report.* New York.

WMO (World Meteorological Organization). 1997. 'An Evaluation of the Current Status and Trends – Hydrological observing stations based on INFOHYDRO statistics for 1997'. In: *The World's Hydrological Networks.* Geneva.

World Bank. 2002. *World Development Indicators.* Washington DC.

——. 2001. *World Bank Atlas 2001.* Washington DC.

——. 1998. 'Knowledge for Development'. In: *World Development Report 1998/99.* Washington DC, Oxford University Press.

Note on Web Sites

By its very nature the Knowledge Base chapter is all-encompassing and vast. Without criteria agreed to and endorsed by all the partner agencies, it did not seem appropriate to single out some data sources as being more important than any others. For example, UNESCO, one of the two co-authors of this chapter, has data available across its fields of competence, in education, science, culture and communication. So how to choose? We have preferred simply to refer the reader to the more specialized lists proposed by the authors at the end of each challenge area.

We would also like to mention that initiatives such as the World Water Portal (see box 14.10) are intended as a response to this very problem of how to organize the mass of material available so that people can access data and information in a useful way. Developing such a resource in a coherent way requires a long-term effort. WWAP is already engaged in this and will continue to do its share.

15

Governing Water Wisely for Sustainable Development

By: UNDESA (United Nations Department of Economic and Social Affairs)/UNDP (United Nations Development Programme)/UNECE (United Nations Economic Commission for Europe)

Table of contents

PARADISE ISLAND
MAURITIUS
NY DIAN'NY "PAPA JEAN PAUL II"
TETO MADAGASIKARA
AN - TSARY

Never doubt that a small group of thoughtful, committed citizens can change the world. In fact, it is the only thing that ever has.

Margaret Mead

As the world is changing, so must our ideas about water problems and their place in the broader picture of human development and needs. Perhaps the greatest shift has occurred in our emerging appreciation of the key role that governance plays. Yes, water is unevenly distributed, with some areas having too much and others not enough. Yes, there are issues of water quality and competition among users. But as this chapter illustrates, good management – that includes integrated and participatory approaches – can go a long way towards attenuating the water crisis. Many encouraging examples are given. These suggest that dynamic new partnerships are forming, preventive actions are being taken and effective institutional and regulatory frameworks are being put in place at levels ranging from the local community on up to central government.

ALTHOUGH WATER GOVERNANCE AND HOLISTIC AND INTEGRATED APPROACHES to water resources management feature strongly in the international water agenda, in many countries water governance is in a state of confusion. The specific water governance issues vary. In some countries there is a total lack of water institutions, and others display fragmented institutional structures (sector-by-sector approach) and overlapping and/or conflicting decision-making structures. In many places conflicting upstream and downstream interests regarding riparian rights and access to water resources are pressing issues that need immediate attention; in many other cases there are strong tendencies to divert public resources for personal gain, or unpredictability in the use of laws and regulations and licensing practices, which impede markets and voluntary action and encourage corruption and other forms of rent-seeking behaviour.

Over the past decades, there has been increasing competition for the available water resources, and increasing water pollution. Consequently, water shortages, water quality degradation and destruction of the aquatic ecosystem are seriously affecting prospects for economic and social development, political stability, as well as ecosystem integrity. In developing countries, scarcity and degradation of water resources may have a severely limiting impact on development options, especially for poor people. In order to meet basic human and ecological needs and services, societies need to address and solve several serious water challenges and must come to terms with dwindling water resources, their uneven geographic and seasonal distribution, and inadequate and inequitable allocation of water services.

The water crisis is essentially a crisis of governance and societies are facing a number of social, economic and political challenges on how to govern water more effectively. The way in which societies organize their water resource affairs is critical for promoting and supporting sustainable development as an integral part of a poverty-focused development strategy. Sustainable development challenges are, at their core, a question of both governance and of how societies can balance economic and social development with ecosystem integrity. Sound and effective governance of water resources and related services are paramount to facilitating and supporting an enabling environment for Integrated Water Resources Management (IWRM). If we do not change the way in which water is governed, negative development impacts will be even more widely felt. It is also important to note that much wider governance issues and policies outside the water sector affect water resource issues. In effect, the challenges facing the sector are systemic in nature and inextricably linked to broader social, political and economic issues of water governance. For example, agricultural and industrial policies, covered in chapters 8 and 9, may have substantial impacts on the water sector.

This chapter focuses on how societies are attempting to govern water in more effective ways. It also contains a discussion on water governance, some of its components and how it can improve water management and service delivery.

Water Governance and the International Water Agenda

The world has changed since Agenda 21 was endorsed in Rio. The end of the Cold War has opened up borders, and globalization and economic and political liberalization have become socio-economic forces that all countries must deal with in order to reap their benefits or avoid their negative impacts. Current understanding is that water governance is a complex issue and very variable. Those who govern must be able to function in situations of rapid change, and often need to become agents for positive change. They also have to deal with competing demands for the resource. There is an ever-growing disparity between those who adapt quickly and easily and those who do not, created in part by the complexity, unpredictability and pace of events in our world. Weaknesses in governance systems are one of the major reasons behind the difficulties encountered in both following a more robust sustainable development pathway and balancing socio-economic needs with environmental sustainability. There is thus a strong need for improved institutions and social arrangements.

Agenda 21 set out a number of challenges for various areas of sustainable development and, in general, there have been huge difficulties converting principles into concrete actions. Although governance of water was not explicit as a programme area in Chapter 18 of Agenda 21, it was represented within most programme areas for water. It envisaged, *inter alia:*

- national comprehensive policies for water resources management, which are holistic, integrated and environmentally sound;
- Institutional strengthening and reform in conjunction with reform of water laws; and
- IWRM based on dynamic, interactive, iterative and multi-sectoral approaches. Its evolution would embrace spatial and temporal integration and all water users, and would be integral to socio-economic planning.

Agenda 21 set a specific target: that by 2000, national action programmes, appropriate institutional structures and legal instruments would be implemented, with water use attaining sustainable patterns. This target remains unfulfilled. It was also stated that subsectoral targets of all freshwater programme areas would be achieved by 2025. National reports to the Commission on Sustainable Development (CSD) were anticipated in order to report on progress towards target implementation, but few national reports contain any such information: a global or regional overview of the formulation of national water policies has therefore yet to emerge. Nevertheless, monitoring progress in relation to water governance is an essential tool for informed decision-making and development of future water governance requirements. Currently, there are very few indicators that can be applied and it is essential to develop the appropriate tools and mechanisms for collecting data at the national level.

Based on experience since Rio, some contextual aspects are important in understanding progress related to more effective water governance. One is the preoccupation that many governments have shown about debt and deficit reduction. During the past decade, these governments have significantly reduced their expenditures on environment-related infrastructure and services, which has generally had a serious negative impact on agencies responsible for water. Many more governments have been steadily backing away from concern for, or commitment to, environmental issues, and instead have emphasized strategies for economic growth based on a neo-liberal ideology and strategy. As a result of a newly emerging political economy in many countries, governments have devolved responsibilities for water and other services to lower levels of government that frequently have not had the human and institutional capacities or financial resources to maintain levels of services. Additionally, governments have been commercializing or privatizing such services. Increasingly, modified management processes should reflect a 'business model' in which efficiency, results-based management and tangible products have been emphasized, and less interest has been shown in providing systematic and transparent consultation processes with the public regarding policy development and implementation.

Since Rio, significant international water goals relating to governance have been set. The Second World Water Forum in The Hague in 2000 identified water governance as one of the highest priorities for action and expressed the need to govern water wisely through the involvement of the public and in the interests of all stakeholders. At the United Nations (UN) Millennium Assembly in 2000, heads of state emphasized conservation and stewardship in protecting our common environment and focused in particular on preventing unsustainable exploitation of water resources through the development of water management strategies at all levels, promoting equitable access and adequate supplies.

Although water-related objectives in Agenda 21 remain unfulfilled, progress has been made in the areas of water governance and management. There now exists a much better global awareness and understanding of the role water plays in ecosystem conservation and the overall cultural, social and economic value of water. The increasing focus on water governance, IWRM and demand-driven approaches marks an important shift in how water is being governed in terms of equitable distribution and efficiency. In general, progress has been made in the following three areas.

- The increasing recognition of water governance and required reforms of policies and institutions as the key to sustainable water development, of which the adoption of appropriate legislation, policies and institutions is only a part of the governance issue: it is the way in which enhanced institutions and policies are being established and implemented that matters. The existence of sufficient rules and regulations means little if they cannot be effectively enforced, due to power politics, vested interests and lack of funds, or the public's absence from the decision-making process.

- Reform of water institutions and policies is now taking place in many countries to address incoherent water property rights, fragmented institutional structures, inadequate policies, lack of incentives for increased partnerships and participation and various other aspects of water governance. However, progress has so far been too slow and too limited.

- Integrated approaches are widely accepted as the main vehicle or instrument to manage water in more effective ways, and the international community has made considerable efforts and progress in increasing awareness of water resources and their management. However, their implementation remains incomplete in both developed and developing countries.

What Is Water Governance?

Governance refers to relationships that can be manifested in various types of partnerships and networks. A number of different actors with different objectives are involved, such as government and civil society institutions and transnational and national private sector interests. An important shift in governance thinking is that development is now increasingly seen as a task that involves society as a whole and not the exclusive domain of governments (Pierre, 2000).

The notion of water governance and its meanings are still evolving and there is no agreed definition. Its ethical implications and political dimensions are all under discussion. Different people use the notion differently, relating it to different cultural contexts.

Some may see governance as essentially preoccupied with questions of financial accountability and administrative efficiency. Others may focus on broader political concerns related to democracy, human rights and participatory processes. There are those who look at governance with a focus on the relationship between the political-administrative and the ecological systems. Other approaches see governance entirely in terms of management, and the operation and maintenance of infrastructure and services. The United Nations Development Programme (UNDP) defines governance as the exercise of economic, political and administrative authority to manage a country's affairs at all levels. It comprises the mechanisms, processes and institutions through which citizens and groups articulate their interests, exercise their legal rights, meet their obligations and mediate their differences.

In this particular context, governance refers essentially to the manner in which power and authority are exercised and distributed in society, how decisions are made and to what extent citizens can participate in decision-making processes. As such, it relates to the broader social system of governing, as opposed to the narrower perspective of government as the main decision-making political entity. Governance of water is perceived in its broadest sense as comprising all social, political and economic organizations and institutions, and their relationships, insofar as these are related to water development and management.

Governance is concerned with how institutions rule and how regulations affect political action and the prospect of solving given societal problems, such as efficient and equitable allocation of water resources. The rules may be formal (codified and legally adopted) or informal (traditionally, locally agreed and non-codified). Sound and effective water governance systems are crucial to pursuing various sustainable water development and management goals.

In essence, water governance refers to the range of political, social, economic and administrative systems that are in place to develop and manage water resources and the delivery of water services, at different levels of society.

Water governance issues are also dependent on properly functioning legal and judicial systems and electoral processes. For example, legislative bodies made up of freely and fairly elected members and representing different parties are important to popular participation and accountability. It is essential that legal and judicial systems protect the rule of law and human rights. Open electoral processes help build political legitimacy. Water reforms that, for example, include decentralization and increased democratization may require constitutional, legal and administrative reforms that enhance the legitimacy and authority of the judiciary and legislative bodies and executing agencies.

Box 15.1: Examples of water governance issues

The water governance issues that need to be addressed and reflected in water policy, law, institutions and management include the following:

- Basic principles such as equity and efficiency in water distribution and allocation, water administration based on catchments, the need for holistic and integrated management approaches, the need to balance water use between socio-economic uses and uses to maintain ecosystem integrity, etc.

- Clarification of the roles of the government, civil society and the private sector and their responsibilities regarding ownership, management and administration of water resources. Under this heading the following issues will be included:
  - absence of or conflicting water rights legislation;
  - lack of effective mechanisms for intersectoral dialogue;
  - lack of economic incentives;
  - fragmentation of water management and administration;
  - lack of mechanisms for the participation of the community or other stakeholders;
  - the role of women in water management;
  - the effects of vested interest;
  - the absence of water quantity and quality standards; and
  - the absence of mechanisms for coordination and conflict resolution.

- Issues related to IWRM, including:
  - inappropriate price regulation and subsidies to resource users and polluters;
  - inappropriate tax incentives and credits;
  - overregulation or underregulation;
  - bureaucratic obstacles or inertia and corruption;
  - conflicting or absent regulatory regimes;
  - mechanisms to incorporate upstream and downstream externalities (environmental, economic and social) in water-planning processes; and
  - mechanisms to resolve disputes.

Some criteria for effective water governance

Governance affects economic, social and environmental outcomes. Water resource institutions regulate who gets what, when they will get it and how much of it they will get. Adequate governance can decrease political and social risks, as well as institutional failures and rigidity. It can also improve capacities to cope with shared problems. Research suggests that there is a strong causal relationship between improved governance and improved development outcomes such as higher per capita incomes, lower infant mortality and higher literacy (Kaufman et al., 1999).

Defining the various components required for effective water governance is a complicated task. In general, we are more familiar with failures than with effective water governance. What makes governance effective can differ from context to context and depends on cultural, economic, social and political settings. More effective governance systems need to be designed and created to deal with governance shortcomings and to increase the development potential of civil society agencies, local communities and the private sector. Box 15.1 presents some of the water governance issues that need to be addressed and reflected in water policy, law, institutions and management. Many of these issues are serious challenges to the development of wise governance.

Effective governance of water resources requires the combined commitment and effort of governments and various civil society actors, particularly at local/community levels, as well as the private sector. Policies must deliver what is needed on the basis of clear objectives and informed decision-making, which should occur at the appropriate level. Policies should also provide clear economic and social gains for society as a whole. Given the complexities of water use within society, managing it effectively and equitably entails ensuring that the disparate voices are heard and engaged in decisions concerning the waters in which they have an interest. Water governance can be said to be effective when there is equitable, environmentally sustainable and efficient use of water resources and its benefits. Such efficient use includes minimizing transaction costs and making the best use of a resource. Although there is no single model for effective governance, the following basic attributes are likely to represent some of its features.

- Participation: all citizens, both men and women, should have a voice – directly or through intermediate organizations representing their interests – throughout processes of policy- and decision-making. Broad participation hinges upon national and local governments following an inclusive approach.

- Transparency: information should flow freely within a society. The various processes and decisions should be transparent and open for scrutiny by the public.

- Equity: all groups in society, both men and women, should have opportunities to improve their well-being.

- Accountability: governments, the private sector and civil society organizations should be accountable to the public or the interests they are representing.

- Coherency: the increasing complexity of water resource issues, appropriate policies and actions must be taken into account so that they become coherent, consistent and easily understood.

- Responsiveness: institutions and processes should serve all stakeholders[1] and respond properly to changes in demand and preferences, or other new circumstances.

- Integrative: water governance should enhance and promote integrated and holistic approaches.

- Ethical considerations: water governance has to be based on the ethical principles of the societies in which it functions, for example by respecting traditional water rights.

These attributes are examples and represent ideal situations, which may not all be found in any single country. Through wide participation and consensus-building, societies should aim at identifying those attributes and actions that are most relevant to them. In this regard, inclusive dialogues at national and local levels are important to identify the appropriate challenges and actions for a given context. An example of public participation in the processes of water governance in Greater Tokyo is given in box 15.2.

Who owns the water?

Property laws often determine who owns or has the right to control, regulate and access water resources. Water rights are often complicated by the variable nature of the resource. Additionally, there are economic, social and environmental values attached to water rights, and any effective water governance structure will need to address this complexity.

There is increasing pressure to recognize and formalize water rights. This is happening in many countries, although it raises complex questions about the multiplicity of claims and water uses, and it may not be sufficient to secure equitable access to water resources. The

1. Stakeholders are sometimes defined as individuals or groups who have a legal responsibility or mandate relative to a decision, and who will be directly or indirectly affected by a decision. The concept of stakeholders has increasingly been used to highlight that while it is not reasonable to involve everyone in every decision, it is important to ensure that those who have legal responsibilities or could be directly or indirectly affected by decisions are represented when decisions are taken.

Box 15.2: Japan promotes public participation

In 1997, the River Law in Japan was revised, and a clause targeting improvement and conservation of the river environment was added. A planning system designed to incorporate the opinions of local residents was also introduced with the aim of establishing a river administration system for flood control, water use and environmental conservation. This system aims to make river areas healthier, while at the same time challenging the public to become more involved in the process.

In order to satisfy the people's need for improvement and conservation of the river environment, and to base such improvement on riverine and regional characteristics such as climate, landscape and culture, it is essential to cooperate closely with local communities. The river improvement plan is twofold: one part deals with matters constituting the fundamental river management policy, and the other deals with the river improvement and conservation plan. The new planning system includes procedures for incorporating the opinions of the local government and residents.

The Tamagawa River System Improvement Plan was implemented in March 2001, the first such plan in Greater Tokyo and the second in Japan. Discussion groups are formed to set up both the planning process and the river basin committee as prescribed by the River Law. The Tamagawa River Basin Discussion Groups (which include the local basin communities, scientific experts, companies, relevant local government authorities and river administrators) are engaged in an ongoing exchange of opinions and information relating to the development of the Tamagawa River and the river basin environment. This exchange enables them to build mutual trust and deepen their cooperation. These meetings are organized to foster a gradual consensus towards creating a healthy river and town.

Another river improvement plan is being implemented, meanwhile, in the Yodogawa River basin, which comprises the cities of Kyoto and Osaka. The River Law provides for setting and implementing goals, but step-by-step consensus-building is also an integral part of the process.

Without a thorough understanding of the current situation and of the problems facing the basin, a consensus cannot be built, and without a consensus, it is impossible to discuss future steps and take action. The entire process, although very lengthy, therefore rests on public involvement, and this trial should provide a basis for partnerships between river administrators and the public.

Source: Prepared for the World Water Assessment Programme (WWAP) by the Ministry of Land, Infrastructure and Transport (MLIT) of Japan, 2002.

process of formalization is all too often biased in favour of the rich and powerful who may abuse the system. In many developing countries, local regulations, customary laws and traditional rights assign rights and responsibilities that differ from state regulations. It is therefore important for formal rights to consider traditional practices.

For formal and informal rights to be meaningful, it is essential that they retain the capacity to protect against competing water users. Due to the nature of water resources, illegal abstractions are generally easy and commonplace. They can be difficult to resolve since the transaction costs for controlling and excluding non-members or owners, particularly in irrigated agriculture, can be very high. Excessive illegal use threatens to break down property rights and established institutions, as well as depleting water resources.

Water can be seen as a common resource system.[2] All water use creates positive or negative externalities (social, economic and/or environmental). The effective governance of water requires that water rights and obligations be clearly defined. For some definitions of property rights, see box 15.3. Such rights and obligations stipulate who is entitled to what quantity and quality of water, and when they are entitled to it. Water entitlements may also include obligations, such as respecting the rights of downstream water users and the discharge of wastewater (Lundqvist, 2000).

Although the state will normally legislate on the issues of property rights, many of the current problems of water governance derive from hierarchical and centralized control by the state and its inability to provide sufficient water-related services or to enforce regulations. It is often held that the local community, together with water users' organizations, can govern common resources in equitable and efficient ways (Bromley, 1992; Ostrom, 1990).

2. The use of common property resources is here seen in similarity with Ostrom's terminology 'common pool resources', which refers to 'a natural or man-made resource system that is sufficiently large as to make it costly (but not impossible) to exclude potential beneficiaries from obtaining benefits from its use' (Ostrom, 1990). Ostrom distinguishes between resource systems and resource units. The former contains groundwater basins, physical infrastructure such as sewerage lines and roads, water basins, etc. The latter is what can be used from the resource systems, e.g. fish, quantity of water withdrawn from a lake or a river.

Box 15.3: Property rights

- Open access property: There is no defined group of users or owners and the water resource is open to anyone.
- Common property: The group in charge of the resource, such as a local community or a particular user group, has a right to exclude non-members from uses and benefits. Members of the management group have both rights and obligations with respect to use and maintenance of the water resource.
- State property: Water users and citizens in general have an obligation to observe use and access rules determined by the controlling government agencies.
- Private property: Within the existing institutional framework the owner has the right to decide on water access and uses. Those without rights or financial means to acquire water are excluded from consumption.

Although rights may be defined on paper, water resources may in practice be considered free-for-all. In many instances, particularly in agriculture, water rights are closely linked to land rights: any reform in water rights has therefore also to address land rights and vice versa. This is being addressed in South Africa's water policy reform (see box 15.5) where land and water rights are being disconnected and the riparian principle may thus not necessarily apply.

Advocates of free market policies are likely to favour private and transferable water rights and pricing that reflects the growing scarcity of the resource. They suggest that this will lead to efficient and equitable allocation of resources and will provide the greatest incentives to avoid wasteful practices. Private property rights imply that the owner can exclude those without rights or those who cannot afford the good. A legitimate concern with privatization and increased commercialization is that such a policy may exclude poorer segments of society from reasonable access to water.

Whose water governance?

It is important to consider to what extent the processes of institutional reform and devolution of water rights serve society, both in its entirety and its component groups. Currently, poor people in both rural and urban areas tend to be disadvantaged in accessing water and sanitation services and in accessing water for food production. If the water resource is managed primarily through private markets, only those having property or sufficient income may have easy access. If public authorities manage water, it is still not certain that poor, isolated or socially immobilized elements will gain improved access. Consensus on public policies in governing water is a problematic issue and raises many questions. Any water governance reform should aim for social and political stability. Mechanisms to compensate those members of society who lose out in the short term may be difficult to establish, or may be omitted if they are few and not politically strong. However, robust and flexible governance structures should be able to cope with such problems.

Changes in water rights and uses can be a very controversial issue. For example, Sri Lanka is preparing a new water act to foster decentralized management of water through river basin organizations representative of the basin-level stakeholders. The new river basin organizations will become responsible for planning, implementing and regulating water allocations between water users in each basin. A National Water Resource Agency will oversee local implementation of the planning and allocation processes. However, the water resource management concepts that underpin the act, including water rights, have proved to be contentious, largely due to fears over possible new water charges and loss of traditional water usage rights. These fears have delayed presentation of the act to Parliament. For more details, see chapter 18, the Ruhuna basins case study.

The Russian Federation is an example of a very large country that has also implemented management units in river basins (see box 15.4). Special consideration must be given to large river basins, particularly where they cross national boundaries. In such situations, the state must be responsible for issuing clear regulations and limiting the rights of local communities where this is necessary to protect downstream users. Such regulations should, where appropriate, reflect international agreements. These issues are discussed in chapter 12 on the sharing of water resources.

Water Governance and Water Management

Governance and management are interdependent. Effective governance systems should enable the more practical management tools to be applied correctly. Public-private partnerships, public participation, economic, regulatory or other instruments will not be

Box 15.4: Governing water wisely – a Russian basin-level approach

The basin-level approach to water management that is underscored in the Water Framework Directive of the European Union (EU) is also the basis for Russian water management policy. The Russian Federation is divided into seventeen big water basins, managed by the specially appointed Water Basin Administrations under the Ministry of Natural Resources. These administrations are responsible for protecting water and managing it in a sustainable manner. Where river basins are shared by several users within the Federation, basin agreements are signed, defining the rights and responsibilities of all regions with regard to water quality. These agreements also form the basis for joint environmental monitoring and data collection needed for joint water management. According to the legislation, agreements should be accompanied by the creation of basin councils representing all main stakeholders. So-called Schemes of Complex Use of Water Resources (which resemble river basin management plans) should be created for each river or lake basin.

However, these measures are often not as efficient as they might be, for the following reasons:

- There is an absence of legislative framework for the work of basin councils.
- Basin agreements have been signed, but their implementation is difficult due to problems with financing water protection measures.
- The Schemes of Complex Use of Water Resources remain undeveloped because of economic constraints.

Source: Prepared for the World Water Assessment Programme (WWAP) by the government of the Russian Federation, 2002.

effective unless the political will exists and broader administrative systems are in place. For example, the polluter pays principle is a management tool specifically designed to decrease water pollution. However, before such a principle can be enforced, it is essential that appropriate rules and regulations, clear mandates for different agencies and transparent financial arrangements be implemented and communicated.

An integrated approach

There is a wide acceptance of IWRM as the appropriate management tool for sustainable use of our water resources and for improved delivery of water services. IWRM promotes participatory approaches, demand and catchment-area management, partnerships, subsidiarity and decentralization, the need to strike a gender balance, the environmental, economic and social value of water and basin or

Figure 15.1: Framework for moving towards IWRM

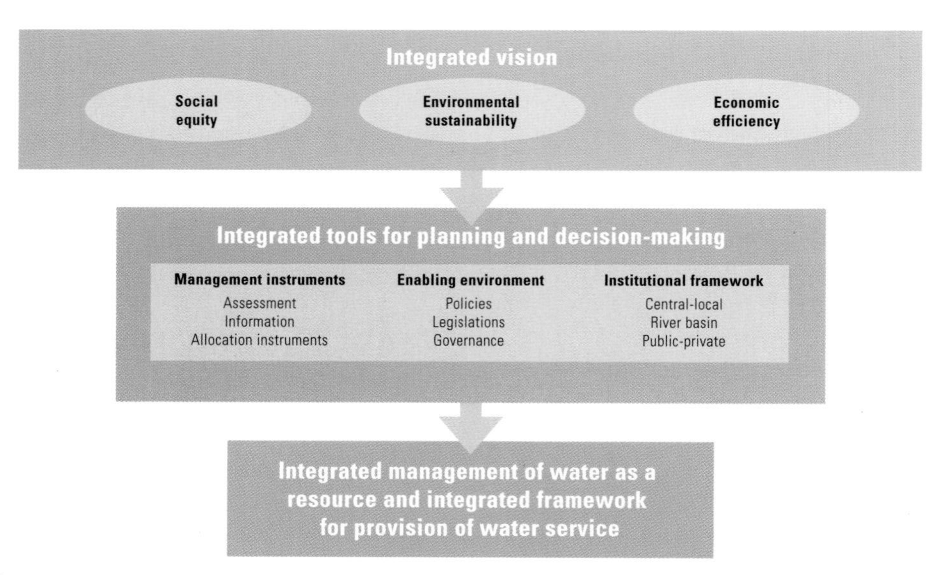

Source: Based on GWP, 2002b.

catchment management (GWP, 2000). It replaces the traditional, fragmented sectoral approach to water management that has led to poor services and unsustainable resource use.

IWRM is based on the understanding that water resources are an integral component of the ecosystem, a natural resource and a social and economic good. Physical processes, such as the naturally occurring interplay between the hydrological cycle, land, flora and fauna, take place in an integrated manner. The challenge is to create governance systems, institutions and management instruments that take into account and reflect such physical complexities in planning, decision-making and implementation processes, while at the same time balancing them with social, economic and environmental needs and objectives.

The Global Water Partnership (GWP) Technical Committee has proposed a simple framework as the starting point for IWRM as illustrated by figure 15.1. Concurrent development and strengthening of three elements is needed: an enabling environment, appropriate institutional roles and practical management instruments.

The enabling environment comprises national, provincial and local policies and legislation. These constitute the 'rules of the game', which allow all stakeholders to play their respective roles. The 'rules' should promote both top-down and bottom-up participation of all stakeholders, from the national level down to the village or municipality, or from the level of a catchment or watershed up to the river basin level.

The government's role should be that of activator and facilitator, rather than top-down manager. Important aspects of the government's role include formulating national water policies and legislation, enacting and enforcing the legislation, and encouraging and scrutinizing the private sector.

In the area of governance and institutional roles, development, financial and human resources, traditional norms and other circumstances will play a large part in determining what is most appropriate. Nevertheless, institutional development is critical everywhere to the formulation and implementation of IWRM policies. Clear demarcation of responsibilities between actors, separation of regulation from service provision functions, adequate coordination mechanisms, filling jurisdictional gaps and eliminating overlaps and matching responsibilities to authority and to capacities for action are all parts of institutional development.

Finally, practical management instruments should be developed to help water managers. The art of IWRM lies in selecting, adjusting and applying the right mix of these tools for a given situation. Five categories deserve special attention.

- Water resource assessment: comprising data collection networks, environmental impact assessment techniques and risk management tools, e.g. for floods and droughts.

- Communication and information: raising awareness is often a potent instrument for improving management, particularly when accompanied by opportunities for informed stakeholder participation.

- Tools for water allocation and conflict resolution: allocation could be done through a mix of regulatory and market instruments based on valuation of costs and benefits; and conflict resolution tools could provide guidance in issues of upstream versus downstream, sector versus sector and human versus nature.

- Regulatory instruments: including direct controls such as land use plans and utility regulation, as well as economic instruments (prices, tariffs, subsidies and others) and encouragement of self-regulation, for example by transparent benchmarking.

- Technology: both new and traditional technologies might provide scope for progress, within the water sector as well as in others that affect water demand.

Integrated management will need to tackle sectoral agencies protecting their traditional roles and responsibilities as well as the problems of overlapping or conflicting legal mandates and responsibilities. The limited array of senior and powerful advocates for the concept of IWRM make it difficult to alter the well-entrenched existing water governance systems, which tend to reflect sectoral approaches

As water-related services are extended to promote public health and food production, uncoordinated institutions can be confusing and lead to water resource depletion. In the Zambian village of Mbala, people received no less then three different pieces of advice from government agencies on how to protect a local water source: to uproot trees, to plant trees around the water source, and to uproot trees and replant them with orange trees to protect the water source (Visscher et al., 1999).

It is equally alarming that in many countries, a large number of water supply and sanitation projects and water management polices continue to be developed in isolation from each other (Visscher et al., 1999).

During the last decade many country initiatives have been taken, ranging from relatively simple changes (creation of interagency coordinating groups) to fundamental reallocations of power and changes in basic values or principles. Some examples of implementation of integrated approaches, although limited, can be found in river basin management in France and in new water laws that encourage cross-sectoral management in South Africa and Zimbabwe. These latter reforms show many similarities in issues of ownership, catchment-based management and the need to obtain a permit for any water use (see box 15.5).

Box 15.5: Water reform in South Africa

The political changes in South Africa and the emergence of a democratic system have allowed for reform of the water sector as regards policy, organizational structure, water rights and legislation. This water reform is often cited as a very comprehensive and innovative approach to water management.

The new water law sets out to meet the objective of managing water quantity and quality to achieve optimum long-term environmentally sustainable social and economic benefits for society, while ensuring that all people have access to sufficient water. Water is considered a national resource vested in the state. The law provides for nineteen catchment management agencies, which have to prepare a management plan, issue water licences, actively promote community participation and perform other functions for implementation of the water law.

In many areas, water services have expanded rapidly. However, in some cases decentralization of service provision and responsibilities for some other areas of the water law has been difficult due to limited human and institutional capacities as well as a shortage of financial resources.

Improving water governance will help to address government, market and system failures. In Latin America there has recently been a move to address aspects of market failure. For example, the Chilean water reforms have placed major emphasis on the correct pricing of water to reflect opportunity costs over and above the tariff. Similar attempts are underway in Costa Rica and Ecuador where downstream users pay the watershed owners and managers for watershed services. The Chilean experience is instructive but the context may be location-specific, since there was a major commitment to developing the entire economy based on an export-oriented open economy. There have been many frictions due to such changes; issues of openness, transparency, participation and ecosystem concerns are now being tackled (Rogers and Hall, 2002). The water reform work in Chile is a case in point of the need to sequence reforms to meet the most urgent requirements, and, along with the example of South Africa, illustrates that water reform is often triggered from outside the water sector (for example, by political and economic liberalization). Experience in the United States suggests that reduced water demand appears to be largely due to reductions in water use subsequent to changes in the energy and agricultural sectors as well as enforcement of federal instream water requirements for ecosystem maintenance (Rogers and Hall, 2002).

Decentralization and participation

Effective water governance requires changes in attitudes and behaviour among individuals, institutions, professionals, decision-makers – in short, among all involved. Participation by the public or stakeholders is an important tool in implementing such changes as it facilitates more informed decision-making and eases conflict resolution. It can also guarantee that voices of relatively powerless groups, such as women and indigenous people, are heard. Participation offers people the opportunity to meet their responsibilities, as well as the opportunity to claim their rights.

Key aspects of sustainability include empowerment of local people, self-reliance and social justice. These reflect concern about principles of equity, accountability and transparency. One way to incorporate these principles into real-life management is to move away from conventional forms of water governance, which have usually been dominated by a top-down approach, and professional experts in the government and private sector and move towards the bottom-up approach, which combines the experience, knowledge and understanding of various local groups and people. An important lesson during the 1990s was recognizing the benefits of combining expert knowledge with local knowledge. The self-help Orangi Pilot Project (OPP), which provided low-cost sanitation to the urban poor, is a good example of the bottom-up approach. The entire project was managed and financed by the local population, clearly illustrating that water governance is an important issue even at local levels (see box 7.6 in chapter 7 for more details on this scheme). Local participation can also be a powerful tool for conflict resolution. An enlightening case is provided by the Taiz region in Yemen (see box 15.6), where social and political conflicts, arising from competing demands on scarce water resources, have started to be resolved by engaging local stakeholders in a continuous dialogue. However, the anticipated end result of this particular case has still to be achieved.

The actual progress in participatory approaches has been modest and uneven. Many governments have a very instrumental view of local communities and related community-based organizations, and their active involvement is normally sought only for implementation of water projects. Participation in a truer sense would entail involvement throughout the whole policy or project cycle. Progress has also been uneven in overcoming the gender gap. Increased attention to gender can enhance project effectiveness as well as provide support for equity issues: it is encouraging that in some places such as Burkina Faso and Bangladesh, thinking and experience have moved beyond women and development to gender and development. In effect, in Burkina Faso, women and men each have their own forms of organization with their own rights to water and land for agriculture: the women in the river valleys, the men on higher ground. When the state took over the land for irrigation, it only gave out plots and water rights to male heads of households

Box 15.6: Taiz water management planning – possibilities for rural/urban conflict resolution

In recent years, efforts have been made by the National Water Resources Authority (NWRA) in Yemen to minimize social and political conflicts. This was done by implementing a system of water transfers from rural to urban communities in the Taiz region within the context of IWRM. Key features of this system included both demand management measures (such as input taxation and raising of public awareness) and social measures (through definition of a regime of tradable water rights). It was felt that the demand management measures would only make meaningful contributions towards achieving the objective of sustainable water resource management if adopted in conjunction with the social measures.

Defining a system for rural/urban water transfers called for detailed consultations with the local rural communities, especially farmers, who did not seem to have much faith in the institutions engaged in the consultative process. Discussions often led to heated arguments. Nevertheless, the process continued over more than three years. The process was followed seriously as a confidence-building opportunity and special efforts were made not to let the dialogue break down at any stage. There were many rounds of discussion, sometimes with large groups of farmers, at other times only with influential community leaders. Each round of discussion built upon the issues and concerns raised in the previous round.

The end result was that communities agreed to the following main principles for rural/urban water transfers:

- There should be clearly defined rights, taking into account ethical considerations such as priority for drinking water needs.
- Water should be allocated through market-like processes, with the exception of water needed for drinking and basic needs.
- Water rights should be tradable, and, to the extent possible, there should be direct compensation of individuals willing to transfer their water rights to others, commensurate with the rights transferred.
- Water transfers should be verifiable. Those who agree to transfer their water rights must reduce their water use accordingly.
- The local communities should participate in designing the rules and mechanisms to govern rural/urban transfers, including a mechanism for monitoring compliance and punishing violators.
- NWRA should have an oversight role in rural/urban transfers to ensure resource sustainability and equity.

Source: Based on a project by UNDESA, 2002. Prepared for the World Water Assessment Programme (WWAP).

and created only male water users' groups. The women lost their production and harvest rights, their traditional organization went unrecognized and they lost their motivation for agriculture. When the government realized this, new plots were also given out to the women and productivity, as well as operation and maintenance of the watercourses, improved.

In Bangladesh, with an abundance of groundwater, large-scale farmers were the first to benefit from state subsidies to install deep wells with mechanized pumps. When shallow wells and smaller pumps became available, this irrigation technology was placed within the reach of the smaller-scale farmers. Out of necessity, they used water more efficiently than the large-scale farmers, and so accumulated a surplus of water that they sold to landless farmers and women, who united and bought pumps to sell water for agriculture. In Bangladesh agriculture, men have access to water technology and land; it is they who mobilize labour, arrange inputs and have the ultimate say over the harvest. Continuing exclusion of women from the developments in water technology has widened the gap. But as water vendors, women have found other opportunities to benefit from the new technology (Van Koppen, 1997).

Institutional reforms have, at least in part, been justified by the principle of subsidiarity (management at the lowest appropriate level). Many national and state governments have delegated responsibility for water and other environmental services to lower levels of existing government, to new institutions created specifically to take on responsibilities at lower tiers of governments. Not all delegation has been within-government: the new Water Law of Zimbabwe, for example, delegates catchment management responsibilities and day-to-day duties of water rights allocation and administration to stakeholder-elected catchment councils. Each catchment council is composed of sub-catchment councils, composed of local water user groups and associations. However, recent political instability in Zimbabwe is seriously threatening attempts to reform the water sector.

The catchment is increasingly accepted as the appropriate scale for water resource management. However, for it to be more useful would require overcoming certain obstacles. Strong sectoral or local interests may secure water first. River basins do not always match existing administrative boundaries, which can make it more difficult for riparians to solve common problems. Many local communities and civil society organizations are facing problems in mobilizing resources and the required human and institutional capacities. It is important that decentralization of water responsibilities to local communities or new catchment-based organizations be done in a transparent and participatory way to prevent powerful local groups from claiming the entire water resource, further marginalizing poor people, women and other politically weak groups.

It is further necessary that local catchment-based management groups respect the rights of other basin users downstream and, where appropriate, international river basin agreements.

Public-private partnerships

The ways in which various government agencies, civil society organizations, private firms and the market relate to each other is crucial for effective public-private partnerships. Governance draws explicit attention to these relationships. Partnership formation can bring about substantial benefits. In cases where less public funding is available for water-related initiatives, partners outside government have sometimes contributed, through money or voluntary action, to expediting activities that would otherwise have been difficult to support. In this manner, partnership arrangements have shown that they can help to maintain or to improve water services.

The Ministerial Declaration at Bonn, 2001, encouraged private sector participation. It also noted that this does not imply private ownership of water resources and that water service providers should be subject to effective regulation and monitoring. Private sector involvement in water may take many forms and is not new. At the most basic level, water service providers have always bought in goods and services from the private sector, and governments have enlisted the private sector to assist in assessing and monitoring water resources, for example in groundwater investigations. In recent years, the trend has been to give the private sector a larger role in managing, operating and maintaining water and wastewater systems. These may be broadly divided into the following:

Box 15.7: Public-private water partnership in France

In order to meet their responsibilities in terms of water services, French communities are often organized into inter-municipality drinking water associations (67 percent of the population) and, more rarely, sanitation associations (16 percent of the population for water collection). They also make use of public/private partnerships by delegating operation, maintenance and development of public potable water and sanitation services to private companies (85 percent of the population for potable water, 36 percent for sanitation). However, they retain ownership of the system and the private service provider must return the network in proper working condition at the end of the contract period. This system allows a clear delineation of roles and exchanges of experience, as private operating companies manage the water services of many different communities. Delegation is also favourable to efficiency, because of the technical expertise and the economic constraints of the private companies.

Source: Based on the Seine-Normandy Basin Agency (AESN), 2002. Prepared for the World Water Assessment Programme (WWAP).

- Divestiture of assets: this model has been used in England and Wales. The private sector owns the infrastructure and is responsible for planning and financing its development, as well as for its operation and maintenance. The driver for privatization of the water industry in England and Wales was the need for investment, and the key to its implementation, a strong regulatory framework. The water and wastewater companies are regulated by an economic regulator, the Office of Water Services (OFWAT), which has limited prices, the Environment Agency, which controls water abstractions and wastewater discharges, and the Drinking Water Inspectorate, which controls the quality of water supplied.

- Concessions: these are granted for the management, operation and development of systems for a limited period (usually about twenty-five years), but ownership of the infrastructure remains with the government. This is the dominant system in France, where there is no regulator, but the interests of consumers are represented by the contract between the service provider and the local government, which owns the assets (see box 15.7). Shorter contracts with minimal investment by the operator (leases) are sometimes employed.

- Build-Operate-Transfer (BOT) and the Build-Own-Operate-Transfer (BOOT) schemes: these involve the private sector in the financing, construction and operation of works. They are usually used for treatment plants and the private investor makes a return on his investment from the revenue for water sold or fees for treated wastewater. The private sector is again controlled through the terms of the contract by a local government or a public utility.

- Service contracts: many utilities use service contracts, that is, they will buy in some goods and services from the private sector. In recent years, some utilities have contracted out substantial parts of their operations, e.g. billing and revenue collection, which would have previously been regarded as a responsibility of the public utility itself.

Pressure from international funding agencies has led to the increased involvement of the private sector in developing countries, largely through concession contracts for the major European companies in the field. In Macao, privatization resulted in an improved level of service. In Buenos Aires, private sector involvement has resulted in increased coverage of services and more reliable water supplies. However, it has been criticized because of lack of transparency in the renegotiating of contracts and tariff increases and decisions to disconnect customers whose payments are late. In all cases involving concession contracts and private sector involvement, success appears to rely on the presence of effective regulation by central or local government agencies. A problem in many developing countries is the lack of capacity and experience to develop an adequate regulatory system.

There is considerable potential for increased private intervention in the near future in providing services to more affluent urban areas of developing countries. However, participation in the extension of service to the urban and rural poor remains more problematic, as this hinges on pricing and cross-subsidy policies that would enable private utilities to generate a fair return on their investments.

A further problem in developing countries is the lack of the necessary skills within the private sector to operate, maintain and develop water and wastewater systems. In this case, private sector participation often implies foreign companies taking over utilities. However, the use of smaller service contracts for specific activities could employ indigenous private companies, thus encouraging the development of greater skills within these companies and enabling local governments to gain more experience in preparing and managing contracts.

The rights to abstract water and discharge wastewater are important to all service providers and are normally controlled by public authorities. However, some economists argue that active trading in water rights promotes water use efficiency as market mechanisms allocate water to the highest valued use. Trading of water rights takes place in parts of some developed countries, such as the United States and Australia.

An alternative to government provision of services to rural and poor urban communities is community-based service delivery. It is often claimed that various civil society organizations are capable of delivering services more effectively than government agencies. Community-based organizations, water users' associations and non-governmental organizations (NGOs) can play important roles, independently or in partnership with government agencies. In addition to delivering services, they can act as a link between state and community, be directly responsible for natural resource management, or act as a 'watchdog'. Much of the competence of civil society organizations is found in their knowledge about and links in the local context, which are important in choosing appropriate solutions. Local knowledge can form a basis for flexible, innovative and dynamic institutional frameworks for sustainable water development. However, many civil society organizations have limited funds and membership, and rely on voluntary work and charismatic leadership. In many cases, the NGOs and other civil society organizations have been inconsistent in their work and have faced difficulties in maintaining and expanding their activities (Tropp, 1998). As previously mentioned, government agencies tend to perceive civil society organizations in a rather narrow instrumental manner and their involvement is normally sought only for project implementation.

Partnership practices have illustrated that there is no blueprint to determine the appropriate model to use. It is obvious that many different kinds of partnerships are needed, ranging from personal or informal to voluntary or legally binding arrangements. They may be short-term and project-specific or long-term and broad in scope. They may involve sharing of work or financial costs, or the sharing only of information. Experience suggests that key ingredients in successful partnerships are a shared vision, compatibility, equitable representation, legitimacy, communication, adaptability, mutual trust and understanding, perseverance, fixed formal or informal rules and transparency. In many parts of the world there is a huge distrust between the state, civil society and the market, which does not render the formation of partnerships any easier.

Water governance and financing

In terms of financing, governance is essentially about creating a favourable environment to increase water investments and to ensure that investment is used correctly. Governance is also concerned with how capital is being spent and how more can be done with existing resources, or even with less. The economic rationale behind governance is that effective water governance is supposed to lower

Box 15.8: Financing water development in Africa

- Water should be explicitly included in Poverty Reduction Strategy Papers (PRSP).
- In most African countries, water management is dispersed between other sectors (agriculture, health, energy, etc.) and is not the responsibility of a specific ministry or authority.
- A fixed percentage of African government budgets (for example 5 percent) could be devoted to water resources development and management.
- Bilateral and multilateral aid could be earmarked as matching funds to African governments' budgetary commitments.
- Urban revenue could be transferred for rural water supply development and human and institutional capacity-building efforts.
- Private finance and public-private partnerships may be best suitable for urban areas. The role of private sector involvement in the African water sector is subject to debate.
- No amount of financial resources can solve Africa's water challenges without firm commitment by its political leaders and decision-makers. Efficient utilization of financial resources can only be achieved when a basic system of effective governance, including transparency, accountability and subsidiarity, is in place to guide public functions.

Source: UNECA, 2002.

transaction costs by preventing corruption and increasing financial efficiency. A fundamental insight is that countries cannot 'construct' themselves out of water problems and capital-intensive infrastructure development must go hand-in-hand with developments in governance of water financing.

It is evident that the water sector is underfinanced and governments have not reached the financial targets set out in Chapter 18 of Agenda 21. However, the limited funds in many water development endeavours should not paralyse action. Currently, the main cost for water-related services in developing countries is carried by governments through taxation and service charges and, to a lesser degree, by donor assistance. The private sector is only modestly involved in water-related services. Governments of developing countries have not been able to raise adequate funds through taxation or the application of water tariffs for enhanced cost recovery. A recent report on financing water development in Africa pointed to some specific sources for additional funding (see box 15.8). But, most importantly, it acknowledged the interdependence between effective water governance, increased funding and efficient utilization of existing resources. The challenging task of raising additional funds should also render decision-makers aware of the need to complement capital-intensive investments with alternative low-cost technology, especially in the sanitation sector.

High levels of corruption and other financial mismanagement reduce the rate of economic growth. Corruption has a pervasive and troubling impact on poor people since it distorts allocation of water resources and related services in favour of the wealthy and powerful. Thus, poor people will receive a lower level of services and infrastructure investment will be biased against projects that serve the poor (UNDP, 1997). The introduction of more effective governance systems with a strong autonomous regulatory authority and transparent and accountable processes would attract new financing. Improving capacity to prepare and manage contracts would also reduce bad utility practices, both public and private.

Throughout the past decade, many developing countries have sought to reduce debts and deficits. This has resulted in large reductions in infrastructure and services expenditure, with serious negative impacts on agencies responsible for water. Policy objectives of debt and deficit reduction have led to significant withdrawal of human and financial resources in supporting environmental services, including water. However, the Heavily Indebted Poor Countries (HIPC) initiative is attempting to reverse this trend. Debt relief is being linked to poverty reduction and, therefore, not only are more funds being made available for the provision of basic services, but countries are being actively encouraged to spend more on these. It may be expected that this will lead to an expansion of funds for water supplies and sanitation services for the poor in both rural and urban areas.

The heavy dependence on public funding and unclear financing mechanisms, institutions and policies are some of the investment characteristics in many developing countries. These issues have to be addressed together with the need for increased financing. The government plays an important role in providing incentives to private finance by establishing clear regulatory and institutional frameworks. Governments should also ensure that poor people are served and can afford water-related services. Countries' economies and prospects for economic growth remain highly dependent on water and other natural resources. There is a growing need to adequately reflect the use of water and other natural resources in national income accounts. Additionally, there is an increasing demand for policies and institutional frameworks that can correct market failures and the economic and social undervaluation of water resources.

Conclusions

The water crisis is essentially about how we as a society and as individuals perceive and govern water resources and services. Although progress in water governance and related management areas has been incredibly slow and uneven, there are encouraging signs that water governance reform is taking place in many countries, promoting and facilitating coherent policy frameworks and institutional integration instead of fragmentation, partnerships and participation.

Water governance will be improved by raising the political will to overcome obstacles and implement water-related commitments made at Rio and afterwards. Although water reforms are evolving in many countries, much remains to be done to achieve the objectives of integrated approaches, sustainable development of water resources and the delivery of adequate water services.

Water resource issues are complex and transcend the water sector itself: indeed, there is an urgent need to broaden the horizon of water issues outside of the water sector. Macro-economic development, population growth and other demographic changes have greater impacts on water demands than water policy. This emphasizes the importance for water professionals to increase their understanding of broader social, economic and political context, while politicians and other key decision-makers need to be better informed about water resource issues. Otherwise water will continue to be an area for political rhetoric and lofty promises instead of implementation of sorely needed actions.

Progress since Rio at a glance

Agreed action	Progress since Rio
Establish by 2000 national action programmes for IWRM	
Emphasize beneficiaries' involvement in all aspects of water resource management and development	
Ensure that interests of all stakeholders are included in the management of water resources	
Establish appropriate institutional structures and network of institutions for IWRM	
Devise legal instruments for equitable sharing of water resources and for the implementation of IWRM	
Establish subsectoral targets for all freshwater programme areas	
Initiate effective programmes for institutional and human capacity-building for IWRM	
Effective mobilization of financial resources held by various stakeholders	

Unsaturated scale: **Unsatisfactory** — **Moderate** — **Satisfactory**

References

Bromley, D.-W. (ed.). 1992. *Making the Commons Work: Theory, Practice and Policy.* San Francisco, Institute for Contemporary Studies.

Federal Ministry for the Environment, Nature Conservation and Nuclear Safety, and Federal Ministry for Economic Co-operation and Development. 2001. Ministerial Declaration, Bonn Keys, and Bonn Recommendation for Action. Official outcomes of the International Conference on Freshwater, 3–7 December 2001, Bonn.

GWP (Global Water Partnership). 2000a. *Integrated Water Resources Management.* TAC Background Paper, No. 4.

——. 2000b. *Towards Water Security: A Framework for Action.* Stockholm.

Jordans, E. and Zwartveen, M. 1997. *A Well of One's Own, Gender Analysis of an Irrigation Program in Bangladesh.* Bangladesh, International Irrigation Management Institute and Grameen Krishi Foundations, International Water Management Institute.

Kaufman, D.; Kraay, A.; Zoido-Lobaton, P. 1999. *Governance Matters.* Policy Research Working Paper, No. 2196. Washington DC, World Bank.

Kooiman, J. (ed.). 1993. *Modern Governance, New Government – Society Interactions.* London, SAGE.

Lundqvist, J. 2000. 'Rules and Roles in Water Policy and Management: Need for Clarification of Rights and Obligations'. *Water International*, Vol. 25, No. 2.

Ministerial Declaration of The Hague on Water Security in the 21st Century. 2000. Official outcome of the Second World Water Forum, 3–7 December 2001, The Hague.

Ostrom, E. 1990. *Governing the Commons: The Evolution of Institutions for Collective Action.* Cambridge, Cambridge University Press.

Pierre, J. (ed.). 2000. *Debating Governance.* London, Oxford University Press.

Rogers, P. and Hall, A.-W. 2002. *Effective Water Governance.* GWP TAC Background Paper.

Tropp, H. 1998. *Patronage, Politics and Pollution – Precarious NGO-State Relationships: Urban Environmental Issues in South India.* Motala, Linköping Studies in Arts and Science 182.

UN (United Nations). 1992. Agenda 21. Programme of Action for Sustainable Development. Official outcome of the United Nations Conference on Environment and Development (UNCED), 3–14 June 1992, Rio de Janeiro.

UNDP (United Nations Development Programme). 1997. *Corruption and Good Governance.* MDGD Discussion Paper 3. New York.

UNECA (United Nations Economic Commission for Africa). 2002. 'Development Challenges of Water Resources management in Africa'. A Briefing Note.

Van Koppen, B. 1997. *Waterbeheer en armoedeverlichting.* Wageningen, Department of Irrigation and Soil Conservation.

Visscher, J.-T.; Bury, P.; Gould, T.; Moriarty, P. 1999. *Integrated Water Resources Management in Water and Sanitation Projects: Lessons from Projects in Africa, Asia and South America.* Occasional Paper Series OP31E, Delft, International Water and Sanitation Centre.

Some Useful Web Sites*

Global Water Partnership (GWP)

http://www.gwpforum.org/

A working partnership among all those involved in water management.

International Water Management Institute (IWMI)

http://www.cgiar.org/iwmi/

Deals with issues related to water management and food security: water for agriculture; groundwater; poverty; rural developments; policy and institutions; health and environment.

United Nations Development Programme (UNDP)

http://www.undp.org/

UN's global development network, advocating for change and connecting countries to knowledge, experience and resources to help people build a better life.

World Bank, Law Library

http://www4.worldbank.org/legal/lawlibrary.html

Organized database of links and tools on international organizations, laws, treaties and laws of nations with links to their constitutions, legislation.

* These sites were last accessed on 7 January 2003.

V

Part V: Pilot Case Studies: A Focus on Real-World Examples

We all know that the view from the top of a tall mountain is different from that down below. As we climb, the horizon expands, stretching as far as the eye can see. People and settlements recede until they are indistinguishable from the landscape itself, a patchwork of shapes and textures.

The view from space is even more startling: whole continents stretch like fabric across the curved surface of the planet. The oceans, river systems, deserts, mountain ranges and even city lights emerge with dramatic clarity against the vastness of space. Yet, how tiny and vulnerable it appears.

From the ground, things look very different: we can see how over-crowded many places are, and how noisy and how polluted they can be. We see the individual people going about their daily lives. We see the effects of population pressures on food and health, we see the competition between different sectors of society for access to services and resources. It is all a question of scale.

The World Water Assessment Programme (WWAP) is both global and local in scale, for it must 'check' the accuracy of the big picture on the basis of snapshots of water in the field. The scale is global, but the solutions and data must be local.

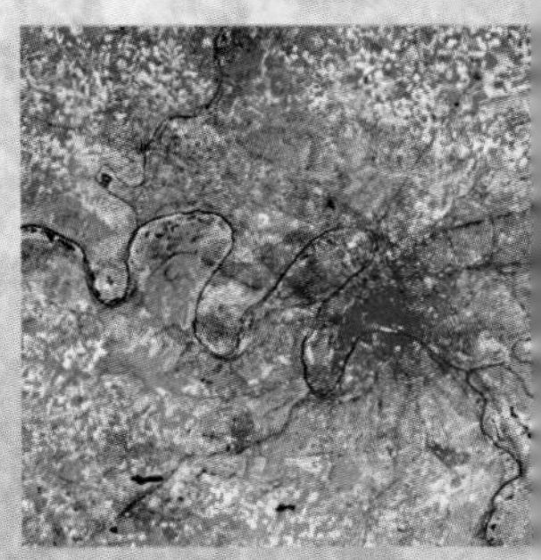

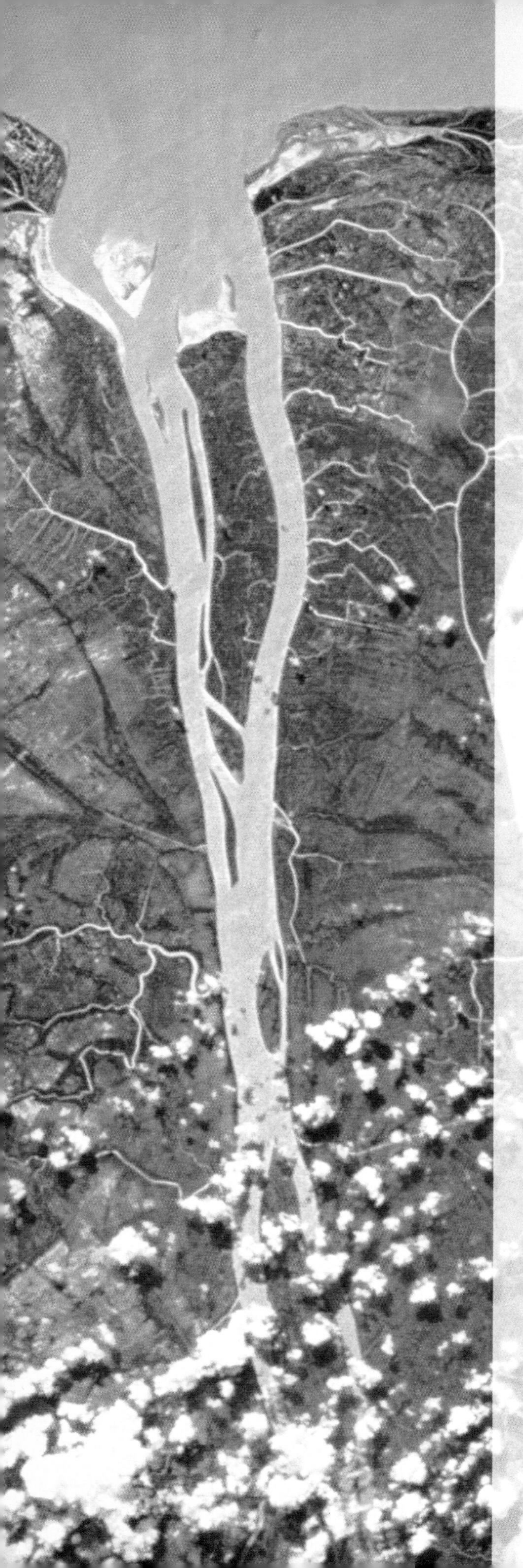

Case studies are, by definition, specific to time and place. The river basin, the city or the aquifer are the best places to observe the effectiveness of different approaches to integrated management and to test our indicators for measuring progress. They offer real-world examples and lessons, as well as providing the on-the-ground picture of actual water resource problems that is essential for filling in the gaps in our planetary overview and correcting our global models.

Taken together, the seven pilot case studies included here are representative of all the challenges that need to be met if water crises are to be addressed and averted. Taken individually, they each tell a story from which we can learn much. The chapters on Tokyo and Chao Phraya examine the problems of river basins in large metropolitan areas in Asia. The study of the Ruhuna basins looks at the pressures on people and institutions in a rapidly changing society where water holds an honoured place in cultural traditions. The Senegal River basin, Lake Titicaca and Lake Peipsi/Chudskoe are all examples of transboundary water bodies – but in very different environmental and socio-economic settings. Finally, the Seine-Normandy chapter provides the example of a European river basin that must both supply the capital city (Paris) and accommodate major industrial and agricultural interests.

The view from the rice paddy is different from the satellite view, but both are necessary and useful. A global exercise such as the WWAP tries to fit the different perspectives together, prodding us to see how each and every one of us has a role to play in determining the well-being of people and planet.

Chao Phraya River Basin, Thailand

By: The Working Group of the Office of Natural Water Resources Committee (ONWRC) of Thailand*

Table of contents

*For further information, contact wres@thaigov.go.th

Chao Phraya River Basin

Scoop up the water, the moon is in your hands.

Zen saying by Kido Chigu (1185–1269)

THAILAND'S CHAO PHRAYA RIVER BASIN supplies a major metropolitan region. It covers 160,000 square kilometres (km^2), representing 30 percent of the country's total area, and is home to 23 million people. Of these, some 8 million live in the capital city of Bangkok. Unlike Japan and France, however, the country has been slow to adopt a comprehensive approach to reform and legislation. It also cannot afford high-tech solutions to such critical water problems as floods, droughts and pollution. When drought conditions lead to water shortages in Bangkok, the result is overpumping of groundwater and subsequent land subsidence and more flooding. Deforestation in the basin's rural areas leads to flash floods, erosion and landslides. There is hope that the newly created River Basin Committees will bring about a more equitable sharing of the resource and that participatory approaches will lead to wiser governance.

THE CHAO PHRAYA RIVER BASIN, traditionally the centre of rice production, is in transition from water richness to water scarcity due to the increasing demands on this limited resource. A more systematic and comprehensive approach to water management is needed to achieve three purposes: equitability (among the diverse stakeholders), sustainability (for the basin's rapidly deteriorating aquatic environment) and efficiency (for international competition). These cannot be accomplished, however, without first carrying out an accurate and up-to-date analysis of the basin's water situation. Assessment tools are therefore essential, and foremost among these are indicators measuring the various conditions in the basin. This chapter is a first attempt to describe the development of such indicators for the Chao Phraya River basin.

Map 16.1: Locator map

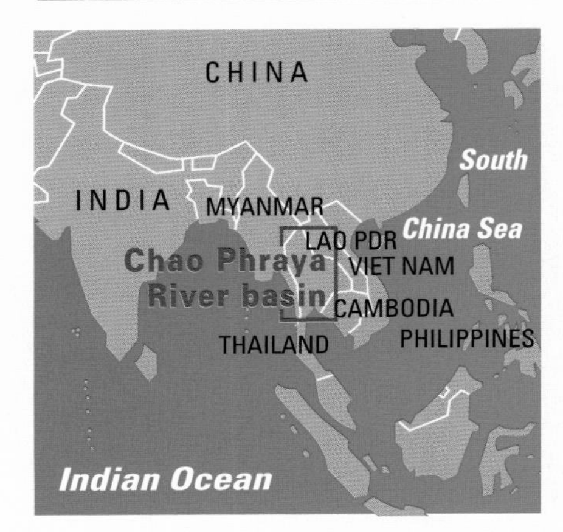

Source: Prepared for the World Water Assessment Programme (WWAP) by AFDEC, 2002.

General Context

Both initiatives, national and international, are overseen by Thailand's National Water Resources Committee (NWRC) through its secretariat, the Office of National Water Resources Committee (ONWRC). To coordinate these efforts, the NWRC set up a subcommittee. As Thailand's bureaucracy is being reformed[1] to increase popular participation and self-governance, as required by its constitution, a working group represented by involved government agencies was created under the subcommittee. This working group's responsibility is to develop, test and, eventually, transfer the indicators and their uses to River Basin Committees. The ONWRC is currently working to establish these committees, and, at the time of writing, all eight sub-basins of the Chao Phraya River basin have been equipped with River Basin Committees. The legislative framework of water management is also undergoing reform. The ONWRC is drafting a Water Resources Law, updating old laws and facilitating the country's systematic water management.

Location

The Chao Phraya River basin covers an area of approximately 160,000 km^2 and is located entirely within Thailand. The basin drains into the gulf of Thailand, part of the South China Sea and the Pacific Ocean. Bangkok, a city of more than 8 million people, is located near the mouth of the Chao Phraya River. Bangkok is not only Thailand's official capital city, but also the capital city for trade, government and South-East Asia air transportation, as well as being the gateway to Indochina and south China.

Major physical characteristics

The Chao Phraya basin is mountainous with agriculturally productive valleys found in the upper region. The lower region contains alluvial plains that are highly productive for agriculture. The Chao Phraya River drains from north to south. Monsoon weather dominates, with a rainy season lasting from May to October and supplementary rain from occasional westward storm depressions originating in the Pacific. Temperatures range from 15°C in December to 40°C in April except in high altitude locations. The whole basin can be classified as a tropical rainforest with high biodiversity. The lower part has extensive irrigation networks and hence intensive rice paddy cultivation. In recent years, however, encroachment of people into forest area in the upper part of the basin and its conversion to agricultural use has become problematic.

Table 16.1: Hydrological characteristics of the Chao Phraya basin

Surface area of the basin	159,283 km^2
Annual precipitation	1,179 mm/year
Annual discharge	196 m^3/s
Annual potential evapotranspiration	1,538 mm/year

1. As a result, ONWRC together with a few other agencies was combined to become the Department of Water Resources under the new Ministry of Natural Resources and Environment. This new Department became effective in October 2002 and acts as Thailand's Apex Body in water management planning and implementation.

Map 16.2: Basin map

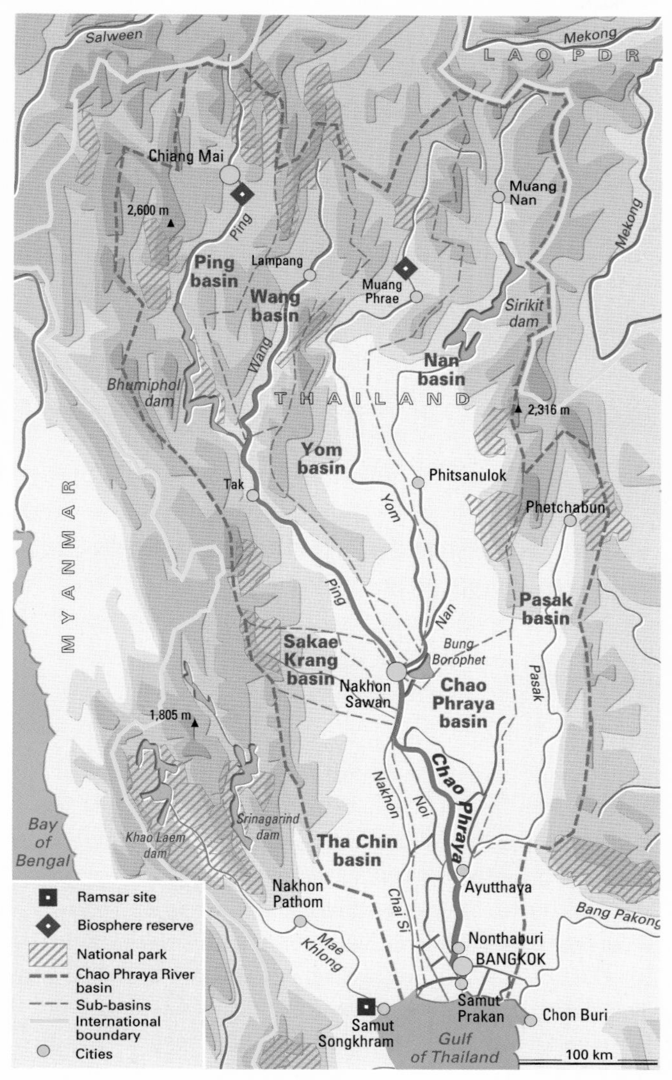

Source: Prepared for the World Water Assessment Programme (WWAP) by AFDEC, 2002.

Major socio-economic characteristics

Population characteristics

The Chao Phraya basin is the most important basin in Thailand. The basin covers 30 percent of Thailand's land area, is home to some 40 percent of the country's population, employs 78 percent of its work force and generates 66 percent of its Gross Domestic Product (GDP). In 1996, the total population of the Chao Phraya basin was 23 million inhabitants. The basin can be divided into eight sub-basins based on the natural distribution of its river system. About half the population (11.5 million) resides in the Lower Chao Phraya basin, in which the highly populated areas of Bangkok Metropolitan Area (BMA) and its environs of Samut Prakan, Nonthaburi and Pathum Thani are located. Similarly, there is a large concentration of population in the Upper Ping where Chiang Mai (the second largest city in Thailand) is located. Overall, about 68 percent of the total population of the basin is rural, but there is considerable variation with over 90 percent of the rural population in the Upper Chao Phraya basin compared to 45 percent in the Lower Chao Phraya basin. It is projected that in the next decade, the rural population will decrease by an annual rate of about 1.31 percent and that the population growth rate will remain low at about 1 percent per annum. These trends should ultimately induce aggregation of farming land with a consequential increase in household incomes. The average population density is 136 inhabitants/km^2, but varies greatly from 44 inhabitants/km^2 in the Nan sub-basin to 533 inhabitants/km^2 in the Chao Phraya sub-basin. Bangkok and its vicinity have the highest population density, with 1,497 inhabitants/km^2.

Economic activities

As Bangkok is located in the Chao Phraya sub-basin, this is economically the most important sub-basin, contributing 78.2 percent of the total GDP of the basin. The sub-basins can be divided into three groups based on their economic growth rate:

- Tha Chin, Chao Phraya and Upper Ping, which have higher growth rates than the national average;
- Pasak and Wang, at about the national average; and
- the Lower Ping, Upper Yom, Lower Yom, Upper Nan, Lower Nan and Sakae Krang sub-basins, all of which have lower growth rates.

The division of the basin into a prosperous north and south and a poor middle is reflected in the socio-economic conditions (see table 16.2). Formal employment and social services, such as health and education, are similarly concentrated in the Bangkok Metropolitan Region (BMR) and the Upper Ping, and generally have a higher per capita provision. Considerable variation exists in the Gross Provincial Product (GPP) and in the economic growth rates of provinces depending on the relative industrial and agricultural shares of sub-basin economies: high-growth sub-basins are industrial while low-growth ones are agricultural.

Agriculture accounts for 5 percent of the GDP in the basin, manufacturing for 33 percent, and wholesale and retail trade for 17 percent. The small overall share of the agricultural sector is due to the large influence of the GDP of the Lower Chao Phraya basin where urbanization and industrialization are intense.

Table 16.2: Population, per capita income and Gross Provincial Product (GDP) in the sub-basins in 1996

Sub-basin	Population	Per capita income in Baht (and US$)	Gross Provincial Product (GDP) in million Baht (and US$)
Upper Chao Phraya basin			
Ping	2,384,946	50,744 (US$1,223)	121,022 (US$2,916)
Wang	717,928	43,419 (US$1,046)	31,172 (US$751)
Yom	1,711,112	31,226 (US$752)	53,431 (US$1,287)
Nan	2,354,766	34,778 (US$838)	82,546 (US$1,989)
Pasak	1,340,559	52,923 (US$1,275)	70,946 (US$1,710)
Sakae Krang	461,542	38,578 (US$930)	17,806 (US$429)
Lower Chao Phraya basin			
Chao Phraya	11,477,193	193,388 (US$4,660)	2,105,979 (US$50,746)
Tha Chin	2,572,201	82,135 (US$1,979)	210,641 (US$5,076)
Total	23,020,247		2,693,543 (US$64,926)

Based on US$1 = approx. 41.5 baht. Considerable variation exists in the GPP and per capita income of the sub-basins: high-growth sub-basins are industrial and low-growth ones are agricultural.

The majority of people in Thailand are of Buddhist background. In the past, when population density was low, the area was rich in water, and the traditional way of life in the lower part of the basin often revolved around both water and agriculture. For example, the famous tourist attraction, the Songkran Festival (a traditional Thai New Year celebration in mid-April where water is splashed among spectators, passers-by and tourists), marks the beginning of the rice-planting season for Thai farmers. In the high-water month of November, another festival, Loi Kratong, involves people floating away their offering in small pontoons placed on the river.[2]

The political system is democratic with a long history of intermittent coups d'état until about ten years ago. The present Constitution, effective as of five years ago and obtained through an elaborate process of popular participation, paves the way for a promising future. With respect to water management in Thailand, the Constitution requires decentralization of water management (as a component of natural resources management) from government agencies to local governments.

2. For more details, see: http://sunsite.au.ac.th/thailand/special_event/songkran/, and http://mcucity.tripod.com/wat3.htm.

Water Resources: Hydrology and Human Impacts

Surface water

Riverine resources

The headwaters of the Chao Phraya River originate in the mountainous terrain of the northern part of the country and consist of four large tributaries: the Ping, Wang, Yom and Nan Rivers. The main river system passes through or close to many of the major population centres of the country including Bangkok, which is situated at its downstream end. The four upstream tributaries flow southward to meet at Nakhon Sawan and form the Chao Phraya River. The river flows southward through a large alluvial plain, called the delta area, splitting into four channels: the Tha Chin (also called the Suphan and Nakhon Chai Si further downstream), the Noi, the Lop Buri and the Chao Phraya Rivers (see table 16.3 for some river data).

Surface water storage

Since 1950, the government has constructed some 3,000 dams to store the monsoon flows for release in the dry season. This enables them to exploit the basin's vast agricultural potential and to meet the growing demands of industrial and urban users. The two largest dams constructed are the Bhumiphol and Sirikit dams, whose purpose is to supply stored water for electricity generation, irrigation, and domestic and industrial water use. Together these two dams control the runoff from 22 percent of the entire basin area. Bhumiphol dam on the Ping River has a live storage capacity of 9.7 billion cubic metres (bm^3), compared to the average annual inflow of 6.6 bm^3 from a drainage basin

Table 16.3: Annual average runoff in the sub-basins

Sub-basin	Catchment area (km²)	Total volume (Mm³)
Ping	35,535	9,073
Wang	11,084	1,624
Yom	19,516	3,684
Nan	32,854	11,936
Sakae Krang	5,020	1,096
Pasak	15,647	2,823
Tha Chin	18,105	2,449
Chao Phraya main stream	21,521	4,435
Total	159,283	37,120

Runoff is fairly constant all over the sub-basins; the Wang and Tha Chin sub-basins show a higher runoff.

of 26,400 km^2. The installed hydroelectricity generation capacity is 713 megawatts (MW). The dam was completed in 1963, and filled for the first time in 1970. Sirikit dam on the Nan River was completed in 1972 and has a live storage capacity of 6 bm^3 compared to the average annual inflow of 5.9 bm^3. The installed hydroelectricity generation capacity is 500 MW. Several other large dams (Kiew Lom, Mae Ngat, Mae Kuang, Mae Chang, Thap Salao and Kra Sieo) have also been built during the last twenty years to increase the total surface water storage in the basin, while another, the Pasak dam, was commissioned in 2000. Table 16.4 presents some information on selected dams in the basin.

Barrages

A number of barrages have been constructed in the Lower Chao Phraya basin to control and divert the water in the canal systems that provide irrigation water to some 1 million hectares in this area. The most important barrages in the delta area are the Rama VI barrage, completed in 1924, and the Chao Phraya (Chainat) diversion dam on the Chao Phraya River, built in 1957. Although the Rama IV barrage was constructed about seventy-five years ago, it still maintains its structure and remains operational. These barrages divert water to a complex system of interconnecting canals serving the Lower Chao Phraya basin irrigation system. Barrages have also been constructed above the Chao Phraya diversion dam. The Naresuan dam across Nan River at Phitsanulok was completed in 1985 and diverts water to the whole area of the Phitsanulok Phase I Project. The water released from the Sirikit dam meets the requirements of the Phitsanulok Irrigation Project as well as part of the demand for the Chao Phraya Delta area.

Groundwater

Aquifer distribution

Hydrogeologically, the Chao Phraya River basin is comprised of seven groundwater sub-basins: Chiangmai-Lampoon basin, Lampang basin, Payao basin, Prae basin, Nan basin, Upper Chao Phraya basin and Lower Chao Phraya basin. Within these groundwater sub-basins, water is held in either confined or unconfined aquifers. Eight separate confined aquifers are located in the Upper Tertiary to Quaternary strata of the Bangkok area. The natural groundwater within this succession of aquifers is highly confined, creating artesian conditions in each. Ease of exploitation, as well as the high chemical quality, are the main reasons for the original development of this source. Groundwater storage and renewable resources have been estimated for each groundwater sub-basin, as shown in table 16.5.

Recharge, flow and discharge

There are few existing estimates available of groundwater recharge on a regional basis. In an artesian aquifer, as that beneath Bangkok, storage depletion is seen through a decline in piezometric pressure and a reduction in the area in which

Table 16.4: Characteristics of major reservoirs

Reservoir name	Sub-basin	Maximum retention (Mm³)	Normal retention (Mm³)	Minimum retention (Mm³)	Effective storage (Mm³)
Bhumiphol	Ping	13,456	13,462	3,800	9,662
Sirikit	Nan	10,640	9,510	2,850	6,660
Kiew Lom	Wang	112	112	4	108
Mae Ngat	Ping	325	265	10	255
Mae Kuang	Ping	263	263	14	249
Mae Chang	Wang	108	NA	NA	NA
Thap Salao	Sakae Krang	198	160	8	152
Kra Sieo	Tha Chin	363	240	40	201
Pasak	Pasak	960	NA	NA	785

The two largest dams of the Chao Phraya River basin are Bumiphol and Sirikit; together, they control the runoff from 22 percent of the entire basin.

Table 16.5: Groundwater storage and renewable water resource of the sub-basins

Groundwater basin	Groundwater storage (million m³)	Renewable water resources (million m³)
Chiangmai-Lampoon	485	97
Lampang	295	59
Chiangrai-Payao	212	42
Prae	160	32
Nan	200	40
Upper Chao Phraya	6,400	1,280
Lower Chao Phraya	6,470	1,294
Total	14,222	2,844

The calculation is based on the assumption that the amount of groundwater stored depends on the change of water level, the area of the aquifer and the storage characteristics, which vary with the geology of each area – unconfined, confined or semi-confined. The Upper and Lower Chao Phraya groundwater basins are by far the largest ones: they store about 90 percent of the overall groundwater resources of the case study area. The table assumes that only 5 percent is renewable each year, a very small amount of the total in the basins.

artesian conditions exist. The continued decline in levels indicates that abstraction is not in balance with recharge. In unconfined aquifers, abstraction of resources in excess of natural recharge normally leads to a much lower rate of water level decline than that of confined aquifers.

Water quality

Surface water quality

A study in 1997 (Binnie & Partners, 1997) reviewed the water quality data for the Central River basin, which is routinely monitored by the National Environmental Board (NEB), the Pollution Control Department (PCD) and the Ministry of Public Health (MOPH). Results indicated that of the major rivers in the Lower basin there was evidence of heavy pollution in both the Chao Phraya and Tha Chin Rivers, while overall water quality was acceptable in the Pasak and Sakae Krang Rivers. The Chao Phraya River exhibited serious organic and bacterial pollution that was a threat to many species of aquatic life. Similarly, water quality in the Tha Chin River was heavily degraded, due to the combined discharges of industrial, domestic and rural inflow.

Groundwater quality

The main chemical constituents affecting groundwater quality are sodium and chloride. The average salinity of the groundwater in the unconfined aquifers shows a general increase in the downstream direction, with the exception of the Ping catchment whose lowest salinity level is comparatively high for its upper catchment location. The groundwater with the lowest salinity comes from the Wang catchment. Nitrate concentrations are almost invariably low in all catchments. The extent to which chemical quality is affected by contamination is not known, except in some specific areas.

Rainfall variation

The basin climate consists of long, distinct dry and rainy seasons. These give rise to typical water problems: water shortages, floods and water pollution. These problems can become the source of environmental and socio-economic problems: for example, water shortage in Bangkok leads to overpumping of groundwater and subsequent land subsidence and flooding. In the basin's rural areas, encroachment on forest land leads to excessive erosion, extreme floods and landslides. Water pollution is widespread in urbanized sections of the Chao Phraya River and its tributaries. Recently, due to encroachment on forested areas, flash floods and landslides have also become frequent.

Flooding

Floods are a natural phenomenon in the Chao Phraya River basin and while residents have adapted their lifestyle to deal with annual flood occurrences, they cause significant economic losses. The major causes have been the decline of flood retention areas and the confinement of flood plains due to increasing development, the rapid urbanization around Bangkok, the growth of provincial cities and the intensification of agriculture.

In recent years, the government has had some success in reducing flooding through the construction of multi-purpose reservoirs, dikes, and other flood control infrastructures. The containment has, however, resulted in a higher overall flood risk as elevation levels are reached more quickly.

Human impacts on water resources

In general, the basin is entering a critical period where small changes in the hydrologic conditions can create large socio-economic disturbances. Due to the increase in population, it is unavoidable that new settlements will arise in areas where water management is difficult. The human impact on water resources, and vice versa, is visible throughout the basin. Native plants acting as surface cover on native land are being destroyed at an alarming rate, causing flash floods, erosion and landslides. The construction of dams and diversions requires the resettlement of people, usually in unclaimed and infertile areas. Highly populated areas are producing solid waste and wastewater that pollute streams and water bodies, and as a result, many native species are disappearing. In the lower part of the basin where intensive irrigation networks exist, rice paddy is grown continuously throughout the year. Thirty years ago, the same area grew a single rice paddy a year. Since then, the number has doubled, then tripled, and today there is continuous cultivation. Clearly, the land is heavily used, with no time to be revitalized. The government now has mobile land doctor units helping farmers diagnose and remedy land degradation problems.

Data and information on water resources

The collection of data and information on water resources is complicated by the high number of agencies involved. This gives rise to such problems as repetition and inconsistency of data. In general, fundamental data on water resources are collected but are difficult to use due to data management problems.

Challenges to Life and Well-Being

Water for basic needs

Water for domestic purposes is provided by water service facilities in urban areas and by wells in rural areas. The majority of piped schemes for farm households are operated and managed by village communities. Eighty-five percent of the population has access to safe and adequate drinking water, 90 percent to latrines and 60 percent to rural pipe water systems.

Water for food

Surface water irrigation systems

Irrigation systems were developed in the Chao Phraya basin as early as the 1890s in the southern Chao Phraya plain. The area was subject to deep and prolonged flooding, and the approach then was to construct canals to provide access to large land areas for the cultivation of flood-dependent rice. The canals also helped spread the floods more evenly and promote drainage at the end of the flood season. Beginning in 1904, with the founding of the Royal Irrigation Department (RID) as the agency responsible for water resource development in Thailand, and through the 1930s, this mode of development was applied to more than 500,000 hectares. Today, irrigation area coverage totals 29 percent.

The irrigation systems in the northern part of the Chao Phraya basin consist of discrete irrigation systems served by their own reservoirs. Many small pump irrigation schemes have also been developed. However, several large dams were built to provide the Lower Chao Phraya basin with irrigation water. It is essential to appreciate that developing the irrigation systems in the Lower Chao Phraya basin was done progressively, often building and expanding on existing systems. The Lower Chao Phraya irrigation scheme, the largest in the country, was initially constructed to provide supplementary irrigation during the monsoon season, but was little by little required to provide ever-increasing amounts of irrigation water during the dry season.

The lower basin (delta) irrigation system is complex and consists of some twenty-six interconnected schemes below the Chao Phraya diversion dam. This barrage diverts the flow of the Chao Phraya River into a distribution network and at the same time releases the required water flow for river maintenance and downstream water needs. The water conveyance and regulation facilities of this system, whose operation is a major concern in efficient water management, are now quite old, and the canals are unlined earth constructions. The whole system is managed by the RID.

Groundwater irrigation systems

In agriculture, groundwater is mainly used to supplement surface water supplies. Groundwater consumption is more intense for land preparation during the dry season and in drought years, for crop needs in the early part of the wet season, and as a supplementary source of water for farms located at the tail end of distribution canals. Pumped irrigation schemes are being implemented by the Department of Energy and Energy Promotion to secure adequate irrigation water throughout the year in the middle-basin area. The potential exists to further develop the use of groundwater for irrigation purposes, but this should be undertaken after a thorough analysis of the sustainable yields of the relevant aquifer.

Water and ecosystems

Public recognition of water requirements for ecosystems is not widespread. It is known only within a limited circle of academics and technical agencies. Efforts are being made to prioritize water requirements for ecosystems in the forthcoming Water Resources Law, described further on. Environmental Law does exist, based on the polluter pays principle, but it is not effectively enforced. The Water Quality Index value for the basin is 59 percent. The index represents a mix of nine water quality items with higher value indicating better water quality.

Water and industry

Industrial growth across the region has been greatest in Bangkok, and pressure on existing infrastructure has led to the creation of new enterprises in the surrounding provinces where land, labour and other resources are more readily available and infrastructure is less congested. Future growth is expected to be greatest in these areas although there are also initiatives to encourage industrial expansion in a number of provincial centres. The amount of water used for industrial purposes in the whole basin is uncertain. Estimates for groundwater use are only available for the Chao Phraya sub-basin. Considerable inconsistencies are found in the data collected, and it is unknown whether figures also take account of inefficiencies in the system. A study (Binnie & Partneis, 1997) of industrial water use for the eight sub-basins estimated an industrial water demand of about 758 Mm^3 in 1996, of which 94 percent was attributed to use in the Chao Phraya/Tha Chin sub-basins. Surface water supplies are less important to industry with about 75 percent of water use derived from groundwater resources. Pipe water use, which is a more economical method of transporting water because of low evaporation and better water quality, represents 70 percent of water used. Indicators show that the polluter pays principle applies to 10 percent, while 50 percent of the water is reused.

Water and energy

Hydropower in the Chao Phraya basin is managed by the Electricity Generating Authority of Thailand (EGAT). Currently there are only two major hydropower installations, at Bhumiphol (713,000 kilowatts [kW]) and Sirikit (500,000 kW), with a smaller installation at Mae Ngat (9,000 kW) in the Upper Ping basin. At present, EGAT is not actively pursuing new hydropower projects in the basin. The construction of further large reservoirs with hydropower potential would involve large-scale resettlement, making such projects problematic. New reservoir construction in the upper basin has also encountered increasing opposition on environmental grounds. The economic value is not used in water allocation for hydroelectricity generation, but there is an Environmental Impact Assessment (EIA) for all hydroelectricity infrastructure.

Water for cities

Across Thailand, potable water supplies are generally provided by two agencies: the Metropolitan Waterworks Authority (MWA) and the Provincial Waterworks Authority (PWA). The MWA engages in production and distribution of potable water in the Bangkok metropolitan region while the PWA is responsible for all provinces in Thailand. The PWA is also responsible for water resource development, conveyance, pumping, treatment, storage and distribution facilities from all urban and rural communities in the provinces. Total domestic water requirements in 1993 were estimated at 3,194 Mm^3 per year. In contrast to industrial water supply, it was estimated that only 12 percent of domestic supply was being met from groundwater sources. There is no cross-subsidy from richer beneficiaries downstream to pay for the expense of protecting the upper watershed. Economic incentives are used in about 70 percent of the municipal areas, and the polluter pays principle is applied in an additional 10 percent of municipal areas.

Navigation

Since early times, the Chao Phraya River has been a major navigation route far into the central part of the basin. Ships and barges have provided a very important means of transport for commercial goods. However, the increasing diversion of river flow for irrigation has reduced minimum flow in critical reaches of the river, and navigation is now restricted during the dry season for vessels over a certain size. A study done for the Harbours Department in 1996 proposed construction of two barrages to restore navigation capacity in the main Chao Phraya River and into the lower reaches of the Nan River. Minimum river flow is also required for navigation in other waterways. Although generally restricted to a smaller and declining number of commercial craft, the Pasak (below Rama VI barrage) and the Tha Chin Rivers are still important waterways as are a number of the RID's supply canals in the lower part of the basin. Allocation of the basin's water supplies must take into account the need to maintain sufficient flow for river transport.

Management Challenges: Stewardship and Governance

Ownership and responsibility

The Chao Phraya basin has a long history of informal water allocation. In some northern areas, informal systems have been successfully implemented for over 200 years. The water allocation systems in the northern parts of the basin differ from those in the middle and lower areas. This is primarily due to topography. In the north, the river valleys are small and socially more defined. In the middle and lower sections, there was usually enough water for a rice crop in the wet season, which was sufficient to sustain a farm family and allow for home consumption throughout the year.

The water allocation system in the lower basin is as follows: EGAT controls the water release from both reservoirs. RID routinely checks the downstream water level at the Chainat diversion dam. If this level is lower than what it should be, considering the quantity released by EGAT, then too much water has been drawn off by users between the reservoirs and the Chainat diversion dam. This phenomenon has led the working group to design a rotation system to conserve water. Organizational and individual users in each province along the

Chao Phraya River basin are assigned a number of days on which water withdrawal is permitted. Provincial authorities are responsible for enforcing this, and are coordinated by the Ministry of the Interior. Compliance with the water allocation plan is good among the agencies represented on the working group but not among farmers, the central reason for this being that farmers can earn more income by planting a second rice crop in defiance of the plan. There is no enforcement because disobeying the water allocation plan is not illegal. Accordingly, efficiency and equity are low.

Institutions

Within the Prime Minister's Office there are six boards and committees responsible for policy planning and coordination of water resources at a national level. In addition, a plethora of government agencies are involved in managing water resource development, use and delivery in Thailand. The three dominant ministries, related to water management, include Agriculture and Cooperatives (MOAC), Science Technology and Environment (MOSTE), and Industry (MOI). The regional offices of government line agencies, provincial governments and local bodies are also involved in water resource development, use and management, while in some areas, farmers have formed strong water user associations to operate and manage water within local irrigation schemes.

National boards and committees

The principal boards and committees responsible for developing policies concerning water resource development, management and conservation are the National Economic and Social Development Board (NESDB), the NEB and the NWRC.

The presence of three bodies having similar functions leads to confusion and indecision in implementing water policies, particularly as their relative importance cannot easily be determined from their functions and powers. Based on the mandates of the three bodies, the NWRC's responsibility relates more directly to water resources than that of the other two boards, with NEB's tasks being more relevant to water resources than those of the NESDB. Since NESDB plans are usually quite broad, more detailed policies and plans are required for each individual sector. For example, when considering natural resources and the environment, the NEB elaborates detailed policies and plans within the framework determined by the NESDB plan. By the same token, since the NEB is not able to address all detailed policy issues regarding natural resources and environment – in particular, water resources – the NWRC should prepare more detailed water policies and plans. As a result, the NESDB plan should be considered as a framework for water policies formulated by NEB and NWEC, respectively. It is apparent that there is some confusion, duplication and lack of clarity and integration in the development and operation of national resource policies and strategies.

Regional offices, provincial government and local bodies

At the provincial level, the Provincial Administration and District Administration offices (and similar agencies at the local government level) have an operational role in supplying local domestic and industrial water, but in reality have little role in water resource planning and management insofar as basin-wide issues are concerned. RID regional offices perhaps contribute most to water management at the provincial and local levels. These offices work closely with water user groups and conduct training programmes in irrigation maintenance and other related issues.

The need for greater coordination of water management at the basin level has been recognized through the recent creation of basin subcommittees. Three such committees have been established in the Chao Phraya basin for the upper and lower Ping River tributary basins, and also for the Pasak tributary basin. These committees are in the formative stage, but have been given wide-ranging advisory roles covering most aspects of Integrated Water Resources Management (IWRM). The ultimate intention is to form such committees in all of the twenty-five designated river basins in Thailand, as they are an important initiative for creating a more appropriate institutional structure for river basin management in the Chao Phraya basin. It remains to be seen whether the committees can be empowered with the necessary knowledge and commitment to fulfil their potential. They must also be viewed as water resource managers and not water development committees if they are to effect good IWRM.

Water user organizations

Since the thirteenth century, the members of water user organizations have developed small-scale irrigation programmes in the Ping River basin without any assistance from the government. These irrigation systems (*Muang Fai*) had their own laws and regulations, agreed upon among the water users who had to pay a water fee either in cash or in kind as stipulated in the agreement. They also had to pay the maintenance fee, contributing to tools and equipment for weir repair and dredging the irrigation canals. The amount of labour and accompanying tools depended on the amount of individual cultivated land and water use in the system.

In general, there are very few successful water user organizations in Thailand. This stems from a number of issues.

The status of such organizations is ambiguous and the necessary legislation is lacking for them to obtain proper legal status. Furthermore, the idea of establishing water user organizations was not initiated by the water users themselves and historically there has been a lack of active farmer participation in all phases of project development as well as inadequate support from the concerned agencies. The successful creation and functioning of water user organizations depend on establishing obvious advantages for the water provider. The draft Water Bill encourages the creation of such organizations: in effect, Section D.3, item ii, states that Basin Committees would have to accommodate, advise and assist water users in forming organizations for the benefit of conserving, developing and utilizing water resources. The effectiveness of Basin Committees depends largely on the involvement of representatives of civil society and local water users in drawing up plans for water management and use.

Legislation

The 1997 Constitution

The enactment of the new Constitution in Thailand in 1997 enabled changes in the way the government, its agencies and local communities manage the country's natural resources. The Constitution is intended to impact on the government's natural resources and environmental policies, the implementation and operation of government projects, and the interpretation of relevant laws and regulations. In particular, the new Constitution provides:

- an increased requirement for the state to encourage citizens to participate in preserving, conserving and using natural resources and biodiversity in a sustainable manner;
- a greater decentralization of government responsibility for the subdistrict (*tambon*) level to manage resource use;
- a more direct participation by civil society in planning, managing and utilizing natural resources, and in developing and enacting laws; and
- a greater citizen access to information.

These provisions favour the concept of IWRM, which demands a high level of community/stakeholder awareness and participation, local-level planning and transparency.

The overall scope of the new Constitution establishes not only a climate of open management, but also an obligation for government administrations to implement this approach. This is particularly relevant to the function and operation of organizational units to be set up under the project to improve water sector management in the Chao Phraya basin. There can be no excuse for not including the basin community/stakeholders in future water resource management decisions: it is required by the new Constitution, and it should unleash a new level of commitment to achieving a balance of sustainable water resource management in Thailand.

Existing water laws

Thailand has at least thirty water-related laws, administered by over thirty departments overseeing water issues in eight ministries.[3] Like water policies, the mass of water laws, codes and instructions have all been framed for particular and usually singular purposes. There is no umbrella legislation linking these laws and codes, and consequently there is no legislative backing for an organization to undertake IWRM. In practice, this results in ad hoc and often erratic relationships among all agencies, as the agencies pursue their narrow objectives and mandates and seem more interested in enhancing water supply to meet the demands of politically powerful groups. Water permits are not issued and bulk water use for irrigation, hydropower, domestic, town water and industry is not properly controlled. New developments arise (though many of these are small and operate locally) and the adverse cumulative impacts of such implantations on the equitable distribution of water and on the health of the aquatic environment are significant. The absence of a modern, comprehensive water resource law is probably the most significant factor inhibiting IWRM in Thailand.

Draft water resources law

The inadequacies in the many water-related laws in Thailand have led to the drafting of a more comprehensive and integrative Water Bill. However, one fundamental flaw in the draft Water Bill is that it does not provide a mandate for any one agency to act as the national water resource manager. This should be at the core of any water resources law. Although it is specified in the draft document that the ONWRC is to be the water resource management agency, the functions specified do not allow the ONWRC to undertake comprehensive water resource management. The draft Water Bill covers water distribution and allocates this role to new River Basin Committees, who are to act as coordinating bodies concentrating largely on strategic natural resources or water resource planning, with a responsibility for

3. Ministry of Agriculture and Cooperatives, Ministry of Transport and Communication, Ministry of Industry, Ministry of the Interior, Ministry of Public Health, Ministry of Labour and Social Welfare, Ministry of Education, Ministry of Defence, and Ministry of Science, Technology and Environment.

broad or bulk allocation of water between water users within geographic zones. Under the new proposed law, they can be innovative and play an important part in achieving good river basin management, providing a critical link between communities and stakeholders at the tributary basin level and a higher regional or national authority.

In summary, the existing laws, policies and strategies relating to water resource management do not clarify roles and functions, and are too discrete to allow for IWRM. In its current form, the draft Water Bill would not provide a suitable basis for a comprehensive approach to river basin management. Clearly, considerable further work has to be undertaken to make the draft water law fulfil all the needs of IWRM throughout the country.

Groundwater

At present, the only direct law related to the regulation of groundwater is the Groundwater Act of 1977 (amended in 1992). The Groundwater Act was put in place with the intention of regulating groundwater use after the government realized the adverse effects of its uncontrolled use. The MOI regulates groundwater through the issue of permits for well drilling. The Groundwater Control Division under the MOI is responsible for reviewing well-drilling requests and for issuing well permits within the BMA and its surrounding provinces, namely Pathum Thani, Nonthaburi, Samut Prakan, Samut Sakhon and Nakhon Pathom. The Groundwater Control Division also takes responsibility for collecting the groundwater fee payment from private sector users, in accordance with the readings from well meters attached to the permitted wells.

Although some acts and regulations are outdated, legislation is in place to adequately control water use. The major problem in controlling excessive and illegal water use and implementing water charges is not the inadequacies of regulations, but lack of political will and the failure of responsible authorities to implement laws and regulations.

Finances

In the past, funding of the basin's water resources emphasized development rather than management. The government budget is allocated on a project-by-project basis, rather than applied to the water sector as a whole. Such a fragmented funding approach is a major cause of inefficiency. As of the next fiscal year, an improved system for allocating budgets will be applied. It is now too early to assess its impact.

Management approaches

Water-related risks such as drought and floods are presently managed on an ad hoc rather than systematic basis. Water valuation is less present in agriculture than in other sectors. Overall, economic tools are not yet fully applied. Water sharing at all levels (basin to basin, upstream to downstream, sector to sector) exists but without clear-cut policy or directions. IWRM and demand management have just been introduced and are not yet practiced. Stakeholder participation, required in the Constitution and the present National Development Plan, is practiced in establishing River Basin Committees. Public-Private Partnership (PPP) is already implemented in domestic and industrial pipe water supply, but not yet in agricultural water supply. Planning and development, done by government agencies in the past, are slowly changing. The newly established River Basin Committees are now expected to perform these duties, while agencies provide technical advice. A Unified Water Resources Information System will be created to ensure the knowledge base. The main features of the system will include public accessibility, and the sharing and linking of all existing agency information bases between themselves and to each individual River Basin Committee's information base.

Policy and policy implementation

Government development plans

Growth under successive five-year plans has been sustained and rapid, except during the economic crisis in Thailand, which started in mid-1997. It is, however, recognized that economic activity and prosperity have remained concentrated in Bangkok and the adjacent provinces. Recent five-year plans have attempted to restore the current imbalance between the more prosperous central region and the poorer rural areas, but with mixed success. One concern has been that the more developed areas are using a higher proportion of the country's natural resources, including water sometimes to the detriment of rural areas.

Water resource development

As previously stated, past government policies have been directed towards water resource development rather than management, with the specific aim of securing more supplies. Guiding principles for national water resource management are outlined in the National Economic and Social Development Plans, but they only indicate very broad objectives. For example, they encourage a holistic approach to basin development (Sixth National Plan) and sustainable development (Eighth National Plan.) A national water policy was announced by the Cabinet in October 2001, which is a starting point for budget allocation in the water sector under the present unified budgeting system.

Water quality

The National Policy regarding water quality is stipulated in both the National Economic and Social Development Plan and in the

National Policy and Plan for Natural Resources and Environment Management, which set long-term (twenty-year) goals, standards and strategies. In the Eighth Plan, the national goal is to maintain the quality of surface water at the 1996 level. Further planning is in process. An action plan for community pollution control will have to be completed for the twenty-five river basins in Thailand. In addition, an emergency plan to prevent and mitigate toxic water pollution has to be undertaken.

Conclusions

From the macro perspective, the critical problem in the basin is the inability to manage water so as to ensure optimum and equitable use and balance benefits (and burden) among the basin's stakeholders. In both its institutions and the dissemination of information, the water management system is disunified and ineffective, while water rights remain unspecified. Demand in all sectors keeps increasing as the economy expands, but water intake and economic output are unbalanced: while agriculture represents about 80 percent of water use, it only contributes to 30 percent of the GDP; industry uses 10 percent of the available water resources and contributes to 60 percent of the GDP. In a basin where one third of the population lives in Bangkok, urban inhabitants pay less for piped water than the production cost, even when the raw water cost, the cost of maintaining upper watershed areas and that of wastewater treatment are not taken into account.

However, these are now recognized problems and important reforms are in process. The bureaucracy is being reformed, and a coordinating water management body is emerging (Apex Body). The Constitution has recognized the need to involve local people in water management, as demonstrated by the establishment of River Basin Committees. The water resources law is being drafted, the water budget system is becoming more holistic, and, if not yet already practiced, IWRM is being introduced.

Box 16.1: Development of indicators

It is assumed in this report that assessment can be made only after indicators have been developed. For this case study, preliminary indicators have been developed but not included. Following is a summary of assessment results.

- Identification of indicators and their values is subject to ongoing improvement.
- Indicators and their values are area-specific. Some indicators are applicable to the whole basin, some are not.
- Each indicator has three components: a name, a target value and an actual value. The target value is the desired value, while the actual value is obtained from real-world situation. To properly assess the water situation at any level – global, regional, basin or other – the target values must be compared to the actual values.

An index for each challenge area may be developed by combining indicators under the same challenge area. Each indicator can be assigned a different value. From WWAP's eleven predetermined challenge areas, the Chao Phraya River basin's priorities relate to health, cities, water sharing, governance and risk management. It should be noted that these priorities are preliminary and apply to the basin as a whole. Priorities for each sub-basin may be different from these and from each other.

References

Lohani, B.-N. et. al. 1978. *Mathematical Optimization Model for Regional Water Quality Management: A Case Study for Chao Phraya River (phase II).* Consultant study by the Thailand Environmental Institute (TEI) for the Pollution Control Department (PCD) of the Ministry of Science, Technology and the Environment (MOSTE). Final Report No. 120. Bangkok, National Environmental Board of Thailand.

Binnie & Partners (Overseas) Ltd. 1997. *Chao Phraya Basin – Water Management Strategy.* Bangkok.

Bureau of the Crown Property. 2000. 'Coordination Framework for Water Resource Management and Development in Chao Phraya Basin'. Working paper. Bangkok.

FAO (Food and Agriculture Organization). 2000. *Thailand, Natural Resources Management Project.* Draft preparation report, No. 00/84/CP-THA. Rome.

RID (Royal Irrigation Department). 2000. *Final Report on Chao Phraya Basin Water Management Project.* Bangkok.

17

Lake Peipsi/Chudskoe-Pskovskoe, Estonia and the Russian Federation

By: The Ministry of the Environment of Estonia and the Ministry of Natural Resources of the Russian Federation

Table of contents

Every seed knows its time.

Russian proverb

LAKE PEIPSI/CHUDSKOE-PSKOVSKOE, a transboundary lake that discharges into the Baltic Sea, is shared by Estonia and the Russian Federation, both of which are responsible for its management. There is abundant water to meet the needs of an ageing, essentially rural, local population. However, uncontrolled fishing, eutrophication and pollution from untreated wastewater and power plant emissions have put the environment under increasing pressure. A joint water commission was created in 1997, but so far the economic potential of the area is underexploited, and more could be done to engage local and regional authorities, stakeholders and private businesses in solving common problems. Many changes will have to occur in 2004/5, as Estonia must adopt new standards in order to gain entry to the European Union.

LAKE PEIPSI/CHUDSKOE-PSKOVSKOE, sometimes called Peipus, is the fourth largest and the biggest transboundary lake in Europe. Its three names originate from the three languages historically used in the region – Estonian, Russian and German. Lake Peipsi belongs to the watershed of the Narva River, a 77 kilometre (km)-long watercourse that connects Lake Peipsi to the Gulf of Finland in the Baltic Sea.

Map 17.1: Locator map

Source: Prepared for the World Water Assessment Programme (WWAP) by AFDEC, 2002.

General Context

The lake consists of three unequal parts: the biggest, northern Lake Peipsi *sensu stricto* (*s.s.*)/Chudskoe, the southern Lake Pskovskoe/Pskovskoe and the narrow, straits-like Lake Lämmijärv/Teploe, which connects Lake Peipsi *s.s.* and Lake Pskovskoe. Lake Peipsi is located on the border of the Republic of Estonia and the Russian Federation between Lake Peipsi lowland (eastern border of Estonia) and the East European Plain (the Russian Federation). The lake lies at 30 metres above sea level.

Major physical characteristics

The catchment area of Lake Peipsi is approximately 160 km in width and 370 km in length. It has a surface area of 47,800 square kilometres (km^2), of which 16,623 km^2 lie in Estonia, 27,917 km^2 in the Russian Federation and 3,650 km^2 in Latvia. The catchment area is a gently undulating glaciolacustrine or till-covered plain, belonging to the forest zone of the East Baltic Geobotanical Subprovince.

The formation of the lake basin is due mainly to Pleistocene glaciers. In the north, some topographical features originated more than 380 million years ago, and have been slightly modified by the glaciers. On the northern part of the lake, scarp, sandy beaches prevail. The flat shores, found mainly on the western part of the lake, are usually swampy, populated with reeds and bulrushes.

Table 17.2 provides some morphometric data on Lake Peipsi.

Table 17.1: Hydrological characteristics of the Lake Peipsi basin

Surface area of the basin	47,800 km^2
Annual precipitation	575 mm/year
Annual discharge	329 m^3/s
Annual inflow to lake	324 m^3/s
Annual outflow from lake	329 m^3/s

Table 17.2: Morphometric data on Lake Peipsi

	Lake Peipsi s.s./ Chudskoe	Lake Lämmijärv/ Teploe	Lake Pskovskoe/ Pskovskoe	Lake Peipsi/ Chudskoe-Pskovskoe
Total basin area, km^2	2,611	236	708	3,555
Distribution of the waters between Estonia and the Russian Federation, km^2	1,387/1,224	118/118	25/683	1,564/1,991
Distribution of the waters between Estonia and the Russian Federation, %	55/44	50/50	1/99	44/56
Percentage of surface area	73	7	20	100
Volume, km^3	21.79	0.60	2.68	25.07
Percentage of total volume	87	2	11	100
Medium depth, m	8.3	2.5	3.8	7.1
Maximum depth, m	12.9	15.3	5.3	15.3
Length, km	81	30	41	152
Medium width, km	32	7.9	17	23
Maximum width, km	47	8.1	20	47
Length of shoreline, km	260	83	177	520
Percentage of total length, km	50	16	34	100

Source: Jaani, 2001.

Map 17.2: Basin map

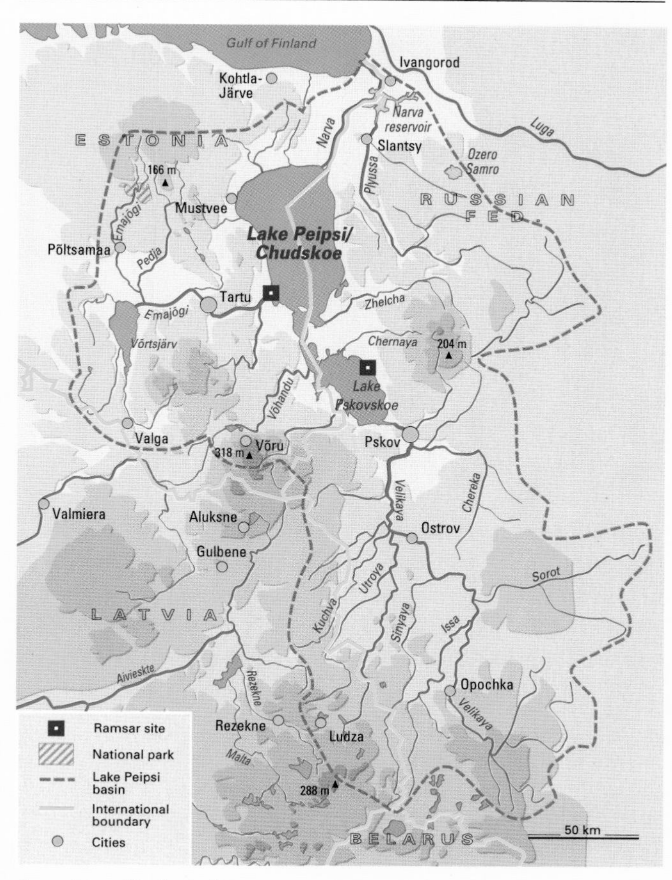

Source: Prepared for the World Water Assessment Programme (WWAP) by AFDEC, 2002.

Climate

Lake Peipsi has a moderate continental climate, rather wet, softened by the relative closeness of the Atlantic Ocean. Positioned between marine and continental climates, it is subject to unstable weather conditions all year. Summer is comparatively warm and wet, with a mild winter. The continental character increases to the east, where winters are longer and summers warmer. The watershed belongs to an area with high activity of low-pressure weather systems: an average of 130 depressions per year are registered, in other words, almost one every three days.

Between 1923 and 1998, the annual average temperature recorded at Tiirikoja, the meteorological station on the west coast of Lake Peipsi, was 4.6°C, with an average precipitation of 575 millimetres (mm) per year (Keevallik et al., 2001). The eastern coast generally has a more continental climate: at the station of Gdov for the same period, the average temperature in July was 17 to 18°C and, in February, -7°C to -8°C. The average annual precipitation near the Narva River is between 700 and 750 mm. The minimum temperature is -39°C on the lake's shores, but can reach -43°C inland.

Major socio-economic characteristics

Population and activities

The total population of the basin is approximately 1 million, but the population density differs in various parts: 24 inhabitants/km^2 in the Estonian and Pskov regions, compared to 11 inhabitants/km^2 on the sparsely populated eastern shore.

There are only two large towns in the basin: Pskov in the Russian Federation, with 204,000 inhabitants, and Tartu in Estonia, with 98,000 inhabitants. The majority of the basin population lives in small settlements. Only 27,000 people live in the local municipalities bordering the lake on the Estonian side. On the Russian side, the Leningrad region had 60,600 inhabitants in 2001, 87 percent of them urban dwellers. Of the 427,000 residents of the Pskov region, almost half live in Pskov itself.

The greatest problems are the ageing population and the departure of the younger generation for bigger towns. An additional problem on the Russian side concerns the very low incomes in the Pskov and Leningrad regions (shown in figure 17.2). More than half the population have incomes that do not cover the cost of living. Comparisons between 1997, 1998 and 1999 indicate that the situation is worsening, in part due to the high rate of unemployment in these regions.

The Lake Peipsi basin can be divided into three regions (southern, central and northern) from the points of view of economic and social development, cultural composition of the population and the type of human impact on the lake.

The southern part of the basin is a rural, sparsely populated area with forestry and agriculture as the main livelihoods. Since agriculture is no longer profitable, many farmers live by cutting down forest from their own lands and selling it. This, and non-point source pollution from agriculture, comprise the main environmental problems of the area. Figures 17.1 and 17.2 show the unemployment rates in Estonian counties bordering Lake Peipsi and in the Pskov region.

Tartu and Pskov are located in the central part of the basin, and the area's economy is defined by the existence of these two population centres. In this region, rural settlements prevail, although on the Russian side the region is only sparsely populated. Pig farms and poultry factories dominate the agricultural sector. On the Estonian side, the culturally mixed rural communities are located along the lake's shores, among them Russian Old Believers, who are renowned for growing cucumbers and onions.

Figure 17.1: Unemployment and average income per month in Estonian counties bordering Lake Peipsi

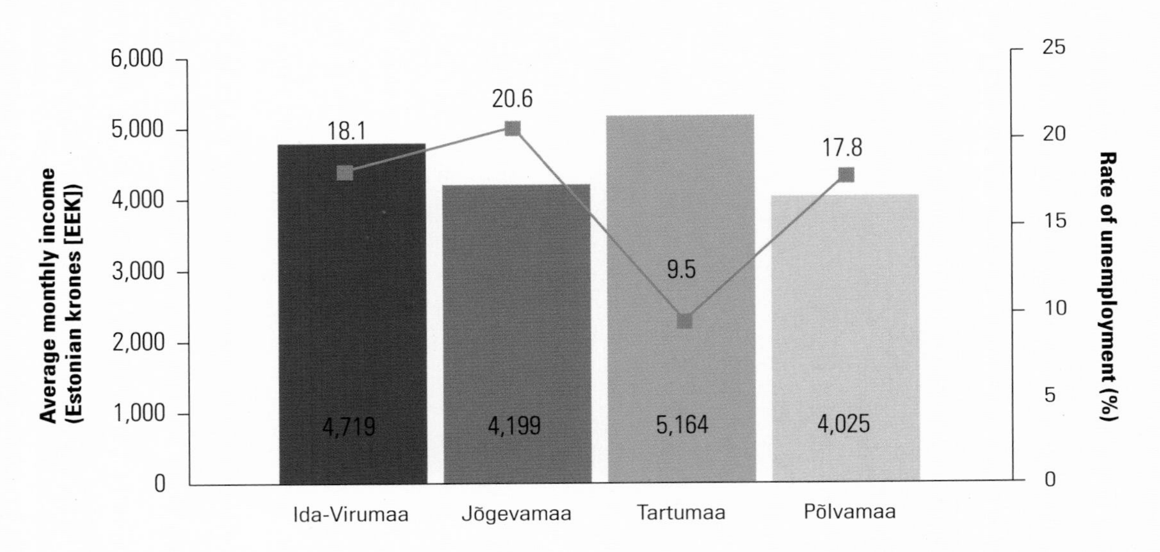

17EEK = US$1. There is a very high rate of unemployment in Estonian counties surrounding the lake. There are, however, disparities: whereas in Tartumaa, unemployment is at 9.5 percent, in Jõgevamaa it reaches 20.6 percent. Meanwhile, the average monthly income is more steady, at around US$267.

Source: Taken from the web site of the Ministry of Labour of the Russian Federation, Department of Population Incomes and Standard of Life, 2002 (http://www.chelt.ru/income/3.html).

Figure 17.2: Unemployment in the Pskov region between 1997 and 2001

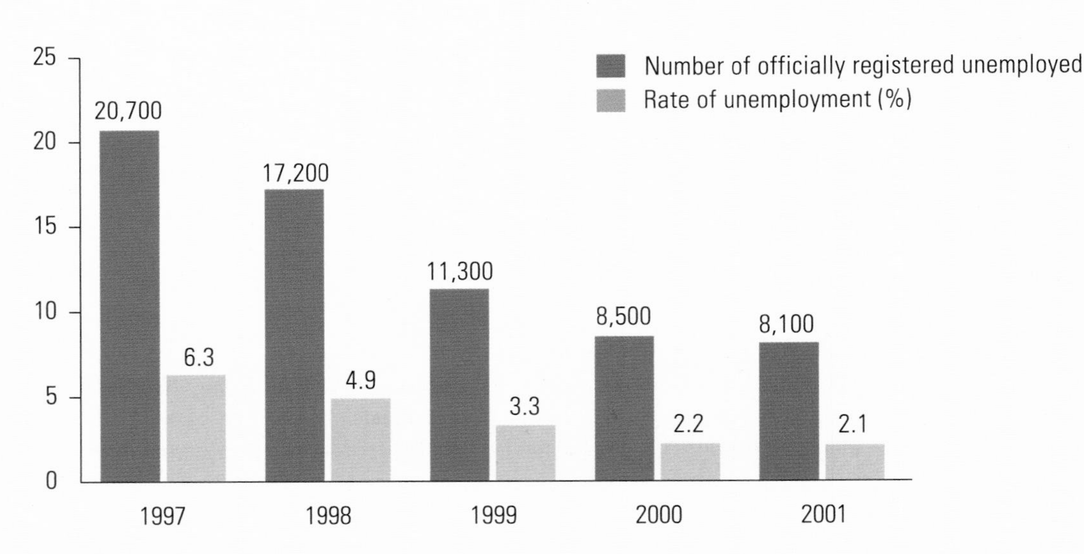

This figure shows both the number of registered unemployed people and the percentage that this represents compared to the total population. Unemployment in the Pskov region is steadily improving, with 2.1 percent unemployment in 2001 compared to 6.3 percent in 1997.

Source: Taken from the official site of the Administration of the Pskov region, 2001 (http://www.pskov.ru/region/invest/news/ 030401.html).

Commercial and small-scale fishing is currently an important source of income, particularly as many small enterprises have been shut down alongside access to the Russian produce market since the end of the Soviet era.

The northern part of the basin is the most industrial and is intrinsically linked to oil shale, the main natural resource of the area. In 2001, the Estonian Oil Shale Ltd. mines produced 11.4 million tons (Mt) of oil shale. The oil shale of the Baltic basin is one of the world's most unique for both its composition and high quality. Eighty percent of the oil shale mined is used to produce energy; the remaining 20 percent is used as raw material for chemical enterprises in the towns such as Kohtla-Järve (Estonia) and Slantsy (the Russian Federation).

Aside from oil shale, the small deposits of construction materials and the sands of the northern and western shores, the other basin resources (such as fish stock, forest, peat) are renewable. The main sources of air pollution are power plants and chemical enterprises. The Russian side of the basin produces construction materials and contains fuel and chemical industries associated with oil shale mining. Peat deposits are used for agricultural needs.

Figure 17.3 shows the distribution of land use in the Lake Peipsi basin.

Figure 17.3: Land use in the Lake Peipsi basin

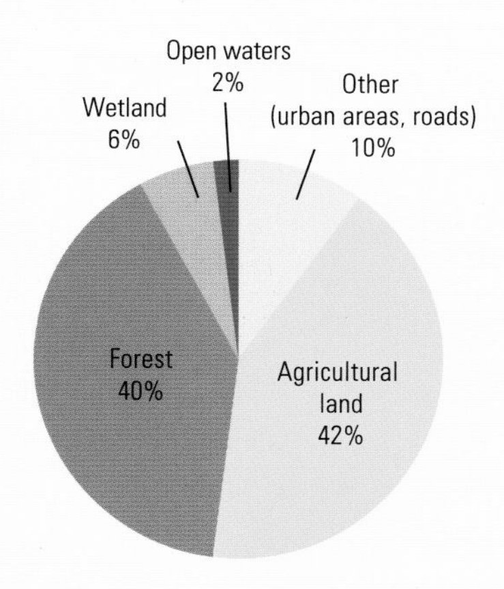

Natural areas (forests, wetlands, lakes) cover almost 50 percent of the basin, but agricultural land use is quite important, at over 40 percent. Urbanized areas, however, cover only 10 percent of the basin.

Source: Taken from the Estonian Department of Statistics web site, 2002 (http://www.stat.ee).

History

Archaeological finds show that permanent agricultural settlements were founded in the first millennium AD. Permanent population centres of this kind existed in the vicinity of Alatskivi and Gdov on the coast of Lake Peipsi, and around Räpina on the coast of Lake Pskovskoe. At the end of the first millennium, Slavs reached Lake Pskovskoe and the eastern coast of Lake Peipsi. The first town-type settlements arose near the Pskov stronghold. In the fourteenth century, strongholds were also built at Gdov and Vasknarva. Beginning in the fifteenth and sixteenth centuries, fishing villages appeared on the coasts of the lake, and in time, the fisher population increased in the area west of Lake Peipsi as well. In the second half of the nineteenth century and the beginning of the twentieth century, the inhabitants of coastal villages grew vegetables and developed many handicrafts.

National minorities

Among the Russian populations dominating the western and northern shores on the Estonian side are the Old Believers. They first settled on the Estonian shore of Lake Peipsi in the eighteenth century to escape the reforms taking place in the Russian Orthodox Church. Since then, they have lived in a separate community by the lake and do not mix with Estonians. The world's largest concentration of them, however, is in the local municipality of Peipsääre, situated north of the mouth of the Emajõgi River. Of the 1,000 inhabitants, almost 900 are Old Believers.

Although the south-western shore of the lake is mainly peopled with Estonians, the eastern and southern coasts of the lake are home to a small Estonian minority group, the Setu. Unlike the majority of Estonians, Setu are orthodox. They have their own regional Ugric language, used by only 1,000 native speakers in south-eastern Estonia and on the Russian side of the border. They live in south-western Estonia and in the county of Pechory in the Pskov region. At present the border divides their living area. At the birth of the Estonian Republic in 1920, the Tartu Peace Treaty attributed the entire Setu area to the Estonian Republic. After the Second World War, this area was split between the Soviet Socialist Republics (SSR) and the Russian Soviet Federated Socialist Republic (RSFSR). The border problem arose during the restoration of the Republic of Estonia, and the Setu area is still halved, a situation that is endangering the Setu culture.

Water Resources

Hydrology

In addition to Lakes Peipsi and Pskovskoe, the region holds more than 4,000 lakes, the largest of which is Lake Vortsjärv (270 km^2). There are also a number of small lakes, with surface areas ranging from 0.1 to 10 km^2. These lakes, excluding Lake Peipsi, form 2 percent of the basin's area, with Lake Peipsi accounting for a further 5 percent.

There are about 240 inlets into Lake Peipsi. The largest rivers are the Velikaya (with a catchment area of 25,600 km^2), the Emajõgi (9,745 km^2), the Vohandu (1,423 km^2) and the Zhelcha (1,220 km^2). They account for about 80 percent of the whole catchment area of Lake Peipsi and of the total inflow into the lake. The only outlet is Narva River, which has a mean annual water discharge of 12.6 km^3 into the Gulf of Finland, about 50 percent of the average volume of Lake Peipsi.

There are considerable lake level fluctuations. Water level changes are characterized by a spring flood that lasts for one and a half months or longer, and is then followed by a low water level that lasts four to five months. A short-term rise occurs in the autumn. Extensive coastal areas are sometimes inundated, and long-term studies have revealed a distinct pattern in the water level fluctuations.

Double currents are commonplace in Lake Peipsi. In Lake Lämmijärv, the velocity of streams may exceed 0.5 metres per second. Owing to a considerable amount of solar radiation accumulated in summer, Lake Peipsi freezes over relatively late, the ice cover usually formed by the end of December. In severe winters, ice thickness may reach 70 to 80 centimetres (cm). In especially warm winters, it reaches about 18 to 20 cm, while in the middle of the lake, the ice cover may be unstable or even absent. Usually the lake thaws in April or at the beginning of May. The total inflow into the lake is 324 cubic metres per second (m^3/s), the total outflow 329 m^3/s, and the residence time is two years.

A comparison between the three parts of Lake Peipsi reveals very different concentrations of phosphorus, nitrogen and chlorophyll. In effect, Lake Peipsi is an eutrophic lake, whereas Lake Pskovskoe is close to being hypertrophic.

Human impacts on water resources

One of the main problems with water protection is the eutrophication of surface waters caused by the increased load of nutrients of anthropogenic origin. Lake Peipsi receives pollution mainly through river water and precipitation directly into the lake. The nutrient content in the basin's rivers was high at the end of the 1980s, causing eutrophication of water bodies. At the beginning of the 1990s, with the dissolution of all collective agricultural farms on the Estonian side and an economic depression on the Russian side where collective farms no longer received subsidies to use herbicides or to keep large cattle stocks, the nutrient load to the lake decreased considerably. Research results indicate that the nitrogen and phosphorus loads decreased by 53 percent and 44 percent respectively during that time.

In Lake Pskovskoe, pollution occurs predominately in the southern part of the lake. In 1999 water samples showed values of up to ten times Russian norms for copper, manganese, oil products, ferrous, nitrate and cadmium. Average biological oxygen demand (BOD) and chemical oxygen demand (COD) levels were also above the limits. The same pollutants also contaminated the eastern part of Lake Peipsi but to a lesser extent. In some samples here also, excessive BOD and COD levels were found. These figures are partially caused by the pollution brought by the rivers – for example, the Velikaya River carries all these pollutants in high concentrations.

The waters flowing into Lake Peipsi are classified as hydrocarbonated calcium-rich. The oxygen content in most of the rivers is quite high as there are no big industrial polluters affecting oxygen conditions in the basin. As the oxygen levels also depend on humic substances of natural origin, the lower oxygen saturation level is caused not only by human impact but also by bog waters carried into rivers. The pH value and alkalinity levels in the rivers of the Lake Peipsi basin are relatively high, indicating an excellent buffering capacity in all the catchment areas. The present BOD level in most rivers of the basin is quite low compared with that of the 1980s when the amounts of wastewater discharged were at their highest.

Figures 17.4 and 17.5 show the ratio of phosphorus and nitrogen pollution loads distributed by country and source.

The majority of phosphorus and nitrogen compounds are carried into the lake by the Estonian Emajõgi River and the Velikaya River in the Russian Federation. These two rivers account for 80 percent of the total nitrogen load and almost 85 percent of the phosphorus load in Lake Peipsi. The first carries biologically treated sewage from Pskov, the latter transports wastewater from Tartu – wastewater that went untreated until the treatment plant opened in 1998.

Studies done in the middle of the 1980s and 1990s show a great decrease in pollution loads, particularly those caused by agriculture. Annual inputs of nitrogen and phosphorus between 1989 and 1998 are described in figures 17.6 and 17.7.

North-east Estonia is one of the most industrially developed regions of the country, where oil-shale industry dominates the sector. The wastewaters and gaseous emissions, including toxic sulphur and nitrogen oxides from power stations and pulverized oil shale, impact on the chemical composition of water in Lake Peipsi.

Figure 17.4: Ratio of phosphorus pollution load by country and source

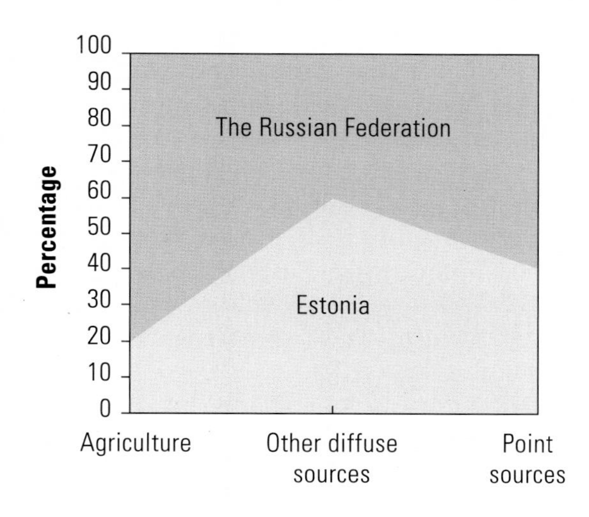

Figure 17.5: Ratio of nitrogen pollution load by country and source

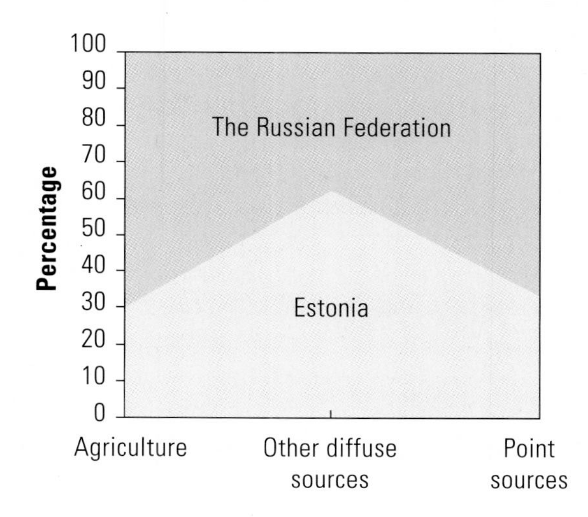

The Russian Federation is the main source of agricultural phosphorus and nitrogen pollution, contributing 80 and 70 percent respectively. It is also the major emitter of other point-source pollution. Estonia, however, emits much more phosphorus and nitrogen from diffuse sources, with 60 percent of the total loads.

Source: Stålnacke et al., 2001.

Figure 17.6: Comparison of annual phosphorus loads by source between 1989 and 1998

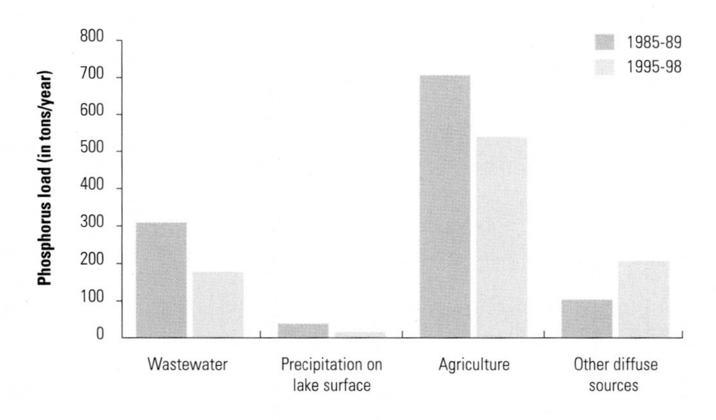

There has been a significant decrease in the amount of phosphorus pollution contributed by wastewater and agriculture between 1989 and 1998, but that contributed by other diffuse sources has almost doubled in the same period.

Source: Stålnacke et al., 2001.

Figure 17.7: Comparison of annual nitrogen loads by source between 1989 and 1998

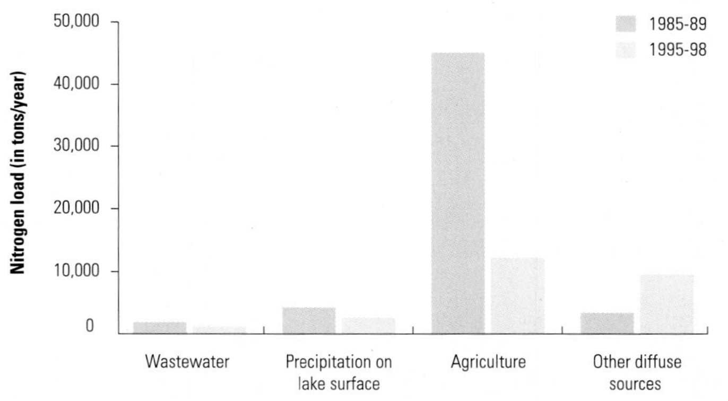

The nitrogen load from agriculture has decreased dramatically. The load from other diffuse sources has, however, nearly tripled and is set to reach the same load as that of agricultural pollution.

Source: Stålnacke et al., 2001.

Pollution loads from big cities, Pskov (mainly to Lake Pskovskoe) and Tartu (to the southern part of Lake Peipsi and to Lake Lämmijärv/Teploe) decrease from south to north, while water transparency and alkalinity caused by the impact of the oil-shale mines and their sediment have the opposite trend.

Data and information on water resources

Each country collects and analyses environmental information on the lake basin in its own manner. Nevertheless, these data are to be exchanged during meetings of the Estonian-Russian joint commission and its working groups. Moreover, it has been agreed that existing programmes and sampling activities will be coordinated to enable joint monitoring activities. However, the joint assessment of the lake is complicated by the countries' different approaches to environmental assessment. Estonia and the Russian Federation have very different monitoring methods and equipment, as well as different norms and standards, so data comparability and credibility remain problematic.

Challenges to Life and Well-Being

Water for fish

The lake itself is not used for irrigation purposes, thus its main use for food production is fishery. According to present data, one species of lamprey and thirty-three fish species permanently inhabit Lake Peipsi or the lower reaches of its tributaries. In the catches, the main commercial fish are lake smelt, perch, pikeperch, ruffe, roach, bream, pike and, until the 1990s, also vendace. The second-rate commercial fish are burbot, whitefish and white bream. The annual catch usually totals 7,000–8,000 tons. Although fishery classifications have categorized the lake as a smelt-bream-type water body, since the second half of the 1980s it has acquired some qualities of a pikeperch lake.

In general, the management of fish resources is regulated by the bilateral intergovernmental Estonian-Russian agreement concluded on 4 May 1994 concerning cooperation in the use and protection of fish resources. In 1995 following this agreement, the Intergovernmental Estonian-Russian Commission on Fishery was formed. It aims, *inter alia*, to:

- adjust management requirements of fishery;
- coordinate scientific research and catches by either party in the other's territorial waters;
- legislate quota exchange; and
- establish catchment limits.

Water for ecosystems

The Lake Peipsi watershed is rich in wetland areas and contains two Ramsar sites: Emajõe Suursoo (Estonia) and Remdovsky (the Russian Federation). Bogs and marshes occupy about 15 percent of the lake basin, while wet areas in general are spread over 35 percent of the territory.

Emajõe Suursoo is located at the mouth of the Emajõgi River on the western shore of the lake. Its surface area is 255 km^2, 180 km^2 of which is covered by the protected area. The total surface area of the Ramsar site is 320 km^2 and also includes the island of Piirissaar. The site is a habitat for several globally endangered birds such as the corncrake, the lesser-spotted eagle and the white-tailed eagle.

The second Ramsar site, the Remdovsky reserve, located on the eastern shore of Lake Peipsi/Chudskoe and Lake Lämmijarv/Teploe, was founded in 1985 and covers approximately 65,000 hectares. In 1996 it was incorporated into the 'Lake Peipsi Lowlands', which is a Ramsar site. It aims to preserve the biodiversity of the region and has international importance due to the presence of endangered species (fifty-eight species of vegetation from the Russian Federation's Red Book, which lists endangered plant and animal species, seven from the Estonian Red Book, and eleven and fifteen species of birds, respectively).

Water for cities

Wastewater from the two main cities in the region – Pskov and Tartu – is partly responsible for lake eutrophication. Smaller towns also contribute to the problem but their populations are much smaller. The biological departments of the sewage treatment plant in Tartu, commenced in 1999, and the plant at Pskov are to be completed in the coming years in order to combat lake eutrophication. Water quality is important, as the inhabitants of Narva use water from the Narva River for drinking purposes. There is, however, sufficient clean groundwater and surface water in the Lake Peipsi watershed to meet the basic needs of the population.

Water for energy

Narva has two waterfalls, Omuti and Narva, stretching to about 7 metres in height and 125 metres in width. Unfortunately, the Narva waterfalls are presently dry, having been drained by the Narva hydropower station. Water that once flowed to the river is now stored in the Narva reservoir so as to guarantee a steady water flow to the hydropower station. The reservoir is also used by two thermal power plants (Baltic and Estonian) to refine oil shale. The biggest industrial user of the lake's water is the Baltic thermal power plant, which uses the water of the Narva River for cooling purposes, with an average annual demand of 470 million m^3. Cooling water accounts for almost 75 percent of Estonia's annual water withdrawals.

Two of the world's largest thermal power plants working on oil shale are located on the Estonian side, both of which are major energy producers. As previously noted, both the power plants and chemical enterprises are the main sources of air pollution.

Management Challenges: Stewardship and Governance

Political set-up and border issues

Both the Republic of Estonia and the Russian Federation are responsible for managing and monitoring Lake Peipsi. The total length of the Estonian-Russian border is 333.8 km, approximately two thirds of which run along Lake Peipsi and the Narva River.

The border treaty between the two countries has not been signed yet, so officially there is no border, but a control line. There are five international crossing points, but none of them are found on the lake, which means that the distance between the southern and northern crossing points is about 200 km.

Five east-Estonian administrative regions (Isa-Virumaa, Jõgevamaa, Tartumaa, Leningrad and Võrumaa) have a common border with the Russian Federation. On the Russian side, the Leningrad and Pskov regions border the Republic of Estonia. The lake is a natural border between people who settled around Lake Peipsi and has resulted in the watershed's large array of cultures.

Governance

There are three levels to Russian water management: federal, regional and territorial. The main state agency responsible for water is the Ministry of Natural Resources (MNR) of the Russian Federation, although other ministries and committees have supportive tasks. The Ministry itself works on the federal level and is represented by Water Basin Administrations on the regional (basin) level, and by Committees of Natural Resources at the territorial level.

In Estonia, water management is coordinated by the Ministry of the Environment and its fifteen country Environmental Departments. Lake Peipsi belongs to the Peipsi sub-basin, where the Tartu Environmental Department is responsible for implementing water policy at a regional/sub-basin level.

Policy issues

As Lake Peipsi is a relatively new transboundary water basin (the control line between Estonia and the Russian Federation was formed when Estonia separated from the Soviet Union in 1991), the procedures of international coordination of water management have still to be elaborated. As noted previously, this challenge will become particularly acute during Estonia's application to join the European Union (EU) in 2004 when it will have to adopt EU standards and norms, which are different from those used in the Russian Federation. Developing cooperative Integrated Water Resources Management (IWRM) is a long process, and especially complicated in the context of an international lake shared by transition countries.

The Water Framework Directive (WFD) of the EU is mandatory only for member countries, and is recommended for potential members. Nevertheless, it could provide a framework for decisions relating to water management in the Russian Federation as well.

Legislation

Each country has its own legislation for water management, but in order to use the water resources responsibly and in a sustainable fashion, several bilateral agreements were signed between the Estonian and Russian governments. These include the following.

- Treaty between the government of Estonia and the Russian Federation on the conservation and use of fish stocks in Lake Peipsi, Lake Lämmijärv and Lake Pskovskoe, concluded on 4 May 1994 in Moscow;
- Agreement between the two governments on cooperation in the field of the environment, concluded on 11 January 1996 in Pskov;
- Agreement between the two governments on cooperation in the field of protection and sustainable use of transboundary watercourses, concluded on 20 August 1997 in Moscow.

Institutions and infrastructure

The Estonian-Russian Transboundary Water Commission was established in 1997 after an intergovernmental agreement was signed aiming to protect and ensure sustainable use of transboundary water bodies between the two countries. The Commission is the main actor in managing Lake Peipsi.

- It organizes the exchange of monitoring data between the parties, in accordance with the agreed monitoring programme.
- It defines priority directions and programmes of scientific studies on protection and sustainable use of transboundary waters.
- It agrees on common indicators of quality for transboundary waters, methods of testing and conducting analyses.

- It facilitates cooperation between agencies of executive power, local governments, scientific and public interest organizations as well as other institutions in the field of sustainable development and protection of transboundary waters.

- It ensures publicity of questions related to the use and protection of the transboundary waters.

Over the last few years, the commission has received considerable support from the Swedish Environmental Protection Agency (EPA) for a project to raise its institutional capacity.

The other commission in existence, previously mentioned, is the Fishery Commission, which works on a constant basis and meets once a year (see section on water for fish).

Management approaches

Managing risks

Risk management is becoming more important in the basin. The Estonian-Russian Transboundary Water Commission established criteria for emergency situations in the whole Narva River basin. These criteria include: accidents on hydrotechnical constructions and transport vehicles, and negative impacts on water bodies (such as accidents at wastewater treatment facilities, extremely high or low water levels, radioactive pollution, heavy pollution and mass death of living organisms). In addition to these criteria, the process for sharing information in each situation was elaborated. Its main objectives are to ensure that each side informs the other in case of extraordinary situations, and to organize assistance and mutual aid. Each side should be informed of any event having transboundary impact (Estonian-Russian Transboundary Water Commission, 2001).

Valuing water

Water prices for home users and water taxes vary greatly in the basin. In Estonian counties, 1 m^3 of supplied drinking water costs between US$0.73 and US$1.37. Current water taxes for surface water amount to 150 krones (about US$9.5) per 1,000 m^3 and 440 krones (US$28.1) per 1,000 m^3 for groundwater.

In the Russian Federation, water prices range from 3 to 7 rubles (between US$0.1 and US$0.22) per cubic metre depending on the region and the type of drinking water (surface or groundwater). In the north-west region, 1,000 m^3 of surface water cost 148 to 172 rubles (US$5 to US$6), whereas the same quantity of groundwater costs 200 to 232 rubles (US$7 to US$8). The Federal Law 'On the payment for use of water bodies' established the base costs, and changes were last made in 2001.

Sharing the resource

As the whole Lake Peipsi and Narva River basin is shared between two countries, cooperation and collaboration in the field of water management is essential. In general, however, the basin has sufficient water so there is no serious competition between industries and local population.

Governing water wisely

One of the main characteristics of the Estonian and Russian water policy is the basin approach. By taking the basin as the main hydrological unit, it closely resembles European water management policy. Thus, the international basin of Lake Peipsi is governed in accordance with both Russian legislation and the EU WFD.

The Estonian-Russian Commission established a formal mechanism for developing cooperation with local authorities, non-governmental organizations (NGOs) and stakeholders, who can communicate their issues and interests directly to the intergovernmental commission. However, very few regional NGOs are involved in the work of the commission and most local NGOs and stakeholders cannot afford greater involvement. External financial support is necessary to develop their capacity and to enable their involvement in managing transboundary waters shared by countries in transition.

Regional NGOs, such as the Peipsi Center for Transboundary Cooperation (Peipsi CTC) and the Council for Cooperation of Border Regions, cooperate with local authorities and stakeholders on regional development projects as well as on educational, research and social projects in the region. Peipsi CTC is also actively involved in the work of the Estonian-Russian Transboundary Water Commission.

Local and regional authorities and businessmen in the Lake Peipsi region also promote their agenda for transboundary economic cooperation in the form of investments in the construction of roads, in water transportation and in tourism infrastructure, as well as international promotion of the region.

Ensuring the knowledge base

Estonian and Russian research cooperation was interrupted by the border re-establishment at the beginning of the 1990s. Estonian environmental experts, together with their Russian colleagues, published a comprehensive monograph on Lake Peipsi in 1999. The Swedish EPA, Danish Ministry of Foreign Affairs and Danish EPA, and the Norwegian Ministry of Foreign Affairs all supported regional studies and the development of strategies for reducing and preventing pollution, as well as regional development in the basin. Reports have also been published in English, Estonian and Russian and have been widely disseminated (see box 14.8 in chapter 14 for further details).

Conclusions

The Lake Peipsi basin faces various challenges, including serious eutrophication affecting both human and fish populations, pollution from power stations, large towns, oil shale industries and mining activities, and the difficulties inherent to transboundary water management. Although the anticipated economic growth in the region is likely to increase the pollution reaching the lake, measures are being taken such as the building of new wastewater treatment facilities to ensure that the pace of eutrophication is slowed.

The region also faces economic challenges. Although fishing has long been one of the major activities of the area, the lake's fish resources are now under severe pressure, further complicated by economic difficulties and high rates of unemployment. The depleted fish populations can no longer provide a living for so many and there is therefore an urgent need to diversify the economy in the basin.

Many bilateral agreements have been implemented concerning various aspects of joint management of the lake, but real coordination and cooperation remain problematic due to the absence of a comprehensive water programme, insufficient public participation and coordination in the lake basin, especially as regards environmental monitoring, and complicated border issues, which have so far impeded effective collaboration. There is, however, a developed legislative base and desire for greater efficiency on both sides upon which to build.

From an environmental point of view, the lake basin can be considered an extensive ecosystem in its own right. It is therefore essential to maintain it in as healthy a condition as possible, and to remember that the loss or change of any one component could have serious consequences on the entire system.

The most important planning and development issue in Lake Peipsi is preparation for the Lake Peipsi Management Plan. In cooperation with Estonian and Russian governments, regional and local authorities, private companies and the public, this is to be completed by 2007, and should provide a starting point for addressing the various challenges encountered in the basin in a more integrated manner.

References

Andersen, J.-M.; Sults, U.; Jaani, A.; Alekand, P.; Roll, G.; Sedova, A.; Nefedova, J.; Gorelov, P.; Kazmina, M. 2001. *Strategy for Wastewater Treatment in the Lake Peipsi Basin.* Tartu, Peipsi Center for Transboundary Cooperation.

Estonian-Russian Transboundary Water Commission. 2001. *Materials and Protocols.*

Jaani, A. 2001. 'The Location, Size and General Characterization of Lake Peipsi'. In: T. Nõges (ed.), *Lake Peipsi. Hydrology, Meteorology, Hydrochemistry.* Tallinn, Sulemees Publishers.

Keevallik, S.; Loitjärv, K.; Rajasalu, R.; Russak, V. 2001. 'Meteorological Regime'. In: T. Nõges (ed.), *Lake Peipsi. Hydrology, Meteorology, Hydrochemistry.* Tallinn, Sulemees Publishers.

Laugaste, R.; Nõges, T.; Nõges, P.; Yastremskij, V.; Milius, A.; Ott, I. 2001. 'Algae'. In: E. Pihu and J. Haberman (eds.), *Lake Peipsi. Flora and Fauna.* Tallinn, Sulemees Publishers.

Leisk, Ü. and Loigu, E. 2001. 'Water quality and pollution load of the rivers of Lake Peipsi basin.' In: T. Nõges (ed.), *Lake Peipsi. Meteorology, Hydrology, Hydrochemistry.* Tallinn, Sulemees Publishers.

Nõges, T., Haberman, J., Jaani, A., Laugaste, R., Lokk, S., Mäemets, A., Nõges, P., Pihu, E., Starast, H., Timm, T.; Virro, T. 1996. 'General Description of Lake Peipsi-Pihkva'. *Hydrobiologia,* No. 338, pp. 1–9.

Peipsi CTC (Center for Transboundary Cooperation). 2001. *Lake Peipsi Business Profile.* Tartu.

Pihu, E. and Kangur, A. 2001. 'Fishes and Fisheries Management'. In: E. Pihu and J. Haberman (eds.), *Lake Peipsi. Flora and Fauna.* Tallinn, Sulemees Publishers.

Russian Scientific Research Institute of Water Management. 2000. *Assessment of Transboundary Water Pollution from Emissions of Baltic and Estonian Thermal Power Stations.* St. Petersburg.

——. 1999. *Analysis of Hydrological and Environmental State of Narva River Basin Including Lake Chudsko Basin and Creation of General Report.* St. Petersburg.

Stålnacke, P. 1999. *Nutrient Loads to the Lake Peipsi – Experiences from a Joint Swedish/Estonian/Russian Project.* Tartu, Peipsi Center for Transboundary Cooperation.

Stålnacke, P.; Sults, Ü.; Vasiliev, A.; Skakalsky, B.; Botina, A.; Roll, G.; Pachel, K.; Maltsman, T. 2001. 'Nutrient Loads to Lake Peipsi'. *Jordforsk Report, No. 4/01.*

St. Petersburg Center of Independent Environmental Assessment of the Russian Academy of Sciences. 2001. *Management of Water Resources and Protection of Narva River Basin including Lake Pskovskoe-Chudskoe Basin.* St. Petersburg.

'A great challenge lies ahead in preserving the environment while building a better life for the Ruhuna basin's population, a good part of whom continues to suffer from malnutrition and poverty. Improving the access to relevant data, implementing more integrated management approaches and continuing the fight against poverty could go a long way to providing a more sustainable future for the basins.'

Ruhuna Basins, Sri Lanka

By: The Ministry of Irrigation and Water Management of Sri Lanka*

Table of contents

*A workshop was held on 7–8 April 2002 in the Koggala Beach Hotel in preparation for the case study report. The authors who presented papers to the workshop made the major input to this paper. Guidance provided by the Secretary of the Ministry of Irrigation and Water Management, the Secretary of the Ministry of Irrigation and the Secretary of the Ministry of Water Management, and the comments made by a large number of Ministries and stakeholder agencies are much appreciated. Special thanks are due to Mr. Ian W. Makin and Dr. Peter Droogers of the International Water Management Institute (IWMI) for their contribution to the study in general and assistance in editing the document, in particular.

Let not a simple drop of water that falls on the land go into the sea without serving the people.

Parakkama-Bahu I, King of Sri Lanka (1153-1186)

THIS CASE STUDY OFFERS THE PICTURE of a rural society that values water in traditional ways, but is changing rapidly. There is little coordination among the many agencies dealing with water. Water resources in Ruhuna are of good – though declining – quality and are highly regulated to support hydropower production and irrigation. However, the basin is currently experiencing water stress due to great seasonal variation in rainfall and a drought in the area. Paddies account for a whopping 95 percent of the total water consumption. Rural poverty is increasing and farmers are often perceived as mere beneficiaries rather than key partners in the management of irrigation and water resources. There is hope that the recent formation of Water Resources Councils can address the need for co-ordination, integrate the concerns of different users, and lead the way to a better life.

THE RUHUNA BASINS IN SOUTHERN SRI LANKA will face major changes over the next two decades. Ambitious development plans indicate that the dominant role now played by agriculture is geared to switch to much more industrial and service-oriented activities. Obviously, these changes will have an enormous impact on society as well as on natural resources and require the inclusion of issues of water management. Currently, almost all water resources that are diverted are used for irrigation with only a small percent used for industry and drinking water. Most recent development plans show that the use of water for urban areas and industry will increase from less than 10 million cubic metres (Mm^3) currently to 100–150 Mm^3 by the year 2025.

Map 18.1: Locator map

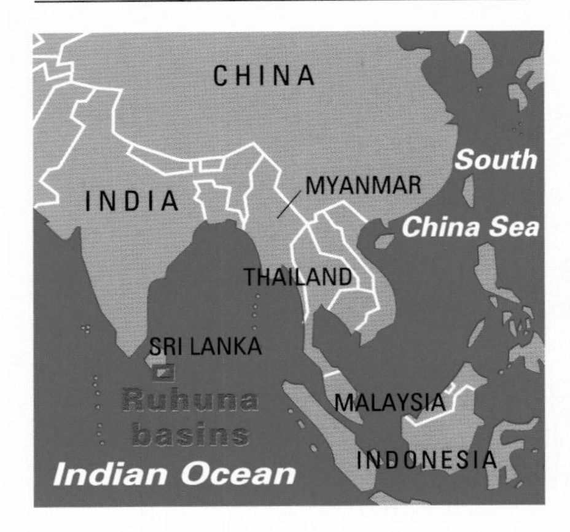

Source: Prepared for the World Water Assessment Programme (WWAP) by AFDEC, 2002.

General Context

Already today, the Ruhuna basins are important in the broader Sri Lanka context: they are the location of a major hydropower plant, and irrigation schemes that make a significant contribution to food production and important nature reserves. However, even before the envisaged development plans are implemented, the basins are already experiencing major water resource problems, clearly demonstrated by the recent drought leading to reduced water for irrigated agriculture, insufficient supply of domestic water and nationwide power cuts for up to eight hours a day.

The Ruhuna basins area includes three main rivers and several smaller basins, and forms part of the hydrologic system of the ancient kingdom of Ruhuna.

Sri Lanka has a surface area of 65,500 square kilometres (km^2) and a population of 19 million inhabitants. It is renowned for its hydraulic civilization in which natural resources have been managed over thousands of years. The country has a mean annual rainfall of about 1,900 millimetres (mm) but this ranges from below 900 mm in the driest parts of the dry zone to over 5,000 mm in the central hills. There are 103 distinct river basins in the island, ranging in size from 9 km^2 to 10,450 km^2. The Ruhuna basins cover 8 percent of the Sri Lanka landmass.

Table 18.1: Hydrological characteristics of the Ruhuna basins

Surface area of the basin	5,578 km^2
Annual precipitation	1,574 mm/year
Annual runoff	78 m^3/s
Annual potential evapotranspiration	1,700 mm/year
Upper catchment	1,458 mm/year
Lower catchment	1,914 mm/year

Major physical characteristics

Geography

The Ruhuna basins are mountainous and relatively wet. Several catchment areas are poorly developed, but there are downstream flat areas with developed water resources. These lowlands consist of rolling plains dotted with a few isolated hills. The rivers originate from the southern slopes of the central highland massif at elevations of up to 2,000 metres. A large part of the basins is made up of many types of composed rock, such as granite, migmatite and quartzites (Panabokke et al., 2002).

Climate

The only source of water is rainfall. Monsoonal rains, which fall from November through March and May through September, contribute a major part of the annual rainfall, which is supplemented by inter-monsoonal rains. The mean annual rainfall for the basin is 1,574 mm, the depth of which decreases from the upper to lower reaches and from west to east. The recent rainfall records at selected stations show a trend of decreasing annual rainfall since 1970. The decrease is not uniform or highly significant in statistical terms. The ambient temperatures in the lowlands range from 25 to 28°C, and in the upper elevations from 23 to 25°C.

Major socio-economic characteristics

The Ruhunu basins include parts of the Ratnapura, Badulla, Moneragala and Hambantota districts with population densities of 307, 291, 71 and 217 inhabitants/km^2, respectively. The total population in the basins is approximately 1.1 million (Jayatillake, 2002b).

The average monthly income per household in the Badulla district is the lowest in the country. For the year 2000 the national average Gross Domestic Product (GDP) per capita was US$850, but for the Ruhuna basins, per capita GDP is estimated at about

Map 18.2: Basin map

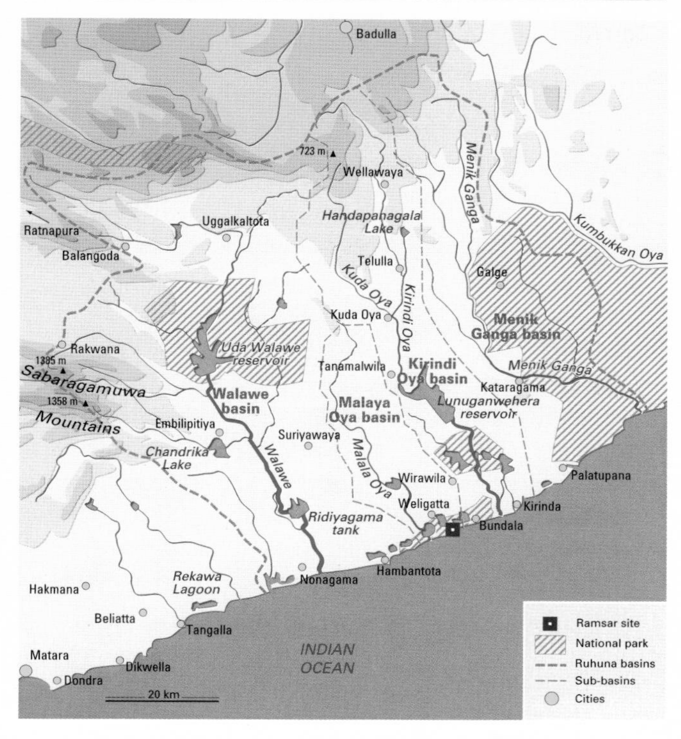

Source: Prepared for the World Water Assessment Programme (WWAP) by AFDEC, 2002.

US$600. The percentage of households receiving food and economic support was measured at 60 percent compared to the national average of 39 percent (Department of Census and Statistics, 2000).

Tea and rubber are grown as commercial crops in the upper basin area, and rice is the major crop in the plains. The major land uses in the basins are: forests (29 percent), scrublands (26 percent), shifting cultivation or chena (23 percent), home gardens (12 percent) and paddy (10 percent). However, there are significant differences in land use among the main sub-basins (Panabokke et al., 2002): for example in the Menik Ganga basin, forests account for 57 percent of land use, compared to only 17 percent in the Walawe basin (see map 18.3).

Water Resources

Surface water

Based on the hydrologic characteristics of basins (see table 18.2) the annual supply of surface water per capita is estimated at 2,291 m^3. Kirindi Oya basin is the most water-stressed of the Ruhuna basins, as it has the lowest per capita surface runoff, relatively high flow requirements for environmental purposes, and farmers in the area are facing various water problems. The potential for further development of water resources within the Kirindi Oya basin is therefore negligible and several studies have been undertaken to explore the options for interbasin transfer.

Groundwater

Many people in the basins depend on shallow groundwater for their domestic needs. Studies indicate that seepage losses from canals and reservoirs have been indispensable for maintaining water levels in shallow wells. Deep groundwater is concentrated in the fractured and weathered aquifers in hard rock areas and alluvial aquifers. Available information indicates that 7 to 10 percent of rainfall contributes to groundwater recharge in the hard rock terrain, and 40 percent in the sandy alluvial aquifers (Panabokke et al., 2002).

Groundwater accounts for 3 percent of total water withdrawals (Jayawardane, 2002 and Jayatillake, 2002b) and there is a high vulnerability to declining groundwater levels and saltwater intrusion in the lower reaches of the basins.

Water quality

To date, the upstream areas of the basins have not faced major water quality problems, as industrial activities are limited. However, studies have indicated that although surface waters are relatively free of fluoride, toxicity problems such as faecal coliform bacteria and organic waste contamination do exist. Data indicate that water

Map 18.3: Major land covers and uses in the Ruhuna basins

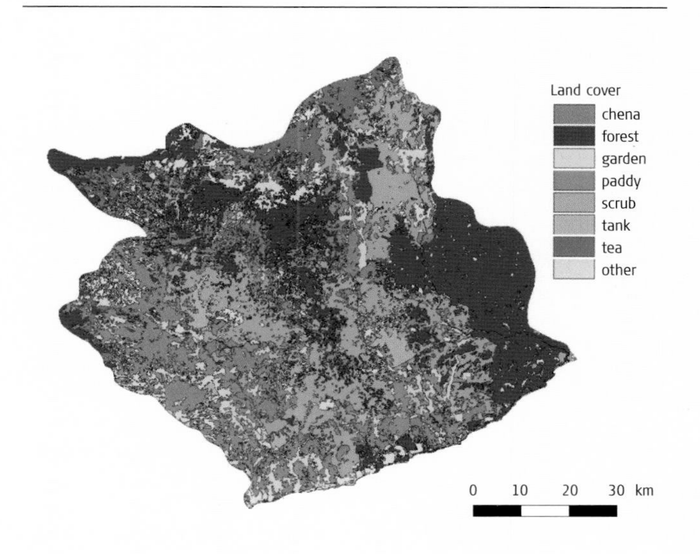

The Ruhuna basins are largely composed of forest, followed closely by scrublands and chena, or shifting, cultivation. Other cultivations include paddy, gardens and tea crops. There are also several large reservoirs (tanks) in the area.

Source: Drawn for the World Water Assessment Programme (WWAP) by the Survey Department of Sri Lanka, 2002.

Table 18.2: Comparison of basic hydrological parameters of sub-basins over a thirty-year time span

Basin	Catchment area (km^2)	Annual precipitation (Mm^3)	Annual surface runoff (Mm^3)	Annual discharge to the sea (Mm^3)
Walawe Ganga	2,471	4,596	1,524	525
Kirindi Oya	1,165	1,713	469	74
Menik Ganga	1,287	2,009	352	326
Malala Oya[1]	402	441	74	NA
Other[1]	253	252	42[2]	NA
Total	5,578	9,011	2,461	

[1]Ungauged rivers and streams [2]Assumed value

Source: Jayatillake, 2002b.

quality is poor in the coastal lagoons and river estuaries and the quality of water appears to deteriorate further especially during the dry season (Handawela, 2002).

There are two national agencies responsible for recommending water quality standards for drinking, bathing, irrigation and other uses: the Sri Lanka Standards Institute and the Central Environmental Agency (CEA).

Human impacts on water resources

Water resources in the basins are highly regulated to support hydropower generation and irrigation. There are twenty large reservoirs (three of which have a capacity exceeding 100 Mm^3) and about 280 smaller reservoirs, giving a total storage capacity of about 900 Mm^3. There are numerous river diversion systems, mainly for irrigation supplies, including eleven large and about 610 small cross-river structures. Storage volume in the three main basins ranges from 57 percent of annual (surface) water resources for the Kirindi Oya basin, to 40 percent in the Walawe basin and almost zero in the Menik Ganga basin.

Changes in land cover will affect water resources, of which about 2,720 km^2 (about 50 percent) is under forest and scrubland. In the Walawe and Kirindi Oya basins, substantial development works have been carried out and the forest and scrub cover have decreased by 30 percent and 23 percent respectively over the past forty years, which is a higher rate of land use change than the national average.

Data and information on water resources

The hydrometric network includes sixteen rainfall stations, seven agro-meteorological stations and six water level stations. The water level observation network is clearly inadequate to provide sound information on water resources in the basins.

The lack of sufficient regular flow observations have led to a large number of sporadic attempts by agencies involved with water resources to collect data, mostly in response to the internal requirements for development projects. The frequency of data collection intensifies during project studies but generally the frequency and quality of data observation diminish once the project is completed.

Integrated Water Resources Management (IWRM) in the basins clearly suffers from this lack of a consistent, continuous and accurate hydrometric network. Moreover, data access and data sharing between the different agencies remain limited, restricting the benefits obtained from even the existing data network.

Challenges to Life and Well-Being

Water for food

The major food crop in the basins is paddy, cultivated over approximately 52,000 hectares of which 90 percent is irrigated. Normally two crops can be grown during the year, during the two monsoon seasons. The water used by major irrigation systems, defined as irrigation duty at the reservoir outlet, is about 1,500 mm in rainy seasons and 1,800 mm in dry. However, a proportion of these releases is used indirectly by small irrigation systems and for domestic use, so actual irrigation applications are lower. Withdrawals for irrigation account for more than 95 percent of the total.

In general large irrigation schemes achieve cropping intensities in excess of 150 percent, while most minor schemes are close to 100 percent. Average paddy yields in major irrigation, minor irrigation and rainfed systems are 4,600, 3,600 and 2,900 kilograms (kg) per hectare, respectively.

In terms of water productivity, the amount of water used to produce 1 kg of crop has been shown in a study in the Kirindi Oya basin to be 0.29, 0.16 and 0.14 kg/m^3 of evapotranspiration, net irrigation and gross irrigation requirement, respectively.

The basins produce large amounts of marine and inland fish and cow and buffalo milk (Weerasinghe et al., 2002). Inland fisheries are becoming popular, are receiving support from the government and are an important source of protein for the rural population. No detailed information is available on the productivity of inland fishery at the basin level.

Water for basic needs

Sixty percent of the basin population has access to safe water and 71 percent to adequate sanitation (Ministry of Plan Implementation, 2001). These figures are slightly lower than the national average, which rests at 75 and 73 percent respectively (Shanmugarajah, 2002). National targets for water supply in Sri Lanka are ambitious: the target for access to safe drinking water is set at 85 percent of the population by 2010 and 100 percent by 2025. Similarly, the target for adequate sanitation is 100 percent by 2035 (Wickramage, 2002).

Some of the major towns in the basins obtain domestic water from irrigation reservoirs, while others abstract water from the river itself. Although the return flow from agricultural lands helps to maintain minimum flow requirements during the dry months in the Walawe sub-basin, it causes water pollution because of the presence of agrochemicals.

Water for ecosystems

'The paddy fields in Ruhuna basins are the most valuable wetlands I've ever seen', claimed a famous American ecologist. As well as these agricultural wetlands, the basins consist of several other ecologically important reserves, including the Ruhuna, Uda Walawe and Bundala National Parks, the lagoon systems adjacent to Bundala park and a large number of man-made reservoirs.

Sri Lanka's first Ramsar site, Bundala National Park, is spread over an area of 6,216 hectares. This area is designated as a sanctuary under the Flora and Fauna Protection Ordinance. Four shallow, brackish lagoons form the major part of the park. Bundala is the most important wintering area in southern Sri Lanka for migratory shorebirds, sometimes accommodating up to 20,000 birds. Elephants and leopards are also found in Bundala (CEA/Arcadis Euroconsult, 1999).

Ruhuna National Park is one of the largest in the country, covering about 126,000 hectares, some of which lie outside the basin. Most of the park's wetlands are well protected. The total land area protected under relevant legislation is about 1,200 km^2, which represents some 21 percent of the basin area. Water use for ecosystems is highest in the Menik River, which flows through the Ruhuna National Park. Concerns about the protection of downstream lagoons have been raised, as the minimum base-flows of the rivers are becoming inadequate to meet ecosystem requirements.

Water for industry

The basin does not currently house any major industrial activities. Some smaller factories – mainly for garment and paper – and hotels and rest-houses for the tourist industry do exist, but industrial water withdrawals are estimated at being less than 1 percent of total withdrawals.

Source protection measures are adopted by these industries only when they use their own water supply systems and when the source is located within their property. Where the industries are extracting water from public water supply systems, there is no specific contribution towards source protection, apart from tariff payment for the consumption of water (Senaratne, 2002).

However, major changes are expected when the proposed Ruhunupura City development plan, which will include an airport, industrial and commercial areas and a commercial harbour, is implemented. The water requirements of Ruhunupura are estimated at between 100 and 150 Mm^3 per year. Studies presently underway are exploring the reservoir options to retain floodwater from rivers in order to meet the greatly expanded water demand.

Water for energy

In 2001/2, Sri Lanka faced a major power crisis resulting in nationwide power cuts for up to eight hours a day for several months. The drought, in combination with hydropower's large share of total power generation (65 percent), was the main cause. The national targets relevant to power generation specify that a reliable electricity grid should be provided to at least 80 percent of the population at affordable prices and that the share of hydropower is to be reduced to about 32 percent by 2013.

The hydropower generation facilities in the Ruhuna basins are restricted to the Walawe sub-basin. Uda Walawe reservoir, constructed in the 1960s, has an installed hydropower capacity of 6 megawatts per hour (MWh). The installed capacity of the hydropower plant at Samanalawewa is 120 MWh, which is about 10 percent of the total installed capacity in Sri Lanka. Based on the records at Samanalawewa, about 1.3 Mm^3 of water was required to produce 1 gigawatt per hour (GWh) of energy (Somatilake, 2002). Water used to generate energy is recaptured downstream for irrigated agriculture, power generation and other uses.

Management Challenges: Stewardship and Governance

Cultural background and the value of water

The Ruhuna, along with being a subkingdom, served as a safe-haven for people fleeing foreign invasions. In ancient times, agriculture played a major role in the economy as well as in national security. Efforts to develop water resources in the area, as in other parts of the country, focused on irrigation. As rainfall is concentrated in the two monsoon seasons, and because of the large interannual variability of rainfall, a good number of reservoirs have been built. Water has been used for recreation, sanitation and hygiene for thousands of years, and as such has been given a very high value in the community.

This historical background has influenced the perception that water is a public good, and has maintained agriculture's role as both a tradition and a major component of the livelihood of the population. Traditions also emphasize that water is a valuable resource that is not to be wasted. A local management structure for water resources was developed that included provisions for cost recovery and regulation. These provisions enabled a self-sustaining rural agrarian society to exist in the villages.

Water-related cultural practices among the rural society emphasized the optimum use of water. However, this management structure was disturbed during the period of colonial occupations. An increase in rural poverty made people more dependent on state subsidies. To some extent this dependency continues under modern irrigated agriculture where, especially in major irrigation areas, there is a heavy emphasis on state control. Farmers are often perceived as mere beneficiaries rather than key partners in management of irrigation and water resources.

The broad recognition of centuries-old water traditions in Sri Lanka, and the considerable number of people still living below the poverty line, enhance the importance of water's social, environmental, cultural and economic values. For example, the numerous minor irrigation systems provide water for domestic use, livestock, wildlife, recharge of groundwater and also for enhancing the village environment. These multiple dimensions that make up the value of water must be considered equitably in planning, developing and managing water resources.

The economic value of water, however, has been the subject of intense discussion in the recent past. A draft policy document that made reference to water as an economic good was rejected following strong pressure from the public and media. Leading politicians have made statements implying that water is set to remain a free good for the foreseeable future.

In agriculture, still the main water use sector, farmers contribute to the maintenance costs of the irrigation network. But, in general, they pay no water supply or service fees. Operation and maintenance costs are minimal and farmers mostly pay through provision of labour for cleaning canals. This is similar to ancient practice. However, in some modern irrigation systems, a minimal service fee is levied.

Political set-up: institutions and legislation

Sri Lanka is a parliamentary democracy divided into eight provinces and twenty-four districts. Rivers flowing through more than one province, and irrigation systems that are served from these channels come under the purview of the central government. The Provincial Councils, which constitute the provincial government, manage smaller rivers, village and provincial irrigation, and environmental issues. The Ruhuna basins belong to the Southern, Uva and Sabaragamuwa provinces.

Water management responsibilities in the basins lie with institutions at the national and local levels: approximately forty agencies exist with responsibility or interest in water. These include the sector agencies dealing with domestic water supply, health and sanitation, agricultural and irrigation services, hydropower generation, groundwater development and ecosystem management. In addition, Provincial Councils established after the thirteenth amendment to the constitution in 1987 have devolved powers for water-related functions. The chief secretary of the province, district secretary and divisional secretary are key government officers that make decisions regarding the management of water resources at the respective levels. At the district and scheme levels, the District Coordinating Committee, District Agricultural Committee and Project Management Committee are also decision-makers.

Such a multitude of institutions requires effective coordination at different levels. At the national level, the Central Coordinating Committee on Irrigation Management provides a forum for policy issues in irrigation management. A similar forum is the Steering Committee on Water Supply and Sanitation. The recent formation of a Water Resources Council (WRC) addresses the need for coordination of water resource issues. Moreover, the proposed formation of river basin committees would rectify the existing inadequacy in dealing with issues of IWRM.

Non-governmental organizations (NGOs) make a significant contribution to water resource management. Several NGOs have invested in minor irrigation, and significant numbers play a vital role in protecting ecosystems and rainwater-harvesting activities. Better coordination of NGOs and government agency actions is essential and could be of great benefit.

Over fifty Acts of Parliament have been established to manage water resources (Ratnayake, 2002). A large number of agencies, each dealing with different aspects (such as irrigation, water supply, sanitation, industries and environment), are charged with implementing these acts. The proposed Water Resources Act, expected to be operational in 2003, should address gaps and implementation problems in the existing legislation.

Finances

Public investment in water resources focused on developing irrigation infrastructure from 1950 until the 1980s. The emphasis has since shifted towards investments in rehabilitation of existing infrastructure and improvement of water management. In the year 2000, national investments in agriculture and irrigation remained at about 8.5 percent of the total capital expenditure. The corresponding figure for the energy and water supply sectors was about 16.5 percent (Central Bank, 2001).

The main investors in urban water supply and sanitation have been the public sector, including the central government, the National Water Supply and Drainage Board (NWSDB), Provincial Councils and local authorities. Investments by community-based organizations and private individuals are significant in rural areas.

Additionally, a substantial portion of irrigation and water supply projects are foreign-funded. Two major irrigation rehabilitation projects and one comprehensive groundwater assessment project are ongoing in the basins and are funded by independent donors.

There have been several attempts to recover the cost of operation and maintenance of irrigation services, but these have so far proved unsuccessful. An ongoing programme consisting of turning irrigation systems over to the users has resulted in substantial contributions from farmers in system management, and in operations and maintenance costs.

Cost recovery mechanisms for urban water supply focus on recovery of operation and maintenance costs of the services. The level of recovery is lower in water supply schemes managed by local authorities. Private investment by individual families for the construction of protected wells and latrines is considerable (Wickramage, 2002).

Differing management approaches

Several management approaches exist in the Ruhuna basins.

- IWRM: although Sri Lanka has been implementing IWRM principles for several years, it is yet to be recognized as state policy. As an example of the attention given to IWRM principles, irrigation development projects in environmentally sensitive areas are submitted to an Environmental Impact Assessment (EIA) and must obtain its approval before they can begin. Also, the National Environmental Act was enacted in 1980 and a ministry was set up in 1991 to deal with specific environmental issues. A ministry dealing with water resource management and a Water Resources Secretariat, charged with formulating a National Water Resources Policy and relevant Water Resources Legislation, were established in 2000 and 1996 respectively.

- Demand management: demand management has been given special attention in recent government policies. In the domestic water supply sector, targets include minimizing unaccounted water and introducing demand management measures. In irrigated agriculture, there is a heavy focus on micro-irrigation methods and improved monitoring of agricultural operations. Optimal water use is a major focus for the ongoing irrigation rehabilitation projects in the basins. In the energy sector, there are campaigns to reduce power consumption.

- Public participation: in 1988, following a number of pilot experiments started in the late 1970s, Sri Lanka adopted participatory management in irrigated agriculture as state policy. A programme of irrigation management turnover (IMT) to farmer organizations is ongoing. Although no systems have been completely turned over, there has been a significant increase in the role of farmers in managing irrigation systems over the past two decades. Most major irrigation systems in the basin are partially turned over. Farmers have traditionally managed minor irrigation systems (command areas of less than 80 hectares). Attempts have been made to encourage community participation in environmental protection; however these efforts are still at an early stage. The Upper Watershed Management Project, implemented by the Ministry of Forestry and Natural Resources, is actively promoting participatory forestry in a focus area that includes the upper catchment of the Walawe basin.

- Public-private partnerships: the concept of farmer companies is being tested at two irrigation systems in the country. One site, the Chandrikawewa Farmer Company, is in the Walawe basin and promotes agricultural production and other rural business activities, while the public sector manages the irrigation system. As yet, a full evaluation of this pilot project has not been undertaken. The other pilot Farmer Company, outside the case study area, is involved in the operation and maintenance of irrigation systems in addition to agricultural input and output marketing. This other company may form a useful model for future public-private partnerships.

Managing risks

Most of the basin's area is located in so-called dry zones and receives less than 1,250 mm of rain annually. The major natural hazard, therefore, is drought. The area is also subject to occasional localized floods, but is at little risk from landslides, coastal erosion, cyclones and earthquakes.

Of the administrative districts of the basin, only Hambantota is classified as drought-prone: during the rainy season drought probability is 28 percent, the highest in Sri Lanka. During the dry season, it increases to 32 percent. Though not classified as drought-prone, some areas of the dry zone in the Moneragala district were seriously affected during the recent drought, as was much of the region. The government initiated a range of measures to mitigate the impacts of future occurrences of droughts. These include short-term emergency measures, such as development of groundwater for emergency domestic supplies, medium-term interventions such as introducing better water management practices; and longer-term studies on the possibility of interbasin water transfers.

Decisions on drought management in agriculture are taken in the seasonal cultivation meeting, attended by farmers and officers. Typical decisions include cultivating a reduced proportion of the command area and sharing the land. In general, domestic water needs are given the highest priority during droughts, a policy that will be formalized with the proposed National Water Resources Policy.

Another concern in some parts of the basins, especially the Moneragala district, is the incidence of malaria. This vector-borne disease can be a predisposing cause of anaemia and malnutrition (for details, see chapter 5 on water and health). Research by the International Water

Management Institute (IWMI) and others have shown that effective water management practices can contribute to eliminating the vector that causes malaria. Initiatives by the Ministry of Health, assisted by the World Health Organization (WHO) and the World Bank, in partnership with national agencies, have addressed the problem considerably.

Sharing the resource

Water delivered through irrigation facilities is a common source of domestic supply for major towns and villages in the basin. Thus, during periods of water shortages, allocating water among different uses becomes an issue.

In one instance, the challenge of sharing water between sectors has been further complicated by the prior water rights of Kaltota farmers, threatened by the Samanalawewa hydropower station. Constructed upstream of the irrigation scheme, the station extracts water from the river to generate energy. As a result, farmers at the Kaltota scheme have had to face occasional water shortages. After a period of intense negotiations and bargaining, a consensus is being built among the authorities dealing with irrigation and hydropower and the farmers.

Evaluating the knowledge base

There is a considerable range of data and knowledge about water and natural resources in the Ruhuna basins and indeed in Sri Lanka in general. However, the available data and information are scattered amongst the different agencies. The World Water Assessment Programme (WWAP) has instigated recognition of the need for greater access to the available information and sharing of knowledge resources between the involved agencies. To this end, a comprehensive database is in the initial stages of design and implementation. The database will be structured to enable monitoring of the WWAP challenge areas and, when complete, will allow improved sharing of data among the agencies.

Identifying Critical Problems and Opportunities

Challenges related to the nature of the resource

Analysis of the meteorological and hydrological data confirms the high temporal variability of rainfall and river flows. Of the three main rivers, a large proportion of the available Kirindi Oya water resource is already developed and there is little scope for further exploitation of the in-stream flows. In comparison, the water resources in the Menik Ganga basin are largely undeveloped, but concerns about the impacts of abstraction on nature and wildlife are currently restricting development plans.

Investigations show that groundwater quality is poor in the lower reaches of the basins, i.e. in the dry zone. In several places, fluoride, hardness, chlorides, sulphate and alkalinity contents are reported to be high, and shallow groundwater, in areas not recharged by irrigation, is falling in some locations due to increased use of agro-wells, that is, wells supplying water to agriculture.

Several water resource development projects are ongoing in the Walawe sub-basin, and other proposals are being studied. With scientific investigations and planning, there is a potential for further development of groundwater. Potential investors in the proposed Ruhunupura industrial complex have indicated their willingness to invest in desalination of seawater to the Ministry of Southern Development.

Challenges related to uses

The main water user in the basins is agriculture, principally for flood-irrigated paddy, the predominant staple-food crop. Paddy cultivation on highly permeable soils has contributed to high water use and thus water shortages and environmental problems. Studies carried out in selected irrigation schemes indicate that many farmers overapply pesticides, herbicides and nitrogen fertilizers (Renwick, 2001). The coastal lagoon system, which forms an important segment of the basin's ecosystem, receives a large volume of drainage flow during the irrigation seasons. This carries the residual wash-off of these agrochemicals. Reduction of dry-weather base flows by irrigation abstractions and increased groundwater use are reported to have aggravated the impacts of drought on the lagoon systems.

In addition to environmental problems from agrochemical use, minor industries in the basin are an occasional source of pollution, such as the paper mill and sugar factories in the Walawe sub-basin.

Several opportunities exist for improving the water use efficiency in agriculture. Increased reuse of irrigation return flows, diversification of crop patterns to include a higher proportion of more water-efficient crops, and the improvement of conveyance systems, better canal operations and field application methodologies are being introduced in the basins. Two donor-funded rehabilitation projects are also supporting initiatives to increase productivity and water use efficiency of the basins' irrigation systems.

It is argued that the village tanks, or dams, should not be considered merely as water sources and production systems but rather as a central part of the socio-economic and cultural system of the rural area. The widespread distribution of minor irrigation systems supports equitable access to water resources. However, transfer of technology innovations to minor irrigation and rainfed farming is less intensive than in the major irrigation schemes. Opportunities do exist to promote higher productivity in rice and other traditional cereal crops in these areas and contribute to sustainable water resource development. Ongoing work for new irrigation facilities has taken these aspects into consideration. Further advances in the use of small-scale groundwater resources (for example through agro-wells) and modern irrigation techniques, including drip, trickle and sprinklers, are supporting increased areas of market-oriented crops.

The reservoirs in the basin allow the impacts of river flow variations to be regulated. However, although farmers using irrigation have reported adverse impacts of other water resource developments, for example hydropower, good coordination and cooperation between sectors have provided more reliable water services to agriculture. Improving agricultural practices and coordination is likely to further ease pressures on basin ecosystems. The proposed river basin organizations will provide a stronger institutional framework to better integrate the concerns of different users in resources planning and management.

Challenges related to management

Several management problems have been identified in the basins. These include inadequate policies and poor coordination among agencies who deal with water, deficiencies in regulating mechanisms and lack of forums to discuss IWRM issues at the basin level. The major issues to be confronted in developing policy and implementing strategies to address the water-related problems include poverty, multisectoral use of water and the coordination of a large number of agencies. Major water resource development projects have focused on agriculture, and intersectoral reallocation is a sensitive issue. There are many legal enactments that deal with safety of infrastructure, water allocation and watershed management, but their implementation is split between several agencies.

Participation of women in decision-making by farmer organizations is low. Only an estimated 22 percent of participants in farmer organizations are female (Atukorala, 2002). As women are concerned with health, sanitation, domestic water and food supplies, their increased participation in water management could positively impact on the issues of sharing water.

Main policy issues related to water supply and sanitation include cost recovery and service coverage for the poor. Present cost recovery levels in urban water supply systems cover only the recurrent operation and maintenance costs and a small part of the capital cost. With government plans to reduce investments in these services, the service providers must, in future, recover the operation, maintenance and capital cost through water charges for urban water supply and sanitation. However, close to 40 percent of the urban population is considered to be poor and increased water charges would not only adversely affect their access to safe water and sanitation, but also hinder attainment of the national targets for service provision.

The government's plans for industrial development should increase employment opportunities and enhance the rural economy. High tariffs on industrial water supplies, the need for more efficient water use by industry, including reuse, recycling and pollution control are important policy issues.

Inadequate use of traditional knowledge about conservation practices has been cited as a cause for environmental degradation (Handawela, 2002). However, access to traditional technology has been found to be unavailable in most irrigation settlements as these are relatively recent developments (Jinapala and Somaratne, 2002). The state has concentrated on better management of irrigated agriculture with little attention directed towards rainfed farming. Even in the irrigated sector, there are administrative demarcations in the management of minor and major irrigation schemes, even though the systems are often linked and are therefore difficult to manage in isolation from one another.

Despite the problems identified, the basins do have great opportunities to establish sustainable water management systems. These opportunities are based on three key characteristics of the basins and the local populations:

- high literacy, which allows for effective communication;
- potential to expand the role of existing farmer institutions to contribute to integrated management of water and other resources; and
- minimal industrial reuse of water, allowing for development potential in that sector. Similarly, there is a possibility to improve reuse of return flows in agriculture.

Conclusions

The Ruhuna basins system is threatened on various fronts, including increasing competition for a scarce resource, agricultural pollution and plans for major industrial development in the future, which is set to raise the stakes even higher. One of the main challenges concerns the lack of IWRM. In effect, with a plethora of government agencies and an even larger number of legislative documents all dealing with water, there is an urgent need to bring these together in a more comprehensive manner. The establishment of river basin organizations has been proposed in order to counter this lack of coordination in both administration and monitoring facilities.

Many efforts have been made in the basins to ensure sustainable use of water resources, and the planned industrial developments are set to provide a great economic boost to the region, for example by creating employment opportunities.

A great challenge lies ahead in preserving the environment while building a better life for the basin's population, a good part of whom continues to suffer from malnutrition and poverty. Improving the access to relevant data, implementing more integrated management approaches and continuing the fight against poverty could go a long way to providing a more sustainable future for the basins.

Box 18.1: Development of indicators

Indicators are the cornerstones to effective management of water resources in the basins. A set of primary indicators has been identified as effective in describing the basins with only a few key numbers. The following indicators are the result of several stakeholder meetings, discussions amongst experts and analysis of available data.

Challenge areas	Sri Lanka indicators
SURFACE WATER INDICATORS	• Annual Water Resources (AWR) = 2,460 million m^3; annual rainfall = 9,010 million m^3 • Precipitation: 1,524 mm (basin average); evapotranspiration: 1,700 mm
WATER QUALITY	• Water quality measured at specific locations; measurements are not continuous • Poor quality experienced in estuaries and coastal lagoons
GROUNDWATER	• A comprehensive groundwater assessment is being carried out • Groundwater recharge = 7 to 10% in hard rock terrain • Groundwater reliance = annual groundwater withdrawals/total annual withdrawals = 3%
PROMOTING HEALTH	• Access to safe water = 60% (the available data do not indicate whether there is sufficient water) • Access to adequate sanitation = 71% • Hours of water supply/day as an indicator has been suggested, but data are not available yet
PROTECTING ECOSYSTEMS	• Total protected land area = 1,200 km^2 = 21% of basins • Forest cover = 1,418 km^2 = 25% • 1 Ramsar site
WATER AND CITIES	• No major cities at present; however, water for cities is a major issue for the future • With the planned industrial development, water needs for domestic and industry expected to increase to 100–150 million m^3
SECURING THE FOOD SUPPLY	• Average paddy yields: major irrigation (> 80 ha) = 4.6 T/ha • Minor irrigation (< 80 ha) = 3.6 T/ha • Rainfed = 2.9 T/ha • Water productivity (selected scheme) = 0.14 kg/m^3 (for paddy)
WATER AND INDUSTRY	• Annual water use (estimated)= 4 million m^3; water reuse is not very common; effluent discharge to open drains observed
WATER AND ENERGY	• Total installed capacity is about 126 MW • About 1.3 million m^3 is used for generating 1 GWh • Basin produces about 245 GWh annually
MANAGING RISK	• Dry season drought probability = 32% • Wet Season drought probability = 28.4% • Risk of malaria and other water-related diseases exist
SHARING WATER	• Agriculture is the major user; estimated use is about 95–97% • Industry use is about 1%. • Domestic water use is about 1–2% • In drought situations domestic water is the priority • Formal policy is being prepared
VALUING WATER	• Economic, social environmental and cultural values are considered in planning water resources • Pipe borne water supply is charged (more information is being collected) • Irrigation water is not directly charged, but part of the management cost is recovered through participatory management
ENSURING KNOWLEDGE	• Hydrometric network consists of 19 rainfall stations, 25 agro-meteorological stations and 6 water level stations • The existing coverage is considered inadequate for comprehensive water resources assessment • A considerable amount of data is collected, but not readily accessible
GOVERNING WATER	• Participatory management in irrigated agriculture became national policy in 1988 • A large number of agencies are involved in water management and related activities • Coordination effected at various levels • Participation by women considered inadequate • Farmer Companies concept is being promoted

References

Atapattu, N.-K. 2002. 'Economic Valuing of Water.' Proceedings of the workshop on the WWAP Sri Lanka Case Study, 7–8 April, 2002, Sri Lanka. Colombo, International Water Management Institute.

Atukorala, K. 2002. 'Gender Gaps, Governance Gaps – a View of Sri Lankan Water Management.' Proceedings of the workshop on the WWAP Sri Lanka Case Study, 7–8 April, 2002, Sri Lanka. Colombo, International Water Management Institute.

CEA (Central Environmental Authority of Sri Lanka)/Arcadis Euroconsult. 1999. *Wetland Atlas of Sri Lanka.* Colombo.

Central Bank of Sri Lanka. 2001. *Annual Report of the Monetary Board to the Hon. Minister of Finance for the Year 2000.* Colombo, Central Bank of Sri Lanka.

Department of Census and Statistics. 2000. *Household Income and Expenditure Survey 1995/96.* Colombo.

Handawela, J. 2002. 'Use of Water for Protecting Ecosystems.' Proceedings of the workshop on the WWAP Sri Lanka Case Study, 7–8 April, 2002, Sri Lanka. Colombo, International Water Management Institute.

Jayatillake, H.M. 2002a. 'Managing Risks.' Proceedings of the workshop on the WWAP Sri Lanka Case Study, 7–8 April, 2002, Sri Lanka. Colombo, International Water Management Institute.

——. 2002b. 'Surface Water Resources of Ruhuna Basins'. Proceedings of the workshop on the WWAP Sri Lanka Case Study, 7–8 April, 2002, Sri Lanka. Colombo, International Water Management Institute.

Jayawardane, D.-S. 2002. 'Groundwater Indicators.' Proceedings of the workshop on the WWAP Sri Lanka Case Study, 7–8 April, 2002, Sri Lanka. Colombo, International Water Management Institute.

Jinapala, K. and Somaratne, P.-G. 2002. 'Relevance of Cultural Knowledge and Practices for Efficient Water Management in Today's Context'. Proceedings of the workshop on the WWAP Sri Lanka Case Study, 7–8 April, 2002, Sri Lanka. Colombo, International Water Management Institute.

Ministry of Plan Implementation. 2001. *Annual Performance Report 1999 and 2000.* Battaramulla, Ministry of Plan Implementation, Regional Development Division.

Panabokke, C.-R.; Kodituwakku, K.-A.; Karunaratne, G.-R; Pathirana, S.-R. 2002. 'Groundwater Resources in Ruhunu River Basins of Sri Lanka'. An unpublished report prepared for the World Water Assessment Programme.

Ratnayake, R. 2002. 'Governing Water.' Proceedings of the workshop on the WWAP Sri Lanka Case Study, 7–8 April, 2002, Sri Lanka. Colombo, International Water Management Institute.

Renwick, M.E. 2001. *Valuing Water in Irrigated Agriculture and Reservoir Fisheries: A Multiple Use Irrigation System in Sri Lanka.* Research Report 51. Colombo, International Water Management Institute.

Senaratne, S. 2002. 'Water for Industry'. Proceedings of the workshop on the WWAP Sri Lanka Case Study, 7–8 April, 2002, Sri Lanka. Colombo, International Water Management Institute.

Shanmugarajah, C.K. 2002. 'Health Water and Sanitation'. Proceedings of the workshop on the WWAP Sri Lanka Case Study, 7–8 April, 2002, Sri Lanka. Colombo, International Water Management Institute.

Somitilake, H.S. 2002. 'Water for Energy'. Proceedings of the workshop on the WWAP Sri Lanka Case Study, 7–8 April, 2002, Sri Lanka. Colombo, International Water Management Institute.

Weerasinghe, K.D., Jayasinghe, A.; Abeysinghe, A.M. 2002. 'Food Security of the Ruhuna Basins'. Proceedings of the workshop on the WWAP Sri Lanka Case Study, 7–8 April, 2002, Sri Lanka. Colombo, International Water Management Institute.

Wickramage, M. 2002. 'Meeting Basic Needs: Domestic Water Supply'. Proceedings of the workshop on the WWAP Sri Lanka Case Study, 7–8 April, 2002, Sri Lanka. Colombo, International Water Management Institute.

'Improving water quality is still the major concern of the Seine-Normandy basin, despite real progress made over the last thirty-five years. Storm runoff during periods of heavy rainfall continues to create problems, causing wastewater to be discharged directly into rivers or overloading wastewater treatment plants, thereby decreasing their efficiency. Non-point source pollution from farming and urban areas is still a major problem as nitrate, pesticide and heavy metal concentrations continue to increase.'

19 Seine-Normandy Basin, France

By: The Seine-Normandy Water Agency (AESN, Agence de l'eau Seine-Normandie)

Table of contents

Seine-Normandy Basin, France

Under the Pont Mirabeau the Seine
Flows with our loves
Must I recall again?
Joy always used to follow after pain.

Guillaume Apollinaire

IT WAS ONLY FORTY YEARS AGO that the Seine River was declared 'dead'. Levels of pollution from industry and agriculture were dangerously high. Native fish had disappeared, plant life was dying, and the water was unsafe for swimming. Today, however, the river and its surroundings have been rehabilitated. The city of Paris even organizes fishing contests on summer afternoons. This dramatic change began with the recognition in 1964 of six river basins as the natural hydrographic units in France and the creation of six water agencies to manage them accordingly. Problems remain, especially nitrate pollution from fertilizers and the continuing disappearance of wetlands, but the case study presented here shows that the application of modern technology, a sound tax base and political will on several levels can go a long way towards reversing some of the neglect of the past.

THE SEINE-NORMANDY BASIN district in the north-west of France covers an area of about 97,000 square kilometres (km^2), nearly 18 percent of the country's total surface area. It is made up of the drainage basins of the Seine River and its tributaries, the Oise, Marne and Yonne, and those of Normandy's coastal rivers.

Map 19.1: Locator map

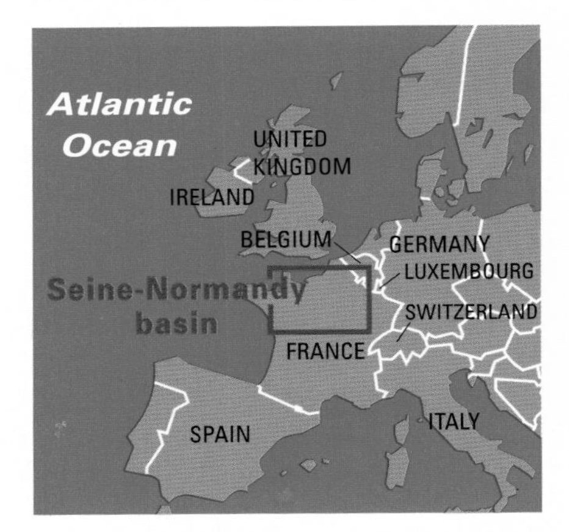

Source: Prepared for the World Water Assessment Programme (WWAP) by AFDEC, 2002.

General Context

The land is relatively flat with altitudes generally under 500 metres. The climate is oceanic and temperate with an average annual rainfall of 750 millimetres (mm) and an average annual potential evapotranspiration of 500 mm. Annual rainfall varies between 300 and 1,600 mm, depending on the area. In Paris, it varies from 400 to more than 800 mm from year to year. The average monthly temperature in Paris is between 2.5°C (in January) and 24.6°C (in August). Periods of freezing temperatures are short along the coast in the west, but lengthen as we move towards the eastern edge of the basin.

Geology

The Seine-Normandy basin includes a large portion of the sedimentary Paris basin. The geological structure of the Paris basin resembles a stack of 'saucers' with the most recent layers (Tertiary) outcropping in the centre and the oldest layers (Mesozoic) outcropping on the outer edges of the basin. These layers overlie the Hercynian bedrock (Palaeozoic) that outcrops in the western part of the basin. This type of geological structure contains numerous aquifers of extremely varying size and structure (alluvial, sedimentary and fractured aquifers). In the Paris basin, about ten of these aquifer formations are very important in terms of usage.

Leached brown soil covers the western part of the basin. There is a thinner layer of acid brown soil, eutrophic brown soil and calcareous soil on the eastern edge of the Paris basin and in Lower Normandy. The rendzina soil found in many places at the base of hills is used for vine growing in the Champagne region.

Population density

The Seine-Normandy basin has an estimated population of 17.5 million people, 80 percent of whom live in urban areas, most of which are located along the basin's rivers and in the Paris region (located approximately in the centre of the basin). Population density in the basin varies greatly. The Paris suburbs are now spreading into Upper Normandy. The Ile-de-France region around Paris is the most popular tourist destination in France and has, for example, 35 million foreign-tourist-nights per year. The populations of some of the departments along the Normandy coast are also subject to very high seasonal variations.

Table 19.1: Hydrological characteristics of the Seine-Normandy basin

Surface area of the basin	97,000	km^2
Annual precipitation	750	mm/year
Annual potential evapotranspiration	500	mm/year
Average discharge in the coastal rivers of Normandy	125.8	m^3/s
Average discharge in the Seine River	460	m^3/s

Economics

Economic activity in the Seine-Normandy basin is dynamic. Industrial production in the basin alone accounts for 40 percent of domestic production, and includes 60 percent of France's automobile industry and 37 percent of its oil refining industry. There are agro-food industries located throughout the basin, whereas heavy industry (chemical, oil, paper, metallurgy) is concentrated in the Lower Seine Valley and the Oise Valley. There is a wide range of economic activity, in terms of both size and diversity, in the Paris region alone. The trades, service and commercial sectors, an integral part of the urban fabric, also flourish due to the high population density.

The basin also has a prosperous farming industry, with extensive farming on vast plains and the renowned wine-producing regions of Champagne and Burgundy. Sixty percent of the surface area of the basin is used for agriculture, and 80 percent of France's sugar, 75 percent of its oil and protein seed crops and 27 percent of its bread grain comes from this region. Since 1970, farming practices in the basin have followed the global trend towards large industrial crops with high added value (sugar beets, rapeseed, potatoes) with a concentration of cereal crops in the south-west and livestock production on the edges of the basin.

Map 19.2: Basin map

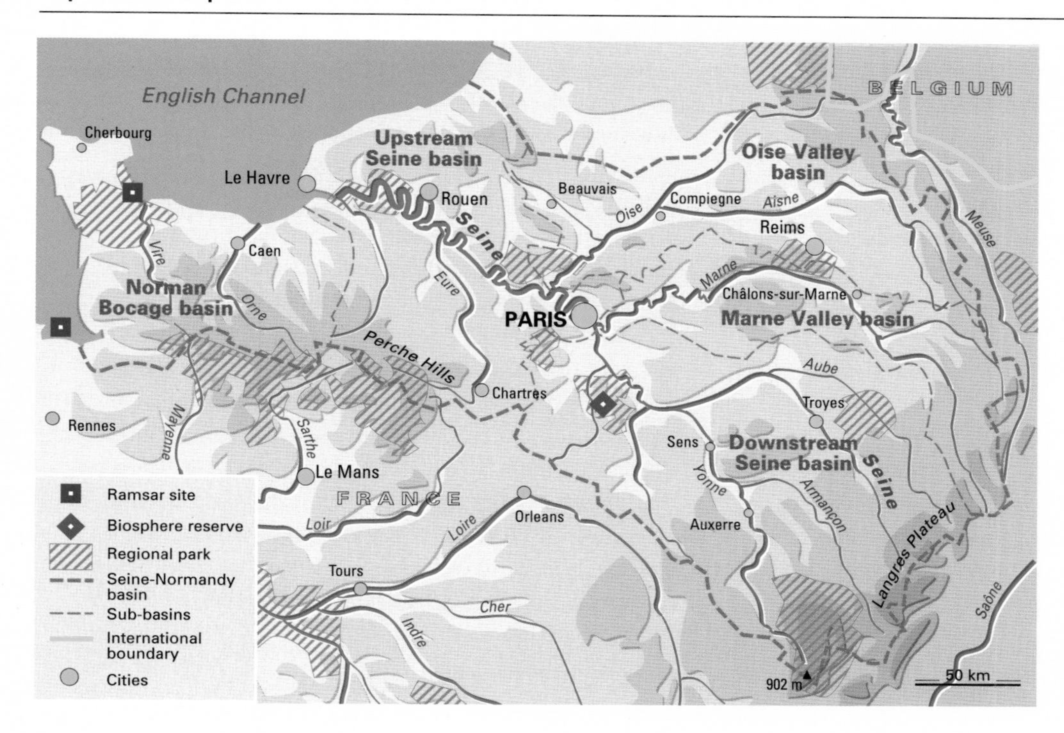

Source: Prepared for the World Water Assessment Programme (WWAP) by AFDEC, 2002.

Water Resources

Increasing anthropogenic pressure on hydromorphology

The Seine-Normandy basin has 55,000 km of water courses. The Seine, fed by the Oise, Marne and Yonne Rivers, is the basin's central artery. The rivers have gentle slopes due to the flat terrain. During flood periods, river water overflows into floodplains that are, in places, more than 10 km wide.

Flooding is indeed a major concern in the basin. Runoff has increased as more of the basin has been rendered impermeable (1,600 km^2 of impermeable ground out of a total surface area of 100,000 km^2, though these areas are concentrated).[1] Flow is often disrupted by the overdeepening of riverbeds, dredging and gravel pits. Dams in the Seine-Normandy basin, partly meant for levelling off of peak flow, often have a minimal effect on floods due to their distance from the large urban areas and their limited capacity compared to the volumes of the exceptional floods.[2] However, these large dams do regulate low flow and without them the rivers upstream from Paris would be dry during the summer due to the large amount of water withdrawn by the Paris region.

Human development also harms the biology of the rivers: migratory fish cannot get past 60 percent of the hydroelectric power plants and less than 20 percent of the dams are equipped with fish-passes. Modifications to the basin's major rivers, particularly for navigation purposes (1,427 km of navigable waterways, 550 of which have large or medium clearance), are the principal cause of decline in the population of migratory fish species.

The quantitative effects of other anthropogenic pressures on the basin are not major issues. Out of about forty aquifer formations, only three have temporarily dropped below their water stress thresholds. In rivers, withdrawal mainly affects quality. The water in some rivers is now simply outflow from wastewater treatment plants.

Water quality: a mixed balance sheet

Despite greater human activity, which produces oxidizable waste, the dissolved oxygen concentration in the basin's rivers has increased over the last thirty years, after becoming seriously depleted in the 1960s.[3] The nitrate situation is, however, worsening. Since 1965, the nitrate concentration in the lower Seine has

1. Additional runoff due to the waterproofing of surfaces is estimated to be 760 million cubic metres per year, based on a runoff coefficient of 100 percent on impermeable surfaces and 20 percent on permeable surfaces, and an average annual rainfall of 600 mm.
2. While the large dams on the Seine, upstream from Paris, can hold 800 million m^3 of water, almost 4 billion m^3 flowed through Paris during the 1910 flood.
3. The average oxygen content at the Poses measuring station (lower Seine), has increased, on average, by 0.9 percent/year over the last twenty-five years, which therefore reflects the efforts made throughout the Seine basin.

increased significantly, even if the rate of progression has slowed since 1989. The same concentrations are also occurring in groundwater.[4] Today, some 25 percent of the basin's groundwater measurement points show more than 40 milligrams (mg) of nitrate per litre; 12 percent show more than 50 mg/litre. But these nitrate rates still are under the groundwater threshold for producing drinking water (which is 100 mg/litre, while it is 50 mg/litre for surface water). Nitrate is also the third biggest cause of coastal and seawater pollution. When combined with phosphate, it can cause eutrophication and the proliferation of toxic microalgae. Phosphate input from continental sources, which is the main cause of eutrophication in freshwater, has decreased considerably but is still too high.[5]

Metal concentrations are also decreasing.[6] While naturally present in small quantities in the aquatic environment, metals also come from insufficiently treated wastewater and from surface runoff in urban areas. While the accidental discharge of highly toxic contaminants that kill fish is now rare, high concentrations are still measured near contaminated sites.

On the other hand, polychlorinated biphenyl (PCB) concentrations are still alarming, even though they are decreasing.[7] Although PCBs have not been manufactured since 1987, they are still produced by the incineration of domestic waste and the manufacture of paints and lubricants. Along with metals and hydrocarbons, PCBs are the second biggest cause of coastal and seawater contamination (after microbiological pollution). Groundwater is, however, little affected by organic micropollutants other than pesticides.

Pesticides, used not only in agriculture but also along railways and roads and in gardens, are a serious problem in the Seine-Normandy basin. Triazines – highly soluble, mobile and persistent organo-nitrogenous compounds – are the most prominent. They are present in surface water (with peaks in spring), coastal waters and, above all, in groundwater.[8]

In general, organic micropollutants are a major water management problem because the concentrations that must be measured are extremely low and new synthetic molecules, which must also be detected, are constantly appearing. Map 19.3 shows the basin's physico-chemical fitness between 1997 and 1999.

4. In 2000, out of 407 wells, 37 percent had nitrate contents between 20 and 40 milligrams per litre and 14 percent had very poor quality water (>50 mg/litre).
5. The flux entering the Seine estuary decreased from 60 to 39 t/day between 1974 and 1999.
6. Since 1976, there has been a decrease at Poses (lower Seine) in the dissolved metal loads of cadmium (a tenfold decrease over the last thirty years), cobalt, chromium, mercury, nickel, lead and zinc. Those of other metals, such as copper, titanium, vanadium, iron and manganese are also decreasing, but less markedly.
7. Their concentration has been decreasing since 1978, but is still five to six times greater than the national average.
8. Half of the 371 wells in the monitoring network were contaminated in 1999. Forty percent had triazine concentrations of over 0.1 micrograms per litre.

Biodiversity on the upturn

Out of a total of thirty-three fish species that have been identified as belonging to the local ecosystem, twenty-six are still commonly found today, a considerable improvement over the 1960s when the diversity and number of fish had declined due to heavy water pollution (Belliard, 2001). The Seine-Normandy basin's Hydrobiological and Piscicultural Network for monitoring fish populations now has 143 stations. Three or four times a year, electrical fishing techniques are used to study fish populations. The live fish are then returned to the river. While conditions near the edges of the basin are generally favourable to fish life, this is not the case in its centre (in the Seine River, in particular). In small rivers, non-point source pollution and the silting up of riverbeds are the main causes of the decrease in fish populations. In large rivers, the causes are mainly physical barriers and discharge from urban areas. Along with the negative impact of anthropogenic pressure (seven species are no longer present), about twenty new species have been introduced by humans,[9] who have also brought in invasive plant species, such as Japanese knotweed.[10]

Readily available water data

The Seine-Normandy Water Agency (AESN, Agence de l'Eau Seine-Normandie) and French government services, together with other public institutions, manage several measurement networks that gather quantitative and qualitative data on surface water and groundwater. For example, common surface water quality parameters were measured at 441 points in the basin in 2000. Of these, 171 were also analysed for metals and 120 for micropollutants. These points are sampled six to forty-eight times a year, for determination of more than 150 parameters (for a total of nearly 2 million data items/year), which vary in time and space with field conditions. The groundwater quality network uses 402 wells and piezometers to monitor the basin's ten major aquifers. Samples are taken twelve to forty-eight times a year for determination of more than 250 parameters (nearly 3 million data items/year). Networks have also been set up to monitor coastal waters. In 2000, water samples from 130 sites were analysed for swimming quality, twenty-two sites for shellfish and eleven sites for marine sediments. In estuaries, coastal rivers, swimming areas and at discharge points, the principal tests carried out are bacteriological analyses (*Escherichia coli*, Enterococci), backed up by chemical analyses

9. Examples of species that have been introduced are ruff and hotu (*Chondrostoma nasus*), and those that have disappeared include sturgeon, sea trout (or squeteague), salmon and sea lamprey.
10. The *phytophora fungus*, part of whose life cycle is in rivers, severely damages alders, which are of major ecological and silvicultural importance. The Dreissene is an invasive mollusc in rivers. *Cladophora* and *Vaucheria* are green algae that thrive in eutrophic environments and have been introduced by humans from aquariums. *Cyanophyta* are also disturbing in freshwater environments for health reasons.

Map 19.3: Water quality in the Seine-Normandy basin

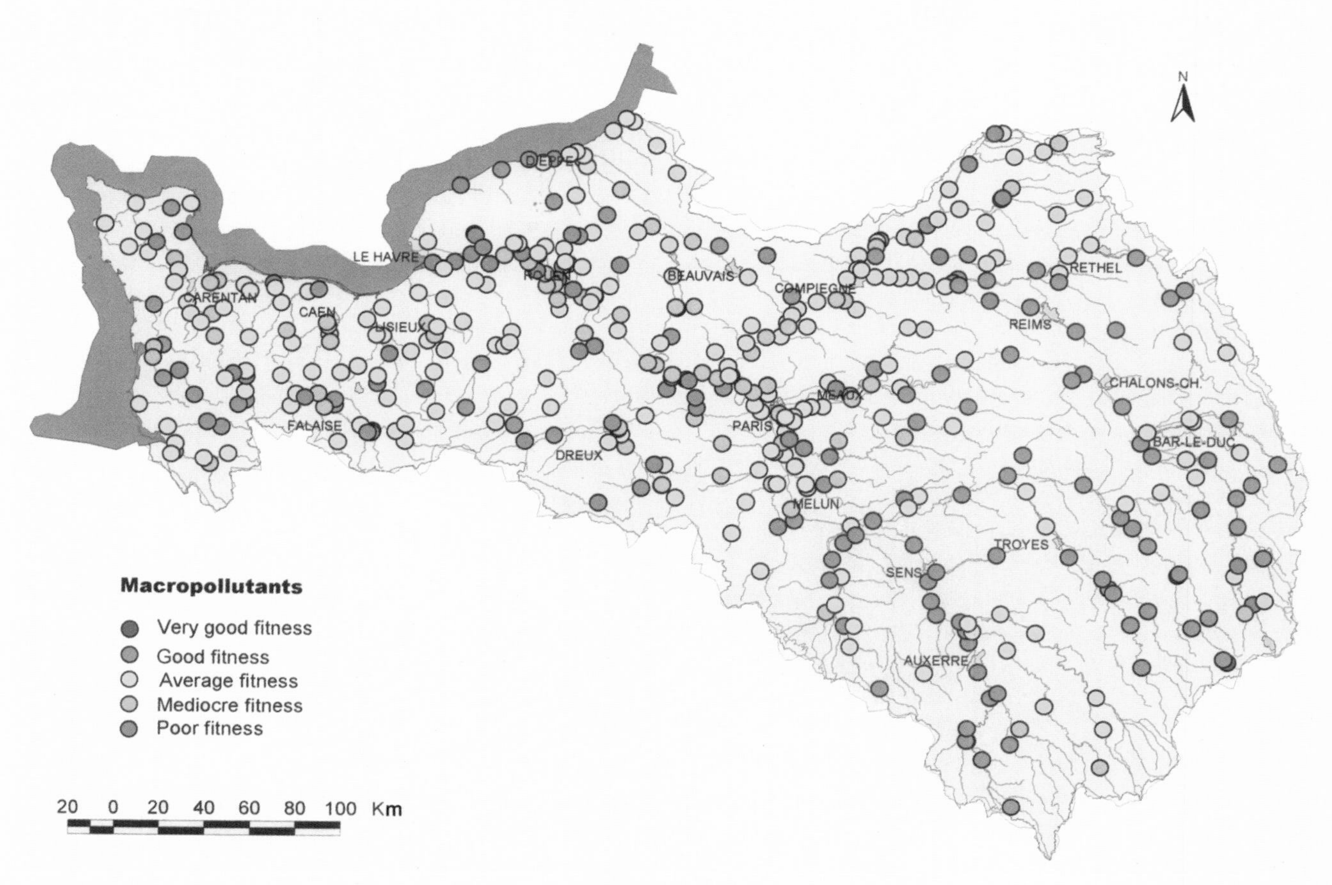

Quantitative and qualitative data on surface water and groundwater are processed using a Quality Evaluation System (SEQ-Eau) based on indicator sets and use requirements.

Sources: AESN, 2001b. IGN-BD Cartho 94.

(suspended particulate matter, oxidizability, nitrate, chloride). Shellfish are analysed for bacteria (total coliform count, Streptococci, Salmonella), metals and radioactivity. Marine sediments are analysed for radioactivity.

The resulting data are processed using a Quality Evaluation System (SEQ-Eau) based on both the notion of indicators (groups of similar parameters, such as 'metal' or 'nitrate') and the requirements of the various uses (drinking water supply, irrigation, swimming). This system is very flexible and enables water quality to be evaluated according to the most relevant criteria for a given use.

Quantitative data are measured at 174 rain gauges, 214 piezometers and 418 hydrometric stations. These enable monitoring of flood risk and the possible effect of flooding on water quality. Water level records are stored in a database and are available to the public via web site.[11] Most of this data is also available to the public on the AESN web site,[12] which uses dedicated software to produce summaries by measuring point and parameter, upon request.

In France, the public is very aware of the seriousness of environmental problems, especially where human health may be affected. People living in the Seine-Normandy basin are much more concerned about water quality than they are about water shortages. They know that pesticides and chemical fertilizers are a major problem and understand that water has to be 'cleaned up' before it is discharged. They therefore consider it normal to pay for this service although they object when they think the price is too high for the service provided or that the costs are not being equitably shared among stakeholders. Interestingly, water consumption has dropped recently, but studies show that price is not directly responsible for this trend.

11. http://infoterre.brgm.fr/
12. www.eau-seine-normandie.fr

Challenges to Life and Well-Being

Stringent health control

The quality of drinking water is much better now than it was thirty-five years ago. Standards are higher and treatment techniques are much more efficient. Drinking water must satisfy criteria based on Maximum Permissible Doses (MPD). The European Water Framework Directive (WFD) requires that forty-eight parameters be taken into account, including microbiology, toxic substances and 'undesirables'. The drinking water of more than half of the basin's population is supplied by groundwater. With groundwater, biological standards can be met simply by protecting wells and slightly disinfecting the pumped water (except when the water is turbid, which can occur during periods of heavy rainfall in karst regions, thus depriving about 500,000 people of water for several days each year). In and around Paris, where drinking water comes mainly from rivers, the treatment required depends on the quality of the raw water. Surface water is ranked in three categories according to quality, each category requiring increasingly stringent treatment. The most polluted water may not be used as a drinking water resource. However, there are few instances of below-grade classification in the Seine-Normandy basin. On the other hand, the MPDs for atrazine (a chemical weed killer) pose a problem, especially since the limits set by the European Union (EU) are even more stringent than those of the World Health Organization (WHO) at 0.5 micrograms (μg) per litre for total organic micropollutants and 0.1 μg/litre for each individual substance. For nitrates, the EU recommends 25 mg/litre, but the current standard for both Europe and WHO is 50 mg/litre. Trends in the basin indicate that the MPDs for nitrate will also soon pose a problem. High lead content in drinking water is frequently a problem in old houses, due mainly to the state of privately owned pipes.

Bathing in rivers is still often restricted due to poor bacteriological quality. The basin's coastline is the site of flourishing tourist activity and a dynamic shellfish farming industry specialized in mussels and oysters, both of which require very high-quality seawater. Microbiological pollution from sewage systems, surface runoff and coastal rivers is the main threat, and has very harmful effects on economic activities. The situation has improved considerably since 1990, but isolated incidents still occur during periods of heavy rainfall. Areas used for shellfish culture are ranked in four quality categories, each of which requires increasingly stringent treatment in order to ensure that the commercialized products meet standards.

Drinking water supply and wastewater treatment

In 1999, 1,564 million cubic metres (Mm^3) of water were pumped in the Seine-Normandy basin for the drinking water supply. This corresponds to a distributed volume of about 1,240 Mm^3, given network losses (estimated to be about 20 percent), providing an average daily consumption of about 190 litres per inhabitant per day or 70 m^3/year.

Concerning wastewater treatment, reduced nitrogen discharge is considered to be the determining factor for river water quality. This depends on the capacity of the receiving environment and the efficiency of wastewater treatment plants.[13] Each city in the basin with a population of more than 10,000 inhabitants has a treatment plant. In 1999, 88 percent of the basin's population was hooked up to municipal wastewater collection networks (the rest of the population lives in isolated dwellings with private sewage disposal systems), and the basin's wastewater treatment plants had a total capacity equivalent to waste from 20.7 million inhabitants. They are usually very effective in removing suspended particulate matter (85 percent) and oxidizable matter (78 percent), but not as good for nitrogenous matter and compounds containing phosphorous (Seine-Normandy Basin Committee, 2000. While these results are adequate, they must be interpreted with caution because they do not include the 60 Mm^3 that flow directly into rivers each year during periods of heavy rainfall when storm runoff exceeds the capacity of the drainage network and/or the treatment plant. Figure 19.1 shows the improvement in water quality following the commissioning of the Saint Dizier treatment plant.

During storms, urban runoff is sometimes passed through wastewater treatment plants, depending on the storage capacity and treatment available, and on the storm's intensity. At present there is inadequate provision for treating contaminated and high-risk runoff. However, the city of Paris is planning to build new stormwater storage with a capacity of 1,6 Mm^3. Hazardous wastes, both solid and liquid, are deposited in seventy-seven private landfills within the basin. Five of these are used for dangerous materials; their impact on water is low and they are well monitored. The remaining seventy-two landfills are for normal waste. Taxes are paid on each ton of waste brought to the landfill.

Domestic users are also a source of non-point source pollution. Contaminants are carried off by surface runoff, which is rarely treated and flows directly into rivers. Waste from public areas and animal droppings in towns and cities are also major sources of contamination.

The disposal of treatment plant sludge (190,000 tons of dry matter/year) is also problematic. Most of it is recycled by farmers, which in turn poses the problem of heavy metals spreading on agricultural land.

13. The effects of an urban area on its drainage river can be expressed by an indicator calculated by relating the discharge capacity of the treatment plant, A, to the five-year minimum flow of the receiving river. Thus, the ratio A/Q_{mna5} enables us to calculate the maximum ammonium concentration in the river at the point of discharge.

Figure 19.1: Improvement of water quality of the Marne River

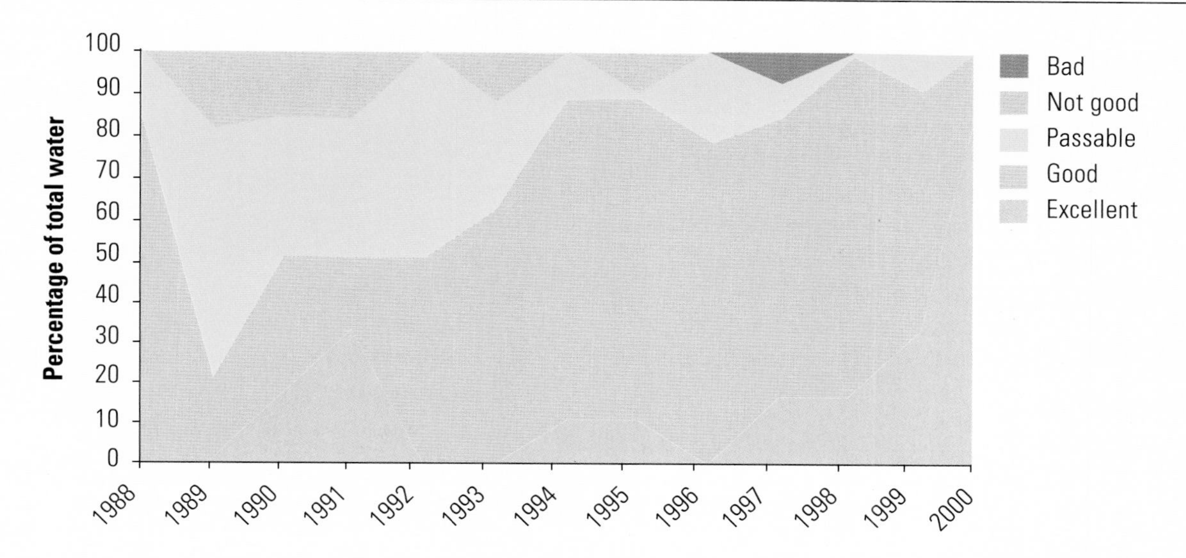

Thanks to the implementation of a wastewater treatment plant in Saint Dizier in 1995, the water quality of the Marne River has considerably improved: more than 80 percent of the water was considered of excellent quality in 2000, compared to 10 percent in 1995.

Source: AESN, 2002.

Agriculture

In the Seine-Normandy basin, irrigation is used to increase crop yield, to improve the quality of the produce, to regulate production and to grow crops that are very sensitive to water shortages (for example, potatoes for the industrial production of chips). At present, 394,000 hectares (7 percent of the usable farm area) can be irrigated: this has nearly doubled since 1988. The quality of the water withdrawn, 90 percent of which is groundwater, is generally very good. In spite of this sharp increase, irrigation still has little quantitative impact on the resource, except for occasional cases of overpumping that have been resolved by regulating demand.[14] Irrigation does, however, have an indirect impact on quality because it favours intensive farming techniques and spring crops, which leave the soil bare for long periods of the year and increase the chemical load in the rivers by leaching and draining.

Non-degradable substances from, or excessive use of, fertilizers, pesticides, liquid manure and other substances spread on crops or coming from livestock activities end up in rivers and groundwater. This has a harmful effect on both the environment and other water uses. Pollution is increasing as meadows are ploughed up (the total surface area of grassland decreased 22 percent between 1988 and 2000) and off-soil production becomes increasingly widespread, leading to problems of effluent management. Soil erosion, which causes water turbidity, is also closely linked to farming practices.

Industry

Industry in the Seine-Normandy basin consumed 1,612 Mm^3 of water in 1999, most of which was pumped directly from rivers, and most of which was used by power plants. The chemical and oil-refining industries also use large amounts of surface water. One third of the water withdrawn from the resource during low water periods comes from rivers. The volume of water withdrawn for industrial purposes, with the exception of power plants, has decreased by about 3 percent a year over the last ten years. The chemical and agro-food industries prefer to use groundwater, and often treat it before use.

Despite the high production of oxidizable matter (20 to 30 tons/day produced by the largest industrial units), treatment reduces average unit fluxes to 300 kilograms (kg) per day. Average purification rates are lower for nitrogenous matter, of which industry produces several hundred kg/day. Some industries are hooked up to municipal wastewater sewage systems, thus increasing the burden on wastewater treatment plants and the heavy metal load in the sludge, thereby limiting its use in agriculture.

Despite the large amounts of hazardous waste generated by industry (ten times more than domestic waste in terms of volume),[15]

14. For example, the water level in the Beauce aquifer dropped sharply in 1992 and 1997, causing water use conflicts.

15. The list of polluted industrial sites is available on the Internet at http://basol.environnement.gouv.fr.

industrial pressure on aquatic environments has been considerably reduced because the basin is very well equipped with waste treatment and disposal facilities.[16] Reducing pollution at the source by using clean processes and recycling polluted materials further reduces the industrial pressure on the resource. Remedial measures include improving the yield of wastewater treatment plants using denitrification processes. AESN is promoting this effort.

Another major issue concerns discharge from wine-producing activities and from the numerous small services and trades businesses that are an integral part of the urban fabric.

About thirty accidental pollution incidents from industrial sources occur each year. In more than half the cases reported, fish died and the contamination spread over more than 3 km.

Aquatic environments for biodiversity and tourism

In the Seine-Normandy basin, two major challenges must be met – protecting wetlands and combating eutrophication – if water is to act as a reservoir of biodiversity and an attractive and healthy environment for outdoor recreation.

The basin's wetlands (about 580,970 hectares) are capable of decreasing the levels of nitrogen and phosphorous in wastewater by 60 to 95 percent. They also help to reduce severe flooding by absorbing groundwater and providing room for rivers to expand. They are also of strategic interest for many water birds: 74 percent of the water birds that nest on a regular basis in France and 81 percent of the overwintering species find shelter in the basin. Six of the ten major migration routes that cross France pass over the basin. Unfortunately, twelve of the seventy-eight nesting species and fifteen of the ninety-four overwintering species are now in decline due to the deterioration of the environment. Indeed, half of the wetlands have disappeared over the last thirty years due to anthropogenic pressures, in particular, drainage for agriculture (1,400 hectares in 1999) and major navigation works, hydroelectricity schemes, and roads/railway lines.

Water is also a major tourist attraction and both rivers and beaches are threatened by eutrophication. In the summer, some beaches are invaded by 'green tides' that can result from a high phosphate and nitrate content in the water.

16. Twelve hazardous waste wastewater treatment plants that can generally destroy or trap more than 95 percent of the pollutants, five centres for burial of ultimate waste using efficient confinement techniques, and seventy-two centres for burial of common industrial waste.

Management Challenges: Stewardship and Governance

The 1964 and 1992 Water Laws and the European Union Water Framework Directive (WFD)

The first French Water Law laid the foundations for the French water management system. This law resulted from the growing need to coordinate the numerous local water uses and responsibilities when, in the early 1960s, the country was faced with both an increase in pollution as a result of urban and industrial growth and a sharp increase in the demand for water. The 1964 Water Law created the novel concept of Water Agencies,[17] each with its own 'water parliament', or Basin Committee. The decentralization of water management was reinforced by the second Water Law, which, in 1992, increased the role of the Water Agencies and created a Master Plan for Water Management (SDAGE, Schéma directeur d'aménagement et de gestion des eaux), guidelines for balanced water management on a river basin scale, to be drawn up by a Basin Committee.[18] The SDAGE reports on the state of the basin (see indicators in box 19.1 at the end of the chapter), with the approval of the various stakeholders, and determines long-term strategic objectives (ten to fifteen years). In 2000, the EU issued its WFD, which outlines the principles for Integrated Water Resources Management (IWRM) at the river basin level and requires that member states achieve 'good status' for all of its water bodies by 2015, using any methods they should choose. From an institutional point of view, the WFD follows the French system. The new French Water Law, which will come into force in 2003, transposes the WFD into French law.

Delineated water management roles

Distinctive aspects of the French water management system are the high degree of local responsibility, public-private sector partnerships, coordination on a river basin level and a taking into account of all water uses.

Municipalities (from small villages to big cities) have been responsible for all services associated with water since the nineteenth century. Today, they are not only responsible for initiating water works and operating facilities, but are also legally responsible for the quality of services and rates charged to users in their community. Towns, therefore, often set up intercommunity

17. The water agencies are public bodies administered jointly by the Ministries of the Environment and Finance. There are six in France, one for each of the country's major catchment basins: Seine-Normandy, Artois-Picardy, Loire-Britanny, Rhône-Mediterranean-Corsica, Adour-Garonne and Rhine-Meuse.
18. The Seine-Normandy SDAGE was approved in 1996; thematic data sheets concerning the SDAGE are available on the Internet at www.environnement.gouv.fr/ile-de-france.

associations to operate drinking water supply (an approach that concerns 67 percent of the basin's population) and wastewater treatment (16 percent of the basin's population) networks. They also create joint public-private partnerships by subcontracting water supply and treatment services to private companies under various types of contracts (85 percent of the basin's population for water supply, 36 percent for wastewater treatment). They are still, however, responsible for the system and the private service provider must return the network in proper working order at the end of the contract period.

In addition to water supply and treatment services, water management involves many responsibilities that are sometimes rather vague. An example of this is the management of privately owned rivers. Their maintenance is, theoretically, the responsibility of riparian owners but, in practice, is often undertaken by intercommunity volunteer groups.

All water users must comply with standards set by the water laws, and compliance is monitored by local representatives of state agencies. The state, therefore, remains the 'guardian' of the resource. It is also responsible for maintaining public rivers, a task largely delegated to the French Navigable Waterways Authority (Voies Navigables de France).

Faced with this complex allocation of responsibility, the role of the Water Agency at the river basin level is to promote measures undertaken to ensure a balance between water resources and needs. Its role is mainly financial. It allocates funds – in the form of loans or subsidies – for projects that correspond to the objectives of the Water Agency programme. Thus, by evaluating proposals and monitoring funded projects, it also plays an advisory and consultant role that is widely recognized by its partners. The money that it distributes comes from users in the form of taxes or fees based on the quantities of water consumed and the amount of pollution discharged. It collects all the water taxes from its river basin. The acknowledged neutrality of the Water Agency also enables it to act as a mediator.

Undeniable but limited public participation

The Basin Committee is an advisory and decision-making body made up of three representative groups – elected officials, water users and people appointed by the state. After studying the situation within the basin, the Committee recommends water tax bases and rates, based on the five-year plans drawn up by the Water Agency and its Board of Trustees (which is made up in the same way as the Basin Committee). Water Agency programmes must follow the SDAGE guidelines, which, in accordance with Water Law requirements, are also the result of extensive collaboration. The state is, therefore, but one of many participants involved in the planning stages (see figure 19.2). It participates in discussions concerning policies that are made and financed by those directly involved. It has, however, administrative control over all of the actions carried out. At a local level (water course, groundwater), very decentralized mechanisms enable the allocation of local responsibilities. Cooperation is achieved through interdepartmental agreements, intercommunity associations and, in particular, local participatory initiatives such as local water management plans (SAGEs, Schémas d'aménagement de gestion des eaux), which are drawn up for sub-river basin catchment areas along the lines of the river basin level SDAGE, and 'rural contracts' created by the AESN.

The composition of the Basin Committee and the steering committees of the local participatory initiatives guarantee, in principle, that water management is, to a certain extent, open to the public. In practice, however, this attempt to be open is sometimes ineffective. User participation in debates is often minimal, local input often being limited to financial-level rather than project-level planning. Faced with this situation, which is closely related to the general timidity of civil society in France, it is clear that the Water Agency needs to encourage more public participation, in particular when the WFD goes into effect.

Payment of water services, financial aid and resource management

The water bill paid by domestic and industrial users hooked up to the municipal water supply network covers the cost of drinking water distribution and wastewater collection and treatment. The price of water varies according to its treatment, management, supply conditions, and wastewater discharge. The bill also includes a pollution tax and a resource withdrawal tax levied by the Water Agency. These taxes represent only a small proportion of the total water bill. Their revenues are redistributed by the Water Agency in the form of interest-free loans or subsidies, in accordance with a five-year programme drawn up jointly by representatives of water users within the framework of the basin's SDAGE. This financial aid is meant to incite users to reduce the impact they have on the resource through investments or improved techniques. The amount of financial aid allocated to the various categories of users is roughly equivalent to the taxes they pay. Funds are, however, shifted somewhat between the basin's various categories of users and geographic areas according to a principle of 'basin solidarity'. For example, domestic users in Paris pay, on average, more in pollution and withdrawal taxes than they get back in aid. This is understandable since their water is pumped upstream and they contribute significantly to pollution downstream. The WFD recommends that the actual cost of water be fully paid by users, that an 'actual cost' indicator be measured, and that appropriate rates be charged to improve quality. In the Seine-Normandy basin, and elsewhere in France, consumers are billed for the cost of

Figure 19.2: Water legislation in France

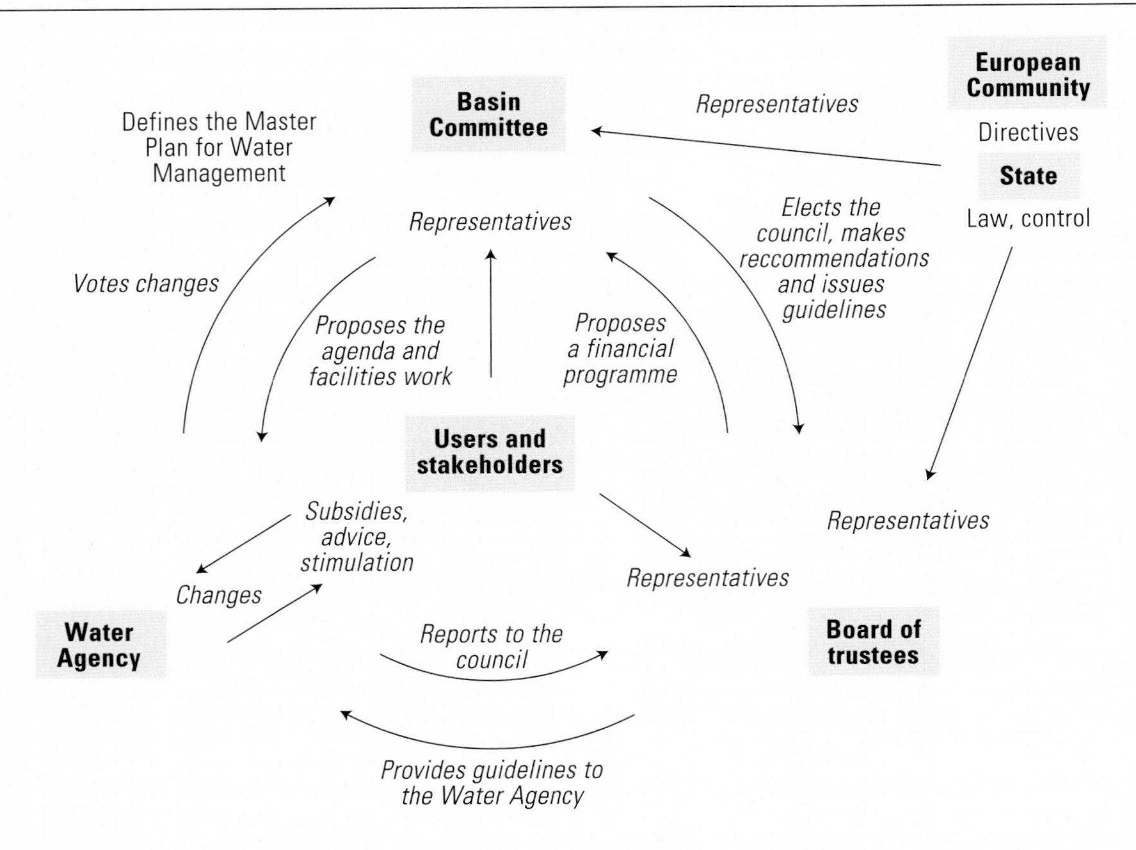

Many stakeholders are involved in the planning stages; the Basin Committee, the advisory and decision-making body, guarantees – to a certain extent – user participation.

distribution and treatment. But users do not pay for environmental damages due, for example, to non-point source pollution, in particular from agriculture. The basin must be more transparent about how it shifts funds between the various categories of users, the state and public institutions, and must take into account costs engendered by the environmental impact of uses.

The calculation of the Water Agency pollution tax is based on the actual quantity of polluted water. For instance, a water treatment allocation has been set up for industries, which is based on water treatment efficiency and the wastewater destination, so that they are taxed only for the actual amount of pollution that they release into the environment and into the local sewage systems. Water management in the Seine-Normandy basin is, therefore, in accordance with the polluter pays principle.

All Water Agency revenues are spent on supporting pollution reduction and clean-up actions. The nuisances caused by one specific water use over another (negative externalities) still have to be measured and accounted for in the rate setting. This WFD recommendation should be met by 2010.

In 1999, the average price of water in the Seine-Normandy basin was 2.74 €/m^3 (about US\$2.8/m^3). The annual bill for domestic users was 126 €/inhabitant (about US\$129/inhabitant), around 20 percent of which were taxes. The average amount spent on water per household was 1.03 percent of total income, and 4 percent of housing costs. In light of these figures, the socio-economic weight of the water sector, which employs 18,700 people in the basin (for water supply and treatment), can be estimated: it represents an annual investment of about 60 €/inhabitant/year (around US\$61/inhabitant/year). Existing supply and treatment facilities were assessed, in 1999, at 9.360 €/household (around US\$9.6/household). The annual expenditure per household in the water sector is therefore 0.5 per thousand of the Gross Domestic Product (GDP).

Achievements of this water management system

Decontamination projects

The Water Agency's first financial aid measures in the 1960s incited many municipalities that had been hesitant until then to launch costly water treatment programmes. During the Water Agency's first five-year programme, the number of wastewater treatment plants in the basin increased threefold. Between 1972 and 1976, financial incentives were created to increase the efficiency of wastewater treatment plants (which has since increased from 40 to 70 percent). Since 1971, SATESE units (technical support for wastewater treatment plants) employ people to monitor the proper operation of wastewater treatment plants. In 1976, inhabitants were made responsible for the pollution they generated and they now pay a unit price for pollution. As a result, wastewater collection and treatment networks have been renovated. At the same time, the Water Agency also began funding private sewage disposal systems. Between 1977 and 1981, Water Agency efforts focused on restoring river water quality by launching numerous joint actions at a local level that were to evolve into river development plans. Diagnostic studies of wastewater collection systems were funded, and the training of SATESE personnel was intensified. Between 1982 and 1986, priority was still given to improving wastewater collection systems. The Water Agency then created 'Reinforced Action Zones' where a higher pollution tax was levied (plus 70 percent) in exchange for an increase in financial aid. Between 1987 and 1991, the fifth Water Agency programme focused on 'black spots', highly polluted areas requiring remediation, and made wide use of pluri-annual contractual agreements to incite urban areas to undertake long-term waste treatment works. Today, the Seine-Normandy basin has 2,100 wastewater treatment plants, which collect wastewater from 3,200 municipalities or intercommunity associations. Current efforts focus on developing more efficient wastewater treatment methods, particularly in terms of nitrogen pollution that take into account surface runoff, and treatment methods that are better suited to rural areas (lagooning, spreading, sand filters, garden filters). The Water Agency is now encouraging small polluters to reduce the spread of pollution.

The Water Agency also participates in the construction of drinking water plants. At the present time, it is financing new filtration techniques.

Preventive measures

The pollution tax has incited industries to make decontamination practices both systematic and highly efficient. An increasing number of companies, supported by the Water Agency, are now using clean processes in order to reduce pollution at the source.

The recent implementation of rural contracts enables local actors to work together on local issues, particularly with respect to combating non-point source pollution. For the moment, these contracts play an important role in terms of raising awareness and inciting action. An encouraging sign in the Seine-Normandy basin is that policies are being developed jointly at the local and regional level to prevent pollution, particularly non-point source agricultural pollution.

In order to reduce this, the Water Agency is financing a pilot project to develop non-polluting farming techniques (soil cover in winter, better use of pesticides, more efficient matching of soils and crops) and to help stockbreeders bring their facilities into compliance with water protection regulations (by means of a new animal-farming tax). There are rules governing the storage of agricultural discharge and most stockbreeders pay an animal-raising tax that corresponds to the impact of discharges into the environment. Farmers using water for irrigation pay a fixed water tax based on their declared area of irrigation. If they own a meter, they pay lower taxes. The installations of meters is subsidized by the Water Agency. The Agency also provides financial assistance to stockbreeders in order to improve their farming practice. They are encouraged to bring manure pits into compliance with nitrate controls and to put down protective groundcover to avoid leaching. The Ministry of Agriculture prohibited the sale of atrazine after September 2002, and will forbid its use after June 2003.

Regulation of and changes in certain chemical products have also had beneficial effects. The increased use of phosphate-free washing powders, combined with particular efforts made by industry, have resulted in a significant decrease in phosphate levels. Cadmium concentrations in Seine estuary sediments have decreased during the last five years since by-products from the manufacture of phosphate fertilizers are no longer discharged into the environment.

Environmental protection and remediation

Other preventive measures specifically concern the environment. One of these is the protection of wetlands. While only 11 percent of the basin's wetlands are protected by regulatory measures, 55 percent are classified under international labels (Ramsar sites, United Nations Educational, Scientific and Cultural Organization [UNESCO] biosphere reserves). The Water Agency also buys wetlands outright (643 hectares were acquired in 2001, almost ten times more than in 1999, and 1,262 hectares have been acquired over the last five years). In addition, it participates in studies and the employment of local wardens and technical personnel. Moreover, for seven years now, the Water Agency has encouraged efforts to restore wetlands by awarding prizes at an annual competition. The Water Agency invested million US$1.53 (1.6 million €) in wetlands in 2000, more than twice the amount allocated in 1998.

The Water Agency has set up technical support units for river maintenance jointly with local authorities and fishermen's federations, and encourages the drawing up of river contracts. Some of these measures are already bearing fruit. Trout have been reintroduced in the Touques River in Normandy, for example, and its river banks have been improved for walkers and anglers. By 2002, some 200 dams were to be equipped with fish passes.

Risk management focused on water shortage, flood and health

The principal risks in the Seine-Normandy basin are those of flooding, severe low water levels and contamination of the drinking water resource. Twenty-two percent of the flood-prone communities have flood risk prevention plans. The local population is informed of the risk of flooding and has Internet access to relevant information.[19] A detailed flood risk map of the region is near completion. Permeable road surfaces are now being used to limit the adverse effects of impermeability on flooding. Since dams do little to control flooding, civil engineering works focus mainly on creating spreading basins. The Water Agency has been able to do little in this area up until now since it has never collected taxes for flood risk management. While the probability of losing life in a flood is extremely low in the basin, property damage is another matter. It is estimated that another flood like the one in 1910 would cost more than 4 billion € (US$4.1 billion).

The large dams on the Seine, the construction of which was subsidized by the Water Agency, ensure that the Marne, the Yonne and the Seine do not dry up in the summer as a result of withdrawals for the Paris region. Using hydrodynamic models, specialists can now study the principal aquifers, particularly as regards the risk of their drying up. Risk thresholds have been set, and specific measures stipulated in management plans are taken if these are exceeded (spreading basins that reduce the flow rate have been defined). Aquifer contracts between groundwater users ensure that, in the event of a crisis, the shortage is shared by all of the users according to a priority scheme (for example, irrigation quotas in the Beauce aquifer, see box 12.4 in chapter 12 for more details).

In order to decrease the risk of contamination of the Paris basin drinking water supply, a project to drill very deep wells to tap the Albien aquifer, 700 metres deep and extensive under the basin, is currently being studied.

Conclusions

Improving water quality is still the major concern of the basin, despite real progress made over the last thirty-five years. Storm runoff during periods of heavy rainfall continues to create problems, causing wastewater to be discharged directly into rivers or overloading wastewater treatment plants, thereby decreasing their efficiency. Non-point source pollution from farming and urban areas is still a major problem as nitrate, pesticide and heavy metal concentrations continue to increase. There are rules for farm waste, just as there are for industries, but only concerning stockbreeding.

Municipal and industrial wastewater treatment still needs to be improved by increasing the efficiency of wastewater treatment plants, particularly with respect to nitrate and phosphate. The wastewater collection system must be improved, and pollution discharged by small businesses and artisans must be controlled. Erosion is another source of non-point source pollution, especially in karst regions where domestic users are still occasionally deprived of drinking water due to turbidity. Combating floods and eutrophication, protecting wetlands, and the spread of treatment plant sludge are also major issues in the Seine-Normandy basin.

All of these problems are on the Water Agency's agenda. Taxes will be levied to more efficiently combat nitrate pollution (a nitrate tax, proportional to the agricultural wastewater discharge) and flooding (a water regime modification tax, based on the impermeable surface area, structures that impede river flow and barriers to flood expansion). An ecology tax has been levied on certain chemical products in order to reduce their use (pesticides, phosphate washing powders).

The European WFD not only confirms French water management principles (management at the scale of major hydrographic basins, direct involvement by those concerned), but also marks a major turning point by setting an ambitious goal to improve the quality of water resources and to achieve 'good status' over the coming fifteen years. Thus, specific prescriptive policies (such as determining discharge standards) will need to be broadened to cover all uses and assess overall environment impact. For those involved in water management, this will mean passing from an obligation of means (doing what is required by law, regardless of results) to an obligation of results (doing whatever is necessary to meet quality objectives required by law). WFD requirements will force the French water management system to increase its public participation and transparency, a task already undertaken by the creation of Basin Committees.

19. www.environnement.gouv.fr/ile-de-france

Box 19.1: Development of indicators

Since the Seine-Normandy SDAGE was implemented, an operating report has been published each year. Progress towards reaching the objectives set by the Basin Committee can thus be monitored. Monitoring is done with indicators that are well suited to the specific context of the basin and focus on Water Agency activities.

Theme	Indicators
MANAGEMENT OF AQUATIC ENVIRONMENTS	• Six indicators of river functionality (fish passes, financial aid for river maintenance) • Three indicators of wetlands preservation (surface areas drained, regulatory protective measures) • Two indicators of decreased dredging of gravel • One indicator of runoff and erosion control
WATER QUALITY MANAGEMENT	• Four indicators of general quality (polluted sites, Seine river water quality) • Five indicators of municipal and industrial discharge • Four indicators of decreased agricultural pollution (demarcation of vulnerable zones, controlling effluents from stockbreeding) • Two indicators of coastal pollution control • Two indicators of drinking water supply (water quality and well protection) • One indicator of pipes and major works
CRISIS MANAGEMENT	• Four indicators of extreme low water level management (expansion zones, warning zones, etc.) • Three indicators of flood control (risk prevention plans, improved forecasting)
INTEGRATED MANAGEMENT	• One indicator of SDAGE progress (local water development and management plans) • Two indicators of contracts
KNOW-HOW AND COMMUNICATION	• Three indicators of knowledge development (research programmes, inventories) • Three indicators of aquatic environment monitoring (measurement network, databases)
STATE OF THE ENVIRONMENT	• Six maps that indicate the state of the aquatic environment: groundwater levels, physico-chemical quality of surface water, pesticide content in surface water, quality of fish populations, quality of coastal waters, maximum nitrate and pesticide concentrations in groundwater

These indicators have been sufficient for monitoring changes in the environment and the management system under the current SDAGE. The EU WFD objectives for achieving 'good status' and covering actual costs will undoubtedly require additional monitoring indicators.

Within the framework of the World Water Assessment Programme (WWAP), the AESN recommends that the following indicators of the environment, governance and financial aspects also be taken into consideration.

Theme	Indicators
ENVIRONMENT: QUALITY (ADAPTABLE TO GROUNDWATER, WATER BODIES AND COASTAL WATERS)	• Maps made using a Quality Evaluation System, based, at least, on the following indicators: BOD5, NH4+, NO3-, P total, suspended particulate matter, pH, conductivity, colour, thermotolerant/ faecal coliform organisms, total chromium, mercury, lead, DDT op', DDT pp', lindane, endrine, dieldrine, aldrine
ENVIRONMENT: WETLANDS	• Surface area and evolution: protected wetlands and RAMSAR sites, basin wetlands, especially those in flood plains, drained areas. • Maps of urban areas and density of industrial fabric. • Changes in the pressures from intensive farming, industry and urban development in the basin.
ECONOMY	• Average price of drinking water/m^3, annual amount paid by inhabitants for drinking water, proportion of the price of water used to protect the resource, the value of distribution and treatment facility assets/household, funds earmarked for water/GDP, volume consumed annually/ inhabitant, annual water bill per household/ annual income. Cost recovery index. Pricing of services and financial autonomy of water works budgets.
GOVERNANCE	• Decentralization, involvement of public in water policy decisions, transparency, allocation of roles, openness to civil society, mobilization of know-how, sharing of knowledge, management system evaluation, public-private partnership (equity and efficiency).

Box 19.1: *continued*

If we attempt to assign global scores for sustainable resource management we run the risk of ending up with meaningless figures. The details on which the ratings are based must be preserved and accompanied by the individual scores for each indicator. The methods used to obtain and calculate indicators must also be described in detail. It might be difficult to assign a precise score to some indicators. The state of the environment, for example, is generally shown with a map. Weighting, which favours certain themes or indicators to the detriment of others, should also be carefully considered. For example, different weighting methods would result in different scores, each of which could be used to more accurately reflect a specific area of water management (shortages, governance, environment).

Data for some of the indicators proposed by WWAP are now being gathered at the Seine-Normandy basin level. Data gathering methods for others are still being developed, while for still others data cannot be gathered at present, or the indicators are too vague.

Challenge area	Seine-Normandy basin indicator
SURFACE WATER	• Withdrawals: 2,044 bm^3/year • Precipitation: 750 mm • Evapotranspiration: 500 mm
WATER QUALITY	• Yearly maps for monitoring the quality of water courses using indicators • SEQ tool (Quality Evaluation System) is used, which covers groups of similar parameters • French Environmental Institute (IFEN) quality indicators
GROUNDWATER	• 10 major water tables • Withdrawals: 1,213 bm^3/year • Volume of underground resources has not yet been accurately assessed • Piezometric monitoring of groundwater tables; three tables have overrun the hydric stress threshold over the last ten years, but have been filled up again due to recent heavy rainfall
PROTECTING HEALTH	• Incidence of water borne diseases is low • Virtually all households have access to good quality potable water (conformity level of potable water analyses > 99% for the sixty-two parameters) • 0.03% of the population is deprived of potable water several days per year due to turbidity in periods of heavy rainfall in certain zones • 88% of the population has access to collective sanitation and 10% use individual sanitation measures (in rural areas) • Right to water is legally recognized • Public drinking fountains, washing places or baths in each town or village • Temporary social aid for the poor in paying their water bills; water cuts very rare and theoretically forbidden for poor people
PROTECTING ECOSYSTEMS	• Present estimation of the surface area of wetlands: 580,969 hectares, 6% of the basin area), of which 2% is protected by national regulations and 9% by international regulations • In 2004, 31% of wetlands will be classified as sites of European importance. • Three Ramsar sites • 1.6% of the surface area in the basin has been rendered impermeable • Around 600,000 ha drained between 1974 and 1999 (in other words 6.2% of the basin area) • 33 fish species listed in the Seine
WATER AND CITIES	• Withdrawals of potable water: 1.6 bm^3 • 20% leaks and is used for cleaning the network; the inhabitants consume 70 m^3/year. • 100% have access to potable water • Access to sanitation in towns currently being evaluated • 100% of communities with more than 10,000 inhabitant equivalents have a wastewater treatment plant • The responsibility (penal) for water and sanitation services is incumbent on the local authority
SECURING THE FOOD	• Indicator that is not very relevant in the Seine-Normandy basin (no problems of securing food)
SUPPLY WATER AND INDUSTRY	• Annual use of water by industry: 95 m^3/inhabitant (of which 2/3 is used for cooling thermal power plants) for 188 m^3/inhabitant of the water withdrawn annually • Pollution from industries not connected to the sewage system: 147 t/day of oxidizable matter; 19 t/day of nitrogenous matter: 3.2 milliequivalents (Meq)/day of inhibiting matter; 2.9 t/day of toxic metals.

Box 19.1: *continued*

Challenge area	Seine-Normandy basin indicator
WATER AND ENERGY	• Annual use of water for cooling of thermal power plants 831 Mm^3/year
MANAGING RISKS	• 2,239 communities are at risk to flooding; 22% of them have a Risk Prevention Plan. Flooding map for the whole basin
SHARING WATER	• Welfare measures have been put into place to ensure that the poor have access to water (cutting off the water supply is prohibited) • 49% of withdrawals are for domestic water requirements, 27% for industry, 3% for farming, 23% for electricity and 5% for other uses (cleaning of roads, etc.) • If there is any conflict between sectors in the use of the resource, domestic supply water is treated as a priority • Water counters exist
VALUING WATER	• Price of potable water is on average 2.74€/m^3 (US$2.80/ m^3), which is 126€/inhabitant/year (US$120/inhabitant/year) • Tariff varies from 0.15 to 5.35€ (US$0.14 to 5.11), depending on the size of the community, the complexity of treatment required, and the specific management set-up • All consumers are billed for water and sanitation services • On average, households spend 1.03% of their income per year on water and sanitation • Pricing system is controlled by the state • On average, the fees represent 20% of the price of potable water • Annual water expenditure (potable water and sanitation) is 0.5% per thousand GDP
ENSURING KNOWLEDGE	• For surface water 150 parameters are measured 6 to 48 times per year, on 441 observation points, relating to 15,000 km of principal drains and 17 million inhabitants • For groundwater: 402 boreholes, measured 12 to 48 times per year, on 250 parameters; relating to 10 or so main aquifers, a total surface area of 97,000 km^2 and 17 million inhabitants
ENSURING KNOWLEDGE	• For quality of coastal waters: 130 'swimming' sites (ten or so parameters), 22 'shellfish' sites (around 5 parameters), 11 'sediment' sites (at least radioactivity, heavy metals) relating to 600 km of coast • In quantitative terms there are: 174 rain gauges, 214 piezometers and 418 hydrometric stations • In total around 5 million sets of data per year, a large proportion of which is available to the general public on the Internet • The quality data is analyzed using the SEQ (Quality Evaluation System), which allows maps to be produced by type of indicator (grouping together of similar parameters)
GOVERNING WATER	• Local authorities responsible for water and sanitation; programming, coordination on a catchment area level • Management interventions adapted to the scale of the problem (river contracts, management schemes on the small catchment area scale) • Effective delineation of roles and balanced public / private partnerships (delegation of water and sanitation services) • Insufficient participation by civil partnerships and vagueness for the responsibility for the upkeep of rivers • Transparency in the price of water and water budget autonomy, fees monitored and well discussed • Problems of imbalance in expertise and insufficient evaluation of management actions • Solidarity funds for rural zones to promote equity, basin solidarity between zones and end users • Delegating water and sanitation services to the private sector (46% of cases) • Close correlation between the public and water policy in small communities, representatives of associations in the Basin Committee, but these openings are not sufficient

References

AESN (Seine-Normandy Water Agency). 2002. *Comment évolue la qualité des eaux depuis trente ans?* Paris.

——. 2001a. *Contribution de l'Agence de l'Eau Seine-Normandie à l'état des lieux* (The Contribution of the Seine-Normandy Water Agency). Inventory according to the European Water Framework Directive. Working paper. Paris.

——. 2001b. *Quelle eau fait-il dans le bassin Seine-Normandie? La qualité des eaux superficielles, souterraines et littorales, synthèse 2001* (How's the Water in the Seine-Normandy Basin? The Quality of Surface, Ground and Coastal Waters, Synthesis 2001). Press kit. Paris.

——. 2000. *Les Forêts Alluviales du Bassin Seine-Normandie. Un Patrimoine à Protéger* (Alluvial Forests in the Seine-Normandy Basin. A Heritage that Must be Protected). Paris.

——. 1999a. *L'eau dans le Bassin Seine-Normandie – trente-cinq ans d'action.* (Water in the Seine-Normandy Basin – Thirty-five Years of Action). Paris.

——. 1999b. *Enquête statistique sur le prix de l'eau du bassin* (Statistical Study of the Price of Water in the Basin). Report. Ecodécision. Paris.

——. 1997a. *Les oiseaux d'eau du bassin Seine-Normandie. Un patrimoine à protéger* (Water Birds in the Seine-Normandy Basin. A Heritage that Must be Protected). Paris.

——. 1997b. *Eléments de sociologie de l'environnement et de l'eau en France, résumé et synthèse de sept etudes et Enquêtes d'opinion* (Sociology of the Environment and of Water in France, Summary of Seven Studies and Opinion Surveys). Paris.

AREA, 2001. *Barrages, entraves à la dynamique biologique des rivières. Recensement des problèmes majeurs en Seine-Normandie, corrections et remèdes possibles* (Dams, Obstructing the Biological Dynamics of Rivers. Inventory of Major Problems in the Seine-Normandy Basin, Possible Solutions). Preliminary version. Paris, Direction Régionale de l'Environnement d'Ile de France.

Belliard, J. 2001. 'Historique du peuplement de poissons dans la Seine' (History of Fish Populations in the Seine River). *Eaux Libres*, September.

Bouleau, G. 2001. *Acteurs et circuits financiers de l'eau en France* (Actors and Financial Channels Involved in Water Management in France). Paris, French Institute of Forestry, Agricultural and Environmental Engineering.

Meybeck, M.; de Marsily, G.; Fustec, E. 1998. *La Seine en son Bassin. Fonctionnement écologique d'un système fluvial anthropisé* (The Seine River and its Basin. Ecological Aspects of an Anthropized River System). Paris, Elsevier.

Seine-Normandy Basin Committee. 2000. *Tableau de bord. Suivi des orientations du Schéma Directeur d'Aménagement et de Gestion des Eaux du bassin Seine-Normandie. Bilan de l'année 2000* (Implementation of Water Development and Management Guidelines in the Seine-Normandy Basin. Annual Report 2000). Paris, Seine-Normandy Water Agency, Direction Régionale des Affaires Sanitaires et Sociales, Direction Régionale de l'Environnement.

——. 1999. *Tableau de bord. Suivi des orientations du Schéma Directeur d'Aménagement et de Gestion des Eaux du bassin Seine-Normandie. Bilan de l'année 1999* (Implementation of Water Development and Management Guidelines in the Seine-Normandy Basin. Annual Report 1999). Paris, Seine-Normandy Water Agency, Direction Régionale des Affaires Sanitaires et Sociales, Direction Régionale de l'Environnement.

——. 1998. *Tableau de bord. Suivi des orientations du Schéma Directeur d'Aménagement et de Gestion des Eaux du bassin Seine-Normandie. Bilan de l'année 1998* (Implementation of Water Development and Management Guidelines in the Seine-Normandy Basin. Annual Report 1998). Paris, Seine-Normandy Water Agency, Direction Régionale des Affaires Sanitaires et Sociales, Direction Régionale de l'Environnement.

Smets, H. 2002. *Le droit à l'eau* (The Right to Water). Paris, Académie de l'Eau, Conseil Européen du Droit de l'Environnement.

Senegal River Basin, Guinea, Mali, Mauritania, Senegal

By: The Organization for the Development of the Senegal River (OMVS, Organisation pour la mise en valeur du fleuve Sénégal)

Table of contents

The chameleon changes colour to match the earth,
the earth doesn't change to match the chameleon.

African proverb

THE SENEGAL RIVER represents a 1,800-kilometre lifeline for a multi-ethnic, multicultural population where livestock outnumber people. It runs through sub-Saharan Africa in a mostly desert region characterized by water scarcity and subsistence economies. The Organization for the Development of the Senegal River (OMVS) was created in 1972 with a mandate to ensure food security and harmony among all riparian users. Thanks to the construction of two main dams providing energy, irrigated agriculture and year-round navigation – and to an original management approach based on a concept of 'optimal distribution among users' rather than volumetric water withdrawals – the area is gradually developing. Ironically, dam construction has brought problems as well as benefits, and the major concern is water-related diseases.

THE SENEGAL RIVER BASIN is located in West Africa, between latitudes 10°30 and 17°30 N and longitudes 7°30 and 16°30 W. It is drained by the 1,800 kilometre (km)-long Senegal River, the second longest river of West Africa, and its main tributaries, the Bafing, Bakoye and Faleme Rivers, which have their source in the Fouta Djallon Mountains (Guinea) or in Mali.

General Context

Most of the Senegal River basin has a sub-Saharan desert climate, aggravated by more or less long periods of drought during the 1970s. Access to sufficient quantities of good-quality water is therefore a particularly sensitive issue and absolutely crucial for the health of the population and the economy.

Physical characteristics

The Senegal River basin covers a surface area of about 300,000 square kilometres (km^2). The high plateaux in northern Guinea represent 31,000 km^2 (11 percent of the basin), 155,000 km^2 are situated in western Mali (53 percent of the basin), 75,500 km^2 are in southern Mauritania (26 percent of the basin) and 27,500 km^2 are in northern Senegal (10 percent of the basin). The basin has three distinct parts: the upper basin, which is mountainous, the valley (itself divided into high, middle and lower) and the delta, which is a source of biological diversity and wetlands (see map 20.2). Topographical, hydrographic and climatic conditions are very different in these three regions and seasonal temperature variations are extensive.

Map 20.1: Locator map

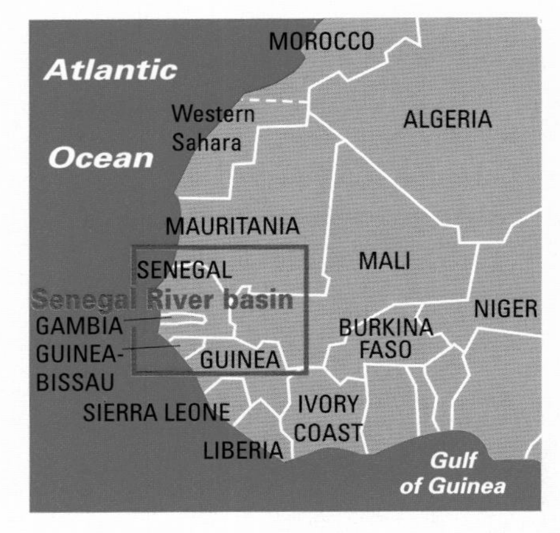

Source: Prepared for the World Water Assessment Programme (WWAP) by AFDEC, 2002.

Table 20.1: Hydrological characteristics of the Senegal River basin

Surface area of the basin	300,000 km^2
Annual precipitation	660 mm/year
Annual runoff (Bakel station)	
Before 1985	698 m^3/s
After 1985	412 m^3/s
Annual discharge (Bakel station)	
Before 1985	863 m^3/s
After 1985	416 m^3/s

Socio-economic characteristics

Population

The Senegal River basin has a total population of around 3,500,000 inhabitants, 85 percent of whom live near the river. This figure includes approximately 16 percent of the total populations of the three OMVS member states – Mali, Mauritania and Senegal – plus the population of the Guinean portion of the upper basin.

The population within the basin is increasing at a rate of about 3 percent per year, which is slightly higher than the individual averages for the three member states.

A large ethnic diversity also characterizes the basin's population, with, among others, Peuls, Toucouleurs, Soninkes, Malinkes, Bambaras, Wolofs and Moors. However, there is massive emigration among the young towards the major cities and to Europe.

Table 20.2: Summary of physical data

		Mali	Mauritania	Senegal	Guinea
Surface area (km^2)	National	1,248,574	1,030,700	197,000	245,857
	Basin	155,000	75,500	27,500	31,000
	(% of basin)	53	26	10	11
Rainfall	National	850	290	800	2,200
annual average (mm)	Basin	300 to 700	80 to 400	150 to 450	1,200 to 2,000
Temperature	National average	29	28	29	26
	Basin min. and max.	15 to 42	18 to 43	17 to 40	10 to 33

Half of the basin is located in Mali, but the main input in terms of water resources comes from the upper basin in Guinea with an average of 1,600 mm of precipitation.

Map 20.2: Basin map

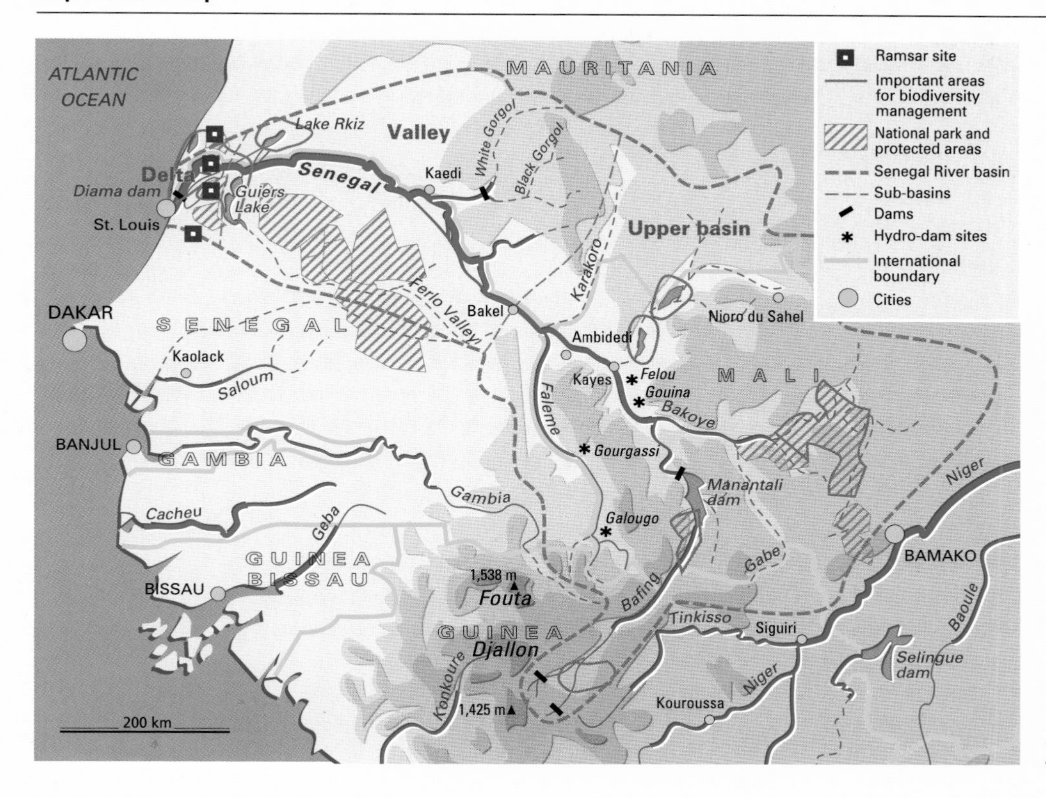

Source: Prepared for the World Water Assessment Programme (WWAP) by AFDEC, 2002.

Table 20.3: Summary of socio-economic data in the OMVS member states

	Senegal River basin	Mali	Mauritania	Senegal
Population (million inhabitants)	3.5	11	3	10
Annual growth rate (%)	3	2.97	2.9	2.8
Urbanization rate (%)	NA	41	53	51
Farmland (ha)	823,000	NA	NA	NA
Irrigated land (ha) – national total	NA	78,630	49,200	71,400
Part in basin		4,000	44,449	67,830
Cattle (x1,000 units)	2,700	6,427	1,394	2,927
Sheep and goats (x1,000 units)	4,500	15,986	10,850	8,330
Fish catch (t/year)	26,000 to 47,000	100,000	620,000	395,000

Population figures have been updated, based on growth rates in each country. Irrigation is the motor of development of the basin, especially in the valley and the delta, and livestock raising has always been a major activity. After agriculture, fishing is the second largest economic activity of the basin.

The financial support received from these people is very important to the livelihood of families remaining in the villages, especially during periods of difficulty, as with droughts or floods. Some of these emigrants return to their villages during the rainy period for seasonal work.

The enormous socio-economic potential of the Senegal River basin was identified long ago by colonialists and some resources were already being developed long before the countries gained their independence in the 1960s. Table 20.3 gives a summary of socio-economic data today in the OMVS member states.

Agriculture

The first major attempts to control the Senegal River discharge were made in the 1940s in order to grow rice in the delta (at Richard-Toll in Senegal). But it was not until 1973 that the State Company for Agricultural Development (SAED, Société d'Aménagement et d'Exploitation des Terres du Delta du Fleuve Sénégal) decided to increase this activity by building dikes around 10,000 hectares of flood land and created, in 1975, an irrigated area of 650 hectares. Thereafter, small irrigated areas were rapidly created as a means to combat the drought cycles in the 1970s that made it almost impossible to grow rainfed and flood recession crops. On the left bank, the surface area of community-based irrigated fields grew from 20 hectares in 1974 to 7,335 hectares in 1983 and 12,978 hectares in 1986. Irrigated agriculture rapidly expanded after the new dams were filled (between 1986 and 1988). Today, irrigation is still the motor of development in the basin, notably in the valley and in the delta, due not only to improved technology, but also to the wider variety of produce grown (rice, onions, tomatoes, potatoes, sweet potatoes). About 100,000 hectares of land are now cultivated in the basin: 60,000 hectares during the rainy season (June-September) and 20,000 hectares during the dry season (March-June).

Livestock-raising has also always been a major economic activity in the basin. Due to the existence of rather high-potential pasture land, combined with the carrying capacity of the grasslands and the flood plains, the riparian populations, and even those living elsewhere, practice transhumance and extensive cattle-, sheep- and goat-raising. These activities are generally profitable.

Fishing, in terms of the income of the work force that it employs, is undoubtedly the largest economic activity in the basin after agriculture, notably for populations living near the river in the valley and the delta. Today, however, the future of this sector is in question because for several years now there has been a steady drop in the tonnage caught throughout the OMVS region. Some observers link this to the river development projects (dams, dikes) and to their impact on the environment (significant decrease in salinity, proliferation of floating water weeds and eutrophication). A recent OMVS study of fish resources indicates, however, that while it is true that some old species have disappeared, new fish species have appeared. It would even seem that the invasive aquatic plants are breeding grounds, which at the same time does not prevent them from seriously hindering the mobility of fishermen. The problem, therefore, needs to be studied in more depth to be able to objectively determine the actual impact of the dams on the fishing sector.

Mining sector

Before independence, mineral exploration carried out by the Geological Survey and Bureau of Mines (BRGM, Bureau de Recherches Géologiques et Minières) had already enabled the French to begin mining several economically profitable ores, notably gold in the Faleme River. The activity later became marginal and today only a few people still pan for gold in the Malian and Senegalese parts of the upper basin. Nevertheless, in light of the mining potential in the sector, with the energy provided by the Manantali dam, since September 2001, and the near completion of the river navigation project, mining will undoubtedly become one of the major development poles in the basin again .

Industry

The industrial sector is not sufficiently developed. The Senegalese Sugar Company (CSS, Compagnie Sucrière Sénégalaise) is the largest agro-industrial unit in operation in the basin. It has a production potential of more than 8,000 hectares of sugar cane in Richard-Toll, using water from both the Senegal River and Guiers Lake. Its two largest subsidiaries are the International Design Industry Services (IDIS), which manufactures polyvinyl chloride (PVC) pipes, and the Senal Cotton and Food Products Commerce and Industry Corp., which produces livestock feed. There are two other, smaller, companies in the Delta – the SOCAS canning plant near Ross Bethio and the SNTI, specialized in industrial tomato processing, in Dagana, Senegal. There are also industrial and private rice paddies managed by the SAED and a public rural development enterprise (SONADER, Société Nationale de Développement Rural) in Mauritania.

Energy

The hydroelectric power plant in Manantali has been in operation since September 2001. The initial objectives of this project were to produce 200 megawatts (MW), to furnish an average of 800 gigawatts per hour per year (GWh/year) to electricity companies in the three OMVS member states. The projected electricity production figures used to estimate profitability were, however, based on hydrological data from 1950 to 1974. New simulations done with data from between 1974 and 1994 when flow coefficients were low, which might correspond better to current conditions, predict an energy production of only 547 GWh. As a result of this decrease in the

Table 20.4: Water use by sector within the OMVS area (in millions m^3)

Sector	Mali	Mauritania	Senegal
Agriculture	1,319	1,499	1,251
Domestic use	27	101	68
Industry	14	29	41
Total	1,360	1,630	1,360
Per capita (m^3/year)	161	923	201

Reference year is 1987, except Mauritania (1985). Agriculture is by far the most important water user in the OMVS area.

Table 20.5: Seasonal changes in Senegal River discharge since 1951

Period	Rainy season (m^3/s)	Dry season (m^3/s)
1951–1999	1,538	138
1951–1971	2,247	172
1972–1999	1,007	112
1972–1990	993	71
1991–1999	1,036	201

This table clearly shows the benefit of the Manantali dam construction in the flow regulation (beginning 1991): the discharge never falls below 200 m^3/s.

energy capability of the Manantali power plant, the expected savings in OMVS member states expenditures for energy would unfortunately drop from 22 to 17 percent.

Navigation

Navigation on the Senegal River is today very limited. The OMVS is aware of the strategic importance of its development over the short term, and a navigation project is under study. Like the exploitation of mineral resources, the ability to transport heavy goods at a lower cost and, especially, access to the Atlantic Ocean for Mali could give a new impetus to the region's economy. Table 20.4 summarizes the water use distribution between the different sectors.

Water Resources

Hydrology

Rainfall

The river's flow regime depends, for the most part, on rain that falls in the upper basin in Guinea (about 2,000 mm/year). In the valley and the delta, rainfall is generally low and there is rarely more than 500 mm/year. During the 1970s (drought years), there was significantly less. This greatly accentuated the interannual irregularity of floods, which, before the dams were built, could vary six-fold between the wettest and driest years. The climatic regime in the basin can be divided into three seasons: a rainy season from June to September, a cold, dry off season from October to February, and a hot, dry off season from March to June. In the river, this creates a high-water period or flood stage between July and October, and a low-water period between November and May to June.

Surface water

The three main tributaries of the Senegal River produce together over 80 percent of its flow. The Bafing alone contributes about half of the river's flow at Bakel. The two largest tributaries on the right bank, above Bakel, the Gorgol and the Oued Gharfa, supply only 3 percent of the water in the Senegal River that flows into the Atlantic Ocean at Saint Louis. At Bakel, considered to be the reference station on the Senegal River due to its location below the confluence with the last major tributary (the Faleme), the average annual discharge is about 690 cubic metres per second (m^3/s), which corresponds to an annual input of around 22 billion cubic metres (bm^3). The annual discharge ranges between a minimum of 6.9 bm^3 and a maximum of 41.5 bm^3. Table 20.5 provides some discharge data.

The total capacity of the Manantali dam, built on the Bafing River, is 11.5 bm^3 of water for a useful volume of around 8 bm^3: it is the largest in the basin. Its purpose is to attenuate extreme floods, generate electric power and store water in the wet season to augment dry-season flow for the benefit of irrigation and navigation.

The Diama dam, located 23 km from Saint Louis near the mouth of the Senegal River in the delta, sits astride the territories of Mauritania and Senegal. Its threefold purpose is:

- to block seawater intrusion and thereby protect existing or future water and irrigation wells;
- to raise the level of the upstream water body, creating reserves to enable irrigation and double cropping of around 42,000 hectares at an altitude of 1.5 metres above sea level (m.a.s.l). and 100,000 hectares at an altitude of 2.5 m.a.s.l.; and
- to facilitate the filling of Guiers Lake in Senegal and Lake Rkiz and the Aftout-es-Saheli depression in Mauritania.

Groundwater

The deep aquifers are, for the most part, represented by the Maestrichian fossil formation and the Continental Terminal formation. The alluvial aquifer is the principal shallow aquifer. It is present in all of the flood plain at various depths, generally less than 2 metres, and has an average thickness of about 25 metres. This aquifer communicates in places with a discontinuous network of lenticular aquifers in the permeable strata interbedded in the alluvium (see map 4.3 on the world's groundwater resources in chapter 4, and map 12.4 on the aquifers in northern Africa in chapter 12).

These aquifers are recharged by the river and by all of the tributaries, distributaries, ponds and lakes in the flood plain. On the edge of the valley, the aquifers tend to deepen, usually with a steep slope, but this is highly variable from one place to another.

The water level in the alluvial aquifer varies with the seasons and river level, along with the general hydrological regime in the valley. Since the dams were filled, both the volume and duration of floods and the geographic distribution of the flooded areas have been disrupted, significantly modifying groundwater recharge and the piezometric surface. Reducing the volume of the floods and building dikes significantly reduces the area of natural recharge zones (infiltration ponds). On the other hand, flow regulation during low water periods (maintenance of a minimum flow) and irrigation of large surfaces, rice paddies in particular, increases groundwater recharge during part of the dry season in some areas.

Water quality

In the Senegal River basin there is, at present, no database for water quality similar to those that exist for quantity and discharge, both of which have been monitored since 1904. There are, however, time series and locally monitored data, generally collected by national water supply companies in Mali, Mauritania and Senegal and within the framework of research carried out by universities, training institutes, cooperative agencies, etc. These data indicate, in some places, a degradation of surface water quality. This deterioration would be caused primarily by eutrophication due to a reduction of the flow velocity and oxygenation of the water caused by the new dams and dikes, the proliferation of water weeds, and chemical and biological pollution related to the discharge of wastewater and pesticides into the river. Furthermore, even if there are, as yet, no figures to confirm it, small alluvial gold-washing activities in the upper basin are a threat to water quality because of the products used (such as mercury and flashlight batteries).

Groundwater in the Senegal River basin is generally salty in areas where there used to be seawater intrusion before the Diama dam was built. The alluvial aquifer has a relatively homogeneous salinity, whereas the lenticular, fluvio-deltaic aquifer formations have a slightly more heterogeneous salinity. As a result, there are large and sometimes abrupt variations, with concentrations rising from 1 or 2 grams (g) per litre to more than 150 g/litre. On average, salinity decreases as one moves away from the centre of the delta (more than 10 g/litre) towards the edge (10 to 0.15 g/litre). The aquifers have a higher load in high areas (with an average of 30 g/litre) than in depressions, which are regularly flooded (13 g/litre). However, the saltiest water is found in depressions such as the Aftout-es-Saheli sebkhas in Mauritania and the Gandiolais lagoons and Ndiael wetlands in Senegal. The pH values also vary (but not with salinity), with a high acidity in and depressions influenced by the acid sulphate deposits of the ancient mangroves. The Sodium Absorption Ratio (SAR)[1] of the aquifers is generally high, which means that there is a risk of alcalinization of soil horizons in contact with these aquifers.

Impact of development on the population and on natural resources

More than ten years after the dams were filled in 1986 and 1987 and the structures (dikes, irrigation systems) associated with the implementation of the OMVS development programme were built, several studies have been carried out, making it possible to conclude that these interventions are having both positive and negative effects on the basin's population and natural resources. Most important, the Senegal basin's flood plain ecology has changed from a salty and brackish aquatic environment with marked seasonal changes to a low-flow perennial freshwater ecology. There is, of course, a cause-effect relationship between the impact of human activities on the physical environment (water resources, soil, vegetation) and the impact of the restoration or degradation of this environment on people because the link between them is fundamental.

Principal negative effects

- Displacement of populations living in the areas where the dams were built;
- proliferation of water-borne diseases (bilharzia [schistosomiasis], malaria, diarrhoea) due to changes in ecological conditions as a result of the blocking of seawater intrusion, with the Diama dam;
- water pollution caused by the development of irrigated agriculture and the agro-industry (CSS, SAED in Senegal, SONADER in Mauritania);
- proliferation of water weeds in the valley and delta, clogging water courses and contributing substantially to making the ecosystem more uniform;
- degradation of the fish population available for independent fishermen (quantity and quality);
- reduction of pasture land;
- riverbank erosion, especially in the upper basin where the topography is much more rugged;

1. Water is generally ranked according to the risk of alcalinization: low hazard (2 to 10), medium hazard (7–18), high hazard (11–26), very high hazard above that (Fetter, 1994).

- degradation of cultivated land;
- modification of the hydrodynamic characteristics of the estuary with the reduction of the 'flushing' phenomenon; and
- disappearance of wetland areas.

The installation of the dams has been accomplished without taking due account of other important aspects of planning.

- Top-down planning has occurred without relationship to the local needs of the beneficiaries.
- The large schemes for groundnuts, cotton and irrigation have been less than successful due to application of inappropriate technologies, lack of markets or access to markets, and lack of local capacity.

Principal positive effects

- Year-round availability of freshwater in sufficient quantities (for agriculture, domestic uses, agro-industry, groundwater recharge), accompanied by reverse immigration of people who had left to find employment in the cities;
- development of irrigated agriculture in the valley (with double cropping);
- partial opening up and stimulation of exchanges between areas where dams have been built and the rest of the subregion due to road construction;
- access to healthcare for several villages near the dams with the construction of dispensaries and health clinics;
- access to drinking water installations for populations living near the dams;
- development of fishing activities for populations living near the Manantali dam;
- reappearance of local fauna and regeneration of vegetation
- flow regulation to decrease or eliminate flooding;
- cheaper electricity in the three member states thanks to the Manantali power plant.

Other positive effects expected over the short term concern:

- the electrification of villages near the dams (a study has been completed and funding obtained for the first phase); and
- navigation on the river between Saint Louis and Kayes (a study is underway).

Water resources database and information

OMVS has abundant quantitative data thanks to a discharge monitoring network set up in 1904, with updated records stored in a database of the OMVS High Commission.

The Technical Department of the High Commission also publishes a monthly hydrological bulletin for the technical services of the member states and other actors (producers, development partners, NGOs, industrial projects) carrying out activities in the basin.

Major studies carried out by the French Research Institute for Development (IRD, Institut de Recherche pour le Développement) and the OMVS have also made it possible to estimate withdrawal and losses during low flow stages. The results are the following: evaporation during low water stages is estimated to be at 65.4 m^3/s, withdrawal for human and industrial consumption at 2.6 m^3/s, and withdrawal for irrigated agriculture during the off-season at 19 m^3/s. The average total water needs downstream from Bakel (reference station) are, therefore, at 87 m^3/s during low water stages.

These studies have also made it possible to develop suitable resource management tools based on analysis of the hydrological behaviour of the river in relation to needs. Software (SIMULSEN) has been developed to evaluate the effects of the various Manantali dam management practices on the degree of satisfaction of demands such as hydroelectric production, flow regulation, flow at Bakel as a function of needs downstream. Specific studies have also been carried out on how flooding is related to the functioning of basins, providing important information concerning their filling and emptying, and the volumes of water potentially available during this period.

Data on water quality, health, livestock-raising, agriculture, fishing, climate and the environment do exist, but are dispersed in various government services, laboratories, universities and research institutes, or even in cooperation institutions, such as the IRD, the United States Agency for International Development (USAID), the United Nations Development Programme (UNDP), and the World Bank.

Data have been collected for many projects, but the resulting databases are incompatible or have simply been lost or abandoned upon completion of projects. The most acute need is in the upper basin, including Guinea, where the lack of data is a concern not only for the government of Guinea but for the whole basin.

Data gaps have long been a handicap for the OMVS. Thus, the High Commission set up an Observatory of the Environment in

November 2000 to create a network of all of the producers/possessors of thematic data and hook them up to a general database that would be managed by the Observatory's Coordination Bureau. Agreement protocols will soon be drawn up by these organizations and the OMVS to formally define the roles and responsibilities of each of the actors in the data collection, processing and storage procedures, on the one hand, and data development, dissemination and sharing, on the other.

Challenges to Life and Well-Being

A difficult context

Before the dams were filled in the mid-1980s, activities of the local inhabitants depended directly on rainfall (rain crops) or on floods (flood recession crops), in particular in the upper basin in Guinea (Fouta Djallon Mountains). But the dramatic and continuous drop in rainfall during the 1960s and 1970s led to the degradation of almost the entire base of natural resources (soil erosion, disappearance of vegetation, drying up of surface water, salinity 200 km upstream from the mouth of the river, drop in the groundwater level, degradation and disappearance of pasture land). Under these conditions, the local inhabitants could not produce enough to survive and the only alternative was emigration. Each year, a large percentage of the population, in particular young people, left the valley and the delta for capital cities in the subregion (Abidjan, Bamako, Dakar, Libreville, Nouakchott) or Europe (usually France or Italy).

The filling of dams

In response to these difficulties, a dam-building project was implemented, in order to partially or totally control river flow and, consequently, enable the development of large areas of land for agriculture to contribute to food security. In addition, the dams built to regulate flow could also be used for hydroelectric power plants, to solve the problem of the low supply and high cost of electricity, and to maintain a sufficient flow depth in the river for fluvio-maritime navigation to relieve Mali's isolation by giving it access to the Atlantic Ocean and lower the cost of transporting heavy goods (making it possible to exploit the basin's mineral resources). It is in this context that the OMVS programme was created.

After the dams were filled, sufficient quantities of water became available year-round, enabling local inhabitants to engage in various highly profitable activities. These new opportunities incited the young men who had left to try their luck elsewhere, without much success, to return home. People from the agrobusiness world also began coming into the area to invest in channels to market or to create small factories to transform the crops grown in the valley and delta.

Preliminary studies showed that the irrigation system would re-establish the basis of profitable production. Flow regularization would guarantee a minimum discharge of 300 m^3/s at Bakel (reference station), and the storage capacity of the Manantali and Diama dams and the Guiers and Rkiz Lakes could be used to irrigate a surface area of 375,000 hectares, three times the surface area cultivated before 1986.

Unfortunately, this initial enthusiasm diminished when, between the sixth and tenth year after the dams were filled, new problems arose. Two in particular, the degradation of ecosystems and the proliferation of water-borne diseases, very rapidly reached severe endemic proportions. These problems are described in detail further on in this chapter.

Management Challenges: Stewardship and Governance

The OMVS river basin organization was established about three decades ago by three out of the four riparian states. Mali's principal interests are the maintenance of river levels so as to obtain navigable access to the sea and energy produced by the Manantali dam. Mauritanian and Senegalese interests converge in power production and irrigation, while Senegal seeks improved livelihoods for local populations. These varied interests are typical of a transboundary water management situation. The Manantali dam, although located in Mali, belongs to all the members of the OMVS authority.

Legal and regulatory framework and governance

The first institutions to develop the Senegal River valley were created during the colonial period. On 25 July 1963, very soon after independence, Guinea, Mali, Mauritania and Senegal signed the Bamako Convention for the Development of the Senegal River Basin. This convention declared the Senegal River to be an 'International River' and created an 'Interstate Committee' to oversee its development. The Bamako Convention was supplemented by the Dakar Convention, signed on 7 February 1964, concerning the status of the Senegal River. The Interstate Committee laid the foundation for subregional cooperation in development of the Senegal River basin. On 26 May 1968, the Labé Convention created the Organization of Boundary States of the Senegal River (OERS, Organisation des Etats Riverains du Sénégal) to replace the Interstate Committee, broadening the field of subregional cooperation. Indeed, OERS objectives were not limited to the valorization of the basin but aimed at the economic and political integration of its four member states. After Guinea withdrew from the OERS, Mali, Mauritania and Senegal decided, in 1972, to set up the OMVS, which pursues the same objectives.

The OMVS has since created a flexible and functional legal framework enabling collaboration and a co-management of the basin. The principal legal texts governing OMVS are:

- the Convention concerning the status of the Senegal River (Convention relative au statut du fleuve Sénégal), 11 March 1972. By this convention, the Senegal River and its tributaries were declared an 'International Watercourse', guaranteeing freedom of navigation and the equal treatment of users;

- the Convention creating the OMVS (Convention portant création de l'Organisation pour la Mise en Valeur du Fleuve Sénégal), 11 March 1972;

- the Convention concerning the Legal Status of Jointly-owned Structures (Convention relative au statut juridique des ouvrages communs), 12 December 1978, supplemented by the Convention concerning the Financing of Jointly Owned Structures (Convention relative aux financements des ouvrages communs), 12 March 1982. These declare that:

  - all structures are the joint and indivisible property of the member states;

  - each co-owner state has an individual right to an indivisible share and a collective right to the use and administration of the joint property;

  - the investment costs and operating expenses are distributed between the co-owner states on the basis of benefits each co-owner state draws from the exploitation of structures. This distribution can be revised on a regular basis, depending on profits;

  - each co-owner state guarantees the repayment of loans extended to the OMVS for the construction of structures;

  - two entities manage the jointly-owned structures for the OMVS: one dedicated to the management and development of the Diama dam (SOGED, Société de gestion et d'exploitation du barrage de Diama), the other to the Manantali dam (SOGEM, Société de gestion de l'énergie de Manantali), both created in 1997.

- in 1992, signature of a framework cooperation agreement between Guinea and the OMVS (Protocole d'accord-cadre de coopération entre la République de Guinée et l'OMVS), creating a framework for cooperation in actions of mutual interest concerning the Senegal River and its basin, including a provision allowing Guinea to attend OMVS meetings as an observer;

- the Senegal River Water Charter, May 2002 (Charte des Eaux du Fleuve Sénégal) whose purpose is to:

  - set the principles and procedures for allocating water between the various use sectors;

  - define procedures for the examination and acceptance of new water use projects;

  - determine regulations for environmental preservation and protection; and

  - define the framework and procedures for water user participation in resource management decision-making processes.

The OMVS functions with the following management bodies:

- Permanent bodies;

- Conference of Heads of State and Government (CCEG, Conférence des Chefs d'Etat et du Gouvernement);

- Council of Ministers (CM, Conseil des Ministres);

- High Commission (HC, Haut Commissariat), executive body;

- Permanent Water Commission (CPE, Commission Permanente des Eaux) made up of representatives of the organization's member states, and which defines the principles of and procedures for the allotment of Senegal River water between member states and use sectors. The CPE advises the Council of Ministers;

- Non-permanent bodies;

- An OMVS national coordination committee in each member state;

- Local coordination committees;

- Regional Planning Committees (CRP, Comités Régionaux de Planification);

- Consultative Committee (CC, Comité Consultatif).

This organizational framework, statutorily strong but flexible on the operational level, enables all of the actors and stakeholders to participate effectively in the efficient management of both the basin's natural resources and its other economic potentials. For more than thirty years now, they have been able to find suitable solutions to all of the technical, social, political and other problems linked to the development of the Senegal River basin's water resources.

Finances

Two types of funding are used to finance the development of the Senegal River basin. The first one covers the operating costs of the various OMVS bodies, and comes from the three member states; each of them pays one third of the total in January of every year. To finance the jointly owned structures and other development activities, funds are sought in the form of loans extended either to the states or directly to the OMVS. In this case, the member states must guarantee the loans. Each member state ensures the reimbursement of its share of the loans.

The apportionment of costs and debts is done according to an accepted formula, subject to revision, as stipulated in the conventions. The underlying principle of cost recovery is that the users pay, but economic conditions are also taken into consideration. Taxes paid to the organization are used to cover operating expenses.

Managing multiple uses: an original approach

Due to potential conflicts between power generation and the other uses of the Senegal River, the three governments have embarked through OMVS on the implementation of an environmental impact alleviation and follow-up programme (PASIE, Plan d'Atténuation et de Suivi des Impacts sur l'Environnement). It is an environmental programme specifically designed to address, monitor and mitigate the environmental issues raised by (or related to) the development and distribution of power from the Manantali power plant.

The OMVS's fundamental conventions of 1972 and the Senegal River Water Charter signed in May 2002, which establish its legal and regulatory framework, clearly state that river water must be allocated to the various use sectors. The resource is not allocated to riparian states in terms of volumes of water to be withdrawn, but rather to uses as a function of possibilities. The various uses can be for agriculture, inland fishing, livestock raising, fish farming, tree farming, fauna and flora, hydroelectric energy production, urban and rural drinking water supply, health, industry, navigation and the environment.

The principles and procedures for the allocation of water were drawn up and a Permanent Water Commission (PWC) was set up to serve as an advisory body to the OMVS's Council of Ministers that makes decisions and asks the High Commission to oversee their application. The OMVS's process for managing needs has four steps.

- First, an inventory of needs is taken by the OMVS National Committees under the Ministries in charge of water in each country. The 'state of needs' is then sent to the OMVS High Commission.

- The High Commission centralizes all of the needs, writes a synthesis report and convenes a meeting of the Permanent Water Commission to vote on recommendations. It then draws up a record of the proceedings with precise recommendations for the Council of Ministers.

- The Council of Ministers makes decisions based on the information provided by the Permanent Water Commission, either in a formal meeting or by informal telephone consultation. The High Commission receives instructions from the Council of Ministers and transmits to member states and other actors the procedures for carrying out the measures adopted by consensus by the member states in the Council of Ministers.

- The work of the Permanent Water Commission and the criteria used by the ministers for decision-making are based on the following general principles:

  - reasonable and fair use of the river water;

  - obligation to preserve the basin's environment;

  - obligation to negotiate in cases of water use disagreement/conflict; and

  - obligation of each riparian state to inform the others before undertaking any action or project that could affect water availability.

The objective of the OMVS method of water allocation is to ensure that local populations benefit fully from the resource, while ensuring the safety of people and structures, respecting the fundamental human right to clean water and working towards the sustainable development of the Senegal River basin.

Approach/procedure

The construction of the basic first-generation infrastructures (the Diama anti-salt dam and the Manantali multipurpose hydroelectric dam) marks the partial conclusion of a major phase, based on a development approach.

Today, the OMVS is attempting to redefine medium- and long-term development strategy for the entire basin, associating development with inextricably integrated and sustainable management. In March 2002, the OMVS began drafting a Master Plan for Development and Management (SDAGE, Schéma Directeur d'Aménagement et de Gestion des Eaux) of the Senegal River basin. This procedure enables progress to be made in the following areas:

- education, by favouring collaboration between stakeholders;
- technology, by reassessing the situation (diagnosis of the entire basin) and defining the strategic orientations and measures required to establish sustainable resource management practices;
- legislation, by ensuring that regulatory actions carried out in all of the member states are coherent; and
- finances by focusing funding on future OMVS programmes.

Moreover, the May 2002 effective date of the Senegal River Water Charter and the start-up of environmental monitoring by the Observatory represent golden opportunities for increasing the involvement of representatives of the various stakeholders in the resource management decision making process. This participatory approach will be reinforced by the launching of the Master Plan next year.

Identifying Main Problems

Degradation of ecosystems

The flood plain ecosystems have been most affected by dam construction. In less than ten years, the degradation of these environments and the consequences on the health of the local population have been spectacular.

Upstream of Diama, the functioning of regularly flooded wetlands, lakes and ponds, such as the Djoudj, Guiers Lake and Lake Diawling, has been seriously disrupted. After 1986, Diama dam blocked seawater intrusion. The water above the dam is now fresh year-round, creating ecological conditions favouring the proliferation of freshwater plants (*Typhas*, *Pistia startioles*, *Salvinia molesta* and various alga species). These are very invasive and eutrophication has begun at some places in the valley and the delta. Downstream of the Diama dam, perturbations in the functioning of ecosystems takes the form of an increase in salinity and/or a drying-up during part of the year (Ndiael wetlands) due to the reduction of flooding or the destruction of water inflow channels during construction of the development works (dikes, irrigated areas). Anthropogenic pollution is caused by the discharge of industrial and agricultural chemicals into these environments.

Other problems arise from increased competition for agricultural land and firewood. As marginal land on slopes and river banks is cleared, there is increased erosion. In addition, large areas of the basin have been denuded due to overgrazing. As was shown in table 20.3, a big percentage of the population is pastoral and therefore must compete for land, increasing competition between agriculture and pastoralism.

Public health

The degradation of the basin's ecosystems has affected the riparian population to various degrees. For example, there has been a drop in productivity in some economic areas (agriculture, fishing, livestock-raising) compared to productivity during the first years after the dams were filled, which has led to a decrease in income and, therefore, a decrease in the standard of living.

The most serious problem that the basin has had to face since 1993/4, however, is the impact of the dams on public health. There has been not only a rapid increase in the prevalence of water-borne diseases that were already present in the area (malaria, urinary schistosomiasis, diarrhoea, intestinal parasitic diseases), but also the appearance of intestinal schistosomiasis, a much more dangerous form of the disease.

Malaria

Malaria is a major public health problem in the basin. Indeed, it is the primary reason for consultations in health clinics and the primary cause of death. It causes 90 percent of the cases of fever. It is caused by *Plasmodium falciparum*, the most deadly species of *Plasmodium*, carried by the anopheles mosquito *Anopheles gambiae*.

Diarrhoea

Surveys done recently by the OMVS in the three member states indicate a diarrhoea prevalence of 15 to 30 percent around Podor (Senegal). This rate is estimated to be about 25 percent in Mauritania and 15 percent in Mali. Many factors favour the appearance of diarrhoea, but in the Senegal River basin, aside from general hygiene, the primary cause is the abusive use of agricultural fertilizers and pesticides, which, at the end of the hydrological cycle, end up in the human food chain.

Schistosomiasis (bilharzia): urinary and intestinal

In the OMVS area of the basin, there are two types of human schistosomiasis: urinary and intestinal. Intestinal schistosomiasis was unknown in the region before the dams were built, but today is rampant in the valley and delta. The blocking of saltwater intrusion upstream has allowed the snails that host the parasite (*Schistosoma mansoni*) to proliferate in the desalinated river and irrigation canals. Humans are contaminated when the parasite penetrates the victim's skin.

Surveys conducted in 2000 by OMVS health services in the member states indicated a prevalence of urinary schistosomiasis disease in about 50 percent of the region around Saint Louis. In Mauritania, in the Trarza, the average infestation is estimated to be at 25 percent, with places where the increase in prevalence is quite spectacular. For example, in Lexeiba, the infestation rate increased from 8 to 50 percent in only a few years. In Mali, urinary schistosomiasis is highly present, with a rate of 64 percent.

The development of intestinal schistosomiasis demonstrates even more clearly the impact of development on the health of the region. Unknown in Mauritania before the dams were filled, the first cases were reported in 1993. One year later, a survey showed that the population of school children in Rosso had an overall prevalence of 32.2 percent. In Senegal the situation is even worse, with a 44 percent rate of infestation in the Walo flood plain, 72 percent in the area around Guiers Lake where more than 90 percent of the villages are affected. In Mali, this form of schistosomiasis is still present in specific areas, with an infestation rate of 3.34 percent in 1997, but the situation calls for close monitoring.

Conclusions

The OMVS has demonstrated its effectiveness. It has been tested for more than thirty years now, and was recently improved by the adoption in May 2002 of the Senegal River Water Charter. This framework enables a collaborative management approach, with effective involvement of local actors/stakeholders, recognized and accepted by all of the riparian states including Guinea, who has signed cooperation agreements with the OMVS prior to being reintegrated into the organization.

It also establishes the principles and terms of water sharing between the different usage sectors, based on the original concept of 'water distribution' among the users and riparian states in which sharing the water resource is not a matter of withdrawals, but rather one of optimal satisfaction of usage requirements. The OMVS looks for an equitable managing and distribution of water resources among multicultural ethnic groups who are gathered around the river basin and its water resources. The new dams and institutional framework have brought greater prosperity and economic revenue and replaced a situation of water scarcity and conflict among users as that which prevailed before the 1980s.

However, these undeniable achievements cannot hide new difficulties emerging with the dams' implementation: displacement of populations, water-borne diseases, proliferation of invasive aquatic vegetation, degradation of cultivated land and water pollution are the major issues the basin will have to cope with in the near future.

Box 20.1: Development of indicators

At the OMVS, the insufficiency or even total lack of temporal and spatial data for several sectors has made it almost impossible to correlate the increased availability of the water due to developments and the environmental and health problems that this has caused and their direct and indirect impact on the living conditions of local populations. Therefore, to eliminate this information control constraint, and to better understand the evolution of development in the basin, the OMVS is being reorganized so that it can collect, process and store all the data needed to monitor development project performance indicators from the upper basin to the mouth of the river. The Environmental Observatory was created by the High Commission in November 2000 for this purpose. Between November 2000 and December 2001, indicators were defined and strategies set up to gather, process and store data that will enable OMVS to correlate water availability, public health, the state of the environment and socio-economic development. These indicators concern:

- the productivity of activity sectors (agriculture, livestock raising, fishing, mining);
- the market rate of crops grown in the basin;
- the percentage of participation of women in economic activities;
- the impact of the involvement of women by activity sector;
- the quality and the quantity of domestic water use;
- the rate of access to drinking water of the populations living along the river;
- the prevalence of water-borne diseases (human and animal);
- the state of the environment (degradation of soil, forests, water bodies);
- the quantitative estimation of the degradation of ecosystems by sector of activity;
- the quantitative estimation of the health situation in each sector of activity;
- the rate of immigration and emigration in the zone; and
- the quantitative estimation of the corrective measures taken to eliminate the negative impact of developments.

References

Adams, A. 2000. *Fleuve Sénégal: gestion de la crue et avenir de la vallée*. London, International Institute for Environment and Development.

Bonavita, D. 2000. *Manantali entre espoir et désillusion*. Paris, Le Figuier.

Coyne and Bellier. 2000. 'Etude pour la mise en place d'un Observatoire de l'Environnement de l'OMVS'. Gennevilliers, France.

Crousse, B.; Mathieu, P.; Seck, S.-M. 1991. *La vallée du fleuve sénégal. Evaluations et perspectives d'une décennie d'aménagements (1980–1990)*. Paris, Kartala.

Fetter, C.-W., Jr. 1994. *Applied Hydrogeology*. 3rd ed. New York, Prentice-Hall Publisher Co.

Gnoumou, Y. 2000. *Prise en compte des pratiques et connaissances du milieu dans la mise en place d'un réseau décentralisé des ressources environnementales et de l'eau du bassin du fleuve Sénégal*. Thesis, published in Dakar.

IRD (Institut de Recherche pour le Développement). 2001a. *Programme d'optimisation de la gestion des réservoirs – Manuel de gestion du barrage de Manantali*. Organization for the Development of the Senegal River.

——. 2001b. *Programme d'optimisation de la gestion des réservoirs – Phase III – Manuel de gestion du barrage de Manantali*. Organization for the Development of the Senegal River.

——. 2001c. *Programme d'optimisation de la gestion des réservoirs – Phase III – Crue artificielle et culture de décrue. Synthèse finale*. Organization for the Development of the Senegal River.

OMVS (Organization for the Development of the Senegal River). 1994. 'L'OMVS, pour un développement économique intégré de la sous-région'. *Bulletin d'Information* (Bulletin Report), No. 3.

——. 1979. *Bulletin d'Information* (Bulletin Report), No. 1.

Salem-Murdock, M. and Niasse, M. 1996. *Conflits de l'eau dans la vallée du fleuve Sénégal. Implications d'un scénario 'zéro inondation'*. London, International Institute for Environment and Development.

SCP (Société Canal de Provence); Coyne and Bellier; Senagrosol. 2002. *Cost-Benefit Analysis to Develop an Optimal Dam Management Scheme*. Dakar.

USAID (United States Agency for International Development). 2002. *Avenir du bassin du fleuve Sénégal – Prendre les bonnes décisions maintenant*. Internet publication.

'There is an urgent need to intensify the fight against poverty and to drastically improve the health system in the [Lake Titicaca] region. Great benefits could be achieved through the provision of better services for waste disposal and treatment, the promotion of environmental education and the continuation of the water regulation works already in progress. Most importantly, however, there needs to be far greater investment in public health services in order to ensure a better quality of life for the population.'

21

Lake Titicaca Basin, Bolivia and Peru

By: The Binational Autonomous Authority of Lake Titicaca (ALT, Autoridad Binacional del Lago Titicaca Bolivia-Perú)

Table of contents

The frog does not drink up the pond in which it lives.

Inca proverb

SITUATED AT AN ALTITUDE OF 3,600 to 4,500 metres in the highest plateaux of the Andes, Lake Titicaca straddles the border of Bolivia and Peru and comprises a basin network of four distinct lakes. The surrounding environment is fragile, subject to flooding and, increasingly, pollution. A unique feature of this pilot case study is the presence of indigenous, pre-Hispanic peoples who continue to follow their ancient cultural traditions and resist assimilation into Western-style societies. These people are extremely poor, and only about 20 percent have access to water and sanitation. The major challenges for the Binational Autonomous Authority of Lake Titicaca, therefore, are to find ways to promote land tenure reform, adopt appropriate farming and irrigation techniques, and develop legislation that will provide an enabling environment in which culturally sensitive development and resource-sharing may occur.

LAKE TITICACA IS A REGION OF MYSTERY AND LEGEND. Originally inhabited by the Urus, a race today extinct, it was dominated successively by Aymara warlords, Quechuas of the Inca empire and Spanish conquerors. Along its banks flourished the Tihuanacu culture (1500 AC) that left behind immense megalithic constructions and complex agricultural systems redolent of an advanced civilization. Before it mysteriously disappeared, its art, culture and religion had spread throughout the entire Andean region.

Map 21.1: Locator map

Source: Prepared for the World Water Assessment Programme (WWAP) by AFDEC, 2002.

General Context

Location

At 14 degrees south, the Andean ridge is divided in two branches: the eastern and western ridges. Between them is a closed hydrological system of approximately 140,000 square kilometres (km^2) located between 3,600 and 4,500 metres above sea level (m.a.s.l.). Within that system lie four major basins (see table 21.2): Lake Titicaca (T), Desaguadero River (D), Lake Poopó (P) and Coipasa Salt Lake (S). Desaguadero River is Titicaca's only outlet and flows into Lake Poopó, the overflow from which in turn gives rise to Coipasa Salt Lake. These four basins form the TDPS system of which the main element, Lake Titicaca, is the largest in South America, the highest navigable lake in the world, and, according to Inca cosmology, the origin of human life.

Major physical characteristics

Topography

Three geographical units can be distinguished in the system:

- the mountain ridge, with altitudes greater than 4,200 m.a.s.l.;
- slopes and intermediate areas, ranging in altitude between 4,000 and 4,200 m.a.s.l. with moderate to steep slopes and a dense hydrographic network; and
- the high plateau, from 3,657 to 4,000 m.a.s.l., in which Lake Titicaca is located. The surrounding area, the most densely populated of the system, varies in height from 3,812 to 3,900 m.a.s.l. Between Titicaca and Poopó, there is a ridge rising to 1,000 metres above the level of the plateau, which is split from west to east by the Desaguadero River. Along the far western edge is a narrow strip

Table 21.1: Hydrological characteristics of the Lake Titicaca basin

Surface area of the basin	57,293 km^2
Annual precipitation	702 mm/year
Annual discharge	281 m^3/s
Annual potential evapotranspiration	652 mm/year

Map 21.2: Basin map

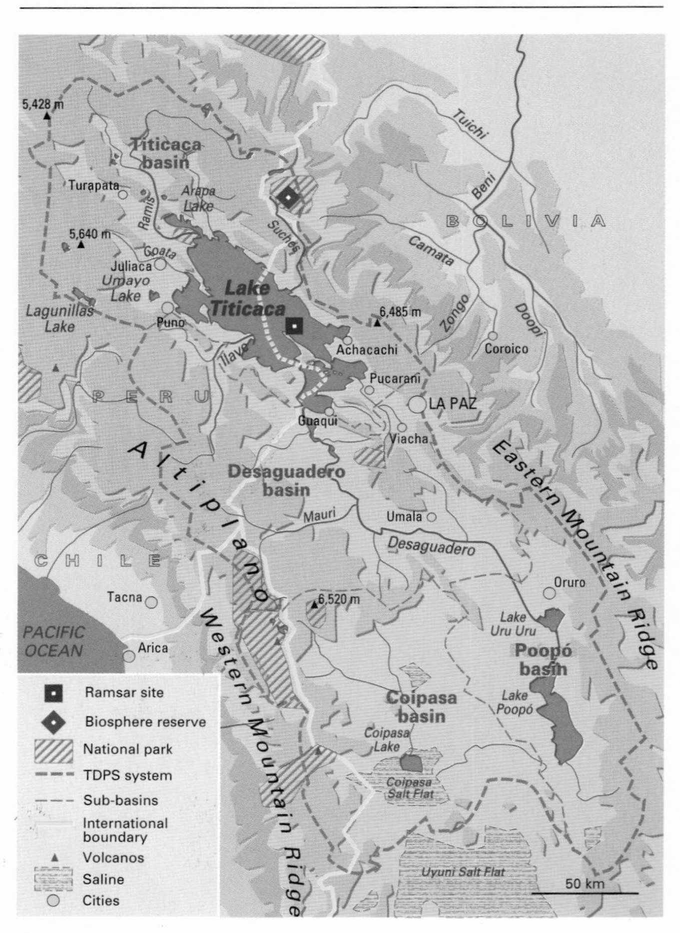

Source: Prepared for the World Water Assessment Programme (WWAP) by AFDEC, 2002.

Table 21.2: TDPS system size

Basin	Area (km^2)
Lake Titicaca	56,300
Desaguadero River	29,800
Lake Poopó	24,800
Coipasa Salt Lake	33,000
TDPS System	143,900
Lake Titicaca	
Average area	8,400 km^2
Average altitude	3,810 m.a.s.l.
Average volume	930 km^3
Maximum length	176 km
Maximum width	70 km
Maximum depth	283 m
Desaguadero River	
Length	398 km
Average flow	70 m^3/s
Average gradient	45 cm/km
Lake Poopó	
Average area	3,191 km^2
Average altitude	3,686 m.a.s.l.
Coipasa Salt Lake	
Average area	2,225 km^2
Average altitude	3,657 m.a.s.l.

of desert that runs along the Pacific coast, and to the east are the Amazon plains that extend to the Atlantic Ocean. The system is located in the southern part of Peru and the north-west of Bolivia. The source that feeds the lake, situated to the north, belongs mostly to Peru. Of the five major rivers flowing into the lake, four run through Peruvian territory. The southern part of the system, which belongs to Bolivia, is drier and ends in the Coipasa Salt Lake, which is formed by the evaporation of overflow from Lake Poopó.

Climate

The climate within the TDPS system is that of a high mountain region with a tropical hydrological regime of great interannual irregularity. In the surrounding area, Lake Titicaca exercises a moderating influence on temperatures and rainfall.

Precipitation varies between 200 and 1,400 millimetres (mm), with maximum value of 800 to 1,400 mm at the centre of the lake. The system shows zones of diminishing humidity from north to south, going from humid around Lake Titicaca, to semi-arid in Lake Poopó, to arid in the Coipasa Salt Lake. There are great seasonal variations, as the area usually has wet summers and dry winters, with a rainy period from December through March and a dry period from May through August. The air temperature varies within the system depending on latitude, longitude, altitude and proximity to the lake, with minimums of -10 to -7°C and maximums of 19 to 23°C. Humidity is low throughout the system, with an average of 54 percent and variations depending on latitude and season. The area also receives strong solar radiation with an annual yield of 533 calories per square centimetre (cm^2) per day: this high radiation explains the intense evaporation that occurs in Lake Titicaca.

Land types

There are four types of land in the Lake Titicaca basin, classified as follows.

- Arable land: Due to climatic conditions and the altitude of the high plateau, special agricultural practices are required. Most of the soils have organic matter and nitrogen deficiencies. Only 33.9 percent of the TDPS land area is arable land. It covers 44,692 km^2.
- Non-arable land: Such land requires special practices to maintain permanent plant cover. Non-arable surface covers 28,063 km^2 or 21.3 percent of the total.
- Marginal lands: These are characterized by moderate to strong erosion processes, but with potential use for extensive grazing of llamas and alpacas. The total area is 40,844 km^2 or 31 percent of the whole basin.
- Badlands: Although unsuitable for agriculture or grazing, such areas can be used for wetlands, recreation and mining. Badlands cover 18,178 km^2, representing 13.8 percent of the system.

Major socio-economic characteristics

Population

The pre-Hispanic ethnic groups on both sides of the lake maintain ancestral cultural patterns which are unlike those of Western culture. The annual economic growth rate for the system is very low, with a declining tendency in the rural areas. This is due mainly to extensive poverty, which results in high infant mortality and rural-to-urban migration. A diminishing soil fertility rate can also be observed. Tables 21.3 and 21.4 give an overview of the system's populations.

Poverty is the most critical social problem in the TDPS system, affecting both rural and urban populations. Families have to devote all their energies to meeting basic needs, and locally available resources are too limited to offer much hope of improved living conditions. Extreme poverty and a lack of opportunity compel the rural population, especially young people, to migrate to the cities,

Table 21.3: TDPS population data

	Peru	Bolivia
Population	1,079,849	1,158,937
% of total	48.2	51.8
Average population density (inh/km^2)	17.6	15.56
Maximum density (inh/km^2)	215	245
Minimum density (inh/km^2)	2.0	2.3
Rural population (%)	60.8	47.9
Urban population (%)	39.2	52.1
Growth rate (%)	1.6	from -1.6 to 9.2
Population trends	generally decreasing	rural decreasing
Population in poverty situation (%)	73.5	69.8

Peru and Bolivia show comparable population situations in terms of numbers, density and the high percentage of people living in poverty. There are, however, more rural populations in Peru than in Bolivia.

Table 21.4: Population in main cities

Main cities	Population	% of country population
Puno (Peru)	91,877	4.10
Juliaca (Peru)	142,576	6.37
El Alto (Bolivia)	405,492	18.11
Oruro (Bolivia)	183,422	8.19

where they crowd into degraded districts. In 1993 Bolivia's urban population grew by 4.3 percent while the rural rate was negative at -0.4 percent. In the same year, the Peruvian side of the TDPS registered 3.4 percent annual growth in population, while the rural population grew at only 0.7 percent.

Education

The conditions of structural poverty in the zone are such that the struggle to survive takes precedence over anything else. Education is therefore not a priority. The illiteracy rate is 22 percent and is differentiated by area and gender. It is higher in rural areas than in cities, and within rural areas, it is higher for females. Among the problems affecting the quality of education are dispersion of the rural population and the existence of non-Spanish mother tongues. The Bolivian Educational Reform Programme has been trying to address both situations for eight years.

Health

Health problems in the TDPS system are clearly related to endemic poverty and, by extension, to such attendant problems as poor nutrition, lack of clean water and sanitation, a fragile environment and the absence of leverage to help people improve their lives or livelihoods. In several cases, the problems are compounded by the existence of strong and persisting ancient cultural traditions; child vaccination, for example, has only recently been adopted by local populations because of legal enforcement. The main health indicators in the region include:

- high rates of morbidity and mortality, mainly in children;
- low life expectancy at birth (lower than the national average);
- high incidence of infectious, respiratory and gastrointestinal diseases;
- high incidence of diseases linked to conditions such as water quality (gastrointestinal diseases) and climate (respiratory diseases);
- deficient nutritional levels in general, both in quantity and quality; and
- health services that are generally poor and mostly concentrated in urban areas.

Economic activities

Bolivia, with a human development index of 0.648, and Peru, with 0.743, are both in the middle human development range. The Gross National Product (GNP) is US$116.6 x 10^3 million Purchasing Power Parity (PPP) in Peru and 19.2 x 10^3 million PPP in Bolivia. GNP per capita is US$4,622 and US$2,355 PPP respectively. However, in 1993 it was estimated that GNP per capita in the Bolivian sector of the TDPS amounted to 35 percent of the national value.

Table 21.5: Health data

	Peru	Bolivia
Life expectancy (years)	60.6	58
Available hospital beds/1000 inhabitants	1.1	1.3
Number of physicians in the area	212	1,128
Infant mortality rate/1000 children < 1 year	81	121
Children suffering chronic malnutrition (%)	71	84
Morbidity (children < 1 year)		
respiratory diseases (%)	39.6	27
nutritional deficiencies (%)	18.5	18
diarrhoea and gastrointestinal diseases (%)	18.7	13
other (%)	23.2	24
Morbidity		
respiratory diseases (%)	20	22
nutritional deficiencies (%)	14	15
diarrhoea and gastrointestinal diseases (%)	7.6	6
other (%)	66	57

The health challenges facing the basin are significant in both countries.

Figure 21.1: Distribution of the active working population of Bolivia and Peru

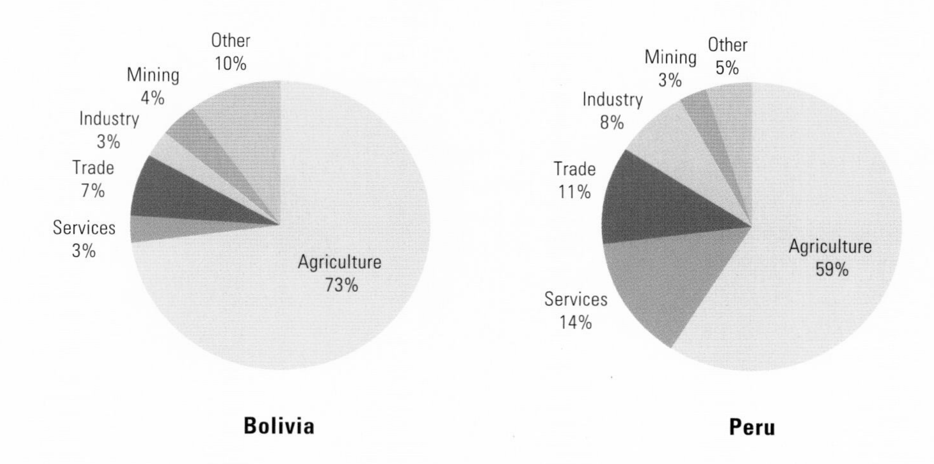

Although Bolivia has a more active agricultural sector than Peru, with a 73 percent working population, both countries are dominated by this activity. Industry is a far less important sector, accounting for only 3 percent of the active working population in Bolivia and 8 percent in Peru.

Agriculture and cattle-raising activities, both focused on food production, are the main economic activities. The staple crops are: quinoa, potato and other tubers, fodder and some leguminous species and vegetables. In general, yields are low because of the limited use of sown seeds, fertilizers and machinery. Drought, floods and frost events are also significant factors (see figures 21.3 and 21.4 further on).

By the middle of the twentieth century, both Bolivia and Peru had independently begun reform processes directed at modifying land ownership. In both countries, land was formerly concentrated among a few owners of large land holdings. But the reform effort resulted in a decrease of agricultural production and an excessive fragmentation of the rural property, particularly in Bolivia. Land holdings on the Peruvian side vary from 0.5 to 20 hectares. Average rural holdings on the Bolivian side of TDPS are small, in extreme cases perhaps no more than a few square metres. In such conditions only small-scale subsistence agriculture is possible. In Bolivia the Political Constitution of the state declares rural property (specifically in the TDPS area) inalienable, meaning that it cannot be sold or used as collateral for a loan, and in fact no market for rural property exists.

Other industries are also present within the system. Agro-industry is underdeveloped and small scale. Forestry is poor although there is potential for increasing production. Trading of farm products is inefficient. Credit is limited and selected, particularly because of the difficulties the native population encounters in trying to understand the banking process. There is a growing trend towards cooperative credit. Although of potential importance, fishery is not dynamic: the fish biomass of Lake Titicaca has been estimated at some 91,000 tons, while extraction fluctuates between 4,000 and 7,000 tons. Figure 21.1 shows the distribution of the active working population in the two countries.

Cultural background

Lake Titicaca is known by the name of Lago Sagrado (Sacred lake) among the Aymaras and constitutes the central element of Inca mythology. The high-plateau population practices its own cultural traditions, which prevail in spite of four centuries of Spanish colonization. The cultural pattern is agrocentric whereby all human activities take agriculture as a central reference point and the value of work becomes the unifying social force as well as the only source of wealth. From this assigned value there are related values pertaining to reciprocity, redistribution and communal democracy.

These patterns and the ways of customary law co-exist with Western patterns and in many cases result in underdevelopment and social exclusion. Because cultural tradition plays such an important role in the lives of local people, it is necessary to understand and take account of the prevailing value system before attempting to introduce changes. Among the many elements to consider, the most important are the following.

- Indigenous wealth: Indigenous peoples in the Titicaca area seek to minimize risk rather than maximize production.
- Property: There exists a complex system in which communal property is superimposed over individual holdings and territory.
- Water resources: The traditional property pattern that exists among the upstream communities sets certain conditions for determining how the resource is shared.

Water Resources

Hydrology

Surface water

As described earlier, four main hydrographic basins form the TDPS system: Lake Titicaca is the main element. Its principal tributaries are located in Peruvian territory: Ramis and Huancané to the north, Coata and Illpa to the west, and Ilave and Zapatilla to the south-west. In Bolivia, the most important affluents are: Huayco, Suchez and Keka to the north and east, and Catari and Tiwuanacu to the south. The Ramis River is the most important as it represents 26 percent of the tributary basin. Its flow is about 76 cubic metres per second (m^3/s). Annual flow in ten stations of the system is shown in table 21.6.

Overflow from Lake Titicaca, observed at the source of the Desaguadero River, amounts to 35 m^3/s. This flow represents only 19 percent of the inflow of the five main tributaries, demonstrating the great volume lost through evaporation.

Between 1914 and 1992, historic levels of variation took place during different cycles: a major cycle lasted twenty-seven to twenty-nine years while an intermediate one lasted twelve to sixteen years. Lake-level oscillation in this period had an amplitude of 6.37 metres compared to the average annual oscillation of about 1 metre.

Groundwater

As can be seen in map 21.3, the main aquifers are located in the middle and lower basins of the Ramis and Coata Rivers, in the lower basin of Ilave River and in a strip that extends from Lake Titicaca to

Map 21.3: Distribution of groundwater in the TDPS system

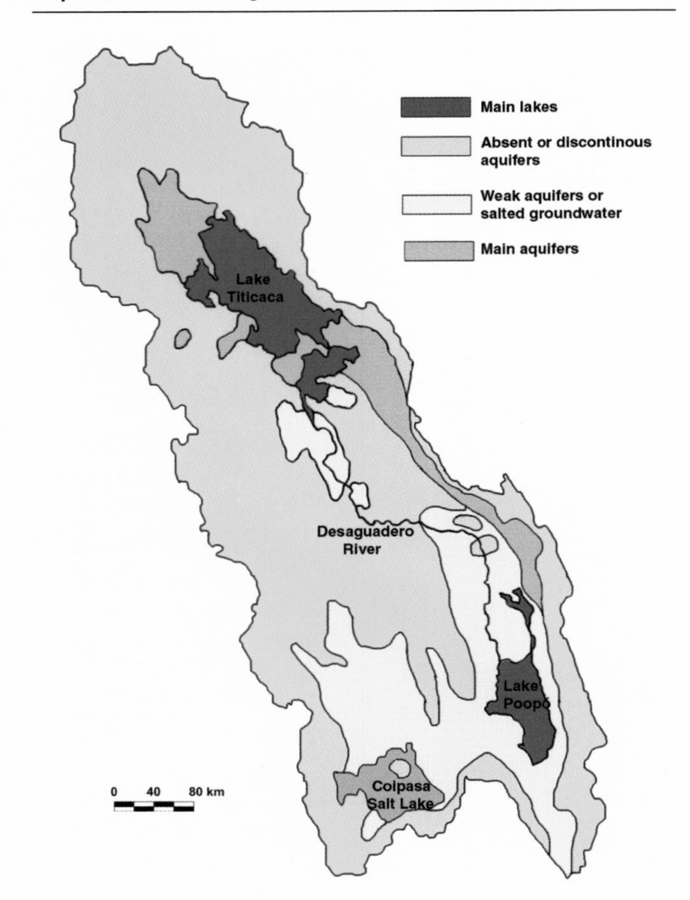

The main aquifers are located in the middle and lower basins of the Ramis and Coata Rivers, in the lower basin of Ilave River and in a strip that extends from Lake Titicaca to Oruro, bordering the eastern ridge.

Table 21.6: Annual flow in ten control stations of Lake Titicaca and Desaguadero River

River	Station	Average (m^3/s)	Maximum (m^3/s)	Minimum (m^3/s)
Ramis	Ramis	75.6	130.4	24.4
Huancané	Huancané	20.0	38.8	6.9
Suchez	Escoma	10.6	18.9	4.0
Coata	Maravilla	41.5	75.5	2.4
Ilave	Ilave	38.5	96.6	5.0
Desaguadero	International	35.5	186.5	–3.5
Desaguadero	Calacoto	51.9	231.6	6.2
Mauri	Abaroa	4.9	9.8	2.3
Caquena	Abaroa	2.8	5.6	0.9
Mauri	Calacoto	18.6	31.8	5.7
Desaguadero	Ulloma	77.1	282.7	19.7
Desaguadero	Chuquiña	89.0	319.3	20.0

The main tributary of the Lake Titicaca basin is the Desaguadero River, with an average annual discharge of 89 m^3/s and a maximum of 319 m^3/s.

Oruro, bordering the eastern ridge. The approximate total volume of groundwater that goes into the system is 4 m^3/s. Most of this groundwater comes from tube wells used to supply water to cities. Such is the case of El Alto, Oruro and several other small towns.

Water table morphology shows that groundwater flows follow the direction of water reservoirs, the location of recharge areas and their base levels. The water tables of Huancané, Ramis, Coata and Parco River basins on the Peruvian side, and Tiwanacu and Catari River basins on the Bolivian side, drain into Lake Titicaca with average hydraulic gradients of 1 to 0.1 percent.

The optimum yield of aquifers and capacity in the Peruvian sector range from 1 to greater than 100 litres per second, and from 0.3 to 5 litres per second respectively. In the Bolivian sector, optimum yield ranges from 2 to 75 litres per second and specific capacity from 0.3 to 4 litres per second.

Water quality

There are higher levels of salinity found in the south of the TDPS system, a result of greater rainfall in the northern part of the system that reduces the concentrations of dissolved salt. On the other hand, evaporation, which is greater in the southern part of the system, increases the concentration levels of dissolved salt. Thus, there is a progressive increase of electric conductivity from north to south. Likewise, it is not uncommon to find tertiary and quaternary formations, which are mainly present on the TDPS system, with parental material formed by rocks containing gypsum and salt.

In general, Lake Titicaca and its tributaries have normal values of water salinity (less than 1,000 milligrams [mg] per litre). Desaguadero River has values between 1 and 2 mg/litre, but downstream values are greater than 2 mg/litre. Lake Poopó has salinity values above 2,000 mg/litre due to natural conditions and mining activity in the surrounding area. Maximum salinity values were found in Coipasa Salt Lake where evaporation is high and rainfall is only 200 mm per year.

Mining activity is the principal cause of heavy metal contamination, and is mostly found in the southern part of the TDPS system. High concentrations of arsenic are found at La Joya, in the western arm of the Desaguadero River. Lake Poopó as well as Coipasa Salt Lake present high levels of lead, cadmium, nickel, cobalt, manganese and chromium.

High values of faecal-coliform (1,000 parts per million [ppm]) and organic matter throughout the Puno Bay are a good indicator that pathogens are present in the water. Those high values are mainly the result of wastewater from the Puno City sewer system. This contamination has generated a process of eutrophication and the growth of aquatic lentils in the bay.

Both water and fish from Lake Titicaca reveal high parasite levels, probably due to inappropriate disposal of wastewater in the cities of Puno and Juliaca in Peru, and Copacabana in Bolivia, as well as animal-raising and agricultural activities in areas surrounding the lake. The parasites infect humans as well, hence the high incidence of gastrointestinal diseases.

Extreme events

Most extreme events in the TDPS system are related to flood risk conditions around Lake Titicaca, drought in the central and southern parts of the system and the incidence of hail and frost throughout (see maps 21.4 and 21.5).

Human impacts on water resources

Surface cover

Until approximately the year 1000, the high plateau was covered with a native tree forest (*Polylepis sp*). Around the year 1100, a severe eighty-year drought changed the surface cover and the forest disappeared. After 1500, inappropriate agricultural practices and imported livestock permanently modified the conditions of the surface cover. Over the last century, human activities have not had a significant impact on the surface cover of the system in large part because of the arid environment and lack of vegetation.

Dams and diversions

The 6.37 metre variation between maximum and minimum registered lake levels produced historical flood events in the lake and surrounding areas. The Master Plan of the TDPS system (see further on for details) has required regulation works to be built that maintain the lake level at the minimum of 3,808 m.a.s.l. with a maximum of 3,811 m.a.s.l. during a normal hydrological cycle.

Pollutants

Organic and bacteriological contamination is caused by human activity, in particular urban wastes and mining. Poor waste disposal is the central cause of organic contamination in all the important urban centres in the basin. The most polluted areas affected by sewage discharge are Puno's interior bay (undergoing a moderate eutrophic process), the lower course of the Coata River, because of the discharge from the city of Juliaca, and Lake Uru Uru, due to discharge from the city of Oruro.

Heavy metal contamination is the result of mining activities in the zone. Although there is not enough available information on this subject, mercury and arsenic concentrations of 0.4 ppm have been found in mackerel captured in Puno Bay.

Non-native species

Non-native fish species with high economic value such as trout and mackerel were introduced into Lake Titicaca around 1930. Since then some native species such as karachi (*Orestia sp*) and boga (*Trichomicterus sp*) have decreased, and their populations are considered vulnerable and endangered.

Map 21.4: Incidence of extreme events – frost

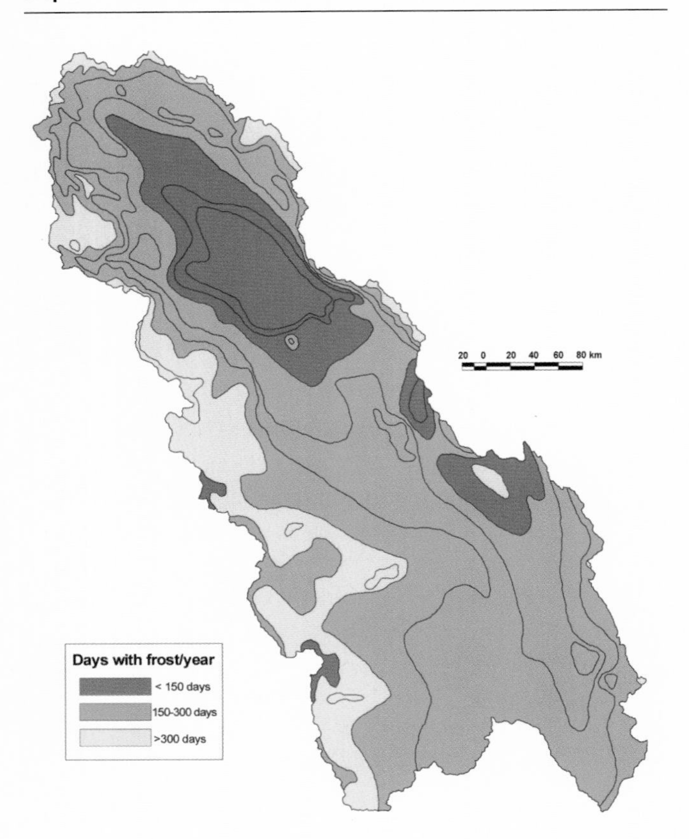

Frost occurs less frequently in the centre of the lake system, where it occurs less than 150 days per year. Most of the system suffers from frost 150 to 300 days per year, while the outer rims of the basin can have frost on more than 300 days.

Map 21.5: Incidence of extreme events – precipitation during periods of drought

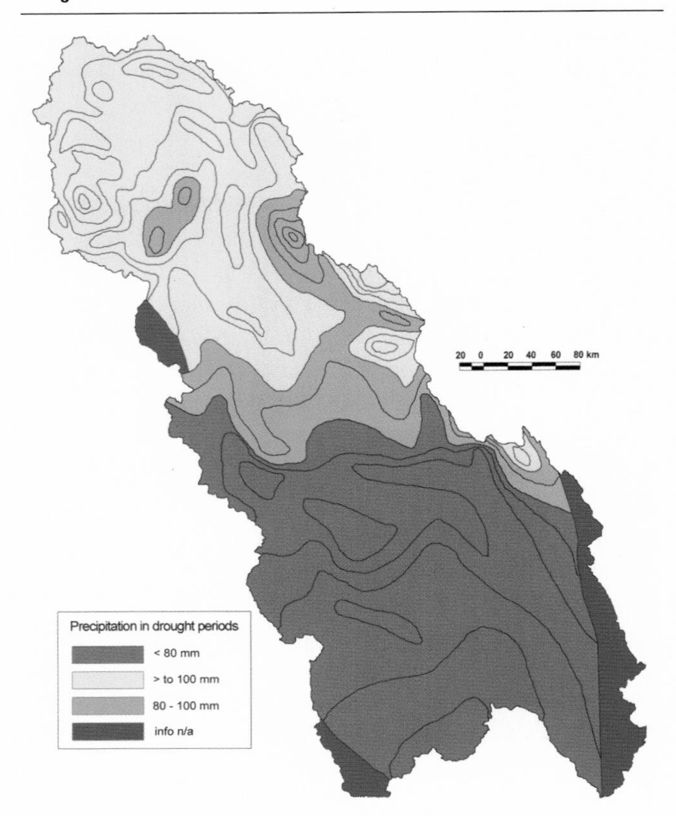

The northern part of the lake system receives more precipitation (more than 100 mm) during drought periods than the southern part (less than 80 mm).

Overharvesting

Land areas suitable for crops and pasture are comparatively lower than the occupied land, with the percentage of areas exploited above their capacity reaching 35.2 percent. This overexploitation is one of the serious environmental problems affecting the high plateau, especially considering the area's very low productivity and the rudimentary technology used to exploit it.

Data and information on water resources

Since the creation of the Binational Autonomous Authority of Lake Titicaca (ALT) in 1993, several efforts have been made to consolidate the available information on water resources in the TDPS system. Most of the information was scattered among different institutions in Bolivia and Peru. Early in the 1990s, international consulting firms prepared a Master Plan that compiled most of the available information about the TDPS system.

The creation of ALT and the elaboration of the Master Plan allowed data and information from different sources to be systematized and it is now possible for Bolivia and Peru to share this information through ALT. However, it is still necessary to improve data collection and dissemination and to standardize the information generated by different institutions.

Table 21.7: Water use in the TDPS system

Use and sector	Surface water (litres/second)	Groundwater (litres/second)	Total use (litres/second)	Net consumption (litres/second)
Domestic	**1,210**	**912**	**2,122**	**424**
Bolivia	361	761	1,122	224
El Alto	51	382	433	86
Oruro	34	379	413	82
Other urban	133		133	27
Rural	143		143	29
Peru	849	151	1,000	200
Puno	25	151	176	35
Juliaca	300		300	60
Other urban	334		334	67
Rural	190		190	38
Irrigation	**7,294**	**85**	**7,379**	**5,534**
Bolivia	4,494		4,494	3,370
Peru	2,800	85	2,885	2,164
Other	1,000		1,000	200
Bolivia	590		590	118
Peru	410		410	82
Total	**9,504**	**997**	**10,501**	**6,158**
Bolivia	5,445	761	6,206	3,712
Peru	4,059	236	4,295	2,446

Irrigation is by far the greatest water user in the TDPS system: it accounts for 75 percent of total withdrawals.

Challenges to Life and Well-Being

Water needs, uses and demands in the TDPS system are mainly directed at covering basic needs and irrigated agricultural production. However, it should be noted that increased water alone cannot improve the local living conditions, which are limited by extreme poverty. Table 21.7 illustrates present water use in the TDPS system.

Water for basic needs

Drinking water and sewage systems are largely deficient throughout the TDPS area. On the Peruvian side, drinking water coverage is between 12 and 30 percent with an average of 19 percent. Sewage system coverage is between 13 and 39 percent with an average of 20 percent. Conditions on the Bolivian side are similar. There is an average drinking water and sewage system coverage of 24 percent and 13 percent respectively.

Bolivian and Peruvian consumption is about 30 and 50 litres per person per day, respectively. However, only 20 percent is taken as a loss to the system because about 80 percent returns to the system as sewage.

Water for food

About 48 percent of the TDPS system area is being used for agriculture; 4.4 percent for crop production, 21.7 percent for grazing, 14.9 percent for grazing-forestry and 7 percent for other uses.

Most of the crop production area is located in the areas surrounding Lake Titicaca. However, only 17 percent of the total area is truly suitable for crop production. Therefore, soil erosion and degradation are a major concern. Excessive property fragmentation is another common problem throughout the system. This fragmentation causes low crop productivity because farmers are not able to use technology to increase their crop yield.

Forestry takes up only 3 percent of the basin. Native brushes cover 2.3 percent of this area and 0.7 percent is formed by modified forest comprising native trees called Kiwiña (*Polylepis sp*).

Irrigated land represents about 19,444 hectares, of which 10,960 lie on the Bolivian side and 8,484 lie in Peru. Taking into consideration the combined needs of water for irrigation projects, crops and potential irrigation areas, the water available for irrigation purposes has been estimated at 7,379 litres per second, mainly taken from surface resources.

In contrast with water for basic needs, water for irrigation represents a significant loss for the system. Most of the water used for irrigation goes to the atmosphere through evaporation and transpiration processes, and only 25 percent returns to the system.

Water for ecosystems

The TDPS has a broad variety of flora and fauna. Biodiversity in the basin area is in a precarious position due to the overexploitation of some species. Both the Bolivian and Peruvian governments have responded by establishing protected areas to help preserve these living resources.

Aquatic flora

Green algae and diatoms are the main components of Lake Titicaca plankton, but there are also cyanobacteria with the capacity to fix nitrogen, as well as large populations of chlorophyll and clorococales. Algae populations are found down to depths of 80 and 100 metres, while nitrogen from agricultural runoff seems to be the main constraint on algal development.

Macroscopic plants – macrophytes – are represented in the system by about fifteen species and are located in shallow zones. Some macrophytes such as Totora and Llachu are important elements for animal feed, especially for cattle. They also help absorb heavy metals such as arsenic, zinc and lead. In the same way, Totora is of central importance for many human activities such as building boats and thatching roofs.

A new algae called Carophiceas (Charas) has been observed in deeper sectors and is important for biomass in the lake systems, as it appears to have a good resistance to high levels of salinity.

Aquatic fauna

Zooplankton, benthic fauna, fish and frogs make up the principal aquatic organisms living in the TDPS system. Among the main groups that form the zooplankton, copepods are broadly dominant over the clodoceros population, and their reproduction occurs throughout the year. Globally, more than 95 percent of the benthic population in Lake Titicaca is found within the first 15 metres of the Minor Lake and in the first 25 metres of the Major Lake.

The TDPS system is rich in fish species, distributed in different hydrological units of the system. Lakes Titicaca and Poopó have the major concentration of commercial fish species.

The frog population is grouped into four genera, among them the *Telmatobius* genus, which includes one of the biggest aquatic species in the world.

Terrestrial plants and animals

Plants and animals living in the TDPS system have adapted to the ecological conditions of the region. The system can be divided into two main regions: Puna (lower than 4,400 m.a.s.l.) and the Altoandina (higher than 4,400 m.a.s.l.). There are different species of bushes and trees in each, including grasslands and native trees. In addition, Lake Titicaca itself was declared a Ramsar site in 1998. It is the only Ramsar site in the TDPS System (see box 6.5 in chapter 6 for further details).

Water and cities

Access to sanitation is low, at only 17.2 percent throughout the basin. One city, El Alto, has a drinking water system administered by a private company (Aguas del Illimani). The others cities are managed by the municipalities. Average coverage in major cities reaches about 60 percent. Average tariffs for drinking water consumption vary from US$0.13 per litre in Puno and Juliaca to US$0.22 per litre in El Alto. The smaller urban centres have small drinking water systems that are managed by the community.

El Alto is the only city that has a wastewater treatment system. The other main cities in the TDPS (Oruro, Puno and Juliaca) do not have appropriate systems and their sewage disposals are a cause of water contamination.

Water and industry

Water demand for mining and industrial activities is not a major problem inside the TDPS system because there are few industries and their water consumption is very low. Water use for mining has not been measured, but is considered insignificant. Conversely, mining is an important source of water contamination.

Industrial pollution of waterways results from inappropriate discharge of wastewater and drainage from mines and mineral processing systems. Water from mines is very acidic and highly contaminated with heavy metals, particularly in Oruro where materials have appreciable amounts of pyrite that produces sulphuric acid when it comes in contact with water. This acid leaches metals such as arsenic, cadmium, cobalt and nickel, producing contaminated water that eventually flows into the basin.

Water and energy

Energy production is not a major activity in the TDPS system. Although there is a lack of information with regards to energy, it has been observed that energy consumption in the area is low and the principal source of energy is biomass (about 70 percent). Only 21 percent of homes on the Peruvian side and 29.8 percent in Bolivia, mainly in urban areas, have electricity. This electricity is generated outside the system and water is not used for hydraulic energy production. There is a small-scale use of liquefied petroleum gas, limited to urban areas due to the difficulties inherent in its transportation.

Other uses

Although recreation and transportation are not considered activities that can affect the water balance or water quality, they are significant activities when viewed as potential new alternatives for developing the TDPS area.

Transportation is of fundamental importance in Lake Titicaca. The lake includes twelve islands where the local population rely on boats and ships for travel. Likewise, Copacabana City, located in the

Manco Kapac Province, is one of Bolivia's most important tourist sites, and getting there necessitates crossing the lake through the Tiquina Strait. Recreation activities are being extended in response to increased demand from international as well as national tourism.

Management Challenges: Stewardship and Governance

The surface of Lake Titicaca is evenly distributed between Bolivia and Peru, countries that exercise an 'exclusive and indivisible joint ownership' over its waters. In fact the joint ownership model not only applies to the water of Lake Titicaca, but also to the watershed, as a way of ensuring integrated management of the water system.

Institutions

Three institutions operate in the TDPS system area with clearly defined roles.

- The Ministry of Sustainable Development and Planning, in Bolivia. By law, this ministry is the supreme national water authority in the country, in charge of designing, planning and enforcing policy, strategies and development initiatives.
- The Peruvian Development Institute (Peru). This institute has the equivalent functions of the Bolivian Ministry of Sustainable Development.
- The Binational Autonomous Authority of Lake Titicaca (ALT). This entity of public international law was created in May 1996. It has two national operative units and its general function is to promote and conduct programmes and projects, and to decide, implement and enforce the regulations on management, control and protection of the system's water resources within the framework of the adopted Master Plan. ALT's political functioning is associated with the Peruvian and Bolivian Ministries of Foreign Affairs. ALT has units for administration, planning and coordination within each government. Two national projects were established for Bolivia and Peru and both technically depend on ALT. The Bolivian project is the Unidad Operativa Boliviana (Bolivian Operative Unit, UOB), located in La Paz, and the Peruvian project is called the Proyecto Especial del Lago Titicaca (Lake Titicaca Special Project, PELT), located in Puno. The units are responsible for coordinating actions with national governments and for centralizing information. The Hydrological Resources Unit and Master Plan Management Unit are in charge of monitoring the water resources and tracking development of the Master Plan, respectively. Figure 21.2 illustrates the structure of the ALT.

Figure 21.2: ALT structure

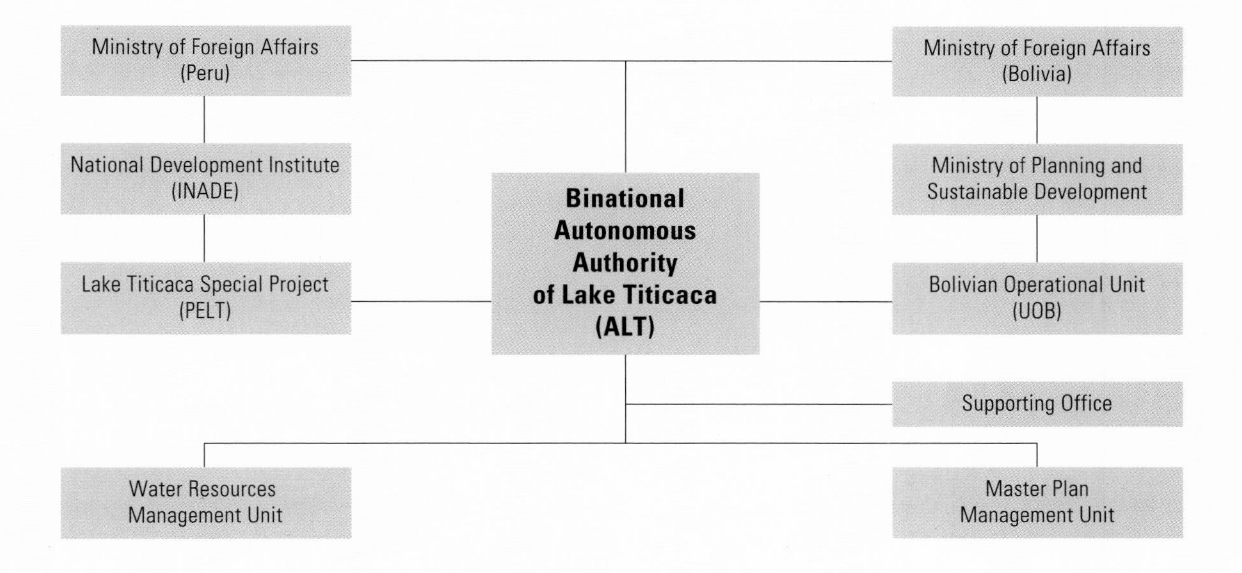

Legislation

Current legislation is incomplete and outdated, both in Bolivia and in Peru. The main legal bodies are:

- 1906 Water Law (Bolivia);
- Bolivian Civil Code;
- Decree No. 03464 on Land Reform and the 1953 Water Regime (Bolivia);
- 1969 General Water Law (Peru);
- 1990 Sanitation Code (Peru);
- 1990 Environment and Natural Resources Code (Peru);
- 1992 Environment Law (Bolivia);
- 1999 Basic Sanitation and Drinking Water Law (Bolivia).

Finances

Investment in the TDPS system comes from the Bolivian and Peruvian governments, international agencies and non-governmental organizations (NGOs). ALT is an autonomous organization with an annual budget of US$250,000 based on equal contributions by the Bolivian and Peruvian governments. In addition, ALT acts to facilitate external donations. In this framework the ALT has developed and is carrying out a number of projects, notably the regulatory works of Lake Titicaca (US$7,000,000), the dredging of the Desaguadero River bed (US$800,000) with ten-year projections (US$25,000,000), a biodiversity conservation project (US$920,000) and other projects oriented to research and validation of Inca and pre-Inca agricultural technologies.

Management approaches

Risk management

TDPS system regulation works allow, under normal hydrological conditions, the maintenance of Lake Titicaca's waters within an average level of 3,809.5 m.a.s.l. with a variation of 1.5 metres. This regulation scheme, based on technical and statistical data, has diminished the flood risk.

Valuing water

The value assigned to water varies according to the rural or urban nature of the water supply systems and the number of served inhabitants. Tariffs for the most populated cities fluctuate between US$0.135/m^3 (Puno and Juliaca) in Peru and US$0.22/m^3 (El Alto and Oruro) in Bolivia. Single tariffs and the concept of non-quantified water are applied in water supply systems for small towns.

The economic value of water is not fully recognized, particularly in rural areas. There is no water use rate and use of water for irrigation is defined by customary practice.

Sharing the resource

The two main uses of water in the system, human consumption and irrigation, are not in conflict at the present time. There is, however, a potential conflict between upstream and downstream users, notably with respect to water for irrigation. The model provided by customary use and the way in which communities have traditionally related to each other play an important role in determining distribution patterns and claims. Upstream communities consider that they have priority over downstream communities through a complex system of retributions and favours.

Governing water wisely

Water management between Bolivia and Peru has been established in terms of the joint ownership of the Lake Titicaca and the entire catchment area. In this way, the ALT has become the proper administrative entity for resolving any such conflicts that may arise.

ALT administration is based on Integrated Water Resources Management (IWRM). The general model promotes coordinated management and development of water, land and related resources, although certain border conditions do not yet permit a complete implementation of the model. Nevertheless, two aspects are coming along slowly: valuing water as an economic good and an improved level of community participation in water management issues.

The Master Plan, developed with the cooperation of the European Community, was drawn up between 1991 and 1993 under the title *Master Plan for Flood Prevention and the Usage of Water Resources of the TDPS System*. This plan constitutes the basic reference and twenty-year framework for the future development of the system. To date, the general scheme proposed by the Master Plan has been implemented.

Ensuring the knowledge base

Implementation of the Master Plan implies the development of a broad knowledge base. In addition to the hydrological knowledge needed for management of the resource, projects for restoration and rescue of ancestral agricultural techniques have been carried out and show a high degree of productivity.

Policy implementation

The following elements are identified as implicit in the Master Plan:

- focusing actions in a framework of sustainable use of natural resources, with these resources as the central element;
- recovering the system's ecological integrity in terms of protecting endangered species, replenishing fish populations and mitigating human impact on the system; and
- promoting human development within the basins.

Lake management shows a good degree of adjustment to the first two points. However, promotion of human development within the system has had a low level of success due to the difficulty of overcoming basic poverty in the area.

Identifying Critical Problems

Challenges related to uncertainty and variability of the resource

Agriculture is the principal economic activity in the TDPS system. As such, and given the general poverty in the area, it is particularly vulnerable to such extreme events as drought and flood. The farmers' survival strategy is thus to diversify their crops in the hope of minimizing risks derived from resource variability.

Although water regulation works in Lake Titicaca have brought a degree of protection against floods, they are effective only during a relatively normal hydrological cycle. Floods have a significant impact on the economy of the area. Although there is no risk of human losses, due to the slow rise of flood waters, the economic damage over a ten-year period has been estimated at US$890,000 for agriculture, animal raising and infrastructure. In addition, flood losses over a twenty-five-year period are estimated at US$1,506,000. These are huge amounts given the region's extreme poverty.

Vulnerability to drought is high and in addition to the economic losses associated with an extremely dry year, drought also causes loss of genetic diversity in native varieties, and thus farmers are forced to buy imported seeds for the following years. There is no available information about losses due to drought events.

Hail and frost cause significant agricultural losses. In the case of frost, Lake Titicaca acts as thermal insulation. Away from the lake, however, and at higher latitudes, frost increases can cover over 300 days a year.

Water salinity increases to the south. It severely limits the soil's agricultural capacity. At the extreme south of the system, the highly saline soils have formed the Coipasa Salt Lake.

Challenges associated with needs, uses and demands

As previously mentioned, sewage discharge from the urban centres in the basin has resulted in organic contamination of the area. The tropical situation of Lake Titicaca, the high levels of solar radiation and the high rate of evaporation make the system very vulnerable, particularly to pollution problems. In contrast, the size of the water body helps to maintain pollution at acceptable levels, although there are some eutrophication problems close to the coastal villages. In addition, there are also problems relating to heavy metal contamination, resulting from mining activities in the zone.

Management-related problems

Although the ALT provides a regulatory framework to both countries, each nation has specific approaches to water management, with no provision for the disparities between them. This lack discourages private investment in the sector, while encouraging poor valuation of the water resource. The nature of the resource as an economic, social or mixed asset has not as yet been defined at the regional or national levels. Because of this, it becomes impossible to allocate installation, maintenance or treatment costs for water systems of any kind. In May 2002, the Interinstitutional Water Council (CONIAG) was created in Bolivia with the mission of reforming the legal, institutional and technical framework of the water sector.

There is, however, work ahead to attain a better integration of community organizations into the management model with the aim of ensuring that the ALT will have higher levels of representation. At present, due to the social and economic instability in both countries, there is no appropriate political climate in which to reach community consensus.

Problems affecting ecosystems

Mining activity, overharvesting and pollution from urban centres all put pressure on the natural resources of the TDPS system.

During the 1980s a regional economic depression diminished mining activity in the area and the level of poverty increased. It was exacerbated by drought and floods, and resulted in increased rural to urban migration. The subsequent depopulation of rural areas, together with a stagnating mining sector have all greatly relieved the pressures on natural resources. However, pollution in urban centres has visibly increased. It is difficult to predict the future with regard to environmental stress and ecosystem health, as so much depends on the general levels of poverty in both countries.

Other problems

Differences between indigenous and Western cultural patterns make it difficult to adopt new agricultural technology, improve production and establish efficient market systems. Associated with these difficulties is the prevailing land ownership model. At present, rural property is fragmented into numerous small plots that are then divided through inheritance into still smaller plots. This model prevents farmers from making the transition to a more efficient and agricultural production and, combined with current land legislation, makes the existence of a real estate market virtually impossible, all of which adds to the structural poverty.

Achievements

Water resources assessment in the TDPS system

In the mid-1950s many initiatives were taken at different levels to make use of the waters of Lake Titicaca, one example being the proposal for bringing irrigation water to the arid zone of Chile. These initiatives involved the extraction of given flows from the lake, but without any evaluation study of the real hydrological potential. Out of concern for possible negative impacts, the Bolivian and Peruvian governments took the first steps to establish the foundation of a binational management system, signing an agreement to study the issue.

Binational Master Plan

The subsequent result of this investigation was the formulation of a Binational Master Plan to provide the guidelines and framework for the future development of the system. One of the conclusions of this study estimates that as little as 20 m^3/s of water could be extracted from the basin for economically productive uses, a much lower volume than proposed in the original. In this way, a possible ecological disaster similar to that of the Aral Sea in the former Soviet Union was averted, and the groundwork was laid for a harmonious and technically effective binational model of administration.

Within the framework of the Master Plan, the following documents were developed:

- a development strategy for irrigation and drainage;
- a strategy for hydraulic regulation of the system;
- an environmental survey and analysis; and
- a conservation plan.

Hydraulic regulation works

Taking into account the fragility of the system with regard to flood protection and prevention, a series of flow regulation works have been defined at basin level and in the system in general, for an amount of US$38 million. In 2001 the first dam was finished, close to the international bridge over the Desaguadero River. The main objective of the dam is to prevent, or at least protect, the surrounding area from floods, according to a rational and planned handling of the lake level when it rises above 3,810 m.a.s.l. Other benefits of this dam include protection of the vast fish populations and aquatic vegetation, provision of 50,000 hectares of secure irrigation to Peru and 15,000 hectares yearly to Bolivia up to a maximum potential of 35,000 hectares, and flood protection for 6,000 to 10,000 hectares on both sides of the lake.

The ALT model

One of the most successful aspects of ALT is its very smooth operation. As indicated earlier, the process of creating the Binational Authority has gone through four stages: conducting an evaluation of resources, designing a legal framework, building a management model and implementing a master plan. These stages represent a scale model that can be replicated nationally as a guide in the process of regulating hydrological resources.

Conclusions

Problems in the TDPS system are mainly of a structural nature. Local poverty undermines every effort to prioritize solutions to problems, thus causing a negative feedback loop. Figure 21.3 represents the cause-effect relationships of the main problems.

There is an urgent need to intensify the fight against poverty and to drastically improve the health system in the region. Great benefits could be achieved through the provision of better services for waste disposal and treatment, the promotion of environmental education and the continuation of the water regulation works already in progress. Most importantly, however, there needs to be far greater investment in public health services in order to ensure a better quality of life for the population.

Other possible means of development include increasing the irrigated surface area of the basin. Much of the agriculture is currently not irrigated. This is due in part to the land fragmentation caused by traditional property rights. The result is a decrease in productivity, which helps to perpetuate the region's poverty. Developing the irrigation potential would require a more detailed evaluation of water reserves and development of the country's extensive natural gas reserves for cheap energy to power the scheme. One possible way to implement more efficient agricultural technologies is through the establishment of legislation that would promote associations of producers managing land surfaces of a more adequate size.

Although there is a legislative body in place and many efforts have been made to develop the area, more can be done to ensure the birth of a more equitable way of life for the inhabitants of the Lake Titicaca basin.

Figure 21.3: Chain of causality

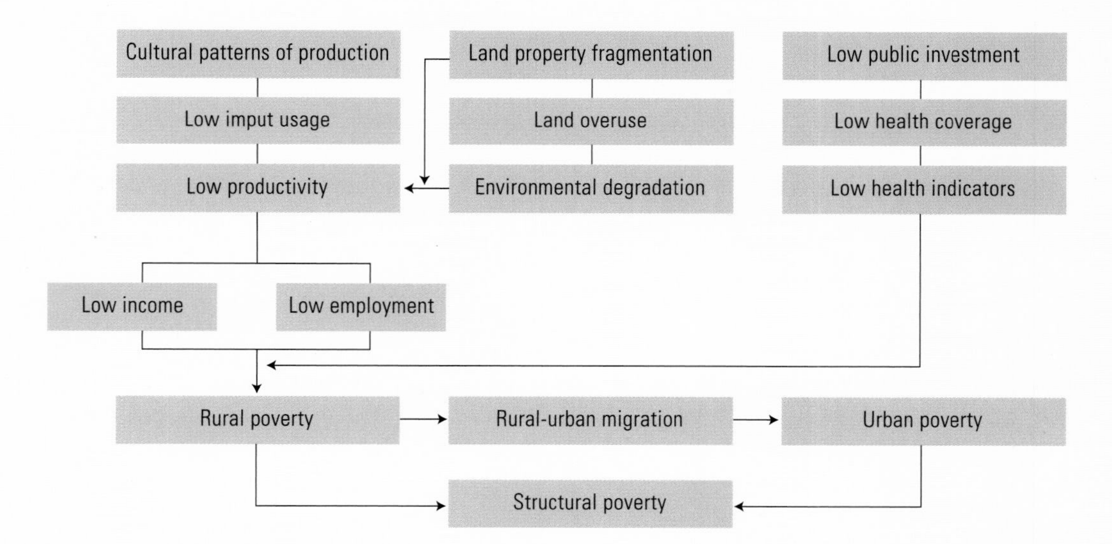

This figure shows the various causes of structural poverty in the TDPS system. Among these causes are fragmented land holdings and inefficient health services, both of which contribute to low unemployment and high poverty in the area.

Source: Prepared for the World Water Assessment Programme (WWAP) by A. Crespo Milliet, 2002.

Box 21.1: Development of indicators

Challenge areas	Lake Titicaca indicators
SURFACE WATER	• Water use (U) = 416.3 m^3 10^6/year • U/A = 0.60 • Rainfall varies between 200 mm and 1,400 mm (normal year)
WATER QUALITY	• Coliforms load 46 unt/100 ml • Puno water system abduction • 1,000 unt/100 ml (Puno Bay) • Turbidity = 1.27 parts per million
GROUNDWATER	• Net recharge = 4 m^3/s; 0.89% of total (see groundwater map for more details)
PROMOTING HEALTH	• Child mortality 14% • 77,663 households with sanitation = 17.2%: Peru = 19%; Bolivia = 24% • Health investment: Peru = 278 Purchasing Power Parity (PPP)/year; Bolivia = 150 PPA/year
PROTECTING ECOSYSTEMS	• Six protected areas 922 km^2 = 0.62% • One Ramsar site: Lake Titicaca 8,400 km^2 = 5.6% of total area
WATER AND CITIES	• 60% of population in major urban centres • 750 tons/day solid waste; 140,000 m^3 wastewater • Only one major city (El Alto) where water system is administrated by a private company
SECURING THE FOOD SUPPLY	• Irrigated land = 19,444 hectares = 0.43% of total productive land
WATER AND INDUSTRY	• No water recycling in the system
WATER AND ENERGY	• Electricity is not generated in the lake area • No water recycling in the system
MANAGING RISK	• Flood area = 90,000 hectares controlled, out of 120,000 hectares in Puno and 10,000 hectares in Desaguadero
SHARING WATER	• Agriculture = 90%; Drinking water = 7%; Other = 3%
VALUING WATER	• Varies between US$0.135 per m^2 (Puno and Juliaca) in Peru and US$ 0.220 per m^3 in Bolivia (El Alto and Oruro) • No demand-responsive water development policies • No demand-responsive water development institutions
ENSURING KNOWLEDGE	• Effective hydrometeorological data collection system exists in both Peru and Bolivia • Data can be collected by requesting institutions • Between Bolivia and Peru there is one main authority (ALT) and eight water scientists
GOVERNING WISELY	• Decentralization process in course; roles are 60% defined • The basin authority has the central and very autonomous responsibility to plan and manage the lake

References

Comité Ad-Hoc de Transición de la Autoridad Autónoma Binacional del Sistema TDPS. 1996. *Diagnóstico ambiental del Sistema TDPS.* Bolivia, United Nations Environmental Programme.

GWP (Global Water Partnership). 2000. *Manejo Integrado de Recursos Hídricos.* (Integrated Management of Hydric Resources). TEC Background Papers No. 4. Stockholm.

Ministry of Sustainable Development and Planning (Ministerio de Desarrollo Sostenible y Planificación). 1999. *Bolivia: Atlas Estadístico de Municipios.* (Bolivia: Statistical Atlas of Municalities). La Paz, Centro de Informacion para el Desarrollo.

——. 1998. *Zonificación Agroecológica y Socioeconómica del la Cuenca del Altiplano del Departamento de La Paz.* (Mapping the Agro-ecological and Socio-economic Aspects of the High Basin of the Department of La Paz). Bolivia, Proyecto ZONISIG.

National Institute of Statistics. 1992. *Population and Housing Census.* Bolivia.

Republics of Bolivia and Peru. 1993. *Master Plan for Flood Prevention and the Usage of Water Resources of the TDPS System.* La Paz, Autoridad Binacional Autonoma del Sistema Hidrico TDPS.

UNDP (United Nations Development Programme). 2002. *Informe de Desarrollo Humano en Bolivia.* (Human Development Report of Bolivia). Bolivia.

UNESCO (United Nations Educational, Scientific and Cultural Organization). 1996. *Mapa Hidrogeológico de America del Sur.* (Hydrogeological Map of South America). Brazil.

——. 1994. *Efficient Water Use.* Uruguay.

22

Greater Tokyo, Japan

Coordinated by: The National Institute for Land and Infrastructure Management – Ministry of Land, Infrastructure and Transport of Japan (NILIM-MLIT)

Table of contents

Greater Tokyo, Japan

Ah, what a pleasure
to cross a stream in summer
– sandals in hand.

Buson (1716-1783)

THIS CASE STUDY PRESENTS the example of river basins that serve one of the world's most populous areas, a region of 27 million people. In addition to its extremely high density, Tokyo metropolitan is subject to seasonal floods and other hazards such as droughts and earthquakes. However, because it is a rich and industrialized country, Japan has the means – and the skills – to manage these risks using infrastructure such as dams, levees and underground floodways. There is also great emphasis placed on public awareness and disaster preparedness. The authorities have developed early warning systems that rely on the Internet, Geographic Information Systems (GIS) and hazard mapping, and there are shelters where people can take refuge. Such continuous efforts have ensured that one of the world's largest economic developments has been able to safeguard its population in the high-risks region. Other concerns include a degraded natural environment and pollution of groundwater, and many efforts, such as river restoration works, are being implemented with wide public participation.

THE GREATER TOKYO REGION (hereafter Greater Tokyo), with its densely populated megacities, includes five river basins covering an area of about 22,600 square kilometres (km^2), with a total population of 27 million and property value assets totalling about US$2.9 trillion. Due to human and industrial activities, various water-related problems have developed, and the need is increasing for better water quality, diversification, protection and improvement of the environment.

General Context

The enormous water resources needed to supply the cities and maintain the safety degree against drought are difficult to manage. In addition, groundwater withdrawal is still causing land subsidence.

About 13.25 million people and 170 trillion yen (approximately US$1.36 trillion) worth of assets in property value are concentrated in 4,800 km^2 of the region's alluvial plains, often the site of substantial flood damage, accentuated by the fact that Greater Tokyo suffers the severe weather conditions of the Asia monsoon period. Changes in land use and increase in rainfall have also enhanced the danger of flood in recent years.

Map 22.1: Locator map

Source: Prepared for the World Water Assessment Programme (WWAP) by AFDEC, 2002.

Map 22.2: Basin map

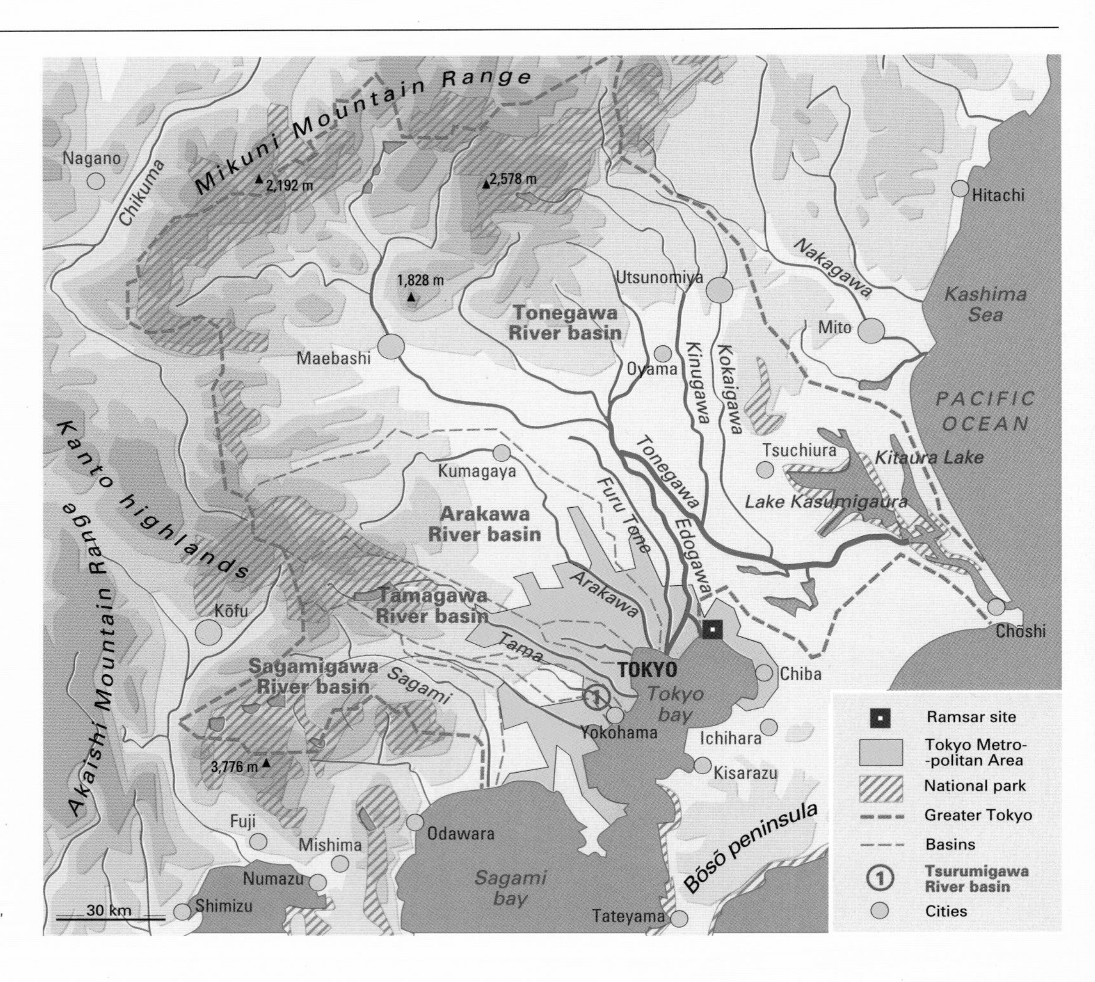

Source: Prepared for the World Water Assessment Programme (WWAP) by AFDEC, 2002.

Table 22.1: Hydrological characteristics of the five river basins in Greater Tokyo

	Tonegawa (1)	Arakawa (2)	Tamagawa (3)	Sagamigawa (4)	Tsurumigawa (5)
Location	Central Honshu, Japan N 35° 32′–37° 5′ E 138° 24′–140° 51′	Central Honshu, Japan N 35° 39′–36° 10′ E 138° 43′–139° 52′	Central Honshu, Japan N 35° 32′–37° 51′ E 138° 494′–139° 46′	Central Honshu, Japan N 35° 32′–37° 5′ E 138° 24′	Central Honshu, Japan N 35° 28′–35° 35′ E 139° 24′–139
Area	16,840 km^2	2941.5 km^2	1,241 km^2	1,668 km^2	239 km
Origin	Mt. Ohminakami	Mt. Kobushi-gatake (2,475 m)	Mt. Kasatori	Mt. Fuji	Tanaka Yato (valley) (Machida, Tokyo)
Outlet	Pacific Ocean	Tokyo bay, Pacific Ocean	Tokyo bay, Pacific Ocean	Sagami bay, Pacific Ocean	Tokyo bay, Pacific Ocean
Length of main stream	322 km	169 km	138 km	108 km	42.5 km
Highest point	1,834 m (trunk of Tonegawa)	Mt. Kobushi-gatake (2,475 m)	1,953 m (trunk of Tamagawa)	3,776 m (trunk of Sagamigawa)	164 m (trunk of Tsurumigawa)
Lowest point	River mouth (0 m)	River mouth (0 m)	River mouth (0 m)	River mouth (0 m)	River mouth (0 m)
Main geological features	Mountain area: sandstone, slate, limestone from the Paleozoic and Mesozoic eras, and volcanic rocks. Plain area: Pleistocene and alluvium.	(Upper basin) Paleozoic, Tertiary; (Lower basin) Quaternary (alluvialand diluvial)	Upper reaches: the Chichibu Paleozoic and Mesozoic strata. The hilly and flat area: the loamy layer of the Kanto District. The low-lying parts: sediments from the delta and coastal sedimentation.	Mountain area: lava bed, conglomerate and volcanic ash. Plain area: igneous rock, mudstone. The low-lying parts: sediments (consisting of lithified clay, sand, silt and conglomerate).	The hilly and flat area: the loamy layer of the Kanto District. The low-lying parts: sediments from the delta and coastal sedimentation.
Main tributaries	Katashinagawa (676.1 km^2) Agatsumagawa (1,355.2 km^2), Kanna-gawa (417.6 km^2), Kaburagawa (632.4 km^2), Karasugawa (759.1 km^2), Watarasegawa (2,621.4 km^2), Kokaigawa (1,043.1 km^2), Kinugawa (1,760.6 km^2)	Akabiragawa (250.0 km^2), Irumagawa (737.3 km^2), Shingashigawa (392.3 km^2), Sumidagawa (243.9 km^2)	Asakawa (154 km^2), Hirasegawa (13 km^2), Nogawa (68 km^2), Hiraigawa (38 km^2), Akigawa (170 km^2)	Doshigawa (152 km^2), Nakatsugawa (140 km^2), Sasagogawa (93 km^2), Kuzunogawa (115 km^2), Mekujirigawa (34 km^2)	Yagamigawa (28 km^2), Hayabuchigawa (20 km^2), Toriyamagawa (11 km^2), Ondagawa (31 km^2)
Main lakes	Kasumigaura, Kitaura, Chuzenji, Imba, Tegan, Ushiku	None	None	Kawaguchi, Yamanaka	None
Main reservoirs	Yagisawa (115.5x10^6 m^3, 1967), Naramata (85.0x10^6 m^3, 1991), Hujiwara (31.0x10^6 m^3, 1958), Aimata (20.0x10^6 m^3, 1959), Sonohara (13.2x10^6 m^3, 1966), Shimokubo (120.0x10^6 m^3, 1969), Ikari (32.0x10^6 m^3, 1956), Kawamata (73.1x10^6 m^3, 1965), Kawaji (76.0x10^6 m^3, 1983)	Futase (26.9x10^6 m^3, 1984), Arima (7.6x10^6 m^3, 1984)	Ogouchi (water supply and power only 185.4x10^6 m^3, 1957)	Sagami (no flood control dam 48.2x$10^6$$m^3$, 1947), Shiroyama (54.7x10^6 m^3, 1964), Miyagase (183x10^6 m^3, 2000)	Tsurumigawa multi-purpose retarding basin (39x10^6 m^3, 2002)
Mean annual precipitation	1,162.6 mm at Maebashi, 1,580.1 mm at Choshi (1971–2000)	1,367 mm (1951–1980), at Chichibu	1,385 mm at Ogouchi (1985–2001)	1,658 mm at Saito bridge (1985–2001)	1,616.5 mm at Tsurukawa, 1,628.5 mm at Tsunashima (1990–2000)

Table 22.1: *continued*

	Tonegawa (1)	Arakawa (2)	Tamagawa (3)	Sagamigawa (4)	Tsurumigawa (5)
Mean annual runoff	165.2 m^3/sec at Yattajima, 220.6 m^3/sec at Kurihashi (1960–2000)	26.4 m^3/s at Yorii (927km^2) (1952–1985)	30.2 m^3/sec at Ishikawa (1991–2000)	50.0 m^3/sec at Sagamiohashi (1991–2000)	8.69 m^3/sec at Kamenoko bridge (1990–1999), 83.4 m^3/sec at Sueyoshi bridge (1983)
Population	About 12,000,000	9,046,643 (1985)	About 3,571,000 (1995)	About 1,284,000	About 1,840,000
Land use	Forest (45.5%), paddy field (18.2%), cropland (11/2%), orchard (3.3%), urban (3.7%), residential area (6.4%), water surface (5.1%), other (6.6%)	Forest (48.2%), paddy field (5.1%), agriculture (6.5%), water surface (4.0%), urban (26.5%) (1985)	Forest (59.6%), paddy field (0.7%), cropland (1.8%), orchard (0.1%), urban (31.3%), other (incl. water surface) (6.5%) (1997)	Forest (78.2%), paddy field (2.5%), cropland, etc. (2.7%), urban (9.0%), other (incl. water surface) (7.6%) (1997)	Paddy field and cropland (10%), forest (5%), urban (85%)
Main cities	Maebashi, Takasaki, Saitama, Tsukuba, Utsunomiya	Tokyo, Omiya, Urawa, Kawagoe, Chichibu	Kawasaki, Chofu, Tachikawa, Tokyo (Ota-ku, Setagayaku)	Hiratsuka, Chigasaki, Zama, Atsugi, Sagamihara	Yokohama, Kawasaki, Machida

Sources: (1) UNESCO-IHP; (2) UNESCO-IHP; (3) Kanto Regional Development Bureau, MLIT; (4) Kanto Regional Development Bureau, MLIT; (5) Kanto Regional Development Bureau, MLIT.

Furthermore, due to intense urbanization, the quality of water in Greater Tokyo has deteriorated. Action was taken to reduce the discharge load, such as drainage regulation and sewage maintenance, and the quality of water started to improve in major rivers. However, the concentration level is still high in some tributaries, lakes and marshes, and new Environmental Endocrine Disruptors have been problematic. In addition, the increase of imported non-native species of fish and plants is becoming a serious ecological problem.

Water Resources

Hydrology

The average precipitation in Greater Tokyo has been 1,551 millimetres (mm) per year for the past thirty years. During periods of drought, the average is 1,213 mm/year, which is 20 percent less than in normal years. Over the last one hundred years, overall precipitation has been decreasing. However, recently, there have been more and more rainfalls of over 100 mm/day (Water Resources Department, MLIT, 2002). Water resources in seven prefectures, including Greater Tokyo, average 374 billion cubic metres (bm^3) per year. During droughts, which occur once every ten years, the average is 247 bm^3, which is 30 percent less than normal (Water Resources Department, MLIT, 2002).

As for the water quality in Greater Tokyo, following are some biological oxygen demand (BOD) measurements taken in 1998: more than 8 milligrams (mg) per litre in Tsurumigawa River, about 4 mg/litre in Arakawa River, less than 2 mg/litre in Tamagawa River and Sagamigawa River. The BOD levels in the Tsurumigawa and Arakawa Rivers tend to increase while those in the other rivers remain stable. In the last ten years, the worst levels were found in Tsurumigawa and Arakawa Rivers. In 1999, the chemical oxygen demand (COD) concentration level was about 7.5 mg/litre (MLIT, 2001a). Generally, the water quality in rivers and lakes has improved; however, the presence of Environmental Endocrine Disruptors, chemical substances thought to affect the endocrine system, was recently discovered in some rivers. It is not yet known if these disrupters have an influence on health or the ecosystem (Ministry of the Environment, 2002). Figure 22.1 shows the variation in BOD levels over a number of years.

Human impacts on water resources

In seven prefectures, including Greater Tokyo, the proportion of lands used for building has increased from 13 percent in 1974 to 20.2 percent in 2000, while that of agricultural lands has diminished, going from 46.3 percent to 39.1 percent. In order to preserve the proportion of lands used for forests and woods, which has been kept stable (although only private forests have so far been counted), only agricultural lands are used for building.[1]

In April 2001, Greater Tokyo counted 183 dams with a total water storage capacity of about 2.5 bm^3. The dams are built for flood control, water supply and electricity generation purposes (Japan Dam Foundation, 2002).

1. *Kotei-shisan no Kakaku-tou no Gaiyou Chousa* (Outline of Protocols, such as Price of Fixed Assets). Taken from the web site of the Ministry of Home Affairs at http://www.soumu.go.jp/czaisei/shiryo.html.

Figure 22.1: Variation in river water quality – biological oxygen demand (BOD) levels

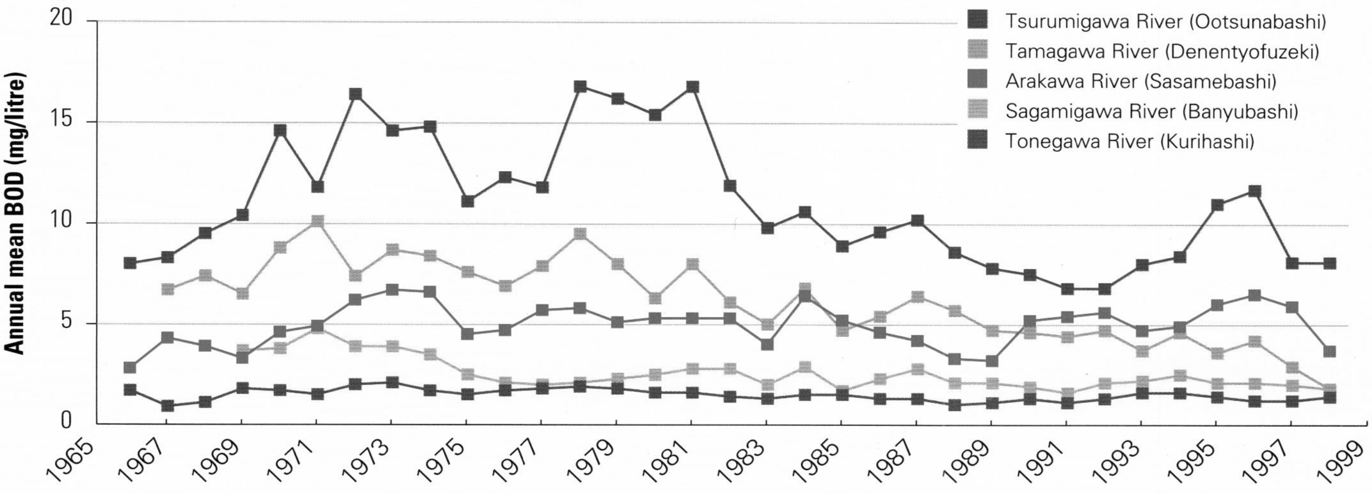

The BOD levels have decreased since the 1970s in all five rivers of Greater Tokyo. However, the level remains high in the Tsurumigawa and Arakawa Rivers.

Source: River Bureau, MLIT, 2001a.

Figure 22.2: Variation in the chemical oxygen demand (COD) discharge load by sector in Tokyo bay

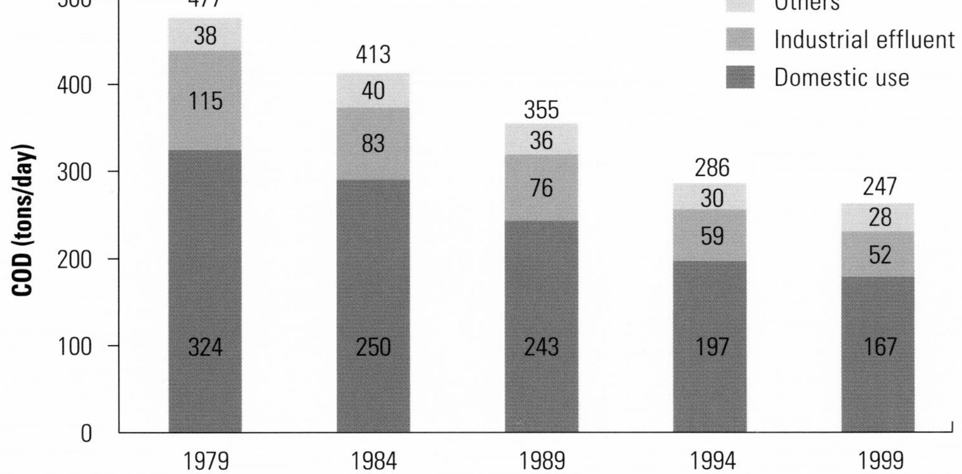

The majority of discharge load in Greater Tokyo is from domestic use, accounting for close to 70 percent of the total amount. The discharge load has drastically decreased, to about half of what it was twenty years ago.

Source: Ministry of the Environment, 2001.

Most of the effluent load of four prefectures (Tokyo, Chiba, Saitama and Kanagawa) is emptied into Tokyo bay, where the COD level in 1999 reached 247 tons per day. Seventy percent of this load is from domestic use (see figure 22.2). Due to drainage regulations, the discharge load has drastically decreased, to about half of what it was twenty years ago (Ministry of the Environment, 2002).

Groundwater resources are widely used in the region. One tenth of the water used to supply Tokyo metropolis is from groundwater resources. Analyses of the water quality in 1998 show results lower than the standard values, except for tetrachloroethylene (a product known to carry severe health risks), which exceeded environmental standards in three out of eighty-seven measurement points (Environmental Bureau of the Tokyo Metropolitan Government, 2000).

A survey in 1999 of non-native species in Japan's rivers yielded the following results: fish represented 6.1 percent, benthos 2.2 percent, plants 11.0 percent, birds 2.4 percent, amphibians 5.3 percent, reptiles 7.7 percent, mammals 18.4 percent and insects 0.7 percent (Foundation for Riverfront Improvement and Restoration, 1999). Black bass and blue gill, which are non-native fish, were found in 40 percent and 30 percent of dams, respectively (Water Resources Environment Technology Center, 2001).

Challenges to Life and Well-Being

To meet the needs and demands of the large human and industrial activities, vast water resources and policy implementation are necessary. The following is a summary of water uses in the region. The water volumes given below represent the total seven prefectures, not the five river basins.

Water use in industry and cities

In 1998, the total volume of water used in Greater Tokyo was 163.5 bm^3. Thirty-four percent of this water is used for households, 14 percent for industrial activities and 52 percent for agricultural activities. The volume of water used in agriculture in Tokyo region is relatively low compared to the 66 percent used country-wide. The volume of water for households has slightly increased in recent years, while that of water used for agriculture has not changed. As for the volume of water used in industry, there is only a very slight increase, the result of massive water recycling (Water Resources Department, MLIT, 2001).

In Greater Tokyo, 44 percent of total water resources are used during normal years, and 66 percent during drought years. This percentage is twice that of the whole country (Water Resources Department, MLIT, 2001).

Groundwater resources make up 22.8 percent of total water use inland, and 13.1 percent in seaside areas. Some 45 percent of households and industries, considerably more than agricultures, rely on groundwater. So as to prevent land subsidence, groundwater withdrawals have been regulated. Limits on withdrawals of groundwater come in the form of two laws: the Industrial Water Law, which targets groundwater used for industrial purposes, and the Law Concerning the Regulation of Pumping-up of Groundwater for Use in Buildings, which targets groundwater used for cooling and other building-related purposes. Groundwater withdrawal in the northern part of the Kanto plain has decreased from 13.1 bm^3 in 1985 to 9.6 bm^3 in 1999 (Water Resources Department, MLIT, 2001). As a result, the rate of land subsidence has stabilized (see figure 22.3).

Figure 22.3: Rate of ground subsidence in cm/year

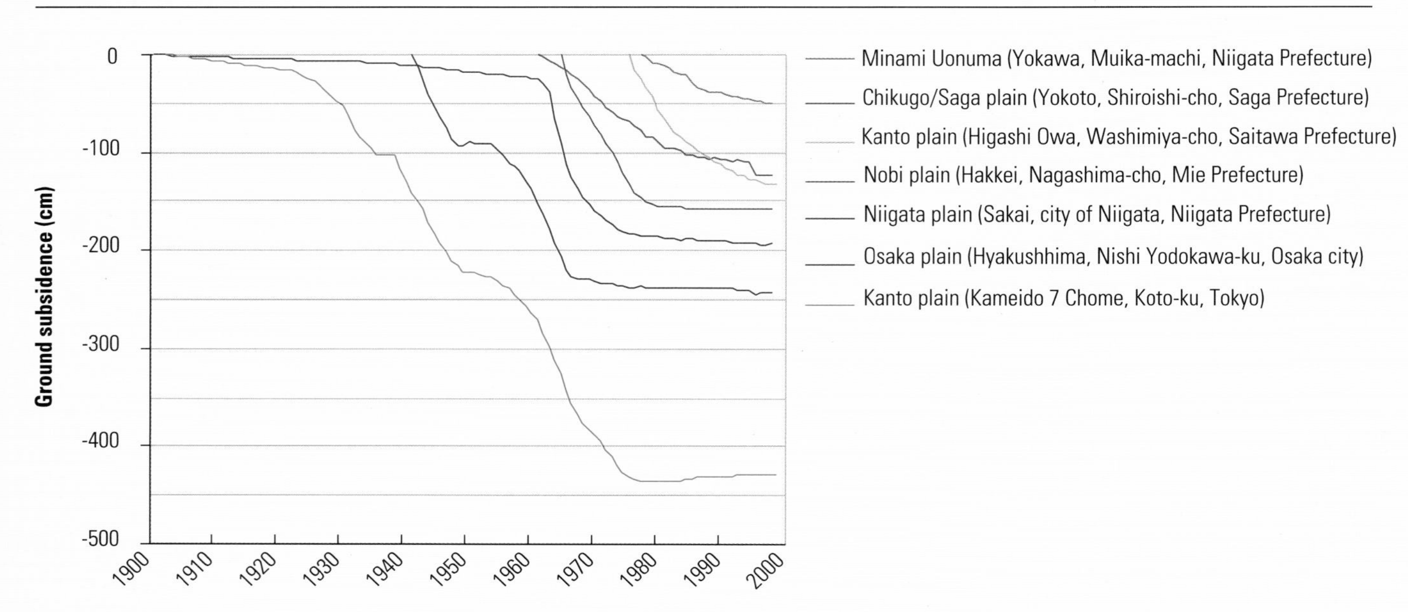

There has been a dramatic increase in ground subsidence in the Greater Tokyo region in the past century. Since the 1980s, this subsidence has stabilized. The Kanto plain has suffered particularly significant ground subsidence, and is now over 4 metres lower than it was in 1900.

Source: Water Resources Department, MLIT, 2001.

Securing the food supply

In 1998, agricultural water use in Greater Tokyo represented about 85.7 bm^3, of which 52 percent was spent on food production. Consumption of rice has decreased, while livestock products and consumption of oils and fats have increased. After 1998, though it had been steadily declining, the food production rate in Japan became stable. In addition, Japan's cereal production has been rising: this is mainly due to an increase in domestic wheat production and feed crops combined with a decreasing demand for livestock (Ministry of the Environment, 2002 and Ministry of Agriculture, Forestry and Fisheries).[2]

Protecting ecosystems

According to a 1998 survey, people's expectations for natural environment and beautiful landscapes are higher than what they were in 1990. About one hundred years ago, wetlands represented about 0.3 percent of the area of seven prefectures, including Greater Tokyo. In 1999, they represented only 0.11 percent, which is a drop of about 60 percent (Geographical Survey Institute, MLIT).[3]

Management Challenges: Stewardship and Governance

The population and property in Greater Tokyo are concentrated in alluvial plains, mostly below flood level, making flood damage potentially serious. Many years of flood control efforts have reduced the total inundated area and made the alluvial plains available for residential, industrial and agricultural use. Since the 1960s, increased land use has required enormous development of water resources especially for households and industry. There was a significant increase in river water use for generating hydroelectric power and for industrial and household water supply. To meet the new demands, a systematic framework for flood control and water use was established. In 1964, the institutional framework was improved by introducing an integrated river management system. A long-term plan for the development of water resources all over the country was also established. However, the increase in population and assets led to environmental problems such as the aggravation of the river water quality and changes in the ecosystem, making environmental conservation an important issue. In addition, with changes in economic and social conditions, the water management system was expected to not only fulfil flood control and water resource purposes, but also provide for recreational use and habitat diversity. As people's concerns about water and the environment grow, public involvement and consensus-building with proper information have become indispensable. In light of these changes, the River Law, on which the river administration is based in Japan, was revised. The revision established a comprehensive river administration system for flood control, water use and environmental conservation.

Water governance

There are several agencies in charge of governing water resources in Japan. Table 22.2 lays out the various water organizations.

Some examples of water management include some of the following elements.

- Water supply: in principle, municipal governments are responsible for water supply, but private sectors can participate upon authorizations from local municipal governments. Since April 2002, outside organizations, including the private sector, have been qualified to conduct technical works on the maintenance of water supply systems.
- Water for agriculture: 'Land Improvement Districts' are responsible for agricultural water use under the Land Improvement Act. The Land Improvement Districts are farmers' associations, and in principle each farmer pays maintenance fees.
- River administration: under the River Law, official river administrators own all rivers for comprehensive river management. The national government (Ministry of Land, Infrastructure and Transport) manages major parts of the 109 river systems. Local governments manage the rest. The river administrators are responsible for flood control, proper use of rivers and conservation of the fluvial environment.

Water rights

The water of Japan's rivers has historically been dominated by a large number of river users with vested water rights, conferring the right to use water, both public and private. With the development of human society and disputes arising over water rights, the need arose for a legal system to regulate water use. The Japanese water rights system was then altered, and the River Law, established in 1896 and which prescribes river management, was revised in 1964. Until then, it had focused only on flood control, and the revision provided a systematic framework for both flood control and general water use. Under this law, an official river administrator manages a basin under a unified and consistent system, and with it water becomes a public resource. In order to better cope with the expanded River Law, it has been necessary to seek conciliation with the prior vested interests: the 1964 revision attempts to modify the system while at the same time taking into account and maintaining

2. See web site at http://www.maff.go.jp/.
3. See web site at http://www1.gsi.go.jp/ch2www/marsh/part/list_4.html.

Table 22.2: Water governance in Japan

Affair	Organization	Sub-section	Main laws
Water supply	Ministry of Health, Labour and Welfare	Water Supply Division, Health Service Bureau	Waterworks Law Law on Execution of Preservation Project of Water for Water Supply
Water use for agriculture	Ministry of Agriculture, Forestry and Fisheries	Rural Development Bureau	Land Improvement Act
Water conservation forest		Forestry Agency	Forest Law
Industrial water supply	Ministry of Economy, Trade and Industry	Industrial Facilities Division, Economic and Industrial Policy Bureau	Industrial Water Law Industrial Water Supply Business Law
Hydropower		Agency of Natural Resources and Energy	Electric Power Development Promotion Law
Sewerage	Ministry of Land, Infrastructure and Transport	Sewerage and Wastewater Management Department, City and Regional Development Bureau	Sewerage Law
Rivers, water resource facilities		River Bureau	River Law Specified Multipurpose Dams Law
Comprehensive and basic policies for water supply and demand, reservoir area		Water Resources Department, Land and Water Bureau	Water Resources Development Promotion Law Water Resources Development Public Corporation Law Law Concerning Special Measures for Reservoir Areas
Water quality, environmental conservation	Ministry of the Environment	Water Environment Department, Environmental Management Bureau	The Basic Environment Law Water Pollution Control Law

the permission systems of the past (Water Use Coordination Sub-Division, Water Administration Division, River Bureau, MLIT, 1995).

Developing water resources

Water resource policies should be carefully promoted from a long-term and comprehensive viewpoint. The new 'National Comprehensive Water Resources Plan' ('Water Plan 21') clarifying the basic direction for the development, conservation and utilization of water resources was settled on in June 1999. This plan provides guidelines for examining various measures concerning water resources for the target years of 2010 through 2015, forecasting water supply and demand for that period. It cites measures against disasters such as the 1995 Hyogoken-Nambu earthquake, as well as the development of policies for conservation of and improvement in the water environment, and for the restoration and nurturing of water-related culture (National Land Agency, 1999). Under the national plan, basic plans for water resources development are established in major river basins, their aim being to reach a water volume of about 258 cubic metres per second (m^3/s) in the Tonegawa and Arakawa Rivers. In 2001, 64 percent of this target had been achieved (Water Resources Department, MLIT, 2001), but more is still needed to meet prospective water demand.

Dams are one of the major tools for water resource development: meeting the demand through river and groundwater intake alone became impossible. As previously mentioned, dams are built for several purposes including household water supply, industrial and agricultural water use, electricity generation and flood control. In April 2001, after improving the water resources development facilities, the total water storage capacity reached about 2.5 bm^3 (Japan Dam Foundation, 2002).

Compensation measures for upstream inhabitants

Before proceeding with the construction of water resource development facilities such as dams, it is important to reach an arrangement with inhabitants in reservoir areas that may suffer significant effects as a result of dam construction. Various measures were designed to mitigate negative effects suffered in reservoir areas and invigorate local communities, and a Law Concerning Special Measures for Reservoir Areas has been created for this purpose (Water Resources Department, MLIT, 2001).

Drought measures
Drought conciliation councils have been established in the Tonegawa, Arakawa, Tamagawa and Sagamigawa Rivers. The conciliation takes place among the water users themselves, while the river administrator offers necessary information at the initial stage of the process, presents drought conciliation proposals and facilitates the process. For example, the drought conciliation in Tonegawa River is characterized by integrated reservoir operation. The efficient operation of several dams as one and the same water system requires consistent management of all dams (River Bureau, MOC, 1997). The Tonegawa Drought Conciliation Council was established in 1970, and comprises the Ministry of Land, Infrastructure and Transport (MLIT), six prefectures and the Water Resource Development Public Corporation (Kanto Regional Development Bureau, MLIT).[4]

Using water effectively
Effective use of water resources does not generally require new large-scale facilities to relieve demand-supply gaps, and is also important in attenuating the effects of drought. One of the examples established in Greater Tokyo is the use of water such as treated sewage, recycled industrial wastewater, rainwater and other types of non-standard water resources. These are lower in quality, but provide for such purposes as toilet-flushing, refrigeration and cooling, and sprinkling (Water Resources Department, MLIT, 2001). Actions such as the construction of water-saving residences are promoted (MOC, 2000).

River improvement measures

Levee maintenance
Between 1991 and 2000, losses resulting from flood damage in Greater Tokyo amounted to about 900 billion yen (US$7.22 billion). The years 1991 and 1998 registered the highest losses, with more than 200 billion yen and a little under 200 billion yen, respectively (equivalent to US$1.6 billion). So far, levee maintenance and dam construction have been adopted as river improvement countermeasures. The levee improvement ratio went from 34.8 percent in 1985 to 45.9 percent in 1999 and, due to such improvements, flood adjustment capacity also increased, from 325 million m^3 (Mm^3) in 1985 to 685 Mm^3 in 2001 (Japan Dam Foundation, 2002; River Bureau, MLIT, 2001b; Japan River Association, 1986-2000).

Non-structural measures
Although the embankment works improved the degree of flood control safety, large flood damage was still a threat, as most of the population and assets are concentrated in the basins. Other measures were therefore necessary to reduce potential flood damage. Non-structural measures include flood warnings, announcement of flood protection measures and the preparation of inhabitant refuges (Arakawa Lower River Work Office, MLIT).[5] Moreover, flood hazard maps are made public to help inhabitants rapidly and efficiently find refuge. The hazard map shows each city's danger zone and the location of the refuge area, specifying access routes (MOC, 2000). With the changes in the rainfall pattern increasing flood risk, the Flood Protection Law was revised in June 2001. Measures such as flood forecasting and the securing of accessible refuges in danger zones were legally supported (MLIT, 2001b).

Managing the environment
With the urbanization of Greater Tokyo, environmental problems arose, such as deterioration of the river water quality, changes in the ecosystem and variations in the landscape. Investigations into water quality and the ecosystem were carried out so as to monitor the actual condition of the river environment.

Monitoring the ecosystem
Local governments (prefectures and designated cities under the Water Pollution Control Law) carry out regular quality surveys of public waters. The waters covered by this survey include those to which Environmental Quality Standards (EQS) are applied. EQS for water pollutants are target levels for water quality to be achieved and maintained in public waters under the Basic Environmental Law (Ministry of the Environment, 2001). The 'National Census of Rivers and Watersides' is another one of the tools for monitoring the condition of the ecosystem. This census examines living conditions for fish and shellfish, benthic organisms, plants, etc., and human activities in rivers and waterside. The census started in 1990 in one hundred and nine class-A rivers (those rivers managed by the national government) and eighty dam reservoirs, and major class-B rivers (those managed by the local government) were added in 1993. The status of non-native species in rivers and dam reservoirs was also investigated (MOC, 2000). As mentioned previously, the River Law was revised in 1997. As well as the usual flood control and water use purposes, the concept 'maintenance and preservation of river environment' was included (see figure 22.4), (MLIT, 2002).[6]

Evaluating the environmental impact
In 1999, the Environmental Impact Assessment Law was implemented, and screening and scoping procedures[7] were

4. See web site at http://www.ktr.mlit.go.jp/kyoku/.

5. See web site at http://www.ara.or.jp/arage/.
6. See web site at http://www.mlit.go.jp/river/.
7. Scoping is the process by which a research plan is developed under the Environmental Impact Assessment Law. Screening is the process which determines the necessity of an Environmental Impact Assessment.

Figure 22.4: Revision of the River Law

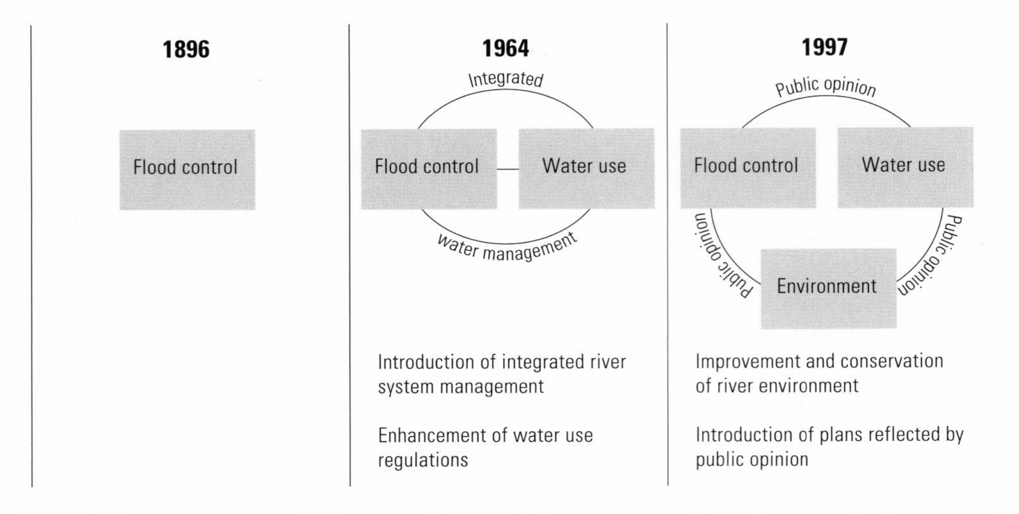

Source: River Bureau, MLIT, 2001c.

introduced. The law also called for 'proper consideration of environmental preservation', including ecosystem and communication activities. Under this law, suitable environment preservation countermeasures are taken for the development of water resources and flood control facilities (Water Resources Environment Technology Center, 2001).

Improving the water quality

The Water Pollution Control Law establishes national effluent standards and authorizes more stringent prefectural standards to regulate wastewater discharged from factories and business establishments into public water bodies. The reinforced factory regulations have been effective in improving water quality, but problems with domestic effluent remain, especially in enclosed or semi-enclosed water bodies such as Tokyo bay (Ministry of the Environment, 2001). Sewage systems are one of the essential components for ensuring the quality of public waters. In 1999, 77 percent of people had access to sewage systems, and 97 percent received clean water through water service systems (Japan Sewage Works Association, 1973–2000 and Ministry of Health and Welfare, 1966–2000).

Public involvement in the river improvement plan

In addition to the preservation measures taken for the river environment, the river-planning system was radically reconsidered in the 1997 River Law revision because of rising concerns about environmental and regional needs for river improvement. By providing comprehensible information to the population and respecting their opinions, the planning process was effectively opened to public participation (MLIT, 2002).[8] In March 2001, the river improvement plan was applied first in the Tamagawa River basin, through various discussions with the inhabitants to gather their opinions, which were also collected through the Internet (Keihin Work Office, MLIT, 2002).[9] Such activities are actively conducted in other basins. For more details on the Tamagawa River Improvement Plan, see box 15.2 in chapter 15 on water governance.

Sharing information

Some database systems exist in Japan, notably the yearly report on water quality in public waters by the National Institute for Environmental Studies[10] and the water information system by the Ministry of Land, Infrastructure and Transport.[11] There are also several paper-based databases, such as *Water Resources of Japan* (Water Resources Department, MLIT, 2002), *Water Service Statistics* (Ministry of Health, Labour and Welfare, 1966–2000) and the *Report on Industrial Statistics (Industrial Land and Water)*, (Ministry of Economy, Trade and Industry, 2002).

Greater Tokyo's catch-phrase, 'the water-aware country', refers to land in which any water-related information is collected, shared with the public and used in a practical way, with consideration given to different geographical settings. It contributes to water resource management, flood control and management of the environment.

8. See web site at http://www.mlit.go.jp/river/.
9. See web site at http://www.keihin.ktr.mlit.go.jp/.
10. See web site at http://www.nies.go.jp/index-j.html.
11. See web site at http://www1.river.go.jp/.

For watershed management, water information such as drainage and basin groundwater needs to be kept updated. People need higher quality services for river administration: downsizing administrative services, making river administrative services more efficient with proper information updating, privatizing some parts of services and simplifying contract procedures. In addition, river information, such as water quality and ecosystem data, has become more useful in education systems.

In 'the water-aware country', information technology is used to manage and share information efficiently. For example, when an area's flood safety degree can be calculated more precisely by information technology, human losses will be reduced through the combination of people reaching suitable refuge under proper guidance, and the river administrator's efficient operational facilities. River GIS also attempt to deal rapidly with water-related disasters, through such measures as structure maps, waterway figures, maps showing the placement of dangerous objects and a water database. To these ends, super-high-speed, big-capacity optical fibre networks are connected to the related organs in order to make the latest data in emergency situations available (Sato, 2002).

Fitting the Pieces Together

In densely populated Greater Tokyo, the water management policy can be considered a success in providing the population and industries with water, and many years of flood control efforts have reduced the total inundated area. Water resource development has focused on stabilizing river flow and meeting new water demands. However, the high concentration of people and industry makes this success fragile and risky. Flood damage has only slightly decreased, due to a higher concentration of population and property: access to a stable water supply has accelerated the rate of concentration and created a new demand for water. The need for water resource management has become more diversified. People are more concerned with the environment and nature. In addition to the water resources policy, it is necessary to establish more integrated river management suitable for sharing risk information and coping with this risk. The concept of risk management is consistent with the revised 1997 River Law, which calls for public participation and environmental consideration. The various policies relating to efficient water use will be evaluated, easily comprehensible water quality indicators developed, and a commitment will be made and enforced to restore the natural environment and to make information public. These are highly related to the participation of citizens and non-governmental organizations (NGOs) in water policy.

Managing risks

As a countermeasure to the high flood damage in Greater Tokyo, an easily comprehensible indicator showing the degree of risk of flood damage was developed and made public (Yasuda and Murase, 2002). The safety degree against flood damage can be expressed through a combination of the flood frequency and the inundation level. This two-dimensional aspect of floods renders it difficult to develop a single-dimensional indicator by which the risk can be expressed.

Figure 22.5 shows that the frequency of floods and the inundation level can be expressed by a colour and a height, respectively. The green represents low flood frequency while the red represents high flood frequency. Comparing the flood levels with the height of people and houses directly indicates a degree of safety against flood damage. Inter-temporal changes in the safety degree can also be expressed, as shown in figure 22.6.

Based on this, flood risk expression has been developed in Tokyo region. The index in the legend is the Flood Risk Indicator (FRICAT) employed in Japan for policy evaluation. The FRICAT represents how often the expected annual damage by flood is higher than that of fire. The average expected annual fire damage in Japan between 1998 and 2001 was 1,165 yen (approximately US$9.3) per person.

Improving water resource management

The tight water resources in Greater Tokyo must be carefully managed, as they have to meet the competing demands of a large population. It has become difficult to develop new water resources facilities such as dams. To further improve the water resource management system, an evaluation of the various policies for efficient use of the limited water resources is being carried out.

Integrated Water Resources Management (IWRM)

The combination of an increased concentration of population and industries, the expansion of urban areas, changes in industrial and social structures and changes in the climate, have all given rise to a variety of water-related problems in Japan. These include water shortages in rivers and groundwater and deterioration of the quality of water, as well as an increase in urban flood damage. The problems originate from changes to the hydrological cycle, such as a lack of infiltration or continuity between surface water and groundwater. Water authorities in Japan are divided among several institutions. In 1998, they reached an agreement on fundamental policy for restoring a healthy hydrological cycle. The policy advocates adopting an integrated water basin approach and sharing knowledge about the hydrological cycle, and encourages efforts to improve the situation in each basin. In addition, case studies were conducted in some rivers around Tokyo and Osaka to monitor and analyse problems.

Figure 22.5: Expression of the safety degree

The frequency of floods and the inundation level can be expressed by a colour and a height, respectively. The green represents low flood frequency while the red represents high flood frequency. Comparing the flood levels with the height of people and houses directly indicates a degree of safety against flood damage.

Source: Yasuda and Murase, 2002.

Figure 22.6: Expression of the safety degree with a time variable

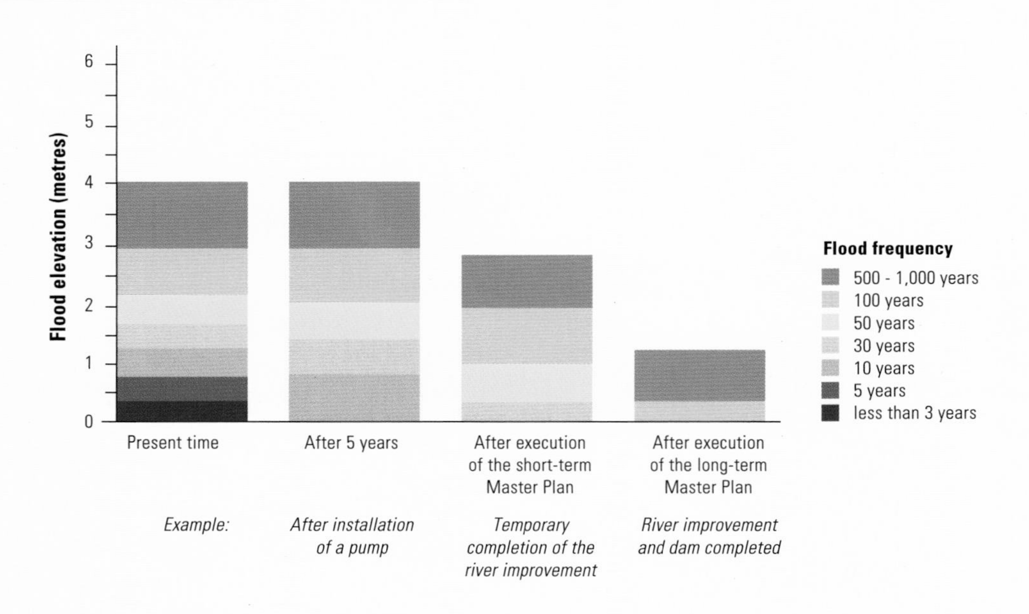

This figure shows how wise management can improve the safety degree against floods.

Source: Yasuda and Murase, 2002.

Use of existing facilities
As the construction of new water resource development facilities is becoming difficult due to the lack of suitable grounds, it is essential to use the already existent facilities as efficiently as possible (MOC, 1995). The Reorganization of Dam Groups is designed to redistribute the storage capacities of the existing dams by taking into consideration the particular features of each type of dam. This redistribution seeks to improve flood control and water use functions, that is, seeks to mitigate flood risk by enhancing the flood-regulating effect, and to improve the riparian environment by restoring the river's flow. The first project for this trial will start in the upper Tonegawa River (Improvement and Management Division, Bureau River, MLIT, 2001).

Upstream/downstream cooperation
Many Japanese cities have developed in downstream areas whereas water facilities were built upstream. People living in upstream areas worry about changes in their lives and jobs, and gain no benefit from the construction of dams, unlike downstream inhabitants. To deal with this problem, measures for the reservoir area development were taken, based on the Act on Special Measures for Reservoir Area Development implemented in 1974. In September 1999 however, a meeting about measures taken for this development highlighted the need for good management of water resources, and for the cooperation between authorities in upstream and downstream areas. In this meeting, the importance of basin management was also raised (MOC, 1999).

Coping with diverse needs
Protecting the natural environment and quality of water has become more important in water management. With people more involved in the water management process, it is necessary to make the process of improving the natural environment and water quality as transparent as possible, with full disclosure of all information. Also, with urbanization, many people have begun to search more actively for places to enjoy nature. More effort must be made to create communities where the places in which people work and live are integrated with a natural river environment.

Developing a water quality indicator
Existing indicators such as BOD levels cannot fully describe the present water condition. In 1998, a study was conducted in the five rivers of Greater Tokyo for the development of easily comprehensible indicators to monitor the water quality conditions, and new indicators are being developed (Kanto Regional Development Bureau, MLIT, 2002). The river administrator proposed indicators through the Internet and collected opinions from the public (see map 22.3). Comprehensibility of indicators is considered important, and the study emphasized and proposed indicators based on three aspects: people's relationship with water, rich biodiversity and drinking water.

Map 22.3: Proposed new water quality indicator for recreational use

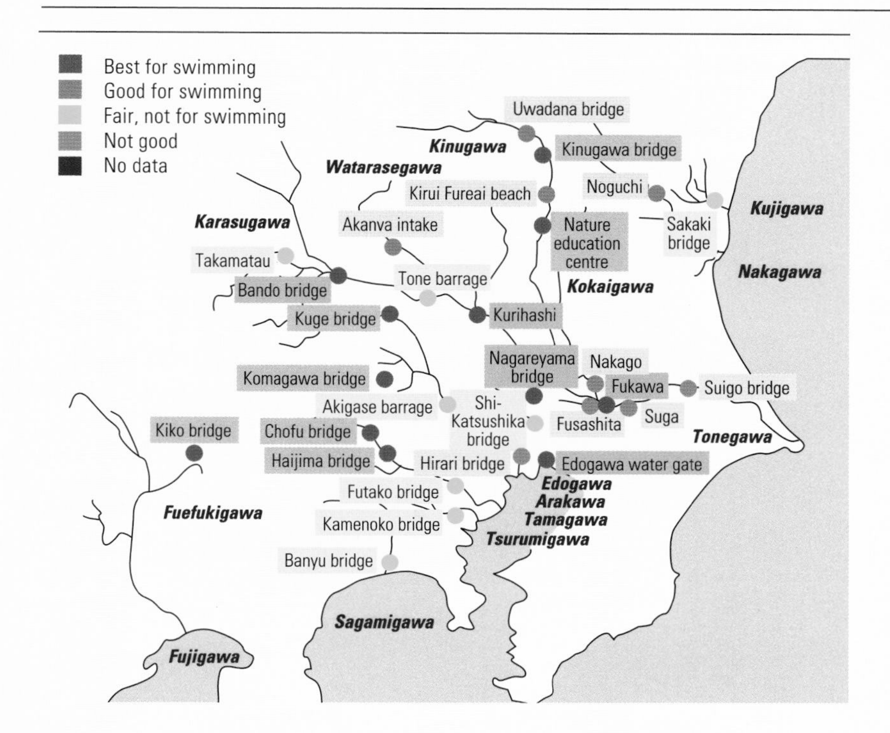

This proposed water quality indicator clearly and efficiently shows the public which areas in Greater Tokyo are suitable for recreational use.

Source: Kanto Regional Development Bureau, MLIT, 2002.

Project for nature restoration

To cater to the population's diverse needs, especially with regards to the natural environment, projects for nature restoration have been implemented. These include the restoration of river meandering, improvement of riverside woods and the restoration of wetlands by frequent flooding, for better habitats along rivers. Such projects are being applied all over the country (Prime Minister of Japan and His Cabinet, 2001). The restoration project of dried wetlands started in the Watarase retarding basin of Tonegawa River (River Bureau, MLIT, 2001c).

Non-structural measures of risk mitigation

Water resource management includes 'hard' and 'soft' management. A hazard map is one of the tools for soft management; it enables people to prepare for disasters and evacuate promptly. Figure 22.7 shows the effect of hazard maps. There is a clear difference between people who watched the map in advance and those who did not. Efficient and comprehensive soft management measures, such as hazard maps, are ensured by the information systems. In Japan, the river administrator processes and provides information to governmental agencies and local residents so that appropriate river management and flood defence measures can be taken.

Conclusions

Greater Tokyo houses one of the world's largest populations and industrial spreads, much of which is set on flood plains and other high-risk areas. In order to mitigate flood damage and to ensure a better quality and security of life in the metropolis, the government of Japan has invested large amounts of money in new technologies, structural measures and greater public information services. Development of comprehensible indicators has also been one of the main priorities in the region. Water is of varying quality in the five rivers that make up Greater Tokyo, with BOD levels sometimes significantly above recommended standards. However, as both this and other environmental matters become increasingly important in the public eye, more and more efforts are being made to control any potential problems.

Greater Tokyo is an area where the public has found a voice in matters of the environment, as the population is frequently consulted and involved through discussions, meetings and media such as the Internet. Although the region faces many challenges, including floods and droughts in a great many urban areas with millions of yen worth of assets, the development of such initiatives as have already been implemented feasibly meets and overcomes many of the challenges that lie ahead.

Figure 22.7: Effect of hazard maps on public safety

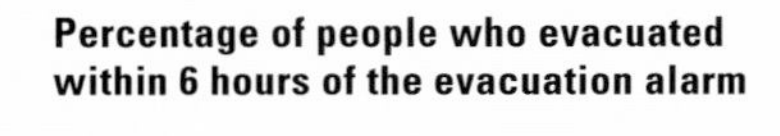

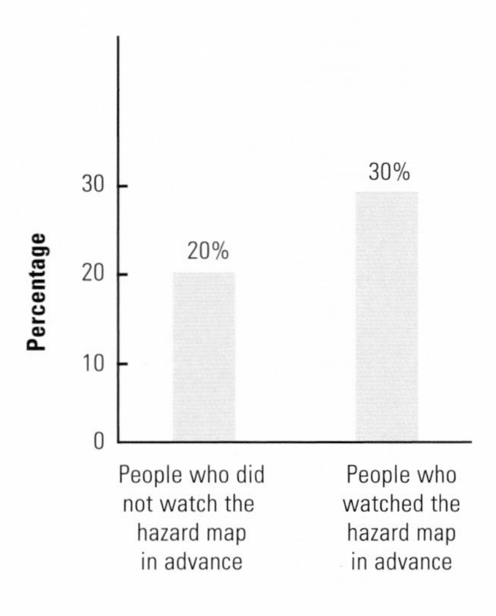

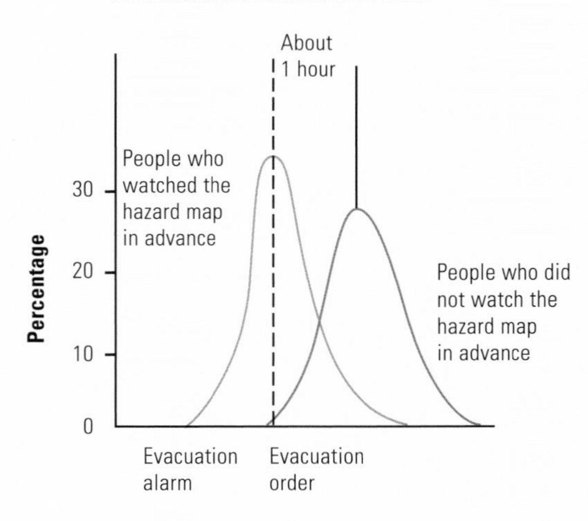

Public awareness of the hazard maps significantly increases safety during floods: 50 percent more people made it to the refuge areas after having watched the map.

Source: Katada Laboratory, 2001.

Box 22.1: Development of indicators

Establishing indicators based on clear criteria is essential for future assessment. The Greater Tokyo case study team proposed the following six criteria: relevance, cost, comprehensibility, clarity, continuity and social benefit (Yasuda and Murase, 2002). The indicators proposed by the World Water Assessment Programme (WWAP) are discussed in the following table. Some indicators are too vague to be calculated.

Challenge areas	Greater Tokyo indicators
SURFACE WATER	• Average water use/water resources normal years: 44%; drought years: 66% • Precipitation inland parts: 1,549 mm/year Precipitation seaside parts: 1,535 mm/year
WATER QUALITY	• Pollutant load in Tokyo bay (not whole basin) was 286 tons/day in 1994. • Biological oxygen demand (BOD): Tonegawa: 1.8 mg /litre Arakawa: 4.4 mg/litre Tamagawa: 2.6 mg/litre Tsurumigawa: 10.4 mg/litre Sagamigawa: 2.1 mg/litre
GROUNDWATER	• Dependence on groundwater inland parts: 22.8% seaside parts: 13.1% • Groundwater withdrawal: northern part of Kanto plain: 960 million m^3 • Degree of dependence of households on underground water: 18.7%
PROMOTING HEALTH	• Number of houses with sewage: 12,052,059 • Percentage of houses with water supply: 97.3% • Investment in water supply: 778,098 million yen • Investment in sanitation: 2,732,671 million yen
PROTECTING ECOSYSTEMS	• Urbanization rate: 13.0% (1974), 20.2% (2000) • Percentage of wetlands: 0.11% • 1 Ramsar site (Japan has 11 sites)
WATER AND CITIES	• Water supply: 97.3%; sanitation (sewage): 76.9% • The basic total pollutant load control system is being applied to Tokyo bay, not the whole Tokyo region. • Pollutant load from household wastewater in Tokyo bay was 167 tons/day in 1999. • Comprehensive flood control measures are implemented through the establishment of the Council for comprehensive flood control measures for individual river basins and through the formulation of basin development plans that include improvement of the environment.
SECURING THE FOOD SUPPLY	• Self-sufficiency ratio in food (all Japan data for 1999) • Self-sufficiency ratio in calorie supply: 39 • Self-sufficiency ratio in staple food cereals: 58 • Self-sufficiency ratio in cereals: 26 • No data at regional level. • Volume of virtual water import: 43.86 bm^3/year.
WATER AND INDUSTRY	• Industrial use of water by total developed water in Tonegawa and Arakawa basins: 25.4% (2000) • Amount of manufactured goods shipped (yen)/amount of water used (m^3): 6,092 yen/m^3 • Ratio of recycled water use: 85.4% • Basic total pollutant load control system is being applied to Tokyo bay, not the whole Tokyo region. • Pollutant loads from industrial wastewater in Tokyo bay was 52 tons/day for 1999.
WATER AND ENERGY	• Amount of annual output/capacity of dam plant for generation of electricity: 9.0 kWh/year/m^3 • The amount of water use for cooling/ amount of water used for industry: 78.9%
MANAGING RISK	• Number of people living within 100-year flood area in Arakawa basin: 2,148,360 people
SHARING WATER	• Industry: 15.4% • Agriculture: 66.1% • Households: 18.5% • Formal policy exists. In addition, drought conciliation are held to conduct a constant water supply during drought.

Box 22.1: *continued*

Challenge areas	Greater Tokyo indicators
VALUING WATER	• Water rate per household: 2,316 yen/month (1999, all Japan) • Water rate per capita: 861 yen/month (all Japan) • No available data on water use by amount of income. • Percentage of water charge/average income: 0.4% • Percentage of water charge/average consumption expenditure: 0.7% • Effective storage capacity of dam/population: 60.3 m^3/capita • Effective storage capacity of dam –flood control storage/population: 49.2 m^3/capita
ENSURING KNOWLEDGE	• Effective computerized system of hydrometeorological data collection exists. • Most data are maintained on local government bases, not on basin-wide. • For transparency, a disclosure law was enacted in 1999, and effective database systems have been developed.
GOVERNING WISELY	• Amount of water resources investment/ population: 3,334 yen/year (2000, all Japan) • Various comprehensive policies such as basin management exist. • River Law exists, revised in 1997. • Institutional cooperation for healthy hydrological cycle started in 1998 (e.g. the River Improvement Plan)

References

Environmental Bureau of the Tokyo Metropolitan Government. 2000. *Tokyo-to Kankyo Hakusyo* (Tokyo Environment White Paper). Tokyo.

Foundation for Riverfront Improvement and Restoration. 1999. *Mizube no Kokusei Chousa Kekka* (Result of the National Census of the Waterside). Tokyo.

Improvement and Management Division, Bureau River, MLIT (Ministry of Land, Infrastructure and Transport). 2001. 'Kizon Damu-Sutokku no Tettei Katsuyou' (Practical Use of Existing Dams). *River.* Tokyo, Japan River Association.

Japan Dam Foundation. 2002. *Dam Nenkan* (Dam's Yearbook). Tokyo.

Japan River Association. 1986–2000. *Kasen Bin-ran* (River Manual). Tokyo.

Japan Sewage Works Association. 1973–2000. *Gesuidou Toukei* (Sewage Statistics). Tokyo.

Kanto Regional Development Bureau. 2002. *Atarashi Suishitsu Sihyo no Tei-an.* (The Proposal of New Water Quality Indicators). Japan, Ministry of Land, Infrastructure and Transport.

Katada Laboratory, Department of Engineering. 2001. 'Survey of People's Behaviors During the Flood in the End of August 1998'. In: MLIT (Ministry of Land, Infrastructure and Transport). *Kokudo Koutsu Hakusyo* (Country Traffic White Paper). Tokyo.

Ministry of Economy, Trade and Industry. 2002. *Report on Industrial Statistics* (Industrial Land and Water). Tokyo.

Ministry of Health and Welfare. 1966–2000. *Suidou Toukei* (Water Service Statistics). Tokyo.

Ministry of the Environment. 2001. *Water Environment Management in Japan.* Tokyo.

——. 2002. *Kankyo Hakusyo* (Environmental White Paper). Tokyo.

MLIT (Ministry of Land, Infrastructure and Transport). 2001a. *Syutoken Hakusyo* (Metropolitan Area White Paper). Tokyo, Printing Bureau, Ministry of Finance.

——. 2001b. *Kokudo Koutsu Hakusyo* (Country Traffic White Paper). Tokyo, Gyosei Corp.

MOC (Ministry of Construction). 2000. *Kensetsu Hakusyo* (Construction White Paper). Japan, Infrastructure Development Institute.

——. 1999. *21-seiki no Suigenchi Vision* (The Source-of-a-stream Vision in the 21st Century). Japan, Infrastructure Development Institute.

——. 1995. *Medias Data.* Japan, Infrastructure Development Institute.

National Land Agency. 1999. *Atarashii Zenkoku Sougou Mizushigen Keikaku* (New National Comprehensive Water Plan). Tokyo, Printing Bureau, Ministry of Finance.

Prime Minister of Japan and His Cabinet. 2001. *21 – seki no 'Wa no Kumi Zukusi'* (Meeting on the Country's Policy in the 21st Century).

River Bureau, MLIT (Ministry of Land, Infrastructure and Transport). 2001a. *1998 Kasen Suishitsu Nenkan* (River Water Quality Yearbook). Tokyo. Ministry of Land, Infrastructure and Transport. (MLIT).

——. 2001b. *Suigai Toukei* (Flood Damage Statistics). Tokyo. Ministry of Land, Infrastructure and Transport.

——. 2001c. 'Shizen Kasen Wetland no Saisei-nado Shizen Kyousei Sesaku no Suishin' (Promotion of a Natural Symbioses Measures, such as Reproduction of Natural River and Wetland). *River.* Tokyo, Japan River Association.

River Bureau, MOC (Ministry of Construction). 1997. *Drought Conciliation and Water Rights.* Tokyo, Tokyo University, Infrastructure Development Institute.

Sato, H. 2002. *Mizu-Joho-Kokudo no Kouchiku ni Mukete* (A Strategy for 'The Land with Water Information'). Japan, Japan River Association.

Water Resources Department, MLIT (Ministry of Land, Infrastructure and Transport). 2001. *Nihon no Mizu-shigen* (Water Resources of Japan). Tokyo, Printing Bureau, Ministry of Finance.

Water Resources Environment Technology Center. 2001. *Damu no Kankyo* (*Environment of Dams).* Tokyo.

Water Use Coordination Sub-Division, Water Administration Division, River Bureau, MLIT. 1995. 'Suiriken toha Nanika' (What are Water Rights?). *Ministry of Construction Monthly,* No. 554, p. 29. Tokyo.

Yasuda, G. and Murase, M. 2002. 'Flood Risk Indicators'. Paper presented at the Workshop on Indicator Development for World Water Development Report held in February 2002, in Rome, Italy.

VI

Part VI: Fitting the Pieces Together

Ten years have passed since the nations of the world met in Rio de Janeiro for the Earth Summit and drew up an action plan for the twenty-first century – Agenda 21. Chapter 18 of this blueprint for action focused entirely on the sustainable use of freshwater resources.

Today, the movement towards a more people-oriented and integrated approach to water management and development is well underway. It is time to take stock. Are we achieving the twin goals of serving society, while also ensuring the sustainable use of natural resources? What is missing from the global freshwater picture?

In this first edition of the *World Water Development Report* (WWDR) we have identified a few of the pieces that make up the giant puzzle of factors contributing to today's global water situation. They come in every shape and size, but each part has a role to play. Each piece contributes to our understanding of the whole. However, unlike a conventional puzzle that remains fixed once the pieces are in place, the freshwater picture is perpetually changing. It is complex and multidimensional.

23

The World's Water Crisis Fitting the Pieces Together

With the collaboration of:
GWP (Global Water Partnership)

Table of contents

Only wholeness leads to clarity.

Friedrich Von Schiller

EACH ONE OF THE ELEVEN CHALLENGES addressed by this report is significant and substantive in its own right. Each will require major and focused effort in order to be met and overcome. Having looked at the challenges individually, however, our purpose now is to fit the pieces together within an integrated framework that reveals how the water crisis affects the daily lives of billions of people. Presenting this bigger picture serves different purposes: to expose the multiple dimensions of the water crisis, to create an awareness of the complexity of the water situation and of how the many dimensions are tied into the overeaching goals.

THOSE AFFECTED MOST BY THE WATER CRISIS are the world's poor. It is they who suffer most immediately from unsafe water, lack of sanitation, food insecurity and from the effects of pollution and a degraded environment. Without representation or any voice in social, economic and political affairs, they are often powerless to improve their situation. This position of powerlessness only reinforces the vicious cycle of poverty, poor health, insecure livelihood and vulnerability to risks of every kind.

Water and Poverty

The plight of the poor

This book has highlighted the plight of the poor in many areas: people who are water-poor; people who are sanitation-poor; people who are food-poor; people who are electricity-poor; people who are income-poor; people who are kept in poverty because they live in disaster-prone areas; and people who, being poor, lack a voice in water governance and water management. What we have still to determine are the numbers of people who suffer two or more of the different dimensions of water poverty, as illustrated in figure 23.1. These figures will not be known until integrated water poverty surveys, such as the Human Development Index, are able to capture the multiple burdens of individuals. There are promising initiatives, but the assessment remains elusive. The central point is that there are very many who are poor, and in many ways at the same time. It is more or less the same two billion people who lack three, four or even five of the basic levels of service and security, and who live with the daily consequences.

Figure 23.1: Multiple burdens of water poverty

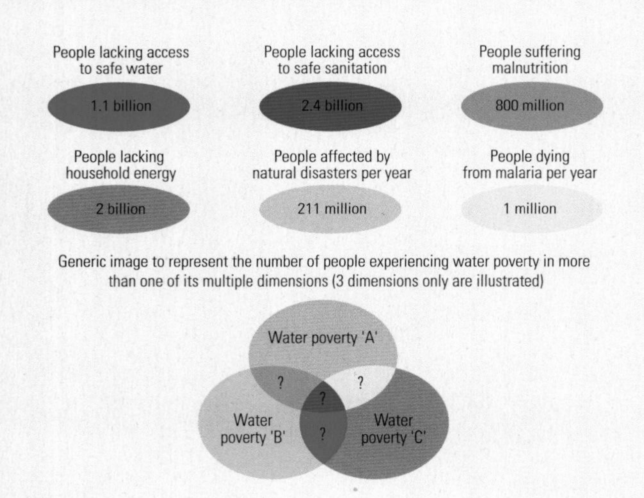

Source: WWAP Secretariat.

Problems of poverty are thus inextricably linked with those of water – its availability, its proximity, its quantity and its quality. They are linked to all the challenge areas, because water problems are people problems. They are sustainable development problems. For this reason, they are the composite parts of the wider goal of attaining water security in the twenty-first century set by The Hague Ministerial Declaration. As such, poverty issues are the compelling element of an overall water agenda. For those concerned with the realization of broad water policy, these issues represent the bigger picture – a picture that is composed both of the broad and the singular, of overarching goals and specific targets, of competing demands and particular interests, of multiple demands on time and finances and of priorities. It is a bigger picture that demands integration, coordination, compromise and trade-offs.

Presenting this bigger picture serves different purposes, including showing how water management, in order to be effective and adopt an integrated approach, must simultaneously address the following three areas:

- the contribution of water to wider economic and social objectives;
- the management of water within its traditional sectoral context; and
- the wise management of water resources in support of all water uses, individually, together and in competition.

Fitting the pieces together actually helps us to gain the clarity needed in order to make difficult strategic choices – to decide where to place our resources so as to have maximum impact. It helps point the way forward.

Linking the Challenge Areas: An Integrated Approach

Global frameworks

The world can be viewed through many different lenses, and there are many different frameworks by which to approach and integrate the challenge areas. Integrated approaches to water resources management have been a recurrent theme throughout this report. It embraces integration across many different dimensions, within both natural and human systems. At the heart of this process is the importance of finding a balance that ensures the sustainability of social, economic and ecosystem goals.

Box 23.1 presents some frameworks resulting from international conferences and ministerial declarations while box 23.2 presents other action frameworks (business, sustainable livelihoods, poverty

Box 23.1: Frameworks arising from international conferences – Millennium Development Goals, Bonn Action Plan and WEHAB

Millennium Development Goals

The Millennium Development Goals, agreed on in 2000 by heads of state, represent a series of specific and urgent improvements in the lives of billions of people, in many different ways. Separate targets have been set related to economic growth, health, agriculture and poverty alleviation. Our discussions of the individual challenges have been very clear about the contribution of water to each. There are also specific targets in water, notably in supply and in sanitation. As already highlighted in Part I of this book, the most significant aspect of the Millennium Development Goals is its embrace of different dimensions as expressed in the following:

> *We resolve further to halve, by the year 2015, the proportion of the world's people whose income is less than one dollar a day and the proportion of people who suffer from hunger and, by the same date, to halve the proportion of people who are unable to reach or to afford safe drinking water.*

By 2015, this Millennium Development Declaration will not have been achieved unless all of the outcomes are achieved. So, the individual challenges in water each need to be resolved, but in concert.

Bonn Action Plan

Ministers at the 2001 Bonn International Conference on Freshwater agreed to a twenty-seven-point Action Plan, further detailed in this chapter. The plan grouped actions within four headings:

- governance;
- mobilizing financial resources;
- capacity-building and sharing knowledge; and
- roles (of different types of institutions).

Many of the specific Bonn actions relate directly to individual challenges. While the twenty-seven individual actions carry greater significance than the headings under which they have been grouped, it is clear that the Bonn conference gave prominence to those water challenges listed above. But its central message was an emphasis on action. It is this emphasis that links the challenges together – only action can overcome the very real challenges in the real world.

WEHAB

The WEHAB initiative, proposed by UN Secretary-General Kofi Annan as a contribution to the World Summit on Sustainable Development (WSSD) held in 2002, targets sustainable development with a stronger emphasis than ever before on development. It provides focus and impetus to action through five key areas: Water (and Sanitation), Energy, Health, Agriculture and Biodiversity (WEHAB). In addition to its own direct contribution, water will also contribute through each of the other areas, as seen in this report. Water may not necessarily dominate the agenda within these other areas, but their respective goals and targets cannot be achieved without actions inclusive of water.

The WEHAB framework establishes the potential conflicts and beneficial interactions between the five development objectives with both positive and negative impacts on water. Such interfaces point to some of the traditional barriers that must now be dismantled in order to move forward.

and action). Each model has the capacity to carry water actions into the future within an integrated approach that both links together the WWDR challenge areas and advances our ability to confront the multidimensional problems of poverty.

The dynamics of competing demands and the delicate balance involved in evaluating trade-offs are best viewed by looking at a real-world example, such as that of dams. This single example includes all of the different challenges and management considerations discussed in the WWDR. It is a microcosm of the water resources dilemma.

Managing dams: a microcosm example of an integrated approach

The water regulated by and stored in dams is considered to be an absolute requirement to meet the water challenges and the development objectives of other sectors in industry, agriculture, energy and risk management. Indeed, it is said that in many parts of the world such challenges cannot be achieved without increased storage, and demand cannot be met by exploitation of natural flow patterns alone. Dams continue to be seen as a solution across all of the use and demand challenges – except those related to the environment. As such, dams remain a controversial issue, with seemingly intractable and divergent standpoints held by actors in different camps.

Many of the critical management issues have been addressed through the recommendations of the World Commission on Dams (WCD). While dams represent a microcosm of the challenges faced as part of the overall water crisis, the best-practice guidelines for dam development and management as proposed by the WCD also have a generic value for all water resource development and management.

Box 23.2: Other frameworks – poverty and action, sustainable livelihoods, business management

Poverty and action framework

The World Bank's *World Development Report 2000/1* placed firm emphasis on the causes of poverty and proposed a framework for action around three equally important areas:

- promoting opportunity;
- facilitating empowerment; and
- enhancing security.

While representing a broad agenda for poverty alleviation in general, it is a framework with strong resonance for the water poverty agenda. The two may be brought together through the specific framing of water challenges within national poverty reduction strategies. Expanding briefly on each of the three areas:

- Opportunity promotion embraces growth as a basis for expanding economic opportunity for poor people – specifically how to achieve rapid, sustainable and pro-poor growth. This hinges around a business environment conducive to private investment and technological innovation, underpinned by political and social stability. Promoting opportunity also embraces economic opportunities for poor people in order to build up their assets, including human capabilities, through health and education. Of importance to the material assets of poor people is ownership of, or access to, natural resources (notably land and water), infrastructure, financial services and social networks.

- Empowerment is seen as enhancing the capacity of poor people to influence the state institutions that affect their lives, removing barriers – political, legal and social – that work against particular groups and building the assets of poor people to enable them to effectively engage in markets. It further embraces state and social institutions that work in the interests of poor people, formal democratic processes and accountability on everyday state or commercial interventions that help or hurt poor people.

- Enhancing security for poor people means reducing their vulnerability to such risks as ill-health, economic shocks and natural disasters, and it means helping them to cope with adverse shocks when these do occur. The links between poverty, insecurity and water are very clear in this regard.

The significance of the individual water challenges very clearly comes together within this broader poverty action framework, as does the fact that its success relies to a great extent on overcoming water challenges.

Box 23.2: *continued*

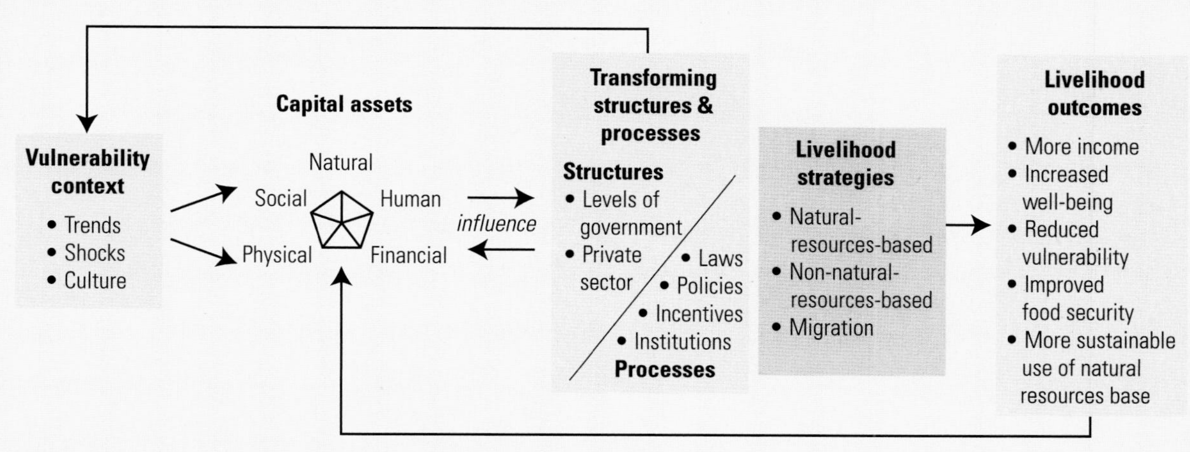

Source: Carney, 1998.

Sustainable livelihoods framework

Short-term survival rather than sustainable management is often the priority of people living in absolute poverty. Securing livelihoods is aimed at enabling individuals and households to secure a regular income, so that goods and services become affordable, and people are better able to find their own ways out of poverty. It is founded on the capabilities, assets and activities that are required for a means of living. A livelihood is sustainable when it can cope with stresses and shocks and can maintain its capabilities and assets both now and in the future, without bankrupting its resource base. A livelihoods framework is consistent with putting people first. Results show up in improved family and household income, more purchasing power and in increasing independence from a subsistence and service provision economy. The associated outcomes include increased well-being, reduced vulnerability and improved security.

A generic framework for sustainable rural livelihoods framework is presented in the figure above. The framework is built around capital assets, but embraces other dimensions that have strong correlation with the challenges in water. It embraces the vulnerability context in which assets exist – the trends, shocks and local cultural practices that affect livelihoods. The framework embraces capital assets upon which people can draw, to which there are five parts: the capital of natural resources (water, land); social resources (the wide institutions of society, membership of groups, networks); the skills, knowledge and ability of human resources; basic infrastructure and production equipment; and financial resources that are available to people (whether savings, supplies of credit or regular remittances, or pensions). The framework also embraces the structures (organizations, from layers of government through to the private sector in all its guises) and processes (policies, laws, rules of the game and incentives) of the institutional setting. A livelihoods perspective provides a framework in which the challenges come together to enable people to make a living.

Business management

Business management represents a framework within which organizations operate on a daily basis. A typical business plan will comprise many areas – business analysis, focused programmes, core functions, technical themes, processes and tools, asset acquisition and management, management maintenance and support of its systems, information flows and human resources, depending on an organization's size and mission. The water challenges come together through the manner in which a business internalizes them within its planning and operations, and ultimately its delivery of services – either voluntarily or through compliance. The Millennium Development targets can only be achieved if organizations orientate their day-to-day business towards such accomplishments.

It is in their multi-purpose use – where stored water in the same dam may be able to support urban water supply, irrigation, power generation and flood storage (implemented through management guidelines addressing issues of environmental protection, value, sharing, risk and governance) that dams bring the different challenges together into a single point of focus. The exact contribution of dams to the resolution of the challenges is argued in different ways in different settings. In many developing countries, and particularly those with high variability in their resource or with cities located far from the resource base, dams appear as an absolute requisite to economic and social development. In other situations, it may be argued that dams are now redundant and can therefore be removed to restore natural ecosystems. Other actors may insist that multi- or bilateral funds should not be directed towards investment in dams where there is a risk of unacceptable social or environmental consequences.

The example of dams clearly reveals the multidimensional nature of sustainable development and its attendant problems for water management. It highlights the need to consider all of the challenges as a practical basis from which to move forward. When an integrated framework is adopted, the actual point of entry matters little: what matters is the convergence of interests and challenges within the broader picture.

The Multidimensional Nature of the World's Water Crisis

Our assessment of individual challenges demonstrates the multidimensional nature of the world's water crisis and the compounding effect of one crisis upon another. It reveals the sheer magnitude of the task faced by those who need to resolve not just one crisis, but several, often simultaneously.

The global overview

Quite simply, sustainable development is not being achieved. It is not being achieved through water supply, sanitation, natural or urban ecosystems, nor through food security, industry, energy or economic and social advancement. The everyday lives of billions of people are not being made more secure. Rather, development has mostly brought additional pressures on water and the environment, and those pressures are set to mount further still.

In 2000, 1.1 billion people were without adequate water supply. More than twice that number – 2.4 billion – were without basic sanitation. In this book, a first estimate has been made of the annual total number of people in ill-health or dying because of deficient water, sanitation and hygiene. Each year, this amounts to 1.7 million deaths (equivalent to 4,740 per day) and the loss of 49.2 million Disability-Adjusted Life Years (DALYs). This does not include the toll of lives ravaged by malaria – over 1 million deaths and 42.3 million life years per year.

Progress has been made over the last ten years in coping with a global population increase of 15 percent. In fact, at a global level, the delivery rates of 'improved' services have stayed slightly ahead of population growth rates. But the gap in water supply, sanitation and health coverage and service has not been closed, and still today it is the poor who remain unserved. The 2000 UN Millennium Summit set a target of halving the proportion of people without sustainable, safe drinking water by 2015. The World Summit on Sustainable Development (WSSD) held in Johannesburg in 2002 agreed on an equivalent sanitation target.

Freshwater ecosystems have been hit hard by reduced and altered flow patterns, by deteriorating water quality, by infrastructure construction and by land conversions. More rivers have been disrupted, fewer rivers retain good ecological status, water quality has declined in many localities, more structures are being built and more lands, including wetlands, are being converted to agriculture. Consequently, biodiversity and commercial fisheries are in global decline as freshwater ecosystems have been more severely affected than the land or the sea. Although many local, and sometimes national, improvements can be seen, these are not yet stemming the global decline in the state of the environment.

Special efforts must be made to provide better water and sanitation services to the world's cities as their populations swell through natural growth and migration. In a little over ten years' time, urban areas will house the majority of the world's population. There is little cause for optimism that development targets will be reached. As chapter 7 explains, the case of African and Asian cities has highlighted that 'improved' coverage is not a question of addressing adequacy of provision in concentrated population centres. Depending on the definitions used, this new evidence could put urban areas particularly – and therefore totals in general – even further behind in coverage, and further increase the quoted numbers of those lacking coverage – perhaps even twofold.

It is not a global food shortage that is causing food insecurity – it is the inability of those in poverty to access the world's available food. Agricultural production has increased steadily in recent decades and the growth of global food supplies has exceeded the demands of an increasing population. The fact that 800 million people are seriously affected by chronic malnutrition is due to the social, economic and political contexts – global and national – that perpetuate, and sometimes cause, unacceptable levels of poverty. Over the next twenty-five years, food will be required for another 2–3 billion people. Within the current demographic context, the global food security outlook is reasonably good. Towards 2050, a stabilized world population could enjoy access to food for all. It is expected that 20 percent of increased crop production in developing countries will come from expansion of agricultural land, and 80 percent from intensification. A 20 percent increase in the extent of land equipped for irrigation is foreseen. Water withdrawals for agriculture, presently at 70 percent of all abstracted water, are expected to increase by 14 percent until 2030, with irrigation water use efficiency improving by an average 4 percent.

Some 22 percent of the world's water withdrawals are for industry, and this is forecast to grow to 24 percent. Globalization – with its accompanying relocation of labour-intensive industries – is creating high water demand in areas that do not have the necessary abundance. High-income countries are deriving more value per unit of water used than lower-income countries. Effluent discharges to water bodies have been reduced in high-income countries, but loads have risen substantially in middle- and low-income countries.

Two billion people have no electricity at all, and 2.5 billion people in developing countries, mainly in rural areas, have little access to commercial energy services. Hydropower has been the most significant source of energy from water, and it contributes 7 percent of total energy production worldwide and one fifth of electricity production. Over the last ten years, the development of new hydropower capacity has kept pace with the overall increase in the energy sector, and it is likely that it will continue to do so for the foreseeable future. Water is likely to contribute to the expansion of electricity coverage through large-scale hydropower and thermal cooling in urban areas. Unless they are brought into supply grids, the greatest potential in rural areas lies in small-scale hydropower.

In response to these challenges to life and well-being, water management continues to seek practical solutions to mitigating risk, sharing water, valuing water, ensuring the knowledge base and governing water effectively. But risks continue to mount and are exerting a heavy toll. Deaths due to natural disasters totalled at least 40,000 in 1998 and 50,000 in 1999, with 97 percent of all these deaths in developing countries. More people are now affected by disasters than ever before – up from an average of 147 million per year (1981–90) to 211 million per year (1991–2000). Economic losses from all disasters amounted to US$70 billion in 1999, compared with US$30 billion in 1990. A dramatic shift is encouraged away from reactive approaches and reliance on failing engineering solutions, towards alternative and more sustainable risk reduction.

A number of common mechanisms have been implemented within countries at an operational level for intersectoral sharing. There has been expanded involvement in the management of international water, including the 1997 UN Convention on the Law of Non-navigational Uses of International Waters. Despite the potential for dispute, the record of cooperation between countries overwhelms the record of acute conflict.

The different ways of valuing water are now recognized, yet this remains a controversial issue. There is a growing acceptance of the need for full cost recovery in water services, but this must be done in a way that safeguards the needs of the poor. Valuing water has become critical to optimizing investment and obtaining viable private sector participation in the efforts to raise the needed projected investments of US$180 billion per year until 2025.

Efforts are being made to extend the knowledge base among the general populace through formal and informal education, public awareness and the media. Recent advances in information and communication technologies (ICT) have made it easier to secure the knowledge base in its many dimensions. But developing countries, with the greatest need for the benefits of ICT in overcoming geographic and economic isolation, are hindered by lack of access. Unless this gap, the so-called digital divide, is narrowed, it will also perpetuate its own vicious cycle of isolation, ignorance and non-participation in development and decision-making.

It is said that the water crisis is a crisis of governance. Specific targets for national action programmes, appropriate institutional structures and legal instruments originally set to be in place by 2000 have not yet been fulfilled in many instances. At the community level, empowerment and self-reliance have shown themselves effective in improving services for the poor. Partnerships with the private sector have also improved services when accompanied by effective regulation. But capacity to regulate is lacking. Remedying the situation requires a long-term commitment to education and training. Stakeholders must be willing and able to participate in decision-making and be recognized as key players in solving water resource problems. Where none exists, an institutional capacity must be created to regulate water questions and provide an enabling framework. The water sector remains seriously underfinanced, although new funding initiatives announced at the WSSD suggest that the situation could improve. Many hope that the private sector can fill this gap and contribute more. Debt relief, in which macrogovernance is proving to be a major determinant, has begun releasing funds in support of development targets.

Regional dimensions

The regional perspective offers another view of the global picture, highlighting certain regional disparities, as shown in figure 23.2. Seen through this prism, the various challenge areas and other pieces in the puzzle reveal a different and sometimes surprising profile. In certain parts of the world the different challenges are actually compounded.

Africa

With only 64 percent of the population with access to improved water supply, Africa has the lowest proportional coverage of any region of the world. The situation is much worse in rural areas, where coverage is 50 percent compared with 86 percent in urban areas. Sanitation coverage in Africa is also poor, although Asia has even lower coverage levels. Currently, only 60 percent of the African population has sanitation coverage, with 80 percent and 48 percent

Figure 23.2: Regional environmental trends

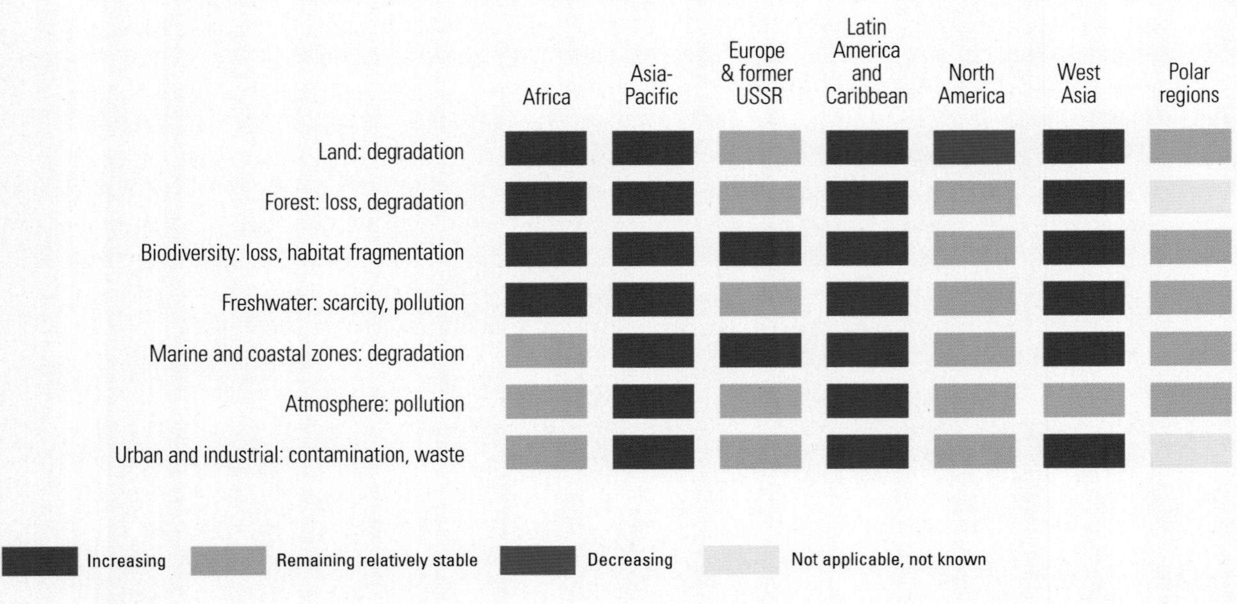

Source: GEO1, 1997.

in urban and rural areas respectively. In global terms, the continent houses 27 percent of the world's population that is without access to improved water supply, and 13 percent without access to improved sanitation. Ten countries have less than 50 percent coverage for both their current national water supply and sanitation coverage. Urban services have remained more or less the same across the 1990s; rural services, however, tell a different story as water supply increased slightly and rural sanitation fell. The African population is expected to increase by 65 percent over the next twenty-five years, with the greatest increases in urban areas. Meeting the 2015 targets will require tripling the rate at which additional people gained access to water between 1990 and 2000 and quadrupling the rate at which they improved sanitation.

The gap between the proportion of urban dwellers with 'improved' provision and provision that is 'safe and sufficient' is evident in many African nations. Whereas 86 percent of urban dwellers have 'improved' supplies, more than half have inadequate provision if the definition is to mean a house connection or yard tap. Many city studies now suggest that the proportion of people with sanitation that is safe and convenient is much lower than the proportion with 'improved' sanitation. In most of the largest cities in Africa, less than 10 percent of their inhabitants have sewer connections. Tens of millions of households, especially in informal settlements, only have access to overused and poorly maintained communal or public toilets. In many African cities, only 10 to 30 percent of all urban households' solid wastes are collected.

For African children under five, the health burden that arises from diarrhoeal disease linked to inadequate water, sanitation and hygiene is up to 240 times higher than in high-income nations. Malaria takes its heaviest toll in Africa south of the Sahara, where about 85.7 percent of the annual global rate of over 1.1 million deaths occur, mainly among children under five. It is the leading cause of death in young children and constitutes 10 percent of the overall disease burden, and slows economic growth in African countries by 1.3 percent a year. Of the estimated 256.7 million people worldwide infected by schistosomiasis (bilharzia), 212.6 million (82.8 percent) cases occur in Africa south of the Sahara. Urban populations in Africa are also among the most affected by lymphatic filariasis.

In the 1990s, 58 percent of sixty-seven urban cities in twenty-nine nations (including most of the continent's largest cities) were using rivers 25 or more kilometres away, and just over half that relied on rivers depended on interbasin transfers. As is common throughout the developing world, very few cities in Africa have rivers flowing through them that are not heavily polluted, and much the same applies to nearby lakes, estuaries and seas.

Per capita food consumption, and associated calorific and nutritional intake, has remained disappointingly low in sub-Saharan Africa over the past forty-five years. The number of undernourished Africans rose steeply during the 1990s. While the total number of undernourished people worldwide has fallen, the proportion of the sub-Saharan Africa population has remained virtually unchanged. Food security actions therefore take on a special urgency in this region.

Africa's dependence on cereal imports is expected to continue to grow, with a widening net trade deficit. When grouped among developing regions generally, 60 percent of food production is from non-irrigated agriculture. A sizeable part of irrigation potential is already used in North Africa and the Near East (where water is the limiting factor), but a large part also remains unused in sub-Saharan Africa. Such low proportions of irrigated land point to underdeveloped infrastructure. From 1998 to 2030, arable land in developing countries is projected to increase by 13 percent, with the bulk of the projected expansion foreseen to take place in sub-Saharan Africa and Latin America. The expansion of irrigation is projected to be strongest in North Africa, as well as in the Near East. By 2030, North Africa will have reached critical thresholds of water availability for agriculture. The proportion of renewable water resources allocated to irrigation in sub-Saharan Africa is likely to remain far below the critical threshold. In this region where no additional land resources are available to exploit, wetlands represent an attractive opportunity for agricultural development and crop production. Yields from inland capture fisheries, highest in Asia in terms of total volume, remain significant in Africa as well.

In many countries across central Africa, hydroelectric power generation – mostly from large-scale schemes – is dominant. The economically exploitable potential is huge. Yet while Asia is likely to have quadrupled its installed capacity by 2010, Africa may achieve a more conservative doubling. Sub-Saharan Africa has the fewest households with access to electricity, and among twenty-three sampled countries, none has more than 40 percent of households served. All countries with more than 25 percent are in West Africa, and in all cases it is the richest households that are served. Outside of West Africa, virtually none outside of the richest 20 percent of the households have electricity. There is potential for small and micro-hydro schemes in sub-Saharan Africa – particularly in support of rural households – where topographic, hydrologic and investment conditions are favourable.

In most African countries, water for industry does not exceed 5 percent of total withdrawals, and industrial value added (in terms of US$ per cubic metre [m^3] of water) is low by world standards. Exceptions are those countries with substantial resources of minerals and precious stones, such as Botswana, Gabon, Namibia, and South Africa where mining adds an order of magnitude to value-added products and on bulk use. Similar orders of value added are achieved by the tourism industry.

Over the past ten years, Africa has experienced nearly one third of all water-related (flood and drought) events that occurred worldwide, with nearly 135 million people affected, 80 percent by drought. In the same time period, deaths have exceeded 10,000, with some 98 percent of these due to flooding. Economic losses – almost always uninsured – have impacted significantly upon national economies, development strategies and households. Economic losses from a poor African household do not have to be high to be a very significant disruption.

The bulk of financing comes from central government, supported by global and regional development banks, multilateral and bilateral institutions and, increasingly, debt relief. Africa, together with most other developing regions, is taking steps towards full cost recovery, but has yet to embrace it fully.

Within catchments where water is already under stress, agriculture is the dominant sector. Africa is the only region worldwide in which this is the case. Intersectoral sharing is thus dominated by agricultural water use. Africa south of the Sahara and the Nile Valley are dominated by shared international river basins. Much of Saharan Africa is dominated by shared international groundwater systems. Unlike many other regions where one country has a single shared river system, many African countries share upwards of two, even as many as nine, international rivers.

Despite the increased enrolment rates in formal education, many countries remain impoverished by poor educational achievement. Four out of every ten primary-aged children in sub-Saharan Africa do not go to school. Gross enrolment at the tertiary level stands at 5 percent compared with 52 percent for developed countries. Up to 30 percent of African scientists are lost to the brain drain. Access to the media remains highly uneven, and the spread of the written press is hampered by lack of financial resources and high illiteracy rates. The digital divide is prevalent throughout the continent.

Asia

The picture in Asia reveals other dimensions of the water crisis. Only 47 percent of the Asian population has improved sanitation coverage, by far the lowest of any region of the world. With just 31 percent coverage, the situation is much worse in rural areas compared with urban areas where coverage is 74 percent. Water supply coverage is at 81 percent, the second lowest after Africa. Like sanitation, coverage is lower in rural areas (73 percent) compared with urban areas (93 percent). Because of the large population sizes in the region notably in China and India, Asia accounts for the vast majority of people in the world without access to improved services. Asia is home to 80 percent of the global population without access to improved sanitation, and almost two thirds do not have access to improved water supply. Currently, approximately two thirds of the Asian population live in rural areas, but the balance is predicted to shift over the coming decades. By the year 2015, the urban population is projected to represent 45 percent of the region's total, and should reach just over one half by 2025. This population growth will place enormous strains on already overburdened urban services. As described for Africa, if adequate provision for sanitation in large cities is taken to mean a toilet connected to a sewer, there is a lack of adequate provision in cities throughout Asia – far greater in fact than the improved coverages suggest. Nearly all urban rivers and nearby water bodies have been seriously polluted. River water quality in the region has seen widespread deterioration to levels that pose significant risks to health standards. Urban populations are among those most affected by lymphatic filariasis. Japanese encephalitis is a burden from east to south Asia, its presence strongly linked to flooded ecosystems.

The number of undernourished people fell steeply in east Asia during the 1990s. Only a small part of the region is expected to contribute substantially to projected expansions in arable land to 2030. The worldwide irrigation expansion is, however, projected to be among the greatest in south and east Asia and, by 2030, south Asia and the Near East will have reached critical thresholds of water availability for agriculture. The proportion of renewable water resources allocated to irrigation in east Asia is likely to remain far below the critical threshold. Asia, and particularly China, has seen the dominant share of aquaculture development, and this growth is expected to continue. Benefiting from irrigation, Asia has had wider diversification in its crop production.

Asian countries are constructing many new hydropower schemes, and the region is set to quadruple its 1995 deployment by 2010, primarily through large hydropower. More than 10 percent of the region's hydropower is generated from small schemes, and micro-hydro installations are widespread, with significant potential for further development.

Withdrawals for industry are low by global standards, and in the absence of high value mineral resources, the industrial value added is low by world standards.

Over the past ten years, Asia has experienced nearly one third of all worldwide water-related (flood and drought) disasters. A total of 1.8 billion people were affected (90 percent of all people affected worldwide). Whereas 80 percent of affected persons in Africa were impacted by drought, in Asia 80 percent of affected persons were impacted by floods. Deaths exceeded 60,000 over the same ten years, 98 percent of which were due to flooding: this represents more than half of all deaths to floods and cyclones worldwide. Economic losses – almost always uninsured – have impacted significantly upon national economies and development strategies, and hugely upon social development, with many events affecting many tens of millions of people on each occasion that they strike.

As in Africa, the central government provides most of the financing. In common with most other developing regions, Asia has yet to adopt full cost recovery.

Within catchments where water is under stress, placing pressures on intersectoral sharing, agriculture is the dominant sector throughout western, central and southern Asia, while industry dominates throughout much of South-East Asia. Most of the countries of western Asia, the northern subcontinent and mainland South-East Asia share international rivers.

Despite the increase in enrolment rates in formal education, educational retention rates and achievement remain poor in many countries in south Asia. Access to the media remains highly uneven, and the spread of the written press is hampered by lack of financial resources and high illiteracy rates, particularly in southern Asia. The digital divide is prevalent throughout Asia.

Europe

In Europe, improved water supply coverage is high, with access provided for 97 percent of the population. One hundred percent of the urban population has coverage, compared with 89 percent of the rural population. In terms of sanitation, 95 percent of the population is totally covered, 99 percent of the urban population and 78 percent of the rural population. Those without access to improved water supply represent 2 percent of the global population, and those without access to improved sanitation represent 1 percent. However, low levels of reporting in some areas suggest that a cautious approach to drawing conclusions is nonetheless warranted. Only four European countries reported not having full water supply and sanitation coverage in 2000, all in eastern Europe (Estonia, Hungary, Romania and the Russian Federation). It is predicted that the European population will begin to decrease, especially in rural areas. The greater need to meet water supply deficiencies rests in eastern Europe.

Improvements have been made in reducing water pollution, mostly through stricter controls on industrial discharges and more sophisticated and comprehensive treatment of sewage and stormwater. But a majority of European rivers, particularly in their middle and lower reaches, are in poor ecological condition due to the impacts of canalization, dams, pollution and altered flow regimes. The European Water Framework Directive (WFD) should accelerate the process of bringing pollution under control. During the twentieth century, hydropower made a dramatic contribution to the electricity sector, and most of the prime sites have now been exploited for big plants. An important role in achieving European renewable energy goals can be played by small hydropower resources, especially if the economic situation for producers improves and environmental constraints decrease. Withdrawals for industry, as a proportion of total water use, are among the highest worldwide, notably in central and eastern Europe. Depending on which particular industries dominate a nation's economy, the industrial value added ranges from among the lowest to the highest, spanning three orders of magnitude (between US$0.26/m^3 in Moldova to US$425/m^3 in the United Kingdom).

About 12 million people have been affected by floods or droughts over the past decade, split about evenly between the two. There have been nearly 2,000 deaths from floods, approximately 0.5 percent of all equivalent deaths worldwide. Most economic losses are covered by insurance and reinsurance, and many personal losses by personal insurance policies. The European reinsurance industry has faced heavy burdens from losses experienced elsewhere around the world.

Many countries have moved close to full cost recovery, and have embraced substantive forms of public-private partnerships, albeit along differing models of implementation.

Within water-stressed catchments, agriculture is the dominant sector in southern Europe, while industry dominates throughout much of central and eastern Europe. The region contains shared international rivers, including the most shared river worldwide – the Danube.

Latin America and the Caribbean

This region has relatively high service levels, but is characterized by large differences from one area to the next. Total water supply coverage is extended to approximately 87 percent of the population, while total sanitation coverage is slightly lower at 78 percent. However, large disparities are apparent between urban and rural areas, with an estimated 86 percent of the urban population with sanitation coverage, compared to only 52 percent of the rural population. With respect to water supply, 94 percent of the urban population enjoys coverage, compared with only 65 percent of the rural population. A total of 68 million people are without access to improved water supply in the region and 116 million people without access to improved sanitation – the vast majority in South America. In the significant majority of countries in the region, more than 75 percent of people enjoy both water supply and sanitation coverage. The countries of the Caribbean have the highest reported coverage levels in the region. The percentage of rural service coverage has increased for both water supply and sanitation. Urban services appear to have changed less, and urban water supply coverage in the region even declined slightly between 1990 and 2000. These trends are strongly affected by Brazil, whose current population represents one third of the regional total. But, as described for Africa, if adequate provision for water supply is taken to be a house connection from a pipe distribution system and sanitation is taken to mean a toilet connected to a sewer, the lack of adequate provision in cities throughout Latin America and the Caribbean is significantly higher than the estimates of improved coverages suggest. As in Africa and Asia, most rivers flowing through cities in Latin America are polluted, as are nearby water bodies.

The generally water-rich region of Latin America shows an aggregated low water use efficiency in agriculture, which is not expected to increase significantly in the future because no other large-scale users compete with agriculture. But where water is locally scarce, high efficiency is obtained. Latin America is likely to see one of the highest regional rates in expansion of arable land, but the proportion of renewable water resources allocated to irrigation is likely to remain far below the critical threshold.

Latin American countries are erecting much new hydropower infrastructure, and the region is set to double its 1995 deployment by 2010.

Withdrawals for industry are mid-range by global standards, and the industrial value added is in the mid to upper range by world standards.

Central America and the Caribbean have experienced about 20 percent of the world's hydrometeorological disasters of the past decade. Although this represents just 1 percent of all people affected worldwide, in the past decade it nonetheless adds up to a total of 36,000 deaths, that is, one third of all deaths worldwide due to flooding.

The central government provides most of the region's financing, with additional help from global and regional development banks, multilateral and bilateral institutions and, increasingly, debt relief. Together with most other developing regions, Latin America is slowly moving towards full cost recovery. Among developing countries, this region is furthest advanced in its engagement with the private sector.

Within catchments where water is under stress, there is no single sector that dominates throughout the region. Approximately one half of the region is composed of shared international river basins, dominated by the Amazon and La Plata basins.

North America

The North American population has the highest reported coverage for any region of the world, at 99.9 percent. Urban coverage of water and sanitation are both reported to be 100 percent. In keeping with world trends, demographic projections for the region suggest that the urban population will continue to grow, while the rural population will decline. As described for Europe, substantial improvements have been made in reducing water pollution. But in 1998, one third of waters in the United States were not clean enough to permit fishing and swimming. The United States is the second largest producer of hydroelectricity in the world, with between 10 and 12 percent of the country's electricity supplied by hydropower. Together with Europe, the United States and Canada have moved close to full cost recovery, and have embraced substantive forms of public-private partnerships. Within water-stressed catchments, agriculture is the dominant sector in the west, while industry dominates in the east.

Oceania

Oceania is the least populated of the six regions described. The current coverage status is relatively good, with 94 percent of the population having access to improved sanitation, and 87 percent to improved water supply. However, these figures are strongly slanted by a well-served Australia and New Zealand. With these two countries excluded, coverage levels are much lower. Based on figures from 1990, to meet the 2015 Millennium targets, an additional 8 million people will need access to improved water supply services and an additional 7.2 million will need access to sanitation during the same period of time. Fiji and Kiribati each report having both water supply and sanitation coverage below 50 percent. Outside of the two countries dominant in population, the region is characterized by problems typically faced by small island states: these include the risks associated with climate change, an abundance of fragile ecosystems and vulnerability to water shortages at certain times of the year.

Country situations

It is beyond the scope of this first *World Water Development Report* to assess the state of water in individual countries. We note only that sources do exist, which may be explored in more depth in future assessment exercises.

First and foremost, there are the offices of national statistics in each country. Also, over 100 national water ministries maintain homepages on the Internet.

Second, there are the publicly available reports that member states have submitted to the United Nations system, and particularly to the Commission on Sustainable Development (CSD). Agenda 21 recommended that countries consider preparing national reports on implementation of Agenda 21 and communicate this information to the CSD. Governments began submitting reports as early as 1993 (see table 23.1).

Third, there is much national information available within the UN system. For example, at the end of each challenge area, we have presented links to web sites where further information may be sourced by those with Internet access. Further, country reports on particular thematic issues have been produced by the UN system in concert with member states. Two series in particular stand out: first, the region reports in support of the Water Supply and Sanitation Sector Assessment, under the auspices of the World Health Organization (WHO), (the generic contents of the country reports are presented in box 23.3), and, second, the regional 'irrigation' reports produced by the Food and Agriculture Organization (FAO). These latter cover geography and population, climate and water resources, irrigation and drainage development, institutional environment, and trends in water resources management for each region.

Looking at the situation as it now stands, the balance sheet is mixed and largely biased towards developed regions of the world in terms of all the challenges that must be faced. Although efforts are being made on all levels and at all scales, how far have we really come? And how far still do we still have to go?

Box 23.3: Generic content of water supply and sanitation sector country profiles

1. Background
1.1 General
1.2 Water resources
1.3 Health and hygiene education
2. Coverage
2.1 Current water supply and sanitation coverage (broken down by urban/rural and type of coverage)
2.2 Operational aspects (water systems providing intermittent supplies, use of disinfectants, rural system functioning, treatment of wastewater from public sewers)
2.3 Water quality standards and effectiveness
2.4 Population projection
2.5 Coverage trends
2.6 Coverage targets
3. Largest city (population, growth, water production, metering, Unaccounted-for-Water and other operational aspects)
4. Costs and investment
4.1 Costs and tariffs (average water production cost, water tariff, sewage tariff)
4.2 User charges (water supply and sanitation)
4.3 Construction costs (water supply and sanitation)
5. Policy, planning and institutions
5.1 Water supply and sanitation policy
5.2 Existence of national water supply and sanitation plan
5.3 Key sector institutions
5.4 External support organizations
6. Collaboration and coordination
7. New approaches
8. Major constraints to sector development
9. Contact point/agency for further information

Table 23.1: National reports submitted by member states through the UN CSD system

Countries	National reports (annual)							National assessment reports	Country profiles	
	1994	1995	1996	1998	1999	2000	2001	WSSD 2002	UNGASS[1] 1997	WSSD
Afghanistan										
Albania								x	x	x
Algeria	x	x			x		x		x	x
Andorra										
Angola										
Antigua and Barbuda				x		x	x			x
Argentina						x		x	x	x
Armenia						x		x	x	x
Australia	x	x	x	x	x	x	x		x	x
Austria	x	x	x			x	x		x	x
Azerbaijan										x
Bahamas		x	x	x					x	x
Bahrain					x				x	x
Bangladesh								x	x	x
Barbados		x	x	x	x	x	x		x	x
Belarus					x			x		x
Belgium	x	x	x	x	x	x	x	x	x	x
Belize										
Benin		x	x	x	x		x		x	x
Bhutan										
Bolivia	x	x		x	x	x		x	x	x
Bosnia and Herzegovina								x		x
Botswana		x		x					x	x
Brazil	x	x	x	x	x	x			x	x
Brunei Darussalam					x					x
Bulgaria			x	x	x	x	x	x	x	x
Burkina Faso	x			x		x	x			x
Burundi								x		
Cambodia										x
Cameroon						x	x		x	x
Canada	x	x		x			x		x	x
Cape Verde										
Central African Republic										
Chad										
Chile	x	x	x		x	x				x
China	x			x		x			x	x
Colombia	x	x	x			x	x		x	x
Comoros								x		
Congo							x			
Costa Rica						x		x	x	x
Côte d'Ivoire				x	x	x			x	x
Croatia					x	x	x	x	x	x
Cuba	x	x	x	x	x	x	x		x	x
Cyprus	x							x		x
Czech Republic	x							x	x	x
Democratic Republic of Congo					x					x
Denmark	x	x	x	x			x		x	x
Djibouti										x
Dominican Republic					x	x		x		x

Table 23.1: *continued*

Countries	National reports (annual)							National assessment reports	Country profiles	
	1994	1995	1996	1998	1999	2000	2001	WSSD 2002	UNGASS[1] 1997	WSSD
East Timor*								x		x
Ecuador	x	x	x	x	x	x			x	x
Egypt	x	x		x					x	x
El Salvador					x			x		x
Equatorial Guinea										
Eritrea										
Estonia	x				x	x		x	x	x
Ethiopia										x
Federated States of Micronesia										
Fiji									x	x
Finland	x	x	x	x	x	x	x	x	x	x
France	x	x	x	x	x	x	x		x	x
Gabon		x							x	
Gambia				x		x	x			x
Georgia					x			x		x
Germany	x	x	x	x	x	x	x		x	x
Ghana		x			x		x			x
Greece		x			x	x	x		x	x
Grenada										
Guatemala										
Guinea			x							
Guinea-Bissau									x	x
Guyana									x	x
Haiti					x				x	
Honduras						x	x	x		x
Hungary	x	x	x	x	x	x		x	x	x
Iceland	x	x	x	x	x	x	x		x	x
India	x	x	x		x	x	x		x	x
Indonesia		x					x		x	x
Iran				x			x			x
Iraq				x	x		x			x
Ireland	x	x	x	x	x				x	x
Israel						x	x		x	x
Italy	x	x							x	x
Jamaica				x		x				x
Japan	x	x		x	x	x	x	x	x	x
Jordan			x					x		x
Kazakhstan						x	x	x	x	x
Kenya		x		x			x			x
Kuwait										
Kyrgyz Republic										
Lao People's Democratic Republic										
Latvia								x		x
Lebanon				x	x		x		x	
Lesotho		x		x						
Liberia										
Libya										
Liechtenstein										x
Lithuania				x	x	x	x	x	x	x
Luxembourg	x								x	x
Madagascar									x	x

Table 23.1: *continued*

Countries	National reports (annual)							National assessment reports	Country profiles	
	1994	1995	1996	1998	1999	2000	2001	WSSD 2002	UNGASS[1] 1997	WSSD
Malawi									x	x
Malaysia	x	x				x			x	x
Maldives										
Mali										
Malta							x			x
Marshall Islands	x									
Mauritania										
Mauritius									x	x
Mexico	x	x	x	x	x	x	x	x	x	x
Monaco				x	x	x			x	x
Mongolia				x	x	x			x	x
Morocco	x	x	x	x	x		x		x	x
Mozambique		x								
Myanmar	x				x	x	x			x
Namibia		x						x		x
Nepal									x	x
Netherlands	x	x	x				x	x	x	x
New Zealand	x	x	x	x	x		x		x	x
Nicaragua					x	x	x	x	x	x
Niger	x							x	x	x
Nigeria					x	x	x		x	x
Norway	x	x	x	x		x	x	x	x	x
Oman						x				
Pakistan	x	x							x	x
Palau										
Panama			x			x	x	x	x	x
Papua New Guinea			x		x	x			x	x
Paraguay								x	x	x
Peru		x	x					x		x
Philippines	x	x		x				x	x	x
Poland	x			x		x	x	x	x	x
Portugal	x	x	x			x			x	x
Qatar									x	x
Republic of Korea	x	x		x	x	x	x		x	x
Republic of Moldova								x	x	x
Romania				x	x		x	x	x	x
Russian Federation				x	x	x	x	x	x	x
Rwanda										
Saint Kitts and Nevis										
Saint Lucia								x		
Saint Vincent and the Grenadines										
Samoa										
San Marino										
Sao Tome and Principe							x		x	x
Saudi Arabia				x		x	x		x	x
Senegal				x	x	x	x	x	x	x
Seychelles										x
Sierra Leone										
Singapore					x	x			x	x
Slovak Republic			x	x		x	x	x	x	x
Slovenia			x	x	x	x		x	x	x
Solomon Islands										
Somalia										

Table 23.1: *continued*

	National reports (annual)							National assessment reports	Country profiles	
Countries	**1994**	**1995**	**1996**	**1998**	**1999**	**2000**	**2001**	**WSSD 2002**	**UNGASS[1] 1997**	**WSSD**
South Africa				x	x	x			x	x
Spain	x	x	x	x	x	x	x		x	x
Sri Lanka	x								x	x
Sudan										
Suriname									x	x
Swaziland									x	x
Sweden	x	x	x	x	x	x			x	x
Switzerland	x	x	x	x	x	x	x	x	x	x
Syrian Arab Republic			x					x	x	x
Tajikistan								x		x
Thailand			x	x	x	x	x		x	x
The FYR of Macedonia					x		x	x	x	x
Togo								x		x
Tonga							x			x
Trinidad and Tobago										
Tunisia	x	x	x				x	x	x	x
Turkey		x	x	x		x		x	x	x
Turkmenistan								x		
Uganda	x	x	x	x				x	x	x
Ukraine		x	x		x			x	x	x
United Arab Emirates										
United Kingdom	x	x	x	x		x	x	x	x	x
United Republic of Tanzania	x	x							x	x
United States	x	x	x			x			x	x
Uruguay	x	x		x					x	x
Uzbekistan					x	x	x	x	x	x
Vanuatu										
Venezuela	x	x	x	x	x		x		x	x
Viet Nam							x		x	x
Yemen										
Yugoslavia				x	x			x		x
Zambia										x
Zimbabwe									x	x

* East Timor only became an independent sovereign state in early 2002.

x Indicates submissions to the CSD. However, this does not indicate the quality of the submitted information.

[1] UNGASS: United Nations General Assembly

Source: Based on information from the National Information Analysis Unit of the United Nations Division for Sustainable Development's web site, http://www.un.org/esa/agenda21/natlinfo/.

What Progress? Looking at the Past and towards the Future

Progress since Rio: where do we stand?

It is always difficult to generalize when drawing conclusions about meeting progress towards targeted goals. There is no single model that describes all situations, as the pilot case studies clearly show. One picture emerges at the global level, but other very different scenarios emerge at ground level within the context of a particular place and specific circumstances. For this reason it is useful to review what individual countries themselves have to report. An evaluation of national progress in implementing Agenda 21 was completed by the CSD immediately prior to the World Summit on Sustainable Development in 2002. It focuses on eleven measures within six themes that were cross-cutting throughout the Programme Areas of Chapter 18, which called for a concerted effort to develop more integrated approaches to water management and for a stronger focus on the needs of the poor. These programme areas are decision-making, programme and projects, education, information, financing and cooperation.

In total, 129 of 190 countries have reported, on at least some of the eleven measures (see figure 23.3). The regional breakdown demonstrates the continental variations in reporting. Participation is strongest in Europe and North America, with 90 percent of countries reporting, and weakest in Africa, with reports from only 50 percent of countries. For our purposes, progress is defined as either action that has been implemented or else as action that is underway. Progress is expressed as number of countries within the region reporting progress as a percentage of the total number of all countries within the region. Countries that have not reported are considered as not having made progress, a conclusion that may not reflect reality in cases where there is a weakness in reporting rather than in implementation.

Progress reported by individual countries against each of the eleven measures is presented in table 23.2, with countries grouped by region. Again, the distinction between 'implemented' and 'in progress of implementation' is retained.

Between a quarter and a third of all countries have reported that implementation has been completed for the significant majority of the eleven measures, an estimate that rises to between one third and one half when implementation and implementation in progress are added together. However, it is not the same countries reporting on progress across all measures. Only 'establishment of a coordinating body on freshwater' is reported as more widely implemented. The figure and table demonstrate the significant variations in reporting, the degree of progress and actions completed or still underway, variations that exist between different regions, between countries and between individual initiatives.

Eight countries report implementation of all eleven actions: Australia, Barbados, Greece, Finland, Republic of Korea, Norway, Singapore and Spain. A further fourteen countries report that all eleven actions have either been implemented or are in process of implementation: Algeria, Belgium, Croatia, Cuba, India, Israel, Jordan, Liechtenstein, Mexico, Monaco, Mongolia, Poland, Slovenia and Venezuela.

Complementary to this evaluation, the report of the Water Action Group of the World Water Council contains many hundreds of individual actions that have been taken since the Second World Water Forum.

Progress towards targets: are we on track?

At Rio in 1992 there was prior agreement that certain actions were to be accomplished by the turn of the century. These included eradication of Guinea worm disease by 1999, access to 40 litres per day of safe water to all urban residents by 2000, national action programmes and water resource assessment services in place by 2000. Among these, best progress has been made in the reduction of Guinea worm infestations. The urban water supply target has not been met, and accomplishment of action programmes and assessment services is confined to a few countries. The Framework for Action, formulated at The Hague in 2000, contains several further actions with impending timeframes. Among them are the following:

- The economic value of water should be recognized and fully reflected in national policies and strategies by 2002.

- The implementation of comprehensive IWRM policies and strategies should be underway in 75 percent of countries by 2005.

- National standards that ensure the integrity of ecosystems should be instituted in all countries by 2005.

Figure 23.3: Comparison of progress in implementing Chapter 18 of Agenda 21 by region

World (190 countries, of which 129 [68%] reported)

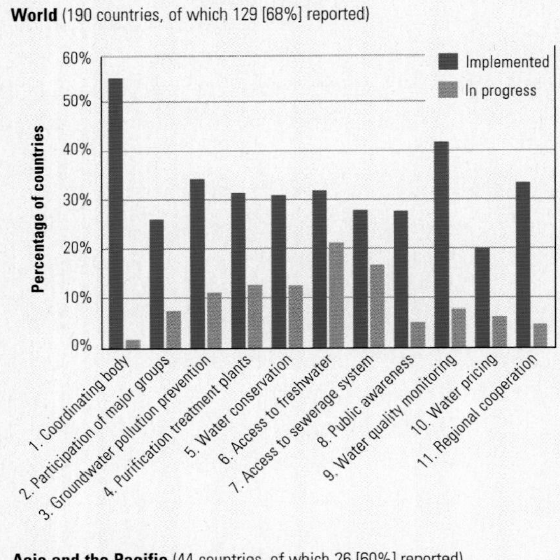

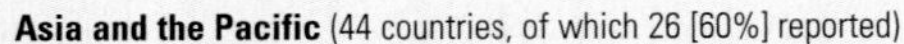

Africa (53 countries, of which 26 [50%] reported)

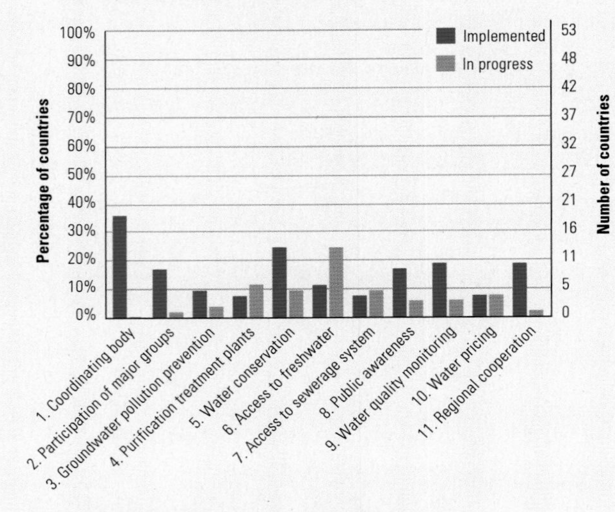

Asia and the Pacific (44 countries, of which 26 [60%] reported)

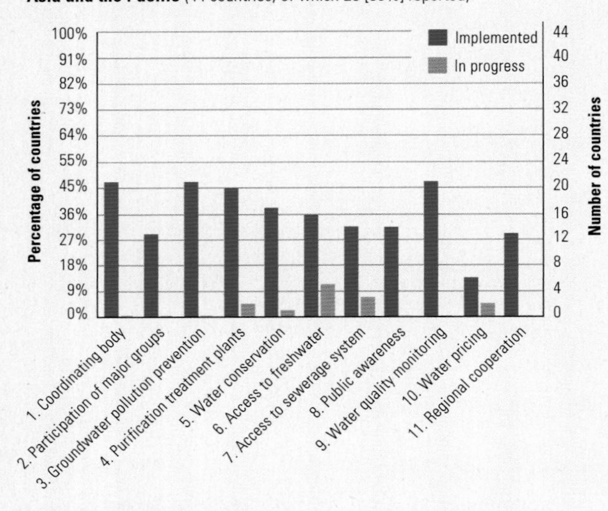

Western Asia (12 countries, of which 9 [75%] reported)

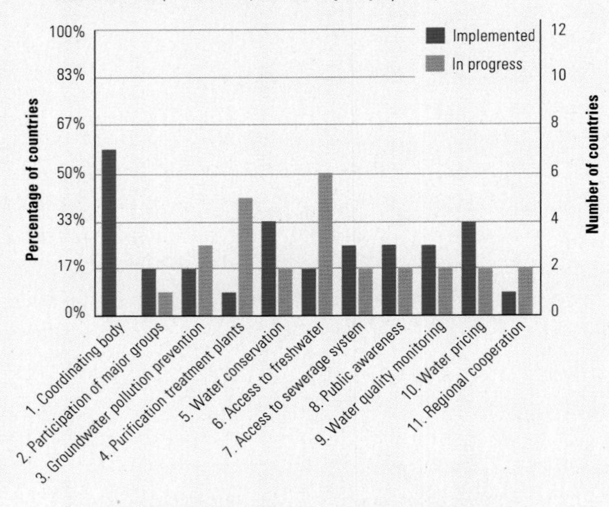

Latin America and the Caribbean (33 countries, of which 25 [75%] reported)

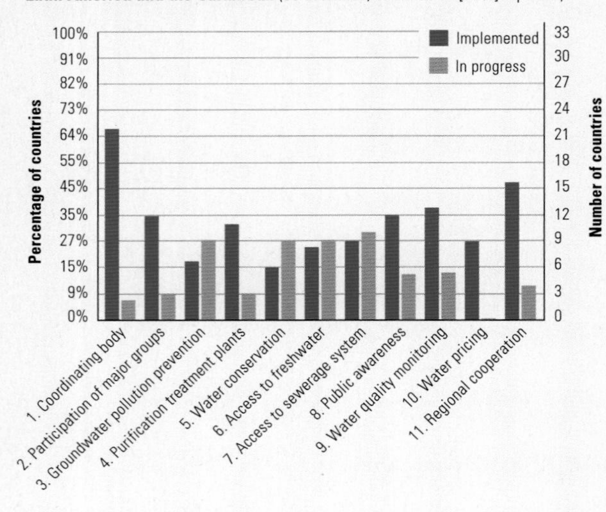

Europe and North America (48 countries, of which 43 [90%] reported)

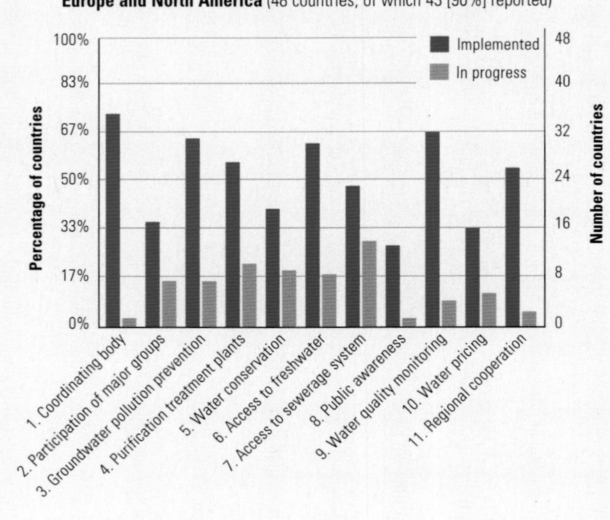

Table 23.2: Summary of national progress in implementing of Chapter 18 of Agenda 21

	Decision-making		Programmes and projects					Education	Information	Financing	Cooperation
AFRICA	1. Coordinating body on freshwater	2. Participation of major groups in decision-making	3. Groundwater pollution prevention	4. Water purification treatment plants	5. Water conservation	6. Access to freshwater	7. Access to sewage systems	8. Public awareness-raising on water conservation	9. Water quality monitoring	10. Water pricing	11. Regional cooperation
1 Algeria	Implemented	Implemented	Implemented	Implemented	Implemented	In progress	Implemented	Implemented	In progress	In progress	Implemented
2 *Angola*											
3 *Benin*	Implemented	Implemented						Implemented	In progress	Implemented	Implemented
4 **Botswana**	Implemented	Implemented		In progress	Implemented	Implemented			Implemented		Implemented
5 **Burkina Faso**											
6 *Burundi*											
7 Cameroon			Implemented		Implemented						
8 *Cape Verde*											
9 *Central African Republic*											
10 *Chad*											
11 *Comoros*											
12 *Congo*											
13 Côte d'Ivoire	Implemented	Implemented	Implemented	In progress	Implemented	In progress	In progress		Implemented	Implemented	In progress
14 *Dem. Rep. of Congo*	Implemented				In progress		In progress	In progress			Implemented
15 *Djibouti*											
16 *Egypt*	Implemented	Implemented	In progress	In progress	Implemented	In progress	Implemented	Implemented	Implemented		Implemented
17 *Equatorial Guinea*											
18 *Eritrea*											
19 *Ethiopia*											
20 *Gabon*											
21 *Gambia*											
22 **Ghana**	Implemented	Implemented			Implemented	Implemented		Implemented	Implemented		Implemented
23 *Guinea*											
24 Guinea-Bissau	Implemented					In progress					
25 Kenya	Implemented	In progress			In progress	In progress			Implemented	In progress	
26 *Lesotho*											
27 *Liberia*											
28 *Libyan Arab Jamahiriya*											
29 *Madagascar*	Implemented			In progress		In progress					
30 *Malawi*	Implemented	Implemented	In progress	In progress	Implemented	In progress	In progress	Implemented	Implemented		Implemented
31 *Mali*											
32 *Mauritania*											
33 *Mauritius*	Implemented					In progress		In progress			
34 **Morocco**	Implemented	Implemented	Implemented	Implemented	Implemented	Implemented	Implemented	Implemented	Implemented	Implemented	
35 *Mozambique*											
36 *Namibia*	Implemented				In progress	Implemented	In progress	Implemented	Implemented		
37 Niger	Implemented					In progress				In progress	Implemented
38 Nigeria	Implemented	Implemented		Implemented	Implemented	In progress		Implemented			
39 *Rwanda*											
40 Sao Tome and Principe											
41 *Senegal*	Implemented				In progress		In progress		Implemented	In progress	
42 *Seychelles*											
43 *Sierra Leone*											
44 *Somalia*											
45 South Africa					Implemented	In progress		Implemented			
46 *Sudan*											
47 Swaziland											
48 *Togo*											
49 *Tunisia*	Implemented			In progress	In progress	Implemented			In progress	Implemented	Implemented
50 **Uganda**				Implemented	Implemented	Implemented	Implemented				Implemented
51 United Rep. of Tanzania			Implemented		Implemented	In progress					
52 *Zambia*											
53 Zimbabwe	Implemented				Implemented	In progress		In progress	Implemented		
Total implemented	19	9	5	4	13	6	4	9	10	4	10
Total in progress	0	1	2	6	5	13	5	3	3	4	1
% implemented	36%	17%	9%	8%	25%	11%	8%	17%	19%	8%	19%
% in progress	0%	2%	4%	11%	9%	25%	9%	6%	6%	8%	2%

Bold = Reporting country submitted 2002 Country Profile
Regular = Reporting country yet to submit 2002 Country Profile
Italics = Non-reporting country

Implemented
In progress

Table 23.2: *continued*

ASIA AND THE PACIFIC	Decision-making		Programmes and projects					Education	Information	Financing	Cooperation
	1. Coordinating body on freshwater	2. Participation of major groups in decision-making	3. Groundwater pollution prevention	4. Water purification treatment plants	5. Water conservation	6. Access to freshwater	7. Access to sewage systems	8. Public awareness-raising on water conservation	9. Water quality monitoring	10. Water pricing	11. Regional cooperation
1 *Afghanistan*											
2 Australia	■	■	■	■	■	■	■	■	■	■	■
3 *Bangladesh*	■	■		▒	■	▒		■	■		
4 *Bhutan*											
5 Brunei Darussalam	■		■	■		■	▒		■		
6 *Cambodia*											
7 China	■		■	■	■	■			■		■
8 *Cyprus*											
9 *Fiji*	■		■	■	▒						
10 India	■	■	■	■	■	■	▒	■	■	▒	■
11 *Indonesia*	■		■	■	■	■		■	■		
12 **Iran**	■		■	■			■		■		■
13 **Japan**	■	■	■	■	■	■	■	■	■	■	
14 **Kazakhstan**	■		■			▒	▒	■	■		■
15 *Kiribati*											
16 *Korea, DPR of*											
17 Kyrgyz Republic											
18 *Lao, People's Dem. Rep.*											
19 Malaysia	■		■	■	■	■	■	■	■		■
20 *Maldives*											
21 *Marshall Islands*											
22 *Micronesia*											
23 Mongolia	■	■	■	■	■	▒	■	■	■	■	■
24 *Myanmar*			■	■	■				■		
25 Nauru											
26 *Nepal*	■	■		■		■	■		■		
27 New Zealand	■	■	■	■	■	■	■		■		■
28 *Pakistan*	■	■	■	■	■	▒	■		■		
29 *Palau*											
30 *Papua New Guinea*											
31 *Philippines*	■		■	▒	■	■	■	■	■	▒	
32 *Republic of Korea*	■	■	■	■	■	■	■	■	■	■	■
33 *Samoa*											
34 *Singapore*	■	■	■	■	■	■	■	■	■	■	■
35 *Solomon Islands*											
36 *Sri Lanka*	■	■	■	■	■	■	■	■	■		■
37 *Tajikistan*											
38 *Thailand*	■	■	■	■	■	■		■	■	■	■
39 Tonga						▒					
40 *Turkmenistan*											
41 *Tuvalu*											
42 *Uzbekistan*	■	■	■	■	■	■	■	■	■		■
43 *Vanuatu*											
44 Viet Nam			■	■		■	■				
Total implemented	21	13	21	20	17	16	14	14	21	6	13
Total in progress	0	0	0	2	1	5	3	0	0	2	0
% implemented	48%	30%	48%	45%	39%	36%	32%	32%	48%	14%	30%
% in progress	0%	0%	0%	5%	2%	11%	7%	0%	0%	5%	0%

Bold = Reporting country submitted 2002 Country Profile
Regular = Reporting country yet to submit 2002 Country Profile
Italics = Non-reporting country

■ Implemented
▒ In progress

Table 23.2: *continued*

Decision-making: 1. Coordinating body on freshwater; 2. Participation of major groups in decision-making
Programmes and projects: 3. Groundwater pollution prevention; 4. Water purification treatment plants; 5. Water conservation; 6. Access to freshwater; 7. Access to sewage systems
Education: 8. Public awareness-raising on water conservation
Information: 9. Water quality monitoring conservation
Financing: 10. Water pricing
Cooperation: 11. Regional cooperation

EUROPE AND NORTH AMERICA

	Country
1	Albania
2	*Andorra*
3	Armenia
4	**Austria**
5	*Azerbaijan*
6	Belarus
7	**Belgium**
8	*Bosnia and Herzegovina*
9	*Bulgaria*
10	Canada
11	*Croatia*
12	**Czech Republic**
13	Denmark
14	Estonia
15	*Finland*
16	*France*
17	Georgia
18	Germany
19	*Greece*
20	**Hungary**
21	**Iceland**
22	Ireland
23	**Italy**
24	Latvia
25	**Liechtenstein**
26	**Lithuania**
27	Luxembourg
28	**Malta**
29	**Monaco**
30	Netherlands
31	**Norway**
32	**Poland**
33	**Portugal**
34	Republic of Moldova
35	**Romania**
36	**Russian Federation**
37	*San Marino*
38	**Slovak Republic**
39	**Slovenia**
40	**Spain**
41	**Sweden**
42	Switzerland
43	The FYR of Macedonia
44	**Turkey**
45	**Ukraine**
46	United Kingdom
47	**United States of America**
48	Republic of Yugoslavia

	1	2	3	4	5	6	7	8	9	10	11
Total implemented	35	17	31	27	19	30	23	13	32	16	26
Total in progress	1	7	7	10	9	8	14	1	4	5	2
% implemented	73%	35%	65%	56%	40%	63%	48%	27%	67%	33%	54%
% in progress	2%	15%	15%	21%	19%	17%	29%	2%	8%	10%	4%

Bold: Reporting country submitted 2002 Country Profile
Regular: Reporting country yet to submit 2002 Country Profile
Italics: Non-reporting country

Implemented
In progress

Table 23.2: *continued*

LATIN AMERICA AND THE CARIBBEAN	Decision-making		Programmes and projects					Education	Information	Financing	Cooperation
	1. Coordinating body on freshwater	2. Participation of major groups in decision-making	3. Groundwater pollution prevention	4. Water purification treatment plants	5. Water conservation	6. Access to freshwater	7. Access to sewage systems	8. Public awareness-raising on water conservation	9. Water quality monitoring	10. Water pricing	11. Regional cooperation
1 Antigua and Barbuda	Implemented		In progress		Implemented	In progress	In progress		Implemented	Implemented	
2 **Argentina**	Implemented	Implemented	Implemented	Implemented			Implemented	Implemented	Implemented		Implemented
3 Bahamas	Implemented		Implemented		In progress	In progress	In progress	Implemented	Implemented	Implemented	Implemented
4 **Barbados**	Implemented	Implemented	Implemented	Implemented	Implemented	Implemented	Implemented	Implemented	Implemented	Implemented	Implemented
5 *Belize*											
6 Bolivia	In progress	Implemented	In progress		In progress	In progress	In progress			Implemented	In progress
7 **Brazil**	Implemented	Implemented	Implemented	Implemented	Implemented	Implemented	Implemented	Implemented	Implemented		Implemented
8 **Chile**	In progress		In progress	In progress	In progress	In progress	In progress	In progress	In progress		
9 **Colombia**	Implemented		In progress	Implemented	In progress	Implemented	Implemented	Implemented	In progress	Implemented	In progress
10 Costa Rica	Implemented	Implemented	Implemented	Implemented	Implemented		Implemented	Implemented	Implemented		Implemented
11 **Cuba**	Implemented	Implemented	In progress	In progress	In progress	In progress	In progress	Implemented	In progress	Implemented	Implemented
12 *Dominica*											
13 Dominican Republic											
14 **Ecuador**	Implemented				In progress						Implemented
15 **El Salvador**	Implemented				Implemented	In progress	In progress	In progress			Implemented
16 *Grenada*											
17 *Guatemala*											
18 Guyana	Implemented	In progress						Implemented	Implemented		
19 **Haiti**	In progress	In progress						In progress	In progress		Implemented
20 **Honduras**	Implemented	In progress	In progress	In progress	In progress	In progress	Implemented	Implemented			In progress
21 Jamaica	Implemented	Implemented	Implemented	Implemented	Implemented			Implemented	Implemented	Implemented	Implemented
22 **Mexico**	Implemented	Implemented	In progress	Implemented	In progress	Implemented	Implemented	Implemented	Implemented	Implemented	Implemented
23 **Nicaragua**	Implemented	Implemented	Implemented			In progress	In progress				Implemented
24 Panama	Implemented	Implemented		Implemented		In progress	In progress		Implemented		In progress
25 *Paraguay*	Implemented				In progress	In progress	In progress		Implemented		Implemented
26 **Peru**	Implemented		In progress	Implemented		Implemented	Implemented		Implemented		Implemented
27 *St Kitts and Nevis*											
28 Saint Lucia											
29 *St Vincent & Grenadines*											
30 Suriname	Implemented					Implemented	In progress				Implemented
31 *Trinidad and Tobago*											
32 Uruguay	Implemented	Implemented	Implemented	Implemented		Implemented	Implemented		Implemented		
33 *Venezuela*	Implemented	Implemented	In progress	Implemented	In progress	Implemented	In progress	Implemented	In progress	Implemented	Implemented
Total implemented	22	12	7	11	6	8	9	12	13	9	16
Total in progress	2	3	9	3	9	9	10	3	5	0	14
% implemented	67%	36%	21%	33%	18%	24%	27%	36%	39%	27%	48%
% in progress	6%	9%	27%	9%	27%	27%	30%	9%	15%	0%	12%

Bold: Reporting country submitted 2002 Country Profile
Regular: Reporting country yet to submit 2002 Country Profile
Italics: Non-reporting country

Implemented
In progress

Table 23.2: *continued*

	Decision-making		Programmes and projects					Education	Information	Financing	Cooperation
WESTERN ASIA	**1. Coordinating body on freshwater**	**2. Participation of major groups in decision-making**	**3. Groundwater pollution prevention**	**4. Water purification treatment plants**	**5. Water conservation**	**6. Access to freshwater**	**7. Access to sewage systems**	**8. Public awareness-raising on water conservation**	**9. Water quality monitoring**	**10. Water pricing**	**11. Regional cooperation**
1 Bahrain											
2 Iraq											
3 Israel											
4 Jordan											
5 *Kuwait*											
6 Lebanon											
7 *Oman*											
8 Qatar											
9 Saudi Arabia											
10 Syrian Arab Republic											
11 *United Arab Emirates*											
12 *Yemen*											
Total implemented	7	2	2	1	4	2	3	3	3	4	1
Total in progress	0	1	3	5	2	6	2	2	2	2	2
% implemented	58%	17%	17%	8%	33%	17%	25%	25%	25%	33%	8%
% in progress	0%	8%	25%	42%	17%	50%	17%	17%	17%	17%	17%

Bold = Reporting country submitted 2002 Country Profile
Regular = Reporting country yet to submit 2002 Country Profile
Italics = Non-reporting country

Implemented
In progress

The definition of these targets was followed in December 2001 by the twenty-seven recommended actions established in Bonn, which do not have set timeframes, but are nonetheless agreed to be critical to the delivery of outcomes. The Bonn recommendations promoted:

- Actions in the field of governance
  1. Secure equitable access to water for all people
  2. Ensure that water infrastructure and services deliver to poor people
  3. Promote gender equity
  4. Appropriately allocate water among competing demands
  5. Share benefits
  6. Promote participatory sharing of benefits from large projects
  7. Improve water management
  8. Protect water quality and ecosystems
  9. Manage risks to cope with variability and climate change
  10. Encourage more efficient service provision
  11. Manage water at the lowest appropriate level
  12. Combat corruption effectively

- Actions in the field of mobilizing financial resources
  13. Ensure significant increase in all types of funding
  14. Strengthen public funding capabilities
  15. Improve economic efficiency to sustain operations and investment
  16. Make water attractive for private investment
  17. Increase development assistance to water

- Actions in the field of capacity-building and sharing knowledge
  18. Focus education and training on water wisdom
  19. Focus research and information management on problem solving
  20. Make water institutions more effective
  21. Share knowledge and innovative technologies
  22. Governments should more actively play their key role in water governance
  23. Empower local communities through social mobilization processes

- Actions to review the role of:
  24. Workers and trade unions
  25. Non-governmental organizations
  26. The private sector
  27. The international community

Next came the Millennial targets that were reaffirmed or revised at the WSSD. But an intermediate reporting on progress is needed to allow us to identify when and how progress is and is not being made, in time to take corrective action. A critical added value to the targets themselves will be reporting on the improvements made in people's daily lives – such as health, mortality and morbidity, time away from tasks, income, school enrolment and educational opportunities. The attainment of the longer-term targets will depend on early reports on the dismantling of barriers by policy actions. The weak links highlighted between progress since Rio and people-centred outcomes argue for a stronger arrangement in which progress in management is reported within a people-first framework.

The *Human Development Report 2002* (UNDP, 2002) summarized the prospects of individual countries attaining the Millennium Development Goals. It does not evaluate all countries, so analysis is limited to those for which information is included. The following subsections detail various elements for several of the Millennium Goals, illustrating why and how assessments such as these are useful.

Millennium Goal 7: a focus on water supply and sanitation

The national prospects of reaching Millennium Goal 7, halving the proportion of people without sustainable, safe drinking water by 2015, are illustrated in table 23.3. In addition to those countries in which the

Table 23.3: Summary of regional progress towards attaining Millennium Development Goal 7

	Achieved	On track	Lagging	Far behind	Slipping back	No data
Sub-Saharan Africa	1	9	4	9	0	21
Arab States	0	8	0	3	0	6
East Asia and the Pacific	0	6	1	4	0	8
South Asia	3	4	0	0	0	1
Latin America and the Caribbean	1	21	1	2	0	8
Central and East Europe and the CIS[1]	0	8	0	0	0	17
Total	**5**	**63**	**7**	**18**	**0**	**75**

The purpose of Millennium Development Goal 7 is to halve the proportion of people without sustainable, safe drinking water. Regions include only Human Development Index countries, whereas total includes all UN member countries excluding high-income OECD members.

[1] CIS: Commonwealth of Independent States

Source: UNDP, 2002.

target has already been reached, and assuming that they do not slip back, it is assessed that a further sixty-three countries with 39 percent of the world's population are on track. Twenty-five countries with 32 percent of the world's population (and possibly up to 100 countries with 42 percent) are not on track, with seven lagging and eighteen far behind. There are four major dimensions to achieving the water supply and sanitation targets in the years to come. First the world must keep pace with a net population growth of more than a billion people over the next fifteen years. Second, the existing coverage and service gap must be closed, and significantly greater outcomes are needed in sanitation coverage. Third, existing and new services have to be sustainable. Fourth, the quality of services needs to be improved. These are not four separate dimensions – they are all part of the same, single challenge. Table 23.3 shows the practical implications of the following points.

- To meet the 2015 targets worldwide, the number of people served by water supply must increase by 1.5 billion, and those served by sanitation by 1.87 billion;
- For water, this means providing services for an additional 100 million people each year, or 274,000 every day, until 2015. Considering that only 901 million people gained access to improved water services during the 1990s, the pace has to be drastically accelerated;
- For sanitation, the challenge is even greater, with services to be provided for an additional 125 million people each year until 2015, or 342,000 every day until 2015. During the 1990s, 1 billion people a year gained access to improved sanitation services.

Rapid urban growth means that more than half of the additional services must be in urban areas, despite the higher current levels of coverage. The lower levels of service in rural areas also mean that nearly half of the improvements will need to take place in rural areas, even though the rural population will grow more slowly than the urban population.

Most of the work will need to be done in Asia, as the absolute needs in Asia outstrip those of Africa and Latin America and the Caribbean combined: the majority of people without access to water supply and sanitation services are in Asia.

Current progress is inadequate to meet the targets. Something will have to change dramatically if the targets are to be met. Unless the pace is increased, the number of people without access will increase sharply. To achieve the 2015 targets, some estimates show that the annual investment in water supply needs to be increased by 31 percent (39 percent for the urban water sector and 19 percent for the rural water sector), and achieving the total sanitation target by 2015 will require that the annual expenditures of the 1990s almost double.

But the magnitude of the challenge may well be even greater than what we have described. The above analysis is based on access to 'improved' sources of water and sanitation. Our analysis of water in cities has demonstrated that 'improved' sources may well be inadequate, unsafe and inconvenient, for present populations who are already served and for future populations. The magnitude of this challenge, in the numbers concerned and the outcomes to be achieved, could be set to rise very significantly if new definitions of coverage become the accepted norm.

Our analysis also assumes that services for those who are already served will be sustained. This is optimistic, as there are still huge constraints affecting sustainability, including funding limitations, insufficient cost recovery and inadequate operation and maintenance. So, in addition to the great demand for constructing new systems, there will also be a need for substantive investments in capacity-building and operation and maintenance.

Millennium Goal 4: reducing child mortality

Water has a massive contribution to make in the attainment of the Millennium Declaration Goal 4, which seeks to reduce child mortality by

Table 23.4: Summary of regional progress towards attaining Millennium Development Goal 4

	Achieved	On track	Lagging	Far behind	Slipping back	No data
Sub-Saharan Africa	0	7	3	24	10	0
Arab States	0	11	1	4	1	0
East Asia and the Pacific	0	13	1	3	1	1
South Asia	0	6	1	1	0	0
Latin America and the Caribbean	0	25	0	8	0	0
Central and East Europe and the CIS[1]	0	10	0	13	2	0
Total	**0**	**85**	**7**	**59**	**15**	**2**

The purpose of Millennium Development Goal 4 is to reduce under-five mortality by two-thirds. Regions include only Human Development Index countries, whereas total includes all UN member countries excluding high-income OECD members.

[1] CIS: Commonwealth of Independent States

Source: UNDP, 2002.

two thirds. In addition to those countries where the target has already been reached, eighty-five countries with 24 percent of the world's population are on track, many of these countries located in Latin America and the Caribbean. Eighty-one countries with 61 percent of the world population are not on track to achieve this goal, with seven lagging and fifty-nine trailing far behind, almost half of these in sub-Saharan Africa, where ten of the fifteen countries that have actually slipped backwards in their progress are also located (see table 23.4)

Millennium Goal 1: eradicating extreme poverty and hunger
Table 23.5 summarizes the prospects of individual countries attaining Millennium Goal 1, the halving by 2015 of the proportion of people suffering hunger, as indicated by levels of malnutrition. Again in addition to those where the target has already been reached, it is estimated that fifty-one countries with 46 percent of the world's population are on track. Twenty-eight countries with 25 percent of the world population are not on track to achieve this goal, with four lagging and twenty-four far behind, many of these again in the sub-Saharan African region. Here as well, an alarming fifteen countries have slipped back in their progress.

Attaining Millennium Goals
Evaluation of national status of progress across all of the relevant Millennium Development Goals is presented in table 23.6, again drawing on content of the *Human Development Report 2002*. The evaluation is preliminary, as it is based on information extrapolated during the 1990s. But such evaluations, ideally from rigorous national analysis, underpin the monitoring of progress towards target attainment. They bring country relevance to the global target. The evaluation shows how certain countries are on track to achieve all of the targets. It also shows how certain countries that are on track to achieve most targets are lagging or slipping back on one or two of them. For many countries, the magnitude of the outstanding effort across the range of development areas is clear. Yet the table also shows that very few countries are behind on all of the targets, and even where progress is behind on most, nearly all countries are on track to meet at least one target.

In Conclusion

There is a water crisis. It is a crisis of governance, and it is a crisis directly impacting life, livelihoods and well-being as these are experienced each day. Despite some action, despite some strategic planning, on a day-to-day basis and from year to year, the everyday lives of an alarming number of people have steadily worsened and will continue to do so if we assume business as usual. The fact remains though that reaching the Millennium targets by 2015 will improve the daily lives of several billion people, yet many countries are still not on track to reach those targets. If they are not reached, or are only partially achieved, many millions of people will subsist in poverty.

We know how difficult the road to attaining these targets will be. What we do not know is whether or not we are even able to attain them. But they are an essential tool for gauging progress and act as incentives to push on. When the mothers of the world see that their children are healthier, that they are better fed, they will know that progress has been made. These are the best indicators.

What is at stake is whether we – the family of nations, countries, local communities and individuals – can honestly say that we have seized every opportunity and mustered every bit of talent and energy to work towards the desired goal. Although it is indeed a daunting task that faces us, every one of us has a role to play in bringing our planet and our people back to health. Ensuring a broader knowledge base and thus empowering everyone to act, is a responsibility we all can – and must – assume. We are all stakeholders in the Earth, and integrating the efforts of governments, institutions, communities and individuals is the only way in which to press forward. To give up is to abandon the Earth and its inhabitants to a world without hope.

Table 23.5: Summary of regional progress towards attaining Millennium Development Goal 1

	Achieved	On track	Lagging	Far behind	Slipping back	No data
Sub-Saharan Africa	2	14	2	11	6	9
Arab States	1	5	0	1	0	10
East Asia and the Pacific	0	6	0	3	1	9
South Asia	0	3	0	3	0	2
Latin America and the Caribbean	3	10	2	5	3	10
Central and East Europe and the CIS[1]	0	11	0	0	1	13
Total	**6**	**51**	**4**	**24**	**15**	**68**

The purpose of Millennium Development Goal 1 is to halve world hunger. Regions include only Human Development Index countries, whereas total includes all UN member countries excluding high-income OECD members.

[1] CIS: Commonwealth of Independent States
Source: UNDP, 2002.

Table 23.6: Summary of national progress towards attainment of relevant Millennium Development Goals

HDI rank		Goal 1 Eradicate extreme poverty and hunger	Goal 2 Achieve universal primary education		Goal 3 Promote gender equality and empower women		Goal 4 Reduce child mortality	Goal 7 Ensure environmental sustainability
		Target: Halve the proportion of people suffering from hunger	Target: Ensure that all children can complete primary education		Target: Eliminate gender disparity in all levels of education[a]		Target: Reduce under-five and infant mortality rates by two-thirds	Target: Halve the proportion of people without access to improved water sources
		Undernourished people (as % of total population)[b]	Net primary enrolment ratio (%)	Children reaching grade 5 (%)	Female gross primary enrolment ratio as % of male ratio	Female gross secondary enrolment ratio as % of male ratio	Under-five mortality rate (per 1,000 live births)	Population using improved water sources (%)
High human development								
22	Israel					On track	On track	
23	Hong Kong, China (SAR)				Achieved	Achieved		
25	Singapore		On track		On track		On track	On track
26	Cyprus		Slipping back	Achieved	Achieved	Achieved	On track	On track
27	Korea, Republic of		On track	On track	Achieved	Achieved	On track	On track
29	Slovenia		On track	Achieved	Achieved	Achieved	On track	On track
30	Malta		Achieved	Achieved	On track	On track	On track	On track
31	Barbados						On track	
32	Brunei Darussalam		On track		On track	Achieved	On track	
33	Czech Republic				On track	Achieved	On track	
34	Argentina		Achieved		On track	Achieved	On track	
35	Hungary		Slipping back		On track	Achieved	On track	On track
36	Slovakia			Achieved	Achieved	Achieved	On track	On track
37	Poland		On track		On track	On track	On track	
38	Chile	Achieved	On track	Achieved	On track	Achieved	On track	On track
39	Bahrain		On track	On track	Achieved	Achieved	On track	
40	Uruguay	Achieved	On track	On track	On track	Achieved	On track	On track
41	Bahamas						On track	On track
42	Estonia	On track	On track		On track	Achieved	Far behind	
43	Costa Rica	On track	On track	On track	On track	Achieved	On track	On track
44	St. Kitts and Nevis						On track	On track
45	Kuwait				On track			
46	United Arab Emirates	Achieved	On track	Achieved	On track	Achieved	On track	
47	Seychelles				On track	Achieved	On track	
48	Croatia		On track	Achieved	On track	Achieved	On track	
49	Lithuania	On track	On track	Achieved	On track	Achieved	Far behind	
50	Trinidad and Tobago	Far behind	Far behind	On track	On track	Achieved	On track	
51	Qatar		Far behind		On track	On track	On track	
52	Antigua and Barbuda						On track	On track

Table 23.6: *continued*

		Goal 1 Eradicate extreme poverty and hunger	Goal 2 Achieve universal primary education		Goal 3 Promote gender equality and empower women		Goal 4 Reduce child mortality	Goal 7 Ensure environmental sustainability
		Target: Halve the proportion of people suffering from hunger	Target: Ensure that all children can complete primary education		Target: Eliminate gender disparity in all levels of education[a]		Target: Reduce under-five and infant mortality rates by two-thirds	Target: Halve the proportion of people without access to improved water sources
HDI rank		Undernourished people (as % of total population)[b]	Net primary enrolment ratio (%)	Children reaching grade 5 (%)	Female gross primary enrolment ratio as % of male ratio	Female gross secondary enrolment ratio as % of male ratio	Under-five mortality rate (per 1,000 live births)	Population using improved water sources (%)
53	Latvia	On track	On track	Achieved	On track	Achieved	Far behind	
Medium human development								
54	Mexico	On track	Achieved	On track	On track	Achieved	On track	On track
55	Cuba	Slipping back	On track		On track	Achieved	On track	On track
56	Belarus			Achieved	On track	Achieved	Far behind	On track
57	Panama	On track					On track	
58	Belize						Far behind	
59	Malaysia		Achieved		Achieved	Achieved	On track	
60	Russian Federation	On track	On track				Far behind	On track
61	Dominica						On track	On track
62	Bulgaria	Slipping back	On track		On track	On track	Far behind	On track
63	Romania		On track	Achieved	On track	On track	On track	
64	Libyan Arab Jamahiriya						On track	Far behind
65	Macedonia, FYR	On track	On track	On track	On track	On track	On track	
66	Saint Lucia						On track	On track
67	Mauritius	On track	On track	On track	Achieved	Achieved	On track	On track
68	Colombia	On track	On track	On track	On track	Achieved	Far behind	On track
69	Venezuela	Slipping back	Far behind	On track	Achieved	Achieved	Far behind	
70	Thailand	On track					On track	On track
71	Saudi Arabia		Far behind	On track	On track	On track	On track	On track
72	Fiji						On track	
73	Brazil	On track					On track	On track
74	Suriname	On track					On track	On track
75	Lebanon				On track	Achieved	Far behind	On track
76	Armenia			Achieved			Far behind	
77	Philippines	Far behind	Achieved		On track	Achieved	On track	Far behind
78	Oman		Far behind	On track	On track	On track	On track	Far behind
79	Kazakhstan			Achieved	Achieved	Achieved	Slipping back	On track
80	Ukraine	On track					Far behind	
81	Georgia			Achieved	On track	On track	Far behind	

Table 23.6: *continued*

82	Peru	Achieved	On track		On track	On track	On track	Lagging
83	Grenada						On track	On track
84	Maldives				On track	Achieved	On track	On track
85	Turkey		On track		On track	Far behind	On track	Lagging
86	Jamaica	On track			On track		Far behind	
87	Turkmenistan	On track					Far behind	
88	Azerbaijan			Achieved	On track	On track	Far behind	
89	Sri Lanka	On track			On track	Achieved	On track	Achieved
90	Paraguay	On track	On track	On track	On track	Achieved	Far behind	On track
91	St. Vincent and the Grenadines						Far behind	On track
92	Albania	On track	Achieved		Achieved	Achieved	On track	
93	Ecuador	On track	On track				On track	
94	Dominican Rep.	Far behind			Achieved	Achieved	On track	Far behind
95	Uzbekistan	On track					Slipping back	
96	China	On track	Achieved	On track	Achieved	On track	Far behind	Far behind
97	Tunisia		Achieved	On track	On track	On track	On track	
98	Iran, Islamic Rep. of	On track	Slipping back		On track	On track	On track	Achieved
99	Jordan	On track					Lagging	On track
100	Cape Verde				On track	Achieved	On track	
101	Samoa (Western)		On track		On track	Achieved	On track	On track
102	Kyrgyzstan	On track	On track		On track	Achieved	On track	
103	Guyana	On track	Slipping back	On track	On track	Achieved	Far behind	On track
104	El Salvador	Far behind	On track		On track	Achieved	On track	
105	Moldova, Rep. of	On track		Achieved	On track	Achieved	Far behind	On track
106	Algeria	On track	On track	On track	On track	On track	Slipping back	On track
107	South Africa		On track		On track	Achieved	Slipping back	
108	Syrian Arab Rep.		On track	On track	On track	On track	On track	
109	Viet Nam	On track			On track	On track	Lagging	Lagging
110	Indonesia	On track	On track	On track	On track	On track	On track	On track
111	Equatorial Guinea						On track	
112	Tajikistan				On track		Far behind	
113	Mongolia	Slipping back		Achieved	Achieved	Achieved	On track	
114	Bolivia	Lagging					On track	On track
115	Egypt	On track	On track		On track	On track	On track	On track
116	Honduras	Far behind					On track	On track
117	Gabon	On track					Far behind	
118	Nicaragua	Far behind	On track	Far behind	Achieved	Achieved	On track	On track
119	Sao Tome and Principe						Far behind	
120	Guatemala	Slipping back			Far behind	On track	On track	Achieved
121	Solomon Islands						On track	
122	Namibia	Far behind	On track		Achieved	Achieved	Far behind	Lagging
123	Morocco	On track	On track	Far behind	On track	On track	On track	On track

Table 23.6: *continued*

		Goal 1 Eradicate extreme poverty and hunger	Goal 2 Achieve universal primary education		Goal 3 Promote gender equality and empower women		Goal 4 Reduce child mortality	Goal 7 Ensure environmental sustainability
		Target: Halve the proportion of people suffering from hunger	Target: Ensure that all children can complete primary education		Target: Eliminate gender disparity in all levels of education[a]		Target: Reduce under-five and infant mortality rates by two-thirds	Target: Halve the proportion of people without access to improved water sources
HDI rank		Undernourished people (as % of total population)[b]	Net primary enrolment ratio (%)	Children reaching grade 5 (%)	Female gross primary enrolment ratio as % of male ratio	Female gross secondary enrolment ratio as % of male ratio	Under-five mortality rate (per 1,000 live births)	Population using improved water sources (%)
124	India	Far behind			On track	Far behind	Lagging	On track
125	Swaziland	Far behind	On track	Far behind	On track	On track	Slipping back	
126	Botswana	Slipping back	Slipping back	On track	Achieved	Achieved	Slipping back	
127	Myanmar	On track					Far behind	Far behind
128	Zimbabwe	Far behind			On track	Far behind	Slipping back	On track
129	Ghana	Achieved					Lagging	On track
130	Cambodia	On track	On track			Lagging	Slipping back	
131	Vanuatu						On track	
132	Lesotho	Lagging	Slipping back		Achieved	Achieved	Far behind	On track
133	Papua New Guinea	Far behind			Far behind	Far behind	Far behind	Far behind
134	Kenya	Far behind			Achieved	On track	Slipping back	Lagging
135	Cameroon	On track					Slipping back	On track
136	Congo	Far behind			On track	Far behind	Far behind	
137	Comoros					On track	On track	Achieved
Low human development								
138	Pakistan	On track					Far behind	On track
139	Sudan	On track			On track	On track	Far behind	On track
140	Bhutan						On track	
141	Togo	On track	On track		Far behind	Far behind	Far behind	Far behind
142	Nepal	Far behind			On track	On track	On track	On track
143	Lao People's Dem. Rep.	Far behind	On track		On track	Far behind	On track	On track
144	Yemen	Far behind					Far behind	Far behind
145	Bangladesh	Far behind					On track	Achieved
146	Haiti	Lagging	On track				Far behind	Far behind
147	Madagascar	Slipping back	Slipping back		On track	Achieved	Far behind	Far behind
148	Nigeria	Achieved					Far behind	Lagging
149	Djibouti		Far behind	Slipping back	Far behind	On track	Far behind	On track
150	Uganda	Far behind			On track	Far behind	Lagging	Far behind
151	Tanzania, U. Rep. of	Slipping back	Far behind	Far behind	On track	On track	Far behind	Far behind
152	Mauritania	On track		Slipping back	On track	Far behind	Far behind	Far behind

Table 23.6: *continued*

153	Zambia	Far behind	Slipping back		On track		Slipping back	On track
154	Senegal	Far behind	On track	On track	On track	Far behind	Far behind	On track
155	Congo, Dem. Rep. of	Slipping back					Far behind	
156	Côte d'Ivoire	On track	Far behind	Far behind	Far behind	Far behind	Slipping back	On track
157	Eritrea		Far behind				On track	
158	Benin	On track	On track		Far behind	Far behind	Far behind	
159	Guinea	On track	Far behind		On track	Far behind	On track	Far behind
160	Gambia	On track			On track	On track	Far behind	
161	Angola	On track					Slipping back	
162	Rwanda	Slipping back					Slipping back	
163	Malawi	On track			On track	On track	Lagging	Lagging
164	Mali	Far behind	Far behind	On track	On track	Slipping back	Far behind	On track
165	Central African Rep.	Far behind					Far behind	Far behind
166	Chad	On track	Far behind	Far behind	Far behind	Far behind	Far behind	
167	Guinea-Bissau						Far behind	
168	Ethiopia		Far behind		Slipping back	Slipping back	Far behind	Far behind
169	Burkina Faso	On track	Far behind		Far behind		Far behind	
170	Mozambique	On track	Slipping back		Far behind	Far behind	Far behind	
171	Burundi	Slipping back			Far behind		Far behind	
172	Niger	Far behind	Far behind	On track	Far behind	On track	Far behind	Far behind
173	Sierra Leone	Lagging					Far behind	
Others								
	Afghanistan	Far behind			Far behind	Slipping back	Far behind	
	Andorra						On track	On track
	Bosnia and Herzegovina	On track					On track	
	Iraq	Slipping back			Far behind	Far behind	Slipping back	
	Kiribati			On track			Lagging	
	Korea, Dem. Rep. of	Slipping back					Far behind	On track
	Liberia	Slipping back					Far behind	
	Liechtenstein						On track	
	Marshall Islands						On track	
	Micronesia, Fed. Sts.						On track	
	Monaco						On track	On track
	Nauru							
	Palau						Far behind	
	San Marino			Achieved			On track	
	Somalia	Slipping back					Far behind	
	Tonga						On track	On track
	Tuvalu						Far behind	On track
	Yugoslavia	On track		Achieved	Achieved	Achieved	On track	

Table 23.6: *continued*

HDI rank	Goal 1 Eradicate extreme poverty and hunger	Goal 2 Achieve universal primary education		Goal 3 Promote gender equality and empower women		Goal 4 Reduce child mortality	Goal 7 Ensure environmental sustainability
	Target: Halve the proportion of people suffering from hunger	Target: Ensure that all children can complete primary education		Target: Eliminate gender disparity in all levels of education[a]		Target: Reduce under-five and infant mortality rates by two-thirds	Target: Halve the proportion of people without access to improved water sources
	Undernourished people (as % of total population)[b]	Net primary enrolment ratio (%)	Children reaching grade 5 (%)	Female gross primary enrolment ratio as % of male ratio	Female gross secondary enrolment ratio as % of male ratio	Under-five mortality rate (per 1,000 live births)	Population using improved water sources (%)
Number of countries in category (% of world population)[c]							
Achieved or on track	57 (49.2)	51 (40.6)	44 (32.2)	90 (63.3)	81 (44.4)	85 (24.4)	68 (43.4)
Lagging, far behind or slipping back	43 (28.0)	24 (5.7)	8 (1.6)	14 (3.4)	20 (22.0)	81 (61.2)	25 (32.1)
No data	68 (8.5)[c]	93 (39.4)	116 (51.9)	64 (19.0)	67 (19.4)	2 (0.1)	75 (10.3)

[a]The goals for gender equality in primary and secondary education are preferably to be achieved by 2005, and by the latest by 2015. Progress towards the goals is assessed here based on a 2015 target.

[b]A complementary indicator for monitoring hunger is the prevalence of underweight children, but very limited trend data are available for that indicator.

[c]Population shares do not sum to 100% because the analysis excludes high-income OECD countries.

The table shows the results of analysis assessing progress towards goals for 2015 based on linear interpolation of trends in the 1990s. Each of the Millennium Development Goals is accompanied by multiple targets. The selection of goals and targets in the table is based principally on data availability. The trend assessment uses two data points at least five years apart. The table includes all UN member countries except high-income OECD countries; it also includes Hong Kong, China (SAR).

Source: UNDP, 2002.

References

Carney, D. 1998. *Sustainable Rural Livelihoods: What Contribution Can We Make?* London, Department for International Development.

Cosgrove, B. and Rijsberman, F.-R. 2000. *World Water Vision: Making Water Everybody's Business.* London, World Water Council, Earthscan Publications Ltd.

Falkenmark, M. and Widstrand, C. 1992. 'Population and Water Resources: A Delicate Balance'. *Population Bulletin.* Population Reference Bureau.

GWP (Global Water Partnership). 2000. *Integrated Water Resources Management.* Stockholm, Technical Advisory Committee Paper No. 4.

Ohlsson, L. 1999. *Environment, Scarcity, and Conflict: A Study of Malthusian Concerns.* University of Göteborg, Sweden, Department of Peace and Development Research.

Raskin, P.; Gleick, P.; Kirshen, P.; Pontius, R.-G. Jr.; Strzepek, K. 1997. 'Water Futures: Assessment of Long-Range Patterns and Problems'. Background document to: Shiklomanov, I.-A. (ed.). *Comprehensive Assessment of the Freshwater Resources of the World.* Stockholm, Stockholm Environment Institute.

Shiklomanov, I.-A. 1997. 'Assessment of Water Resources and Water Availability in the World'. Background document to: Shiklomanov, I.-A. (ed.) *Comprehensive Assessment of the Freshwater Resources of the World.* Stockholm, Stockholm Environment Institute.

UN (United Nations). 2002. *Johannesburg Summit 2002 – National Implementation of Agenda 21 – Main Report.* New York, United Nations Department of Economic and Social Affairs, Division of Sustainable Development, National Information Analysis Unit.

UNDP (United Nations Development Programme). 2002. *Human Development Report 2002: Deepening Democracy in a Fragmented World.* New York, Oxford University Press.

UNEP (United Nations Environment Programme). 1997. *Global Environmental Outlook 1.* Global State of the Environment Report. Nairobi.

WHO/UNICEF (World Health Organization/United Nations Children's Fund). 2000. *Global Water Supply and Sanitation Assessment 2000 Report.* New York.

World Bank/IBRD (International Bank for Reconstruction and Development). 2001. *World Development Report 2000/2001: Attacking Poverty.* New York, Oxford University Press.

Annexes

Acronyms

ADB: Asian Development Bank
AFDEC: Association Française pour le Développement de l'Expression Cartographique
ALT: Binational Autonomous Authority of Lake Titicaca (Autoridad Binacional del Lago Titicaca)
AQUASTAT: Country Information on Water and Agriculture
AESN: Seine-Normandy Water Agency (Agence de l'eau Seine-Normandie)
AWEC: Annual Water Experts Conference
BMA: Bangkok Metropolitan Area
BMR: Bangkok Metropolitan Region
BOD: Biological oxygen demand
BRGM: French Geological Survey and Bureau of Mines (Bureau de recherches géologiques et minières)
BSH: Basic Systems in Hydrology
CAPNET: International Network for Capacity Building for Integrated Water Resources Management
CBH: Capacity-Building in Hydrology and Water Resources
CBO: Community-Based Organization
CDRM: Comprehensive Disaster Risk Management
CEDAW: Convention on the Elimination of All Forms of Discrimination Against Women
CEOP: Coordinated Enhanced Observing Period
CEOS: Committee on Earth Observation Satellites
CESCR: Covenant on Economic, Social and Cultural Rights
CFR: Case Fatality Rate
CONIAG: Interinstitutional Water Council
CRED: Centre for Research on the Epidemiology of Disaster
CSE: Centre for Science and Environment
CMC: Community Multimedia Centres
CO_2: Carbon Dioxide
COD: Chemical oxygen demand
CODI: Community Organizations Development Institute (ex-UCDO: Urban Community Development Office)
CRC: Convention on the Rights of the Child
CSD: Commission on Sustainable Development
C_v: Coefficients of Variation
DAC: Development Assistance Committee (OECD Department)
DALY: Disability-Adjusted Life Years
DANIDA: Danish Agency for Development Assistance
DDT: Dichlorodiphenyltrichloroethane (toxic Chemical)
DHS: Demographic Health Surveys
DIGESA: Ministry of Health's Directorate for Environmental Health of Peru
DFID: Department for International Development (United Kingdom)
DPSIR: Driving Force-Pressure-State-Impact-Resource
DPSEEA: Driving Force-Pressure-State-Exposure-Effect-Action
DRPC: Danube River Protection Convention
DSS: Data Synthesis System
EDC: Endocrine disrupting chemical
EGAT: Electricity Generating Authority of Thailand
EIA: Environmental Impact Assessment
EMS: Environmental Management Systems
EPA: Environmental Protection Agency
EQS: Environmental Quality Standards
ERB: Experimental and Representative Basins
ESDG: Environmentally Sustainable Development Group
EU: European Union
FAH: Forecasting and Applications in Hydrology
FAO: Food and Agriculture Organization
FRICAT: Flood Risk Indicator (Japan)
FRIEND: Flow Regimes from International Experimental and Network Data
GCOS: Global Climate Observing System
GDP: Gross Domestic Product
GEF: Global Environment Facility
GEMS/WATER: Global Environmental Monitoring System, Freshwater Quality Programme
GEO: Global Environment Outlook
GEWEX: Global Energy and Water Cycle Experiment
GHP: GEWEX Hydrometeorology Panel
GIS: Geographic Information Systems
GIWA: Global International Waters Assessment
GNIP: Global Network for Isotopes in Precipitation
GNP: Gross National Product
GPA: Global Plan of Action for the Protection of the Marine Environment from Land Based Activities

GPCC: Global Precipitation Climatology Centre
GPP: Gross Provincial Product
GRDC: Global Runoff Data Center
GRID: Global Resource Information Database
GTN-H: Global Terrestrial Network on Hydrology
GTOS: Global Terrestrial Observing System
GWP: Global Water Partnership
HELP: Hydrology, Environment, Life and Policy
HIA: Health Impact Assessment
HOMS: Hydrological Operational Multipurpose System
HWRP: Hydrology and Water Resources Programme
IAEA: International Atomic Energy Agency
IAH: International Association of Hydrogeologists
IAHS: International Association of Hydrological Sciences
IBA: International Bird Area
IBRD: International Bank for Reconstruction and Development (World Bank Group)
ICESCR: International Covenant on Economic, Social and Cultural Rights
ICJ: International Court of Justice
ICOLD: International Commission on Large Dams
ICPRD: International Commission for the Protection of the Danube River
ICSU: International Council for Science
ICT: Information and Communication Technology
ICZM: Integrated Coastal Zone Management
IDPAD: Indo-Dutch Programme on Alternatives in Development
IDWSSD: International Drinking Water Supply and Sanitation Decade
IETC: International Environmental Technology Centre
IFAD: International Fund for Agriculture and Development
IFRC: International Federation of Red Cross and Red Crescent Societies
IGBP: International Geosphere Biosphere Programme
IHE: Institute for Infrastructural, Hydraulic and Environmental Engineering
IHD: International Hydrological Decade
IHP: International Hydrological Programme
IIED: International Institute for Environment and Development
IISD: International Institute for Sustainable Development
IMF: International Monetary Fund
INBD: International Network of Basin Organizations
IPCC: Intergovernmental Panel on Climate Change
IRBM: Integrated River Basin Management
IRC: International Research Centre for Water Supply and Sanitation
IRD: Research Institute for Development (Institut de Recherche pour le Développement)
ISARM: International Shared Aquifer Resource Management
IUCN: World Conservation Union
IVA: Industrial Value Added
IWMI: International Water Management Institute
IWRM: Integrated Water Resources Management
JMP: Joint Monitoring Programme
KWSB: Karachi Water and Sewerage Board
MAB: Man and the Biosphere
MAP: Mediterranean Action Plan
MAWAC: Managing Water for African Cities
MDGs: Millennium Development Goals
MICS: Multiple Indicator Cluster Surveys
MLIT: Ministry of Land, Infrastructure and Transport
MOAC: Ministry of Agriculture and Cooperatives (Thailand)
MOC: Ministry of Construction
MOI: Ministry of Industry (Thailand)
MOSTE: Ministry of Science Technology and Environment (Thailand)
MOPH: Ministry of Public Health (Thailand)
MPD: Maximum Permissible Doses
MWA: Metropolitan Waterworks Authority (Thailand)
NCPC: National Cleaner Production Centres
NEAP: National Environmental Action Plan
NEB: National Environmental Board (Thailand)
NEPAD: New Partnership for Africa's Development
NESDB: National Economic and Social Development Board (Thailand)
NFIP: National Flood Insurance Programme
NGO: Non-governmental Organization
NILIM: National Institute for Land and Infrastructure Management
NSDF: National Slum Dwellers Federation
NWRC: National Water Resources Committee (Thailand)
OECD: Organization for Economic Cooperation and Development
OED: Operation Evaluation Department
OERS: Organization of the Boundary States of the Senegal River (Organisation des États Riverains du Fleuve Sénégal)
OFDA: Office of Foreign Disaster Assistance (United States)
OKACOM: Okavango River Basin Commission
O&M: Operation and Maintenance
OMVS: Organization for the Development of the Senegal River (Organisation pour la mise en valeur du fleuve Sénégal)
ONWRC: Office of National Water Resources Committee (Thailand)
OPP: Orangi Pilot Project
PAH: Polycyclic Aromatic Hydrocarbon
PASIE: Environment Impacts Alleviation and Follow-Up Programme (Plan d'Atténuation et de Suivi des Impacts sur l'Environnement, Senegal)
PCB: Polychlorinated Biphenyl
PCCP: From Potential Conflict to Cooperation Potential
PCD: Pollution Control Department (Thailand)
Peipsi CTC: Peipsi Center for Transboundary Cooperation
PELT: Lake Titicaca Special Project (Proyecto Especial del Lago Titicaca)
PHAST: Participatory Hygiene and Sanitation Transformation
PPA: Power Purchase Agreement
PPP: Public-Private Partnership
PPPUE: Public-Private Partnerships for the Urban Environment
PRSP: Poverty Reduction Strategy Papers
PSR: Pressure State Response
PWA: Provincial Waterworks Authority (Thailand)
PWC: Permanent Water Commission
RBD: River Basin District
RBC: River Basin Committee
RBI: River Basin Initiative
R&D: Research and Development
RID: Royal Irrigation Department (Thailand)
RIVPACS: River Invertebrate Prediction and Classification System
SADC: Southern African Development Community
SAFE: Integrated strategy involving Surgery, Antibiotic treatment, promotion of Facial cleanliness and the initiation of Environmental changes

SAGEs: Local Water Management Plans (Schémas d'aménagement de gestion des eaux)
SAPP: Southern African Power Pool
SAR: Sodium Absorption Ratio
SDAGE: Master Plan for Water Management (Schémas directeurs d'aménagement et de gestion des eaux)
SDC: Swiss Development Cooperation of the Swiss Government
SDX: Swiss Agency for Development and Cooperation
SDW: Sustainable Development of Water Resources
SPARC: Society for the Promotion of Area Resource Centres
SWTR: Surface Water Treatment Rule
TEST: Transfer of Environmentally Sound Technology
TDPS: Titicaca, Desaguadero, Poopó, Coipasa Salt Lake
UCDO: See CODI
UfW: Unaccounted-for-Water
UN: United Nations
UN ACC/SCWR: United Nations Administrative Committee on Coordination Subcommittee on Water Resources
UNCBD: United Nations Secretariat of the Convention on Biological Diversity
UNCCD: United Nations Secretariat of Convention to Combat Desertification
UNCED: United Nations Conference on Environment and Development
UNDESA: United Nations Department of Economic and Social Affairs
UNDP: United Nations Development Programme
UNECA: United Nations Economic Commission for Africa
UNECE: United Nations Economic Commission for Europe
UNECLAC: United Nations Economic Commission for Latin America and the Caribbean
UNEP: United Nations Environment Programme
UNESCAP: United Nations Economic and Social Commission for Asia and the Pacific
UNESCO: United Nations Educational, Scientific and Cultural Organization
UNESCWA: United Nations Economic and Social Commission for Western Asia
UNFCCC: United Nations Secretariat of United Nations Framework Convention on Climate Change
UNFPA: United Nations Population Fund
UNGIWG: UN Geographic Information Working Group
UNHCR: United Nations High Commissioner for Refugees
UNICEF: United Nations Children's Fund
UNIDO: United Nations Industrial Development Organization
UNISDR: United Nations International Strategy for Disaster Reduction
UNSIA: United Nations Special Initiative for Africa
UNSO: Office to Combat Desertification and Drought
UNU: United Nations University
UNU/INWEH: International Network on Water, Environment and Health
UOB: Bolivian Operative Unit
USEPA: United States Environmental Protection Agency
UV: Ultraviolet
VASTRA: Swedish Water Management Research Programme (Vattenstrategiskarforskningsprogrammet)
VIP: Ventilated Pit Latrine
WCD: World Commission on Dams
WCED: World Commission on Environment and Development
WCP: World Climate Programme
WCRP: World Climate Research Group
WDI: World Development Indicators
WEHAB: Water, Energy, Health, Agriculture, Biodiversity
WES: Water, Environment and Sanitation Programme
WFD: Water Framework Directive
WGMS: World Glacier Monitoring Service
WHO: World Health Organization
WHYCOS: World Hydrological Cycle Observing System
WMO: World Meteorological Organization
WPI: Water Poverty Index
WRI: World Resource Institute
WSI: Water Stress Index
WSP: Water Safety Plans
WSSCC: Water Supply and Sanitation Collaborative Council
WSSD: World Summit on Sustainable Development
WSH: Water, Sanitation and Health Programme
WWAP: World Water Assessment Programme
WWC: World Water Council
WWDR: World Water Development Report
WWF: World Wide Fund International
YLD: Healthy years lost due to disability (Years lived with a disability)
YLL: Years lost due to premature mortality

Main Units of Measurement

b: billion
€: euro
G: giga
ha: hectare
k: kilo
l: litre
m: metre
m^2: square metre
m^3: cubic metre
m.a.s.l: metres above sea level
M: mega
M: million
s: second
T: tera
t: ton
US$: United States dollar
W: watt
Wh: watt per hour

Some Other Global Assessment Publications

This annotated list covers some of the more recent global reports in water, water-related and development issues.

Beck, T.; Guendel, S.; Kodsi, E.; Fuhr, N. 2000. *Sustainable Human Development and Good Project Design: An Assessment of Project Formulation and Design in UNDP Agriculture Programming, 1994–1999.* The Food Security and Agriculture Programme, UNDP/BDP/SEED (United Nations Development Programme/ Bureau for Development Policy/Energy and Environment Department).
An assessment of the extent to which the sustainable human development goals have been met in UNDP agriculture programming. The good-practice-for-results review, of which this study is one part, provides an assessment of project document formulation and design, strategic guidance for future programmes and projects, and a set of good-practice tools for use mainly by UNDP and partner staff.
[Cross-cutting]

FAO (Food and Agriculture Organization). Annual publication. *The State of Food and Agriculture.* Rome.
FAO's annual report on current developments and issues in world agriculture. The Organization monitors the global agricultural situation as well as the overall economic environment surrounding world agriculture. The 2002 report calls for increased international financial flows towards agriculture and rural areas. It also examines one of the possible new mechanisms for this financing: the Clean Development Mechanism (CDM) deriving from the Kyoto Protocol to the United Nations Framework Convention on Climate Change. Particular attention is paid to the potential use of the CDM as an instrument for both enhancing carbon sequestration through land use changes and for reducing rural poverty.
[Main challenge area: Securing the food supply]

——. 2000. *World Agriculture: Towards 2015/2030.* Rome.
A summary of FAO projections and messages intended for the general reader. The projections cover supply and demand for the major agricultural commodities and sectors, including fisheries and forestry. This analysis forms the basis for a more detailed examination of other factors, such as nutrition and undernourishment, and the implications for international trade. The report also investigates the implications of future supply and demand for the natural resource base and discusses how technology can contribute to a more sustainable development.
[Main challenge area: Securing the food supply]

Gleick, P. Biennial publication. *The World's Water.* Washington DC, The Pacific Institute for Studies in Development, Environment, and Security, Island Press.
The World's Water is a comprehensive reference series on worldwide freshwater resources and the political, economic, scientific and technological issues associated with them. Published every two years since 1998, it provides both detailed analysis of the most significant trends and events and up-to-date data on water resources and their use.
[Cross-cutting]

Groombridge, B. and Jenkins, M. 1998. *Freshwater Biodiversity: a Preliminary Global Assessment.* Cambridge, United Kingdom, WCMC/UNEP (World Conservation Monitoring Centre/United Nations Environment Programme), World Conservation Press.
This document provides useful information on inland waters and their biodiversity to a wide audience, ranging from those interested in the state of the world environment generally, to those needing an overview of the global and regional context in order to improve planning, management and investment decisions.
[Main challenge area: Protecting ecosystems]

Leisinge, K.-M.; Schmitt, K.; Pandya-Lorch, R. 2002. *Six Billion and Counting: Population Growth and Food Security in the 21st Century.* Baltimore, United States, Johns Hopkins University Press, International Food Policy Research Institute.
More people will inevitably mean greater demand for food, water, education, health care, sanitary infrastructure and jobs, as well as greater pressure on the environment. There must come a point when population growth threatens global food security and the Earth's finite natural resources. But what specific threats does population growth present now and in the coming decades? How can the world achieve sustainable development in the face of an ever-growing population? This book deals with these questions.
[Main challenge area: Securing the food supply]

Pardey, P.-G.; Beintema, N.-M. 2001. *Slow Magic. Agricultural R&D a Century After Mendel.* Washington DC, International Food Policy Research Institute, Agricultural Science and Technology Indicators Initiative.
This report assembles and assesses new and updated evidence regarding investments in agricultural R&D by public and private agencies, contrasting developments in rich and poor countries. It tracks trends in agricultural R&D over the past several decades.
[Main challenge area: Securing the food supply]

Pinstrup-Andersen, P.; Pandya-Lorch, R.; Rosengrant, M.-W. 1999. *World Food Prospects: Critical Issues for The Early Twenty-First Century.* Washington DC, Food Policy Report, International Food Policy Research Institute.
This report provides a summary of the most recent results from the International Food Policy Research Institute projections of the future world food situation. It then identifies and discusses six recent developments and emerging issues that will influence the prospects for global food security. It also discusses new evidence on the opportunities offered by agro-ecological approaches, the potential role of modern biotechnology and the relevance of new information technology and precision farming for small farmers in developing countries.
[Main challenge area: Securing the food supply]

Revenga, C.; Brunner, J.; Henninger, N.; Kassem, K.; Payne, R. 2000. *Pilot Analysis of Global Ecosystems: Freshwater Systems.* Washington DC, World Resource Institute.
This study analyses quantitative and qualitative information and develops selected indicators of the current and future capacity of freshwater systems to continue providing the full range of goods and services needed or valued by humans.
[Main challenge area: Protecting ecosystems]

Revenga, C.; Murray, S.; Abramovitz, J.; Hammond, A. 1998. *Watersheds of the World.* Washington DC, World Resources Institute, Worldwatch Institute.
This report and collection of maps present and analyse a wide range of global data at the watershed level, assessing 145 basins around the world. The analysis is based on fifteen global indicators that characterize watersheds according to their value, current condition and vulnerability to potential degradation.
[Cross-cutting]

Rosengrant, M.-W.; Paisner, M.-S.; Meijer, S.; Witcover, J. 2001. *2020 Global Food Outlook: Trends, Alternatives, and Choices. A Vision for Food, Agriculture, and the Environment Initiative.* Washington DC, International Food Policy Research Institute.
This report shows how, and how much, certain policy decisions and social changes will affect the world's future food security. It projects the likely food situation in 2020 if the world continues on more or less its present course, and it then shows how alternative choices could produce a different future.
[Main challenge area: Securing the food supply]

UNDP/UNEP/World Bank/WRI (United Nations Development Programme/United Nations Environment Programme/World Bank/World Resources Institute). 2000. *World Resources 2000–2001 – People and Ecosystems: The Fraying Web of Life.* Washington DC.
This book is a comprehensive guide to the global environment. This edition provides an assessment of five of the world's major ecosystems: agro-, coastal and marine, forest, freshwater and grassland ecosystems.
[Main challenge area: Protecting ecosystems]

UNDP (United Nations Development Programme). Annual report. *Human Development Report.* New York, Oxford University Press.
The *Human Development Report* (HDR) was first launched in 1990 with the single goal of putting people back at the centre of the development process in terms of economic debate, policy and advocacy. The goal was both massive and simple, with far-ranging implications – going beyond income to assess the level of people's long-term well being. Since the first Report, four new composite indices for human development have been developed – the Human Development Index, the Gender-related Development Index, the Gender Empowerment Measure and the Human Poverty Index. Each Report also focuses on a highly topical theme in the current development debate, providing path-breaking analysis and policy recommendations. The 2002 edition, *Deepening Democracy in a Fragmented World,* examines political participation as a dimension of human development.
[Cross-cutting]

———. 1997. *Energy After Rio: Prospects and Challenges.* New York.
The report focuses on the main linkages between energy and global problems, including: poverty, gender disparity (biases and discrimination against women), population growth, food, water and health, urban air pollution, climate change, acidification, land degradation, investment and foreign exchange requirements, energy imports and security, and nuclear proliferation. The report analyses energy from an end-use and service viewpoint, rather than from the traditional supply-side approach.
[Main challenge area: Water and energy]

UNEP/IETC (United Nations Environment Programme/International Environmental Technology Centre). 1998. *Sourcebook of Alternative Technologies for Freshwater Augmentation.*
A series of source books on alternative technologies for augmenting freshwater, covering Africa, some countries in Asia, East and Central Europe, small island developing states, and Latin America and the Caribbean. Mainly aimed at managers and planners dealing with freshwater resources and the use of environmentally sound technologies, this publication can also be of interest for NGOs and agencies dealing with the subject.
[Governing water wisely]

UNDP/UNDESA/WEC (United Nations Development Programme/United Nations Department of Economic and Social Affairs/World Energy Council). 2000. *World Energy Assessment: Energy and The Challenge of Sustainability.*
The *World Energy Assessment* was undertaken, in part, to build consensus on how we can most effectively use energy as a tool for sustainable development. Its analysis shows that more is needed to promote efficient and renewable energy, and to encourage advanced technologies that offer alternatives for clean and safe energy. Supporting developing countries is also needed to find ways to avoid retracing the wasteful and destructive stages that have characterized industrialization in the past.
[Main challenge area: Water and energy]

UNESCO (United Nations Educational, Scientific and Cultural Organization). 2001. *Monitoring Report on Education for all 2001.* Paris, UNESCO/Education for All.
This document provides insight into the state of the world's education and monitors the progress that countries and Education for All partners have made towards achieving the Dakar goals (adopted by the World Education Forum held in Dakar, Senegal, 26–28 April 2000). The report highlights important trends and findings and points to future actions.
[Main challenge area: Ensuring the knowledge base]

——. 2000. *EFA 2000 Assessment.* Paris, UNESCO/Education for All.
This book documents the progress made in education since 1990. It presents global, regional and national education indicators in six key areas: the demand for education, early childhood programmes, primary education, education finance, teachers and literacy.
[Main challenge area: Ensuring the knowledge base]

——. Biennial publication. *World Communication and Information Report.* Paris.
A comprehensive reference series on communication and information and the political, economic, scientific and technological issues associated with them. Published every two years, the report provides both detailed analysis of the most significant trends and events and up-to-date data on communication and information and their developments worldwide. The latest report, 1999–2000, concentrates on the relation between information and communications technologies (ICTs) and on some of their socio-cultural impact.
[Main challenge area: Ensuring the knowledge base]

——. Biennial publication. *World Science Report.* Paris.
This report provides information on the more important technical development of the last two years with a discussion of the main issues raised in this area by some of the most eminent world specialists. The last release of the report (1998) includes chapters looking at how science is helping to safeguard our two most basic commodities – food and water – in a context of rapid demographic growth and environmental stress.
[Main challenge area: Ensuring the knowledge base]

——. Biennial publication. *World Social Science Report.* Paris.
First released in 1999, the report takes stock of the social sciences as they are, and looks forward to their continuing development in the coming decades. It is divided into two parts. The first, *A Global Picture*, provides an overview of the history (since the eighteenth century), future prospects and current organization, financing and resources of the social sciences. The second takes up three central issues: science and technology in society, development and the environment. A final section reviews two areas of contact between the natural and social-cognitive science and the evolutionary study of human behaviour.
[Main challenge area: Ensuring the knowledge base]

UN-Habitat. 2001. *The State of the World's Cities Report 2001.* Nairobi.
The State of the World's Cities Report 2001 is a first in-depth attempt to monitor, analyse and report on the realities faced by urban populations around the world. The report was produced by UN Habitat to coincide with the Istanbul + 5 Special Session of the United Nations General Assembly. Its central message is that people's processes and initiatives and enabling governing structures must unite to form broad-based partnerships that will promote justice, equity and sustainability in cities.
[Main challenge areas: Water and cities, Meeting basic needs]

——. 2001. *Cities in a Globalizing World: Global Report on Human Settlements 2001.* Nairobi.
The report argues that technology-driven options for growth and development – which spur globalization – have led to divided cities where the lines of stratification between people, places and groups are becoming more magnified. The costs and benefits of globalization are unevenly distributed both within and between cities. In many countries, real incomes have fallen, the costs of living have gone up and the number of poor households has grown, particularly in urban areas. Sixty countries have become steadily poorer since 1980.
[Main challenge areas: Water and cities, Meeting basic needs]

UNICEF (United Nations Children's Fund). 2002. *Children in the New Millennium: Environmental Impact on Health.* Geneva.
By illustrating the link between the environment and the well-being of our children, this joint United Nations Environment Programme (UNEP), UNICEF and World Health Organization (WHO) report raises awareness and deepens our understanding of environmental health issues. It provides an overview of key environmental health threats to children. The report concludes with a series of practical recommendations for action at the local, national, regional and international levels to stimulate discussion and intensify action in the field of children's environmental health.
[Main challenge area: Meeting basic needs]

——. 1997. *A Child's Right to Sustainable Development.* New York.
The report argues that children are our future and thus they will bear the consequences of today's decisions. If sustainable development is to be achieved, it must be from their perspective.
[Main challenge area: Governing water wisely]

WCD (World Commission on Dams). 2000. *Dams and Development: A New Framework for Decision-Making.* London, Earthscan Publications.
Through its global review of the performance of dams, WCD presents an integrated assessment of when, how and why dams succeed or fail in meeting development objectives. This provides the rationale for a fundamental shift in options assessment and in the planning and project cycles for water and energy resources development. The Commission's framework for decision-making is based on five core values: equity, sustainability, efficiency, participatory decision-making and accountability.
[Main challenge area: Water and energy]

WHO/UNICEF (World Health Organization/United Children's Fund). 2000. *Global Water Supply and Sanitation Assessment 2000 Report.* New York.
The WHO and UNICEF Joint Monitoring Programme provides a snapshot of water supply and sanitation worldwide at the turn of the millennium using information available from different sources. The *2000 Assessment*, the fourth in the series, differ from the previous reports, in that it is focusing on users as primary sources of data, rather than on providers. As a result, the assessment provides the baseline and monitoring methodology that will ensure reliable and consistent statistics with which to report progress with confidence in the years to come.
[Main challenge area: Meeting basic needs]

World Bank. *World Development Indicators (WDI).* New York, Oxford University Press.
World Bank's premier annual compilation of data about development. *WDI 2002* includes about 800 indicators in eighty-seven tables, organized in six sections: World View, People, Environment, Economy, States and Markets, and Global Links. The tables cover 152 economies and fourteen country groups-with basic indicators for a further fifty-five economies. This WDI print edition offers the current overview of reliable data from the past few years.
[Cross-cutting]

Worldwatch Institute. Annual publication. *State of the World Series.* Washington.
The Worldwatch Institute's flagship annual is in publication since 1984 and is used by government officials, corporate planners, journalists, development specialists, professors, students and concerned citizens in over 120 countries. Published in more than twenty different languages, it is one of the most widely used resources for analysis. The 2002 special World Summit Edition focuses on the agenda of the World Summit on Sustainable Development.
[Cross-cutting]

——. Annual publication. *Vital Signs Series.* Washington.
Published annually since 1992, this book provides comprehensive, user-friendly information on key trends and includes tables and graphs that help readers assess the developments that are changing their lives for better or worse. The 2002 edition features more than fifty key indicators of long-term trends, from the growth of fish farms and bicycle production, to the increase in solar cell and Internet use, to the decrease in land mine production.
[Cross-cutting]

"The face of water, in time, became a wonderful book . . . And it was not a book to be read once and thrown aside, for it had a new story to tell every day."

Mark Twain

Index

Notes: text in boxes is indicated by **bold page numbers** and in figures and tables by *italic page numbers*; footnotes are denoted by the suffix 'n' – e.g. 'Beauce aquifer 437n(14)' refers to footnote (14) on page 437. 'Millennium Development Goals' is abbreviated in this index to 'Goals' under each country.

Index

D

E

F

Index

Index

N

O

Index

Index

Credits

Photographs

Cover
© UNICEF - S. Noorani

Inside front cover
© Still Pictures – J.J. Alcalay, P. Frischmuth, R. Seitre, F. Suchel, J-C. Muñoz
© UNESCO – CZAP/ASA, F. Gattoni, D. Riffet
© X. Lefèvre
© Still Pictures – A. Bartschi, M. Edwards, J. Etchart, P. Frischmuth (p.I)
© UNESCO – D. Roger (p.I)

Prelims
© UNEP

Part I
© UNESCO – M. Becka (p.1)

Chapter 1
© Still Pictures – E. Cleijne (p.2)
© UNICEF (p.3)
© UNEP (p.4)

Chapter 3
© UNESCO – G. Guit; M. Soler-Roccat; D. Roger (p.31)

Chapter 4
© UNESCO – G.-A.Vicas (pp.64/65)
© Still Pictures – A. Bartschi (p.64)
© Still Pictures – R. Seitre (p.65)
© Y. Arthus-Bertrand (p.66)

Chapter 5
© UNESCO – D. Roger (p.97)
© UNICEF (p.99)
© Swynk (p.100)
© UNESCO – Y. Nagata (pp.100/101)
© Swynk ; © UNESCO – D. Roger (p. 101)

Chapter 6
© UNCHS (p.129)

Chapter 7
© FAO – Bizzari; © UNESCO – Robinson (p.158)
© A. Clayson; © IRD; © Swynk (p.159)

Chapter 8
© IRD (p.189)
© FAO – M. Marzot ; © UNESCO – G. Malempré; © P. Unt (p.190)
© Swynk (pp.190/91)
© FAO – P. Johnson; © Swynk; © UNESCO – D. Riffet; © P. Unt (p.191)
© Swynk; © UNESCO – D. Roger; © X. Lefèvre (p.192)

Chapter 9
© Still Pictures – R. Janke (p.225)
© UNESCO – D. Roger (p.226)

Chapter 10
© UNESCO – A. Testut (p.248)
© A. Clayson ; © FAO – M. Marzot ;
© Swynk; © UNESCO; © UNESCO – G. Malempré (p.249)

Chapter 11
© Swynk (p.268)
© UNICEF – M. Murray-Lee; © WMO – Cincinnati Post (p.269)

Chapter 12
© UNICEF – J. Isaac (p.291)
© Y. Arthus-Bertrand; © UNESCO – D. Riffet (p.292)
© Swynk (pp.292/3, p.293)
© UNESCO – D. Roger (p.293)

Chapter 13
© UNESCO – J. Foy (p.324)
© Swynk (pp.324/5, p.325)

Chapter 14
© UNESCO – D. Roger (p.345)
© Swynk (p.346)
© © IRD; Still Pictures – G. Nicolet; © Swynk; © UNESCO – D. Roger (p. 347)

Chapter 15
© UNESCO – A. Périé (p.367)
© UNESCO – D. Roger (p.368)
© Swynk ; © UNESCO – I. Forbez ;
© C. Fernandez-Jauregui (p.369)

Chapter 16
© Swynk (p.385)
© ONWRC; © UNESCO – G. Malempré (p.388)
© FAO – P. Johnson; © UNESCO – M. Becka (p.389)

Chapter 17
© R. Urbel (p.401)
© P. Unt (p.402, p.403)
© T. Tuul (pp.402/3)

Chapter 18
© UNESCO – P. Coles

Chapter 19
© AESN (p.429)
© UNESCO – D. Chenot (p.430)
© AESN (p.431)

Chapter 20
© OMVS (p.447)
© UNESCO – G. Malempré (p.448)
© Y. Arthus-Bertrand (pp.448/9)
© OMVS ; © UNESCO – D. Roger (p.449)

Chapter 21
© UNESCO, A. Jonquière (p.463, pp.464/5, p.465)
© UNESCO – G. Malempré, P.A.Pittet (p.465)

Chapter 22
© MLIT (p.482, pp. 482/3, p.483)
© UNESCO – D. Roger (p.483)

Chapter 23
© FAO – M.Marzot;
© Still Pictures – F.Suchel;
© Swynk ; © UNICEF – M.Murray-Lee (p.500)
© FAO – P. Johnson, M. Marzot ; © Swynk;
© UNESCO – Y. Nagata;
© UNICEF (p.502)
© FAO – P. Johnson;
© MLIT ; © Swynk; © T. Tul

Inside back cover
© Still Pictures – G. Nicolet, R.Seitre, R.Janke, H.Bloch
© UNICEF– M.Murray-Lee, S.Noorani, N.Toutounji
© UNESCO – G.Fernandez, D.Roger